Encyclopedia of
Plant Physiology

New Series Volume 8

Editors

A. Pirson, Göttingen
M. H. Zimmermann, Harvard

Secondary Plant Products

Edited by
E. A. Bell and B.V. Charlwood

Contributors

B. L. Archer D.V. Banthorpe E. A. Bell B.V. Charlwood
E. E. Conn G. B. Fodor T.W. Goodwin D. Gröger
C. Grunwald J. B. Harborne E. Leete M. Luckner
T. J. Mabry K. Mothes J. G. Roddick T. A. Smith
A. M. Stephen G. A. Thompson, Jr. D. R. Threlfall
E. W. Underhill

Springer-Verlag Berlin Heidelberg New York 1980

Professor E. ARTHUR BELL
Dr. BARRY V. CHARLWOOD
University of London King's College
68 Half Moon Lane
London SE 24 9JF/United Kingdom

With 176 Figures
and Numerous Schemes and Formulas

ISBN-13: 978-3-642-67362-7 e-ISBN-13: 978-3-642-67360-3
DOI: 10.1007/978-3-642-67360-3

Library of Congress Cataloging in Publication Data, Main entry under title Secondary plant products (Encyclopedia of plant physiology, new ser , v 8) Bibliography p Includes index 1 Plant products I Bell, Ernest Arthur, 1926 – II Charlwood, B V , 1945 – III Archer, B L IV Series QK711 2 E5 vol 8 [QK887] 581 1'08s [581 1'92], 79-16289

Softcover reprint of the hardcover 1st edition 1980

Preface

It is impossible in a single volume to deal comprehensively with all classes of secondary plant compounds. In the earlier series of this Encyclopedia emphasis was laid on the isoprenoids and plant phenols. While these compounds have not been neglected in the present volume we have attempted to achieve a more balanced presentation by drawing attention to the importance of nitrogenous secondary metabolites such as the alkaloids, amines and non-protein amino acids.

Most of the compounds or groups of compounds included in Volume 8 are of restricted distribution within the plant kingdom and wherever possible we have provided information concerning their chemistry, biochemistry, taxonomic significance and probable ecological roles. Secondary compounds cannot be defined in terms of restricted distribution, however, nor can they be defined without reference to the plants in which they occur, as it is possible that a given compound occurring in two species may have a primary role in one and not in the other. As our knowledge of biochemistry increases we shall no doubt find it necessary to revise our ideas concerning the roles of a great many of the compounds which are found in plants.

It is a pity indeed that the term "secondary" should have ever been applied to these compounds as the word gives the unfortunate impression that they are all relatively unimportant. By definition they do not play a role in the primary metabolic processes essential for the maintenance of life in an individual plant, nevertheless many may be absolutely essential to the survival of the species as a whole in a given natural habitat. A fuller account of the historic development of ideas concerning "non-primary" compounds is provided by Professor MOTHES in the Introduction.

The isolation of natural products has been facilitated during the last twenty years by the advances which have been made in all branches of chromatography and electrophoresis, while structural elucidation has been greatly simplified by the rapid development of techniques such as mass, ^{1}H and ^{13}C NMR spectroscopy. These latter methods have also been applied to biosynthetic studies and it may be fairly stated that interest in the biosynthesis and ecological significance of secondary compounds has steadily increased throughout the period under review.

As for the future, application of biological techniques (such as tissue culture, selective breeding and the artificial inducement of genetic changes) to the production of abnormally high or low concentrations of secondary metabolites in living systems offers, we believe, the most exciting prospect in this field.

We should like to thank all the authors who have contributed to this volume, also the publishers, and Mrs. J.E. POTTER for invaluable help in preparing the indexes.

London, September 1979

E. A. BELL
B.V. CHARLWOOD

Contents

4.5 Isoprenoid Alkaloids

J.G. RODDICK

5 The Isoprenoids

5.1 The Terpenoids

D.V. BANTHORPE and B.V. CHARLWOOD (With 12 Figures)

5.2 Steroids

C. GRUNWALD

5.4 Polyisoprenoids

5.4.1 Polyprenols and Terpenoid Quinones and Chromanols

D.R. THRELFALL (With 2 Figures)

5.4.2 Polyisoprene

B.L. ARCHER (With 2 Figures)

6 Plant Phenolics

J.B. HARBORNE (With 16 Figures)

List of Contributors

B.L. ARCHER
Malaysian Rubber Producers' Research Association
Brickendonbury,
Hertford, Herts. SG13 8NP/
United Kingdom

D.V. BANTHORPE
Chemistry Department
University College
20 Gordon Street
London WC1H-OAJ/United Kingdom

E.A. BELL
Dept. of Plant Sciences
University of London King's College
68 Half Moon Lane
London SE 24 9JF/United Kingdom

B.V. CHARLWOOD
Dept. of Plant Sciences
University of London King's College
68 Half Moon Lane
London SE 24 9JF/United Kingdom

E.E. CONN
Dept. of Biochemistry and Biophysics
University of California
Davis, California 95166/USA

G.B. FODOR
Dept. of Chemistry
West Virginia University
Morgantown, WV 26506/USA

T.W. GOODWIN
Dept. of Biochemistry
University of Liverpool
P.O. Box 147, Liverpool L69 3BX/
United Kingdom

D. GRÖGER
Akademie der Wissenschaften der DDR
Institut für Biochemie der Pflanzen
Postfach 250
401 Halle/GDR

C. GRUNWALD
Dept. of Botany and Plant Pathology
Illinois Natural History Survey
University of Illinois
Urbana, Illinois 61801/USA

J.B. HARBORNE
Dept. of Botany
The University of Reading
Whiteknights
Reading, RG6 2AS/United Kingdom

E. LEETE
Dept. of Chemistry
University of Minnesota
Minneapolis, Minnesota 55455/USA

M. LUCKNER
Martin-Luther-Universität
Halle-Wittenberg
Sektion Pharmazie
Weinbergweg 15
402 Halle/GDR

T.J. MABRY
Dept. of Botany
The University of Texas at Austin
Austin, Texas 78712/USA

K. MOTHES
Hoher Weg 23
402 Halle/GDR

J.G. RODDICK
Dept. of Biological Sciences
University of Exeter
Washington-Singer Laboratories
Perry Road
Exeter, EX4 4QG/United Kingdom

T.A. SMITH
Research Station
University of Bristol
Long Ashton
Bristol, BS18 9AF/United Kingdom

A.M. Stephen
Dept. of Organic Chemistry
University of Cape Town
Rondebosch, Cape Province 7700/
Republic of South Africa

G.A. Thompson, Jr.
Dept. of Botany
University of Texas
Austin, Texas 78712/USA

D.R. Threlfall
Dept. of Plant Biology
University of Hull
Hull HU6 7RX/United Kingdom

E.W. Underhill
Prairie Regional Laboratory
National Research Council of Canada
Saskatoon, Sask. S7N OW9/Canada

1 Historical Introduction

K. Mothes

The introduction of Schwarze (1958) to volume X of the first Encyclopedia reflects well the earlier state of our knowledge on the biology and biochemistry of secondary products. I should like to add some further remarks which complete the picture and introduce our present knowledge.

Two notable publications by Bonner (1950) and Paech (1950) contributed to the broad application of the term "secondary plant products" although a uniform and unequivocal definition of these compounds, and guide lines for distinguishing them from "primary" products could not be given. Nevertheless, the term "secondary products" has a long history and has proved valuable.

Let us begin with the pioneers of modern plant physiology who considered it to be an experimental and causal science. Although the detailed chemical structures of most secondary products were unknown, their formation was recognized as a plant-specific phenomenon and treated accordingly. In his lectures on plant physiology Sachs (1882) devoted a special chapter to laticiferous ducts and secretory containers. He especially emphasized the lack of phylogenetic continuity which distinguishes their distribution from that of most other plant organs. They are found only in certain plant species and these do not necessarily belong to related phylogenetic groups. He states that lattices may be of different origin and contain not only valuable substances (carbohydrates, starch and proteins) but also waste products, e.g., resins, rubber, polyterpenes, alkaloids, etc. Moreover, Sachs pointed out that a secretory container should rather be called an excretory container. Hence he was aware of a special feature which later was called "storage excretion" in the zoological literature (Wigglesworth, 1972). He also considered lignin, cutin of epidermal cells, oxalate crystals, resins, and essential oils to be excretory products. It must be remembered however, that some metabolically useful substances such as sugars or organic acids in fruits may have a secondary function in attracting animals and thus promoting the dispersion of the species.

A clearer distinction between the physiological role of a compound and its ecological significance was elaborated by Pfeffer (1897) in the first volume of the second edition of his *Plant Physiology*. He wrote that the soluble carbohydrates in nectaries cannot be considered as storage compounds, even if they are reabsorbed and reused in metabolism. It cannot be denied that some metabolic by-products are occasionally degraded whereas others are not. Storage may necessitate compartmentation, e.g., in the vacuole, or conjugation (as of phenolics and tannins with carbohydrates), thereby preventing damage to the protoplast. Hence, beside storage excretion the phenomenon of intracellular metabolic excretion had already been described at the end of the last century. Pfeffer (1897) was convinced that secondary compounds have no physiological function essential for the maintenance of life. He concluded this from the observation that closely related varieties of almonds may or may not contain cyanogenic glycosides. At any rate, many of these compounds, e.g., many pigments (disregarding chlorophyll and similar pigments), alkaloids, ptomaines, essential oils, resins, etc., have only ecological importance and cannot be reused in metabolism. By precipitating tannins of *Spirogyra* with methylene blue and showing that their removal did not affect the growth of this alga, he presented direct evidence that tannins are not required for the life of this organism. Already SACHS had called tannic substances "by-products of metabolism".

With the beginning of this century intensive investigations on secondary products in higher plants were started. In CZAPEK'S *Plant Biochemistry*, the third volume of which appeared in 1921 in the second edition, 183 pages are concerned with respiration and 547 pages with secondary plant products. This indicates that secondary products were considered to be characteristic of plant metabolism. In the headings of the main chapters they are called either "excretory products" or "end products of metabolism". When describing the alkaloids CZAPEK wrote the following: „Vielleicht ist die sporadische Verbreitung dieser Basen, die Inkonstanz ihres Auftretens bei nahe verwandten Pflanzen, ein Zeichen dafür, daß es sich bei der Bildung solcher Stoffe um Prozesse handelt, die nicht jedem Zellplasma eigen sind, sondern mehr sekundären Charakter haben. Doch ist Vorsicht bei solchen Schlüssen geboten.[1]" He pointed out that alkaloids are never found in sieve tubes, that the germination of *Datura* seeds proceeds normally after the alkaloids have been extracted, and that alkaloids of seeds generally are not used as storage compounds. CZAPEK emphasized that current theories about alkaloids as defense substances, although interesting, do not concern the physiological role of these compounds. Although some hints on a possible ecological function have been reported, definite proof is lacking. Metabolic degradation of secondary products is possible, but the peculiarities of regulation determine whether degradation takes place or not. CZAPEK considered it striking that no secondary products had been found in animals. An essential hint at the origin of the term "secondary" may already be found in CZAPEK'S Biochemistry, Vol. I, page 24. He mentions ALBRECHT KOSSEL who in 1891 presented a lecture to the Physiological Society of Berlin entitled: "Über die chemische Zusammensetzung der Zelle" (On the chemical constitution of cells). This lecture was published in Archiv für Physiologie (1891, p. 181): „Ebenso wie die mikroskopische Forschung dahin gelangt ist, daß sie die Zelle alles unwesentlichen Beiwerkes entkleidet hat, daß sie das Gehäuse und die in ihr aufgespeicherten Reservestoffe von den eigentlichen Trägern des Lebens zu trennen weiß, so muß auch die Chemie versuchen, diejenigen Bestandtheile heraus zu sondern, welche in dem entwicklungsfähigen Protoplasma ohne Ausnahme vorhanden sind und die zufälligen oder für das Leben nicht unbedingt nöthigen Zellstoffe als solche zu erkennen. Die Aufsuchung und Beschreibung derjenigen Atomcomplexe, an welche das Leben geknüpft ist, bildet die wichtigste Grundlage für die Erforschung der Lebensprocesse. Ich schlage vor, diese wesentlichen Bestandtheile der Zelle als *primäre* zu bezeichnen, hingegen diejenigen, welche nicht in *jeder* entwicklungsfähigen Zelle gefunden werden als *sekundäre*. Die Ent-

1 The sporadic distribution of these bases, the irregularity of their occurrence in closely related species, perhaps indicate that the production of such substances involves processes not inherent to every cell plasma, but more of a secondary character. Caution is, however, necessary in drawing such conclusions

scheidung, ob ein Stoff zu den primären oder secundären Bestandtheilen gehört, ist in manchen Fällen äußerst schwierig.[2]"

In the next period ending about 1950 interest in natural products increased considerably. Beside the discovery of several hormones in animals and plants evidence was obtained of the relationship between secondary and other groups of natural compounds. The progress of histology and cytology gave deeper insight into the localization of secondary products in cells and tissues. Above all a more dynamic treatment of secondary metabolism began. It soon became apparent that the separation of secondary products from primary ones was difficult, especially when considering hormones or storage compounds. Hence, BONNER and GALSTON (1952), distinguished between "highways and byways of metabolism" thus avoiding the qualification of the *products* of these processes (end products, by-products). The highways include synthesis and degradation of proteins, carbohydrates, and lipids. "But individual species of higher plants are able to make not only many chemical materials characteristic of, and essential to, the functioning of the plant, but also particular chemical substances which are not essential to the economy of the plant and which have no recognizable role in metabolism. It is characteristic of such compounds that they often occur only in scattered species distributed at random throughout the plant kingdom. Among these secondary products of metabolic by-ways are the alkaloids, the terpenes, rubbers, sterols and steroids, the tannins, and many of the other plant materials which contribute to the welfare of mankind."

In *Plant Biochemistry*, edited by BONNER and VARNER (1965), one fifth of the book is concerned with secondary plant products. The distinction between primary and secondary products is much clearer than that provided by PAECH (1950) who, in a special book about secondary products, devoted more than fifty pages to carboxylic acids and fat metabolism.

There was a tremendous increase of work on the organic chemistry of natural plant products in the middle of this century, brought about by the development of new physical and physicochemical methods of separation, purification, and structure elucidation. Simultaneously, the interest of the pharmaceutical industry in these compounds increased. This development is typified by the on-going search for antibiotics initiated by the discovery of penicillin and by the investigations of cancer-inhibiting substances which led to the discovery and structure elucidation of nearly 800 alkaloids of the tryptamine-monoterpene type from the *Gentianales*.

The modern development of natural product chemistry also led to the isolation of secondary products from lower plants (fungi) and even from prokaryotes (bacteria). Among these frequently rather complicated compounds, pigments, toxins and antibiotics were found. The number of the known antibiotics at present exceeds 3000. Many relationships between these compounds and secondary products from

2 Just as microscope research has succeeded in stripping the cell of its non-essential accessories and in separating its casing and the reserves stored in it from the actual life-carriers, so now chemistry must attempt to separate those compounds which are present, without exception, in a protoplasma capable of developing, and to recognise the substances which are either incidental or not absolutely necessary for life. Finding and describing those atom complexes to which life is bound comprises the most important basis for the investigation of life processes. I propose calling these essential components of the cell primary components, and those which are not found in every cell capable of developing, secondary. To decide whether a substance belongs to the primary or secondary components is extremely difficult in certain cases

higher plants have been found. Thus the chromophore of the actinomycins is formed from two molecules of 3-hydroxy-4-methyl-anthranilic acid. In many cases the D-isomers of amino acids are found in peptides. In lower plants and bacteria, as in higher plants, no definite proof of an ecological function of secondary products could be given. Frequently they are true excretory products. The very large number of secondary products already found in microorganisms leads to the conclusion that the range of chemicals that organisms can synthesize has become more restricted during evolution.

More recently it has been shown that insects as well as lower plants are capable of producing secondary compounds. Of particular importance in this area has been the work of BUTENANDT and his co-workers (BUTENANDT et al., 1940; BUTENANDT and KARLSON, 1954; BUTENANDT, 1959; KARLSON and BUTENANDT, 1959; KARLSON, 1963). They investigated tryptophan metabolism in pupae of butterflies, elucidated the chemical structure of insect hormones and showed them to occur in plants as well. They found that certain ecologically important compounds function as inter-organismic signals between insects of the same species (pheromones) and between insects and plants. Stimulated by these experimental results a systematic search for defense substances in insects started, thus initiating a broad and important field of natural products chemistry called chemical ecology.

It should be noted that the chemical structures of secondary animal products are frequently much simpler than their counterparts in plants. Furthermore, secondary products of animals may be derived from the plants on which they feed as illustrated by cardiac glycosides in butterflies and grasshoppers and alkaloids in butterflies (SCHILDKNECHT, 1970; REICHSTEIN, 1967; ROTHSCHILD and REICHSTEIN, 1976). However, cyanogenic compounds and alkaloids like the quinazolones may be synthesized by the insects themselves using biosynthetic pathways similar to those found in higher plants. It is interesting in this respect to learn that certain chrysomelid beetles contain monohydroxylated digitoxigenin glycosides even when the beetles have been fed on plants which are free of cardenolides (PASTEELS and DALOZE, 1977). Further evidence for such insect biosynthesis is that arabinose and xylose constitute the sugar moiety of these particular glycosides. These carbohydrates are rarely found in plant cardiac glycosides.

The occurrence of secondary animal products is by no means restricted to insects though the chemical potential of the higher animals seems less developed. This may be due to their well-organized excretory organs, e.g., kidneys. Evidently most secondary products or even their precursors are rapidly excreted. Some strikingly toxic compounds are produced by certain chordates, e.g., tetrodotoxin of the bowl fish and the Californian newt, a *Taricha* species (RUGGIERI, 1976). The biosynthetic pathway to tetrodotoxin cannot be immediately deduced from its chemical structure. However, considering its isolated occurrence the possibility cannot be excluded that it is taken up from special nutrients. A pertinent example is reported for the purine derivative saxitoxin of some mussel species, which was found also in the Dinoflagellate *Gonyaulax catenella* (SCHANTZ et al., 1975). Both of these extremely active toxins have attained considerable importance in experimental pharmacology and neurology (WITKOP, 1971).

The tremendous increase of the number of known secondary products in all living organisms has been accompanied by a corresponding increase in our knowledge of secondary metabolism stimulated by technical advances in chromatography and tracer techniques. Within less than three decades the main biosynthetic pathways for most secondary plant products have been worked out although the development of the pertinent enzymology has only just started. Cytological problems, i.e., the sites of secondary product formation and their transport and storage, are closely related to cell differentiation – a fascinating experimental field for future work particularly because of its evolutionary significance (LUCKNER et al., 1977). As these products are toxic to the protoplasm their formation has necessarily required the co-evolution of compartments for their segregation (SCHNEPF, 1973). Semipermeable membranes with unidirectional transport play a major role in this respect, while membranes may also participate directly in the synthesis of secondary products.

It is not surprising that the term "secondary product" has been a matter of discussion throughout this period of rapid advance. Boundaries of secondary metabolism are not very distinct. Originally emphasis was put on the secondary importance of the products: compounds which occur in certain species or even in certain chemical races only cannot play a fundamental role in metabolism. However, such negative statements are of limited value because they may be the result of our lack of knowledge. GEISSMAN and CROUT (1969) pointed out that only two decades ago shikimic acid was considered an exotic plant product of *Illicium religiosum*, before it was suddenly detected as a key intermediate of the biosynthesis of aromatic compounds. The same holds true of squalene, and other so-called secondary compounds may yet prove to be key intermediates. Such interesting exceptions do not invalidate the general statement that secondary products are physiologically nonessential compounds. This is best demonstrated by certain cyanogenic compounds which are sometimes excretory products but can also be formed during early developmental stages and afterwards disappear again. Thus, in *Sorghum vulgare* (CONN, 1973) a large part of the nitrogen mobilized from the seed storage proteins is reincorporated via tyrosine into the cyanogenic glycoside dhurrin which evidently plays a role in the embryo development which might otherwise be played by asparagine or glutamine.

Other examples of such secondary compounds are cellulose, chitin, and lignin, which are metabolically inert and have great importance as mechanical constituents of the cell wall (FREY-WYSSLING, 1935, 1970). Storage proteins of seeds and other organs may also be mentioned in this context. These have no physiological activity even if they are enzymes, e.g., urease of seeds. Such proteins are in effect stored amino acids. Hence, the realization of their biological function depends on their degradation (MÜNTZ, 1974, 1977, 1978). Another example is provided by the lectins which may have a function because of their carbohydrate affinity (RÜDIGER, 1978), but do not have a general distribution.

It is reasonable to assume that storage compounds may be changed in their chemical structure, e.g., as a consequence of mutations, without loss of their function. Thus the fat mixture of seeds of *Brassica napus* L. var. *napus* (rapeseed) can be devoid of erucic acid, but not of linolenic acid, which is presumably required for chloroplast biogenesis (THIES, 1971; RÖBBELEN, 1976). Qualitative and quantitative variability are characteristic features of typical secondary products. This is exemplified by the "sweet" lupine of Sengbusch or by the alkaloid- and pigment-free sclerotia of certain wild-type strains of the ergot fungus *Claviceps purpurea*. It should be noted however that the latter strains are rarely found in nature probably because of their lower viability. Strains with exceptionally high alkaloid content are likewise rare. Ergot strains with 0.5%–0.6% alkaloids have proved advantageous for the parasitic production because only these give high and reproducable yields (MOTHES and SILBER, 1954; GRÖGER, 1966).

The hormones of higher plants which we know at the present time are regulators of growth and differentiation. There is not a single autotrophic flowering plant which does not require one of these hormones, although under certain experimental conditions one hormone may occasionally be substituted for another. Recent evidence suggests that these hormones, e.g., auxin, zeatine (cytokinins), and gibberellins are found in bacteria, fungi, and even in animals. In these organisms they have no comparable function, and hence can be considered as true "secondary products". It is a matter of speculation that the acquisition of the ability to synthesize these compounds occurred early in evolutionary history before the development of higher plants. They became essential for higher plants but remained without biological function in other organisms. Decisive steps during evolution were the development of target organs and of physiological responses, thus enabling secondary products to become plant hormones.

The same situation holds true for animal hormones. When shortly after the discovery of sexual hormones of higher animals BUTENANDT (1933) found the same type of compound in plants it created great interest and induced much speculation about the uniformity of sexual processes in all organisms. Today we know that there is no basis for such speculation and quite recently the detection of large amounts of the insect hormone ecdysone in several plant species (KARLSON, 1963, 1976) made this even clearer. Metabolism of cholesterol by demethylation, oxidation, and side chain cleavage results in a broad spectrum of compounds. Therefore, the occurrence of secondary products resembling animal steroid hormones in plants or other organisms is not surprising.

Even with our present limited knowledge of secondary plant and animal products it is clear that living organisms have an enormous chemical ability especially if we remember the remarkable ability of bacteria and fungi to carry out the specific hydroxylation of steroid derivatives (CHARNEY and HERZOG, 1967). Certain strains of these microorganisms gained great importance in the industrial production of adrenocortical hormones. Certain fungi were even found to contain enzymes catalyzing the transformation of thebaine into codeinone, a reaction hitherto only observed in two *Papaver* species. It is at present impossible to give an estimate of how many biochemical reactions, though coded for by the genetic material of the organisms, are only expressed if the particular substrates are offered exogenously (TSUDA, 1964; GRÖGER and SCHMAUDER, 1969).

Considering this modern view of secondary metabolism older attempts to describe the products as waste products ("flotsam on the metabolic beach") or end products of metabolism appear inadequate. The introduction to the book by GEISSMAN and CROUT and the treatise on secondary metabolism by LUCKNER (1967) represent our present knowledge in this respect. It is essential to avoid unclear restrictions of the term "secondary products" but rather follow the idea of BONNER who wrote about "byways of metabolism" when treating this part of biochemistry.

The term secondary products has also been used to mean that these compounds are derivatives of primary products. This is surely valid for most of the true alkaloids whose N- and C-atoms are usually derived from protein amino acids, the basic ring system being modified by methyl, acetyl, isoprene or other groups. This concept is also valid for cyanogenic compounds, tannins, etc. However, it is essential to notice that many secondary products (including some alkaloids) have their own biosynthetic pathways which are not directly related to primary metabolism. Furthermore, not all protein amino acids give rise to specific alkaloids. Amongst those which frequently do, however, are the aromatic amino acids phenylalanine, tyrosine, and tryptophane, the basic amino acids arginine, ornithine, and lysine and the branched chain amino acids leucine and valine. The reason for this preferential incorporation into alkaloids may be explained either by a special reactivity or by the slow rate of metabolism of these primary products.

It is not unusual for the same secondary product to be formed in different plants by two basically different pathways, e.g., nicotinic acid, which may arise from tryptophane degradation, or be synthesized from aspartic acid and glyceraldehyde. As our knowledge, particularly of the biosynthesis of secondary products in animals, increases, further such examples will be reported.

It must be kept in mind that only those secondary products are available for study which do not intoxicate the synthesizing organism. They may be nontoxic to the organism, or alternatively they may be sequestered into compartments surrounded by semipermeable membranes which isolate the toxins from the protoplasm, or they may be detoxicated by methylation or conjugation with carbohydrates, amino acids, or proteins. Hence, the accumulation of these products depends on the co-evolution of additional biochemical or structural systems for their safe storage. Co-evolution does not mean that both characters have evolved simultaneously but rather sequentially, as was pointed out by JERMY (1976). [3]

The close relationship between hormones and secondary products is evident, if we think of the need for nature "to invent mechanisms" for regulating the concentration of plant hormones which may be toxic at too high concentrations. Thus, hormones can be "detoxicated" by conjugation with carbohydrates, amino acids, or proteins. Such conjugated hormones at the same time are hormones ready for transport to other parts of the organism.

It must be assumed that the tremendous number of secondary products would be even larger, if all plants which acquired the ability to produce such compounds during evolution had survived (MOTHES, 1972, 1973; ZÄHNER, 1977). This, however, does not exclude the possibility that some of these compounds have acquired functions which aid the propagation of various species. This, the ecological aspect of the subject, has already been of interest to earlier representatives of classical

---

3 An interesting example of secondary product accumulation in special compartments is described by EISNER et al. (1971). In the defense glands of the grasshopper *Romalea microptera* in addition to a mixture of various phenolic compounds and quinones they detected for the first time a chlorinated aromatic compound, 2.5-dichlorophenol. "This compound which is repellent to ants and therefore defensively useful to the grasshopper, probably stems from herbicide or herbicide derivative ingested by the insect with its diet." The reason for this particular accumulation is to be looked for in the preformed characteristics of the grasshopper membranes which allow unidirectional flow of certain specific compounds. Such membranes existed long before 2.5-D was ever taken up with the nutrient plant material. They prevent 2.5-D from reaching all parts of the body and becoming toxic to the grasshopper itself

plant physiology and plant biochemistry. For most higher plants with colored flowers unpigmented varieties are described, the latter having undiminished viability compared with the original species. But if we assume that only white flowering species are growing on a meadow of the Alps, intraspecific pollination may be decisively impeded by a lack of color markers for the insects. A similar situation holds true for scent substances. Much more difficult to evaluate is the role of secondary products as repellents to predatory animals, especially insects. Though plants of the Brassicaceae species are protected by their mustard oil glycosides, these repellents act simultaneously as attractants to the white cabbage butterflies *Pieris brassicae* and *P. rapae* and to the diamondback moth *Plutella maculipennis*. For these insects mustard oil glycosides do not serve as nutrients, but as compounds signaling an unoccupied niche for egg deposition. Examples of the same compounds acting as repellents and attractants can also be given, however. Multiplicity and versatility in nature do not allow absolute resistance (FRAENKEL, 1959, 1969). We are far from being able to give a complete account of the role which secondary products play in the composition of larger plant societies. For example, precise knowledge of chemical allelopathy is just beginning to emerge. Yet we now know that phytoalexins can be produced by plants after being infected by a fungus, and these phytoalexins control the further growth of the parasite on its host plant (DEVERALL, 1977).

The multiple problems of ecological biochemistry which are fundamental to a deeper understanding of life as a whole, i.e., of life as a constantly interacting community of different species, are treated in several monographs and symposium proceedings, e.g., SONDHEIMER and SIMEONE, 1970; HARBORNE 1972, 1977; LUCKNER et al., 1976; WALLACE and MANSELL, 1976; MARINO-BETTOLO, 1977.

To define the role of secondary products within the framework of general metabolism the term excrete has been used several times. I cannot go into the details of the terms excrete and secrete used in animal physiology (BIEDERMANN, 1917; FREY-WYSSLING, 1935, 1970). Excretory products in plants are frequently formed in large excess. The rubber content of certain species may exceed 20% of the dry weight and the saponin content of seeds may be greater than 15%. This kind of "energy waste" is evidently restricted to photoautotrophic organisms.

Finally I want to point out that syntheses of secondary products in most instances require multistep reactions. The complexity of the whole process called secondary metabolism therefore becomes apparent if we take into account the numerous enzymes or multienzyme complexes involved, the regulated synthesis of these enzymes, and the necessity of their spatial organization. The chemical differentiation of cells or tissues allows an insight into the enormous chemical potential contained in the genetic material. The coordinate gene expression in secondary metabolism under the influence of intrinsic and extrinsic factors reflects a general phenomenon of the developmental biology. This is treated by LUCKNER in a separate chapter of this volume (cf. also NOVER and MOTHES, 1977; NOVER et al., 1978). As the occurrence of particular secondary products may be restricted to few species or even to a single chemical race, as they are sometimes formed at certain stages of the individual's development (so-called idiophase, BU'LOCK and BARR 1968, BU'LOCK et al., 1972), they are excellently suited to the study of the central problems of cell differentiation. The special importance of investigations of the regulation of secondary product formation in developmental biology derives from the fact that some of these compounds play an outstanding role as plant and animal hormones.

Acknowledgment. I thank Dr. L. Nover for the translation of the manuscript into English.

References

Biedermann, W.: Sekretion und Sekrete. Pflüger's Arch. Physiol. *167*, 1–110 (1917)
Bonner, J.: Plant biochemistry. p. 537. New York: Academic Press 1950
Bonner, J., Galston, A.W.: Principles of plant physiology. 1st ed. p. 449. San Francisco: Freeman 1952
Bonner, J., Varner, J.E.: Plant biochemistry. p. 1054 New York: Academic Press 1965
Bu'Lock, J.D., Barr, J.G.: A regulation mechanism linking tryptophan uptake and synthesis with ergot alkaloid synthesis in *Claviceps*. Lloydia *31*, 342–354 (1968)
Bu'Lock, J.D., Drake, D., Winstanley, D.J.: Specificity and transformation of the trisporic acid series of fungal sex hormones. Phytochemistry *11*, 2011–2018 (1972)
Butenandt, A.: Zur Biologie und Chemie der Sexualhormone. Naturwissenschaften *21*, 49–54 (1933)
Butenandt, A.: Wirkstoffe des Insektenreiches. Naturwissenschaften *46*, 461–471 (1959)
Butenandt, A., Jacobi, H.: Über die Darstellung eines krystallisierten pflanzlichen Tokokinins (Thelykinins) und seine Identifizierung mit dem d-Follikelhormon. Hoppe-Seyler's Z. Physiol. Chem. *218*, 104–112 (1933)
Butenandt, A., Karlson, P.: Über die Isolierung eines Metamorphose-Hormons der Insekten in kristallisierter Form. Z. Naturforsch. *9B*, 389–393 (1954)
Butenandt, A., Weidel, W., Becker, E.: α-Oxytryptophan als "Prokynurenin" in der zur Augenpigmentbildung führenden Reaktionskette bei Insekten. Naturwissenschaften *28*, 447–448 (1940)
Butenandt, A., Weidel, W., Weichert, R., von Derjugin, W.: Über Kynurenin. Hoppe-Seyler's Z. Physiol. Chem. *279*, 27–43 (1943)
Charney, W., Herzog, H.L.: Microbial transformations of steroids. New York: Academic Press 1967
Conn, E.E.: Biosynthesis of cyanogenic glycosides. Biochem. Soc. Symp. *38*, 287–302 (1973)
Czapek, F.: Biochemie der Pflanzen, 2. Aufl., 3. Bd. Jena: G. Fischer 1921
Deverall, B.J.: Defense mechanisms of plants. Cambridge University Press 1977
Eisner, Th., Hendry, L.B., Gekall, D.B., Meinwald, J.: 2,5-Dichlorophenol (from ingested herbicide?) in defensive secretion of grasshopper. Science *172*, 277–278 (1971)
Fraenkel, G.S.: The raison d'être of secondary plant substances. Science *129*, 1466–1470 (1959)
Fraenkel, G.S.: Evaluation of our thoughts on secondary plant substances. Entomol. Exp. Appl. *12*, 473–486 (1969)
Frey-Wyssling, A.: Die Stoffausscheidung der höheren Pflanze. p. 378. Berlin: Springer 1935
Frey-Wyssling, A.: Betrachtungen über die pflanzliche Stoffelimination. Ber. Schweiz. Bot. Ges. *80*, 454–466 (1970)
Geissman, T.A., Crout, D.H.G.: Organic chemistry of secondary plant metabolism. p. 592. San Francisco: Freeman, Cooper & Co. 1969
Gröger, D.: Fortschritte der Chemie und Biochemie der Mutterkornalkaloide. Fortschr. Chem. Forsch. *6*, 159–194 (1966)
Gröger, D., Schmauder, H.P.: Mikrobielle Umwandlung von Thebain. Experientia *25*, 95–96 (1969)
Harborne, J.B. (ed.): Phytochemical ecology. p. 277. London: Academic Press 1977a
Harborne, J.B.: Introduction to ecological biochemistry. p. 241. London: Academic Press 1977
Jermy, T. (ed.): The host-plant in relation to insect behaviour and reproduction. p. 322. Symp. Biol. Hung. *16*, New York, London: Plenum Press 1976
Karlson, P.: Insect hormones. In: Comprehensive biochemistry. Vol. XI, pp. 168–178. Amsterdam: Elsevier 1963
Karlson, P.: Animal hormones as secondary plant products and their ecological relevance. Nova Acta Leopold. Suppl. *7*, 423–432 (1976)

Karlson, P., Butenandt, A.: Pheromones (ectohormones) in insects. Annu. Rev. Entomol. *4*, 39–58 (1959)
Luckner, M.: Der Sekundärstoffwechsel in Pflanze und Tier. S. 360. Jena: G. Fischer 1967
Luckner, M., Mothes, K. (eds.): Secondary plant metabolism and co-evolution. Nova Acta Leopold. Suppl. *7*, S. 613. Halle: Deutsche Akademie der Naturforscher Leopoldina 1976
Luckner, M., Nover, L., Böhm, H.: Secondary metabolism and cell differentiation, pp.130 Berlin, Heidelberg, New York: Springer 1977
Marino-Bettolo, G.B. (ed.): Natural products and the protection of plants. Pontif. Acad. Sci. Scripta Varia *41*. Civ. Vat.: Pontif. Acad. Sci. (1977)
Mothes, K.: Über sekundäre Pflanzenstoffe. Abh. Sächs. Akad. Wiss. Math. Naturwiss. Kl. *52*, 1–29 (1972)
Mothes, K.: Pflanze und Tier. Ein Vergleich auf der Ebene des Sekundärstoffwechsels. Sitzungsber. Österr. Akad. Wiss. Math. Naturwiss. Kl. Sonderh. Abt. I. *181*, 1–37 (1973)
Mothes, K., Silber, A.: Über die Variabilität des Mutterkorns. Forsch. Fortschr. *28*, 101–104 (1954)
Müntz, K.: Cell specialization processes during biosynthesis and storage of proteins in plant seeds. In: Regulation of developmental processes in plants. Jena: G. Fischer 1978a
Müntz, K.: Biosynthese und Speicherung von Reserveproteinen während der Reifung pflanzlicher Samen. In: Zelldifferenzierung. Nover, L., Luckner, M., Parthier, B. (eds.) Jena: G. Fischer 1978b
Müntz, K., Scholz, G.: Speicherproteine und Proteinspeicherung in pflanzlichen Samen. Biol. Rundsch. *12*, 225–244 (1974)
Nover, L., Mothes, K. (eds): Cell differentiation in microorganisms, plants and animals. Leopold. Symp. Jena: G. Fischer 1977
Nover, L., Luckner, M., Parthier, B. (eds.): Zelldifferenzierung. S. 582. Jena: G. Fischer 1978
Paech, K.: Biologie und Physiologie der sekundären Pflanzenstoffe. S. 268. Berlin: Springer 1950
Pasteels, J.M., Daloze, D.: Cardiac glycosides in the defense secretion of chrysomelid beetles: evidence for their production by insects. Science *197*, 70–72 (1977)
Pfeffer, W.: Pflanzenphysiologie, 2. Aufl. I. Bd., S. 991. Leipzig: Engelmann 1897
Reichstein, T.: Cardelonide (herzwirksame Glykoside) als Abwehrstoffe bei Insekten. Naturwiss. Rundsch. *20*, 449–511 (1967)
Röbbelen, G.: Züchtung und Erzeugung von Qualitätsraps in Europa. Fette Seifen Anstrichm. *73*, 10–17 (1976)
Rothschild, M., Reichstein, T.: Some problems associated with the storage of cardiac glycosides by insects. Nova Acta Leopold. Suppl. *7*, 507–550 (1976)
Rüdiger, H.: Lectine, pflanzliche zuckerbindende Proteine. Naturwissenschaften *65*, 239–244 (1970)
Ruggieri, G.D.: Drugs from the sea. Science *194*, 491–497 (1976)
Sachs, J.: Vorlesungen über Pflanzenphysiologie. S. 203–222. Leipzig: Engelmann 1882
Schantz, E.J., Chazarossian, V.E., Schnoes, H.K., Strong, F.M., Springer, J.P., Pezzani, J.O., Clardy, J.: The structure of saxitoxin. J. Am. Chem. Soc. *97*, 1238–1239 (1975)
Schildknecht, H.: Die Wehrchemie von Land- und Wasserkäfer. Angew. Chem. *82*, 17–25 (1970)
Schnepf, E.: Sezernierende und exzernierende Zellen in Pflanzen. In: Grundlagen der Cytologie. Hirsch, G.C., Ruska, H., Sitte, P. (eds.), pp. 461–477. Jena: G. Fischer 1973
Schwarze, P.: Einführung zu: Encyclopedia Plant Physiol. Vol X p. 1–23. Berlin: Springer 1958
Sondheimer, E., Simeone, J.B.: Chemical ecology. New York: Academic Press 1970
Thies, W.: Der Einfluss der Chloroplasten auf die Bildung der ungesättigten Fettsäuren in reifenden Rapssamen. Fette Seifen Anstrichm. *73*, 710–715 (1971)
Tsuda, K.: Int. Assoc. Microbiol. Soc. Symp. *6*, 167–193 (1964)
Wallace, J.W., Mansel, R.L. (eds.): Biochemical interaction between plants and insects. Rec. Adv. Phytochem. *10*, 425 (1976)
Wigglesworth, V.B.: The principles of insect physiology. London: Chapman & Hall 1972
Witkop, B.: New directions in the chemistry of natural products: the organic chemist as a pathfinder for biochemistry and medicine. Experientia *27*, 1121–1138 (1971)
Zähner, H.: Einige Aspekte der Antibiotika-Forschung. Angew. Chem. *89*, 696–703 (1977)

2 The Possible Significance of Secondary Compounds in Plants

E.A. Bell

1 Introduction

Significance is used here to mean phylogenetic and ecological significance.

As the phylogenetic significance of various specific groups of secondary compounds are considered in detail elsewhere in this volume I have chosen to emphasise here that the phylogenetic significance of a particular group of compounds may be very different at different taxonomic levels.

The concept of an ecological role for secondary compounds is not accepted by all scientists. In this chapter however I have presented evidence and arguments which strongly suggest that some at least of the secondary compounds do confer a selectionary advantage on the species which synthesise them.

The reader is provided with an introduction to the literature of both subjects.

2 Phylogenetic Significance

Interest in the possible use of chemistry and biochemistry in plant systematics has increased rapidly during the past twenty years; this increase of interest has been due largely to the development of improved analytical methods. These have made it possible to analyse large numbers of plant extracts rapidly and with relative ease. These methods normally involve the extraction of the constituents with suitable solvents, the separation of the extracted components from one another by chromatography or electrophoresis, and the subsequent identification of "unknowns" using a combination of methods such as elementary analysis, UV, IR, NMR, and mass spectrometry. The particular procedures chosen are determined by the nature of the compounds under investigation. With volatile components, gas/liquid chromatography combined with mass spectrometry has proved particularly valuable. Liquid/liquid chromatography on papers or thin layers is perhaps the most generally applicable separation technique however, while ion-exchange chromatography and methods using high and low voltage electrophoresis have been used with success for separating ionisable molecules. For further information concerning these methods the reader is referred to Harborne (1973).

The great advantages which chromatography and electrophoresis offer to the chemotaxonomist are speed and the possibility of determining many different constituents of a complex mixture at the same time. Using 2D paper chromatography it is possible, for example, not only to resolve and identify from their R_f values known constituents of a complex mixture, but also to recognise the presence of unknown components; indeed many new natural products have first been detected as spots occupying unusual positions after developing chromatograms with reagents appropriate to the compounds under investigation. Using gas/liquid chromatography an unusual volatile component may be recognised as an unidentified peak amongst familiar peaks and using electrophoresis an unusual ionised molecule may be recognised as an unusual spot on an electrophoresis paper.

The use of secondary compounds has clear advantages over the use of primary compounds in establishing phylogenetic relationships because differences in the complement of secondary compounds are qualitative differences, whereas differences in the concentrations of primary compounds are quantitative differences, and these are subject to environmental as well as to genetic control. The existence of a common pattern of secondary compounds may indeed provide much clearer evidence of common ancestry than morphological similarities attributable either to common ancestry or to convergent evolution. Provision of qualitative as well as of quantitative evidence concerning differences and similarities between taxa is indeed the prime reason for using secondary compounds in chemosystematic studies. The accumulation of a particular concentration of a primary metabolite in a species may of course be a genetically controlled characteristic of that species. The difficulty arises, however, in deciding whether this really is the case, or whether the particular concentration (high or low) is the result of growing conditions or the availability of nutrients. In some instances the accumulation of a primary metabolite may be so striking and consistent that there can be little doubt of genetic control, but such instances are exceptional. The presence of an unusual secondary compound, however, clearly indicates that the species synthesising it possess an unusual biosynthetic pathway and the unusual gene or combination of genes which control that pathway. This is true whether the unusual compound is present in high or low concentrations.

While the foundations of chemotaxonomy can be traced back to early pharmacological studies in which plants with similar medicinal properties were classed together, chemotaxonomy as we know it today is based on more specific knowledge of the actual constituents found in plants. This modern approach to chemotaxonomy is the subject of a number of volumes which range in content from the detailed compilations of HEGNAUER (1962–1969) and GIBBS (1974) to the more general texts of ALSTON and TURNER (1963a) and SMITH (1976). A number of multi-author collections covering different aspects of chemotaxonomy are also available and these include volumes edited by LEONE (1964), SWAIN (1963, 1966), HAWKES (1968), HARBORNE (1970), SWAIN et al. (1971) and BENDZ and SANTESSON (1974).

In the following chapters the taxonomic significance of a number of different groups of secondary compounds has been reviewed. I shall not attempt to review these reviews, but rather emphasise that different groups of secondary compounds may have systematic significance at different taxonomic levels, as also may individual members of the same group.

At the highest taxonomic level BATE-SMITH (1962) has drawn attention to a wide correlation between the presence of leucoanthocyanins (I) and the woody habit in plants. This correlation, although not perfect, provides evidence of a chemical difference between woody and herbaceous species although few taxonomists other than HUTCHINSON (1959) have regarded such a division as very fundamental.

At the familial level we find that the anthocyanin pigments (II) are replaced by betalain pigments (III) in most families of the Centrospermae, and here we have an example of chemical evidence reinforcing the accepted views of the classical taxonomist while at the same time clarifying, as in the case of the Didiereaceae, relationships which were not previously clear (MABRY, Chap II, this vol).

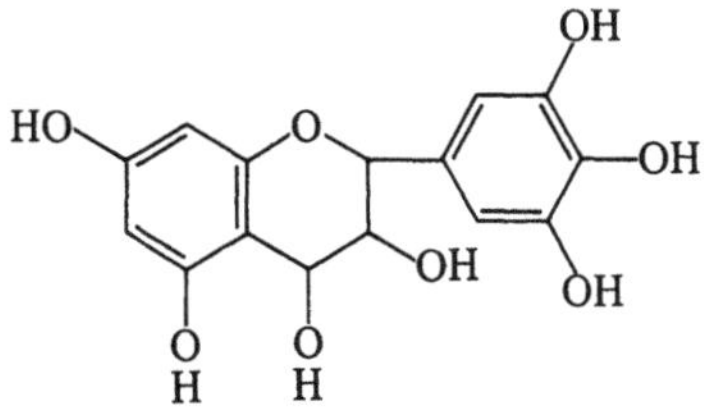

(I)

Leuco–delphinidin: a leucoanthocyanidin

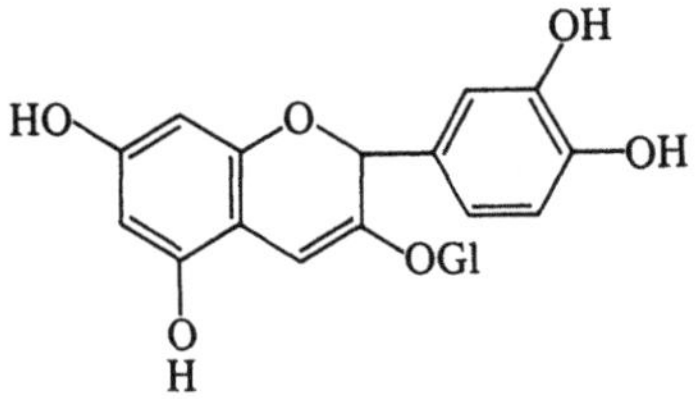

(II)

Cyanidin–3–glucoside: an anthocyanin

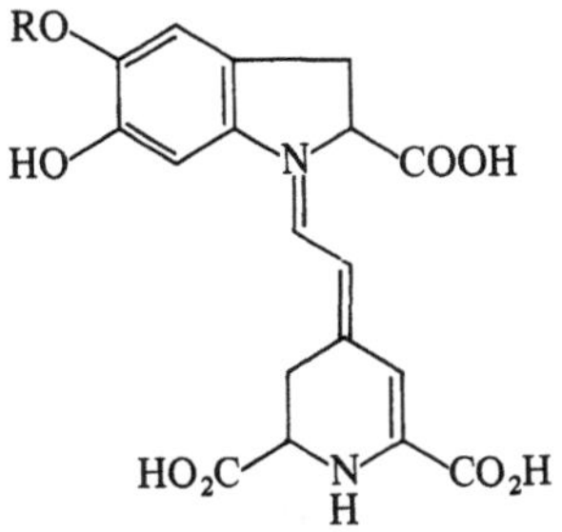

(III)

Betanin: a betalain pigment
R = β–D–glucosyl

Within a single family the non-protein amino acid canavanine (IV) can be cited as an example of a secondary compound which has provided evidence of phylogenetic relationships at the sub-familial and tribal levels. This amino acid is only known in species of one of three sub-families of the Leguminosae – the Papilionoideae. Even within this sub-family canavanine is not found in all genera or in all species. Surveying analytical data from 1200 species representing 240 genera of the Papilionoideae BELL et al. (1978) have concluded that this amino acid probably arose in forebears of the tribe Tephrosieae, and because it conferred a selective advantage on the species which contained it, it has persisted in most of the more advanced tribes derived from the Tephrosioid stock. This conclusion implies that the Australian "Podalyrieae", the Bossiaeeae, and the genus *Hypocalyptus* (which is presently placed in the Liparieae) are derived from the Tephrosioid stock and not from forebears of the Sophoreae as previously supposed.

At the generic level and the subgeneric level non-protein amino acids may also be significant in establishing phylogenetic relationships. The seeds of *Mucuna* species and the seeds of no other species analysed are characterised by the accumulation of high concentrations of L-3,4-dihydroxyphenylalanine (V) (L-DOPA). Similarly high concentrations of 5-hydroxy-L-tryptophan (VI) are restricted to the seeds of *Griffonia* (Bandeiraea) species (BELL et al., 1976) and high concentrations of azetidine-2-carboxylic acid (VII) to seeds of *Bussea* (WATSON and FOWDEN, 1973; EVANS and BELL, 1978). Non-protein amino acids may also however be of significance in establishing discrete biochemically defined sub-genera. Examples of this are found in the legume genera *Lathyrus* and *Vicia* (BELL, 1971) and *Acacia* (SENEVIRATNE and FOWDEN, 1968; EVANS et al. 1977).

$H_2N-C(=NH)-NH-O-CH_2-CH_2-CH(NH_2)-CO_2H$

Canavanine
(IV)

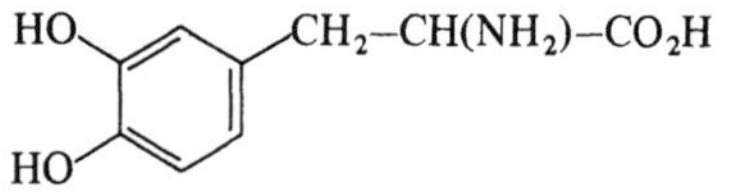

3,4–Dihydroxyphenylalanine
(V)

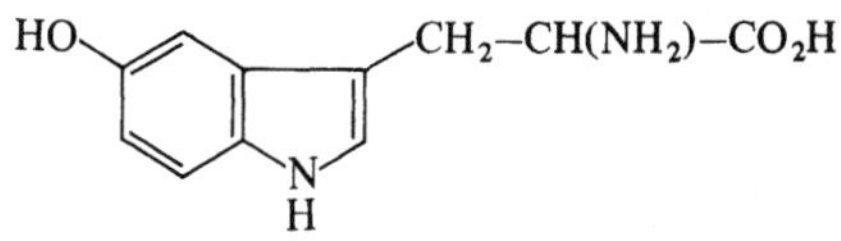

5–Hydroxytryptophan
(VI)

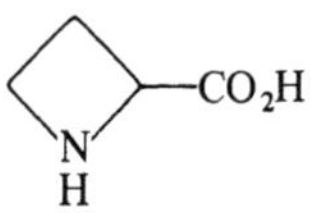

Azetidine–2–carboxylic acid
(VII)

A good example of the use of secondary compounds in the study of infraspecific variability is provided by MABRY (1970) who has established the existence of several distinct chemical races of *Ambrosia psilostachya* which accumulate different sesquiterpene lactones. Two of these chemical races have in fact been found in nearby habitats in Texas. One occupies the mainland gulf coast area, while the other occupies the off-shore chain of islands which lie parallel to, but isolated from, that coast. The mainland race is characterised by the accumulation of sesquiterpene monolactones (VIII) and the island race by the accumulation of sesquiterpene dilactones (IX). MABRY (1970) has suggested that the islands have been colonised from the south to the north, the dilactone race coming from mainland populations in the very south of Texas or in Mexico.

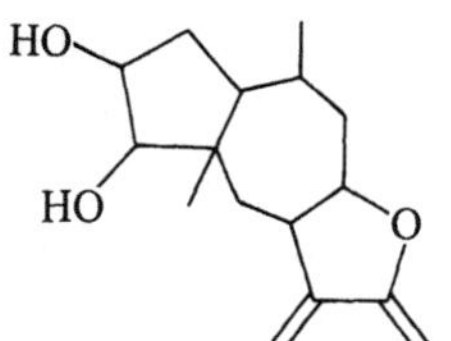

Cumanin, a monolactone
(VIII)

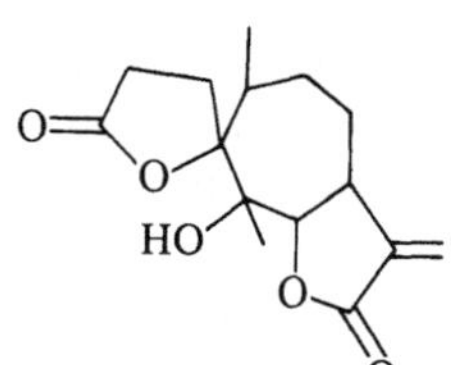

Psilostachin: a dilactone
(IX)

The distribution of secondary compounds may also be of value in identifying hybridisation among sympatric populations of different wild species. One of the best examples of this use of secondary compounds is also one of the earliest. It is that in which ALSTON and TURNER (1963b) used the distinctive flavonoid patterns which characterise the flower colours of different species of *Baptisia* to analyse interbreeding populations of 2, 3, and 4 different species growing together on a number of different sites.

From the few examples presented, it is clear that all secondary compounds are not of equal value in determining relationships at any particular taxonomic level and the choice of appropriate secondary compounds is of paramount importance when undertaking chemotaxonomic studies.

3 Ecological Significance

There is a growing interest in the chemical and biochemical aspects of ecology and these have been the subject of a number of publications in recent years. These include books written by FLORKIN and SCHOFFENIELS (1969), SONDHEIMER and SIMEONE (1970) and HARBORNE (1977a), multi-author volumes edited by HARBORNE (1972), GILBERT and RAVEN (1975), WALLACE and MANSELL (1976), LUCKNER et al. (1976), FRIEND and THRELFALL (1976), MARINI-BETTOLO (1977), HARBORNE (1978), ROSENTHAL and JANZEN (1979) as well as shorter review articles by WHITTAKER and FEENY (1971), LEVIN (1976), SIEGLER (1977), HARBORNE (1977b) and SWAIN (1977).

In this volume we are dealing with secondary plant products, and before going further it is perhaps right to ask whether there is any justification for supposing that these compounds have an ecological role at all. In practice we know so little about most of the secondary compounds that a general answer to this question is not possible. There is however no reason to suppose that the biochemistry of a plant is any less subject to the selectionary pressures exercised by its environment than is the plant's physiology or morphology, nor to suppose that the biochemically inefficient plant is more likely to survive competition than the morphologically inefficient one. I suggest that during the course of evolution millions of secondary products have been synthesised from time to time by different species of plants and when the presence of a particular secondary product has conferred a selectionary advantage on the plant containing it, then the chances of survival of that plant, its offspring, and the secondary product itself will have been enhanced. The great majority of secondary products thrown up during the course of evolution probably proved to be of no advantage to the plants which synthesised them and neither the plants nor the compounds are likely to have survived for our study today. Such random syntheses of new compounds by plants have however provided the variability which is the prerequisite for the operation of natural selection and the advancement of all living organisms.

The environment itself is far from static however, and the living and non-living factors which exercise selectionary pressures on a plant species vary themselves from place to place and from period to period. At any given time individual members of a particular species may be growing in various different habitats and the selectionary pressures operating in one habitat may differ markedly from those operating in another, and these differences may lead to evolutionary divergence within the species. Conditions and selectionary pressures may also alter with time in the same geographical location. This may lead to a situation in which a plant displays certain chemical characteristics which it acquired as a result of environmental pressures which no longer exist, just as certain anatomical features may reflect past rather than present needs. It is also possible that a secondary product may have survived in a plant, not because it conferred an advantage on that plant, but rather because the gene or genes controlling its synthesis were closely associated on the same chromosome with genes determining another character which proved to be of major selective advantage. This advantage could very well have outweighed the disadvantage of synthesising a valueless molecule. BELL (1978) has suggested that plants synthesise a greater array of secondary compounds than animals because plants cannot rely on physical mobility to escape their predators and have therefore evolved a chemical defence against such predators. It is not true, however, that all plants, even higher plants, lack mobility, and although higher plants cannot move bodily from one place to another we are all familiar with the remarkable response of *Mimosa pudica* which closes its leaflets and lowers its leaves in response to physical stimuli. An advantage of this reaction to *M. pudica* may well be in reducing the plant's availability to herbivores and in *M. pudica* we may have an example of a secondary product, a plant hormone, facilitating physical movement rather than serving as a substitute for it (HESSE et al., 1957). SCHILDKNECHT et al. (1978) have identified this hormone as either the D-apio-D-furanosyl or the L-apio-L-furanosyl derivative of gentisic acid (X). SCHILDKNECHT and IYENGAR (1975) have also shown that a compound which inhibits the recovery of closed leaves occurs in *M. pudica* and this they have identified as D-pinitol. (XI).

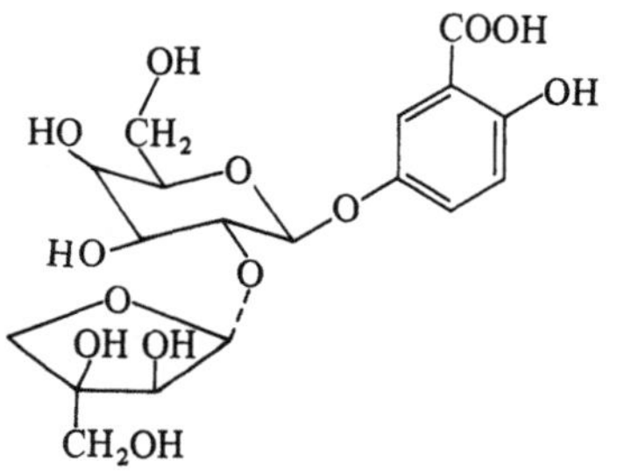

Gentisic acid–5–O–β–
L–apio–L–furanoside:
one of the 2 possible configurations
of Mimosa leaf Movement–
Factor I (M–LMFI)
(X)

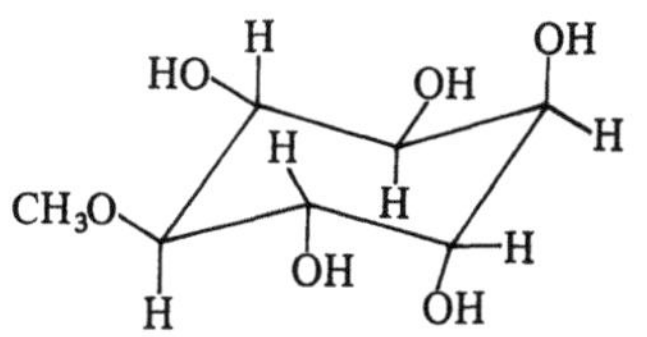

D–Pinitol
(XI)

More usually the role of secondary compounds in defence is related to their irritant, toxic, or unpalatable characteristics. In Europe one of the most obvious examples of a plant with the ability to deter human predators is *Urtica dioica*, the stinging nettle. The nettle in fact provides a good example of a plant in which the evolution of a defence mechanism has involved both morphological and chemical adaptation. The stinging cells of this species have brittle ends which break when touched and the sharp, now open, cells penetrate the skin, acting as hypodermic syringes from which the irritant mixture is injected. The composition of the nettle poison has not been entirely elucidated, but histamine (XII), acetylcholine (XIII), and 5-hydroxytryptamine (XIV) are among the physiologically active compounds reported to be present (SCHILDKNECHT, 1977). It is probable, however, that the extreme pain induced by nettle stings is contributed to by components other than these. 5-Hydroxytryptamine is also held to be responsible, in part at any rate, for the irritant properties of the trichomes which are found on the pods of *Mucuna pruriens*, these trichomes produce red weals on the skin when touched (BOWDEN et al., 1954).

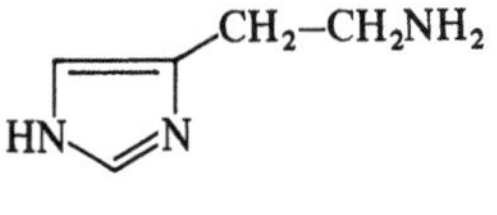

Histamine
(XII)

$$CH_3CO{-}O{-}CH_2{-}CH_2{-}\overset{+}{N}(CH_3)_3\ \overset{-}{O}H$$

Acetylcholine
(XIII)

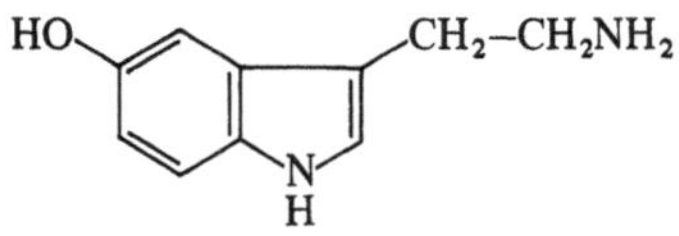

5-Hydroxytryptamine
(XIV)

As well as obvious irritants like those of the stinging nettle, plants synthesise other types of compounds which can provide protection against predators and/or pathogens. Phenolic compounds, for example, are responsible for the resistance of *Phaseolus lunatus* (lima bean) to *Phytophthora phaseoli*, the organism causing downy mildew (VALENTA and SISLER, 1962), while tannins provide resistance to *Verticillium* wilt in a number of species (BELL, 1972). An increase in the tannin content of oak leaves is also responsible for the desertion of these leaves by the larvae of the winter moth as the leaves mature in June (FEENY, 1970).

In certain *Solanum* species alkaloids deter the Colorado beetle (STURCHOW and LOW, 1961) and many non-protein amino acids have been found experimentally to inhibit feeding in the grasshoppers, *Locusta, Chortoicetes* and *Schistocerca* (NAVON and BERNAYS, 1978).

Perhaps the clearest evidence that some at least of the secondary compounds synthesised by plants have a protective role is to be found in the responses which these compounds have elicited in species co-evolving with the plants. The leaves and seeds of *Cycas circinalis*, for example, contain glycosides of the carcinogen methylazoxymethanol (XV), the toxic aglycone is liberated by β-glucosidase en-

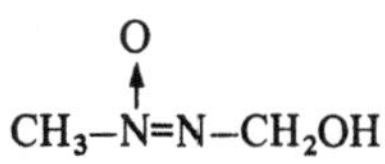

Methylazoxymethanol
(XV)

zymes which occur in the gut of insects and also in the microflora of the mammalian intestine. The leaves of this species are nevertheless the food of the larvae of the moth *Seirarctia echo*. TEAS (1967) has shown that these larvae, although equipped like other insects with a β-glucosidase in the gut wall, are nevertheless able to reglucosylate the aglycone and sequester the toxin as cycasin (methylazoxy-methanol-β-D-glucoside). Similarly, ROSENTHAL et al. (1976) have shown that the larvae of *Caryedes brasiliensis* which eat the canavanine-rich seeds of *Dioclea megacarpa* possess an arginyl-tRNA synthetase which is able to discriminate against canavine and prevent the non-protein amino acid from being incorporated into the insect protein. The same larvae are also able to metabolise canavanine (ROSENTHAL et al., 1977).

The actual use by insects of toxins obtained from plants in their own protection affords further evidence of the complexity of co-evolution and of the pressures which have led to the selection of plants containing such compounds. Protection from birds is afforded to the adult monarch butterfly (*Danaus plexippus*) by cardiac glycosides (XVI) obtained from milkweeds (*Asclepias* spp.) by its larvae and to

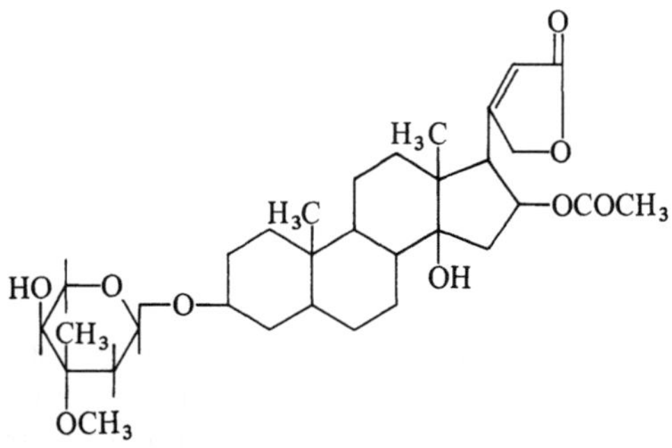

Oleandrin: a cardiac glycoside
(XVI)

the cinnabar moth (*Tyria jacobaeae*) by pyrrolizidine alkaloids obtained by the larvae of that moth from the ragwort (*Senecio jacobea*). These and other examples are discussed by ROTHSCHILD (1972). The presence in plants of compounds which show hormonal activity in insects (WILLIAMS, 1970) suggests yet another way in which plant biochemistry may have responded to selectionary pressures exerted by herbivores.

In addition to their involvement in plant–animal relations, secondary compounds are also concerned in plant–plant relationships. A classical example is that of juglone (XVII) which is liberated as an inactive precursor by the walnut tree. Following oxidation to the active form, this compound inhibits the growth of many other species of plant. MÜLLER and CHOU (1970) have also shown that the final dominance of shrubs in the fire-cycle succession of the Californian chapparal is controlled by the production of secondary compounds by these shrubs which suppress the growth of herbaceous species.

More recently WILSON and BELL (1978) have demonstrated that non-protein amino acids and β-aminopropionitrile produced by particular legume species may inhibit seedling growth in closely related species which do not synthesise these compounds.

The secondary compounds discussed so far are all synthesised by normal healthy plants. The synthesis of a second group of protective secondary compounds, the phytoalexins (XVIII), only takes place under a stimulus such as the invasion of the plant tissue by a fungus. Different groups of plants produce different phytoalexins of very different chemical structures. The majority of phytoalexins produced by species of the Leguminosae are, for example, pterocarpans related to the isoflavonoids, whereas most of the phytoalexins from the Solanaceae are terpenoids. Exceptions to such generalisations are known, however. Wyerone, the phytoalexin of *Vicia faba* (Leguminosae) is the ester of a highly unsaturated straight chain fatty acid. The nature of the phytoalexin response in a plant is characteristic of the species of plant, rather than the inducing pathogen, nevertheless the relative quantities of different phytoalexins produced by a single plant species in response to different pathogens may vary. Little is known as yet of the mode of action of phytoalexins, but the diversity of chemical structures found among these compounds suggests that different phytoalexins probably affect invading pathogens in different ways. For more detailed information on phytoalexins the reader is referred to the reviews by DEVERALL (1972), INGHAM (1972) and CRUICKSHANK (1977). Phytoalexins are also extensively treated in Volume 4 of this Encyclopedia by KUĆ (1976).

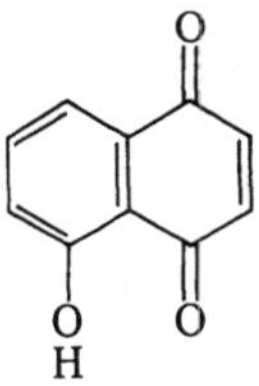

Juglone
(XVII)

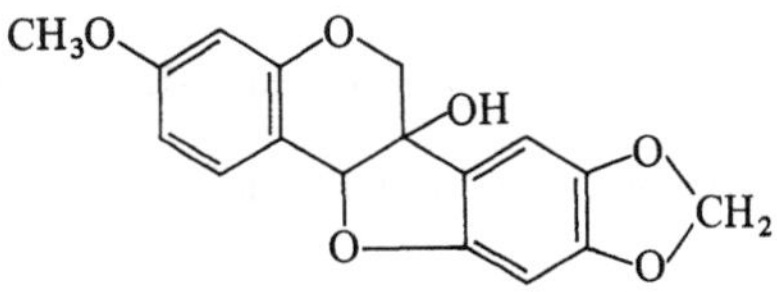

Pisatin: a phytoalexin
(XVIII)

While much work remains to be done in the field of biochemical ecology the examples given here suggest that some at least of the secondary compounds produced by plants may be of prime importance either in protecting plants against attack by micro-organisms and/or animals, or possibly by increasing the ability of one plant species to compete with another in a given habitat.

References

Alston, R.E., Turner, B.L.: Biochemical systematics. New Jersey: Prentice-Hall 1963a

Alston, R.E., Turner, B.L.: Natural hybridisation among four species of *Baptisia* (Leguminosae). Am. J. Bot. *50*, 159–173 (1963b)

Bate-Smith, E.C.: Chromatography and taxonomy in the Rosaceae with special reference to Potentilla and Prunus. J. Linn. Soc. London Bot. *58*, 39–64 (1961)

Bate-Smith, E.C.: The phenolic constituents of plants and their taxonomic significance. Bot. J. Linn. Soc. *58*, 95–173 (1962)

Bell, A.A.: Nature of disease resistance. In: Verticillium wilt of cotton. pp. 34–37. Washington, D.C.: U.S.D.A. Agric. Res. Serv. 1972

Bell, E.A.: Comparative biochemistry of non-protein amino acids. In: Chemotaxonomy of the leguminosae. Harborne, J.B., Boulter, D., Turner, B.L., (eds.), pp. 179–206. London: Academic Press 1971

Bell, E.A.: Toxins in seeds. In: Biochemical aspects of plant-animal co-evolution. Harborne, J.B., (eds.). pp. 143–161. London: Academic Press 1978
Bell, E.A., Fellows, L.E., Qureshi, M.Y.: 5-Hydroxy-L-tryptophan, taxonomic marker and chemical defense in *Griffonia*. Phytochemistry *15*, 823 (1976)
Bell, E.A., Lackey, J., Polhill, R.M.: Systematic significance of canavanine in the Papilionoideae (Faboideae). Biochem. Syst. Ecol. *6*, 201–212 (1978)
Bendz, G., Santesson, J. (eds.): Chemistry in botanical classification. New York: Academic Press 1974
Bowden, K., Brown, B.G., Batty, J.E.: 5-Hydroxytryptamine: its occurrence in cowhage. Nature (London) *174*, 925–926 (1954)
Cruickshank, I.A.M.: A review of the role of phytoalexins in disease resistance mechanisms. In: Natural products and the protection of plants. Scripta Varia. Marini-Bettolo, G.B. (ed.), Vol. 41, pp. 509–561. Vatican City: Pontifical Academy of Science 1977
Deverall, B.J.: Phytoalexins. In: Phytochemical ecology. Harborne, J.B. (ed.), pp. 217–233. London: Academic Press 1972
Evans, C.S., Bell, E.A.: "Uncommon" amino acids in seeds of 64 species of Caesalpinieae. Phytochemistry *17*, 1127–1129 (1978)
Evans, C.S., Qureshi, M.Y., Bell, E.A.: Free amino acids in seeds of *Acacia* species. Phytochemistry *16*, 565–570 (1977)
Feeny, P.P.: Seasonal changes in oak leaf tannins and nutrients as a cause of spring feeding by winter moth caterpillars. Ecology *51*, 565–581 (1970)
Florkin, M., Schoffeniels, E.: Molecular approaches to ecology. New York: Academic Press 1969
Friend, J., Threlfall, D.R. (eds.): Biochemical aspects of plant-parasite relationships. London: Academic Press 1976
Gibbs, R.D.: Chemotaxonomy of flowering plants. Montreal: McGill-Queen's University Press 1974
Gilbert, L.E., Raven, P.H. (eds.): Coevolution of animals and plants. Austin-London: University of Texas Press 1975
Harborne, J.B. (ed.): Phytochemical phylogeny. London: Academic Press 1970
Harborne, J.B. (ed.): Phytochemical ecology. London: Academic Press 1972
Harborne, J.B.: Phytochemical methods. London: Chapman and Hall 1973
Harborne, J.B.: Introduction to ecological biochemistry. London: Academic Press 1977a
Harborne, J.B.: Chemosystematics and evolution. Pure Appl. Chem. *49*, 1403–1421 (1977b)
Harborne, J.B. (ed.): Biochemical aspects of plant and animal co-evolution. London: Academic Press 1978
Hawkes, J.G. (ed.): Chemotaxonomy and serotaxonomy. London: Academic Press 1968
Hegnauer, R.: Chemotaxonomie der Pflanzen. Vols. *1–5*, Basle: Birkhauser 1962–1969
Hesse, G., Banerjee, B., Schildknecht, H.: Die Reizbewegungen der Mimosoideen und ihre Hormone. Experientia *13/1* 13–19 (1957)
Hutchinson, J.: The families of flowering plants. Oxford: Clarendon Press 1959
Ingham, J.L.: Phytoalexins and other natural products as factors in plant disease resistance. Bot. Rev. *38*, 343–424 (1972)
Kuć, J.A.: Phytoalexins. In: Physiological plant pathology. Heitefuss, R., Williams, P.H. (eds.), pp. 632–652. Encyclopedia of Plant Physiol. Vol. 4. Berlin, Heidelberg, New York: Springer 1976
Leone, C.A. (ed.): Taxonomic biochemistry and serology. New York: Ronald Press 1964
Levin, D.A.: The chemical defenses of plants to pathogens and herbivores. Annu. Rev. Ecol. Syst. *7*, 121–159 (1976)
Luckner, M., Mothes, K., Nover, L. (eds.): Secondary metabolism and coevolution. Nova Acta Leopold. Suppl. No. 7. Leipzig: Barth 1976
Mabry, T.J.: Infraspecific variation of sesquiterpene lactones in *Ambrosia* (Compositae). Applications at the population level. In: Phytochemical phylogeny. Harborne, J.B. (ed.), pp. 269–300. London: Academic Press 1970
Marini-Bettolo, G.B. (ed.): Natural products and the protection of plants. Scripta Varia. Vol. *41*, Vatican City Pontifical Academy of Science 1977
Müller, C.H., Chou, C.H.: Phytotoxins: An ecological phase of phytochemistry. In: Phytochemical ecology. Harborne, J.B. (ed.), pp. 201–216. London: Academic Press 1970

Navon, A., Bernays, E.A.: Inhibition of feeding in acridids by non-protein amino acids. Comp. Biochem. Physiol. *59A*, 161–164 (1978)
Rosenthal, G.A., Janzen, D.H. (eds.): Interrelationship of secondary plant constituents with herbivores. New York: Academic Press (in press)
Rosenthal, G.A., Dahlman, D.L., Janzen, D.H.: A novel means for dealing with L-canavanine, a toxic metabolite. Science *192*, 256–258 (1976)
Rosenthal, G.A., Janzen, D.H., Dahlman, D.L.: Degradation and detoxification of canavanine by a specialised seed predator. Science *196*, 658–660 (1977)
Rothschild, M.: Some observations on the relationships between plants, toxic insects and birds. In: Phytochemical ecology. Harborne, J.B. (ed.), pp. 1–12. London: Academic Press 1972
Schildknecht, H.: Protective substances in arthropods and plants. In: Natural products and the protection of plants. Scripta Varia. Marini-Bettolo, G.B. (ed.), Vol. *41*, pp. 59–99. Vatican City: Pontifical Academy of Science 1977
Schildknecht, H., Iyengar, D.S.P.: On the biological chemistry of Mimosaceae. I. D-Pinitol from *Mimosa pudica* L. Naturwissenschaften *62*, 533 (1975)
Schildknecht, H., Tausher, B., Moeschler, H., Edelmann, J.: Detection and structure elucidation of leaf movement factors from Mimosaceae. Proc. 11th Int. Symp. Chem. Nat. Prods. part I. Vol. *4*, pp. 97–111. Sofia: Bulgarian Acad. Sci. 1978
Seigler, D.S.: Primary roles for secondary compounds. Biochem. Syst. Ecol. *5*, 195–199 (1977)
Seneviratne, A.S., Fowden, L.: Amino acids of the genus *Acacia*. Phytochemistry *7*, 1039–1045 (1968)
Sondheimer, E., Simeone, J.B.: Chemical ecology. New York: Academic Press 1970
Smith, P.M.: The chemotaxonomy of plants. London: Arnold 1976
Sturchow, B., Low, I.: Die Wirkung einiger Solanum-alkaloid-glykoside auf dem Kartoffelkäfer, *Lepotinotarsa decemlineata* Say. Entomol. Exp. Appl. *4*, 133–142 (1961)
Swain, T. (ed.): Chemical plant taxonomy. London: Academic Press 1963
Swain, T. (ed.): Comparative phytochemistry. London: Academic Press 1966
Swain, T.: Secondary compounds as protective agents. Annu. Rev. Plant Physiol. *28*, 479–501 (1977)
Swain, T., Boulter, D., Turner, B.L. (eds.): Chemotaxonomy of the Leguminosae. London: Academic Press 1971
Teas, H.J.: Cycasin synthesis in *Seirarctia echo* (Lepidoptera) larvae fed methylazoxymethanol. Biochem. Biophys. Res. Commun. *26*, 686–690 (1967)
Valenta, J.R., Sisler, H.D.: Evidence for the chemical basis of resistance of lima bean plants to downy mildew. Phytopathology *52*, 1020–1037 (1962)
Wallace, J.W., Mansell, R.L. (eds.): Biochemical interaction between plants and insects – recent advances in phytochemistry. Vol. 10. New York: Plenum Press 1976
Watson, R., Fowden, L.: Amino acids of *Caesalpinia tinctoria* and some allied species. Phytochemistry *12*, 617–622 (1973)
Whittaker, R.H., Feeny, P.P.: Allelochemics: chemical interactions between species. Science *171*, 757–770 (1971)
Williams, C.M.: Hormonal interactions between plants and insects. In: Chemical ecology. Sondheimer, E., Simeone, J.B. (eds.), pp. 103–132. New York: Academic Press 1970
Wilson, M.F., Bell, E.A.: Amino acids and β-aminopropionitrile as inhibitors of seed germination and growth. Phytochemistry *17*, 403–405 (1978)

3 Expression and Control of Secondary Metabolism

M. Luckner

1 Introduction: Expression of Secondary Metabolism – a Feature of Cell Specialization

One of the outstanding characteristics of secondary metabolism is its dependence on the developmental stages of the producer organisms. In unicellular organisms, e.g., bacteria, the cells after a phase of rapid division may pass to a state of specialization at which secondary products are formed. In multicellular organisms formation of secondary products is expressed as a specific feature of certain organs or tissues during restricted periods of their development and specialization. This phase dependence was shown in several instances to be due to the phase-dependent formation of the enzymes synthesizing secondary products, demonstrating that expression of secondary metabolism is the result of a differentiation process (Luckner, 1971; Luckner et al., 1977).

Differentiation is one of the basic features of metabolic regulation in all living beings (cf. Nover et al., 1978). The proteins formed as the result of the manifold differentiation processes can be classified with respect to their biological significance for and function within the producing organism as given in Table 1. In accordance with this scheme secondary metabolism may be defined as biosynthesis, transformation, and degradation of endogenously produced compounds by specialization proteins. Formation of secondary products, like other properties of specialized cells, is without direct significance for the synthesizing cell itself, but may be of importance for the producer organism as a whole (Luckner et al., 1977). The following discussion is based on that definition because it focuses attention on the general regulatory principles by which the formation of the different types of secondary products is controlled.

2 Regulation of Enzyme Amount in Secondary Metabolism

The basic regulatory principle in the expression of secondary product formation is the synthesis of secondary metabolic enzymes. Table 2 presents a selection of experimental results demonstrating that in microorganisms and higher plants the enzymatic activities of secondary metabolism appear before or during the phase of secondary products synthesis, and that expression depends on the de novo synthesis of RNA and proteins. However, of the large number of methods which may show the necessity of formation of RNA and proteins (cf. Luckner et al.,

Table 1. Proteins as products of differentiation processes (LUCKNER et al., 1977; NOVER et al., 1978)

A. Proteins of primary metabolism	B. Specialization proteins
Proteins which themselves or the products of which are typical for the cells of all organisms or at least large groups of organisms, and which are of direct importance for the existence and reproduction of the cells themselves. Formation and function of these proteins is of intracellular importance	Proteins that are found in some cells of certain organisms only and which themselves or the products of which are frequently without direct importance for the cell producing them. Formation and function of these proteins may have a vital function for the organism as the whole
1. Enzymes	1. Enzymes
Enzymes involved in gene expression, enzymes of nucleic acid and amino acid metabolism, enzymes of energy supply, of synthesis and degradation of fatty acids, etc.	Enzymes that enable the cell to perform special functions, e.g., enzymes of secondary metabolism, enzymes with a function in morphogenesis, special catabolic enzymes that degrade drugs, special nutrients, etc.
2. Nonenzymatic proteins	2. Nonenzymatic proteins
Structural proteins of the cell membrane and the protoplasm, ribosomal structural proteins, regulatory proteins, e.g., histones, acidic proteins of chromatin, etc.	Proteins that enable the cell to perform special functions within the organism, e.g., hemoglobin, muscle proteins, storage proteins, cilial proteins and exoproteins such as collagen, blood plasma proteins, antibodies, milk proteins, and proteohormones, structural and regulatory proteins required for the specialized function of the cell, etc.

1977) in the investigation of secondary metabolism only the following have been applied so far:

a) The comparison of the rate of synthesis of secondary products in vivo with the activities of the relevant enzymes measured in vitro.
b) The examination of the influence of inhibitors of gene expression.
c) The use of radioactive labeling in the study of enzyme synthesis.
d) The study of mRNA translation in vitro.

Inhibitors of gene expression have been used most frequently. Application of these compounds offers the opportunity to obtain data on regulatory details and on the dynamics of the total process of gene expression with moderate experimental expense. A prerequisite for the correct interpretation of results, however, is a basic knowledge of the inhibitor effects on the organism under investigation, in order to distinguish between primary effects on gene expression and secondary effects frequently evoked by interference with other metabolic processes. In *Penicillium cyclopium*, for instance, in which the influence of inhibitors on the expression of alkaloid metabolism has been extensively investigated (NOVER and MÜLLER, 1975; EL KOUSY et al., 1975; NOVER and LUCKNER, 1976), of many compounds tested only cycloheximide and 5-fluorouracil were active, and under given experimental conditions were without side effects.

Table 2. Selected experimental systems in which differential gene expression in secondary metabolism was examined (LUCKNER et al., 1977)

Organism or system	Secondary products whose synthesis has been investigated	Methods applied[a]	References
Bacteria			
Bacillus sp.	Peptide antibiotics	A	KURAHASHI, 1974
		B	WEINBERG and TONNIS, 1966, 1967
Bacillus megaterium	Dipicolinic acid	A	BACH and GILVARG, 1966
		B	STERLINI and MANDELSTAM, 1969; MANDELSTAM, 1971, 1976
Bacillus subtilis	Sulfolactic acid	B	WOOD, 1971
Streptomyces antibioticus	Actinomycin chromophore	B	MARSHALL et al., 1968
Fungi			
Penicillium cyclopium	Benzodiazepine alkaloids	A	LUCKNER, 1977; VOIGT et al., 1978
		B	NOVER and MÜLLER, 1975; EL KOUSY et al., 1975; NOVER and LUCKNER, 1976
Penicillium patulum	6-Methylsalicylic acid, patulin	A	LIGHT, 1969; 1970
		B	LIGHT, 1967; BU'LOCK et al., 1969
Aspergillus fumigatus	Orsellinic acid and other polyketides	B	PACKTER and COLLINS, 1974; WARD and PACKTER, 1974
Blakeslea trispora	Carotenoids, trisporic acids	B	FEOFILOVA and ARBUZOV, 1975; FEOFILOVA and BEKHTEREVA, 1976
Claviceps sp.	Ergoline alkaloids	A	HEINSTEIN et al., 1971; ERGE et al., 1973; MAIER and GRÖGER, 1976
		B	BU'LOCK and BARR, 1968
Polyporus hispidus	Styrylpyrone	A	VANCE et al., 1974 TOWERS et al., 1974
		B	VANCE et al., 1974; TOWERS et al., 1974
Higher plants			
Sinapis alba	Cinnamic acid and flavane derivatives	A	WELLMANN, 1974
		B	LANGE and MOHR, 1965; RISSLAND and MOHR, 1967
		C	SCHOPFER and HOCK, 1971; ACTON and SCHOPFER, 1975; SCHOPFER et al., 1976
Petroselinum hortense cell cultures	Cinnamic acid and flavane derivatives	A	HAHLBROCK et al., 1971a; GRISEBACH and HAHLBROCK, 1974; HAHLBROCK, 1976; HAHLBROCK et al., 1976; EBEL and HAHLBROCK, 1977
		B	HAHLBROCK and RAGG, 1975; HAHLBROCK, 1976
		C	HAHLBROCK and SCHRÖDER, 1975; WELLMANN and SCHOPFER, 1975; SCHOPFER et al., 1976

Table 2 (continued)

Organism or system	Secondary products whose synthesis has been investigated	Method applied[a]	References
		D	SCHRÖDER et al., 1976; HAHLBROCK, 1977
Phaseolus vulgaris	Phaseollin	B	HESS and HADWIGER, 1971
Amaranthus caudatus, Celosia plumosa	Betalains	B	KÖHLER, 1972, 1975; GIUDICI DE NICOLA et al., 1973
Pisum sativum, Malus silvestris	Ethylene	B	LIEBERMANN and KUNISHI, 1975
Hordeum vulgare	Hordenine	A	MANN et al., 1963
		B	MANN et al., 1963

[a] A: Determination in vitro of an activity increase of enzymes of secondary metabolism before or at the beginning of the phase of secondary product formation.
B: Measurement of the influence of inhibitors of gene expression on the formation of secondary products or on the formation of enzymes involved in secondary metabolism.
C: Isotope-labeling of enzymes catalyzing secondary product formation to demonstrate their de novo formation at certain developmental phases of the producer organism.
D: Translation of mRNA in vitro to show phase-dependent presence of mRNA species coding secondary metabolism enzymes.

By the methods mentioned in Table 2 it has been shown that in many instances the beginning of the formation of secondary products is coupled directly with the synthesis of the corresponding enzymes. One example in this respect is biosynthesis of flavonoids in cell cultures of *Petroselinum hortense* (cf. Fig. 1). Here the regulatory mechanisms by which the activity increase of the enzymes of flavonoid biosynthesis is controlled has been extensively investigated with phenylalanine ammonia-lyase (PAL), the key enzyme of the biosynthetic chain. Experiments with inhibitors of transcription and translation have shown that PAL biosynthesis depends on RNA and protein biosynthesis (HALBROCK and RAGG, 1975; HAHLBROCK, 1976). Labeling experiments have demonstrated that it is caused by an accelerated rate of enzyme synthesis (HAHLBROCK and SCHRÖDER, 1975; WELLMANN and SCHOPFER, 1975; SCHOPFER et al., 1976).

Stimulation of RNA synthesis seems to commence immediately upon UV irradiation while the increase of the rate of protein synthesis begins at the end of the lag period (cf. Fig. 1) which precedes increase of enzyme activity (HAHLBROCK and RAGG, 1975; HAHLBROCK, 1976). In agreement with these results in vitro translation has shown that the amount of PAL-mRNA rapidly increases after the beginning of irradiation, and that the rate of PAL synthesis calculated on the accumulating PAL activity corresponds to the amount of PAL-mRNA. This clearly demonstrates that in the expression of flavonoid metabolism in *Petroselinum hortense* cell cultures, as in the secondary metabolism of most other organisms, the control of transcription is the basic regulatory principle (SCHRÖDER et al., 1976; HAHLBROCK, 1977).

In some experimental systems, however, the control of transcription may not be the only regulatory mechanism. Experiments with actinomycin D for instance,

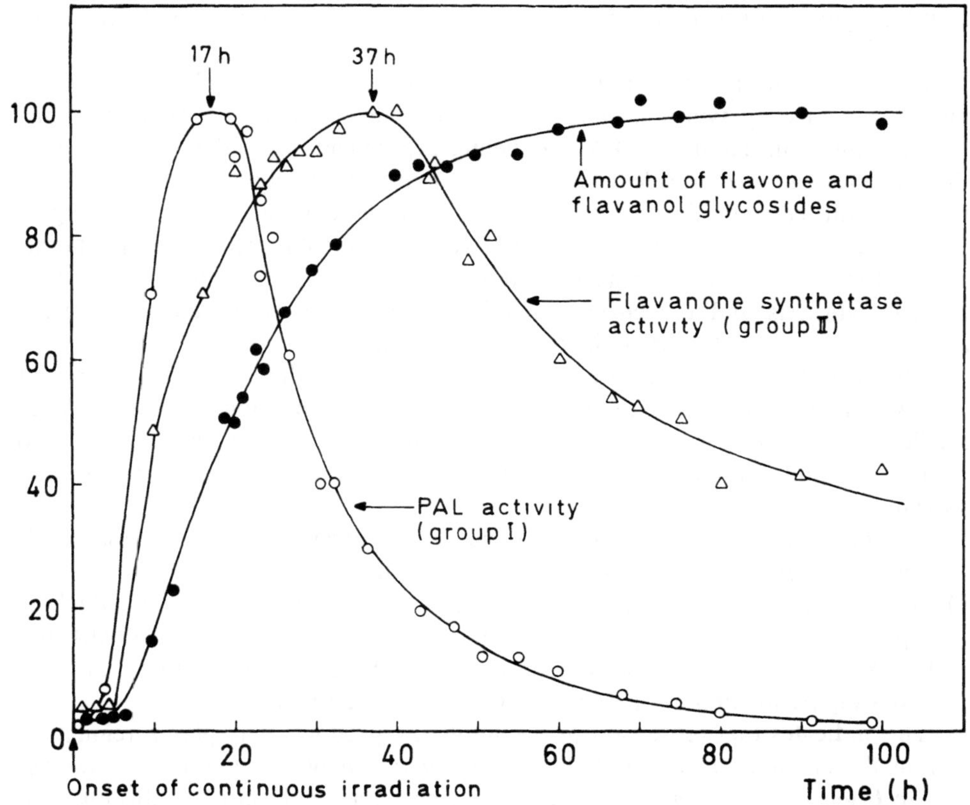

Fig. 1. Time course of light-induced flavonoid accumulation and activities of enzymes of flavonoid biosynthesis in cell cultures of *Petroselinum hortense* (HAHLBROCK et al., 1976). Flavonoid biosynthesis is induced upon radiation with UV-light. With respect to their lag-phases and activity changes the enzymes involved form two groups with coordinated regulation. *Group I*, which includes PAL, comprises enzymes of general cinnamic acid metabolism. *Group II*, to which flavanone synthetase belongs, catalyzes the flavone glycosid pathway (cf. Sect. 5.3). *Ordinate*, relative values

an inhibitor of transcription, on dipicolinic acid formation during bacterial sporulation have revealed a considerable gap between the onset of transcription and the actual beginning of secondary product formation (STERLINI and MANDELSTAM, 1969; MANDELSTAM, 1971). This result suggests a dual control of enzyme synthesis on the transcriptional and on one of the post-transcriptional levels. It is thought that in the bacteria stable mRNA species exist whose translation is also controlled.

A further feature of post-transcriptional control is the processing of proenzymes (enzymogens). Here the increase of enzyme activity is not directly coupled with de novo protein formation. For instance the appearance and increase of cyclopenase activity during conidiospore maturation in *Penicillium cyclopium* cannot be prevented by cycloheximide, an inhibitor of translation, in concentrations which suppress protein biosynthesis (EL KOUSY et al., 1975). The mechanism of cyclopenase activation has so far remained unclear. In a similar case, however, (activation of prochitin synthetase in *Saccharomyces cerevisiae*), it has been shown that the active enzyme is formed by proteolytic processing of a proteinogen (CABIB and ULANE, 1973; HOLZER, 1975).

In some organisms the amount of enzymes in secondary metabolism is determined not only by synthesis, but also by enzyme degradation. Again this has been examined in detail with PAL. Labeling experiments have demonstrated that this enzyme, e.g., in *Petroselinum hortense* cell cultures, undergoes a rapid turnover in the uninduced state. During induction the balance between synthesis and degradation is disturbed by an increased rate of synthesis of PAL-mRNA (HAHLBROCK, 1977). However, as early as about 17 h after beginning of the induction, degradation exceeds synthesis and the enzyme activity starts to decrease, returning within about 80 h almost to the pre-induction level (cf. Fig. 1).

The reasons for the decrease of PAL-mRNA biosynthesis in cell cultures of *Petroselinum hortense* are unknown. Probably product repression is involved. With seedlings of *Cucumis sativus,* in which after illumination (as in the cell cultures) an increase and subsequent decrease of PAL activity was detected, it has been shown that

a) cinnamic acid, the product of PAL action on phenylalanine, represses PAL synthesis if exogenously supplied (cf. Table 3) and
b) the application of α-aminooxyacetic acid, a phenylalanine analog, which is a powerful inhibitor of PAL (AMRHEIN et al., 1976b, cf. also AMRHEIN and GOEDEKE, 1977), causes prolongation of the phase of PAL activity increase and a higher maximum of PAL activity (AMRHEIN, 1979).

This indicates that repression by accumulating products may play a regulatory role in the cinnamic acid metabolism of *Cucumis*. In other organisms, however, e.g., in *Fagopyrum esculentum*, this regulatory principle could not be substantiated, though the PAL activity after illumination showed a similar time course with and without α-aminooxyacetic acid (AMRHEIN et al., 1976b, AMRHEIN, 1979).

3 Regulation of Enzyme Activity

In addition to regulation of the enzyme amount, secondary metabolism may also be controlled by regulation of the activity of the enzymes. Evidence in this respect comes, for instance, from the large discrepancies which have been found in many experimental systems between the relatively high enzyme activities measurable in vitro and the much lower activities in the living cell. In some organisms low in vivo activity has been shown to be due to limited substrate supply. One example in this respect is the low activity of cyclopenase in cultures of *Penicillium cyclopium*. Cyclopenase transforms the benzodiazepine alkaloids cyclopenin and cyclopenol to the quinoline derivatives viridicatin and viridicatol. It is located at the inner side of the plasma membrane of the mature conidiospores (WILSON and LUCKNER, 1975). However, in spite of high activities measurable in vitro the enzyme remains inactive in vivo. Hence in contrast to the formation and excretion of large amounts of cyclopenin and cyclopenol, the synthesis of viridicatin and viridicatol in cultures of *P. cyclopium* is only small. After artificial increase of membrane permeability, e.g., by lowering the cellular ATP level (ROOS and LUCKNER, 1977) the rate of viridicatin-viridicatol biosynthesis greatly increases. The enzyme now is accessible to its substrates inside and outside the spores. The alkaloid metabolism of cultures of *P. cyclopium* under these experimental conditions resembles that of certain strains of *P. viridicatum* which produce the same benzodiazepine and quinoline alkaloids, but in contrast to *P. cyclopium*, after a short period of cyclopenin-cyclopenol excretion take up the benzodiazepine alkaloids from the culture medium and convert them into viridicatin and viridicatol (Roos et al., 1976).

Under certain experimental conditions differences between the in vitro and the in vivo activities have also been demonstrated with PAL. The activity of this enzyme in living cells has been estimated after feeding of L-[3-^{3}H]-phenylalanine and determination of the ^{3}HOH liberated (Amrhein et al., 1976a). By this method it has been shown that the formation of ^{3}HOH by flotation of buckwheat, strawberry, or sunflower leaves on water shows a considerable increase whereas the in vitro measurable PAL activity remains stable (Amrhein and Goedeke, 1976; Amrhein et al., 1976b). Probably this result is due to the decrease of phenylalanine formation in the cells during starvation, which causes a change in the ratio between the fed labeled and the endogenously produced unlabeled phenylalanine. Also in many other plants PAL activity in vivo seems to be regulated by the substrate supply (Margna, 1977).

In some cases, however, the regulation of the in vivo activity of secondary metabolic enzymes is more complicated. In hyphae of *Penicillium cyclopium* for instance, where the activities of the enzymes of benzodiazepine alkaloid biosynthesis

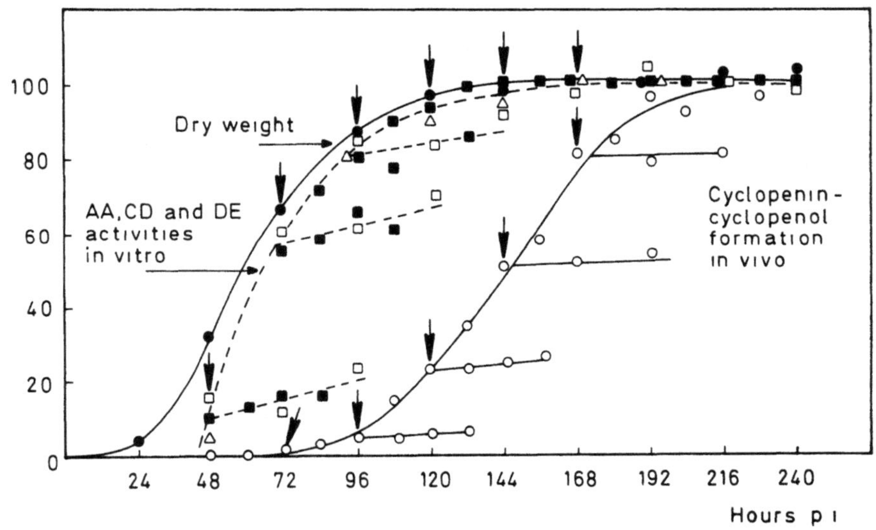

Fig. 2. Time course of the activities of enzymes of alkaloid biosynthesis and of the rates of alkaloid formation in hyphae of *Penicillium cyclopium* (Voigt et al., 1978). Enzymes of alkaloid biosynthesis are induced in a coordinated manner by an internal signal during development of the mould. However, the rates of alkaloid production increase later than enzyme activities. The increase of both processes can be stopped by addition of inhibitors of gene expression, e.g., 5-fluorouracil and cycloheximide (cf. the *arrows*) indicating that alkaloid production is controlled by formation of a protein which is not an enzyme of alkaloid biosynthesis. *Ordinate* relative values *AA* Anthranilic acid activating enzyme (△) *CD* Cyclopeptine dehydrogenase (■) *DE* Dehydrocyclopeptine epoxidase (□); p.i.: after inoculation

Table 3. Examples of product inhibition and product repression in secondary metabolism

Organism or system	Enzyme investigated	Inhibitors/repressors	Inhibition of activity	Repression of synthesis	References
Bacteria					
Actinomyces streptomycini	Streptidine kinase	Streptomycin	+		WALKER and WALKER, 1967
Fungi					
Claviceps purpurea	Dimethylallylpyrophosphate: L-tryptophan dimethylallyltransferase, chanoclavine-I cyclase	Agroclavine, elymoclavine lysergic acid	+	?	ERGE et al., 1973; FLOSS et al., 1974a; HEINSTEIN and FLOSS, 1976; KRUPINSKI et al., 1976; FLOSS et al., 1979 (cf. however, MAIER and GRÖGER, 1976)
Penicillium cyclopium	Dehydrocyclopeptine epoxidase	Cyclopenin, cyclopenol, viridicatin, viridicatol	+		VOIGT and LUCKNER, unpublished
Higher plants					
Different plants	PAL	Cinnamic acid and derivatives		+	JOHNSON et al., 1975; LAMB and RUBERY, 1976a; cf. also ENGELSMA, 1968, 1979; RANJEWA et al., 1977
		Cinnamic acid derivatives, benzoic acids and flavonoids	+		KOUKOL and CONN, 1961; CAMM and TOWERS, 1973; IREDALE and SMITH, 1974; ENGELSMA, 1974, 1979; NARI et al., 1974; KALGHATGI and RAO, 1975; RANJEWA et al., 1977

Solanum tuberosum	Cinnamic acid 4-hydroxylase	p-Coumaric acid		+	Lamb and Rubery, 1976a
Pisum sativum	Cinnamic acid 4-hydroxylase	p-Coumaric acid	+		Russel, 1971
Bambusa sp.	Caffeic acid methylase	5-Hydroxyferulic acid	+		Shimada et al., 1973
Petunia sp.	Caffeic acid CoA ligase	p-Coumaric acid and caffeic acid esters of quinic acid	+		Ranjewa et al., 1977
	Ferulic acid CoA ligase	Naringenin	+		Ranjewa et al., 1977
Petroselinum hortense cell cultures	Flavanone synthetase	Naringenin	+		Kreuzaler and Hahlbrock, 1975a
Phalaris tuberosa	Tryptophan decarboxylase	N,N-Dimethyl-tryptamine	+		Baxter and Slaytor, 1972
Nicotiana tabacum	Ornithine decarboxylase, putrescine N-methyltransferase, N-methylputrescine oxidase	Nicotine		+	Mizusaki et al., 1973

become measurable at a relatively early stage of the mould's development, whereas alkaloid formation is expressed later (Fig. 2), the increase of the rates of alkaloid biosynthesis is immediately stopped by inhibitors of gene expression, e.g., 5-fluorouracil or cycloheximide, even at developmental stages at which the in vitro measurable enzyme activities have already reached their maximum levels. This indicates the existence of a protein limiting the rate of alkaloid formation, which is synthesized much later than the investigated enzymes. Probably this protein does not take part in the biosynthesis of the alkaloid precursors phenylalanine, anthranilic acid, and methionine, because feeding of these compounds to cultures at the beginning of alkaloid production showed no increase of alkaloid biosynthesis. One might thus speculate that intracellular channeling of precursors or co-substrates to the site of alkaloid biosynthesis requires a specific protein which limits the production rate. In accordance with this assumption it has been shown that different pools of phenylalanine exist in the hyphae of *Penicillium cyclopium* from which the amino acid is transported to the sites of protein and alkaloid biosynthesis and that channeling to and between these pools causes pronounced changes in the incorporation rates of phenylalanine into protein and alkaloids under different developmental and experimental conditions (NOVER, 1979; NOVER et al., 1979).

Enzyme activity in secondary metabolism may be influenced not only by the accessibility of precursors but also by the accumulation of products (product inhibition by allosteric interaction or competition with the substrates at the binding sites). Table 3 shows results of in vitro experiments which demonstrate that product inhibition is a property of many secondary metabolic enzymes. In living cells, however, enzymes and secondary products usually are separated by compartmentation (cf. Sect. 4). It is thus difficult to decide whether product inhibition in vivo occurs as frequently as in vitro.

One example where product inhibition obviously plays a role in living cells is ergoline alkaloid formation in *Claviceps*. Table 3 shows that dimethylallylpyrophosphate: L-tryptophan dimethylallyltransferase (DMAT synthetase), the first enzyme of ergoline biosynthesis, undergoes inhibition by certain ergoline alkaloids in vitro. However, also in stationary batch cultures of the mould the intracellular concentration of ergoline alkaloids and the activity of DMAT synthetase measurable in homogenates undergo fluctuations. Thereby the enzyme activity tends to decrease when the intracellular alkaloid concentration is at a maximum. Hence this oscillation may be explained by (a) accumulation of the alkaloids at periods of high enzyme activity where diffusion into the medium is limiting (b) reversible inhibition of the enzyme by the accumulating alkaloids and (c) by decrease in the stored amount by release into the medium and again increase in enzyme activity and alkaloid biosynthesis (FLOSS et al., 1974a; HEINSTEIN and FLOSS, 1976).

4 Compartmentation as Principle of Control in Secondary Metabolism

Spatial organization is one of the most important regulatory features in secondary metabolism. It is brought about by (a) membranes separating different parts of

Table 4. Sites of intracellular location of some secondary metabolic enzymes and secondary products (LUCKNER et al., in press)

Group of compounds	Secondary metabolic enzymes and secondary products	Sites of location	References
Cellulose	β-1,4-Glucan synthetase	Dictyosomes and Golgi vesicles as well as the plasma membrane (M)[a]	WHALEY, 1975; AMELUNXEN et al., 1976; HELSPER et al., 1977
	Cellulose	Dictyosomes and Golgi vesicles, extraplasmic space (F)	SHAFIZADEH and MCGINNIS, 1971; WHALEY, 1975; AMELUNXEN et al., 1976; ALBERSHEIM, 1976
	Cellulase	Plasma membrane, extraplasmic space (M/F)	MATILE, 1975; KOEHLER et al., 1976
Cell wall matrix polysaccharides	β-1,3-Glucan synthetase, xylan synthetase etc.	Dictyosomes and Golgi vesicles (M)	KARR, 1976
	Hemicellulosic and pectic polymers	Dictyosomes and Golgi vesicles, extraplasmic space (F)	BULL and CHESTERS, 1966; SCHNEPF, 1969; WHALEY, 1975; KARR, 1976; ALBERSHEIM, 1976
	β-1,3-Glucanases, xylanases, pectinases	Extraplasmic space (F)	MATILE, 1975; ALBERSHEIM, 1976; DEKKER and RICHARDS, 1976; REXOVÁ-BENKOVÁ and MARKOVIČ, 1976
Chitin	Chitin synthetase	Plasma membrane (M)	DURAN et al., 1975
	Chitin	Extraplasmic space (F)	BARTNICKI-GARCIA, 1968
	Chitinase	Extraplasmic space (F)	MATILE, 1975
Sulfated polysaccharides	Fucoidan synthetase	Dictyosomes and Golgi vesicles	QUATRANO, 1974; CALLOW and EVANS, 1976
	Fucoidan	Extraplasmic space (F)	QUATRANO, 1974; CALLOW and EVANS, 1976; QUATRANO and STEVENS, 1976
Chlorophyll	Enzymes of protoporphyrin biosynthesis	Chloroplasts (F)	GIVAN and HARWOOD, 1976
	Enzymes converting protoporphyrin into chlorophyll	Chloroplasts (M)	REBEIZ and CASTELFRANCO, 1973; GIVAN and HARWOOD, 1976
	Chlorophyll	Chloroplasts (M)	REBEIZ and CASTELFRANCO, 1973; EGGER and SITTE, 1973

Table 4 (continued)

Group of compounds	Secondary metabolic enzymes and secondary products	Sites of location	References
Plastid quinones	Enzymes transforming phenylalanine into benzoic acid derivatives	Chloroplasts (M)	LÖFFELHARDT and KINDL, 1975; 1976
	Plastoquinones, phylloquinones, ubiquinones, tocopherylquinones	Chloroplasts (M; F in plastoglobuli)	LICHTENTHALER and WEINERT, 1970; EGGER and SITTE, 1973
Carotenoids	Enzymes of carotenoid biosynthesis	Chloroplasts (F)	GIVAN and HARWOOD, 1976
	Enzymes of violaxanthin cycle	Chloroplasts (M)	SIEFERMANN and YAMAMOTO, 1975
	Carotenoids	ER, lysosomes, tonoplast, chloroplasts (M; F in spherosomes and plastoglobuli)	SITTE, 1974; COSTES et al., 1976
Cinnamic acid derivatives	PAL, cinnamic acid o- and p-hydroxylases, p-coumaric acid-3-hydroxylase	ER, lysosomes, chloroplasts, mitochondria, plasma membrane (M)	STAFFORD, 1974a, b; GREGOR, 1976; ALIBERT et al., 1977; CZICHI and KINDL, 1977
	Cinnamic acid derivatives	Vacuole, extraplasmic space, chloroplasts (F)	STAFFORD, 1974a; WHITMORE, 1974; RILEY and KOLATTUKUDY, 1975; ALIBERT et al., 1977
Flavonoids	Chalcone-flavanone isomerase, flavonoid hydroxylase	ER, chloroplasts (M)	FRITSCH and GRISEBACH, 1975; WEISSENBÖCK et al., 1976
	Flavonoids	Vacuole, extraplasmic space, chloroplasts (F)	STAFFORD, 1974a; MCCLURE, 1975; WEISSENBÖCK et al., 1976; PLESSER and WEISSENBÖCK, 1977
Alkaloids	Enzymes of morphine synthesis, clavicipitic acid synthetase, cyclopenase	ER, tonoplast, cell membrane (M)	FAIRBAIRN et al., 1974; WILSON and LUCKNER, 1975; SAINI et al., 1976
	Alkaloids with different structures	Vacuole, extraplasmic space, chloroplasts (F)	JANS, 1973; VERZÁR-PETRI, 1973; MATILE, 1976a; NEUMANN, 1976; MÜLLER et al., 1976

Cyanogenic glycosides	Enzymes of p-hydroxymandelonitrile biosynthesis	ER (M)	SHIMADA and CONN, 1977; MØLLER and CONN, 1977
	Cyanogenic glycosides	Vacuole (F)	SAUNDERS et al., 1977
Polypeptide antibiotics	Edeine synthetase	Plasma membrane (M)	KURYLO-BOROWSKA, 1975
	Edeine	Extraplasmic space (F)	KURYLO-BOROWSKA, 1975
Peroxidases		ER, dictyosomes, tonoplast, vacuoles, lysosomes, plasma membrane, extraplasmic space, chloroplasts (M/F)	STAFFORD, 1974a, b; LIU and LAMPORT, 1974; PARISH, 1975; HENRY, 1975a; MÄDER, et al., 1975
Phenoloxidases		ER, lysosomes, extraplasmic space (M/F)	STAFFORD, 1974a, b; HENRY, 1975b; 1976; LIEBEREI and BIEHL, 1976; KATO et al., 1976; BARTSCH et al., 1979

[a] M: Membrane-bound or associated; F: Free.

the cells (b) formation of more or less stable bindings between cell constituents, e.g., between enzyme complexes and their substrates and intermediates, and (c) the influence of the so-called microenvironment, e.g., by weak protein–protein interactions and by the so-called unstirred layer effects (SRERE and MOSBACH, 1974), by mutual influences of enzymes which are located in membranes in proximity to each other, etc.

Table 4 demonstrates that many secondary enzymes are bound to or associated with the plasma membrane or with the membranes of ER, dictyosomes, vacuoles, and microbodies, whereas hydrophilic secondary products accumulate in the non-plasmic compartments separated by these membranes from cytoplasm (e.g., vacuoles, Golgi vesicles, and the extraplasmic space). Lipophilic secondary products may accumulate in the membranes themselves (e.g., carotenoids in nearly all cell membranes; COSTES et al., 1976; and chlorophyll and gibberellic acids in the membranes of chloroplasts; REBEIZ and CASTELFRANCO, 1973; EGGER and SITTE, 1973; COOKE and KENDRICK, 1976; cf. also BROWNING and SAUNDERS, 1977) or are located in the extraplasmic space (e.g. cutin and suberin). In few instances lipophilic secondary products were also shown to be stored in the cytoplasm (e.g., cautchouc particles in the cytoplasm of the latex cells of *Hevea, Papaver,* and *Taraxacum,* SCHULZE et al., 1967; cf. also MATILE, 1976b; essential oils in the cytoplasm of gland cells of *Dictamnus albus* and *Cleome spinosa*; AMELUNXEN and ARBEITER, 1967, 1969). A relatively large group of secondary metabolic enzymes is found in cytoplasm. The occurrence of secondary metabolic enzymes in nonplasmic compartments, however, is rare. The enzymes which there may be found are involved in the formation of impermeable polymers (e.g., melanins or lignins) or in polymer degradation.

The separation of the hydrophilic secondary products from cytoplasm prevents interference of the often toxic compounds with metabolically active proteins. However, this "metabolic excretion" (MOTHES, 1966) seems to be incomplete in some cases. This is indicated by the action of secondary products on the formation of secondary metabolic enzymes (cf. the discussion of sequential gene expression in Sect. 5.4 and of product repression in Sect. 2) as well as by their action on the activity of enzymes (cf. product inhibition in Sect. 3).

Table 4 implies that synthesis and excretion of secondary products in many instances are directly coupled. The enzymes synthesizing secondary products have been found frequently either as constituents of or associated with the membranes separating the cytoplasm from the main storage site of the secondary products (cf. e.g., cellulose synthetase, chitin synthetase, and edeine synthetase) or they are bound to or associated with the membranes of the endoplasmic reticulum and of dictyosomes which secrete secondary compounds into the nonplasmic space of these organelles. From there the products are transported to the storage site by means of vesicles whose membranes fuse with that of the storage compartment under release of their content in the nonplasmic storage space (cf. excretion of hemicelluloses and sulfated polysaccharides). In both cases formation of secondary products is coupled with transport through membranes to the nonplasmic space. Such coupling would be unnecessary only if the secondary compound were both lipophilic and nontoxic. Hence these substances may be found in membranes themselves or in the cytoplasm (see above), or they may penetrate through membranes into nonplasmic compartments by simple diffusion.

Once secreted, secondary products are held back in their storage compartments by (a) lack of solubility, caused e.g., by polymerization, cf. cell wall constituents (b) hydrophilic interaction which prevents penetration through the lipophilic membranes of the storage compartments, cf. many glycosides, (c) lipophilic interaction averting dissolution in the cytoplasm, cf. essential oils, and (d) interaction with non-permeable substances, as is thought, for instance, to be the case with alkaloids which obviously are held back by an ion trap mechanism (JANS, 1973; MATILE, 1976a; MÜLLER et al., 1976). However, this does not mean that further metabolism or degradation of secondary products is impossible. Chlorogenic acid, flavonoids, cyanogenic glycosides and different types of alkaloids, for instance, undergo rapid turnover which may be shown by feeding radioactive labeled precursors in pulse-chase experiments (cf. TAYLOR and ZUCKER, 1966; BOUGH and GANDER, 1971; ROBINSON, 1974; BARZ, 1975; STRACK and REZNIK, 1976). Figure 3 demonstrates this with respect to chlorogenic acid, whose specific activity after administration of the precursor phenylalanine first rapidly increases and then decreases with a half-life of less than 20 h in spite of a slight increase in its absolute amount, indicating on the one hand steady synthesis of the new unlabeled compound and on the other hand continuous degradation.

With some enzymes of secondary metabolism it has been demonstrated that nonidentical enzyme forms (isoenzymes) are located in different compartments.

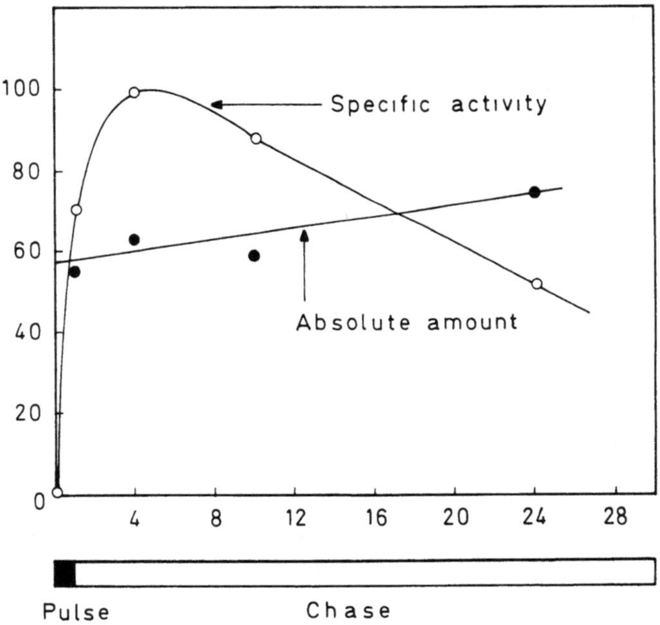

Fig. 3. Turnover of chlorogenic acid in leaf discs of *Xanthium pennsylvanicum* (TAYLOR and ZUCKER, 1966). [U-^{14}C]-L-Phenylalanine was applied to the upper surface of leaf discs for 1 h (pulse). Then the discs were rinsed and placed on moist filter paper exposed to light (chase). At the times indicated the absolute amount and the specific activity of chlorogenic acid were measured. *Ordinate*, relative values

Different forms of peroxidase for instance have been found in the protoplast and in the periplasmic space of higher plants (MÄDER et al., 1975; MÄDER, 1976). Furthermore roots of *Quercus pedunculata* contain two forms of PAL, one in the mitochondria-microbody fraction, the other in the microsome fraction, which differ in their sensitivity to cinnamic acid and benzoic acid derivatives and obviously belong to different metabolic pathways (ALIBERT et al., 1972).

Of special interest is the secondary metabolism of plastids because a great variety of secondary products and secondary metabolic enzymes have been detected in the envelope as well as in the inner membranes and in the nonparticulate fraction of this organelle (cf. LUCKNER et al., in press). Lipophilic secondary products are either found as constituents of the membranes (e.g., gibberellic acids, chlorophylls, phycobilins, and plastid quinones) or as the flavonoid aglycones and excess amounts of plastid quinones, they are accumulated in plastoglobuli, i.e., lipophilic droplets located in the stroma. The site of accumulation of most hydrophilic secondary products is unknown. However, in rooted leaves of *Nicotiana rustica*, in which large amounts of nicotine were accumulated, alkaloid was detected in the nonplasmic space between both membranes of the plastid envelope and in vacuoles located within the stroma, which obviously are formed from the membrane (NEUMANN, 1975).

So far it has not been possible to demonstrate plastidic inheritance in secondary metabolism. However, for instance with chlorophyll, nuclear genes have been identified which control biosynthesis (WETTSTEIN et al., 1974; KAHN et al., 1976). It is therefore expected that the enzymes of secondary metabolism, like many other plastid proteins which are coded by the genome, are formed extraplastidically and penetrate through the plastid envelope (GIVAN and HARWOOD, 1976; but see HAMPP and ZIEGLER, 1975).

Low molecular substances may also cross the membranes of the plastid envelope, especially at the early stages of greening, i.e., when the ability to fix CO_2 in the chloroplast has not yet been developed and the permeability of the envelope is relatively high. At that period large amounts of precursors of secondary products, such as oxaloacetate, succinate, or mevalonate are taken up from the extraplastidic site (HAMPP and WELLBURN, 1976; WELLBURN and HAMPP, 1976a, b; 1977).

On the other hand secondary products synthesized in the plastids may be released into extraplastidic compartments. There are indications that flavonoids, for example, are synthesized in the chloroplasts and transported to their extraplastidic storage sites, namely the vacuoles and the periplasmic space (ZAPROMETOV and BUCKLAEVA, 1967; CHARRIÉRE-LADREIX, 1976) and that during the biosynthesis of methoxylated furanocoumarins intermediates are released from chloroplasts and transformed into the final product in other compartments (DHILLON and BROWN, 1976). The process of release in some cases at least is thoroughly regulated. Hence it has been shown that the movement of gibberellic acids, formed and stored in the chloroplasts, to the cytoplasm is influenced by red and far red radiations and thus is obviously controlled by phytochrome (COOKE and KENDRICK, 1976; EVANS and SMITH, 1976).

A special aspect of compartmentation is the control of secondary metabolism by localization of enzymes belonging to one metabolic pathway in different tissues and organs or by synthesis and storage of secondary products in different cells. One example of this is provided by some Solanaceae. In plants of this family it has been demonstrated that alkaloids formed in the roots and transported to and stored in the aerial parts (cf. MOTHES, 1969) undergo structural modifications during this translocation (cf. the epoxidation and demethylation of hyoscyamine in shoots of *Datura;* ROMEIKE, 1959, 1964). Another example is provided by cell cultures of some *Macleaya* species where, in contrast to most cells which obviously are able to synthesize alkaloids, only few accumulate these compounds. It has been demonstrated that the synthesizing cells release alkaloids into the surroundings, from which they are taken up by the storage cells. Hence the typical alkaloid accumulation in *Macleaya* cell cultures needs the cooperation of and is controlled by two different cell types (BÖHM, 1978).

A simple, but efficient form of compartmentation is brought about by multienzyme complexes. They (a) allow channeling of intermediates and the consequent separation of possibly competing pathways (b) may have kinetic advantages caused by the proximity of the enzymes (microenvironment effect) and (c) simplify the control of the catalytic properties of the combined enzymes by protein–protein interactions (SRERE and MOSBACH, 1974; WELCH and GAERTNER, 1976).

A most efficient form of channeling is brought about by covalent binding of precursors and intermediates to enzyme proteins. In secondary metabolism this has been shown in the biosynthesis of peptide antibiotics and polyketides (Table 5). In vivo experiments indicate that similar complexes are also involved in the biosynthesis of certain fungal peptide alkaloids (FRAMM et al., 1973; FLOSS et al., 1974b; GRÖGER, 1975).

Some of the synthetases of bacterial polypeptide antibiotics, e.g., gramicidin S synthetase, tyrocidine synthetase, edeine synthetase, and bacitracin synthetase, separate easily into two or three complementary enzyme fractions (LEE et al., 1973; LIPMANN, 1973; KURYLO-BOROWSKA and SEDKOWSKA, 1974; FRØYSHOV, 1974; LEE and LIPMANN, 1977). In some cases these fractions, which still show part of the catalytic properties of the total synthetase, may be further decomposed to simple proteins with a molecular weight of about 70,000. Each of these proteins still activates one of the amino acids involved in peptide biosynthesis. In addition one or more subunits with a molecular weight of about 17,000 are found representing the pantetheine-containing carrier protein, that by a shuttle mechanism mediates sequential transpeptidation of the activated amino acids (LIPMANN, 1973; LEE et al., 1973; LEE and LIPMANN, 1977). Hence tyrocidine synthetase for instance is an enzyme complex formed by 12 nonidentical subunits, which catalyze activation and linkage of ten amino acids in a highly controlled manner.

Channeling is also brought about by binding of enzymes to or integration in membranes in proximity to each other. Even if no covalent binding of precursors and intermediates occurs, the microenvironment in these complexes may cause "catalytic facilitation" (GAERTNER et al., 1970), i.e., preference for transformation of intermediates formed by the enzyme complex rather than of externally administered intermediates. Membrane-bound enzyme complexes have been detected in plastids, mitochondria, and ER. They participate in synthesis and metabolism of cinnamic acid derivatives and in the formation of cyanogenic compounds (Table 5). The extent of cooperation of the enzymes involved depends on the integrity of the membranes. It is reduced, for example, by membrane purification or increase of membrane permeability by treatment with ethylene (CZICHI and KINDL, 1975b; 1977; KINDL, 1979).

Table 5. Multienzyme complexes catalyzing reactions of secondary metabolism (LUCKNER et al., in press)

Organism or system	Multienzyme complex	Catalyzed reaction	Intermediates bound		References
			covalently	non-covalently	
Bacteria					
	Alamethicin synthetase	Amino acids→ alamethicin	+		RINDFLEISCH and KLEINKAUF, 1976
	Bacitracin synthetase	Amino acids→ bacitracin	+		FRØYSHOV, 1974; FRØYSHOV and LALAND, 1974; KURAHASHI, 1974
	Edeine synthetase	Amino acids→ edeine	+		KURYLO-BOROWSKA and SEDKOWSKA, 1974; KURAHASHI, 1974
	Gramicidin S synthetase	Amino acids→ gramicidin S	+		LIPMANN, 1971; 1973; LALAND and ZIMMER, 1973; KURAHASHI, 1974; PASS et al., 1974; DUNGEN, VON et al., 1976
	Polymyxin synthetase	Amino acids→ polymyxins	+		KURAHASHI, 1974
	Tyrocidine synthetase	Amino acids→ tyrocidine	+		LIPMANN, 1971; 1973; LEE et al., 1973; LALAND and ZIMMER, 1973; KURAHASHI, 1974
Fungi					
	6-Methyl-salicylic acid synthetase	Acetyl-CoA+ Malonyl-CoA→ 6-Methylsalicylic acid	+		LIGHT and HAGER, 1968; DIMROTH et al., 1970; 1972; 1976; LIGHT, 1970; SCOTT et al., 1974
	Orsellinic acid synthetase	Acetyl-CoA+ Malonyl-CoA→ Orsellinic acid	+		GAUCHER and SHEPHERD, 1968
	Alternariol synthetase	Acetyl-CoA+ Malonyl-CoA→ Alternariol	+		GATENBECK and HERMODSSON, 1965; SJÖLAND and GATENBECK, 1966; LIGHT, 1970
Blue green algae					
Anacystis nidulans	Thylakoid membranes containing PAL and benzoic acid (p-hydroxy-benzoic acid) synthetase	L-Phe→Benzoic acid (p-hydroxy-benzoic acid)		+	LÖFFELHARDT and KINDL, 1976

Table 5 (continued)

Organism or system	Multienzyme complex	Catalyzed reaction	Intermediates bound covalently	Intermediates bound non-covalently	References
Green algae					
Dunaliella marina	Thylakoid membranes containing PAL and cinnamic acid hydroxylases	Phe→o-and p-cinnamic acids		+	CZICHI and KINDL, 1975a
Higher plants					
	Thylakoid membranes containing PAL, cinnamic acid 4-hydroxylase and benzoic acid synthetase	Phe, tyr→ benzoic acid, p-hydroxybenzoic acid		+	LÖFFELHARDT and KINDL, 1975
	Mitochondrial membranes containing benzoic acid synthetase	p-Coumaric acid→ p-hydroxybenzoic acid		+	HAGEL and KINDL, 1975
	Microsomal membranes containing PAL and cinnamic acid hydroxylases	Phe→o- and p-coumaric acids		+	CZICHI and KINDL, 1975b, 1977
	Flavanone synthetase[a]	p-Coumaric acid-CoA + Malonyl-CoA→flavanones	+		KREUZALER and HAHLBROCK, 1975a
Sorghum bicolor	Microsomal membranes containing p-hydroxymandelonitrile synthetase	Tyr→p-hydroxymandelonitrile		+	McFARLANE et al., 1975; SHIMADA and CONN, 1977; MØLLER and CONN, 1977; CONN et al., 1979

[a] Under certain experimental conditions also forming the α-pyrone derivative bis-noryangonin (KREUZALER and HAHLBROCK, 1975b). Genetic experiments indicate that the first product in cyclization is a chalcone (KUHN et al., 1978).

5 Integration of Secondary Metabolism in Programs of Differentiation and Development

The phase-dependent expression of secondary metabolism, mentioned in Section 1 as one of the most important features in the control of secondary product formation, is caused by integration of secondary metabolism into the programs of differentiation and development. These programs, which are characteristic of all living beings, mediate the coexpression of the different fields of cell specialization with the various aspects of secondary product formation e.g., with the synthesis of secondary metabolic enzymes, with the formation of the facilities necessary for channeling of precursors and intermediates and with the compartmentation of secondary metabolic enzymes and products.

An example in which integration of secondary product formation in a differentiation program has been extensively investigated is bacterial sporulation. During formation of bacterial endospores a complex sequence of morphological and biochemical characteristics is expressed which involves at least 30 independently regulated gene groups with more than 100 genes. Besides formation of secondary products (peptide antibiotics, dipicolinic acid, sulfolactic acid and brown pigments of unidentified structure) many other metabolic activities and morphological characters are expressed (HANSON et al., 1970; PIGGOT, 1973; MANDELSTAM, 1976). All events are strictly fixed in their temporal sequence. Based on some outstanding attributes the process of sporulation may be divided into six stages (cf. Table 6).

The mutual regulatory interactions between the processes of sporulation have been investigated by the use of mutants and inhibitors of gene expression. Interruption of spore formation by this means at certain stages prevents the expression of the cytological and biochemical characteristics of the subsequent stages (WAITES et al., 1970; WOOD, 1971; MANDELSTAM, 1976). This demonstrates that the process of sporulation is a sequence of dependent processes (MANDELSTAM, 1976), i.e., it has the character of a program. However, there are obviously steps which are necessary for progress of the program and others which are not. Secondary product formation for instance may be eliminated by mutation without impairing the expression of the subsequent steps of sporulation (KAMBE et al., 1974; HAAVIK and FRØYSHOV, 1975; and the review of MANDELSTAM, 1976). It thus appears that at least in the artificial milieu of the laboratory none of the secondary products shown in Table 6 has an essential function in the process of spore formation.

Neither the molecular organization of bacterial sporulation nor that of most other differentiation programs is known (except in very simple cases, e.g., the programed expression of the viral genetic material in bacterial host cells; SZYBALSKI, 1977), nor is any detailed information available about the molecular basis of the integration into the differentiation programs of secondary product formation. Thus in the following section only some general principles are discussed, which show the interrelations between control of secondary metabolism and that of other metabolic activities of the producer organisms:

5.1 The Influence of Nutrients

Suppression of secondary product formation by excess nutrients, especially by glucose, other easily metabolizable carbon sources, and phosphate, is a common phenomenon in microbial cultures. It has been found for instance in the biosynthesis of gibberellic acids, polyketides, certain antibiotics, and alkaloids (DEMAIN, 1968; 1971, 1972; WEINBERG, 1970; GRISEBACH, 1975; LUCKNER et al., 1977). Usually the suppression of secondary product formation is accompanied by the suppression

Table 6. Morphological and biochemical events associated with sporulation in *Bacillus* sp. (LEE et al., 1975; MANDELSTAM, 1976; cf. also LUCKNER et al., 1977)

	0 h	1.5 h	2.5 h	4.5 h	6 h	7 h	8 h
Stage	0	I	II	III	IV	V	VI
Morphological characteristics	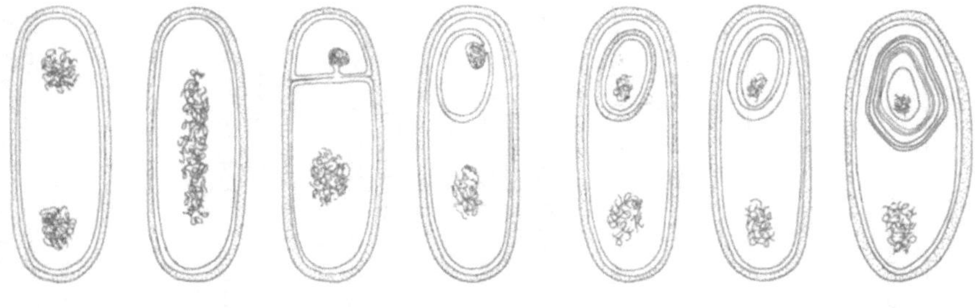						
	Vegetative cell	Chromatin filament	Spore septum	Spore protoplast	Cortex formation	Coat formation	Maturation
Biochemical characteristics		Exoprotease	Alanine dehydrogenase	70,000 dalton RNA binding protein	Uptake of Ca^{2+}		
		Ribonuclease	β-alanine and pantothenic acid uptake	Alkaline phosphatase	Ribosidase	Cysteine incorporation	Alanine racemase
		Protein turnover	Peptide antibiotics	Glucose dehydrogenase	Adenosine deaminase	Octanol resistance	Heat resistance
				Heat-resistant catalase	Sulfolactic acid Dipicolinic acid	Brown pigment	

of other characteristics of cell specialization (in *Penicillium cyclopium* for instance conidiospore formation, LUCKNER et al., 1977), indicating a general influence of nutrient supply on cell specialization.

Investigation of the molecular mechanisms by which nutrients affect secondary metabolism has been basically influenced by the results concerning glucose repression of different groups of catabolic enzymes in bacteria (catabolite repression). These have revealed that glucose acts via cAMP which, as an effector of the so-called catabolite activating protein, is required for the transcription of the glucose-sensitive genes (PASTAN and PERLMAN, 1970). In eukaryotic systems, however, the mode of action of glucose is still unknown. In *Penicillium cyclopium* for instance glucose suppression of alkaloid biosynthesis cannot be overcome by administration of cAMP or dibutyryl cAMP (LUCKNER et al., 1977) and with respect to suppression of β-galactosidase induction, it has been demonstrated that glucose acts at one of the post-transcriptional levels (ININGER and NOVER, 1975).

Table 7. Signals influencing the expression of secondary metabolism (from LUCKNER et al., 1977)

Organism or system	Secondary products or enzymes whose synthesis is influenced	Signal: Name	Signal: Substrate-like	Signal: Nonsubstrate-like	References
Bacteria					
Actinomyces streptomycini	Streptomycin	A-factor		+	KHOKHLOV et al., 1967; 1973; 1976; KLEINER et al., 1976; KHOKHLOV and TOVAROVA, 1979
Fungi					
Aspergillus fumigatus, Claviceps paspali C. fusiformis	Ergolin alkaloids	Tryptophan and analogues	+		RAO and PATEL, 1974; FLOSS et al., 1974, 1979; ROBBERS and FLOOS, 1976, ROBBERS et al., 1978
Penicillium cyclopium	Benzodiazepine alkaloids, cyclopeptine dehydrogenase	Phenylalanine and analogues	+		DUNKEL et al., 1976
		Mycelial factors		+	DUNKEL et al., 1976
Cephalosporium acremonium	Cephalosporin C	Methionine, norleucine	+	?	DREW and DEMAIN, 1973; cf. GRÖGER, 1976; 1977
Blakeslea trispora	Carotenoids	β-Ionone	+		FEOFILOWA and ARBUZOW, 1975; FEOFILOWA and BEKHTEREVA, 1976
Mucorales (*Blakeslea* sp., *Phycomyces* sp., *Mucor* sp.)	Trisporic acid, β-carotene and its oxygenated polymer sporopollenin	Trisporic acids	+		GOODAY, 1974, 1978; BU'LOCK, 1975; RAO and MODI, 1977
Higher plants					
Different plant species and plant cell cultures	Cinnamic acid derivatives, flavanes, lignin and enzymes involved in biosynthesis of these compounds	Auxins		+	CONSTABEL et al., 1971; ALFERMANN and REINHARD, 1971; 1976; ALFERMANN et al., 1975; MCCLURE, 1975
		Cytokinins		+	CONSTABEL et al., 1971; PECKET and HATHOUT

					BASSIM, 1974; MCCLURE, 1975
		Phytochrome		+	ZUCKER, 1972; SMITH, 1972; 1973; CAMM and TOWERS, 1973; MOHR et al., 1974; MCCLURE, 1975; SCHOPFER, 1977; BENVENISTE et al., 1978; TONG and SCHOPFER, 1978; cf. also DRUMM et al., 1975
		Ethylene		+	CRAKER and WETHERBEE, 1973a, b; HYODO and YANG, 1974; MCCLURE, 1975; CRAKER, 1975
		Gibberellic acid		+	LÄCHELE et al., 1974; MCCLURE, 1975; HEINZMANN and SEITZ, 1977
		Cinnamic acid and derivatives	+		HESS, 1967a, b; 1968; LAMB and RUBERY, 1976b; LAMB, 1977; ENGELSMA, 1979 (cf. also Table 3)
Amaranthus tricolor, A. paniculata, Celosia plumosa	Betalains	Phytochrome		+	GUIDICI DE NICOLA et al., 1973; 1975; KÖHLER, 1973a; COLOMAS and BULARD, 1975
		Cytokinins		+	KÖHLER and CONRAD, 1966; PIATTELLI et al., 1971; GUIDICI de NICOLA et al., 1973; KÖHLER, 1973b
		Gibberellic acid		+	COLOMAS et al., 1973; KINSMAN et al., 1975a, b; LALORAYA et al., 1976
Different plant species	Ethylene	Phytochrome		+	CRAKER et al., 1973; BURG, 1973; KANG and BURG, 1973
		Auxins		+	KANG et al., 1971; SAKAI and IMASEKI, 1971; BURG, 1973; LIEBERMANN and KUNISHI, 1975
Morinda citrifolia cell cultures	Anthraquinones	Auxins		+	ZENK et al., 1975
Different plant species	Carotenoids	Phytochrome		+	THOMAS and JEN, 1975; GRUMBACH and LICHTENTHALER, 1975; LICHTENTHALER, 1975b; KLEUDGEN and LICHTENTHALER, 1975
Hordeum vulgare	Prenyl chains	Phytochrome		+	LICHTENTHALER, 1975a

It is of interest that glucose suppression of alkaloid metabolism in *Penicillium cyclopium* can be impeded at least in part by the morphological organization of the mycelium. In emerged cultures, i.e., under growth conditions which are normal in nature, *Penicillium cyclopium* forms a relatively dense mycelial mat, in which cell specialization, as indicated by alkaloid production and conidiospore formation, cannot be prevented by administration of high glucose concentrations (LUCKNER et al., 1977). Obviously this is due to limitation of nutrients within the mat. Cell culture experiments have shown that in higher plants nutrient suppression as a principle of control in secondary metabolism seems not to exist.

5.2 Action of Signals

In many microorganisms and higher plants secondary metabolism is triggered by as yet unknown internal signals produced as part of the programs of differentiation and development (cf. the expression of alkaloid metabolism in *Penicillium cyclopium*, Fig. 2, and of secondary product formation during bacterial sporulation, Table 6). In other organisms expression is influenced by external signals (cf. Table 7). These may be divided into substrate-like and nonsubstrate-like signals (LUCKNER et al., 1977). However, only few substrate-like signals seem to influence secondary metabolism directly. One example is probably cinnamic acid, which represses PAL formation (cf. Sect. 2) and induces hydroxycinnamoyl-CoA: quinate hydroxycinnamoyl transferase, an enzyme of chlorogenic acid biosynthesis (LAMB, 1977). Most other signals show indirect action. The plant hormone gibberellic acid (a nonsubstrate-like signal), reduces amaranthine biosynthesis, for instance, by limitation of the precursor supply, an effect reversed by the administration of tyrosine or DOPA (LALORAYA et al., 1976). Furthermore the plant hormones and most microbial signals affect not only secondary metabolism but also a wide variety of other metabolic activities.

Usually the response of the target cells depends on their state of differentiation. In many cases the phase during which secondary metabolism may be influenced is not linked directly to the stage of secondary product formation. Hence for instance the following microbial signals are active only if added at the beginning of the growth phase, i.e., at a developmental stage long before secondary product formation is actually expressed:

a) the A-factor which triggers streptomycin formation in *Actinomyces streptomycini,*
b) tryptophan which influences ergoline alkaloid biosynthesis in *Claviceps*
c) phenylalanine and the peptides which increase benzodiazepine alkaloid biosynthesis in *Penicillium cyclopium*, and
d) methionine and norleucine, which stimulate cephalosporin formation in *Cephalosporium acremonium.*

Phases of susceptibility in which secondary metabolism in higher plants is influenced by appropriate signals have also been detected. Hence the phytochrome-mediated induction of anthocyanin biosynthesis in the epidermal cells of cotyledons of *Sinapis alba* begins only 27 h after sowing, in contrast to the fact that the phytochrome system per se is fully functional immediately after sowing. Competence with respect to anthocyanin formation is determined endogenously (STEINITZ et al., 1976), i.e., by the state of differentiation of the epidermis cells.

5.3 Coordinated Enzyme Expression

Coordinated regulation of the activity of enzymes which are members of one biosynthetic pathway seem to occur frequently in secondary metabolism (cf. e.g., Figs. 1 and 2). This has been examined in detail in the flavonoid metabolism of higher plants. In cell cultures of *Petroselinum hortense* after light induction the enzymes of flavonoid biosynthesis fall into two groups according to their in vitro measurable activities. Group I comprises the enzymes of general cinnamic acid metabolism: PAL, cinnamate 4-hydroxylase and cinnamate (p-coumarate) CoA ligase which produce the precursors of different groups of secondary phenylpropanoids. Group II is formed from the enzymes of the special pathway leading from the cinnamic acid derivatives to the flavonoids themselves: flavanone synthetase, chalcone-flavanone isomerase, chalcone-flavanone oxidase, malonyl, glucosyl, and apiosyl transferases, etc. Whereas the group I enzymes have a lag phase of about 2 h and reach maximum activity about 17 h after onset of radiation, the lag phase of group II enzymes lasts about 4 h and highest activity is found after about 37 h (HAHLBROCK et al., 1976). The activities of the enzymes of group I also increase in parallel to each other after transfer of the cells to new medium (HAHLBROCK and WELLMANN, 1973) indicating that they really form a regulatory unity in parsley cell cultures.

Experiments with whole plants have demonstrated that in these another pattern of control is established. In cotyledons and leaves of *Petroselinum hortense* seedlings PAL and several group II enzymes show a regulatory behavior which indicates interdependence (HAHLBROCK et al., 1971b) Furthermore in the seedlings PAL is induced by light whereas cinnamate (p-coumarate) CoA ligase, another group I enzyme, is not (MCCLURE and GROSS, 1975). This corresponds with results from primary metabolism which have shown flexibility in the integration of genes into groups with coordinated expression.

5.4 Sequential Gene Expression in Secondary Metabolism

The subsequent induction of enzymes which belong to one metabolic pathway by the products of preceding enzymes (sequential gene expression), is a characteristic of catabolic pathways in bacteria (ORNSTON, 1971). Although in many plants and microorganisms sequential formation of secondary products and secondary metabolic enzymes has been observed (LUCKNER et al., 1977) in very few organisms has the direct influence of secondary products on the expression of enzymes catalyzing their further transformation been demonstrated. One example is probably the light-induced chlorogenic acid biosynthesis in *Solanum tuberosum*. In potato tuber discs radiation causes first an increase of PAL activity, which is followed by an increase in hydroxycinnamoyl-CoA: quinate hydroxycinnamoyl transferase (HQT) activity and the accumulation of chlorogenic acid. It has been demonstrated that feeding of cinnamic acid in potatoes gives rise to an increase of HQT activity, and it therefore seems possible that the cinnamic acid, produced endogenously by PAL, induces the enzymes of chlorogenic acid biosynthesis (LAMB, 1977). In *Fagopyron esculentum,* however, indications for such a feed-forward control in chlorogenate biosynthesis could not be demonstrated (ULBRICH and AMRHEIN, 1978).

5.5 Secondary Metabolism and Cell Specialization

It is characteristic of many microbial cultures that after a phase of rapid cell division in a programed, multistep process the growth rate slows down and formation of secondary products and other features of cell specialization begins (so-called trophophase-idiophase development, BU'LOCK, 1975; LUCKNER et al., 1977). This kind of development has also been shown to occur in higher plants. The terms trophophase and idiophase, however, can be applied only to tissues or organs of higher plants and not to the organism in toto. One example of trophophase-idiophase development in higher plants is formation and ripening of pollen grains where after meiosis, during the process of maturation, formation of carotenoids, peroxidases, and sporopollenin (SHAW, 1971), as well as sequential formation of enzymes and products of flavonoid metabolism proceeds (WIERMANN, 1973; QUAST and WIERMANN, 1973; SÜTFELD and WIERMANN, 1974; STANLEY and LINSKENS, 1974; WIERMANN, 1979).

It should be mentioned, however, that cell division and cell specialization are not mutually exclusive. Conidiospores of *Penicillium cyclopium* for instance contain the enzymes of benzodiazepine alkaloid biosynthesis as constitutive features, whereas in the hyphae these enzymes are formed during maturation of already existing cells (VOIGT et al., 1978). Thus the enzymes present in the spores must have been synthesized during cell division in the conidia-producing cells, the phialides. Conidiation is an example of a so-called quantal cell cycle (HOLTZER et al., 1973; HOLTZER and RUBINSTEIN, 1977) where daughter cells during formation are programed to carry out certain forms of specialization. The conidiospores of *Penicillium cyclopium* for instance in their mature state differ from the phialides by their spherical shape and their thick rigid cell wall, by the typical spore pigments, and by the enzyme cyclopenase (NOVER and LUCKNER, 1974). Quantal cell cycles are frequently found in animals, but seem to be rare in plants and microorganisms.

In summary: the features discussed in Sections 5.1–5.5 demonstrate that the integration in the programs of differentiation and development of producer organisms is one of the most important control mechanisms in secondary metabolism. It causes the formation of secondary products during certain phases and within certain specialized cells only, it accounts for the phases of susceptibility to signals which influence secondary metabolism and the determination of secondary product formation by signals which do not act directly on secondary metabolism, and it connects secondary metabolism with most other biochemical and morphological activities of the producer organism.

6 Some Speculations About the Evolution of the Control Mechanisms in Secondary Metabolism

In the preceding sections it has been seen that secondary metabolism is characterized by a high degree of order, i.e., by the precise regulation of enzyme level and activity in secondary metabolic pathways, the compartmentation and channeling of enzymes, precursors, intermediates, and products as well as the integration of secondary metabolism in the programs of cell specialization of the producer organism. However, characteristics of secondary metabolism are also the bizarre chemical structures of the formed products, the restricted occurrence of the different secondary compounds within the groups of living beings and their usefulness which usually is small or absent (CZAPEK, 1921; KOSTYTSCHEW, 1926; PAECH, 1950), i.e., properties which give secondary metabolism its erratic features.

Both characteristics seem at first sight to be incompatible. However, they may be understood on the basis of the following three evolutionary principles:

1. Either the formation of secondary products (e.g., as a means for the removal of excess or waste compounds), or the secondary products themselves (e.g., as regulatory, physiological or ecological factors) create an advantage for the producer organism, or at least created a superiority at that period of evolution during which the secondary products, unused today, evolved. Only by admitting the present, or past usefulness of secondary compounds can we explain why selection pressure created the regulatory principles mentioned above.

2. In contrast evolution of the details of the chemical structures is comparable with the evolution of some morphological characteristics, e.g., the shape of leaves of higher plants, which undergo a relatively low selective pressure. In none of the organisms investigated could it be demonstrated that the single member of the frequently found groups of closely related secondary products is in fact of advantage or disadvantage to the producer organism even if the group in toto may be useful.

3. Formation of secondary products may be preserved for long periods even if these compounds have no further usefulness in the producer organisms. This may happen because synthesis of most secondary products constitutes only a minor diversion of metabolic resources.

In a number of organisms, however, secondary metabolism shows features which may be signs of decay:

a) On the one hand enzymes or enzyme groups may be disconnected from the common regulation of enzyme activities in metabolic pathways. One example of this is cyclopenase, an enzyme of alkaloid metabolism in *Penicillium cyclopium*, whose occurrence is restricted to the conidiospores, whereas the other enzymes of alkaloid biosynthesis are found in the hyphae and in the conidiospores. Cyclopenase shows a regulation which is independent of that of the other enzymes of alkaloid metabolism present in the spores and in contrast to its very high in vitro activity is inactive in the living cells (NOVER and LUCKNER, 1974; ROOS et al., 1976). A second example has been found in the alkaloid metabolism of *Papaver bracteatum*. With this plant genetic varieties exist which produce the alkaloid alpinigenine only in their youth, but not in their mature state. It has been shown that this behavior is brought about by the disappearance of one or more of the alpinigenine-synthesizing enzymes during aging. Other alkaloids, however, e.g., thebaine, are produced also in mature plants. Hence regulation of the disappearing enzymes of alpinigenine biosynthesis is independent of that of the other enzymes of alkaloid metabolism (BÖHM, 1971; BÖHM, unpublished results; LUCKNER et al., 1977).

b) A second principle seems to be that genes coding enzymes of secondary metabolism under the normal conditions of development are not further expressed or that secondary metabolic enzymes after formation remain inactive. This will explain the synthesis of new secondary compounds under extreme physiological conditions, e.g., in cell cultures (BÖHM, 1977) or after administration of unphysiological amounts of precursors or intermediates (cf. the formation of harmane alkaloids from tryptophan in cell cultures of *Phaseolus vulgaris*, a plant unable to form any indole alkaloids under normal conditions; VELIKY, 1972).

The obviously significant amount of unexpressed genetic material and of unused secondary metabolic enzymes and secondary products form a reservoir of chemical potential which may be used during further evolution of the producer organisms or of the living beings with which they coexist. It allows establishing of new intercellular or interorganismic signals, of new mediators of ecological relations etc.

7 Summary

The following control mechanisms seem to be the most important in regulation of secondary metabolism:

a) Synthesis and degradation of secondary metabolic enzymes, i.e. control of enzyme amount,
b) regulation of enzyme activity,
c) compartmentation and channeling of enzymes, precursors, intermediates and secondary products, and
d) integration of secondary metabolism in the programmes of differentiation and development of the producer organism.

They demonstrate that expression and control in secondary metabolism is subject to the same regulatory mechanisms which are also found in other parts of metabolism. Regulatory abnormalities, e.g., the noncoordinated regulation of single enzymes or small enzyme groups within metabolic pathways, or the nonexpression of genetic material coding for secondary metabolic enzymes, as well as the latency of some enzymes of secondary metabolism under normal physiological conditions, are visualized as the consequence of the low selective pressures acting on the details of secondary metabolic pathways.

References

Acton, G.J., Schopfer, P.: Control over activation or synthesis of phenylalanine ammonia-lyase by phytochrome in mustard (*Sinapis alba* L.). Biochim. Biophys. Acta *404*, 231–242 (1975)

Albersheim, P.: The primary cell wall. In: Plant biochemistry. Bonner, J., Varner, V.E., (eds.), pp. 225–274. New York: Academic Press 1976

Alfermann, A.W., Reinhard, E.: Isolierung anthocyanhaltiger und anthocyanfreier Gewebestämme von *Daucus carota*: Einfluß von Auxinen auf die Anthocyanbildung. Experientia *27*, 353–354 (1971)

Alfermann, A.W., Reinhard, E.: Influence of phytohormones on secondary product formation in plant cell cultures. In: Secondary Metabolism and Coevolution. Luckner, M., Mothes, K., Nover, L., (eds.). Nova Acta Leopold Suppl. *7*, 345–356 (1976)

Alfermann, A.W., Merz, D., Reinhard, E.: Induktion der Anthocyanbiosynthese in Kalluskulturen von *Daucus carota*. Planta Med. Suppl. 70–78 (1975)

Alibert, G., Ranjeva, R., Boudet, A.M.: Recherches sur les enzymes catalysant la formation des acides phénoliques chez *Quercus pedunculata* (EHRH.). II Localisation intracellulaire de la phenylalanine ammoniaque-lyase, de la cinnamate 4-hydroxylase, et de la "benzoate synthase". Biochim. Biophys. Acta *279*, 282–289 (1972)
Alibert, G., Ranjeva, R., Boudet, A.M.: Organisation subcellulaire de voies de synthèse des composés phénoliques. Physiol. Vég. *15*, 279–301 (1977)
Amelunxen, F., Arbeiter, H.: Untersuchungen an den Spritzdrüsen von *Dictamnus albus* L. Z. Pflanzenphysiol. *58*, 49–69 (1967)
Amelunxen, F., Arbeiter, H.: Untersuchungen an den Drüsenhaaren von *Cleome spinosa* L. Z. Pflanzenphysiol. *61*, 73–80 (1969)
Amelunxen, F., Thio Tiang Nio, E., Spiess, E.: Untersuchungen zur Zellwandbildung. I Elektronenmikroskopische Beobachtungen zur Feinstruktur der Ranken von *Cucurbita maxima* L. Cytobiol. *13*, 233–250 (1976)
Amrhein, N.: Novel inhibitors of phenylpropanoid metabolism in higher plants. In: Regulation of secondary product and plant hormone metabolism. Luckner, M., Schreiber, K. (eds.). Proc. 12th FEBS Meet. Vol. 55. Oxford: Pergamon Press, 173–182 (1979)
Amrhein, N., Gödeke, K.H.: The estimation of phenylalanine ammonia-lyase (PAL)-activity in intact cells of higher plant tissue II. Correlations and discrepancies between activities measured in intact cells and cell-free extracts. Planta *131*, 41–45 (1976)
Amrhein, N., Gödeke, K.H.: α-Aminooxy-β-phenylpropionic acid – a potent inhibitor of L-phenylalanine ammonia-lyase in vitro and in vivo. Plant Sci. Lett. *8*, 313–317 (1977)
Amrhein, N., Gödeke, K.H., Gerhardt, J.: The estimation of phenylalanine ammonia-lyase (PAL)-activity in intact cells of higher plant tissue I. Parameters of the assay. Planta *131*, 33–40 (1976a)
Amrhein, N., Gödeke, K.H., Kefeli, V.I.: The estimation of relative intracellular phenylalanine ammonia-lyase (PAL)-activities and the modulation in vivo and in vitro by competitive inhibitors. Ber. Dtsch. Bot. Ges. *89*, 247–259 (1976b)
Bach, M.L., Gilvarg, C.: Biosynthesis of dipicolinic acid in sporulating *Bacillus megaterium*. J. Biol. Chem. *241*, 4563–4564 (1966)
Bartnicki-Garcia, S.: Cell wall chemistry, morphogenesis and taxonomy of fungi. Annu. Rev. Microbiol. *22*, 87–108 (1968)
Bartsch, E., Lerbs, W., Luckner, M.: Phenoloxidase activity and pigment synthesis in conidiospores of *Penicillium cyclopium*. Z. Allg. Mikrobiol. *19*, 75–82 (1979)
Barz, W.: Abbau von Flavonoiden und Isoflavonoiden – ein Überblick. Ber. Dtsch. Bot. Ges. *88*, 71–81 (1975)
Baxter, C., Slaytor, M.: Biosynthesis and turnover of N, N-dimethyltryptamine and 5-methoxy-N,N-dimethyltryptamine in *Phalaris tuberosa*. Phytochemistry *11*, 2767–2773 (1972)
Benveniste, I., Salaün, J.-P., Durst, F.: Phytochrome-mediated regulation of a monooxygenase hydroxylating cinnamic acid in etiolated pea seedlings. Phytochemistry *17*, 359–363 (1978)
Böhm, H.: Über *Papaver bracteatum* Lindl., 8. Mitt.: Erneutes Auftreten von Alpinigenin in Thebain-Typen nach Verfütterung von Tetrahydropalmatin. Biochem. Physiol. Pflanz. *162*, 474–477 (1971)
Böhm, H.: Secondary metabolism in cell cultures of higher plants and problems of differentiation. In: Secondary metabolism and cell differentiation. Luckner, M., Nover, L., Böhm, H. (eds.), pp. 103–123. Berlin, Heidelberg, New York: Springer 1977
Böhm, H.: Regulation of alkaloid production in plant cell cultures. In: Frontiers of plant tissue culture. Thorpe, T.A. (ed.). pp. 201–211. University of Calgary Offset Printing Service, 1978
Bough, W.A., Gander, J.E.: Exogenous L-tyrosine metabolism and dhurrin turnover in Sorghum seedlings. Phytochemistry *10*, 67–77 (1971)
Browning, G., Saunders, P.F.: Membrane localized gibberellins A_9 and A_4 in wheat chloroplasts. Nature (London) *265*, 375–377 (1977)
Bull, A.T., Chesters, C.G.C.: The biochemistry of laminarin and the nature of laminarinase. Adv. Enzymol. *28*, 325–364 (1966)
Bu'Lock, J.D.: The two-faced microbiologist: contributions of pure and applied microbiology to good research. Dev. Ind. Microbiol. *16*, 11–19 (1975)
Bu'Lock, J.D., Barr, J.G.: A regulation mechanism linking tryptophan uptake and synthesis with ergot alkaloid synthesis in *Claviceps*. Lloydia *31*, 342–354 (1968)

Bu'Lock, J.D., Shepherd, D., Winstanley, D.J.: Regulation of 6-methylsalicylate and patulin synthesis in *Penicillium urticae*. Can. J. Microbiol. *15*, 279–285 (1969)
Burg, S.P.: Ethylene in plant growth. Proc. Natl. Acad. Sci. U.S.A. *70*, 591–597 (1973)
Cabib, E., Ulane, R.: Chitin synthetase activating factor from yeast, a protease. Biochem. Biophys. Res. Commun. *50*, 186–191 (1973)
Callow, M.E., Evans, L.V.: Localization of sulphated polysaccharides by X-ray microanalysis in *Laminaria saccharina*. Planta *131*, 155–157 (1976)
Camm, E.L., Towers, G.H.N.: Phenylalanine ammonia-lyase. Phytochemistry *12*, 961–973 (1973)
Charriére-Ladreix, Y.: Repartition intracellulaire du sécrétat flavonique de *Populus nigra* L. Planta *129*, 167–174 (1976)
Colomas, J., Bulard, C.: Irradiations à faible énergie et biosynthèse d'amarantine chez des plantules d'*Amarantus tricolor* L. var. bicolor ruber Hort. Planta *124*, 245–254 (1975)
Colomas, J., Laytou, J.-L., Bulard, C.: Influence de deux gibbérellines GA_3 et GA_4 sur la biosynthèse de l'amaranthine induite par la kinétine chez *Amaranthus tricolor* L. var bicolor ruber Hort. C.R. Acad. Sci. Paris *277 D*, 173–176 (1973)
Constabel, F., Shyluk, J.P., Gamborg, O.L.: The effect of hormones on anthocyanin accumulation in cell cultures of *Haplopappus gracilis*. Planta *96*, 306–316 (1971)
Cooke, R.J., Kendrick, R.E.: Phytochrome controlled gibberellin metabolism in etioplast envelopes. Planta *131*, 303–307 (1976)
Conn, E.E., McFarlane, I.J., Møller, B.L., Shimada, M.: Channelling of intermediates during biosynthesis of cyanogenic glycosides. In: Regulation of secondary product and plant hormone metabolism. Luckner, M., Schreiber, K. (eds.). Proc. 12th FEBS Meet. Vol. 55. Oxford: Pergamon Press, 63–71 (1979)
Costes, C., Burghoffer, C., Carrayol, E., Ducet, G., Diano, M.: Occurence of carotenoids in non-plastidial materials from potato tuber cells. Plant Sci. Lett. *6*, 253–259 (1976)
Craker, L.E.: Effect of ethylene and metabolic inhibitors on anthocyanin biosynthesis. Phytochemistry *14*, 151–153 (1975)
Craker, L.E., Wetherbee, P.J.: Ethylene, light, and anthocyanin synthesis. Plant Physiol. *51*, 436–438 (1973a)
Craker, L.E., Wetherbee, P.J.: Ethylene, carbon dioxide, and anthocyanin synthesis. Plant Physiol. *52*, 177–179 (1973b)
Craker, L.E., Abeles, F.B., Shropshire, W.: Light-induced ethylene production in Sorghum. Plant Physiol. *51*, 1082–1083 (1973)
Czapek, F.: Biochemie der Pflanzen, Vol. III, p. 220, Jena: Fischer Verlag 1920/1921
Czichi, U., Kindl, H.: A model of closely assembled consecutive enzymes of membranes: Formation of hydroxycinnamic acids from L-phenylalanine on thylakoids of *Dunaliella marina*. Hoppe-Seylers Z. Physiol. Chem. *356*, 475–485 (1975a)
Czichi, U., Kindl, H.: Formation of p-coumaric acid and o-coumaric acid from L-phenylalanine by microsomal membrane fractions from potato: Evidence of membrane bound enzyme complexes. Planta *125*, 115–125 (1975b)
Czichi, U., Kindl, H.: Phenylalanine ammonia-lyase and cinnamic acid hydroxylases as assembled consecutive enzymes on microsomal membranes of cucumber cotyledons: Cooperation and subcellular distribution. Planta *134*, 133–143 (1977)
Dekker, R.H.F., Richards, G.N.: Hemicellulases: their occurence, purification, properties and mode of action. Adv. Carbohydr. Chem. Biochem. *32*, 277–352 (1976)
Demain, A.L.: Regulatory mechanisms and the industrial production of microbial metabolites. Lloydia *31*, 395–418 (1968)
Demain, A.L.: Overproduction of microbial metabolites and enzymes due to alteration of regulation. Advan. Biochem. Eng. *1*, 113–142 (1971)
Demain, A.L.: Cellular and environment factors affecting the synthesis and excretion of metabolites. J. Appl. Chem. Biotechnol. *22*, 345–362 (1972)
Dhillon, D.S., Brown, S.A.: Localization, purification and characterization of dimethylallylpyrophosphate: umbelliferone dimethylallyltransferase from *Ruta graveolens*. Arch. Biochem. Biophys. *177*, 74–83 (1976)
Dimroth, P., Walter, H., Lynen, F.: Biosynthese von 6-Methylsalicylsäure. Eur. J. Biochem. *13*, 98–110 (1970)

Dimroth, P., Greull, G., Seyffert, R., Lynen, F.: 6-Methylsalicylic acid synthetase. Hoppe-Seylers Z. Physiol. Chem. *353*, 126 (1972)

Dimroth, P., Ringelmann, E., Lynen, F.: 6-Methylsalicylic acid synthetase from *Penicillium patulum*. Some catalytic properties of the enzyme and its relation to fatty acid synthetase. Eur. J. Biochem. *68*, 591–596 (1976)

Drew, S.W., Demain, A.L.: Methionine control of cephalosporin C formation. Biotechnol. Bioengineering *15*, 743–754 (1973)

Drumm, H., Wildermann, A., Mohr, H.: The "high-irradiance response" in anthocyanin formation as relatec to phytochrome level. Photochem. Photobiol. *21*, 269–273 (1975)

Dungen, von, A., Vater, J., Kleinkauf, H.: Biosynthesis of gramicidin S with the aid of dipeptides by gramicidin S synthetase. Eur. J. Biochem. *66*, 623–626 (1976)

Dunkel, R., Müller, W., Nover, L., Luckner, M.: Stimulation of alkaloid formation in *Penicillium cyclopium* Westling by phenylalanine and mycelial extracts. In: Secondary metabolism and coevolution. Luckner, M., Mothes, K., Nover, L. (eds.). Nova Acta Leopold. (Suppl. *7*), 281–288 (1976)

Durán, A., Bowers, B., Cabib, E.: Chitin synthetase zymogen is attached to the yeast plasma membrane. Proc. Natl. Acad. Sci. U.S.A. *72*, 3952–3955 (1975)

Ebel, J., Hahlbrock, K.: Enzymes of flavone and flavonol-glycoside biosynthesis. Coordinated and selective induction in cell suspension cultures of *Petroselinum hortense*. Eur. J. Biochem. *75*, 201–209 (1977)

Egger, K., Sitte, P.: Plastiden und Photosynthese. In: Grundlagen der Cytologie. Hirsch, C.G., Ruska, H., Sitte, P. (eds.), pp. 345–387. Jena: VEB G. Fischer Verlag 1973

El Kousy, S., Pfeiffer, E., Ininger, G., Roos, W., Nover, L., Luckner, M.: Influence of inhibitors of gene expression on processes of cell specialization during idiophase development of *Penicillium cyclopium* Westling. Biochem. Physiol. Pflanz. *168*, 79–85 (1975)

Engelsma, G.: Photoinduction of phenylalanine deaminase in gherkin seedlings. III. Effects of excision and irradiation on enzyme development in hypocotyl segments. Planta *82*, 355–368 (1968)

Engelsma, G.: On the mechanism of the changes in phenylalanine ammonia-lyase activity induced by ultraviolet and blue light in gherkin hypocotyls. Plant. Physiol. *54*, 702–705 (1974)

Engelsma, G.: Inhibition of phenylalanine ammonia-lyase by cinnamic acid derivatives. In: Regulation of secondary product and plant hormone metabolism. Luckner, M., Schreiber, K. (eds.). Proc. 12th FEBS Meet. Vol. 55. Oxford: Pergamon Press, 163–172 (1979)

Erge, D., Maier, M., Gröger, D.: Untersuchungen über die enzymatische Umwandlung von Chanoclavin-I. Biochem. Physiol. Pflanz. *164*, 234–247 (1973)

Evans, A., Smith, H.: Localization of phytochrome in etioplasts and its regulation in vitro of gibberellin levels. Proc. Natl. Acad. Sci. U.S.A. *73*, 138–142 (1976)

Fairbairn, J.W., Hakim, F., El Kheir, Y.: Alkaloidal storage metabolism and translocation in the vesicles of *Papaver somniferum* latex. Phytochemistry *13*, 1133–1139 (1974)

Feofilova, E.P., Arbuzov, B.A.: Mechanism of action of β-ionone on carotenoid-synthesizing enzymes of *Blakeslea trispora*. Mikrobiologija *44*, 395–399 (1975)

Feofilova, E.P., Bekhtereva, M.N.: Effect of vitamin A on biosynthesis of carotene by *Blakeslea trispora*. Mikrobiologija *45*, 557–558 (1976)

Floss, H.G., Robbers, J.E., Heinstein, P.F.: Regulatory control mechanisms in alkaloid biosynthesis. In: Metabolism and regulation of secondary plant products. Runeckles, V.C., Conn, E.E., (eds.). Recent Adv. Phytochem. *8*, 141–175 (1974a)

Floss, H.G., Tscheng-Lin, M., Kobel, H., Stadler, P.: On the biosynthesis of peptide ergot alkaloids. Experientia *30*, 1369 (1974b)

Floss, H.G., Robbers, J.E., Heinstein, P.F.: Regulation of ergot alkaloid formation: Influence of tryptophan and its analogs and of the endproduct elymoclavine. In: Regulation of Secondary Product and Plant Hormone Metabolism. Luckner, M., Schreiber, K. (eds.). Proc. 12th FEBS Meet. Vol. 55. Oxford: Pergamon Press, 121–131 (1979)

Framm, J., Nover, L., El Azzouny, A., Richter, H., Winter, K., Werner, S., Luckner, M.: Cyclopeptin und Dehydrocyclopeptin, Zwischenprodukte der Biosynthese von Alkaloiden der Cyclopenin-Viridicatin-Gruppe bei *Penicillium cyclopium* Westling. Eur. J. Biochem. *37*, 78–85 (1973)

Fritsch, H., Grisebach, H.: Biosynthesis of cyanidin in cell cultures of *Haplopappus gracilis*. Phytochemistry *14*, 2437–2442 (1975)
Frøyshov, Ø.: Bacitracin biosynthesis by three complementing factors from *Bacillus licheniformis*. FEBS Lett. *44*, 75–78 (1974)
Frøyshov, Ø., Laland, S.G.: On the biosynthesis of bacitracin by a soluble enzyme complex from *Bacillus licheniformis*. Eur. J. Biochem. *46*, 235–242 (1974)
Gaertner, F.H., Ericson, M.C., DeMoss, J.A.: Catalytic facilitation in vitro by two multienzyme complexes from *Neurospora crassa*. J. Biol. Chem. *245*, 595–600 (1970)
Gatenbeck, S., Hermodsson, S.: Enzymatic synthesis of the aromatic product alternariol. Acta Chem. Scand. *19*, 65–71 (1965)
Gaucher, G.M., Shepherd, M.G.: Isolation of orsellinic acid synthase. Biochem. Biophys. Res. Commun. *32*, 664–671 (1968)
Giudici de Nicola, M., Piatelli, M., Amico, V.: Photocontrol of betaxanthin synthesis in *Celosia plumosa* seedlings. Phytochemistry *12*, 353–357 (1973)
Giudici de Nicola, M., Amico, V., Sciuto, S., Piatelli, M.: Light control of amaranthin synthesis in isolated *Amaranthus* cotyledons. Phytochemistry *14*, 479–481 (1975)
Givan, C.V., Harwood, J.L.: Biosynthesis of small molecules in chloroplasts of higher plants. Biol. Rev. *51*, 365–406 (1976)
Gooday, G.W.: Fungal sex hormones. Annu. Rev. Biochem. *43*, 35–49 (1974)
Gooday, G.W.: Functions of trisporic acid. Phil. Trans. R. Soc. Lond. B. *284*, 509–520 (1978)
Gregor, H.D.: Studies on phenylalanine ammonia-lyase from castor bean endosperm. I. Subcellular localization of the enzyme. Z. Pflanzenphysiol. *77*, 454–463 (1976)
Grisebach, H.: Neuere Untersuchungen zur Biosynthese einiger Antibiotica: Planta Med. Suppl. 232–250 (1975)
Grisebach, H., Hahlbrock, K.: Enzymology and regulation of flavonoid and lignin biosynthesis in plants and plant cell suspension cultures. In: Metabolism and regulation of secondary plant products. Runeckles, N.C., Conn, E.E., (eds.). Recent Adv. Phytochem. *8*, 21–52 (1974)
Gröger, D.: Biosynthese des Pepidteils der Mutterkornalkaloide vom Cyclol-typ. Planta Med. *28*, 37–51 (1975)
Gröger, D.: Influence of methionine on β-lactam antibiotic formation by Cephalosporium. In: Secondary metabolism and coevolution. Luckner, M., Mothes, K., Nover, L., (eds.). Nova Acta Leopold. (Suppl.7), 271–279 (1976)
Gröger, D.: Biochemische Aspekte der β-Lactamantibiotica. Pharmazie *32*, 309–317 (1977)
Grumbach, K.H., Lichtenthaler, H.K.: Kinetic of lipoquinone and pigment synthesis during red light-induced thylakoid formation in etiolated barley seedlings. Z. Naturforsch. *30*, 337–341 (1975)
Haavik, H.I., Frøyshov, Ø.: Function of peptide antibiotics in producer organisms. Nature (London) *254*, 79–82 (1975)
Hagel, P., Kindl, H.: p-Hydroxybenzoate synthase: a complex associated with mitochondrial membranes of roots of *Cucumis sativus*. FEBS Lett. *59*, 120–124 (1975)
Hahlbrock, K.: Regulation of phenylalanine ammonia-lyase activity in cell-suspension cultures of *Petroselinum hortense*. Apparent rates of enzyme synthesis and degradation. Eur. J. Biochem. *63*, 137–145 (1976)
Hahlbrock, K.: Coordinated regulation of the enzymes of flavonoid biosynthesis. In: Cell differentiation in microorganisms, plants and animals. Nover, L., Mothes, K., (eds.). pp. 524–537. Jena: VEB G. Fischer Verlag and Elsevier/North-Holland 1977
Hahlbrock, K., Ragg, H.: Light-induced changes of enzyme activities in parsley cell suspension cultures. Effects of inhibitors of RNA and protein biosynthesis. Arch. Biochem. Biophys. *166*, 41–46 (1975)
Hahlbrock, K., Schröder, J.: Light-induced changes of enzyme activities in parsley cell suspension cultures. Increased rate of synthesis of phenylalanine ammonia-lyase. Arch. Biochem. Biophys. *166*, 47–53 (1975)
Hahlbrock, K., Wellmann, E.: Light independent induction of enzymes related to phenylpropanoid metabolism in cell suspension cultures from parsley. Biochim. Biophys. Acta *304*, 702–706 (1973)
Hahlbrock, K., Ebel, J., Ortmann, R., Sutter, A., Wellmann, E., Grisebach, H.: Regulation

of enzyme activities related to the biosynthesis of flavone glycosides in cell suspension cultures of parsley (*Petroselinum hortense*). Biochem. Biophys. Acta *244*, 7–15 (1971a)

Hahlbrock, K., Sutter, A., Wellmann, E., Ortmann, R., Grisebach, H.: Relationship between organ development and activity of enzymes involved in flavone glycoside biosynthesis in young parsley plants. Phytochemistry *10*, 109–116 (1971b)

Hahlbrock, K., Knobloch, K.H., Kreuzaler, F., Potts, J.R.M., Wellmann, E.: Coordinated induction and subsequent activity changes of two groups of metabolically interrelated enzymes. Light-induced synthesis of flavonoid glycosides in cell suspension cultures of *Petroselinum hortense*. Eur. J. Biochem. *61*, 199–206 (1976)

Hampp, R., Wellburn, A.R.: Changes in the permeability of the inner mitochondrial membrane associated with chloroplast development. Planta *131*, 21–26 (1976)

Hampp, R., Ziegler, H.: Lichtabhängige Neusynthese von δ-Aminolaevulinsäure-Dehydratase in isolierten *Avena*-Etioplasten. Planta *124*, 255–260 (1975)

Hanson, R.S., Peterson, J.A., Yousten, A.A.: Unique biochemical events in bacterial sporulation. Annu. Rev. Microbiol. *24*, 53–90 (1970)

Heinstein, P., Floss, H.G.: Feedback inhibition and end product repression in secondary metabolism. In: Secondary metabolism and coevolution. Luckner, M., Mothes, K., Nover, L., (eds.). Nova Acta Leopold. (Suppl. *7*), 299–310 (1976)

Heinstein, P., Lee, S.-L., Floss, H.G.: Isolation of dimethylallylpyrophosphate: tryptophan dimethylallyl transferase from the ergot fungus (*Claviceps* spec.). Biochem. Biophys. Res. Commun. *44*, 1244–1251 (1971)

Heinzmann, U., Seitz, U.: Synthesis of phenylalanine ammonia-lyase in anthocyanin-containing and anthocyanin-free callus cells of *Daucus carota* L. Planta *135*, 63–67 (1977)

Helsper, J.P.F.G., Veerkamp, J.H., Sassen, M.M.A.: β-Glucan synthetase activity in golgi vesicles of *Petunia hybrida*. Planta *133*, 303–308 (1977)

Henry, E.W.: Peroxidases in tobacco abscission zone tissue. III. Ultrastructural localization in thylakoids and membrane-bound bodies of chloroplasts. J. Ultrastruct. Res. *52*, 289–299 (1975a)

Henry, E.W.: Polyphenol oxidase activity in thylakoids and membrane-bound granular components of *Nicotiana tabacum* chloroplasts. J. Microsc. Biol. Cell. *22*, 109–116 (1975b)

Henry, E.W.: The ultrastructural localization of polyphenol oxidase in chloroplasts of *Brassica napus* cv. Zephyr. Z. Pflanzenphysiol. *78*, 446–452 (1976)

Hess, D.: Die Wirkung von Zimtsäuren auf die Anthocyansynthese in isolierten Petalen von *Petunia hybrida*. Z. Pflanzenphysiol. *56*, 12–19 (1967a)

Hess, D.: Substratinduktion bei der Anthocyansynthese von *Petunia*. Naturwissenschaften *54*, 289–290 (1967b)

Hess, D.: Substratinduktion durch Zimtsäuren als physiologischer Mechanismus bei der Einleitung der Anthocyansynthese. Z. Pflanzenphysiol. *59*, 293–296 (1968)

Hess, S.L., Hadwiger, L.A.: The induction of phenylalanine ammonia-lyase and phaseollin by 9-aminoacridine and other deoxyribonucleic acid intercalating compounds. Plant Physiol. *48*, 197–202 (1971)

Holtzer, H., Rubinstein, N.: Binary decisions, quantal cell cycles and cell diversification. In: Cell differentiation in microorganisms, plants and animals. Nover, L., and Mothes, K., (eds.), pp. 424–437. Jena: VEB G. Fischer Verlag and Elsevier/North-Holland 1977

Holtzer, H., Weintraub, H., Biehl, J.: Cell cycle-dependent events during myogenesis, neurogenesis, amd erythrogenesis. In: Biochemistry of cell differentiation. Monroy, A., Tsanev, R., (eds.), pp. 41–53. London: Academic Press 1973

Holzer, H.: Chemistry and biology of macromolecular inhibitors from yeast acting on proteinases A and B, and carboxypeptidase Y. Adv. Enzyme Regul. *13*, 125–134 (1975)

Hyodo, H., Yang, S.F.: The effect of ethylene on the development of phenylalanine ammonia-lyase in potato tuber discs. Z. Pflanzenphysiol. *71*, 76–79 (1974)

Ininger, G., Nover, L.: Regulation of β-galactosidase formation in emerged cultures of *Penicillium cyclopium* Westling. Biochem. Physiol. Pflanz. *167*, 585–595 (1975)

Iredale, S.E., Smith, H.: Properties of phenylalanine ammonia-lyase extracted from *Cucumis sativus* hypocotyls. Phytochemistry *13*, 575–583 (1974)

Jans, B.P.: Untersuchungen am Milchsaft des Schöllkrautes (*Chelidonium majus*). Ber. Schweiz. Bot. Ges. *83*, 306–344 (1973)

Johnson, C., Attridge, T., Smith, H.: Regulation of phenylalanine ammonia-lyase synthesis

by cinnamic acid. Its implication of the light mediated regulation of the enzyme. Biochim. Biophys. Acta *385*, 11–19 (1975)
Kahn, A., Avivi-Bleiser, N., Wettstein, von, D.: Genetic regulation of chlorophyll synthesis analyzed with double mutants in barley. In: Genetics and Biogenesis of chloroplasts and mitochondria. Bücher, T., Neupert, W., Sebald, W., Werner, S., (eds.), pp. 119–131. Amsterdam: Elsevier/North-Holland 1976
Kalghatgi, K.K., Rao, P.V.S.: Microbial L-phenylalanine ammonia-lyase. Purification, subunit structure and kinetic properties of the enzyme from *Rhizoctonia solani*. Biochem. J. *149*, 65–72 (1975)
Kambe, M., Imae, Y., Kurabashi, K.: Biochemical studies on gramicidin S non-producing mutants of *Bacillus brevis* ATCC 9999. J. Biochem. (Tokyo) *75*, 481–493 (1974)
Kang, G., Burg, S.P.: Role of ethylene in phytochrome induced anthocyanin synthesis. Planta *110*, 227–235 (1973)
Kang, G., Newcomb, W., Burg, S.P.: Mechanism of auxin induced ethylene production. Plant Physiol. *47*, 504–509 (1971)
Karr, A.L.: Cell wall biogenesis. In: Plant biochemistry. Bonner, J., Varner, V.E., (eds.), pp. 405–426. New York: Academic Press 1976
Kato, C., Uritani, I., Saiyo, R., Takeo, T.: Cellular localization of particulate-bound polyphenol oxidase in tea leaves. Plant Cell Physiol. (Tokyo) *17*, 1045–1052 (1976)
Khokhlov, A.S., Tovarova, I.I.: Autoregulator from *Streptomyces griseus*. In: Regulation of secondary product and plant hormone metabolism. Luckner, M., Schreiber, K. (eds.). Proc. 12th FEBS Meet. Vol. 55. Oxford: Pergamon Press, 133–145 (1979)
Khokhlov, A.S., Tovarova, I.I., Borisova, L.N., Pliner, S.A., Shevchenko, L.A., Kornitskaja, E.Ya., Ivkina, N.S., Rapoport, I.A.: The A-factor responsible for the biosynthesis of streptomycin in mutant strains of *Actinomyces streptomycini*. Dokl. Akad. Nauk USSR *177*, 232–235 (1967)
Khokhlov, A.S., Anisova, L.N., Tovarova, I.I., Kleiner, E.M., Kovalenko, I.V., Krasilnikova, O.I., Kornitskaya, I.Ya., Pliner, S.A.: Effect of A-factor on the growth of asporogenous mutants of *Streptomyces griseus*, not producing this factor. Z. Allg. Mikrobiol. *13*, 647–655 (1973)
Khokhlov, A.S., Tovarova, I.I., Anisova, L.N.: Regulators of streptomycin biosynthesis and development of *Actinomyces streptomycini*. In: Secondary metabolism and coevolution. Luckner, M., Mothes, K., Nover, L. (eds.). Nova Acta Leopold. (Suppl. *7*), 289–298 (1976)
Kindl, H.: Compartments and microcompartments channeling intermediates in phenylpropanoid metabolism. In: Regulation of secondary product and plant hormone metabolism. Luckner, M., Schreiber, K. (eds.). Proc. 12th FEBS Meet. Vol. 55. Oxford: Pergamon Press, 49–61 (1979)
Kinsman, L.T., Pinfield, N.J., Stobart, A.K.: A gibberellin bioassay based on betacyanin production in *Amaranthus caudatus* seedlings. Planta *127*, 149–152 (1975a)
Kinsman, L.T., Pinfield, N.J., Stobart, A.K.: The hormonal control of amaranthin biosynthesis in *Amaranthus caudatus* seedlings. Planta *127*, 207–212 (1975b)
Kleiner, E.M., Pliner, S.A., Soifer, V.S., Onoprienko, V.V., Balashova, T.A., Rozynov, B.V., Khokhlov, A.S.: The structure of A-factor, a bioregulator from *Streptomyces griseus*. Bioorg. Chim. *2*, 1142–1147 (1976)
Kleudgen, H.K., Lichtenthaler, H.K.: Die Wirkung von Phytochrom auf die Bildung von Einzelkarotinoiden in etiolierten *Hordeum*-Keimlingen. Z. Naturforsch. *30*c, 67–68 (1975)
Koehler, D.E., Leonard, R.I., Vanderwoude, W.J., Linkins, A.E., Lewis, L.N.: Association of latent cellulase activity with plasma membranes from kidney bean abscission zones. Plant. Physiol. *58*, 324–330 (1976)
Köhler, K.H.: Action of inhibitors of protein and nucleic acid synthesis on light-dependent and kinetin-stimulated betacyanin synthesis. Phytochemistry *11*, 127–131 (1972)
Köhler, K.H.: Die Steuerung der Amaranthinbiosynthese durch das Phytochromsystem (einschließlich Beschreibung der Bestrahlungsanlage und der Lichtenergiemessung). Biol. Zentralbl. *92*, 307–336 (1973a)
Köhler, K.H.: Die Betalaine, eine neue Klasse chymochromer Pigmente. Pharmazie *28*, 18–24 (1973b)
Köhler, K.H.: Kinetik der durch Inhibitoren des Protein- und Nukleinsäurestoffwechsels be-

wirkten Hemmung der Amaranthinbildung bei *Amaranthus caudatus* Keimlingen. Biochem. Physiol. Pflanz. *168*, 113–122 (1975)
Köhler, K.H., Conrad, K.: Ein quantıtativer Phytokinin-Test. Biol. Rundsch. *4*, 36–37 (1966)
Kostytschew, S.: Lehrbuch der Pflanzenphysiologie. Berlin: Springer 1926
Koukol, J., Conn, E.E.: The metabolism of aromatic compounds in higher plants. IV. Purification and properties of the phenylalanine deaminase of *Hordeum vulgare*. J. Biol. Chem. *236*, 2692–2698 (1961)
Kreuzaler, F., Hahlbrock, K.: Enzymatic synthesis of an aromatic ring from acetate units. Partial purification and some properties of flavanone synthetase from cell-suspension cultures of *Petroselinum hortense*. Eur. J. Biochem. *56*, 205–213 (1975a)
Kreuzaler, F., Hahlbrock, K.: Enzymatic synthesis of aromatic compounds in higher plants. Formation of bis-noryangonin (4-hydroxy-6[4-hydroxystyryl]2-pyrone) from p-coumaryl-CoA and malonyl CoA. Arch. Biochem. Biophys. *169*, 84–90 (1975b)
Krupinski, V.M., Robbers, J.E., Floss, H.G.: Physiological study of ergot: Induction of alkaloid synthesis by tryptophan at the enzymatic level. J. Bacteriol. *125*, 158–165 (1976)
Kuhn, B., Forkmann, G., Seyffert, W.: Genetic control of chalcone-flavanone isomerase activity in *Callistephus chinensis*. Planta *138*, 199–203 (1978)
Kurahashi, K.: Biosynthesis of small peptıdes. Annu. Rev. Biochem. *43*, 445–459 (1974)
Kurylo-Borowska, Z.: Biosynthesis of edeine. II. Localization of edeine synthetase within *Bacillus brevis* Vm4. Bıochim Biophys. Acta *339*, 31–41 (1975)
Kurylo-Borowska, Z., Sedkowska, J.: Biosynthesis of edeine. Biochim. Biophys. Acta *351*, 42–56 (1974)
Lächele, E., Lehner, F., Seitz, U.: Aufnahme und Verbreitung von ^{14}C-markierten Vorstufen der Anthocyane unter dem Einfluß von Gibberellinsäure A_3 bei Calluskulturem von *Daucus carota*. Z. Pflanzenphysiol. *71*, 375–383 (1974)
Laland, S.G., Zimmer, T.L.: The protein thiotemplate mechanism of synthesis for the peptide antibiotics produced by *Bacillus brevis*. Essays Biochem. *9*, 31–57 (1973)
Laloraya, M.M., Srivastan, H.N., Guruprasad, K.N.: Recovery of gibberellic acid inhibition of betacyanin biosynthesis by pigment precursors. Planta *128*, 275–276 (1976)
Lamb, C.J.: *Trans*-Cinnamic acid as a mediator of the light stimulated increase in hydroxycinnamoyl-CoA: quinate hydroxycinnamoyl transferase. FEBS Lett. *75*, 37–40 (1977)
Lamb, C.J., Rubery, P.H.: Phenylalanine ammonia-lyase and cinnamıc acid 4-hydroxylase: product repression of the level of enzyme activity in potato tuber discs. Planta *130*, 283–290 (1976a)
Lamb, C.J., Rubery, P.H.: Photocontrol of chlorogenic acid biosynthesis in potato tuber discs. Phytochemistry *15*, 665–668 (1976b)
Lange, H., Mohr, H.: Die Hemmung der Phytochrom-induzierten Anthocyansynthese durch Actinomycin D und Puromycin. Planta *67*, 107–121 (1965)
Lee, S.G., Lipmann, F.: Isolation of amino acid activating subunit-pantetheine protein complexes: Their role in chain elongation in tyrocidine synthesis. Proc. Natl. Acad. Sci. U.S.A. *74*, 2343–2347 (1977)
Lee, S.G., Roskoski, R., Bauer, K., Lipmann, F.: Purification of the polyenzymes responsible for tyrocidine synthesis and their dissociation into subunits. Biochemistry *12*, 398–405 (1973)
Lee, S.G., Littau, V., Lipmann, F.: The relation between sporulatıon and the induction of antibiotic synthesis and of amıno acıd uptake in *Bacillus brevis*. J. Cell Biol. *66*, 233–242 (1975)
Lichtenthaler, H.K.: Regulation of prenyl chain synthesis in etiolated *Hordeum* seedlings by far-red and white light. Physiol. Plant. *33*, 241–244 (1975a)
Lichtenthaler, H.K.: Control of light-induced carotenoid synthesis ın *Raphanus* seedlings by phytochrome. Physiol. Plant. *34*, 357–358 (1975b)
Lichtenthaler, H.K., Weinert, H.: Die Beziehungen zwischen Lipochinonsynthese und Plastoglobulibildung in den Chloroplasten von *Ficus elastica* Roxb. Z. Naturforsch. *25b*, 619–623 (1970)
Lieberei, R., Biehl, B.: Freisetzung und Aktivierung von Polyphenoloxydasen aus Thylakoidmembranen der Spinat-Chloroplasten. Ber. Dtsch. Bot. Ges. *89*, 663–676 (1976)
Liebermann, M., Kunishi, A.T.: Ethylene forming systems in etiolated pea seedlings and apple tissue. Plant Physiol. *55*, 1074–1078 (1975)

Light, R.J.: Effects of cycloheximide and amino acid analogues on biosynthesis of 6-methylsalicylic acid in *Penicillium patulum*. Arch. Biochem. Biophys. *122*, 494–500 (1967)
Light, R.J.: 6-Methylsalicylic acid decarboxylase from *Penicillium patulum*. Biochim. Biophys. Acta *191*, 430–438 (1969)
Light, R.J.: Enzymatic studies on polyketide biosynthesis. J. Agric. Food. Chem. *18*, 260–267 (1970)
Light, R.J., Hager, L.P.: Molecular parameters of 6-methylsalicylic acid synthetase from gel filtration and sucrose density gradient centrifugation. Arch. Biochem. Biophys. *125*, 326–333 (1968)
Lipmann, F.: Attempts to map a process evolution of peptide biosynthesis. Science *173*, 875–884 (1971)
Lipmann, F.: Nonribosomal polypeptide synthesis on polyenzyme templates. Acc. Chem. Res. *6*, 361–367 (1973)
Liu, E.H., Lamport, D.T.H.: An accounting of horseradish peroxidase isoenzymes associated with the cell well and evidence that peroxidase does not contain hydroxyproline. Plant. Physiol. *54*, 870–876 (1974)
Löffelhardt, W., Kindl, H.: The conversion of L-phenylalanine into benzoic acid on the thylakoid membranes of higher plants. Hoppe-Seylers Z. Physiol. Chem. *356*, 487–493 (1975)
Löffelhardt, W., Kindl, H.: Formation of benzoic acid and p-hydroxybenzoic acid in the blue green alga *Anacystis nidulans*. A thylakoid-bound enzyme complex analogous to the chloroplast system. Z. Naturforsch. *31c*, 693–699 (1976)
Luckner, M.: Was ist Sekundärstoffwechsel? Pharmazie *26*, 717–724 (1971)
Luckner, M.: The integration of benzodiazepine and quinoline alkaloid formation into the developmental program of *Penicillium cyclopium*. In: Cell differentiation in microorganisms, plants and animals. Nover, L., Mothes, K. (eds.), pp. 538–558. Jena: VEB G. Fischer Verlag and Elsevier/North-Holland 1977
Luckner, M., Nover, L., Böhm, H.: Secondary metabolism and cell differentiation. Berlin, Heidelberg, New York: Springer 1977
Luckner, M., Diettrich, B., Lerbs, W.: Secondary metabolism in microorganisms and higher plants. Cellular compartmentation and channelling. Progress Phytochem., in press (1979)
Mäder, M.: Die Lokalisation der Peroxidase-Isoenzymgruppe G_1 in der Zellwand von Tabak-Geweben. Planta *131*, 11–15 (1976)
Mäder, M., Meyer, Y., Bopp, M.: Lokalisation der Peroxidase-Isoenzyme in Protoplasten und Zellwänden von *Nicotiana tabacum* L. Planta *122*, 259–268 (1975)
Maier, W., Gröger, D.: Dimethylallyltryptophan-Synthetase in *Claviceps*. Biochem. Physiol. Pflanz. *170*, 9–15 (1976)
Mandelstam, J.: Recurring patterns during development in primitive organisms. Symp. Soc. Exp. Biol. *25*, 1–26 (1971)
Mandelstam, J.: Bacterial sporulation: a problem in the biochemistry and genetics of a primitive developmental system. Proc. R. Soc. London Ser. B *193*, 89–106 (1976)
Mann, J.D., Steinhart, C.E., Mudd, S.H.: Alkaloids and plant metabolism. J. Biol. Chem. *238*, 676–681 (1963)
Margna, U.: Control at the level of substrate supply – an alternative in the regulation of phenylpropanoid accumulation in plant cells. Phytochemistry *16*, 419–426 (1977)
Marshall, R., Redfield, B., Katz, E., Weissbach, H.: Changes in phenoxazinone synthetase activity during the growth cycle of *Streptomyces antibioticus*. Arch. Biochem. Biophys. *123*, 317–323 (1968)
Matile, P.: The lytic compartment of plant cells. In: Cell biology monographs. Vol. *1*, Wien: Springer 1975
Matile, P.: Localization of alkaloids and mechanisms of their accumulation in vacuoles of *Chelidonium majus* laticifers. In: Secondary metabolism and coevolution. Luckner, M., Mothes, K., Nover, L. (eds.). Nova Acta Leopold. Suppl. *7*, 139–156 (1976a)
Matile, P.: Vacuoles. In: Plant biochemistry. Bonner, J., Varner, V.E. (eds.), pp. 189–224. New York: Academic Press 1976b
McClure, J.W.: Physiology and function of flavonoids. In: The Flavonoids. Harborne, J.B., Mabry, T.J., Mabry, H. (eds.), pp. 970–1055. London: Chapman and Hall 1975
McClure, J.W., Gross, G.G.: Diverse photoinduction characteristics of hydroxycinnamate

coenzyme A ligase and phenylalanine ammonia-lyase in dicotyledonous seedlings. Z. Pflanzenphysiol. *76*, 51–55 (1975)

McFarlane, I.J., Lees, E.M., Conn, E.E.: The *in vitro* biosynthesis of dhurrin, the cyanogenic glycoside of *Sorghum bicolor*. J. Biol. Chem. *250*, 4708–4713 (1975)

Mizusaki, S., Tanabe, Y., Noguchi, M., Tamaki, E.: Changes in the activities of ornithine decarboxylase, putrescine N-methyltransferase and N-methylputrescine oxidase in tobacco roots in relation to nicotine biosynthesis. Plant Cell Physiol. (Tokyo) *14*, 103–110 (1973)

Mohr, H., Drumm, H., Kasemir, H.: Licht und Farbstoffe. Ber. Dtsch. Bot. Ges. *87*, 49–69 (1974)

Møller, B.L., Conn, E.E.: Biological formation of N-hydroxyamino acids. Proc. West. Pharmacol. Soc. *20*, 103–107 (1977)

Mothes, K.: Zur Problematik der metabolischen Exkretion bei Pflanzen. Naturwissenschaften *53*, 317–323 (1966)

Mothes, K.: Biologie der Alkaloide. In: Biosynthese der Alkaloide. Mothes, K., Schütte, H.R. (eds.), pp. 1–20. Berlin: VEB Deutscher Verlag der Wissenschaften 1969

Müller, E., Neumann, D., Nelles, A., Bräutigam, E.: Alkaloid transport and distribution in *Nicotiana* and *Macleaya*. Diversity and Unity. In: Secondary metabolism and coevolution Luckner, M., Mothes, K., Nover, L. (eds.). Nova Acta Leopold. Suppl. *7*, 133–138 (1976)

Nari, J., Mouttet, C., Fouchier, F., Ricard, J.: Subunit interactions in enzyme catalysis. Kinetic analysis of subunit interactions in the enzyme L-phenylalanine ammonia-lyase. Eur. J. Biochem. *41*, 499–515 (1974)

Neumann, D.: Beiträge zur Physiologie der Alkaloide. VI. Ultrastruktur der Parenchymzellen und Speicherung von Alkaloid in Blattstecklingen von *Nicotiana rustica*. Biochem. Physiol. Pflanz. *168*, 511–518 (1975)

Neumann, D.: Interrelationship between the morphology of alkaloid storage cells and the possible mechanism of alkaloid storage. In: Secondary metabolism and coevolution. Luckner, M., Mothes, K., Nover, L. (eds.). Nova Acta Leopold. Suppl. *7*, 77–81 (1976)

Nover, L.: Phenylalanine compartmentation and alkaloid synthesis in *Penicillium cyclopium* Westling. In: Regulation of secondary product and plant hormone metabolism. Luckner, M., Schreiber, K. (eds.). Proc. 12th FEBS Meet. Vol. 55. Oxford: Pergamon Press, 73–89 (1979)

Nover, L., Luckner, M.: Expression of secondary metabolism as part of the differentiation processes during the idiophase development of *Penicillium cyclopium* Westling. Biochem. Physiol. Pflanz. *166*, 293–305 (1974)

Nover, L., Luckner, M.: Influence of inhibitors of gene expression on processes of cell specialization in *Penicillium cyclopium* Westling. In: Secondary metabolism and coevolution. Luckner, M., Mothes, K., Nover, L. (eds.). Nova Acta Leopold. Suppl. *7*, 229–241 (1976)

Nover, L., Müller, W.: Influence of cycloheximide on the expression of alkaloid metabolism in partially synchronized emerged cultures of *Penicillium cyclopium* Westling. FEBS Lett. *50*, 17–20 (1975)

Nover, L., Luckner, M., Parthier, B. (eds.): Differenzierung – Molekulare Grundlagen, Regulation. Jena: VEB G. Fischer Verlag 1978

Nover, L., Lerbs, W., Müller, W., Luckner, M.: Channelling of exogenous applied phenylalanine to the sites of storage and of alkaloid and protein biosynthesis in *Penicillium cyclopium*. Biochim. Biophys. Acta *584*, 270–283 (1979)

Ornston, L.N.: Regulation of catabolic pathways in *Pseudomonas*. Bacteriol. Rev. *35*, 87–116 (1971)

Packter, N.M., Collins, J.S.: Effect of inhibitors of protein synthesis on the formation of phenols derived from acetate and shikimic acid in *Aspergillus fumigatus*. Eur. J. Biochem. *42*, 291–302 (1974)

Paech, K.: Biochemie und Physiologie der sekundären Pflanzenstoffe. Berlin: Springer 1950

Parish, R.W.: The lysosome-concept in plants. I. Peroxidases associated with subcellular and wall fractions of maize root tips: Implications for vacuole development. Planta *123*, 1–13 (1975)

Pass, L., Zimmer, T.-L., Lalano, S.G.: On the use of affinity chromatography in demonstrating the transfer of thioester-bound D-phenylalanine from the light enzyme of gramicidin S synthetase to an acceptor side for this amino acid on the heavy enzyme. Eur. J. Biochem. *47*, 607–611 (1974)

Pastan, I., Perlman, R.: Cyclic adenosine monophosphate in bacteria. Science *169*, 339–344 (1970)
Pecket, R.C., Hathout Bassim, T.A.: The effect of kinetin in relation to photocontrol of anthocyanin biosynthesis in *Brassica oleracea*. Phytochemistry *13*, 1395–1399 (1974)
Piattelli, M., Guidici De Nicola, M., Castrogiovanni, V.: The effect of kinetin on amaranthin synthesis in *Amaranthus tricolor* in darkness. Phytochemistry *10*, 289–293 (1971)
Piggot, P.J.: Mapping of asporogenous mutations of *Bacillus subtilis*: a minimum estimate of the number of sporulation operons. J. Bacteriol. *114*, 1241–1253 (1973)
Plesser, A., Weissenböck, G.: Untersuchungen zur Lokalisation von Flavonoiden in Plastiden. IV. Flavonoidgehalt intakter Chloroplasten aus *Avena sativa* L. Z. Pflanzenphysiol. *81*, 425–437 (1977)
Quast, L., Wiermann, R.: Über das Vorkommen verschiedener substituierter Chalkone während der Mikrosporogenese bei *Tulipa*. Experientia *29*, 1165–1166 (1973)
Quatrano, R.S.: Developmental biology: Development in marine organisms. In: Experimental marine biology, Chap. 8. pp. 303–346. New York: Academic Press 1974
Quatrano, R.S., Stevens, P.T.: Cell wall assembly in *Fucus zygotes*. Plant. Physiol. *58*, 224–231 (1976)
Ranjeva, R., Boudet, A.M., Alibert, G.: Autorégulation cellulaire de métabolisme des phénylpropanoides. Physiol. Vég. *15*, 303–312 (1977)
Rao, K.K., Patel, V.P.: Effect of tryptophan and related compounds on alkaloid formation in *Aspergillus fumigatus*. Lloydia *37*, 608–610 (1974)
Rao, S., Modi, V.V.: Carotenogenesis: Possible mechanism of action of trisporic acid in *Blakeslea trispora*. Experientia *33*, 31–33 (1977)
Rebeiz, C.A., Castelfranco, P.A.: Protochlorophyll and chlorophyll biosynthesis in cell-free systems from higher plants. Annu. Rev. Plant Physiol. *24*, 129–172 (1973)
Rexová-Benková, L., Markovič, O.: Pectic enzymes. Advan. Carbohydrate Chem. Biochem. *33*, 323–385 (1976)
Riley, R.G., Kolattukudy, P.E.: Evidence for covalently attached p-coumaric acid and ferulic acid in cutins and suberins. Plant Physiol. *56*, 650–654 (1975)
Rindfleisch, H., Kleinkauf, H.: Biosynthesis of Alamethicin. FEBS Lett. *62*, 276–280 (1976)
Rissland, I., Mohr, H.: Phytochrom-induzierte Enzymbildung (Phenylalanindesaminase) ein schnell ablaufender Prozeß. Planta *77*, 239–249 (1967)
Robbers, J.E., Floss, H.G.: Induction by tryptophan of the enzymes of ergot alkaloid biosynthesis. In: Secondary metabolism and coevolution. Luckner, M., Mothes, K., Nover, L. (eds.). Nova Acta Leopold. Suppl. *7*, 243–269 (1976)
Robbers, J.E., Eggert, W.W., Floss, H.G.: Physiological studies on Ergot: Time factor influence on the inhibitory effect of phosphate and the induction effect of tryptophan on alkaloid production. Lloydia *41*, 120–129 (1978)
Robinson, T.: Metabolism and function of alkaloids in plants. Science *184*, 430–435 (1974)
Romeike, A.: Über die Mitwirkung des Sprosses bei der Ausbildung des Alkaloidspektrums. II. Bildungsstätte und Biogenese des Scopolamins in *Datura ferox* L. Flora (Jena) *148*, 306–320 (1959)
Romeike, A.: Über die Mitwirkung des Sprosses bei der Ausbildung des Alkaloidspektrums. III. Entmethylierung des Hyoscyamins in den oberirdischen Organen verschiedener *Datura* Arten. Flora (Jena) *154*, 163–173 (1964)
Roos, W., Luckner, M.: ATP-dependent permeability of membrane barriers in *Penicillium cyclopium* Westling. Biochem. Physiol. Pflanz. *171*, 127–138 (1977)
Roos, W., Fürst, W., Luckner, M.: Significance of membrane permeability for the synthesis and distribution of alkaloids in cultures of *Penicillium cyclopium* Westling. In: Secondary metabolism and coevolution. Luckner, M., Mothes, K., Nover, L. (eds.). Nova Acta Leopold. Suppl. *7*, 175–182 (1976)
Russel, D.W.: The metabolism of aromatic compounds in higher plants. X. Properties of cinnamic acid 4-hydroxylase of pea seedlings and some aspects of its metabolic and developmental control. J. Biol. Chem. *246*, 3870–3878 (1971)
Saini, M.S., Cheng, M., Anderson, J.A.: Conversion of 4-γ,γ-dimethylallyltryptophan to clavicipitic acid. Some properties of the enzyme. Phytochemistry *15*, 1497–1500 (1976)
Sakai, S., Imaseki, H.: Auxin-induced ethylene production by mungbean hypocotyl segments. Plant Cell Physiol. (Tokyo) *12*, 349–359 (1971)

Saunders, J.A., Conn, E.E., Chin Ho Lin, Stocking, C.R.: Subcellular localisation of the cyanogenic glucoside of Sorghum by autoradiography. Plant Physiol. *59*, 647–652 (1977)
Schnepf, E.: Sekretion und Exkretion bei Pflanzen. Protoplasmatologia VIII. Wien: Springer 1969
Schopfer, P.: Phytochrome control of enzymes. Annu. Rev. Plant Physiol. *28*, 223–252 (1977)
Schopfer, P., Hock, B.: Nachweis der phytochrominduzierten de novo-Synthese von Phenylalaninammoniumlyase (Pal, E.C. 4.3.1.5.) in Keimlingen von *Sinapis alba* durch Dichtemarkierung mit Deuterium. Planta *96*, 248–253 (1971)
Schopfer, P., Wellmann, E., Acton, G.J.: Demonstration of phytochrome controlled de novo synthesis of phenylalanine ammonia-lyase by density labelling. In: Secondary metabolism and coevolution. Luckner, M., Mothes, K., Nover, L. (eds.). Nova Acta Leopold. Suppl. *7*, 213–228 (1976)
Schröder, J., Betz, B., Hahlbrock, K.: Light-induced enzyme synthesis in cell suspension cultures of *Petroselinum hortense*. Demonstration in a heterologous cell-free system of rapid changes in the rate of phenylalanine ammonia-lyase synthesis. Eur. J. Biochem. *67*, 527–541 (1976)
Schulze, C., Schnepf, E., Mothes, K.: Über die Lokalisation der Kautschukpartikel in verschiedenen Typen von Milchröhren. Flora (Jena) *158A*, 458–460 (1967)
Scott, A.I., Beadling, L.C., Georgopapadakou, N.H., Subbarayan, C.R.: Biosynthesis of polyketides. Purification and inhibition studies of 6-methylsalicylic acid synthetase. Bioorg. Chem. *3*, 238–248 (1974)
Shafizadeh, F., McGinnis, G.D.: Morphology and Biogenesis of cellulose and plant cell-walls. Adv. Carbohydr. Chem. Biochem. *26*, 297–349 (1971)
Shaw, G.: The chemistry of sporopollenin. In: Sporopollenin. Brooks, J., Grant, P.R., Muir, M., Va Gijzei, P., and Shaw, G. (eds.), pp. 305–348. London: Academic Press, 1971
Shimada, M., Conn, E.E.: The enzymatic conversion of p-hydroxyphenylacetaldoxime to p-hydroxymandelonitrile. Arch. Biochem. Biophys. *180*, 199–207 (1977)
Shimada, M., Kuroda, H., Higuchi, T.: Evidence for the formation of methoxyl groups of ferulic and sinapic in *Bambusa* by the same o-methyltransferase. Phytochemistry *12*, 2783–2875 (1973)
Siefermann, D., Yamamoto, H.Y.: Properties of NADPH and oxygen-dependent zeaxanthin expoxidation in isolated chloroplasts. A transmembrane model for the violaxanthin cycle. Arch. Biochem. Biophys. *171*, 70–77 (1975)
Sitte, P.: Plastiden-Metamorphose und Chromoplasten bei *Chrysosplenium*. Z. Pflanzenphysiol. *73*, 243–265 (1974)
Sjöland, S., Gatenbeck, S.: Studies on the enzyme synthesizing the aromatic product alternariol. Acta Chem. Scand. *20*, 1053–1059 (1966)
Smith, H.: The photocontrol of flavonoid biosynthesis. In: Phytochrome. Mitrakos, K., Shropshire, W. (eds.), pp. 433–481. London: Academic Press 1972
Smith, H.: Regulatory mechanisms in the photocontrol of flavonoid biosynthesis. In: Biosynthesis and its control in plants. Milborrow, B.V. (ed.), pp. 303–321. London: Academic Press 1973
Srere, P.A., Mosbach, K.: Metabolic compartmentation: symbiotic, organellar, multienzymic and multienvironmental. Annu. Rev. Microbiol. *28*, 61–83 (1974)
Stafford, H.A.: Possible multienzyme complexes regulating the formation of C_6–C_3 phenolic compounds and lignins in higher plants. In: Metabolism and regulation of secondary plant products. Runeckles, V.C., Conn, E.E. (eds.). Recent Adv. Phytochem. *8*, 53–57 (1974a)
Stafford, H.A.: The metabolism of aromatic compounds. Annu. Rev. Plant Physiol. *25*, 459–486 (1974b)
Stanley, R.G., Linskens, H.T.: Pollen: biology, biochemistry, management. Berlin, Heidelberg, New York: Springer 1974
Steinitz, B., Drumm, H., Mohr, H.: The appearance of competence for phytochrome-mediated anthocyanin synthesis in the cotyledons of *Sinapsis alba* L. Planta *130*, 23–31 (1976)
Sterlini, J.M., Mandelstam, J.: Commitment to sporulation in *Bacillus subtilis* and its relationship to development in actinomycin resistence. Biochem. J. *113*, 29–37 (1969)
Strack, D., Reznik, H.: Die Dynamik von Flavonolglykosiden während der Keimlingsentwicklung von *Cucurbita maxima* Duchesne. Z. Pflanzenphysiol. *79*, 95–108 (1976)

Sütfeld, R., Wiermann, R.: Die Bildung von Coenzym A-Thiolestern verschieden substiuierter Zimtsäuren durch Enzymextrakte aus Antheren. Z. Pflanzenphysiol. *72*, 163–171 (1974)
Szybalski, W.: A network of developmental controls in coliphage lambda. In: Cell differentiation in microorganisms, plants and animals. Nover, L., Mothes, K. (eds.), pp. 262–316. Jena: VEB G. Fischer Verlag and Elsevier/North-Holland 1977
Taylor, A.O., Zucker, M.: Turnover and metabolism of chlorogenic acid in *Xanthium leaves* and potato tubers. Plant Physiol. *41*, 1350–1359 (1966)
Thomas, R.L., Jen, J.J.: Phytochrome-mediated carotenoid biosynthesis in ripening tomatoes. Plant Physiol. *50*, 452–453 (1975)
Tong, W., Schopfer, P.: Absence of Pfr-destruction in the modulation of phenylalanine ammonia-lyase synthesis of mustard cotyledons. Plant Physiol. *61*, 59–61 (1978)
Towers, G.H.N., Vance, C.P., Nambudiri, A.M.D.: Photoregulation of phenylpropanoid and styrylpyrone biosynthesis in *Polyporus hispidus*. In: Metabolism and regulation of secondary plant products. Runeckles, V.C., Conn, E.E. (eds.). Recent Adv. Phytochem. *8*, 81–94 (1974)
Ulbrich, B., Amrhein, N.: Induction by light of hydroxycinnamoyl-CoA quinate hydroxycinnamoyl transferase in buckwheat (*Fagopyron esculentum* Moench): Absence of feed-forward control by trans cinnamate. Planta *138*, 69–71 (1978)
Vance, C.P., Tregunna, E.B., Nambudiri, A.M.D., Towers, G.H.N.: Styrylpyrone biosynthesis in *Polyporus hispidus*. I. Action spectrum and photoregulation of pigment and enzyme formation. Biochem. Biophys. Acta *343*, 138–147 (1974)
Veliky, I.A.: Synthesis of carboline alkaloids by plant cell cultures. Phytochemistry *11*, 1405–1406 (1972)
Verzár-Petri, G.: Histochemical and histoautoradiographic examination of alkaloid localisation in the vegetative organs of *Datura innoxia* Mill. Acta Bot. Acad. Sci. Hung. *18*, 257–271 (1973)
Voigt, S., El Kousy, S., Schwelle, N., Nover, L., Luckner, M.: Activities in vitro and in vivo of enzymes of benzodiazepine alkaloid biosynthesis during development of *Penicillium cyclopium*. Phytochemistry *17*, 1705–1709 (1978)
Waites, W.M., Kay, D., Dawes, I.W., Wood, D.A., Warren, S.C., Mandelstam, J.: Sporulation in *Bacillus subtilis*. Correlation of biochemical events with morphological changes in asporogenous mutants. Biochem. J. *118*, 667–676 (1970)
Walker, J.N., Walker, M.S.: Streptomycin biosynthesis. Enzymatic synthesis of o-phosphorylstreptidine from streptidine and adenosinetriphosphate. Biochim. Biophys. Acta *148*, 335–341 (1967)
Ward, A.C., Packter, N.M.: Relationship between fatty-acid and phenolsynthesis in *Aspergillus funigatus*. Eur. J. Biochem. *46*, 323–333 (1974)
Weinberg, E.D.: Biosynthesis of secondary metabolites: Role of trace metals. Adv. Microbiol. Physiol. *4*, 1–44 (1970)
Weinberg, E.D., Tonnis, S.M.: Action of chloramphenicol and its isomers on secondary biosynthetic processes of *Bacillus*. Appl. Microbiol. *14*, 850–856 (1966)
Weinberg, E.D., Tonnis, S.M.: Role of manganese in biosynthesis of bacitracin. Can. J. Microbiol. *13*, 614–615 (1967)
Weissenböck, G., Plesser, A., Trinks, K.: Flavonoidgehalt und Enzymaktivität isolierter Haferchloroplasten (*Avena sativa* L.). Ber. Dtsch. Bot. Ges. *89*, 457–472 (1976)
Welch, G.R., Gaertner, F.H.: Coordinate activation of a multienzyme complex by the first substrate. Arch. Biochem. Biophys. *172*, 476–489 (1976)
Wellburn, A.R., Hampp, R.: Movement of labelled metabolites from mitochondria to plastids during development. Planta *131*, 17–20 (1976a)
Wellburn, A.R., Hampp, R.: Uptake of mevalonate and acetate during plastid development. Biochem. J. *158*, 231–233 (1976b)
Wellburn, A.R., Hampp, R.: Changes in plastid and mitochondrial envelopes during greening. Biochem. Soc. Trans. *5*, 91–94 (1977)
Wellmann, E.: Gewebespezifische Kontrolle von Enzymen des Flavonoidstoffwechsels durch Phytochrom in Kotyledonen des Senfkeimlings (*Sinapis alba* L.). Ber. Dtsch. Bot. Ges. *87*, 275–279 (1974)
Wellmann, E., Schopfer, P.: Phytochrome-mediated *de novo* synthesis of phenylalanine ammonia-lyase in cell suspension cultures of parsley. Plant Physiol. *55*, 822–827 (1975)

Wettstein, D.v., Kahn, A., Nielsen, O.F., Gough, S.: Genetic regulation of chlorophyll synthesis with mutants in parsley. Science *184*, 800–802 (1974)
Whaley, W.G.: The Golgi apparatus. In: Cell biology monographs. Vol. 2. Wien: Springer 1975
Whitmore, F.W.: Phenolic acids in wheat coleoptile cell walls. Plant Physiol. *53*, 728–731 (1974)
Wiermann, R.: Über die Beziehungen zwischen flavonolaufbauenden Enzymen, einem flavonolumwandelnden Enzym und der Akkumulation phenylpropanoider Verbindungen während der Antherenentwicklung. Planta *110*, 353–360 (1973)
Wiermann, R.: Stage-specific phenylpropanoid metabolism during pollen development. In: Regulation of secondary product and plant hormone metabolism. Luckner, M., Schreiber, K. (eds.). Prod. 12th FEBS Meet. Vol. 55. Oxford: Pergamon Press, 231–241 (1979)
Wilson, S., Luckner, M.: Cyclopenase, ein Lipoproteid der Protoplasmamembran von Konidiosporen des Pilzes *Penicillium cyclopium* Westling. Z. Allg. Mikrobiol. *15*, 45–51 (1975)
Wood, D.A.: Sporulation in *Bacillus subtilis*. The appearance of sulphoacetic acid as a marker event for sporulation. Biochem. J. *123*, 601–605 (1971)
Zaprometov, M.N., Buklaeva, S.V.: On the biosynthesis of phenolic compounds by chloroplasts. Dokl. Akad. Nauk USSR *176*, 470–473 (1967)
Zenk, M.H., El-Shagi, H., Schulte, U.: Anthraquinone production by cell suspension cultures of *Morinda citrifolia*. Planta Med. Suppl. 79–101 (1975)
Zucker, M.: Light and enzymes. Annu. Rev. Plant Physiol. *23*, 133–156 (1972)

4 The Alkaloids

4.1 Alkaloids Derived from Ornithine, Lysine, and Nicotinic Acid

E. LEETE

1 Alkaloids Derived from Ornithine

1.1 Introduction

Feeding experiments with isotopically labeled ornithine have established this amino acid to be the precursor of the pyrrolidine rings found in the tobacco alkaloids (nicotine, nornicotine), the tropane alkaloids (hyoscyamine, hyoscine, meteloidine), the pyrrolizidine alkaloids (retronecine), and phenanthroindolizidines (tylophorine).

It should be emphasized that not all pyrrolidine rings found in nature are derived from ornithine. Some notable exceptions are shihunine (I) (derived from *o*-succinylbenzoic acid; LEETE and BODEM, 1976), anthramycin (II) (whose pyrroline ring is formed by catabolism of the aromatic ring of tyrosine, HURLEY et al., 1975a, b, 1976a, b), and vasicine (III), also known as peganine, whose pyrrolidine ring is derived from an acetate residue (WAIBLINGER et al., 1972) and aspartic acid (JOHNE et al., 1973). An earlier report (LILJEGREN, 1971) that ornithine was incorpo-

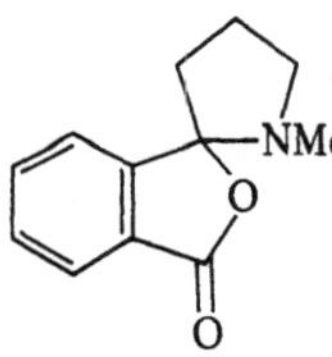

Shihunine
(I)

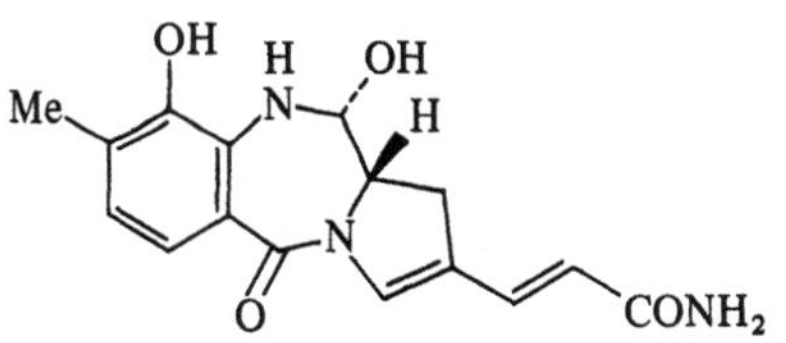

Anthramycin
(II)

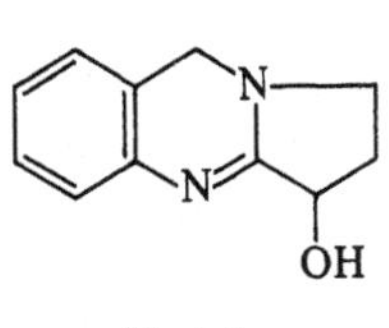

Vasicine
(III)

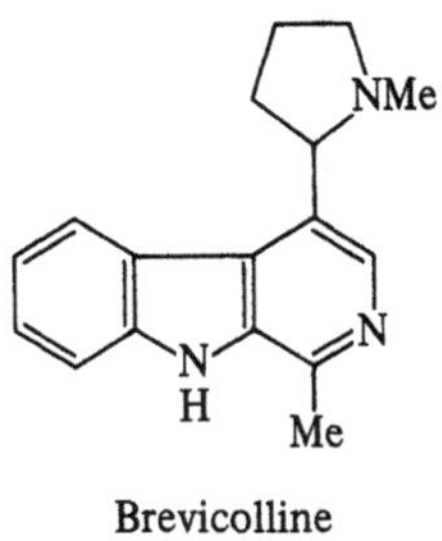

Brevicolline
(IV)

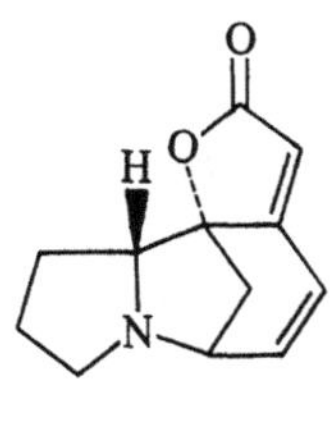

Norsecurinine
(V)

rated into the pyrrolidine moiety of this alkaloid was rationalized (LEETE, 1973a) by postulating that the administered [2-^{14}C] ornithine served as a source of [1-^{14}C] acetate.

Experimental work has not been carried out on several alkaloids which contain a pyrrolidine ring, however, by analogy with related compounds, their derivation from ornithine seems probable, for example brevicolline (IV) (KOMPIS et al., 1969) and norsecurinine (V).

1.2 The Tobacco Alkaloids

It is generally accepted that the immediate precursor of the N-methylpyrrolidine ring of nicotine (XX) is 4-methylaminobutanal (XIV) which exists in acidic solution as the N-methyl-Δ^1-pyrrolinium salt (XVII). This compound is an excellent precursor (18%–33% incorporation) of the pyrrolidine ring of nicotine (KISAKI et al., 1966; LEETE, 1967b; MIZUSAKI et al., 1967). N-Methylputrescine (XI) is a precursor of (XIV) and a partially purified enzyme, called N-methylputrescine oxidase, which catalyzes the formation of (XVII) has been obtained from tobacco roots (MIZUSAKI et al., 1972). This enzyme was not found in tobacco leaves (MIZUSAKI et al., 1973), a result consistent with the earlier observations of DAWSON (1942a, b) and DAWSON and SOLT (1959) who showed that tobacco roots are the main site of nicotine formation. N-Methylputrescine is formed from putrescine (X) and *S*-adenosyl-L-methionine, and the enzyme catalyzing this reaction, called putrescine-N-methyltransferase, has also been obtained from a cell-free extract of tobacco roots (MIZUSAKI et al., 1971). Putrescine is formed by the decarboxylation of ornithine (IX), and ornithine decarboxylase has been detected in tobacco roots (MIZUSAKI et al., 1973). It has also been established that putrescine can be formed from arginine (XV), via agmatine (XII), and N-carbamylputrescine (XIII) (YOSHIDA and MITAKE, 1966). However the direct conversion of ornithine to putrescine seems to be the main biosynthetic pathway.

The feeding of [2-^{14}C]-ornithine or [2-^{14}C]-glutamic acid (VII) (which is converted to ornithine via glutamic-γ-aldehyde (VI)) to tobacco afforded nicotine which was labeled equally at C-2′ and C-5′ (DEWEY et al., 1955; LAMBERTS and BYERRUM, 1958; LAMBERTS et al., 1959; LEETE, 1955, 1958a, 1976b; LEETE and SIEGFRIED, 1957; WU and BYERRUM, 1965). This symmetrical labeling of the pyrrolidine ring is rationalized by postulating the intermediacy of free (i.e. nonenzyme bound) putrescine in this biosynthetic sequence. 5-N-Methylornithine (VIII) was also incorporated into the pyrrolidine ring of nicotine (GILBERTSON and LEETE, 1967). However, it is considered that this is an "aberrant biosynthesis" since this amino acid is not found in tobacco. Apparently a nonspecific decarboxylase catalyzes the formation of N-methylputrescine from this amino acid.

For several years RAPOPORT and his coworkers (LIEBMAN et al., 1965, 1967, RUEPPEL et al., 1974) have been reporting that short-term feeding of $^{14}CO_2$ to tobacco affords nicotine which is unsymmetrically labeled in its pyrrolidine ring. Such a result would not be in accord with the biosynthetic sequence illustrated in Fig. 1. BYERRUM and coworkers (ZIELKE et al., 1968) reported symmetrical labeling of the pyrrolidine ring after a short-term feeding with $^{14}CO_2$. HUTCHINSON et al. (1976) fed $^{13}CO_2$ and attempted to determine the distribution of label by ^{13}C-NMR spectroscopy, however, their results were equivocal. This controversy regarding the origin of the pyrrolidine ring has been reviewed (LEETE, 1976a, 1977b) and the majority of the experimental evidence favors the formation of the pyrrolidine ring from ornithine, via putrescine.

Nornicotine (XVIII) derived from [2-^{14}C]-ornithine was also equally labeled at C-2′ and C-5′ (LEETE, 1976b) in contrast to an earlier report (MIZUSAKI et al., 1965) in which unsymmetrical labeling of the pyrrolidine ring was claimed. Nornicotine seems to be formed irreversibly from nicotine, a reaction which occurs both in the living plant and during the curing and drying of tobacco leaves (LEETE and SLATTERY, 1976; TSO, 1972; TSO and JEFFREY, 1956, 1957; WADA 1956). It was shown by DAWSON (1945a, b, 1951) that demethylation of nicotine takes place in the aerial parts of the plant. This demethylation is not specific for the natural (−)-2′*S*-nicotine. Indeed it was discovered that the (+)-2′*R*-nicotine is demethylated faster than the natural isomer (KISAKI and TAMAKI, 1961a, b, 1966). The production of racemic nornicotine from (−)-nicotine involves the loss of the C-2′ hydrogen (LEETE and CHEDEKEL, 1972). The intermediates involved in the demethylation of nicotine are currently unknown. It has been suggested that nicotine-1′-N-oxide is an intermediate (CRAIG et al., 1964; EGRI, 1957; ILJIN and SEREBROVKAYA, 1958; WENKERT, 1954). However the administration of nicotine-1′-N-oxide to *Nicotiana glutinosa* or *glauca* plants indicated that there was not a direct conversion to nornicotine (ALWORTH et al., 1969; LEETE, 1977e). In animals, evidence has been obtained (NGUYEN et al., 1976) that nicotine is demethylated via the iminium salt (XIX), and it has been suggested (LEETE, 1977e) that racemization could occur by postulating a tautomeric shift of this iminium salt to compound (XVI).

1.3 Tropane Alkaloids

The administration of [2-^{14}C]-ornithine to *Datura stramonium* plants afforded hyoscyamine (XXVII) (Fig. 3) labeled solely at C-1 (LEETE, 1962, 1964; LIEBISCH and SCHÜTTE, 1967). The unsymmetrical incorporation of ornithine is apparently achieved by methylation of the 5-amino group to yield 5-N-methylornithine (VIII) which was incorporated directly into hyoscyamine (AHMAD and LEETE, 1970). Decarboxylation affords N-methylputrescine (XI) which was also shown to be an efficient precursor of the tropane skeleton (LIEBISCH et al., 1966). Oxidation affords the N-methyl-Δ^1-pyrrolinium salt (XVII). Condensation with acetoacetate (derived from two molecules of acetic acid (KACZKOWSKI et al., 1961)) affords hygrine (XXII) presumably via the keto acid (XXI). This reaction occurs in vitro (GALINOVSKY et al., 1951; EL-OLEMY et al., 1966). Hygrine is incorporated into the tropine (XXVI) moiety of hyoscyamine (LIEBISCH et al., 1972; O'DONOVAN and KEOGH, 1969) probably via dehydrohygrine (XXIII) and tropanone (XXIV). Hygrine is also a precursor of cuskhygrine (XXV) (O'DONOVAN and KEOGH, 1969; LIEBISCH et al., 1969) presumably by reaction with another molecule of (XVII). These biosynthetic steps are illustrated in Fig. 2.

Tropine on esterification with tropic acid affords hyoscyamine. The biosynthesis of tropic acid has been reviewed (LEETE, 1973a) and its formation involves an intramolecular rearrangement of the side chain of phenylalanine (LEETE et al., 1975). The acid moiety of littorine (XXX), phenyllactic acid, is also derived from phenylalanine (EVANS and WOOLLEY, 1969). Cochlearine (XXXIII) is another tro-

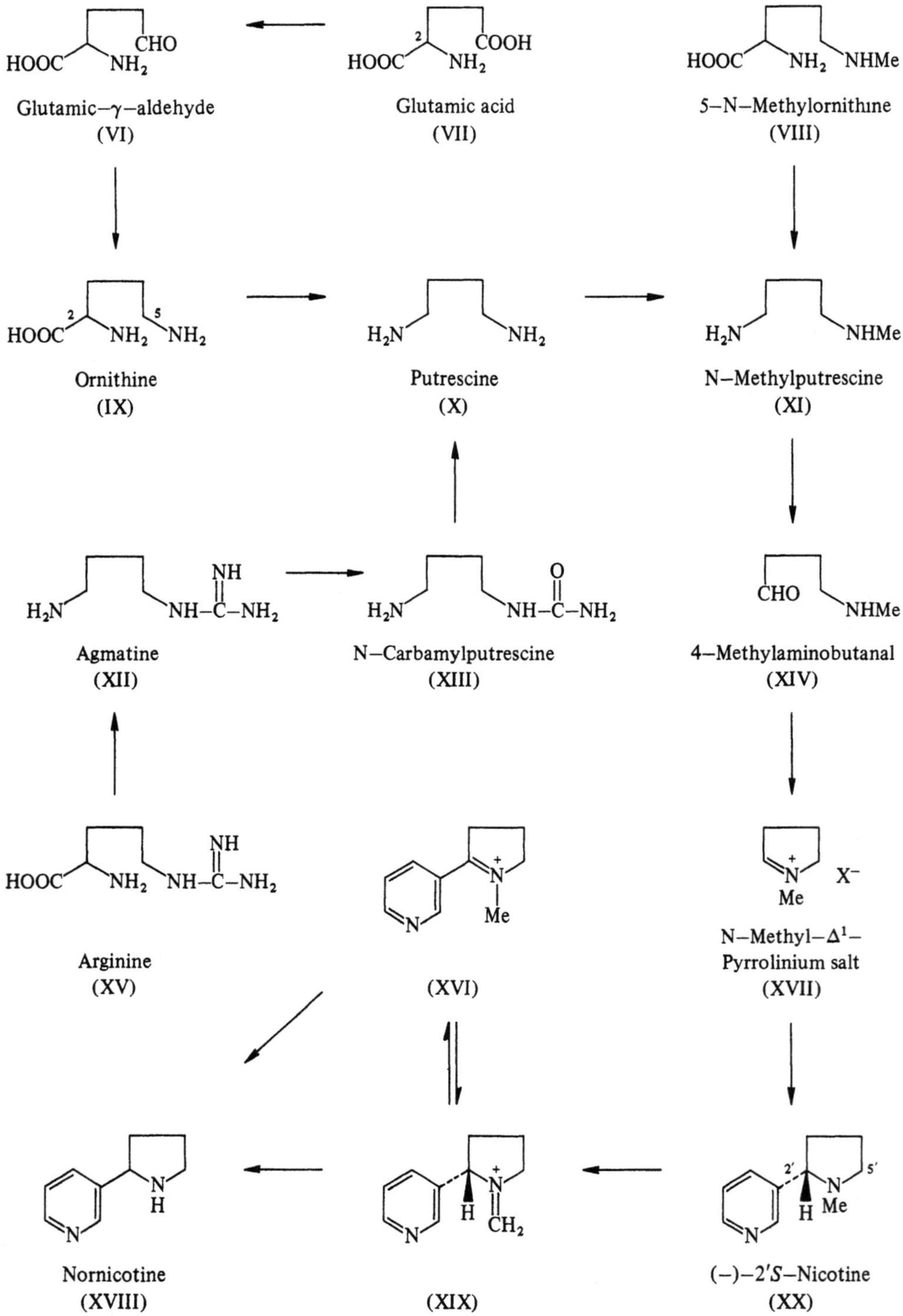

Fig. 1. Precursors of the pyrrolidine ring of nicotine

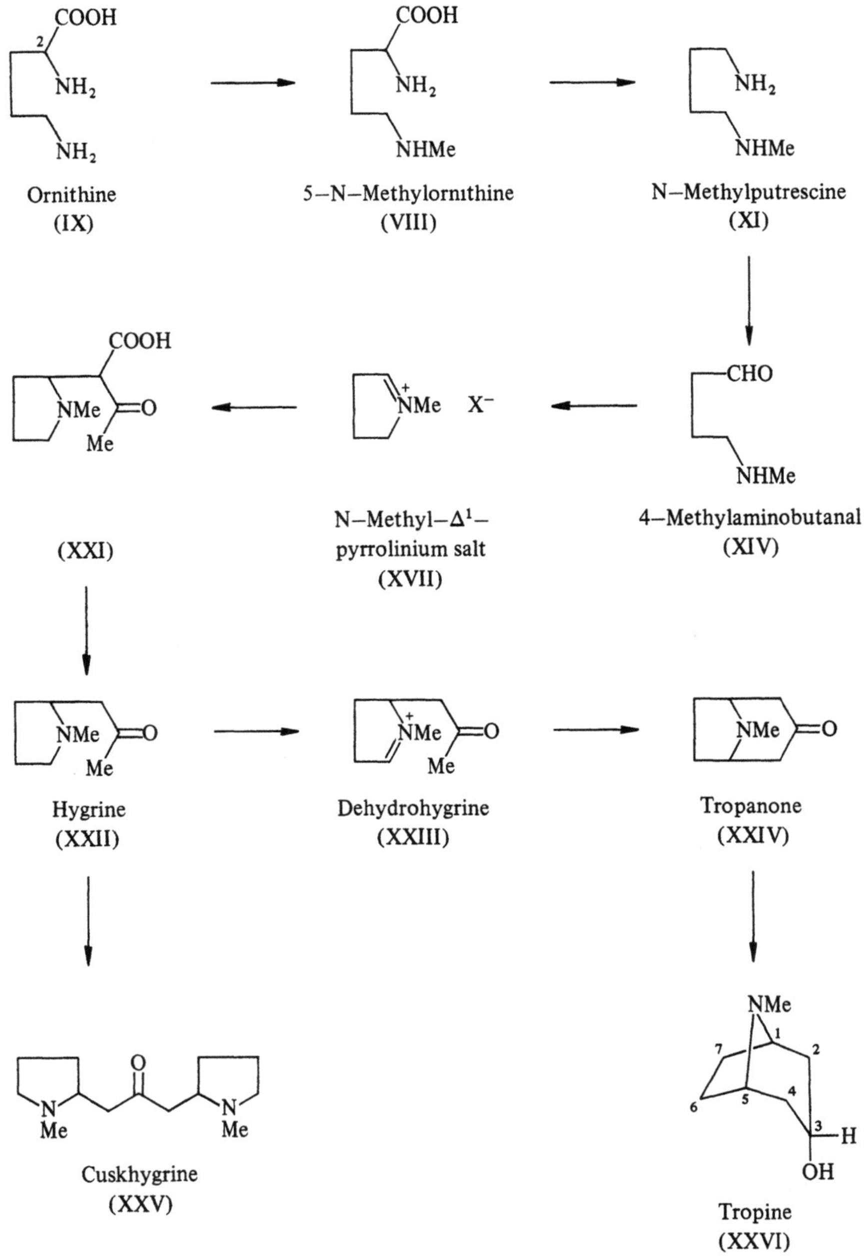

Fig. 2. Biosynthesis of the tropane skeleton

pine ester, and its unusual acid moiety is derived from *m*-hydroxyphenylalanine (Liebisch et al., 1973). Hyoscine (also known as scopolamine) (XXXI) is a metabolite of hyoscyamine, and epoxide formation probably takes place via 6β-hydroxyhyoscyamine and 6,7-dehydrohyoscyamine (XXVIII) (Fodor et al., 1959; Romeike, 1956, 1959, 1962; Romeike and Fodor, 1960). It has been shown (Leete and Lucast, 1976) that the epoxide ring of hyoscine is introduced into the tropine moiety of hyoscyamine with loss of the β-hydrogens at C-6 and C-7.

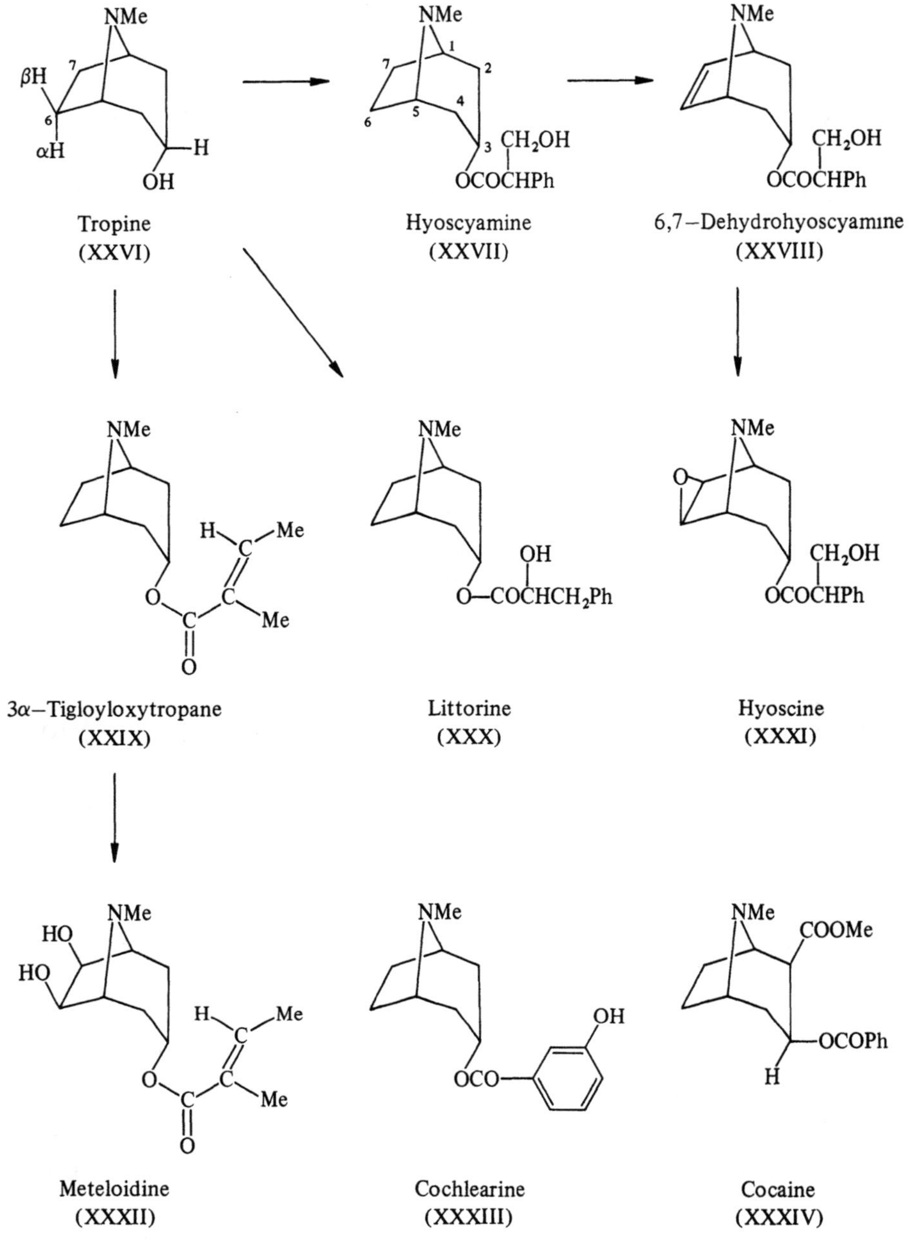

Fig. 3. Alkaloids derived from tropine

Tropine is also a precursor of meteloidine (XXXII) (LEETE, 1972; LEETE and NELSON, 1969) and it is formed via 3α-tigloyloxytropane (XXIX) (BERESFORD and WOOLLEY, 1975a, b; LEETE and LUCAST, 1975). The tiglic acid moiety of these alkaloids is formed from isoleucine (LEETE and MURRILL, 1967; WOOLLEY, 1966) via 2-methylbutanoic acid (BASEY and WOOLLEY, 1973; LEETE, 1973b).

It seems reasonable that cocaine (XXXIV) is related biogenetically to the other tropane alkaloids, probably involving the intermediate (XXI) with retention of

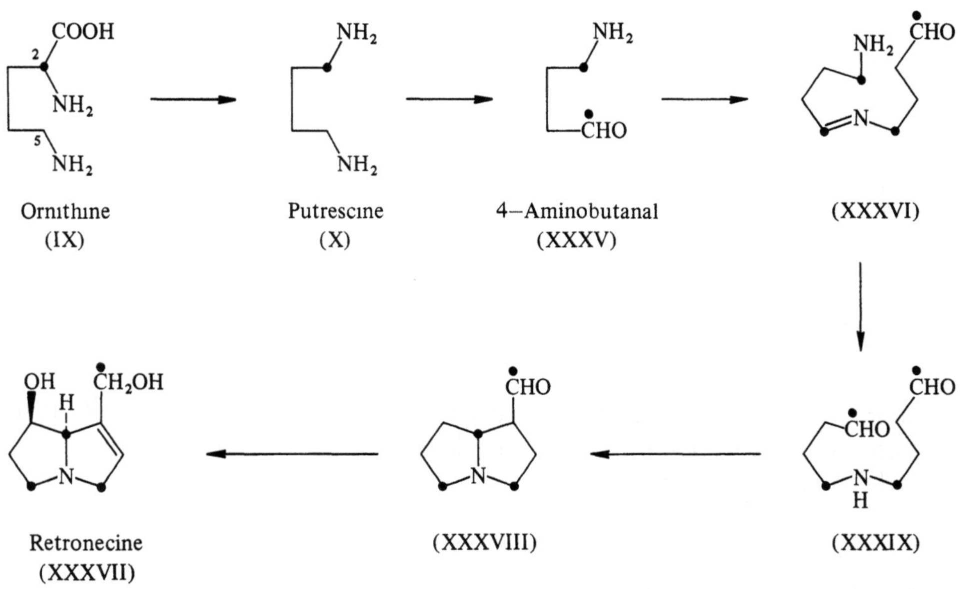

Fig. 4. Hypothetical biosynthesis of retronecine

its carboxyl group in the ultimate alkaloid. However, tracer experiments have been inconclusive (LIEBISCH et al., 1972; TAMPRATEEP et al., 1963). The benzoic acid moiety of cocaine is derived from phenylalanine (GROSS and SCHÜTTE, 1963).

1.4. Pyrrolizidine Alkaloids

Labeled ornithine is incorporated into retronecine (XXXVII; Fig. 4) (BALE and CROUT, 1975; BOTTOMLEY and GEISSMAN, 1964; HUGHES et al., 1964; NOWACKI and BYERRUM, 1962). The pattern of labeling in retronecine derived from [2-^{14}C]- and [5-^{14}C]-ornithine (BOTTOMLEY and GEISSMAN, 1964) is consistent with the biosynthetic scheme illustrated in Fig. 4. Decarboxylation of ornithine affords free putrescine which is oxidized to 4-aminobutanal (XXXV), two molecules of which combine to yield the Schiffs base (XXXVI). Reduction of the aldimine, and oxidation of the primary amino group yields (XXXIX) which undergoes an intramolecular Mannich reaction to afford (XXXVIII), and thence to retronecine by unexceptional steps. The distribution of label in retronecine derived from [2-^{14}C]-ornithine is indicated by means of heavy dots in Fig. 4. The results of HUGHES et al. (1964) are not consistent with the occurrence of a symmetrical intermediate such as (XXXIX), and they suggested that the two rings of retronecine, although both derived from ornithine, are incorporated via different metabolites, namely Δ^1-pyrroline-2-carboxylic acid and 2-keto-glutaric-4-aldehyde. Further work is necessary to resolve these contradictory results.

1.5 Phenanthroindolizidine Alkaloids

The pyrrolidine ring of tylophorine (XL) is derived from ornithine, however, it is not currently known whether it is incorporated via free putrescine with subsequent

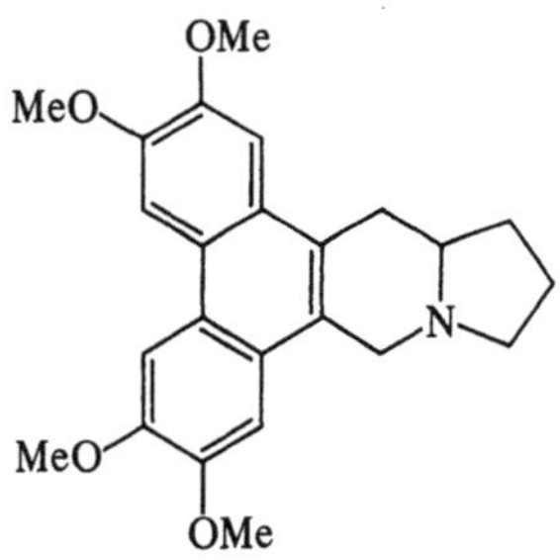

Tylophorine
(XL)

equilibration of the two carbons α to the pyrrolidine nitrogen (HERBERT et al., 1976, MULCHANDANI et al., 1969, 1971, 1976). The rest of the molecule is derived from phenylalanine and tyrosine.

2 Alkaloids Derived from Lysine

2.1 Unsymmetrical Incorporation of Lysine

Lysine (XLI) is the precursor of the piperidine ring of several alkaloids, and an almost universal intermediate en route to these alkaloids seems to be Δ^1-piperideine (LEETE, 1969b, 1971) (XLIV). It has been suggested by LEISTNER and SPENSER (1973) that this compound is formed by the stepwise pathway illustrated in Fig. 5. The Schiffs base of L-lysine and pyridoxal phosphate (XLII) undergoes a decarboxylation, with retention of configuration (LEISTNER and SPENSER, 1975) affording the Schiffs base of cadaverine with pyridoxal phosphate (XLIII). The bound cadaverine then undergoes a stereospecific dehydrogenation, so that the hydrogen which was originally present at C-2 in lysine is retained in the resultant Schiffs base of 5-aminopentanal (XLVI). Hydrolysis yields 5-aminopentanal (XLV) which cyclizes to Δ^1-piperideine. SPENSER considered that an equilibirium exists between the bound form of cadaverine (XLIII) and free cadaverine. In some species such as *Nicotiana glauca*, the equilibrium apparently favors the bound form of cadaverine, and the piperidine ring of anabasine (XLVII) found in this species is labeled solely at C-2′, following the administration of [2-^{14}C]-lysine (GILBERTSON, 1972; GRIFFITH and GRIFFITH, 1966; LEETE, 1956; 1958b; LEETE and CHEDEKEL, 1972; LEETE et al., 1964; LOVKOVA et al., 1974a, b). Other alkaloids which incorporate lysine unsymmetrically are illustrated in Fig. 6, the location of label from [2-^{14}C]-lysine being indicated with a heavy dot. Adenocarpine (XLVIII) is a dimer of Δ^1-piperideine, a condensation which also occurs in vitro (SCHÜTTE et al., 1964b). Two molecules of adenocarpine afford santiaguine (XLIX) (O'DONOVAN and CREEDON, 1974). N-Methylpelletierine (L) results from a condensation between Δ^1-piperideine and acetoacetate, followed by a decarboxylation (GUPTA and SPENSER, 1969; KEOGH and O'DONOVAN, 1970a, LEISTNER et al., 1973). Anaferine (LI) probably arises by reaction of N-methylpelletierine with a second molecule of Δ^1-piperideine (KEOGH and O'DONOVAN, 1970a). The side chain of sedamine (LII)

Stepwise mechanism

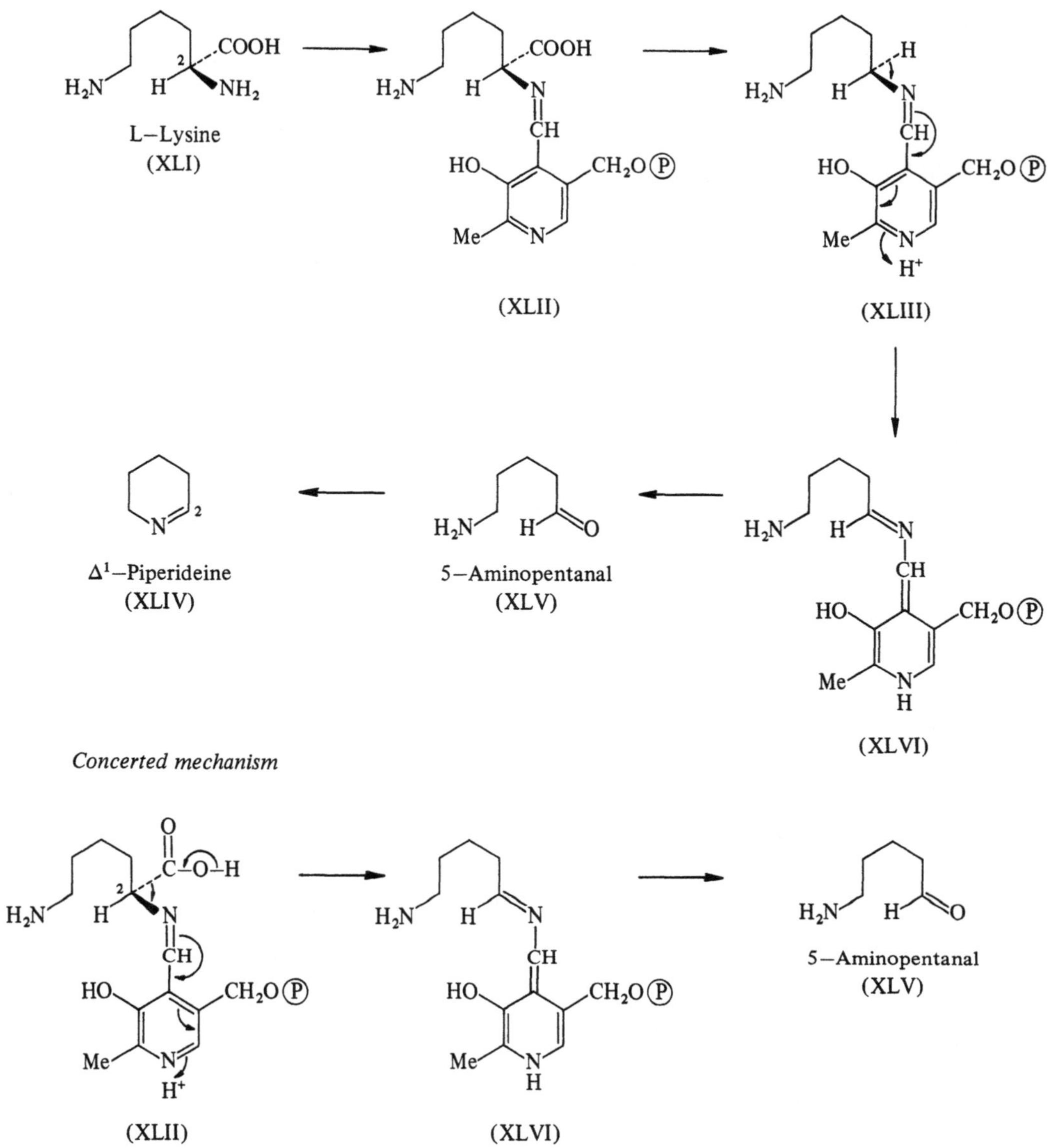

Fig. 5. Hypothetical conversion of lysine to 5-amino-pentanal involving retention of the hydrogen at C-2

is derived from phenylalanine, probably incorporated via benzoylacetic acid (GUPTA and SPENSER, 1967, 1970). The origin of the side chain of halosaline (LIII) is unknown, however, its piperidine ring is formed from lysine (O'DONOVAN and CREEDON, 1971). Lobinaline (LIV) is plausibly formed by dimerization of a compound related to sedamine (GUPTA and SPENSER, 1971). Lobeline (LV) derived from [2-^{14}C]-lysine was labeled on both carbons α to the nitrogen (KEOGH and O'DONOVAN, 1970b), however, it was suggested that this result is a consequence

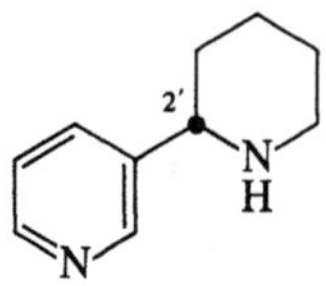

Anabasine
(XLVII)

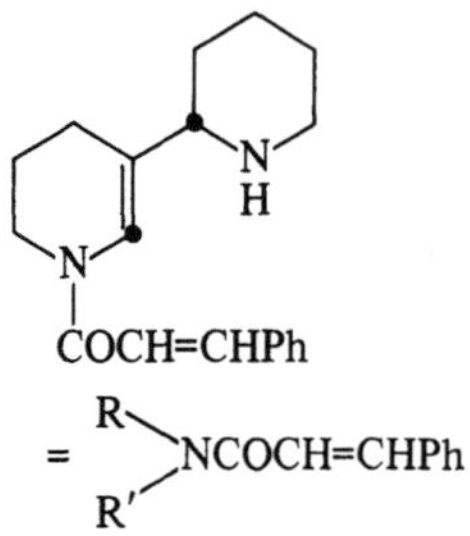

Adenocarpine
(XLVIII)

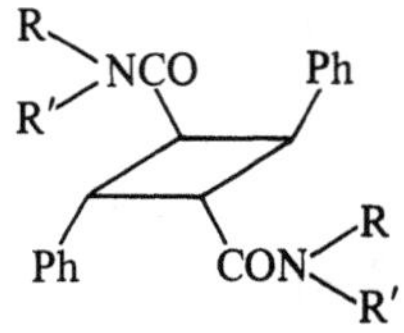

Santiaguine
(XLIX)

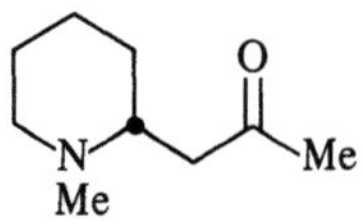

N–Methylpelletierine
(L)

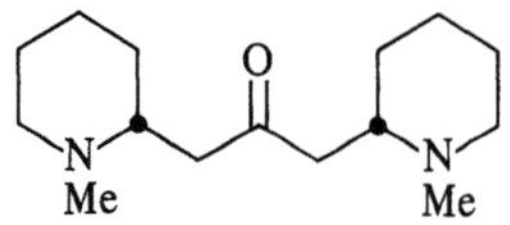

Anaferine
(LI)

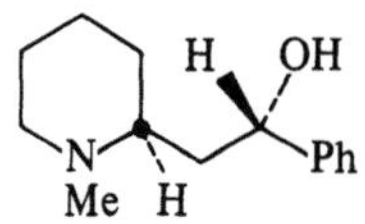

Sedamine
(LII)

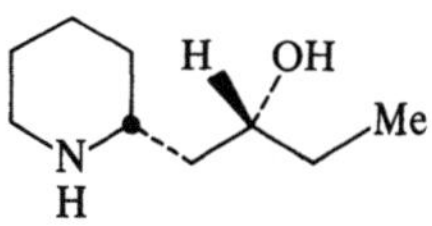

Halosaline
(LIII)

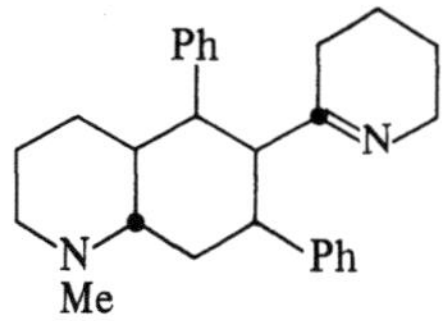

Lobinaline
(LIV)

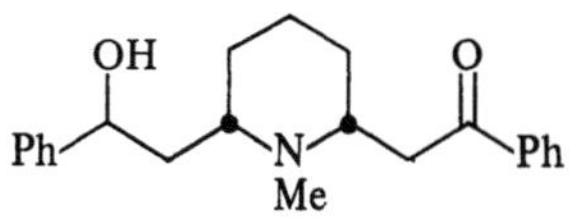

Lobeline
(LV)

Fig. 6. Alkaloids formed from lysine via an unsymmetrical intermediate. The positions labeled after the administration of [2-^{14}C]-lysine indicated with *heavy dots*

of a symmetical precursor, namely the diketone corresponding to lobeline, en route to the final alkaloid (O'DONOVAN et al., 1975).

LEETE and CHEDEKEL (1972) have suggested a concerted oxidative decarboxylation of lysine-pyridoxal (Fig. 5) to yield in one step the Schiffs base of 5-aminopentanal with pyridoxal. The incorporation of cadaverine into these alkaloids does not invalidate either of these hypotheses, and may be rationalized by postulating its oxidation to 5-aminopentanal by a nonspecific oxidase.

2.2 Symmetrical Incorporation of Lysine

Several other piperidine alkaloids are apparently formed via free cadaverine since [2-^{14}C]-lysine (or [6-^{14}C]-lysine) yields alkaloids which have their radioactivity equally distributed between the carbons α to the nitrogen. The quinolizidine alkaloids fall into this class and their biosynthesis has been reviewed by SCHÜTTE (1969), specific examples are lupinine (LVI) (SCHÜTTE and HINDORF, 1964), sparteine (LVII) (SCHÜTTE et al., 1961, 1964a), lupanine (LVIII) (SCHÜTTE et al., 1962b, SCHÜTTE and HINDORF, 1965), matrine (LIX) (ABDUSALAMOV et al., 1969, 1971a, b; KUSCHMURADOV et al., 1972; SCHÜTTE et al., 1962a, 1965; SHIBATA and SANKAWA, 1963; TANKANAEV and ABDUSALAMOV 1971). Only one of the piperidine rings of decodine (LX) is formed from lysine (KOO et al., 1970b), the rest of this molecule, and a related alkaloid cryogenine (LXI) is derived from acetic acid and phenylalanine (KOO et al., 1970a; ROTHER and SCHWARTING, 1969, 1972). These alkaloids are illustrated in Fig. 7.

It seems reasonable to assume that lupinine, the simplest of the quinolizidine alkaloids, is formed by an intramolecular Mannich reaction on the dialdehyde (LXII) produced from cadaverine analogous to the formation of its lower homolog from putrescine (see Sect. 1.4 on the pyrrolizidine alkaloids). GOLEBIEWSKI and SPENSER (1976) have proposed a new biosynthetic scheme for the tetracyclic lupin alkaloids such as sparteine. It is suggested that these alkaloids are modified trimers of Δ^1-piperideine. One of these trimers is isotripiperideine (LXIV), the favored stereochemistry being the all-chair conformation illustrated in Fig. 9. From this compound sparteine and lupanine are obtained via the intermediates (LXIII) and (LXV). The stereochemistry of the ultimate alkaloids is predetermined at all sites except C-11. The stereochemistry at this position is determined by the direction of the intramolecular ring closure of (LXV) to (LXVI). In this figure, isotripiperideine is numbered so that the carbon atoms correspond to those in sparteine. Lupanine derived from [2-^{14}C]- or [6-^{14}C]-Δ^1-piperideine was labeled in accord with this hypothesis.

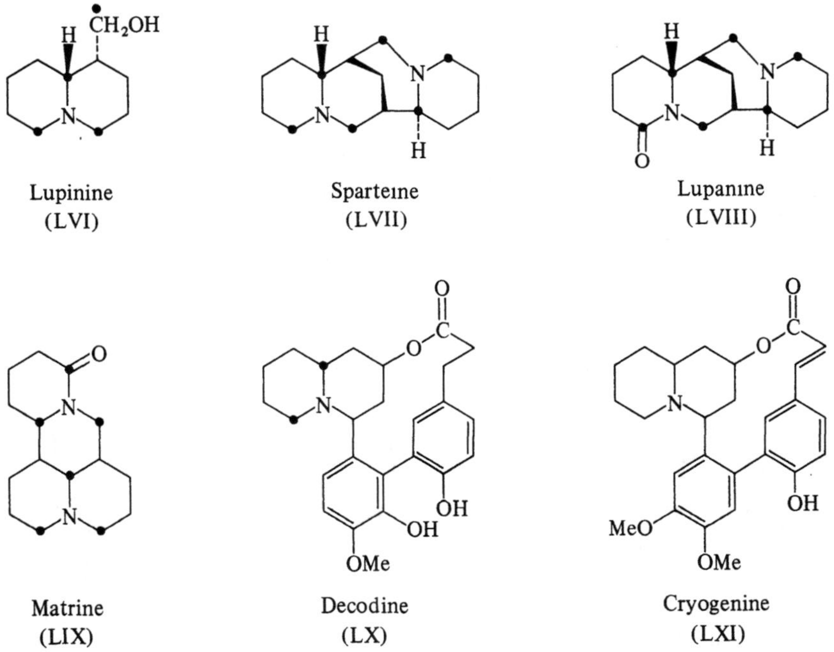

Lupinine (LVI) — Sparteine (LVII) — Lupanine (LVIII)

Matrine (LIX) — Decodine (LX) — Cryogenine (LXI)

Fig. 7. Alkaloids formed from lysine via a symmetrical intermediate. The positions labeled after administration of [2-^{14}C] or [6-^{14}C]-lysine indicated with *heavy dots*

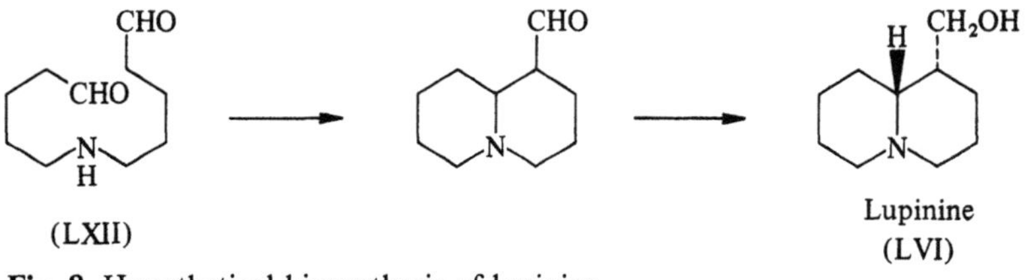

Fig. 8. Hypothetical biosynthesis of lupinine

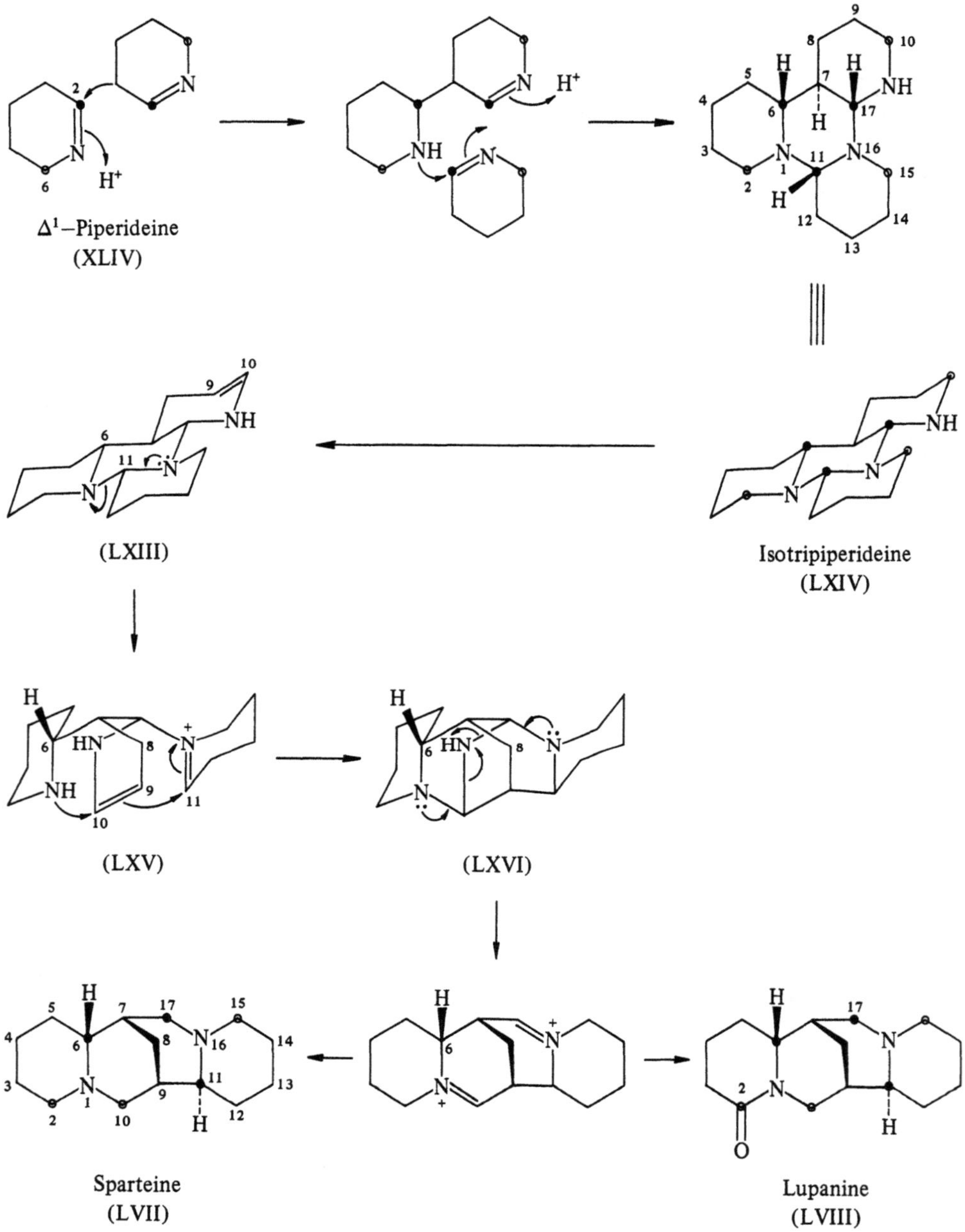

Fig. 9. Spenser's hypothesis for the biosynthesis of the tetracyclic lupine alkaloids. The pattern of labeling expected from Δ^1-piperideine indicated with *open circles* and *heavy dots*

2.3 Lycopodium Alkaloids

Lycopodine (LXIX) is derived from two molecules of Δ^1-piperideine, and one of these is incorporated via pelletierine (LXVII) (BRAEKMAN et al., 1972; CASTILLO

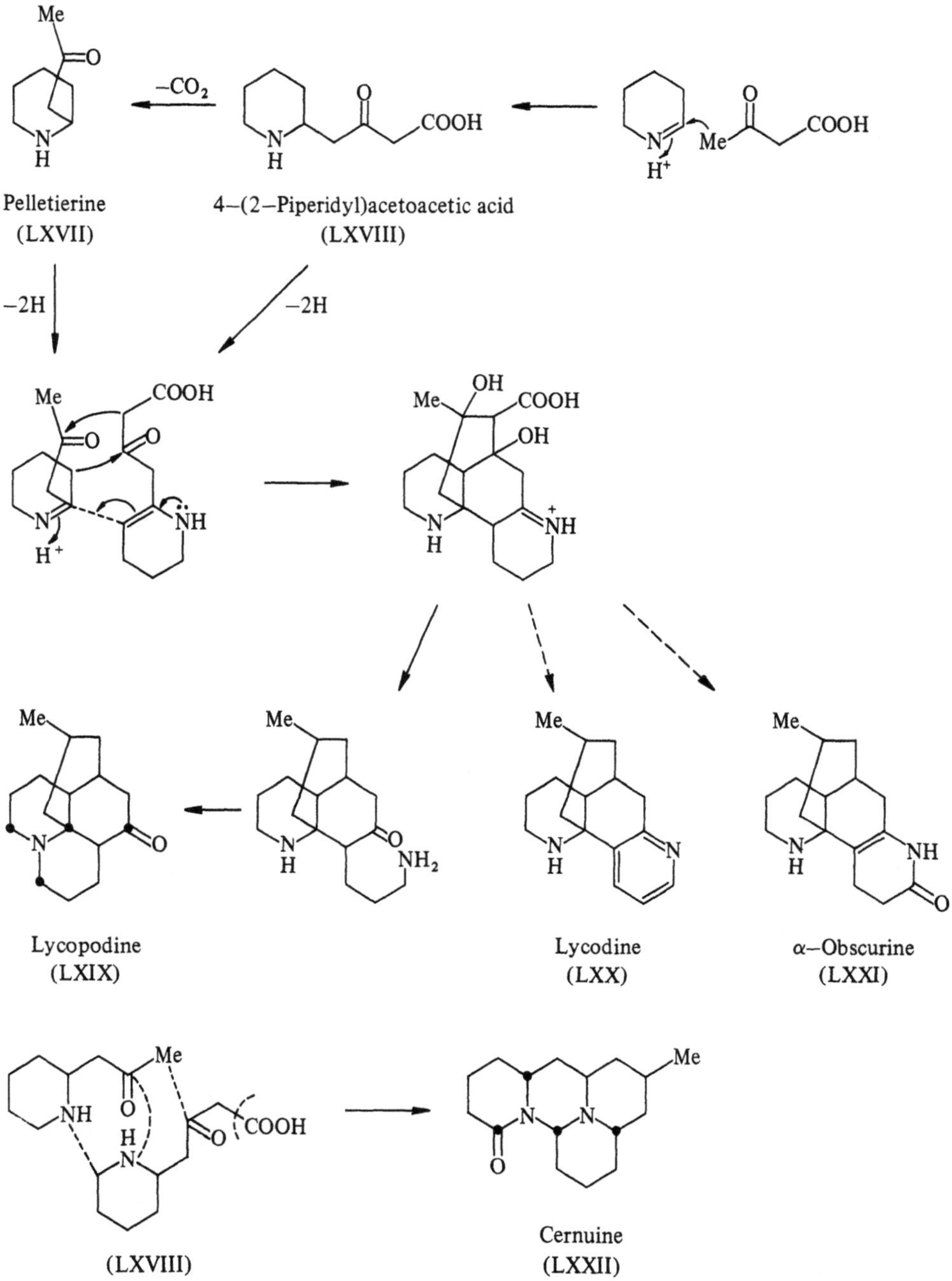

Fig. 10. Biosynthesis of the *Lycopodium* alkaloids. The pattern of labeling in lycopodine and cernuine derived from [2-^{14}C] or [6-^{14}C]-lysine indicated with *heavy dots*

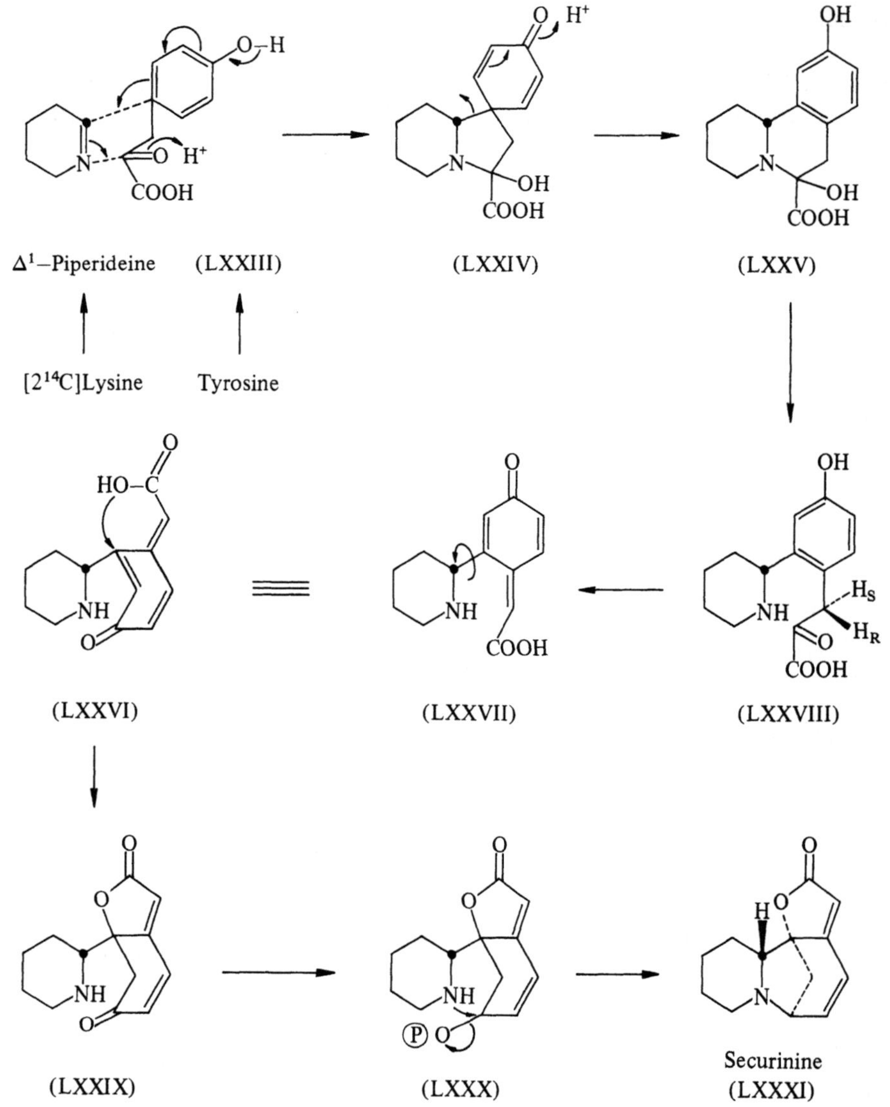

Fig. 11. Hypothetical biosynthesis of securine from lysine and tyrosine. The positions labeled following the feeding of [2-^{14}C] lysine indicated with a *heavy dot*

et al., 1970a, b; GUPTA et al., 1968; MARSHALL et al., 1976). The exact nature of the other half is unknown, however, a plausible biosynthetic sequence (Fig. 10) can be drawn utilizing 4-(2-piperidyl)acetoacetate (LXVIII), formed from Δ^1-piperideine and acetoacetate. This scheme (CASTILLO et al., 1970b) also accomodates other alkaloids found in *Lycopodium* species such as lycodine (LXX), α-obscurine (LXXI), and cernuine (LXXII) (HO et al., 1971). Lysine is incorporated into lycopodine and cernuine via a symmetrical intermediate (cadaverine). Thus [2-^{14}C]-lysine labels these alkaloids at four positions as indicated in Fig. 10.

2.4 Securinine

Securinine (LXXXI) is derived from lysine and tyrosine and the hypothetical scheme illustrated in Fig. 11 (GOLEBIEWSKI et al., 1976) is consistent with feeding experiments involving labeled tyrosine and lysine (PARRY, 1974, 1975; SANKAWA et al., 1974, 1977). Reaction between Δ^1-piperideine (whose nitrogen is derived from the 6-amino group of lysine) and *p*-hydroxyphenylpyruvic acid (LXXIII) (formed from tyrosine by transamination) affords (LXXIV). A dienone-phenol rearrangement yields the quinolizidine (LXXV), which undergoes a ring opening affording (LXXVIII). Loss of the pro-S hydrogen from the position which was originally C-3 of tyrosine by oxidation, and decarboxylation yields the dienone (LXXVII), redrawn in formula (LXXVI). The lactone (LXXIX) results by attack of the carboxyl group on the 2, 3-unsaturated ketone. Securinine is then formed via (LXXX), this final ring closure being plausibly facilitated by phosphorylation of the reduced ketone.

2.5 Mimosine

Mimosine (LXXXII, see also Chap. 7, this vol.), is also derived from lysine (HYLIN, 1964; NOTATION and SPENSER, 1964; TIWARI et al., 1967; TIWARI and SPENSER, 1965) however, the intermediates in this transformation are unknown. Degradations

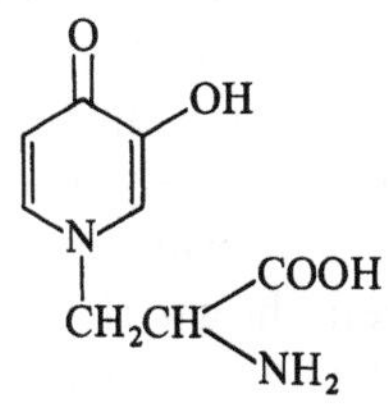

Mimosine
(LXXXII)

were not carried out to determine the distribution of activity in the alkaloid derived from [2-^{14}C] or [6-^{14}C]-lysine.

3 Alkaloids Derived from Nicotinic Acid

3.1 Introduction

It is now well established that nicotinic acid (LXXXIII) is formed in higher plants (tobacco, *Ricinus communis*, corn) from aspartic acid and a three-carbon compound related to glycerol, probably glyceraldehyde-3-phosphate. Extensive tracer work in this area has been reviewed (GROSS, 1969; LEETE, 1967a, 1969a; SPENSER, 1968) and the scheme illustrated in Fig. 12 is consistent with data, although most of the intermediates in this biosynthetic sequence are still hypothetical. Quinolinic acid (LXXXIV) seems to be the immediate precursor of nicotinic acid. More

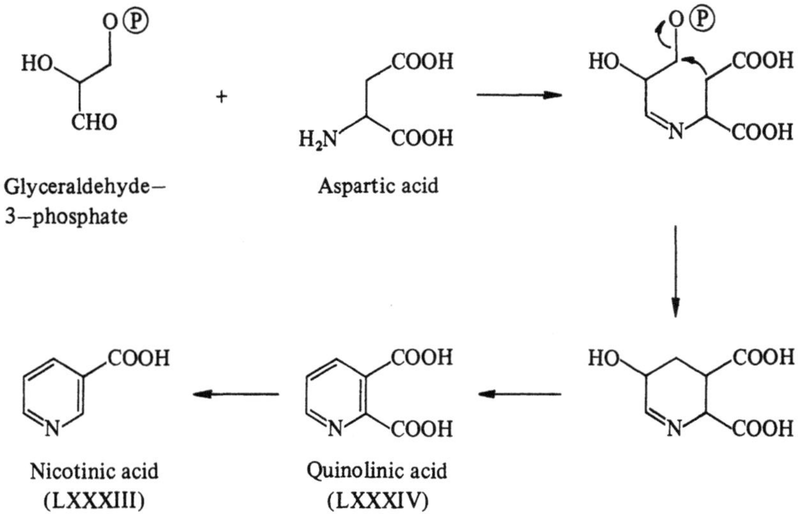

Fig. 12

recent work (LOVKOVA et al., 1970; SCOTT and DEVONSHIRE, 1971) is also consistent with this hypothetical biosynthesis.

Nicotinic acid serves as a precursor of the pyridine rings of the tobacco alkaloids (nicotine, nornicotine, anabasine, anatabine), the pyridone ring of ricinine, and unexpectedly, the isoquinuclidine moiety of dioscorine. It is also found in the sesquiterpenoid alkaloids of *Catha edulis* as a nicotinoyl ester. An example is cathedulin-2 (LXXXV) (BAXTER et al., 1976a, b). Another alkaloid, obviously related biogenetically to this one, is maytoline (LXXXVI) found in *Maytenus serrata* (KUPCHAN and SMITH, 1977; KUPCHAN et al., 1970). It has been shown that the 2-substituted nicotinoyl residue found in wilfordine (LXXXVII) (SHIZURI et al., 1973) is derived from nicotinic acid (LEE and WALLER, 1972).

3.2 The Tobacco Alkaloids

DAWSON and coworkers (1960a, b) showed that nicotinic acid is incorporated into the pyridine ring of nicotine and anabasine (SOLT et al., 1960) with the loss of the carboxyl group. It was also established that the point of attachment of the pyrrolidine ring of nicotine is at the C-3 position from which the carboxyl group is lost (YANG et al., 1965; SCOTT and GLYNN, 1967). It has also been shown that the hydrogen located at C-6 of nicotinic acid is lost during the formation of nicotine (DAWSON et al., 1960a, LEETE and LIU, 1973). Of the various hypotheses (DAWSON, 1962; DAWSON and OSDENE, 1972; LEETE, 1977e; WENKERT, 1968) which have been suggested for the biosynthesis of nicotine, the one illustrated in Fig. 13 is currently favored. Reduction of nicotinic acid yields 3,6-dihydronicotinic acid (LXXXVIII). It is proposed that the hydrogen added stereospecifically at C-6 is the one which is retained in the ultimate nicotine and the stereochemistry of the initial reduction is as shown (LEETE, 1978). This dihydronicotinic acid is then considered to undergo a concerted decarboxylation and reaction with the N-methyl-Δ^1-pyrroli-

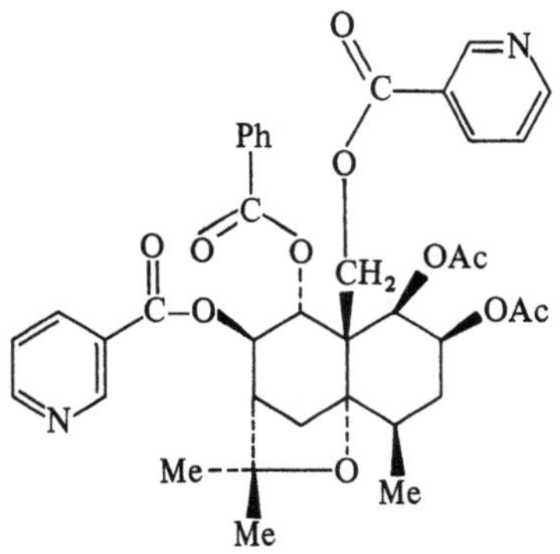

Cathedulin–2
(LXXXV)

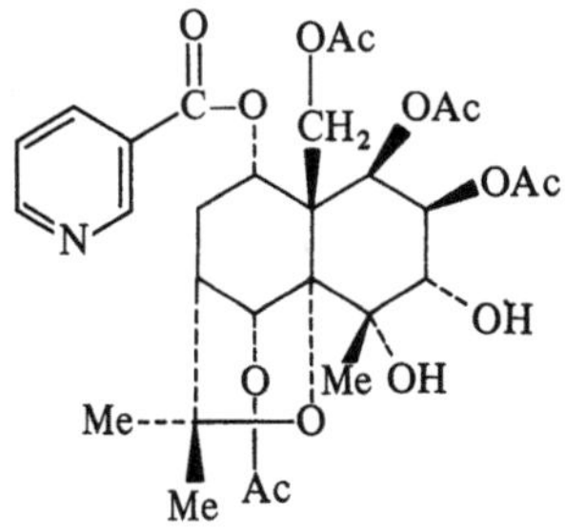

Maytoline
(LXXXVI)

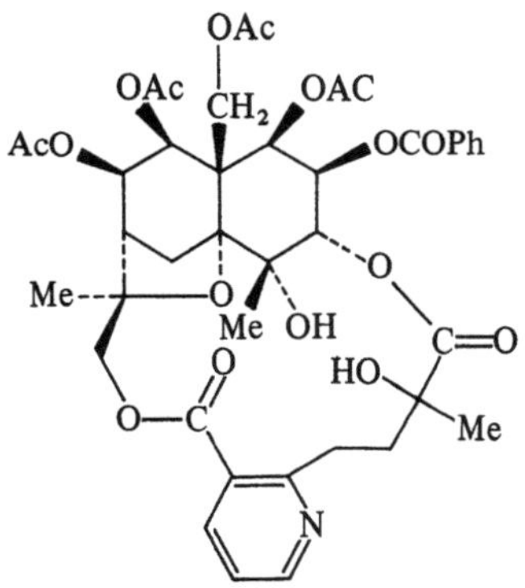

Wilfordine
(LXXXVII)

nium salt (derived from ornithine) or Δ^1-piperideine (derived from lysine). Dehydrogenation of the resultant compounds (LXXXIX) and (XC) yields nicotine and anabasine respectively. It is proposed that this final dehydrogenation is also stereospecific, eliminating the hydrogen originally present at C-6 in nicotinic acid.

Both rings of anatabine (XCIII) are formed from nicotinic acid (LEETE, 1975, 1977a, LEETE and SLATTERY, 1976) and the tracer results are consistent with the scheme illustrated in Fig. 14. Decarboxylation of the 3,6-dihydronicotinic acid (LXXXVIII) yields 2,5-dihydropyridine (XCI) which dimerizes to afford (XCIV) which is converted to anatabine by dehydrogenation. There is no interconversion of anatabine and anabasine in tobacco, however, 2,3-dipyridyl (XCII) may be formed by the oxidation of anatabine, especially in harvested leaves (LEETE and SLATTERY, 1976).

3.3 Dioscorine

This alkaloid, containing an isoquinuclidine nucleus, is derived from nicotinic acid (with loss of its carboxyl group) and acetic acid. Dioscorine (XCIX) derived

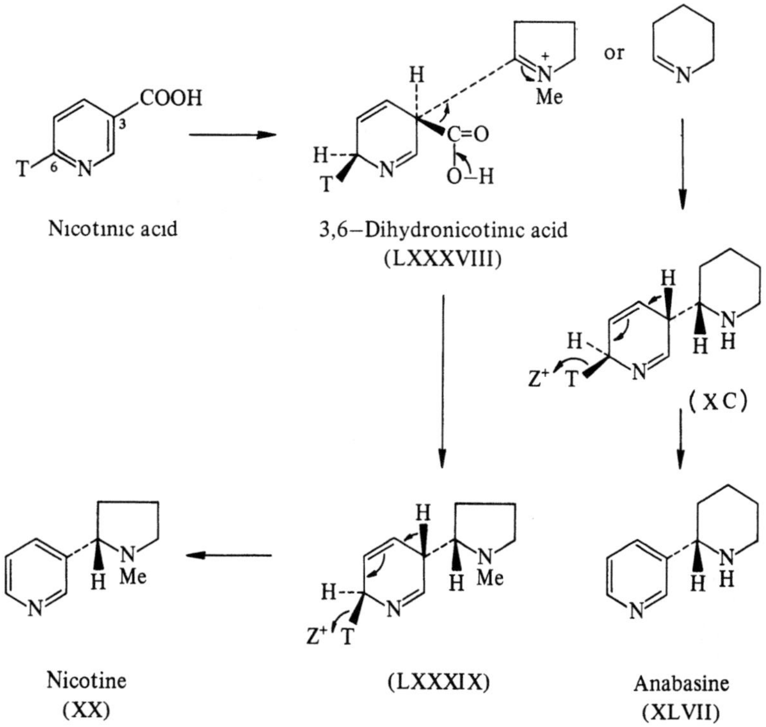

Fig. 13. Biosynthesis of nicotine and anabasine from nicotinic acid. The fate of the hydrogen at C-6 of nicotinic acid is illustrated by means of a tritium label (*T*)

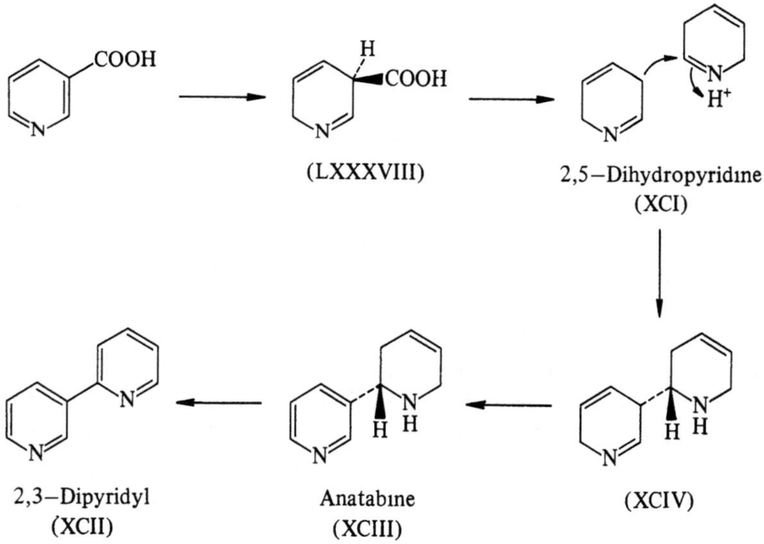

Fig. 14. Biosynthesis of anatabine

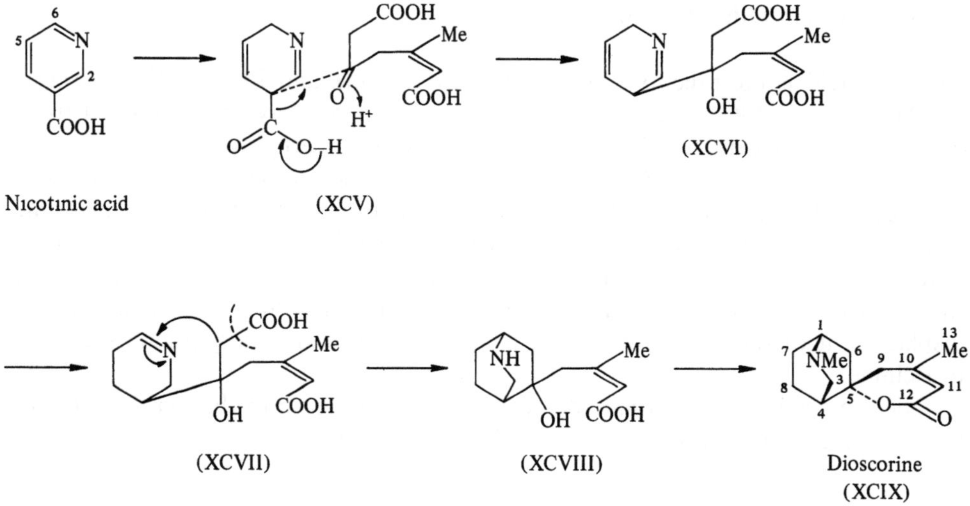

Fig. 15. Biosynthesis of dioscorine

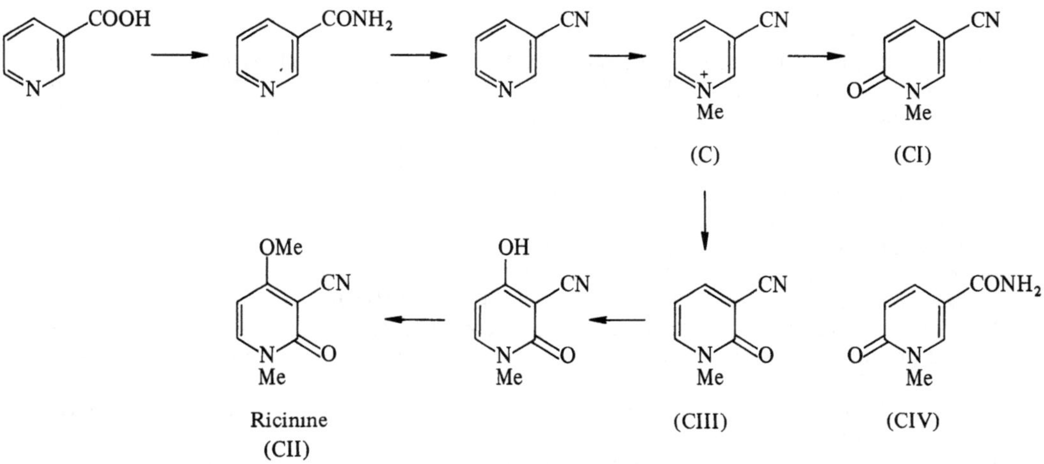

Fig. 16. Biosynthesis of ricinine

from [1-^{14}C]-acetate was labeled on alternate carbons of the unsaturated lactone ring (C-5, 10, 12) (LEETE and PINDER, 1972). [2-^{14}C]-Nicotinic acid yielded dioscorine labeled at C-3 (LEETE, 1977d) whilst [5,6-$^{13}C_2$]-nicotinic acid afforded material labeled at C-1 and C-7, location of the label being determined by means of ^{13}C-NMR (LEETE, 1977c). A hypothesis consistent with these data is illustrated in Fig. 15. As with the tobacco alkaloids it is suggested that the activated form of nicotinic acid is 3,6-dihydronicotinic acid (LXXXVIII). A concerted decarboxylation and condensation with a branched 8-carbon acetate derived unit (XCV) yields compound (XCVI). Shift of a double bond in the dihydropyridine and further reduction of the ring affords compound (XCVII). An aldol condensation, possibly with decarboxylation, leads to (XCVIII), which contains the isoquinuclidine ring system. Dioscorine then arises by lactone formation and N-methylation.

3.4 Ricinine and Other Pyridones

Nicotinic acid is the precursor of ricinine (CII) (LEETE and LEITZ, 1957; WALLER and HENDERSON, 1961) in *Ricinus communis* and enzymes have been isolated from this plant which catalyze the oxidation of 3-cyano-l-methylpyridinium salts (C) to the pyridones (CI) and (CIII) (FU et al., 1972). It is thus suggested that ricinine is produced by the route illustrated in Fig. 16. The pyridine moieties of nicotinic acid mononucleotide, nicotinic acid adenine dinucleotide, and nicotinamide adenine dinucleotide are also converted to ricinine (WALLER et al., 1966). It may be that these nucleotides are the actual precursors of ricinine, and N-methylation, by displacement of the phosphoriboside, is a late stage in the biosynthetic sequence to ricinine.

Nicotinic acid and nicotinamide are also incorporated into the pyridone (CIV) found in *Trewia nudiflora* (SASTRY and WALLER, 1972).

References

Abdusalamov, B.A., Takanaev, A.A., Sadykov, A.S., Aslanov, Kh.A.: Biosynthesis of *Sophora pachycarpa* alkaloids. Uzb. Biol. Zh. *13*, 10–12 (1969) Chem. Abs. *71*, 27925 h (1969)

Abdusalamov, B.A., Takanaev, A.A., Aslanov, Kh.A., Sadykov, AS.: Biosynthesis of alkaloids by *Goebelia pachycarpa*. Biokhimiya *36*, 290–294 (1971 a) Chem. Abs. *75*, 45685s (1971 a)

Abdusalamov, B.A., Takanaev, A.A., Aslanov, Kh.A., Sadykov, A.S.: *Goebelia pachycarpa* alkaloids. Khim. Prir. Soedin. *7*, 127–128 (1971 b)

Ahmad, A., Leete, E.: Biosynthesis of the tropine moiety of hyoscyamine from δ-N-methylornithine. Phytochemistry *9*, 2345–2347 (1970)

Alworth, W.L., Liberman, L., Ruchstahl, J.A.: Metabolism of nicotine-1′-oxide in excised *Nicotiana glutinosa* leaves. Phytochemistry *8*, 1427–1432 (1969)

Bale, N.M., Crout, N.H.G.: Determination of the relative rates of incorporation of arginine and ornithine into retronecine during pyrrolizidine alkaloid biosynthesis. Phytochemistry *14*, 2617–2622 (1975)

Basey, K., Woolley, J.G.: Biosynthesis of the tigloyl esters in *Datura*: The role of 2-methylbutyric acid. Phytochemistry *12*, 2197–2201 (1973)

Baxter, R.L., Crombie, L., Simmons, D.J., Whiting, D.A.: Extractives of *Catha edulis* (Khat): Occurrence of Celastraceaeous alkaloids having mono- and bis-macrolide bridges. J. Chem. Soc. Chem. Commun. 463–464 (1976 a)

Baxter, R.L., Crombie, L., Simmons, D.J., Whiting, D.A.: Structures of cathedulin-2 and cathedulin-8, new sesquiterpene alkaloids from *Catha edulis*. J. Chem. Soc. Chem. Commun. 465–466 (1976 b)

Beresford, P.J., Woolley, J.G.: Biosynthesis of 3α, 6β-ditigloyl-oxytropane and 3α, 6β-ditigloyl-oxytropan-7β-ol in *Datura*. Phytochemistry *14*, 2205–2207 (1975 a)

Beresford, P.J., Woolley, J.G.: Competitive feeding experiments with tropine in *Datura*. Phytochemistry *14*, 2209–2212 (1975 b)

Bottomley, W., Geissman, T.A.: Pyrrolizidine alkaloids. The biosynthesis of retronecine. Phytochemistry *3*, 357–360 (1964)

Braekman, J.C., Gupta, R.N., MacLean, D.B., Spenser, I.D.: Biosynthesis of lycopodine. Pelletierine as an obligatory intermediate. Can. J. Chem. *50*, 2591–2602 (1972)

Castillo, M., Gupta, R.N., MacLean, D.B., Spenser, I.D.: Biosynthesis of lycopodine from lysine and acetate. The pelletierine hypothesis. Can. J. Chem. *48*, 1893–1903 (1970 a)

Castillo, M., Gupta, R.N., Ho, Y.K., MacLean, D.B., Spenser, I.D.: Biosynthesis of lycopodine. Incorporation of Δ^1-piperideine and of pelletierine. Can. J. Chem. *48*, 2911–2918 (1970 b)

Craig, J.C., Mary, N.Y., Goldman, N.L., Wolf, L.: Tertiary amine oxide rearrangements. III. The mechanism of the demethylation of nicotine. J. Am. Chem. Soc. *86*, 3866–3869 (1964)

Dawson, R.F.: Accumulation of nicotine in reciprocal grafts of tomato and tobacco. Am. J. Bot *29*, 66–71 (1942a)

Dawson, R.F.: Nicotine synthesis in excised tobacco roots. Am. J. Bot. *29*, 813–815 (1942b)

Dawson, R.F.: An Experimental analysis of alkaloid production in *Nicotiana*: The origin of nornicotine. Am. J. Bot *32*, 416–423 (1945a)

Dawson, R.F.: On the biogenesis of nornicotine and anabasine. J. Am. Chem. Soc. *67*, 503–504 (1945b)

Dawson, R.F.: Alkaloid biogenesis. III. Specificity of the nicotine-nornicotine conversion. J. Am. Chem. Soc. *73*, 4218–4221 (1951)

Dawson, R.F.: Biosynthesis of the *Nicotiana* alkaloids. In: Science in progress. Brode, W.R. (ed.), Vol. 12, pp. 117–143. New Haven, Conn.: Yale University Press, 1962

Dawson, R.F., Osdene, T.S.: A speculative view of tobacco alkaloid biosynthesis. In: Structural and functional aspects of phytochemistry, Recent advances in phytochemistry. (ed. Runeckles, V.C., Tso, T.C. (eds.), Vol. 5, pp. 317–338. New York: Academic Press 1972

Dawson, R.F., Solt, M.L.: Estimated contributions of root and shoot to the nicotine content of the tobacco plant. Plant Physiol. *34*, 656–661 (1959)

Dawson, R.F., Christman, D.R., Solt, M.L., Wolf, A.P.: The biosynthesis of nicotine from nicotinic acid. Chemical and radiochemical yields. Arch. Biochem. Biophys. *91*, 144–150 (1960a)

Dawson, R.F., Christman, D.R., D'Adamo, A., Solt, M.L., Wolf, A.P.: The biosynthesis of nicotine from isotopically labelled nicotinic acids. J. Am. Chem. Soc. *82*, 2628–2633 (1960b)

Dewey, L.J., Byerrum, R.U., Ball, C.D.: Biosynthesis of the pyrrolidine ring of nicotine. Biochim. Biophys. Acta *18*, 141–142 (1955)

Egri, L.: Investigation of the alkaloids of some tobacco types with special consideration to research on the biosynthesis of nornicotine. Fachliche Mitt. Oesterr. Tabakregie 19–24 (1957)

El-Olemy, M.M., Schwarting, A.E., Kelleher, W.J.: Simulated biosynthesis of hygrine, cuskhygrine, anahygrine, anaferine and isopelletierine. Lloydia *29*, 58–67 (1966)

Evans, W.C., Woolley, V.A.: Biosynthesis of the (+)-2-hydroxy-3-phenylpropionic acid moiety of littorine in *Datura sanguinea* and *Anthocercis littorea*. Phytochemistry *8*, 2183–2187 (1969)

Fodor, G., Romeike, A., Janzsó, G., Koczor, I.: Epoxidation experiments *in vivo* with dehydrohyohyoscyamine and related compounds. Tetrahedron Lett. *7*, 19–22 (1959)

Fu, P., Kobus, J., Robinson, T.: The multiplicity of pyridinium oxidase in *Ricinus communis*. Phytochemistry *11*, 105–112 (1972)

Galinovsky, F., Wagner, A., Weiser, R.: Conversion of l-methyl-2-pyrrolidone and 1-methyl-2-piperidone with $LiAlH_4$ to ω-(methylamino)-aldehydes – Synthesis of hygrine, cuskhygrine, and methylisopelletierine. Monatschrift *82*, 551–559 (1951)

Gilbertson, T.J.: Biosynthesis of the piperidine nucleus: Metabolism of D- and L-lysine-2-^{14}C by *Nicotiana glauca*. Phytochemistry *11*, 1737–1739 (1972)

Gilbertson, T.J., Leete, E.: Biosynthesis of *Nicotiana* alkaloids. XII. The incorporation of α- and δ-N-methylornithine into the pyrrolidine ring of nicotine. J. Am. Chem. Soc. *89*, 7085–7088 (1967)

Golebiewski, W.M., Spenser, I.D.: The biosynthesis of the lupine alkaloids. A reexamination. J. Am. Chem. Soc. *98*, 6726–6728 (1976)

Golebiewski, W.M., Horsewood, P., Spenser, I.D.: Biosynthesis of securinine: The mode of incorporation of lysine. J. Chem. Soc. Chem. Commun. 217–218 (1976)

Griffith, T., Griffith, G.D.: Biosynthesis of piperidine alkaloids – I. Evidence for use of the diaminopimelic acid pathway for lysine synthesis in *Nicotiana glauca*. Phytochemistry *5*, 1175–1179 (1966)

Gross, D.: Pyridine alkaloids. In: Biosynthese der Alkaloide. Mothes, K., Schütte, H.R. (eds.), pp. 215–274. Berlin: VEB Deutscher Verlag der Wissenschaften 1969

Gross, D., Schütte, H.R.: The biosynthesis of tropane alkaloids. III. Phenylalanine as a precursor to tropic acid and benzoic acid. Arch. Pharm. *296*, 1–6 (1963)

Gupta, R.N., Spenser, I.D.: The biosynthesis of sedamine. Can. J. Chem. *45*, 1275–1285 (1967)
Gupta, R.N., Spenser, I.D.: Biosynthesis of N-methylpelletierine. Phytochemistry *8*, 1937–1944 (1969)
Gupta, R.N., Spenser, I.D.: Biosynthesis of the piperidine nucleus: The occurrence of two pathways from lysine. Phytochemistry *9*, 2329–2334 (1970)
Gupta, R.N., Spenser, I.D.: Biosynthesis of lobinaline. Can. J. Chem. *49*, 384–397 (1971)
Gupta, R.N., Castillo, M., MacLean, D.B., Spenser, I.D., Wrobel, J.T.: Biosynthesis of lycopodine. J. Am. Chem. Soc. *90*, 1360–1361 (1968)
Herbert, R.B., Jackson, F.B., Nicholson, I.T.: Biosynthesis of phenanthroindolizidine alkaloids *via* derivatives of 2-phenacylpyrrolidine and benzoylacetic acid. J. Chem. Soc. Chem. Commun., 865–867 (1976)
Ho, Y.K., Gupta, R.N., MacLean, D.B., Spenser, I.D.: Biosynthesis of cernuine. Can. J. Chem. *49*, 3352–3361 (1971)
Hughes, C.A., Letcher, R., Warren, F.L.: The *Senecio* alkaloids, part XVI. The biosynthesis of the "necine" bases from carbon-14 precursors. J. Chem. Soc. 4974–4978 (1964)
Hurley, L.H., Gairola, C., Zmijewski, M.: Biosynthesis of the antiviral antibiotic 11-demethyltomaymycin by *Streptomyces achromogenes*. J. Chem. Soc. Chem. Commun. 120–121 (1975a)
Hurley, L.H., Zmijewski, M., Chang, C.-J.: Biosynthesis of anthramycin. Determination of the labelling pattern by the use of radioactive and stable isotope techniques. J. Am. Chem. Soc. *97*, 4372–4378 (1975b)
Hurley, L.H., Das, N., Gairola, C., Zmijewski, M.: Biosynthetic incorporation of DL-tryptophan-(5-^{3}H) into anthramycin, sibiromycin and tomaymycin: N.I.H. shift produced by actinomycetes. Tetrahedron Lett., 1419–1422 (1976a)
Hurley, L.H., Gairola, C., Das, N.V.: Pyrrolo-[1,4]-benzodiazepine antibiotics. Biosynthesis of the antitumor antibiotic 11-demethyltomaymycin and its biologically inactive metabolite oxotomaymycin by *Streptomyces achromogenes*. Biochemistry *15*, 3760–3769 (1976b)
Hutchinson, C.R., Hsia, M-T.S., Carver, R.A.: Biosynthetic studies with $^{13}CO_2$ of secondary plant metabolites. *Nicotiana* alkaloids 1. Initial experiments. J. Am. Chem. Soc. *98*, 6006–6011 (1976)
Hylin, J.W.: Biosynthesis of mimosine. Phytochemistry *3*, 161–164 (1964)
Iljin, G.S., Serebrovkaya, K.B.: Demethylation of nicotine and absorption of oxygen by tobacco plants. Dokl. Akad. Nauk USSR *118*, 139–141 (1958)
Johne, S., Waiblinger, K., Gröger, D.: Investigation of the biosynthesis of the quinazoline alkaloid peganine in *Adhatoda vasica* Nees. Pharmazie *28*, 403–406 (1973)
Kaczkowski, J., Schütte, H.R., Mothes, K.: The role of acetate in the biosynthesis of the tropane alkaloids. Biochim. Biophys. Acta *46*, 588–594 (1961)
Keogh, M.F., O'Donovan, D.G.: Biosynthesis of some alkaloids of *Punica granatum* and *Withania somnifera*. J. Chem. Soc. (*C*), 1792–1797 (1970a)
Keogh, M.F., O'Donovan, D.G.: Biosynthesis of lobeline. J. Chem. Soc. (*C*), 2470–2472 (1970b)
Kisaki, T., Tamaki, E.: Phytochemical studies on the tobacco alkaloids. I. Optical rotatory power of nornicotine. Arch. Biochem. Biophys. *92*, 351–355 (1961a)
Kisaki, T., Tamaki, E.: Phytochemical studies on the tobacco alkaloids. III. Observations on the interconversion of DL-nicotine and DL-nornicotine in excised tobacco leaves. Arch. Biochem. Biophys. *94*, 252–256 (1961b)
Kisaki, T., Tamaki, E.: Phytochemical studies on the tobacco alkaloids. X. Degradation of the tobacco alkaloids and their optical rotatory changes in tobacco plants. Phytochemistry *5*, 293–300 (1966)
Kisaki, T., Mizusaki, S., Tamaki, E.: γ-Methylaminobutyraldehyde, a new intermediate in nicotine biosynthesis. Arch. Biochem. Biophys. *117*, 677–678 (1966)
Kompis, I., Grossmann, E., Terent'eva, I.V., Lazur'evskii, G.V.: Biosynthesis of brevicolline. Khim. Prir. Soedin., *5*, 39–43 (1969)
Koo, S.H., Comer, F., Spenser, I.D.: Biosynthesis of *Lythraceae* alkaloids: Mode of incorporation of phenylalanine. J. Chem. Soc. Chem. Commun. 897–898 (1970a)
Koo, S.H., Gupta, R.N., Spenser, I.D., Wrobel, J.T.: Biosynthesis of the *Lythraceae* alkaloids: Incorporation of lysine. J. Chem. Soc. Chem. Commun. 396–397 (1970b)

Kupchan, S.M., Smith, R.M.: Maytoline, maytine, and maytolidine, novel nicotinoyl sesquiterpene alkaloids from *Maytenus serrata* (Hochst., ex A. Rich) R. Wilczek. J. Org. Chem. *42*, 115–118 (1977)
Kupchan, S.M., Smith, R.M., Bryan, R.F.: Maytoline, a nicotinoyl sesquiterpene alkaloid prototype from *Maytenus ovatus*. J. Am. Chem. Soc. *92*, 6667–6668 (1970)
Kuschmuradov, J.K., Gross, D., Schütte, H.R.: Biosynthesis and metabolism of the quinolizidine alkaloids of *Sophora alopecuroides*. Phytochemistry *11*, 3441–3445 (1972)
Lamberts, B.L., Byerrum, R.U.: Glutamate as a precursor for the pyrrolidine ring of nicotine. J. Biol. Chem. *233*, 939–943 (1958)
Lamberts, B.L., Dewey, L.J., Byerrum, R.U.: Ornithine as a precursor for the pyrrolidine ring of nicotine. Biochim. Biophys. Acta *33*, 22–26 (1959)
Lee, H.J., Waller, G.R.: Biosynthesis of pyridine alkaloids from *Tripterygium wilfordii*. Phytochemistry *11*, 2233–2240 (1972)
Leete, E.: Biogenesis of nicotine. Chem. Ind. London 537 (1955)
Leete, E.: The biogenesis of nicotine and anabasine. J. Am. Chem. Soc. *78*, 3520–3523 (1956)
Leete, E.: The biogenesis of nicotine. V. New precursors of the pyrrolidine ring. J. Am. Chem. Soc. *80*, 2162–2164 (1958a)
Leete, E.: The biogenesis of the *Nicotiana* alkaloids. VI. The piperidine ring of anabasine. J. Am. Chem. Soc. *80*, 4393–4394 (1958b)
Leete, E.: The stereospecific incorporation of ornithine into the tropine moiety of hyoscyamine. J. Am. Chem. Soc. *84*, 55–57 (1962)
Leete, E.: Biosynthesis of hyoscyamine. Proof that ornithine-2-^{14}C yields tropine labelled at C-1. Tetrahedron Lett. 1619–1622 (1964)
Leete, E.: Alkaloid biogenesis. In: Biogenesis of natural compounds. Bernfeld, P. (ed.), pp. 953–1023. Oxford: Pergamon Press 1967a
Leete, E.: Biosynthesis of *Nicotiana* alkaloids. XI. Investigation of tautomerism in N-methyl-Δ^1-pyrrolinium chloride and its incorporation into nicotine. J. Am. Chem. Soc. *89*, 7081–7084 (1967b)
Leete, E.: Alkaloid biosynthesis. Adv. Enzymol. *32*, 373–422 (1969a)
Leete, E.: Biosynthesis of the *Nicotiana* alkaloids. XIV. The incorporation of Δ^1-piperideine-6-^{14}C into the piperidine ring of anabasine. J. Am. Chem. Soc. *91*, 1697–1700 (1969b)
Leete, E.: Biosynthesis of the hemlock and related piperidine alkaloids. Acc. Chem. Res. *4*, 100–107 (1971)
Leete, E.: Biosynthesis of meteloidine. Phytochemistry *11*, 1713–1716 (1972)
Leete, E.: Biosynthesis of alkaloids. In: Biosynthesis. Geissman, T.A. (ed.), Vol. 2, pp. 106–182. London: The Chemical Society 1973a
Leete, E.: Biosynthetic conversion of α-methylbutyric acid to tiglic acid in *Datura meteloides*. Phytochemistry *12*, 2203–2205 (1973b)
Leete, E.: Biosynthesis of anatabine and anabasine in *Nicotiana glutinosa*. J. Chem. Soc. Chem. Commun. 9–10 (1975)
Leete, E.: Biosynthesis of alkaloids. In: Biosynthesis. Bu'lock, J.D. (ed.), Vol. 4, pp. 97–178. London: The Chemical Society 1976a
Leete, E.: A new systematic degradation of nicotine to determine activity at C-2′ and C-5′. The pattern of labelling in nicotine and nornicotine formed from [2-^{14}C]ornithine in *Nicotiana glutinosa* and in nicotine obtained from *N. tabacum* exposed to [^{14}C, ^{13}C]carbon dioxide. J. Org. Chem. *41*, 3438–3441 (1976b)
Leete, E.: The incorporation of [5,6-$^{13}C_2$]nicotinic acid into the tobacco alkaloids examined by the use of ^{13}C nuclear magnetic resonance. Bioorg. Chem. *6*, 273 (1977a)
Leete, E.: Biosynthesis of alkaloids. In: Biosynthesis. Bu'lock, J.D. (ed.), Vol. 5, pp. 136–239 London: The Chemical Society 1977b
Leete, E.: Biosynthesis of the isoquinuclidine moiety of dioscorine. Incorporation of [5,6-$^{13}C_2$]nicotinic acid established by means of ^{13}C nuclear magnetic resonance. J. Am. Chem. Soc. *99*, 648–650 (1977c)
Leete, E.: Biosynthesis of dioscorine: Incorporation of nicotinic acid into the isoquinuclidine moiety. Phytochemistry *16*, 1705–1709 (1977d)
Leete, E.: Biosynthesis and metabolism of the tobacco alkaloids. Tob. Chem. Symp. Am. Chem. Soc. pp. 365–388. New Orleans (1977e)

Leete, E.: Stereochemistry of the reduction of nicotinic acid when it serves as a precursor of anatabine. J. Chem. Soc. Chem. Commun. 610–611 (1978)
Leete, E., Bodem, G.B.: Biosynthesis of shihunine in *Dendrobium pierardii*. J. Am. Chem. Soc. *98*, 6321–6325 (1976)
Leete, E., Chedekel, M.R.: The aberrant formation of (−)-N-methylanabasine from N-methyl-Δ^1-piperideinium chloride in *Nicotiana tabacum* and *N. glauca*. Phytochemistry *11*, 2751–2756 (1972)
Leete, E., Leitz, F.H.B.: Biogenesis of ricinine. Chem. Ind. London 1572 (1957)
Leete, E., Liu, Y-Y.: Metabolism of 2- and 6-tritium labelled nicotinic acid in intact *Nicotiana tabacum* plants. Phytochemistry *12*, 593–596 (1973) and unpublished work of E. Leete.
Leete, E., Lucast, D.H.: Biosynthesis of meteloidine from 3α-tigloyloxytropane in *Datura innoxia*. Phytochemistry *14*, 2199–2200 (1975)
Leete, E., Lucast, D.H.: Loss of tritium during the biosynthesis of meteloidine and scopolamine from [N-methyl-^{14}C, 6β, 7β-$^3H_2$]tropine. Tetrahedron Lett. 3401–3404 (1976)
Leete, E., Murrill, J.B.: Biosynthesis of the tiglic acid moiety of meteloidine in *Datura meteloides*. Tetrahedron Lett. 1727–1730 (1967)
Leete, E., Nelson, S.J.: Biosynthesis of the teloidine moiety of meteloidine in *Datura meteloides*. Phytochemistry *8*, 413–418 (1969)
Leete, E., Pinder, A.R.: Biosynthesis of dioscorine. Phytochemistry *11*, 3219–3224 (1972)
Leete, E., Siegfried, K.J.: Biogenesis of nicotine. III. Incorporation of ornithine into the pyrrolidine ring. J. Am. Chem. Soc. *79*, 4529–4531 (1957)
Leete, E., Slattery, S.A.: Incorporation of [2-^{14}C]- and [6-^{14}C]-nicotinic acid into the tobacco alkaloids. Biosynthesis of anatabine and α, β-dipyridyl. J. Am. Chem. Soc. *98*, 6326–6330 (1976)
Leete, E., Gros, E.G., Gilbertson, T.J.: The biosynthesis of anabasine. Origin of the nitrogen of the piperidine ring. J. Am. Chem. Soc. *86*, 3907–3908 (1964)
Leete, E., Kowanko, N., Newmark, R.A.: Use of carbon-13 nuclear magnetic resonance to establish that the biosynthesis of tropic acid involves an intramolecular rearrangement of phenylalanine. J. Am. Chem. Soc. *97*, 6826–6830 (1975)
Leistner, E., Spenser, I.D.: Biosynthesis of the piperidine nucleus. Incorporation of chirally labeled [1-^{3}H]-cadaverine. J. Am. Chem. Soc. *95*, 4715–4725 (1973)
Leistner, E., Spenser, I.D.: Stereochemistry of the enzymatic decarboxylation of L-lysine. J. Chem. Soc. Chem. Commun., 378–379 (1975)
Leistner, E., Gupta, R.N., Spenser, I.D.: A general method for the determination of precursor configuration in biosynthetic precursor-product relationships. Derivation of pipecolic acid from D-lysine and of piperidine alkaloids from L-lysine. J. Am. Chem. Soc. *95*, 4040–4047 (1973)
Liebisch, H.W., Schütte, H.R.: Studies on the biosynthesis of the tropane alkaloids. VIII. Precursors of the pyrrolidine moiety. Z. Pflanzenphysiol. *57*, 434–439 (1967)
Liebisch, H.W., Maier, W., Schütte, H.R.: Biosynthesis of tropane alkaloids. VI. N-Methylputrescine as possible precursor of the pyrrolidine ring in hyoscyamine and scopolamine. Tetrahedron Lett. 4079–4082 (1966)
Liebisch, H.W., Radwan, A.S., Schütte, H.R.: Biosynthesis of the tropane alkaloids. X. Formation of cuskhygrine. Ann. Chem. *721*, 163–167 (1969)
Liebisch, H.W., Peisker, K., Radwan, A.S., Schütte, H.R.: Biosynthesis of tropane alkaloids. XI. Formation of the C_3 bridge of tropine. Z. Pflanzenphysiol. *67*, 1–9 (1972)
Liebisch, H.W., Bernasch, H., Schütte, H.R.: The biosynthesis of the tropane alkaloids. XII. The biosynthesis of cochlearine. Z. Chem. *13*, 372–373 (1973)
Liebman, A.A., Morsingh, F., Rapoport, H.: Biosynthesis of nicotine in *Nicotiana glutinosa* from carbon-14 dioxide. Labeling pattern in the pyrrolidine ring. J. Am. Chem. Soc. *87*, 3499 (1965)
Liebman, A.A., Mundy, B.P., Rapoport, H.: The biosynthesis of nicotine in *Nicotiana glutinosa* from carbon-14 dioxide. Labeling pattern in the pyrrolidine ring. J. Am. Chem. Soc. *89*, 664–672 (1967)
Liljegren, D.J.: Biosynthesis of quinazoline alkaloids of *Peganum harmala*. Phytochemistry *10*, 2661–2669 (1971)
Lovkova, M.Y., Ilyin, G.S., Klimentyeva, N.I.: Role of aspartic acid in the biosynthesis of nicotine and anabasine. Fiziol. Rast. *17*, 409–415 (1970)

Lovkova, M.Y., Nurimov, E., Ilyin, G.S.: Role of lysine in the biosynthesis of anabasine and lupinine in *Anabasis aphylla*. Biokhimiya *39*, 388–393 (1974a)
Lovkova, M.Y., Nurimov, E., Ilyin, G.S.: Role of aspartic acid in the biosynthesis of pyridine and piperidine rings of anabasine in *Anabasis aphylla*. Biokhimiya *39*, 523–526 (1974b)
Marshall, W.D., Nguyen, T.T., MacLean, D.B., Spenser, I.D.: Biosynthesis of lycopodine. The question of the intermediacy of piperidine-2-acetic acid. Can. J. Chem. *53*, 41–50 (1976)
Mizusaki, S., Kisaki, T., Tamaki, E.: Phytochemical studies on the tobacco alkaloids. IX. Question for nornicotine as a precursor of nicotine. Agric. Biol. Chem. (Tokyo) *26*, 714–718 (1965)
Mizusaki, S., Kisaki, T., Tamaki, E.: Phytochemical studies on the tobacco alkaloids. XII. Identification of γ-methylaminobutyraldehyde and its precursor role in nicotine biosynthesis. Plant Physiol. *43*, 93–98 (1967)
Mizusaki, S., Tanabe, Y., Noguchi, M., Tamaki, E.: Phytochemical studies on tobacco alkaloids. XIV. Occurrence and properties of putrescine N-methyltransferase in tobacco roots. Plant Cell Physiol. (Tokyo) *12*, 633–640 (1971)
Mizusaki, S., Tanabe, Y., Noguchi, M., Tamaki, E.: Phytochemical studies on the tobacco alkaloids. XV. N-Methylputrescine oxidase from tobacco roots. Phytochemistry *11*, 2757–2762 (1972)
Mizusaki, S., Tanabe, Y., Noguchi, M., Tamaki, E.: Phytochemical studies on the tobacco alkaloids. XVI. Changes in the activities of ornithine decarboxylase, putrescine N-methyltransferase, and N-methylputrescine oxidase in tobacco roots in relation to nicotine biosynthesis. Plant Cell Physiol. (Tokyo) *14*, 103–110 (1973)
Mulchandani, N.B., Iyer, S.S., Badheka, L.P.: Incorporation of tyrosine-2-^{14}C into tylophorine. Phytochemistry *8*, 1931–1935 (1969)
Mulchandani, N.B., Iyer, S.S., Badheka, L.P.: Incorporation of phenylalanine-2-^{14}C into tylophorine. Phytochemistry *10*, 1047–1050 (1971)
Mulchandani, N.B., Iyer, S.S., Badheka, L.P.: Incorporation of cinnamic acid-2-[^{14}C] into tylophorine. Phytochemistry *15*, 1697–1699 (1976)
Nguyen, T-L., Gruenke, L.D., Castagnoli, N.: Metabolic N-demethylation of nicotine. Trapping of a reactive iminium species with caynide ion. J. Med. Chem. *19*, 1168–1169 (1976)
Notation, A.D., Spenser, I.D.: Biosynthesis of mimosine. Incorporation of aspartic acid into the pyridone nucleus. Can J. Biochem. *42*, 1803–1808 (1964)
Nowacki, E., Byerrum, R.U.: A study on the biosynthesis of the *Crotalaria* alkaloids. Life Sci. *5*, 157–161 (1962)
O'Donovan, D.G., Creedon, P.B.: Biosynthesis of the alkaloids of *Haloxylon salicornicum* I. Tetrahedron Lett. 1341–1344 (1971)
O'Donovan, D.G., Creedon, P.B.: Biosynthesis of santiagume in *Adenocarpus foliosus*. Part II. J. Chem. Soc. Perkin I, 2524–2528 (1974)
O'Donovan, D.G., Keogh, M.F.: Role of hygrine in the biosynthesis of cuscohygrine and hyoscyamine. J. Chem. Soc. (*C*), 223–226 (1969)
O'Donovan, D.G., Long, D.J., Forde, E., Geary, P.: The biosynthesis of *Lobelia* alkaloids. Part III. Intermediates in the biosynthesis of lobeline, biosynthesis of 8,10-diethyllobelidione. J. Chem. Soc. Perkin I, 415–419 (1975)
Parry, R.J.: Biosynthesis of the *Securinega* alkaloids I. Specific incorporation of tyrosine. Tetrahedron Lett. 307–310 (1974)
Parry, R.J.: Biosynthesis of the *Securinega* alkaloids. Stereospecificity of hydrogen loss from C-3 of tyrosine. J. Chem. Soc. Chem. Commun. 144–145 (1975)
Romeike, A.: Participation of the shoot in the formation of the alkaloid spectrum. Epoxide formation in hyoscyamine in *Datura ferox*. Flora (Jena) *143*, 67–86 (1956)
Romeike, A.: The contribution of the shoot in the elaboration of the spectrum of alkaloids. II. Site of the formation and biogenesis of scopolamine in *Datura ferox*. Flora (Jena) *148*, 306–320 (1959)
Romeike, A.: The occurrence of 6-hydroxyhyoscyamine in *Datura*. Naturwissenschaften *49*, 281 (1962)
Romeike, A., Fodor, G.: The biogenesis of hyoscine in *Datura stramonium* L. Tetrahedron Lett. *22*, 1–4 (1960)

Rother, A., Schwarting, A.E.: Biosynthesis of cryogenine. J. Chem. Soc. Chem. Commun. 1411–1412 (1969)
Rother, A., Schwarting, A.E.: Phenylalanine as a precursor for cryogenine biosynthesis in *Heimia salicifolia*. Phytochemistry *11*, 2475–2480 (1972)
Rueppel, M.L., Mundy, B.P., Rapoport, H.: Biosynthesis of the pyrrolidine ring of nicotine in *Nicotiana glutinosa*. Phytochemistry *13*, 141–151 (1974)
Sankawa, U., Yamasaki, K., Ebizuka, Y.: Biosynthesis of securinine. Incorporation of radioactive tyrosine, lysine, and cadaverine. Tetrahedron Lett. 1867–1868 (1974)
Sankawa, U., Ebizuka, Y., Yamasaki, K.: Biosynthesis of securinine, the main alkaloid of *Securinega suffruticosa*. Phytochemistry *16*, 561–563 (1977)
Sastry, S.D., Waller, G.R.: Biosynthesis of N-methyl-5-carboxamide-2-pyridone from *Trewia nudiflora*. Phytochemistry *11*, 2241–2245 (1972)
Schütte, H.R.: The quinolizidine alkaloids. In: Biosynthese der Alkaloids. Mothes, K., Schütte, H.R. (eds.), pp. 324–343. Berlin: VEB Deutscher Verlag der Wissenschaften 1969
Schütte, H.R., Hindorf, H.: The biosynthesis of lupinine. Z. Naturforsch. *19b*, 855 (1964)
Schütte, H.R., Hindorf, H.: The biosynthesis of lupanine and hydroxylupanine in *Lupinus angustifolius*. Ann. Chem. *685*, 187–194 (1965)
Schütte, H.R., Bohlmann, F., Reusche, W.: The biosynthesis of sparteine. Arch. Pharm. *294*, 610–615 (1961)
Schütte, H.R., Aslanow, H., Schäfer, C.: Biosynthesis of quinolizidine alkaloids. VI. Biosynthesis of matrine. Arch. Pharm. *295*, 34–36 (1962a)
Schütte, H.R., Nowacki, E., Schäfer, C.: The formation of lupanine and hydroxylupanine from cadaverine-[1,5-^{14}C]-dihydrochloride in *Lupinus angustifolius*. Arch. Pharm. *295*, 20–23 (1962b)
Schütte, H.R., Hindorf, H., Mothes, K., Hübner, G.: Biosynthesis of sparteine in *Lupinus luteus* and *Sarothamnus scoparius*. Ann. Chem. *680*, 93–104 (1964a)
Schütte, H.R., Kelling, K.L., Knofel, D., Mothes, K.: The biosynthesis of adenocarpine in *Adenocarpus viscosus*. Phytochemistry *3*, 249–252 (1964b)
Schütte, H.R., Lehfeldt, J., Hindorf, H.: Biosynthesis of matrine in *Sophora tetraptera*. Ann. Chem. *685*, 194–199 (1965)
Scott, T.A., Devonshire, A.L.: The incorporation of [4-^{14}C]-aspartic acid into nicotinic acid N-glucoside in *Nicotiana tabacum*. Biochem. J. *124*, 949–950 (1971)
Scott, T.A., Glynn, J.P.: The incorporation of nicotinic-2,3,7-^{14}C acid into nicotine by *Nicotiana tabacum*. Phytochemistry *6*, 505–510 (1967)
Shibata, S., Sankawa, U.: Biosynthesis of matrine. Chem. Ind. London 1161–1162 (1963)
Shizuri, Y., Yamada, K., Hirata, Y.: The structures of wilfordine and a new alkaloid alatamine, obtained from *Euonymus alatus* forma *striatus* (Thunb.) Makino. Tetrahedron Lett. 741–744 (1973)
Spenser, I.D.: The biosynthesis of alkaloids and of other nitrogenous secondary metabolites. In: Comprehensive biochemistry. Florkin, M., Stotz, E.H. (eds.), Vol. 20, pp. 231–413. Amsterdam, London, New York: Elsevier 1968
Solt, M.L., Dawson, R.F., Christman, D.R.: Biosynthesis of anabasine and of nicotine by excised root cultures of *Nicotiana glauca*. Plant Physiol. *35*, 887–894 (1960)
Takanaev, A.A., Abdusalamov, B.A.: Biosynthesis of alkaloids of *Sophora pachycarpa*. Uzb. Biol. Zh. *15*, 20–22 (1971) Chem. Abs. *75*, 115946 (1971)
Tamprateep, P., Taylor, E.H., Ramstad, E.: Biosynthesis of the alkaloids of *Erythroxylon novogranatense*. Lloydia *26*, 203 (1963)
Tiwari, H.P., Spenser, I.D.: Precursors of mimosine in *Mimosa pudica*. Can. J. Biochem. *43*, 1687–1691 (1965)
Tiwari, H.P., Penrose, W.R., Spenser, I.D.: Biosynthesis of mimosine: Incorporation of serine and of α-aminoadipic acid. Phytochemistry *6*, 1245–1248 (1967)
Tso, T.C.: Physiology and biochemistry of tobacco plants. Stroudsburg, Pa.: Dowden, Hutchinson and Ross, Inc. 1972
Tso, T.C., Jeffrey, R.N.: Tobacco alkaloids. I. Changes in nicotine and nornicotine content in *Nicotiana*. Plant Physiol. *31*, 433–440 (1956)
Tso, T.C., Jeffrey, R.N.: Tobacco alkaloids. II. Formation of nicotine and nornicotine in tobacco supplied with nitrogen-15. Plant Physiol. *32*, 86–92 (1957)

Wada, E.: Conversion of nicotine to nornicotine in cherry red tobacco during flue curing. Arch. Biochem. Biophys. *62*, 471–475 (1956)
Waiblinger, K., Johne, S., Gröger, D.: Biosynthesis of the pyrrolidine ring of peganine. Phytochemistry *11*, 2263–2265 (1972)
Waller, G.R., Henderson, L.M.: Biosynthesis of the pyridine ring of ricinine from succinate and other labeled compounds. Biochem. Biophys. Res. Commun. *5*, 5–10 (1961)
Waller, G.R., Yang, K.S., Gholson, R.K., Hadwiger, L.A., Chaykin, S.: The pyridine nucleotide cycle and its role in the biosynthesis of ricinine by *Ricinus communis* L. J. Biol. Chem. *241*, 4411–4418 (1966)
Wenkert, E.: The role of oxidation in the biogenesis of alkaloids. Experientia *10*, 346–350 (1954)
Wenkert, E.: Alkaloid synthesis. Acc. Chem. Res. *1*, 78–81 (1968)
Woolley, J.G.: The biosynthesis of tigloyl esters in *Datura*. In: Abh. Dtsch. Akad. Wiss. Berlin Kl. Chem. Gol. Biol. *3*, pp. 531–534 (1966)
Wu, P.L., Byerrum, R.U.: The biosynthesis of the pyrrolidine ring of nicotine. Biochemistry *4*, 1628–1632 (1965)
Yang, K.S., Gholson, R.K., Waller, G.R.: Studies on nicotine biosynthesis. J. Am. Chem. Soc. *87*, 4184–4188 (1965)
Yoshida, D., Mitake, T.: Agmatine and N-carbamylputrescine as intermediates in the formation of nicotine by tobacco plants. Plant Cell Physiol. (Tokyo) *7*, 301–305 (1966)
Zielke, H.R., Byerrum, R.U., O'Neal, R.M., Burns, L.C., Koeppe, R.E.: Incorporation of carbon dioxide and acetate into the pyrrolidine ring of nicotine. J. Biol. Chem. *243*, 4757–4761 (1968)

4.2 Alkaloids Derived from Phenylalanine and Tyrosine

G.B. FODOR

1 Introduction

This short review of a rather complex field aims at giving an insight and overview based on recent isotopic tracer studies – facts and no history.

Since both phenylalanine (I) and tyrosine (II) occur in animal tissues and also contribute to the build-up of a number of plant alkaloids it seemed that (II) would be a product of hydroxylation of (I) and thus an intermediate for a large number of alkaloids. However, it has been shown that only mammalian liver (SOLOWAY, 1966) is able to convert (I) into (II) while a similar phenylalanine hydroxylase is missing from the plant cell and labelled (I) was not converted (LEETE, 1966) into (II). Therefore, the large number of plant alkaloids with a recognisable 2-phenylethylamine moiety produced from (I) and those produced from (II) ought to be discussed separately. On the other hand, some 5,6-dihydroxyindole derivatives such as betalains or gliotoxin are derived from tyrosine (II), via DOPA (III) (Fig. 1) and not from the indolylamino acid tryptophan. Hence, these will be discussed in this chapter. The skeletons of DOPA (III), dopamine (XXIII) and noradrenalin (XXVII) are present in a number of alkaloids inter alia in isoquinolines (Fig. 9).

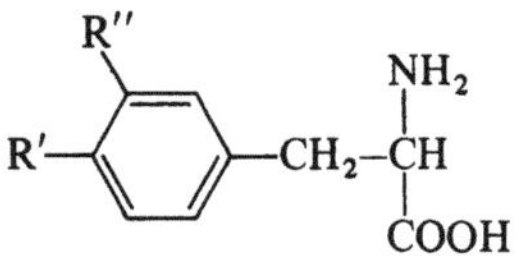

(I) R' = R'' = H
(II) R' = OH; R'' = H
(III) R' = R'' = OH

Fig. 1

2 Alkaloids Derived from Phenylalanine

The most important representatives which are formed from this biogenetic precursor are listed below:

2.1 Ephedra alkaloids: ephedrine, pseudoephedrine
2.2 Microbial metabolites: cytochalasin B and D

2.3 *Penicillium terlikowskii* alkaloid: gliotoxin
2.4 Beauveria alkaloid: tenellin
2.5 Taxus alkaloids: taxin
2.6 Lunaria alkaloids: lunarine and lunaridine
2.7 Lythraceae alkaloids

In addition, deamination of phenylalanine leads to a number of aromatic carboxylic acids which, in turn, are incorporated into alkaloids, e.g., tropic acid into atropine, benzoic acid into the tropane alkaloids tropacocaine and cocaine; cinnamic acid into cinnamoyl ecgonine, etc.

2.1 Ephedra Alkaloids

Ephedra alkaloids, mostly phenylpropanoids have been the subject of extensive biosynthetic investigations with labelled compounds, thus an impressive and convincing scheme (Fig. 2) has been outlined (Shibata and Imaseki, 1956; Shibata

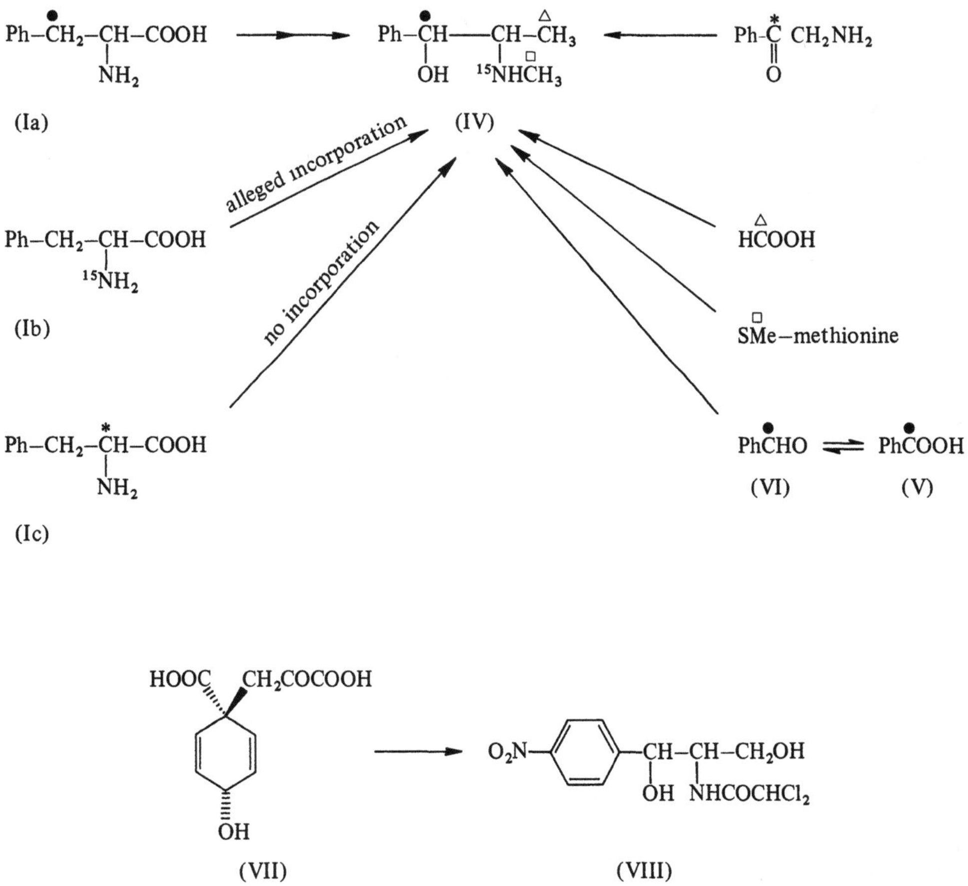

Fig. 2

et al., 1958) for ephedrine in *Ephedra*, and of norpseudoephedrine in *Catha edulis* (LEETE, 1958). The main pillars of that pathway were: incorporation of [3-^{14}C]-phenylalanine (Ia), [carbonyl-^{14}C]-2-aminoacetophenone and of [^{14}C]-formic acid into the carbon skeleton, the alleged incorporation of [^{15}N]-phenylalanine (Ib) to give the nitrogen atom and of S-Me-labelled methionine to afford the N-methyl group of ephedrine (IV). Unfortunately, that scheme has been invalidated by the failure of [2-^{14}C]-phenylalanine (c) to be incorporated into ephedrine and by the latest report (YAMASAKI et al., 1973) that (Ia) and (II) were actually poor precursors. On the other hand, [^{14}C]-benzoic acid (V), and [^{14}C]-benzaldehyde (VI), respectively, gave highly radioactive (IV) (YAMASAKI et al., 1969) which led to the assumption of a C6-C1 rather than a C6-C2 building block. However, benzoic acid could be directly formed from prephenic acid (VII) (ZENK and MÜLLER, 1964; VOLLMER and GRISEBACH, 1964, 1966) therefore no definite scheme seems to be justified at present. To add to the confusion [6-^{14}C]-glucose and [U-^{14}C]-aspartic acid were also incorporated into the C2-N unit of (IV) which is hard to reconcile with the high incorporation of formate into the terminal methyl group of ephedrine.

The phenylpropanolamine antibiotic chloramphenicol (VIII) from *Streptomyces venezuelae*, although not formed from phenylalanine, has been shown to have a very closely related biogenetic pathway diverging either at the stage of chorismic or prephenic acid (VII) (GOTTLIEB et al., 1962) (Fig. 2).

2.2 Cytochalasins

Cytochalasins are microbial metabolites of great biological interest.[1] [^{14}C]-Labelled (I) was incorporated into both cytochalasin B and D (IX, X) (KIECHEL et al.,

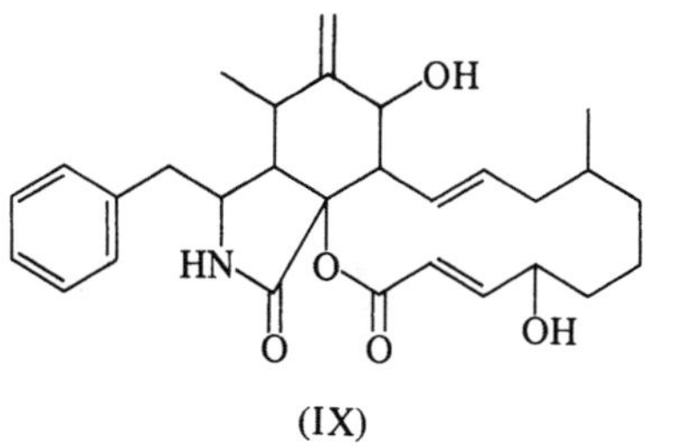

(IX)

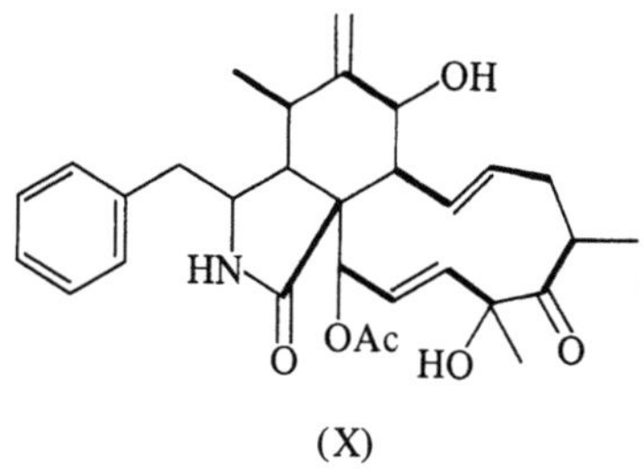

(X)

(I)

(II)

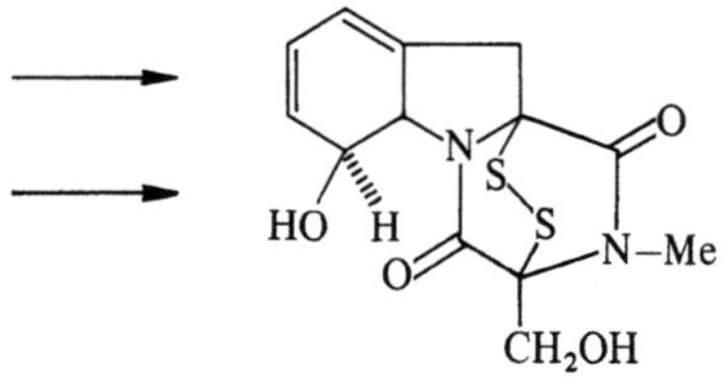

Fig. 3 (XI)

1 See several chapters in Vol. 7 of this Encyclopedia: Plant movements (HAUPT, W., FEINLEIB, M.E., eds.), 1979

1970) while the residual skeleton stems from acetate (BINDER et al., 1970; GRAF et al., 1974; LEBET and TAMM, 1974; VEDERAS et al., 1975) (Fig. 3).

2.3 Gliotoxin

Gliotoxin (XI) a fungal alkaloid from *Penicillium terlikowskii* may be formed from phenylalanine (BU'LOCK and RYLES, 1970), or from tyrosine (MACDONALD and SLATER, 1975) as has been shown by labelling experiments. Stereochemical subtleties were revealed, e.g., that the *pro*-R-β-methylene proton of (I) undergoes exchange (JOHNS et al., 1975), although that process is not part of gliotoxin biosynthesis (BU'LOCK et al., 1972).

2.4 Tenellin

Tenellin (XII), an alkaloid from *Beauvaria tenella* and *B. bassiana* is formed from [1,3-^{13}C]-phenylalanine (MCINNES et al., 1974a, b) via a skeletal rearrangement (VINING et al., 1975) as indicated by a ^{13}C-^{13}C coupling between C-4 and C-5 of this alkaloid (LEETE et al., 1975) (Fig. 4).

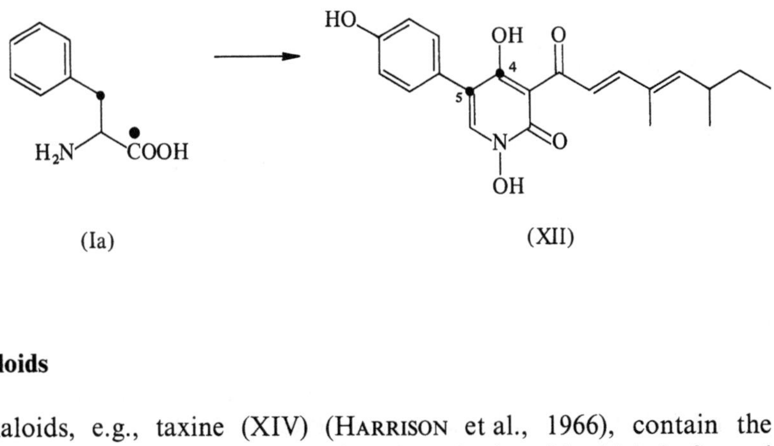

Fig. 4 (Ia) (XII)

2.5 Taxus Alkaloids

Taxus ester-alkaloids, e.g., taxine (XIV) (HARRISON et al., 1966), contain the rearranged amino acid, 2-dimethylamino-3-phenylpropionic acid which is formed via cinnamic acid (XIII), from [3-^{14}C]-phenylalanine, as indicated by the presence of all ^{14}C-label on the benzylic carbon (LEETE and MARION, 1953; LEETE and BODEM, 1966) (Fig. 5).

2.6 The Lunaria Alkaloids

Lunarine and lunaridine are the most interesting representatives. [3-^{14}C]-Phenylalanine was incorporated in *Lunaria biennis* (POUPAT and KUNESCH, 1971) into lunarine (XV) but [3-^{14}C]-tyrosine was not. Therefore p-coumaric acid undergoes oxidative coupling to the dicarboxylic acid which is then condensed with spermidine to give lunarine. [3-^{14}C, 4'-^{3}H]-Phenylalanine, on the other hand, gives lunarine with

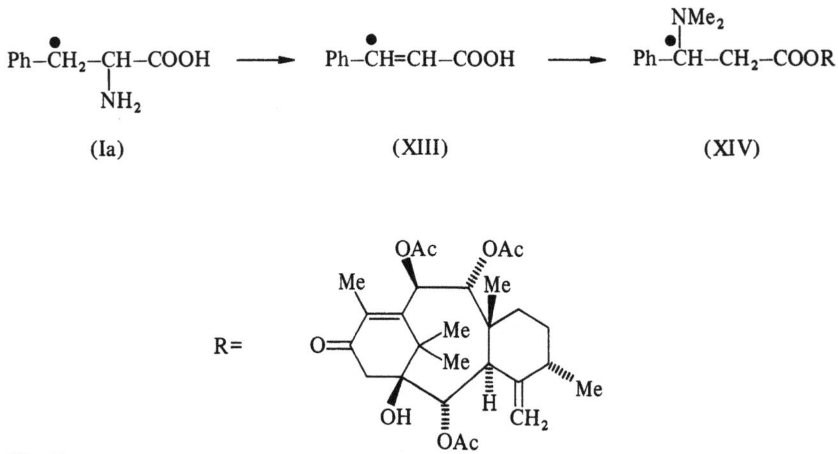

Fig. 5

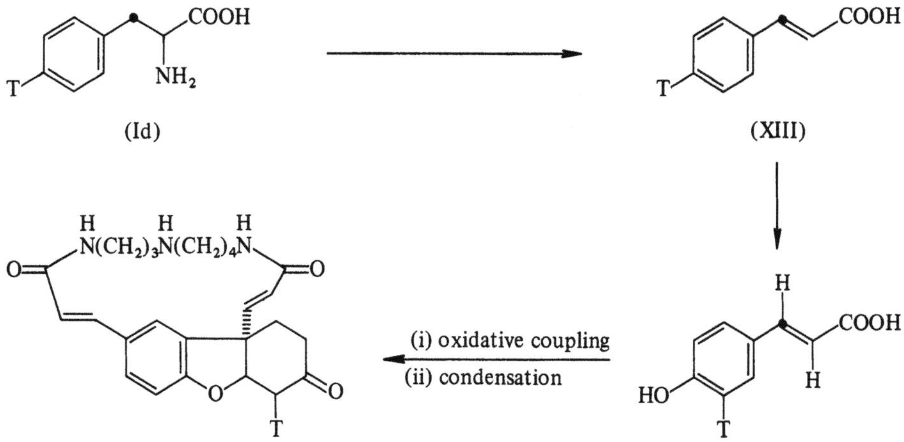

(XV)

Fig. 6

32% tritium retention, owing to some loss occurring via enolization of the carbonyl, i.e., NIH shift occurs (BOWMAN et al., 1969) (Fig. 6).

2.7 The Lythraceae Alkaloids

Three main representatives of this class are cryogenine (XVI), decodine (XVII) and decinine (XVIII) (KOO et al., 1970a, b; ROTHER and SCHWARTING, 1972, 1975) (Fig. 7). [1,3-^{14}C]-Phenylalanine was incorporated into ring B and into the carbon adjacent to ring D (31:61%) of cryogenine in *Heimia salicifolia* (DOBBERSTEIN et al., 1975); the heterocyclic ring arises from pelletierine (XIX), based on incorporation of both [2-^{14}C]-lysine and [6-^{14}C]-lysine into the heterocyclic moiety of decodine and decinine, in *Decodon verticillatus*. The biogenetic sequence shown in Scheme 2 (Fig. 8) has been suggested for the biosynthesis of (XVII) and (XVIII) that also includes phenol oxidative coupling.

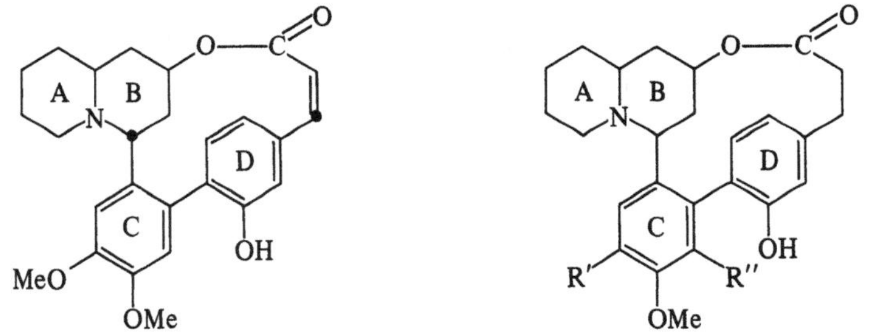

(XVI)

(XVII) R′ = H R″ = OH

(XVIII) R′ = OMe R″ = H

Fig. 7

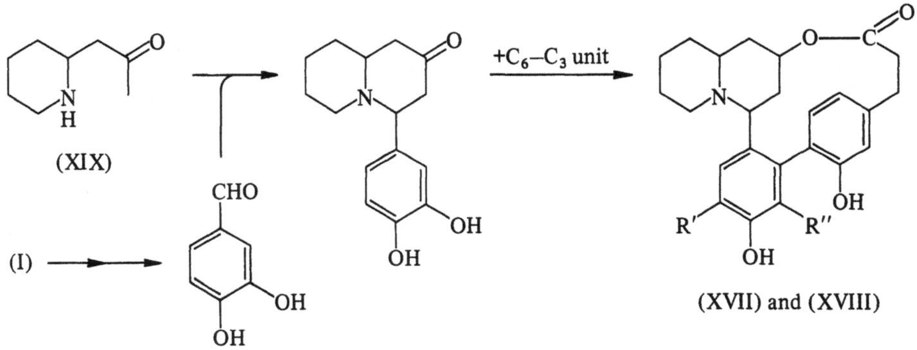

Fig. 8

3 Alkaloids Derived from Tyrosine and/or DOPA

It is a much more complex picture to describe the biogenesis of alkaloids from (II) than to review the alkaloids which are formed from phenylalanine (I). For convenience one may group the large number of the alkaloids which are formed from tyrosine according to the plant families in which they occur, according to their structures or according to their biosynthetic origins.

3.1 Simple Phenylethylamines

Tyramine (XX), N-methyltyramine (XXI) and hordenine (XXII) in the barley plant are obviously derived from (II) (Frank et al., 1956) by decarboxylation (Leete et al., 1952; Leete and Marion, 1953) followed or preceded by N-methylation. Mescaline (XXIV) is derived from DOPA (III) (Liss, 1961) via dopamine (XXIII) (Agurell et al., 1967) by methylation, hydroxylation and a second methylation of (XXV) (Leete, 1959, 1966; Battersby et al., 1967a; Rosenberg et al., 1967). The isomer, (XXVI) is the precursor of isoquinolines. Noradrenalin (XXVII),

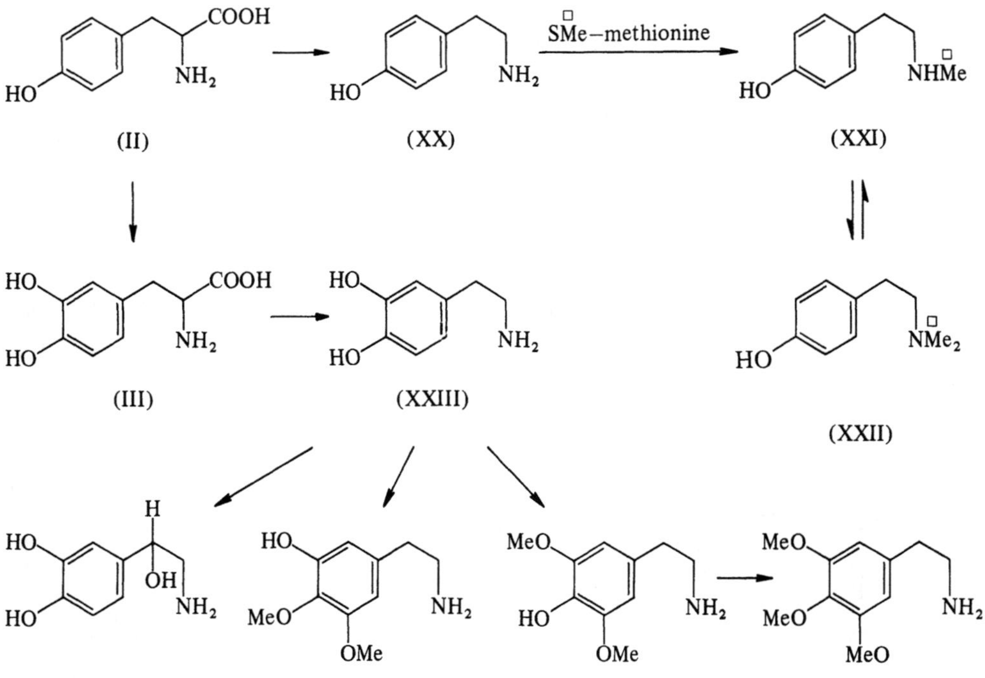

Fig. 9

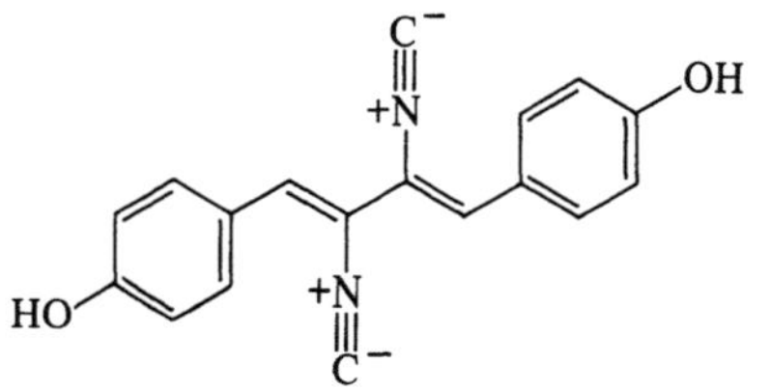

Fig. 10

on the other hand, probably arises from the C-3 hydroxylation (AGURELL, 1969; PAUL et al., 1969) of dopamine (XXIII) (SMITH and KIRSHNER, 1962). Noradrenalin, (XXVII) which has been isolated from *Trichocereus pachanoi* (DOBBINS, 1968), is the precursor of a number of alkaloids, e.g., berberastine (MONKOVIC and SPENSER, 1965a, b) (Fig. 9).

Xanthocillin (XXVIII) is derived from tyrosine, but the origin of the isonitrile carbon(s) is unknown. [1-^{14}C]-Tyrosine was not incorporated into (XXVIII), whilst the product from [2-^{14}C, ^{15}N]-tyrosine showed a sharp increase in the ^{14}C/^{15}N ratio, indicative of the loss of the tyrosine nitrogen (ACHENBACH and GRISEBACH, 1965) (Fig. 10).

3.2 Simple Isoquinoline Alkaloids

[2-^{14}C]-Tyrosine (IIa) is incorporated into anhalonidine (XXIX), (LEETE, 1966), in peyote cactus according to the Scheme (LUNDSTRÖM and AGURELL, 1971); and also into pellotine (XXX), anhalamine (XXXI), anhalidine (XXXII), lophophorine (XXXIII) and anhalonine (XXXIV) (Fig. 11). The most plausible intermediate is 3-hydroxy-4,5-dimethoxyphenylethylamine (XXXV) which has been detected by GLC/MS technique in peyote (POISSON, 1960). On the other hand, 4-hydroxy-3-methoxyphenylethylamine (XXXVI) was detected in *Lophophora williamsii* (DOBBINS, 1968) which suggests that this is the intermediate between tyramine (or dopamine) and, for example, lophocerine (XXXVII). Feeding experiments with [^{14}C]- and [^{3}H]-3-hydroxy-4,5-dimethoxyphenylethylamine (XXXV) resulted in high incorporation of the label into both anhalamine (XXXI) and anhalonidine (XXIX) (LUNDSTRÖM and AGURELL, 1971). Conversion of anhalonidine (XXIX), into anhalonine (XXXIV), and of pellotine (XXX) into lophophorine (XXXIII) most likely follows the general biogenetic pathway of o-methoxyphenols into methylenedioxy-derivatives (BATTERSBY et al., 1967a; 1968a). Only a low incorporation took place into the N-methyl bases, e.g., pellotine (XXX). However, N-[^{14}C]-methyl-[2,3-^{3}H]-3-hydroxy-4,5-dimethoxyphenylethylamine (XXXVIII) was efficiently incorporated into pellotine (XXX) (LUNDSTRÖM, 1971) indicating parallel hydroxylation and cyclization mechanisms of the N-methylated and the *nor*-phenylethylamines. On the other hand, methylation and reverse demethylation of (XXXI) $\rightleftharpoons$ (XXXII) and (XXXIII) $\rightleftharpoons$ (XXXIV) was reported (BATTERSBY et al., 1967a). Salsoline (XXXIX) and carnegine (XLI) are formed, by analogy, from dopamine with (XXXVI) as a plausible intermediate. The "NIH shift" (DALY and JERINA, 1969) i.e., the migration of ring substituents during the course of enzymatic aryl hydroxylation, could account for these changes without assuming an intermediate isomeric with (XXXVI) (Fig. 12). The origin of the 2 "extra" carbons in peyote alkaloids was revealed by feeding methyl-labelled pyruvic acid (LEETE and BRAUNSTEIN, 1969), which was fully incorporated. The intermediate peyoruvic acid (XLII) was indeed isolated from *L. williamsii* (KAPADIA et al., 1970) (Fig. 13). Furthermore, [9-^{14}C]-peyoruvic acid was incorporated into 1,2-dehydroanhalonidine, indicative of an oxidative decarboxylation. Anhalamine (XXXI), on the other hand has its "extra" isoquinoline C-1 carbon from the analogous peyoxylic acid, i.e., from glyoxylic acid. [1,9-^{14}C]-Peyoxylic acid was indeed specifically incorporated into the peyote cactus alkaloids (KAPADIA et al., 1970).

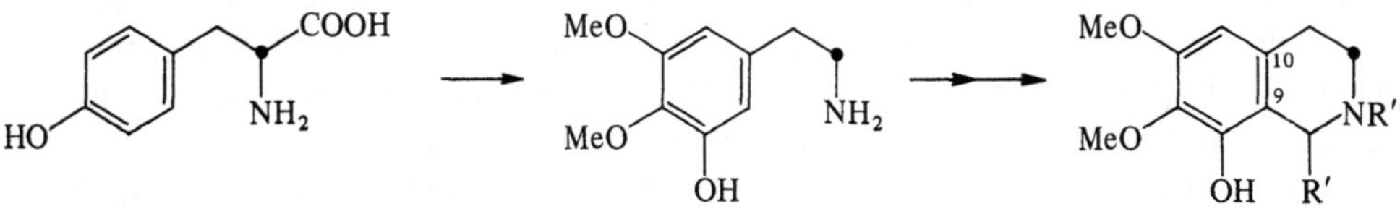

(IIa) (XXXV)

(XXIX) R′ = Me, R″ = H
(XXX) R′ = R″ = Me
(XXXI) R′ = R″ = H
(XXXII) R′ = H, R″ = Me

Fig. 11

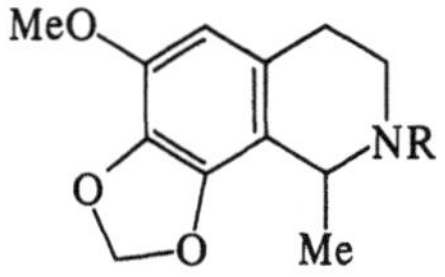

(XXXIII) R = Me
(XXXIV) R = H

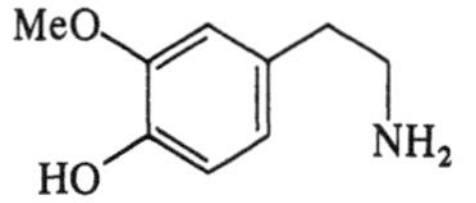

(XXXVI)

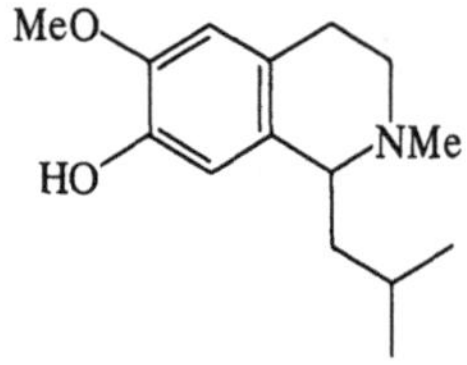

(XXXVII)

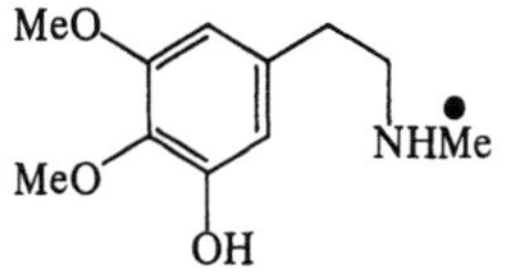

XXXVIII

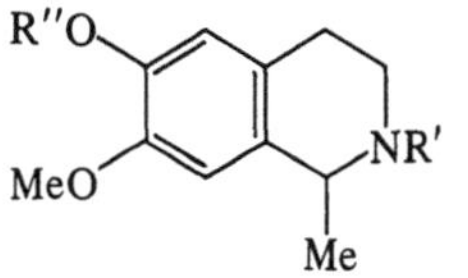

(XXXIX) R′ = R″ = H
(XL) R′ = H, R″ = Me
(XLI) R′ = R″ = Me

Fig. 12

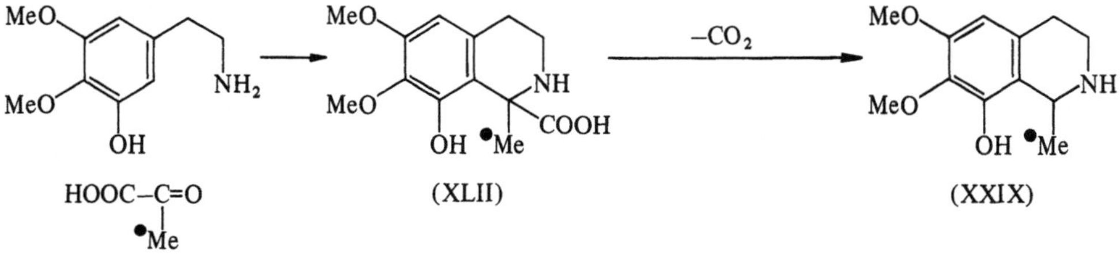

(XLII) (XXIX)

Fig. 13

An essentially similar mechanism for the incorporation of a C2 unit has been outlined for the formation of salsoline (XXXIX) (McFarlane and Slaytor, 1972) and carnegine (XLI). Interestingly, these two α-keto acid intermediates were postulated as early as 1935 (Hahn et al., 1935). The C5 unit in lophocerine (XXXVII), on the other hand, was of isoprenoid origin (Lundström, 1971).

3.3 Benzylisoquinolines and Related Alkaloids Derived from Two Moles of Tyrosine

It was assumed (Wintersteiner and Trier, 1910), and later put in refined terms (Robinson, 1917), that papaverine, morphine and most of the benzylisoquinoline alkaloids are formed from C6-C2 units, e.g., dopamine and 3,4-dihydroxyphenylacetaldehyde (Battersby et al., 1964b). The first common intermediate would be norlaudanosoline (XLIII). Recent tracer work has shown (Wilson and Coscia, 1975; Battersby et al., 1975c) that 3,4-dihydroxypyruvic acid (itself formed by oxidation and transamination of tyrosine), and *not* phenylacetaldehyde, provided C-1 and the benzyl group of the benzylisoquinoline skeleton (Fig. 14). Practically all previous assumptions as to the interconversion of norlaudanosoline into more than 70 alkaloids are compatible with the biogenetic theory if one takes "phenol-coupling" into account (Erdtman and Wachtmeister, 1957; Barton and Cohen,

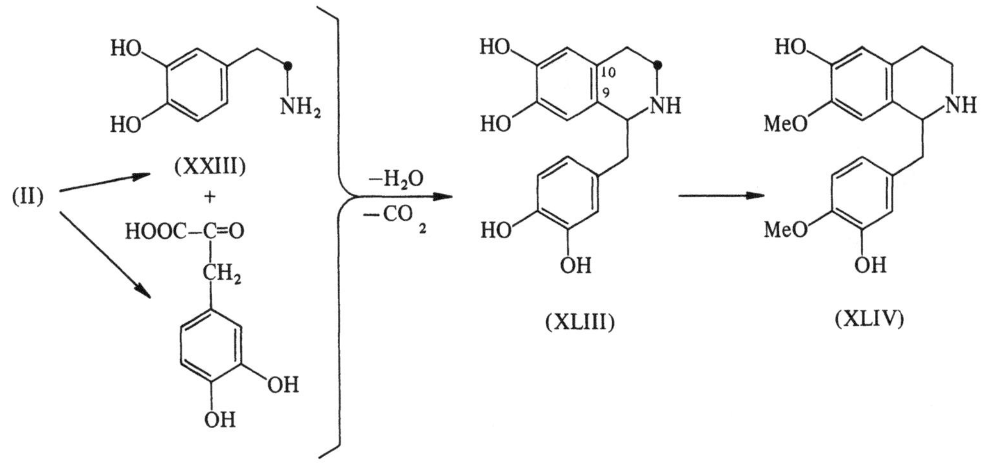

Fig. 14

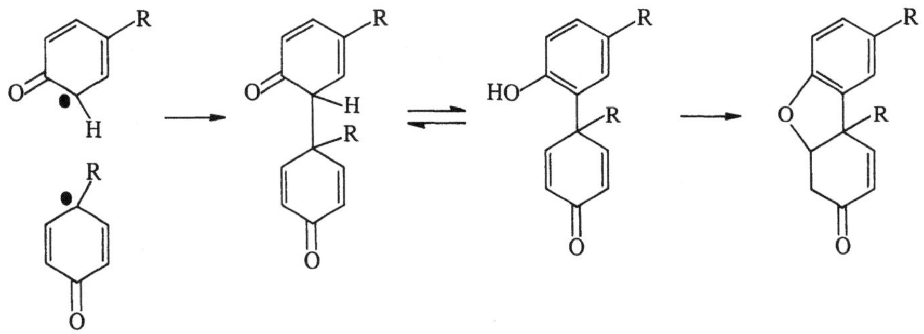

Fig. 15

1957; BARTON, 1963, 1964, 1967). This concept was outlined by ERDTMAN as early as 1933, revived and generalised by BARTON to account for the formation of new C–C bonds in a variety of benzylisoquinoline and related alkaloids. The general mechanistic sequence for an o, p′ coupling is shown in the Scheme (Fig. 15). Similar canonical structures can be written for o,o′ and p,p′ coupling, for the phenolate free radicals written for convenience in the ketonic form.

The major argument in favour of the concept of two different building blocks in providing the norlaudanosoline skeleton is that dopamine gives only N and carbons 3 to 10 of the isoquinoline ring of (XLIII), but not C-1 and the benzyl group, while tyrosine can provide both (GEAR and SPENSER, 1963).

Three partially methylated derivates of (XLIII), i.e., norprotosinomenine (XLIV), orientaline (XLV), and reticuline (XLVI) (Fig. 16) are key intermediates to the benzylisoquinolines. Norprotosinomenine leads to the erythrina group of alkaloids, to the isothebaine group and to morphine, berberine, and protopine. Complete-O-methylation (XLIII), followed by dehydrogenation of the tetrahydroisoquinoline ring gives papaverine. Protoemetine can also be derived from norlaudanosoline.

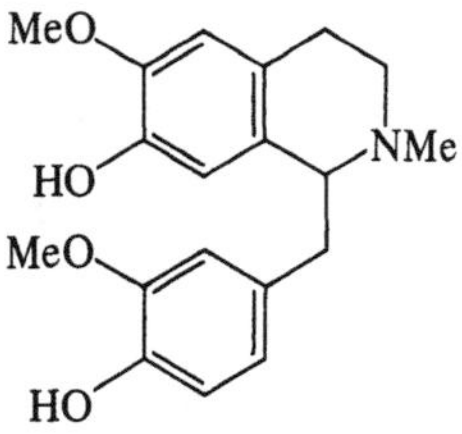

(XLV)

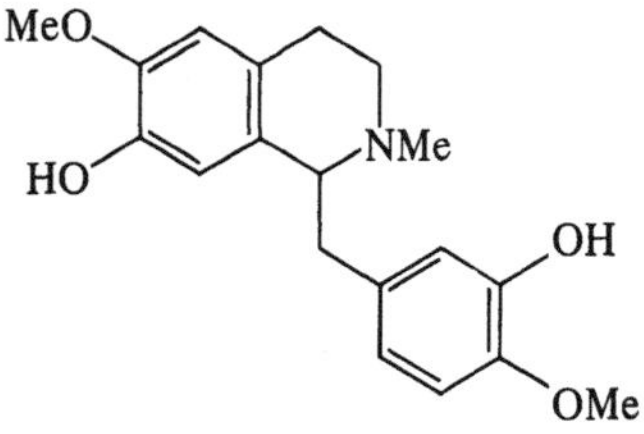

(XLVI)

Fig. 16

3.3.1 Erythrina Alkaloids

(−) Norprotosinomenine (XLIV) is converted by phenolic coupling and rearrangement into (XLVII), (XLVIII) and the chiral, but optically unstable dibenzazonine (XLIX), and then probably, through a quinonoid intermediate, leads via 5*S*-(−)erysodienone (L) to erythraline (LI) and erythroidine (LIV) (BARTON et al., 1970, 1974; BATTERSBY et al., 1967b). Using tritiated erysodine (LII), erysopine (LIII), erythraline (LI) and erysodienone (L) further conversion to the corresponding lactone bases, α- and β-erythroidine (LIV) has been shown (BARTON et al., 1974) (Fig. 17). Similar phenolic coupling leads to homoaporphines (BATTERSBY et al., 1967b).

3.3.2 Alkaloids from Orientaline

(+) Orientaline (XLV) undergoes oxidative o, p′-phenolic coupling to give orientalinone (LV) and this, by reduction to orientalinol and subsequent rearrangement and aromatization (by dehydration) gives rise to isothebaine (LVI), an aporphine alkaloid (BATTERSBY et al., 1965a, 1969). However, the structurally related alkaloid, boldine (LVII) of *Litsea glutinosa*, was only formed from reticuline and norreticuline (BARTON et al., 1967) (Fig. 18).

Isoboldine (LVIII) and magnoflorine (LIX) are minor alkaloids of *Papaver somniferum*. [3-^{14}C]-Reticuline and N-methyl-[^{14}C]-reticuline were incorporated into isoboldine through o, p-phenol coupling, while an o, o′-coupling (BROCHMANN-HANSSEN et al., 1971b, 1972) leads to magnoflorine. (+)-Thalpenine (LX), on the other hand, could be formed from (LVII) by the oxidation of an-OMe to-OCH_2 group and cyclization followed by N-methylation (by S-Me methionine). An NIH shift seems to be excluded due to the loss of 64% ^{14}C-label from (±)-[6-^{14}C-Me, 1-^{3}H]-reticuline (TEWARI et al., 1974) (Fig. 19).

The biosynthesis of the aporphine alkaloids of *Dicentra eximia*, i.e., corydine (LXIII), glaucine (LXV) and dicentrine (LXVI) was shown by labelling experiments to start with norprotosinomenine (XLIV) but neither nororientaline, norreticuline, nor the N-methyl-(XLIV) were incorporated. The methylation pattern seems to be of utmost importance. Scheme (Fig. 20), which assumes a p,o′-phenolic coupling leading to (LXI), is consistent with all these facts. A p,p′-phenolic coupling gives glaucine and dicentrine (BATTERSBY et al., 1971b): Scheme (Fig. 21). There is an apparent NIH shift occurring in the benzenoid part of the isoquinoline ring during the synthesis of corydine.

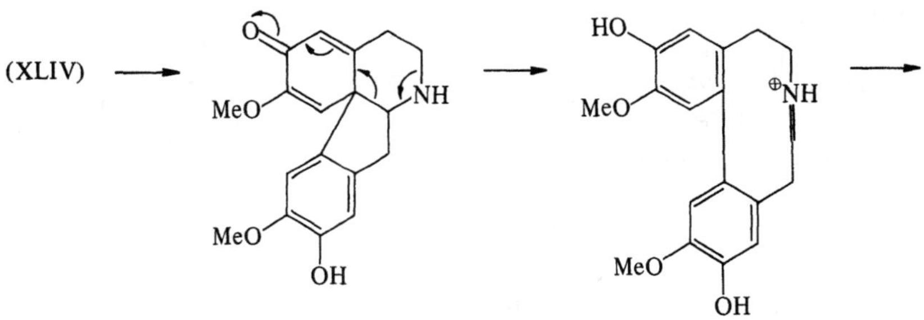

(XLVII) (XLVIII)

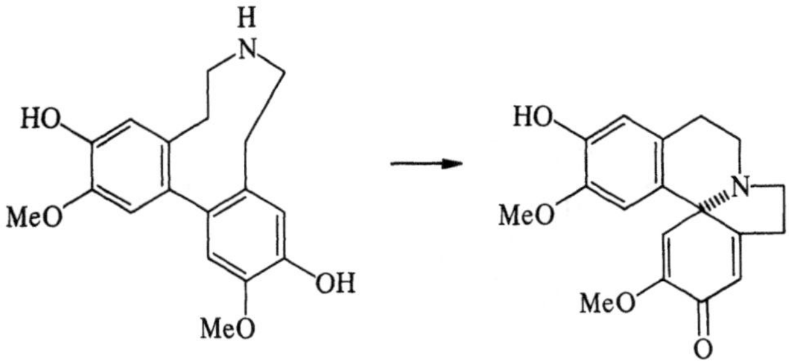

(XLIX) (L)

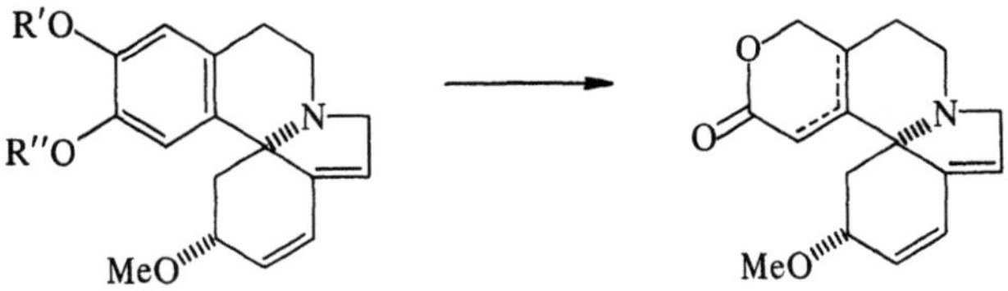

(LI) R′ + R″ = CH_2
(LII) R′ = H, R″ = Me
(LIII) R′ = R″ = H

(LIV)

Fig. 17

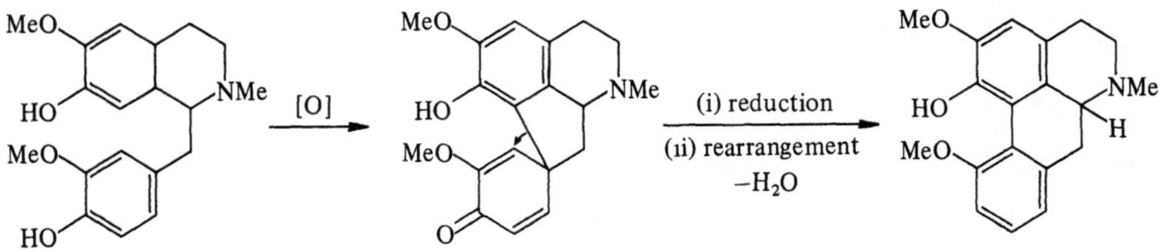

(XLV) (LV) (LVI)

Fig. 18

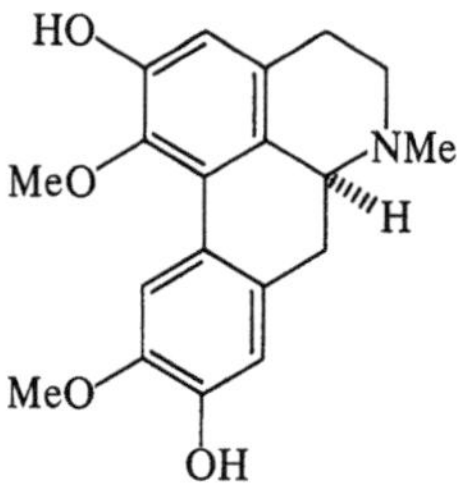

(LVII)

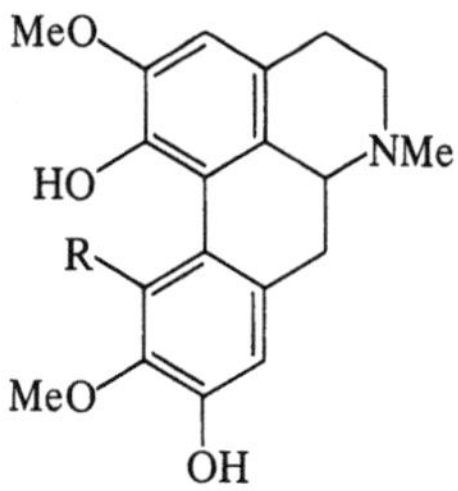

(LVIII) R = H
(LIX) R = OH

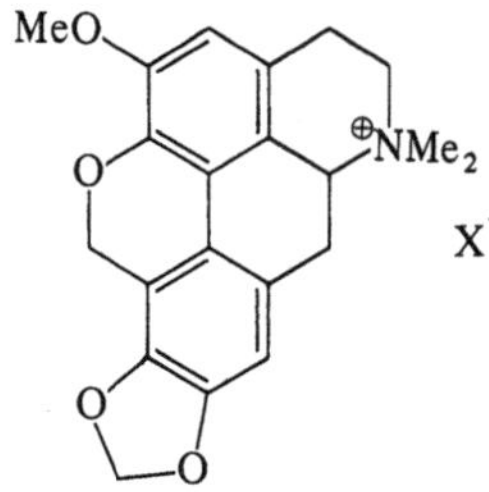

(LX)

Fig. 19

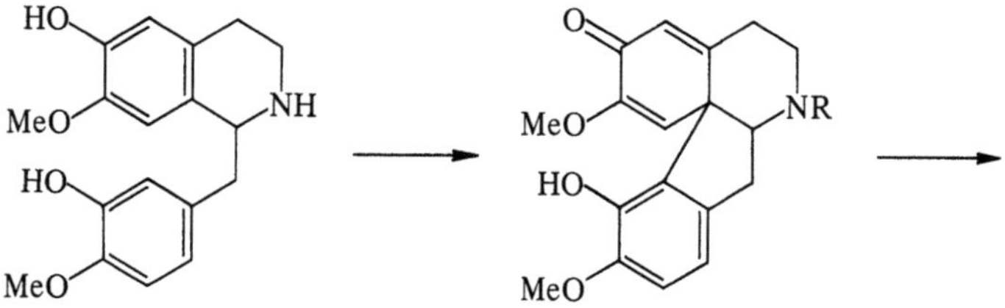

(XLIV) (LXI)

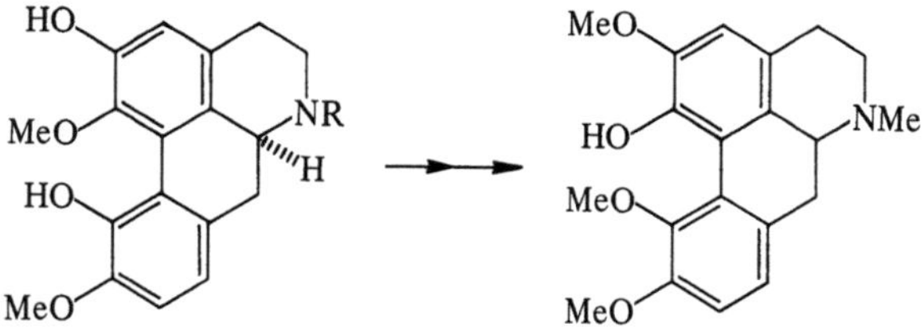

(LXII) (LXIII)

Fig. 20

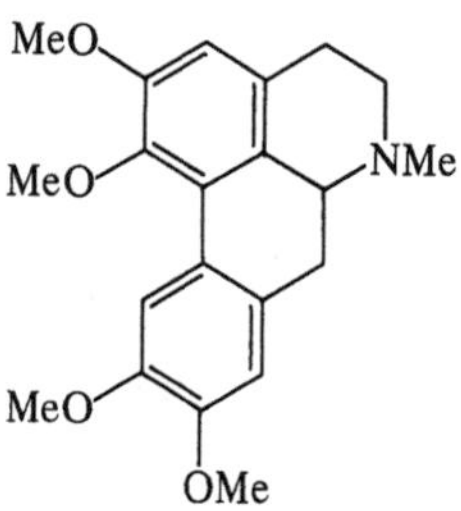

(LXV)

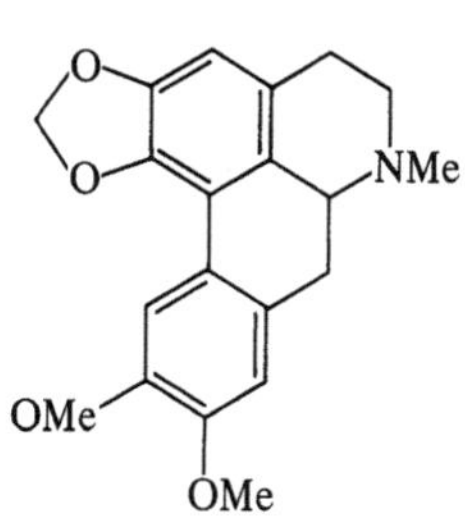

(LXVI)

Fig. 21

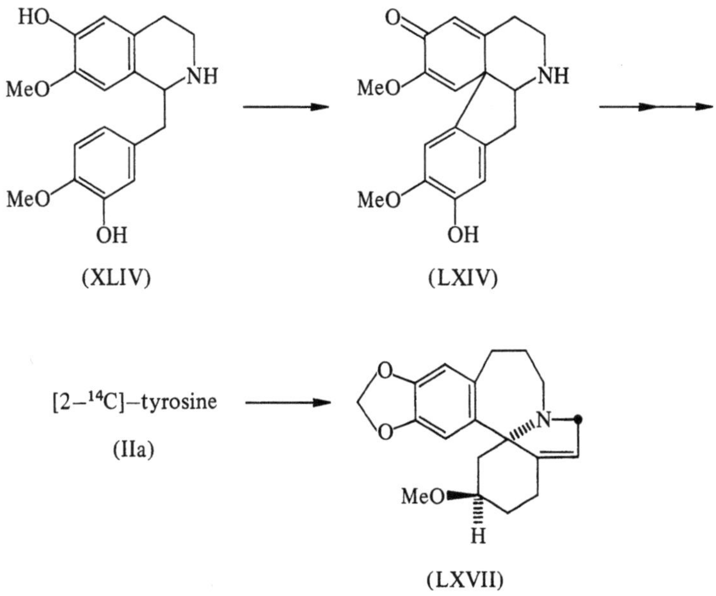

Fig. 22

(LXVIII)

Fig. 23

Schelhammeridine (LXVII) an alkaloid of *Schelhammera pedunculata* is structurally related to erythrina alkaloids. The label from [2-^{14}C]-tyrosine was incorporated into the five-membered ring, next to the nitrogen, but phenylalanine and cinnamic acid were also incorporated into the "bottom part" (Battersby et al., 1974a, 1975d) (Fig. 22). Cephalotaxine (LXVIII) (Fig. 23) from *Cephalotaxus wilsoniana* seems to be closely related to the homoerythrina group. The label from [3-^{14}C]-tyrosine was incorporated into positions 9 and 16 of (LIVIII), only one label from [2-^{14}C]-tyrosine was present at C-7, and none at C-3 or C-2. These facts may be explained if a cleavage of one of the benzene rings is assumed in the Scheme (Fig. 24) (Parry and Schwab, 1975).

3.3.3 Benzylisoquinoline Alkaloids Derived from Reticuline

Reticuline (XLVI) proved to be a focal intermediate in the biosynthesis of all the alkaloids shown in the Scheme (Fig. 25).

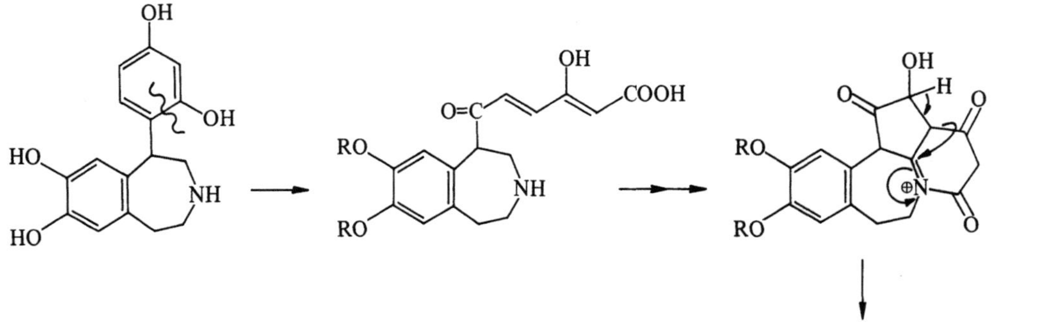

(LXVIII)

Fig. 24

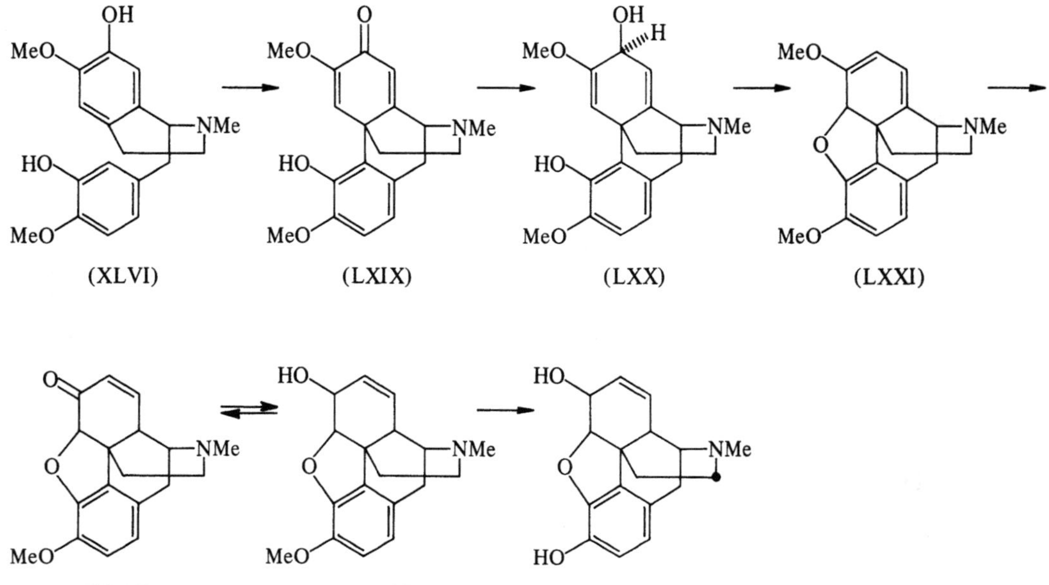

(XLVI) (LXIX) (LXX) (LXXI)

(LXXII) (LXXIII) (LXXIV)

Fig. 25

3.3.3.1 Morphine Alkaloids

Morphine (LXXIV) is derived from two moles of tyrosine (II) (BATTERSBY and HARPER, 1958; BATTERSBY et al., 1962, 1964b) via norlaudanosoline (XLIII) and (−) reticuline (XLVI) (BATTERSBY et al., 1965b); oxidative o,p′ phenolic coupling leads to the known alkaloid salutaridine (LXIX) which by reduction gives salutaridinol (LXX). Allylic type of intramolecular hydroxyl addition gives rise to thebaine (LXXI) (NEUBAUER, 1965), and this compound is converted into codeinone (LXXII) by hydrolysis of the methyl enolether group. Codeine (LXXIII), formed by reduction of codeinone, is in turn converted into morphine by O-demethylation (KIRBY, 1967; KIRBY et al., 1972; STERMITZ and RAPOPORT, 1961). All these steps,

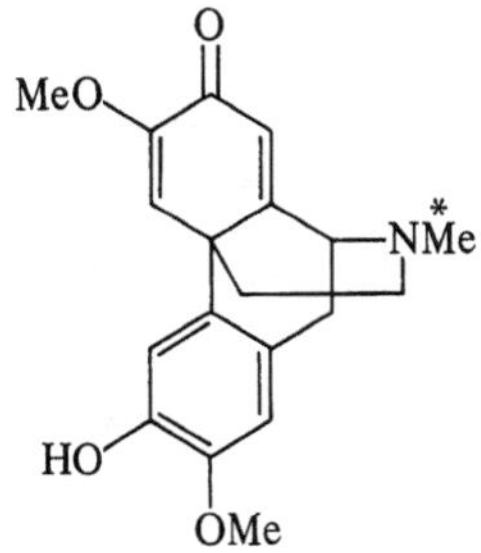

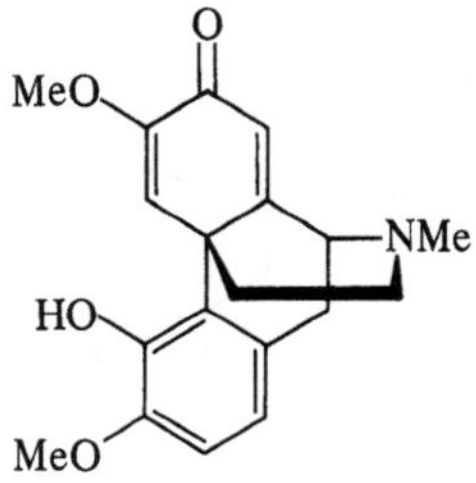

Fig. 26 (LXXV) (LXXVI)

which had been predicted previously (BARTON and COHEN, 1957; BARTON, 1963), have been verified by feeding labelled intermediates (I), (XLVI), (LXIX) and (LXX) to *Papaver somniferum*. For example, [2-^{14}C]-tyrosine provides carbons 10 and 16; [2-^{14}C]-dopamine (XXIII) gives only 16-C labelled morphine (LXXIV); labelled laudanosoline, reticuline (XLVI), salutaridine (LXIX) all gave morphine labelled in the expected positions.

Flavinanthine (LXXV) is derived from [N-methyl-^{14}C]-reticuline (STUART et al., 1969) via p,p′ phenol coupling. Similarly two molecules of phenylalanine (BARTON et al., 1967; STUART and GRAHAM, 1973) give rise to flavinanthine which is a positional isomer of (LXIX) (Fig. 26). Flavinanthine has been found in *Croton flavens*, together with sinoacutine (LXXVI) (STUART and GRAHAM, 1973) which is the precursor for sinomenine.

3.3.3.2 Berberine and Related Alkaloids

[3-^{14}C,N-methyl-^{14}C]-laudanosoline in *Berberis japonica* gave rise to berberine wherein the "bridge" contained practically all the radioactivity of the N-Me group. Furthermore, [N-methyl-^{14}C, O-methoxy-^{14}C]-reticuline (XLVI) was incorporated by *Hydrastis canadensis* into berberine (LXXVII) the bridge and the methylenedioxy-carbons containing most of the label. This can be interpreted in terms of the interconversions shown in the Scheme (BATTERSBY et al., 1963; BARTON et al., 1963a; GEAR and SPENSER, 1963), in which O-methylation precedes N-methylation (Fig. 27).

The study of the biogenesis of the tetrahydroberberines, e.g., stylopine (LXXXI) and chelidonine (LXXXIV) in *Chelidonium majus* showed that multiple labelled *S*-reticuline (XLVI) gave rise to *S*-scoulerine (LXXIX) and this was oxidised via (LXXX) to S-stylopine (LXXXI) (BATTERSBY et al., 1975a, b, e). The latter gives rise to chelidonine (LXXXIV) by N-methylation and decomposition of the quaternary base to an enamine which then undergoes cyclisation (BATTERSBY et al., 1975c, f). Chelidonine forms from tyrosine (LEETE, 1963a) [N-methyl-^{14}C, 8-^{3}H]-Stylopine methochloride gave chelidonine (LXXXIV) and protopine (LXXXV) essentially without change in isotope ratio (cf. TANI and TAGAHARA, 1974) (Fig. 28). By simple ring opening, both *S*-reticuline (XLVI) and *S*-scoulerine (LXXIX) led to protopine (LXXXV) accompanied by the loss of tritium label from C-14 of stylopine (LXXXI). The C-5 and C-8 labelled isocorylpalmine (LXXXVI) gave

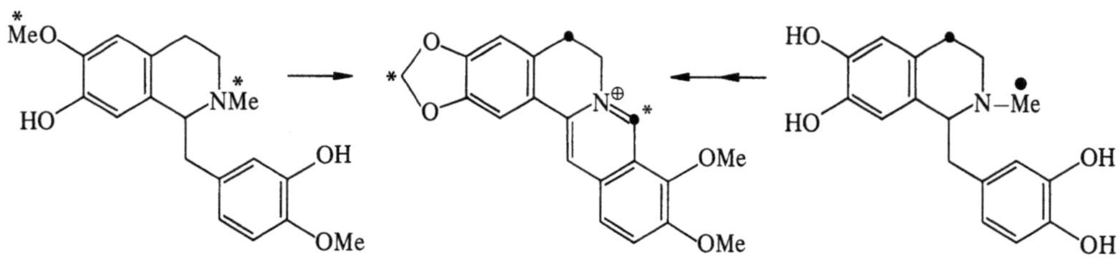

(XLVI) (LXXVII) (LXXVIII)

Fig. 27

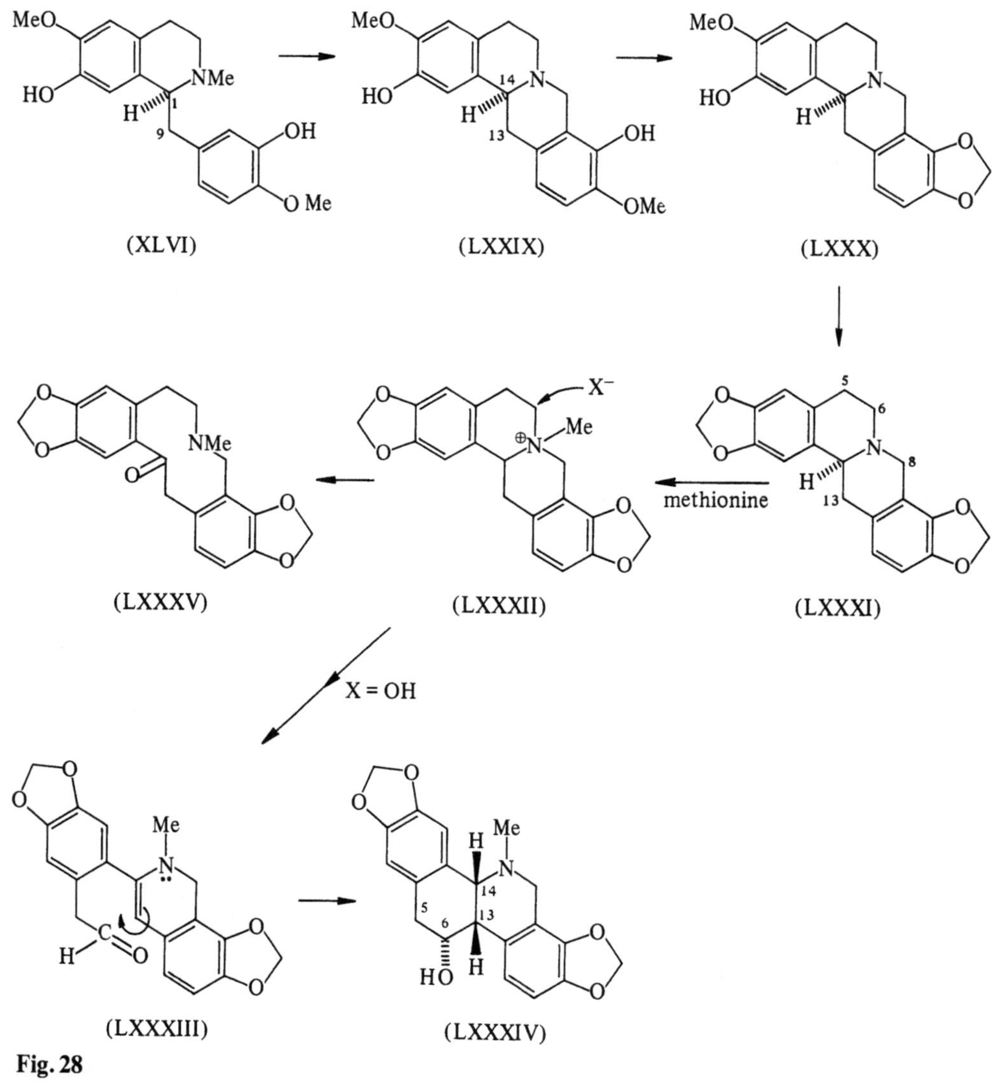

Fig. 28

rise to the correspondingly labelled allocryptopine (LXXXVII) (BARTON et al., 1965) by an analogous mechanism (Fig. 29).

3.3.3.3 *Papaverine and Narcotine*

Papaverine (XCI), i.e., tetradehydro-O-tetramethyllaudanosoline, is formed from norlaudanosoline and (−)-norreticuline (BROCHMANN-HANSSEN, et al., 1971a) via tetrahydropapaverine (UPRETY et al., 1975). Interestingly, only (−)-tetrahydropapaverine acts as substrate (BROCHMANN-HANSSEN et al., 1975).

Narcotine was generated from scoulerine (BATTERSBY et al., 1968b), like papaverine, but now it was shown that labelled isocorylpalmine (LXXXVI) and canadine (LXXXVIII), were also effective precursors of narcotine. Oxidative ring opening, probably of (LXXXIX), followed by N-methylation explain why the label from "bridge" carbon was not present in the N-methyl group of narcotine (XC). The pro-*S* hydrogen was lost from C-13 of canadine during oxidation (BATTERSBY et al., 1975d) (Fig. 29).

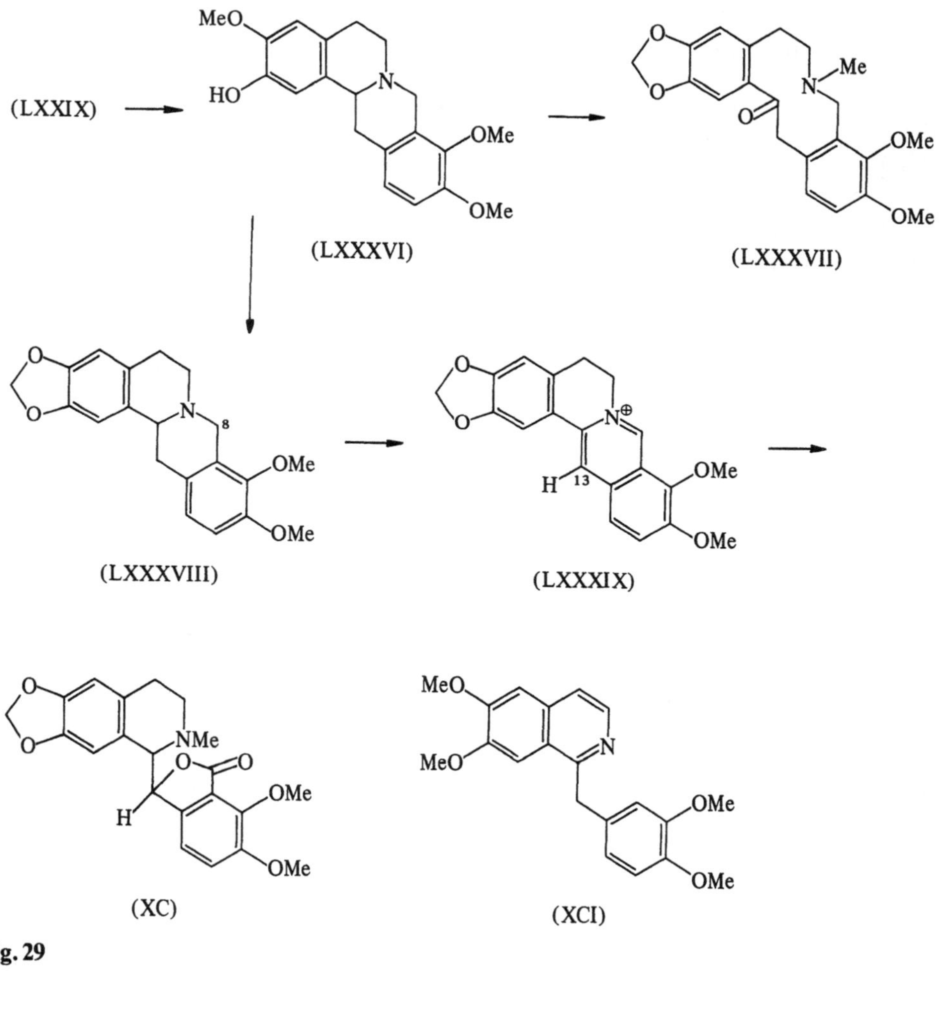

Fig. 29

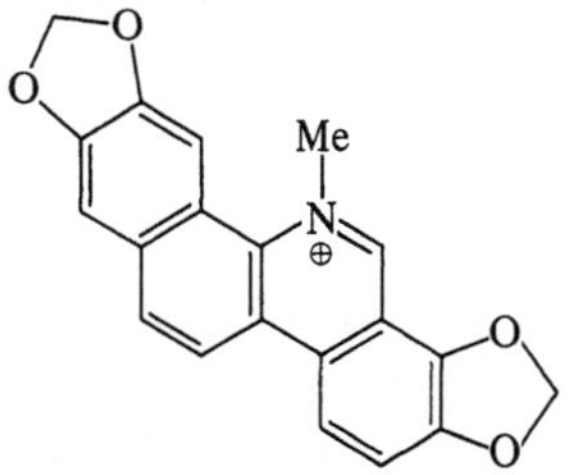

(XCII)

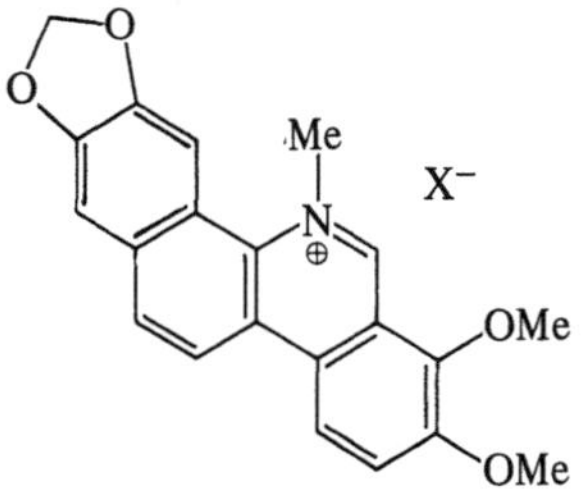

(XCIII)

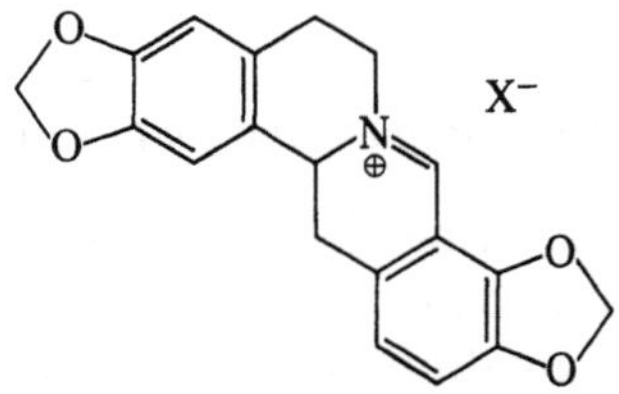

(XCIV)

Fig. 30

Radioactive *S*-scoulerine (LXXIX) is also incorporated into sanguinarine (XCII) and chelerythrine (XCIII); stylopine (LXXXI) on the other hand, gave the skeleton of coptisine (XCIV) (BATTERSBY et al., 1975a, b) (Fig. 30). Allocryptopine (LXXXVII) and protopine (LXXXV) probably are formed via different oxidation processes from scoulerine (Fig. 28).

3.3.4 Phenylisoquinoline Alkaloids

Cryptostyline I (XCVII) from *Cryptostylis erythroglossa* was the first alkaloid with the l-arylisoquinoline system (instead of alkyl, benzyl) of high biogenetic interest. [^{14}C]-Tyrosine and [^{14}C]-DOPA were indeed incorporated into (XCVII) (AGURELL, 1974; AGURELL et al., 1974). On the other hand, only labelled homovanillylamine (XCV) and vanillin (XCVI) (but not homo-isovanillylamine) were incorporated giving rise to the Ph-C moiety of cryptostyline I; (XCVIII) was very well incorporated into cryptostyline, indicative of a reduction *after* ring closure (AGURELL et al., 1974) (Fig. 31).

Corydaline (CIII) forms from [3-^{14}C]-tyrosine (IIb), via reticuline, and [S-^{14}C-methyl]-methionine in *Corydalis*, probably via (XCIX)→(C)→(CII) (BLASCHKE, 1968, 1970).

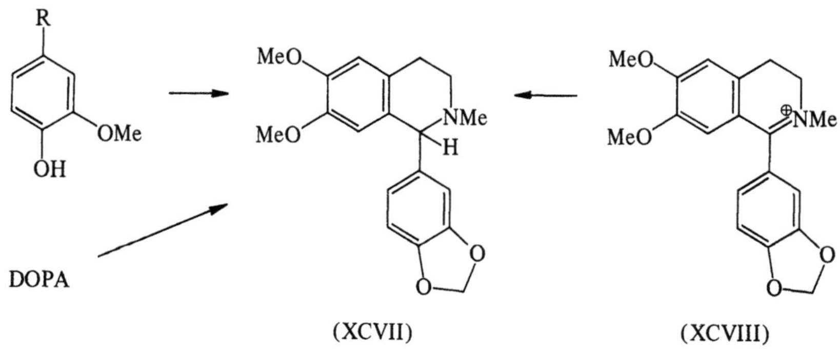

(XCVII) (XCVIII)

(XCV) R = $(CH_2)_2NH_2$
(XCVI) R = CHO

Fig. 31

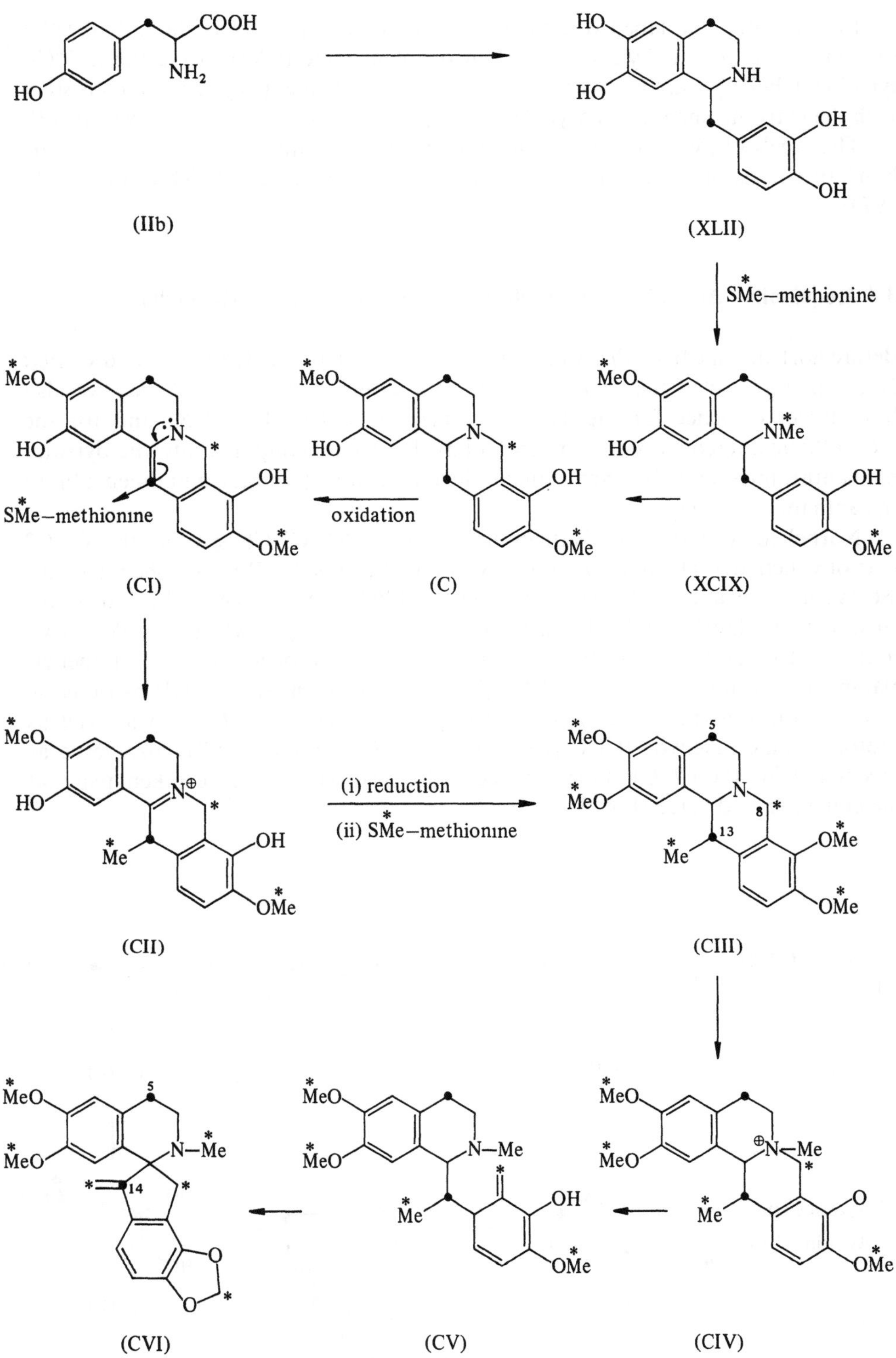
(IIb)
(XLII)
SMe–methionine
(CI)
(C)
(XCIX)
oxidation
(CII)
(i) reduction
(ii) SMe–methionine
(CIII)
(CVI)
(CV)
(CIV)

Fig. 32

Furthermore, the labelled methyl of methionine appeared as the C-9 skeletal carbon and as the methylene at C-14 in ochotensimine (CVI), while the [3-^{14}C]-tyrosine label appeared on carbons 5 and 14. Scheme (Fig. 32) is consistent with the rearrangement of a corydaline-type-protoberberine→(CIV)→(CV)→(CVI).

The labels of [Me-^{14}C, ^{3}H] methionine were incorporated into the "extra carbon" of corydaline, and, partly into the berberine-bridge carbon (HOLLAND et al., 1974).

3.4 Isoquinoline Alkaloids from Norbelladine (Amaryllidaceae Alkaloids)

Before norbelladine from the Amaryllidaceae had been isolated, BARTON and COHEN (1957) had predicted that it must occur as a common precursor. Today we also have tracer evidence for the natural occurrence of belladine (CX) and for the biosynthetic interconversion of norbelladine, by phenol coupling, into the pyrrolophenanthridines, (e.g., lycorine), ethanophenanthridines (crinine), and dibenzofuran (galanthamine) series.

Norbelladine and its derivatives have a C6-C2-N-C1-C6 system, the C6-C2 part of which arises from tyrosine via tyramine, but not DOPA, nor phenylalanine (SUHADOLNIK et al., 1962; BATTERSBY et al., 1964a; WILDMAN et al., 1962a, b). However, the labelled C6-C1 must come from [3-^{14}C]-phenylalanine (WILDMAN et al., 1962b; ARCHER et al., 1968), but not from [2-^{14}C]- or [carboxyl-^{14}C]-phenylalanine nor from tyrosine nor DOPA. [3-^{14}C]-labelled cinnamic (XIII) p-coumaric (CVII) (SUHADOLNIK and ZULALIAN, 1963) and caffeic acids (CVIII) as well as protocatechuic aldehyde (CIX) (ZULALIAN and SUHADOLNIK, 1974) also serve as precursors for the C6-C1 moiety hence hydroxylation must have taken place at the cinnamic acid stage (Fig. 33).

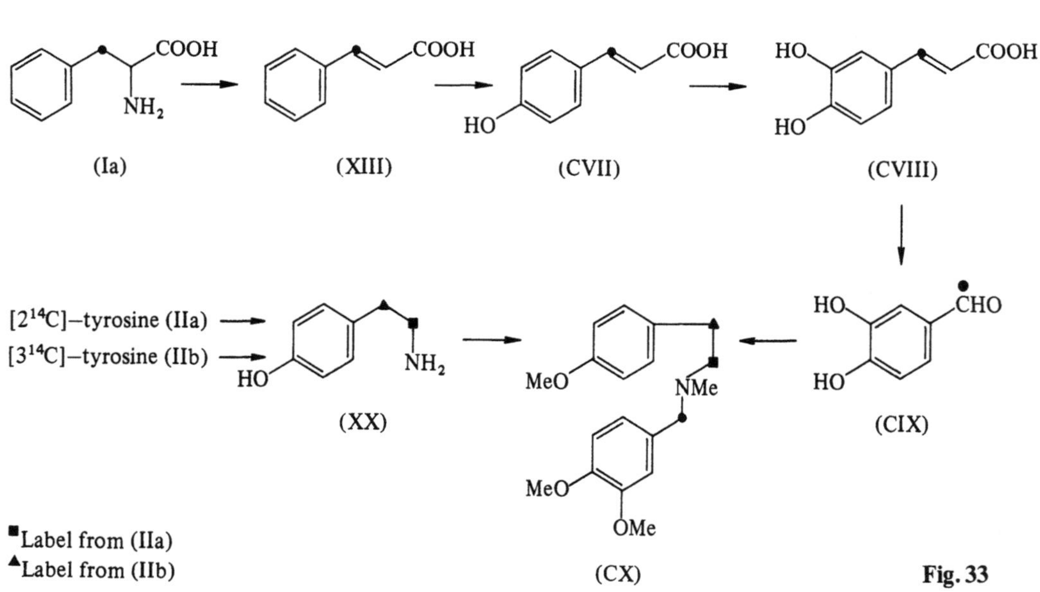

Fig. 33

Labelled norbelladine (CXI) was incorporated into all three major groups of alkaloids, (BATTERSBY et al., 1961a, b; BARTON et al., 1963b) according to the following schemes via different kinds of phenolic couplings (o, p′; p, p′; p, o′).

3.4.1 Norpluviine, Lycorine Type of Alkaloids

The o, p′ phenolic coupling of (CXI) to give (CXII), followed by intramolecular cyclization by the secondary amine nitrogen and subsequent steps (CXII)→(CXIII)→(CXIV) (norpluviine) are self-explanatory, see Scheme (Fig. 34). Caranine is the methylenedioxy derivative of lycorine (CXV) which is formed from (CXIV) by phenolic hydroxylation. It has been shown that [1,1′-$^{14}C_2$]-norbelladine gives lycorine (WILDMAN et al., 1962b) and that [2β-^{3}H]-caranine (CXVI) loses the tritium label during its hydroxylation to lycorine (BRUCE and KIRBY, 1968a, b) (Fig. 35). Conversion of norpluviine into lycorine probably occurs via an epoxide.

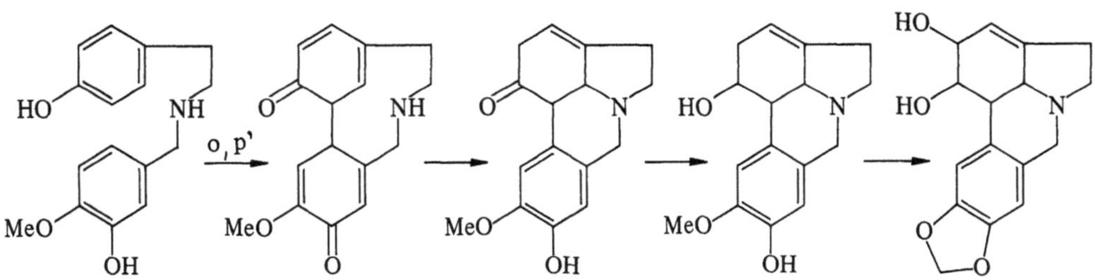

(CXI) (CXII) (CXIII) (CXIV) (CXV)

Fig. 34

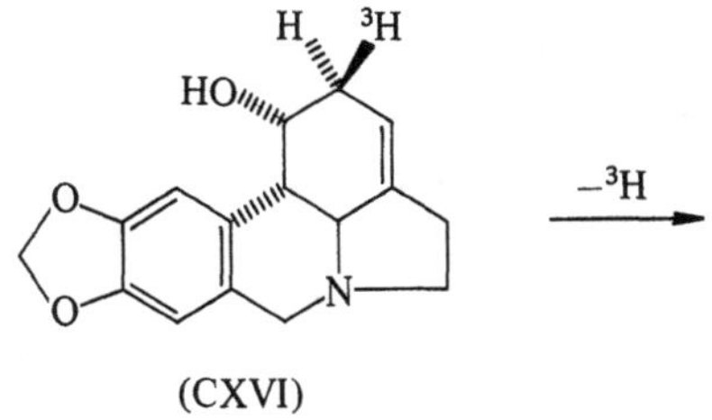

(CXVI) (CXV)

Fig. 35

3.4.2 Crinine, Haemanthamine Type of Alkaloids

The p′-O-methylnorbelladine (CXI) undergoes p, p′ phenol coupling giving rise to (CXVII) or (CXVIII) and in turn, to the tetracyclic compounds. The bridge can form either exo (CXVIII) or endo (CXVII); accordingly two series can arise. The exo one comprises crinine (CXIX). The endo-bridged derivatives vittatine (CXX), crinamine (CXXI), haemanthamine (CXXII) and haemanthidine (CXXIII) are partly hydroxylated on the bridge. The incorporation of the tritium label of haemanthamine into tazettine (CXXIV) unequivocally proved the biogenetic pathway (CXXII)→(CXXIII)→(CXXIV) (FALES and WILDMAN, 1964) (Fig. 36).

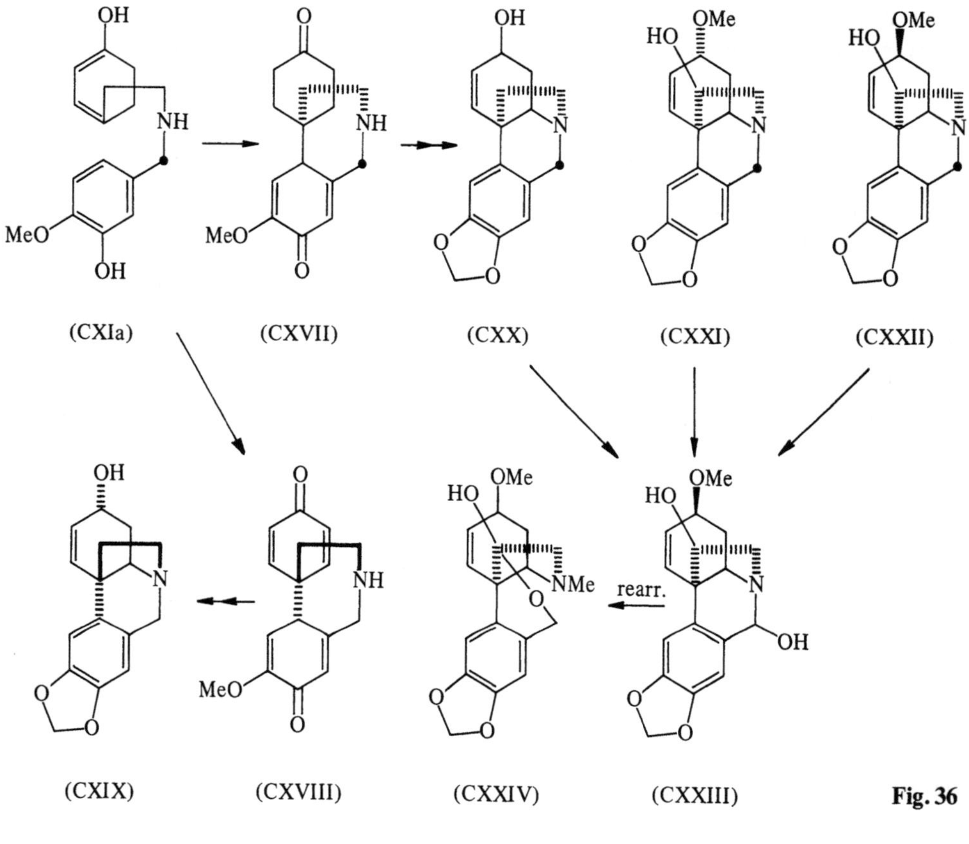

Fig. 36

[1-^{14}C, 2-^{3}H]-o-methylnorbelladine (CXIa) led to radioactive heamanthamine, involving the loss of the pro-R hydrogen from C-2 (BATTERSBY et al., 1971a, 1973). The C6-C1 unit of haemanthamine forms from phenylalanine via cinnamic acid (SCHÜTTE, 1969). Amaryllidaceae have the enzyme phenylalanine lyase (SUHADOLNIK et al., 1963) which converts (I) into trans-cinnamic acid. Stereospecifically labelled [3-^{3}H]-tyrosine lost its H_R during hydroxylation to haemanthamine (KIRBY and MICHAEL, 1971). Tazettine (CXXIV) proved, by tracer work, to be a product of further rearrangement of haemanthidine.

3.4.3 Galanthamine Series

Alkaloids of this class, present in daffodils, have an N-methyl group; their precursor being established as N, p′-O-dimethylnorbelladine (CXXV) and the phenolic coupling involves the p- and o′-carbon atoms to the hydroxyls. The sequence (CXXV)→(CXXVI)→narwedine (CXXVII)→galanthamine (CXXVIII) is somewhat similar to the formation of the other Amaryllidaceae alkaloids, also to the oxide bridge formation in thebaine. The [N-methyl-^{14}C]-label of (CXXV) was fully recovered in the N-methyl group of galanthamine (CXXVIII) (Fig. 37). Triple labelled (CXXV) entered (CXXVIII) without any scrambling of the label (BARTON et al., 1962). It was also shown, by further labelling experiments, that norbelladine was methylated first on the nitrogen.

Narciclasine (CXXXI) was formed from vittatine via the (tritiated) intermediates (CXXIX) and (CXXX) (Fig. 38), by a retro-Prins reaction (FUGANTI, 1973, 1975), involving loss of 2 carbons (BATTERSBY et al., 1971a, b); (FUGANTI and MAZZA, 1971a, b, 1972, 1973; FUGANTI et al., 1971).

Mesembrine (CXXXIV) from the Aizoaceae family is formed by a somewhat more complex biogenetic pathway involving the incorporation of [3-^{14}C]-tyrosine or ring labelled phenylalanine (JEFFS et al., 1967, 1971). Mesembrine is a member of the Amaryllidaceae alkaloids and is formed by the phenol coupling from a N-methyl-norbelladine derivative with the loss of a single carbon. [1′-^{14}C, 2′, 6′-

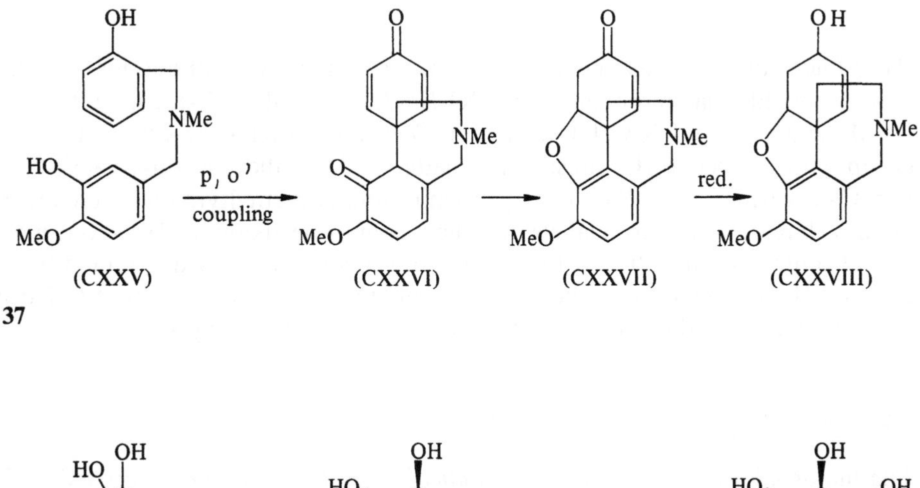

(CXXV) (CXXVI) (CXXVII) (CXXVIII)

Fig. 37

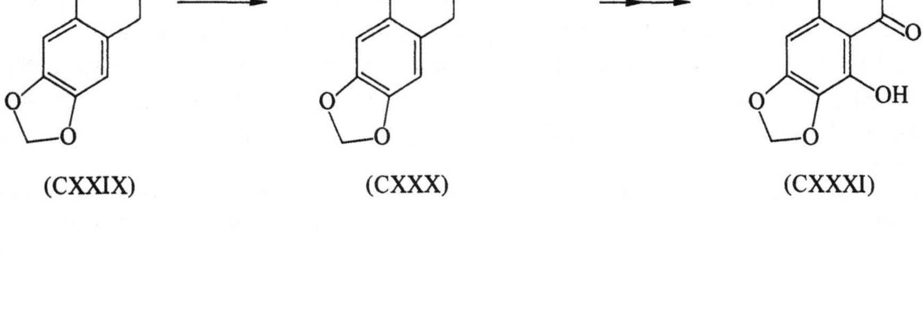

(CXXIX) (CXXX) (CXXXI)

Fig. 38

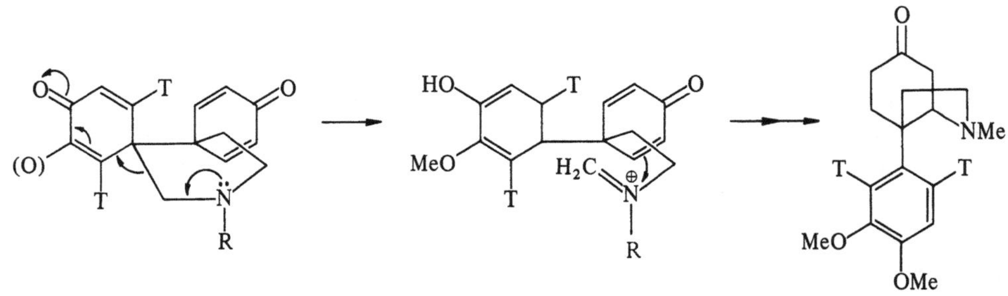

(CXXXII) (CXXXIII) (CXXXIV)

Fig. 39

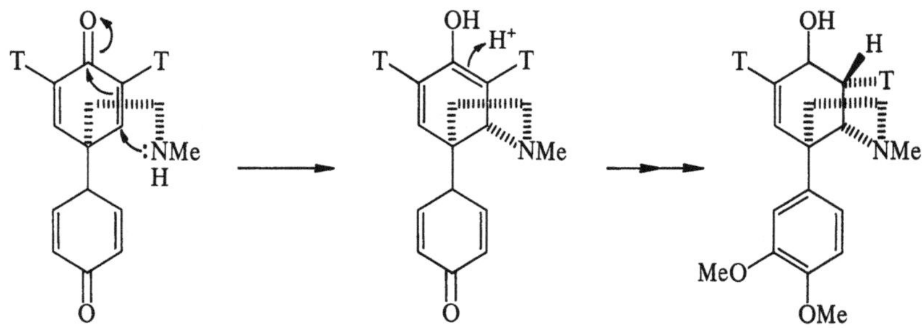

(CXXXV) (CXXXVI) (CXXXVII)

Fig. 40

$^{3}H_2$-]-Phenylalanine was incorporated into mesembrine, without loss of the T atoms, possibly via spirodienone (CXXXIII) (JEFFS et al., 1974) (Fig. 39).

Mesembrenol (CXXXVII), formed in *Sceletium strictum,* resembles the crinine group with respect to the substitution pattern of the alicyclic ring. L-[3′, 5′-$^{3}H_2$, U-^{14}C]-Tyrosine, [3′, 5′-$^{3}H_2$, 1-^{14}C]-tyramine and N-methyltyramine were incorporated with retention of both labels into the mesembrine alkaloids. However, mesembrenol contained only 50% of the tritium and this was equally distributed between H-5 and H-7α. This result indicates that a stereospecific protonation at C-7 occurs from the β-face of the enol (CXXXVI) (JEFFS et al., 1976) (Fig. 40).

3.5 Colchicine [1]

The major alkaloid of *Colchicum autumnale* and *C. byzanthinum* proved to be a modified phenylethylisoquinoline alkaloid. Tracer studies with phenylalanine (LEETE and NEMETH, 1960) and cinnamic acid on the one hand, and labelled tyrosine on the other proved that both α-amino acids are precursors of the peculiar ring system. The carboxyl group of phenylalanine gave rise to C-7 of colchicine (CXLVI), while the C-3 and C-2 carbons became C-6 and C5, respectively (LEETE, 1963b; BATTERSBY et al., 1964c) via cinnamic acid (BATTERSBY et al., 1964c; LEETE, 1965; WIGHTMAN et al., 1972). The incorporation of [3-^{14}C]-tyrosine into the

1 Effects of this alkaloid on cell structures are extensively treated in several chapters of Vol. 7 of this Encyclopedia: Plant movements (HAUPT, W., FEINLEIB, M.E., eds.), 1979

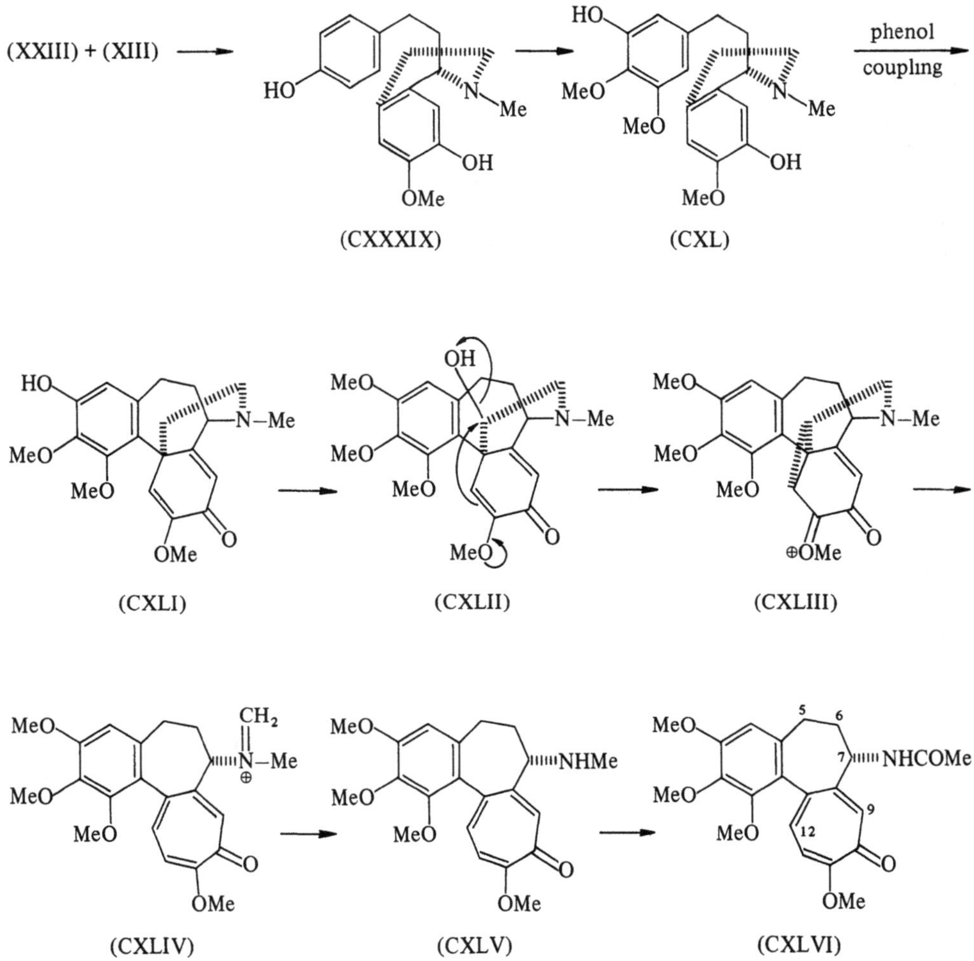

Fig. 41

tropolone ring resulted in labelling at C-12 of (CXLVI), whereas label from [4′-^{14}C]-tyrosine was incorporated into C-9 indicating ring enlargement (Battersby and Herbert, 1964; Battersby et al., 1966). Further support for such a mechanism is provided by the work of Avundzhyan (1967), Barker et al. (1967) and Battersby et al. (1972a, b) who isolated demecolcine (CXLV) and androcymbine (CXLI), the supposed intermediates, and also interconverted ^{3}H-labelled O-Me-androcymbine into colchicine (CXLVI). The ^{14}C/^{15}N ratio of *S*-autumnaline (CXL) remained virtually unchanged after incorporation of (CXL) into colchicine (Battersby et al., 1972a, b). (±)-[1-^{13}C]-Autumnaline (CXL) showed the enhancement of the ^{13}C-nmr signal only at C-7 in (CXLVI) (Battersby et al., 1974b) which supports the Scheme (Fig. 41).

Homoallylic ring expansion (Chapman and Fitton, 1963) is an essential part of this mechanism. It became obvious that the main building block for both parts was not dopamine. Also, [2-^{14}C]-noradrenaline was not incorporated into colchicine while [2-^{14}C]-dopamine gave a radioactive product (Spenser, 1968).

3.6 Emetine and Related Ipecacuanha Alkaloids

[2-^{14}C]-Tyrosine gave rise to label in two of the carbon atoms next to the two nitrogen atoms in emetine (CXLVIII) and cephaeline (CXLVII) (see Scheme in Fig. 42), while protoemetine (CXLIX), a related alkaloid, had radioactivity in isoquinoline carbons 1 and 3 when the plant had been fed with [2-^{14}C]-tyrosine. Therefore, protoemetine is not the precursor of emetine. One C6-C2 unit from tyrosine or DOPA seemed to react with secologanin (CL), itself derived from

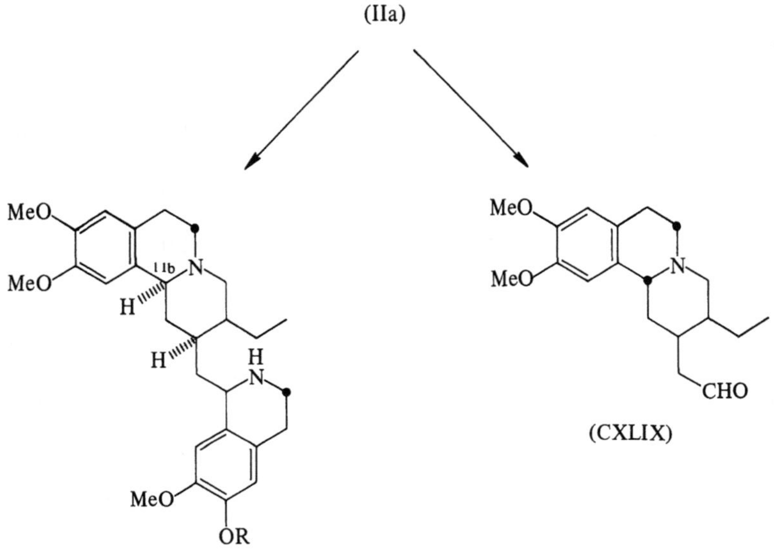

(CXLVII) R = H
(CXLVIII) R = Me

Fig. 42

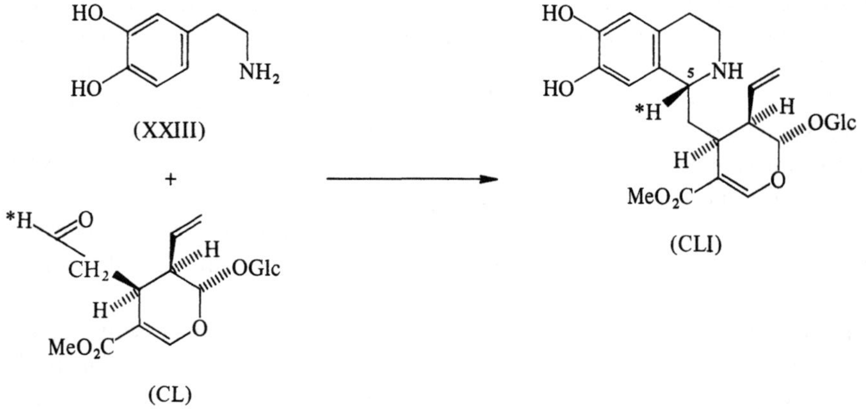

Fig. 43

geraniol to give desacetylipecoside (CLI) (BATTERSBY and GREGORY, 1968). Structural rearrangement of (CLI) followed by condensation with a second molecule of dopamine and methylation gives rise to emetine (CXLVIII) (Fig. 43). Labelled [O-methyl-^{3}H, 6,6-$^3H_2$]-secologanine was indeed efficiently incorporated into (CLI) and (CXLVII) with C-11b in (CXLVII) carrying the radioactive hydrogen (BATTERSBY and PARRY, 1971). Surprisingly, the 5-epimer of (CLI) (desacetylisopecoside) was not incorporated into (CXLVII). Thus, the incorporation of (CLI) to (CXLVII) requires inversion of configuration at the chiral isoquinoline methine carbon (KENNARD et al., 1971).

3.7 Betalains

These dihydroindole pigments of Centrospermae are treated in detail elsewhere in this volume (MABRY, Chap. 11) (Fig. 44). It will therefore suffice here to indicate briefly the biogenetic pathway from tyrosine by analogy with melanin formation. Surprisingly, 90% of label from [1-^{14}C]- and [2-^{14}C]-DOPA was contained in the tetrahydropyridine moiety of betanidine (CLII) after feeding experiments, and only 10% in the cyclo-DOPA unit (WYLER et al., 1963; MINALE et al., 1965). The low incorporation into cyclo-DOPA may be explained by assuming that DOPA-glucoside is a precursor of the dihydroindole part of betanin.

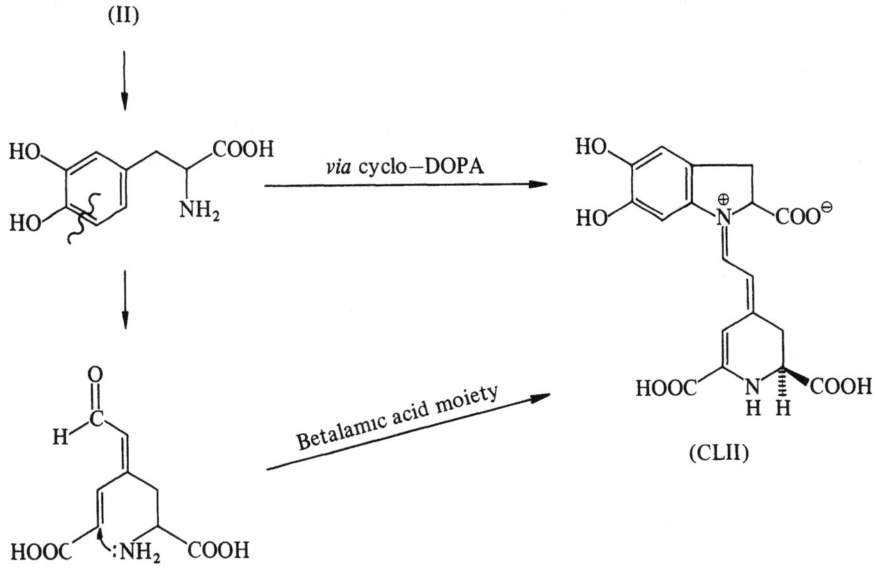

(CLII)

Fig. 44

3.8 Alkaloids Derived from Noradrenaline

Noradrenaline is formed in both mammalian and plant tissues, from tyrosine, via DOPA and dopamine (MEISTER, 1965). Berberastine (CLIII), an alkaloid of

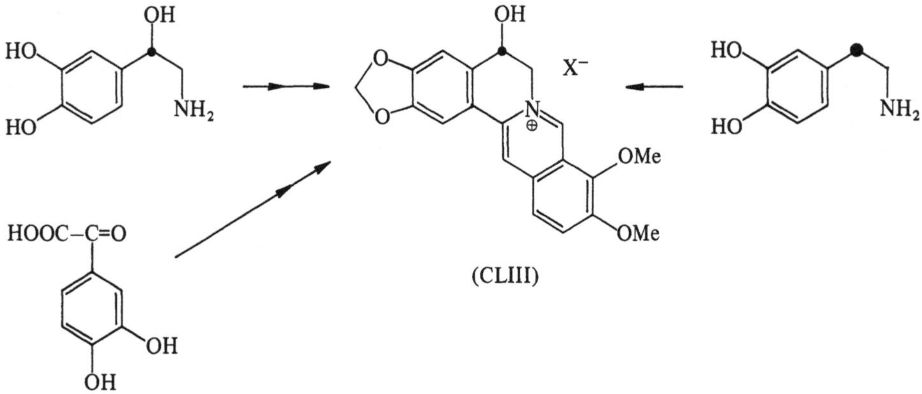

Fig. 45

Hydrastis canadensis, can be labelled from either [1-^{14}C]-dopamine or [2-^{14}C]-noradrenaline (MONKOVIC and SPENSER, 1965a, b) (Fig. 45). At the same time berberine and canadine (tetrahydroberberine) only incorporate DOPA but not noradrenaline. Therefore, the possibility of an ultimate hydroxylation of berberine seems excluded.

3.9 Complex Alkaloids Formed from Tyrosine and a Different Precursor

3.9.1 Securinine

[2-^{14}C]-Tyrosine was incorporated into securinine (CLVII) as was lysine. In Euphorbiaceae the mechanism of the formation of (CLVII) appears to involve a condensa-

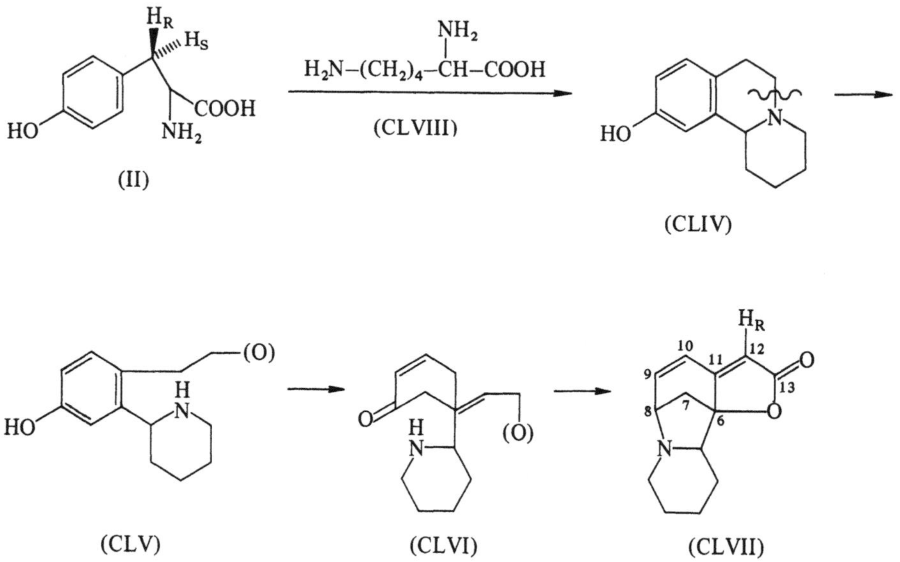

Fig. 46

tion of tyrosine (II) (PARRY, 1974) with lysine (CLVIII) (GOLEBIEWSKI et al., 1976) to give an azaphenanthrene (CLIV). This, upon ring opening to (CLV) and subsequent rearrangement and ring closure (CLV)→(CLVI) gives the alkaloid (CLVII) that has lost the tyrosine nitrogen but retained the tyrosine skeleton in an oxidized form (PARRY, 1974) (Fig. 46).

In order to obtain further elucidation of the mechanism, (3*S*)-[3-^{14}C, 3-^{3}H]-tyrosine (II) was fed to the plant and it was shown that the pro-*S* hydrogen was lost during the formation of (CLVII) (PARRY, 1975).

3.9.2 Anthramycin, Tomaymycin

Tyrosine and/or DOPA are incorporated into the skeleton of 11-demethyltomaymycin (CLIX) and of anthramycin (CLX) (HURLEY and ZRUJEWSKI, 1974; HURLEY et al., 1975) (Fig. 47).

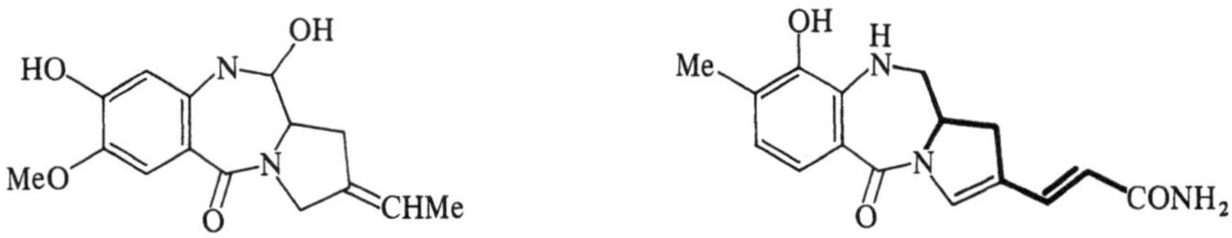

(CLIX) (CLX)

Tyrosine moiety outlined

Fig. 47

References

Achenbach, H., Grisebach, H.: Biosynthesis of xanthocillin. Z. Naturforsch. *20B*, 137–140 (1965)

Agurell, S.: Identification of alkaloid intermediates by gas chromatography – mass spectrometers. Potential mescaline precursors in *Trichocereus* species. Lloydia *32*, 40–45 (1969)

Agurell, S.: Studies on orchidaceae alkaloids. XL. Biosynthetic studies of (−)-cryptostyline I in *Cryptostyles erythroglossa hayata*. Acta Chem. Scand. *28B*, 1175–1179 (1974)

Agurell, S., Lundström, J., Sandberg, F.: Biosynthesis of mescaline in *Peyote*. Tetrahedron Lett. 2433–2435 (1967)

Agurell, S., Granelli, I., Lüning, B., Rosenblom, J.: Studies on orchidaceae alkaloids. XXIX. Isolation of (−) cryptostyline I, II, III and two quaternary salts from *Cryptostylis erythroglossa hayata*, biosynthetic studies of (−) cryptostylines. Acta Chem. Scand. *28B*, 239–243 (1974)

Archer, D.A., Breuer, S.W., Binks, R., Battersby, A.R., Wildman, W.C.: Biosynthesis in the amaryllidaceae· further evidence on the C_6-C_1 unit and the methylenedioxy group. Proc. Chem. Soc. 168 (1963)

Avundzhyan, E.S.: Metabolism of nicotine-type alkaloids in plants. Usp. Sovrem. Biol. *63*, 232–248 (1967)

Barker, A.C., Battersby, A.R., McDonald, E., Ramage, R., Clements, J.H.: Biosynthesis of colchicine: Ring expansion and later stages. Structure of speciosine. J. Chem. Soc. Chem. Commun. 390–392 (1967)

Barton, D.H.R.: The biogenesis of phenolic alkaloids. Proc. Chem. Soc. 293–299 (1963)

Barton, D.H.R.: Phenol oxidation and biogenesis. Proc. Robert A. Welch Found. Conf. Chem. Res. *4*, 165–188 (1961)

Barton, D.H.R., Cohen, T.: Some biogenetic aspects of phenol oxidation. Festschrift A. Stoll. Basel: Birkhäuser 117–143 (1957)
Barton, D.H.R., Kirby, G.W., Taylor, J.B., Thomas, G.M.: Multiple labelling experiments in the biosynthesis of amaryllidaceae alkaloids. Proc. Chem. Soc. 179 (1962)
Barton, D.H.R., Hesse, R.H., Kirby, G.W.: The origin of the berberine carbon. Proc. Chem. Soc. 267–268 (1963a)
Barton, D.H.R., Kirby, G.W., Taylor, J.B., Thomas, G.M.: Phenol oxidation and biosynthesis, Part XI. The biogenesis of amaryllidaceae alkaloids. J. Chem. Soc. 4545–4558 (1963b)
Barton, D.H.R., Hesse, R.H., Kirby, G.W.: Phenol oxidation and biosynthesis, Part VIII. J. Chem. Soc. 6379–6389 (1965)
Barton, D.H.R., Bhakuni, D.S., Chapman, G.M., Kirby, G.W., Haynes, L.J., Stuart, K.L.: Phenol oxidation and biosynthesis, Part XIV. The biosynthesis of crotonosine. J. Chem. Soc. (*C*), 1295–1298 (1967)
Barton, D.H.R., Boar, R.B., Widdowson, D.A.: Phenol oxidation and biosynthesis, Part XXI. The biosynthesis of the erythrina alkaloids. J. Chem. Soc. (*C*) 1213–1218 (1970)
Barton, D.H.R., Bracho, R.D., Potter, C.J., Widdowson, D.A.: Phenol oxidation and biosynthesis, Part XXIV. Origin of chirality in the erythrinan system and derivative of the lactone rings of α and β erythroidine. J. Chem. Soc. Perkin I. 2278–2283 (1974)
Battersby, A.R., Gregory, B.: Biosynthesis of the ipecac alkaloids and ipecoside, a cleaved cyclopentane monoterpene. J. Chem. Soc. Chem. Commun. 134–135 (1968)
Battersby, A.R., Harper, B.J.T.: Biogenesis of morphine. Chem. Ind. (London) 364 (1958)
Battersby, A.R., Herbert, R.B.: Colchicine. Proc. Chem. Soc. 260 (1964)
Battersby, A.R., Parry, R.J.: Biosynthesis of the ipecac alkaloids and ipecoside. J. Chem. Soc. Chem. Commun. 901–902 (1971)
Battersby, A.R., Binks, R., Breuer, W., Fales, H.M., Wildman, W.C.: Biosynthesis in the amaryllidaceae: incorporation of norbelladine into lycorine and norpluviine. Proc. Chem. Soc. 243 (1961a)
Battersby, A.R., Fales, H.M., Wildman, W.C.: Biosynthesis in the amaryllidaceae. Tyrosine and norbelladine as precursors of haemanthamine. J. Am. Chem. Soc. *83*, 4098–4099 (1961b)
Battersby, A.R., Binks, R., Harper, B.J.T.: Alkaloid biosynthesis. Part II. The biosynthesis of morphine. J. Chem. Soc. 3534–3544 (1962)
Battersby, A.R., Francis, R.J., Hirst, M., Staunton, J.: Biosynthesis of the "Berberine Bridge". Proc. Chem. Soc. 268 (1963)
Battersby, A.R., Binks, R., Breuer, S.W., Fales, H.M., Wildman, W.C., Highet, R.J.: Alkaloid biosynthesis, Part III. Amaryllidaceae alkaloids: the biosynthesis of lycorine and its relatives. J. Chem. Soc. 1595–1609 (1964a)
Battersby, A.R., Binks, R., Francis, R.J., McCaldin, D.J., Ramuz, H.: Alkaloid biosynthesis, Part IV. 1-benzylisoquinolines as precursors of thebaine, codeine, and morphine. J. Chem. Soc. 3600–3610 (1964b)
Battersby, A.R., Binks, R., Reynolds, T.J., Yeowell, D.Y.: Alkaloid biosynthesis, Part VI. The biosynthesis of colchicine. J. Chem. Soc. 4257–4268 (1964c)
Battersby, A. R., Brown, R.T., Clements, J.H., Iverach, G.G.: On the biosynthesis of isothebaine. J. Chem. Soc. Chem. Commun. 230–232 (1965a)
Battersby, A.R., Foulkes, D.M., Binks, R.: Alkaloid biosynthesis, Part III. Use of optically active precursors for investigations on the biosynthesis of morphine alkaloids J. Chem. Soc. 3323–3332 (1965b)
Battersby, A.R., Herbert, R.B., McDonald, E., Ramage, R., Clements, J.H.: Biosynthesis of colchicine from a phenylisoquinoline. J. Chem. Soc. Chem. Commun. 603–605 (1966)
Battersby, A.R., Binks, R., Huxtable, R.: Biosynthesis of cactus alkaloids. Tetrahedron Lett. 563–565 (1967a)
Battersby, A.R., McDonald, E., Munro, M.H.G., Ramage, R.: Homoaporphine systems and related dienones: isolation, structure, and synthesis. J. Chem. Soc. Chem. Commun. 934–935 (1967b)
Battersby, A.R., Binks, R., Huxtable, R.: Biosynthesis of pellotine. Tetrahedron Lett. 6111–6115 (1968a)
Battersby, A.R., Hirst, M., McCaldin, D.J., Southgate, R., Staunton, J.: Alkaloid biosynthesis, Part XII. The biosynthesis of narcotine. J. Chem. Soc. (*C*) (1968b)

Battersby, A.R., Brockson, T.J., Ramage, R.: Further studies on the synthesis and biosynthesis of isothebaine. J. Chem. Soc. Chem. Commun. 464–465 (1969)
Battersby, A.R., Kelsey, J., Staunton, J.: Hydroxylation at saturated carbon: haemanthamine. J. Chem. Soc. Chem. Commun. 183–184 (1971a)
Battersby, A.R., McHugh, J.L., Staunton, J., Todd, M.: Biosynthesis of the apparently "directly coupled" aporphine alkaloids. J. Chem. Soc. Chem. Commun. 985–986 (1971b)
Battersby, A.R., Herbert, R.B., McDonald, E., Ramage, R., Clements, J.H.: Alkaloid biosynthesis, Part XVIII. Biosynthesis of colchicine from the 1-phenylisoquinoline system. J. Chem. Soc. Perkin I. 1741–1746 (1972a)
Battersby, A.R., Herbert, R.B., Pijewska, L., Santavy, F., Sedmera, P.: Alkaloid biosynthesis, Part XVII. The structure and chemistry of androcymbine. J. Chem. Soc. Perkin I. 1736–1740 (1972b)
Battersby, A.R., Kelsey, J.E., Staunton, J., Suckline, K.E.: Studies of enzyme-mediated reactions, Part III. Stereoselective labelling at C-2 of tyramine. Stereochemistry of hydroxylation at saturated carbon. J. Chem. Soc. Perkin I. 1609–1615 (1973)
Battersby, A.R., Bohler, P., Munro, M.H.G., Ramage, R.: Biosynthesis, Part XX. Biosynthesis of C-homoaporphines in *Kreysigia multiflora*. J. Chem. Soc. Perkin I. 1399–1402 (1974a)
Battersby, A.R., Sheldrake, P.W., Milner, J.A.: Biosynthesis of colchicine: incorporation of a ^{13}C-labelled precursor in a higher plant. Tetrahedron Lett. 3315–3318 (1974b)
Battersby, A.R., Francis, R.J., Hirst, M., Ruveda, E.A., Staunton, J.: Biosynthesis, Part XXI. Investigations on the biosynthesis of stylopine in *Chelidonium majus*. J. Chem. Soc. Perkin I. 1140–1147 (1975a)
Battersby, A.R., Francis, R.J., Hirst, M., Ruveda, E.A., Staunton, J.: Biosynthesis, Part XXII. The origin of chelidonine and of other alkaloids derived from the tetrahydroprotoberberine skeleton. J. Chem. Soc. Perkin I. 1147–1156 (1975b)
Battersby, A.R., Jones, R.C.F., Kazlauskas, R.: Experiments of the early steps of morphine biosynthesis. Tetrahedron Lett. 1873–1876 (1975c)
Battersby, A.R., McDonald, E., Milner, J.A., Johns, S.R., Lamberton, J.A., Sioumis, A.A.: Biosynthesis of schelhammeridine: mode of specific incorporation of [2-^{14}C] tyrosine. Tetrahedron Lett. 3419–3421 (1975d)
Battersby, A.R., Staunton, J., Wiltshire, H.R., Bircher, B.J., Fuganti, C.: Studies of enzyme-mediated reaction, Part V. Synthesis of (13*S*)- and (13*R*)-[13-^{3}H] scoulerine from stereospecifically labelled (R)- and (S)-[α-$^3H_1$] benzyl alcohols; stereochemistry of enzymic reaction at saturated benzylic carbon. J. Chem. Soc. Perkin I. 1162–1171 (1975e)
Battersby, A.R., Staunton, J., Wiltshire, H.R., Francis, R.J., Southgate, R.: Biosynthesis, Part XXII. The origin of chelidonine and of other alkaloids derived from the tetrahydroprotoberberine skeleton. J. Chem. Soc. Perkin I. 1147–1156 (1975f)
Blaschke, G.: Alkaloid biosynthesis II. Reticuline as biosynthesis precursor of corydaline Arch. Pharm. *301*, 439–443 (1968)
Blaschke, G.: Biosynthesis of alkaloids. 3. Mechanism of diphenyl coupling in the biosynthesis of aporphine alkaloids. Arch. Pharm. *303*, 358–363 (1970)
Bowman, W.R., Bruce, I.T., Kirby, G.W.: Biosynthetic incorporation of [3-^{14}C; 3,5-$^2H_2$; 4-^{3}H] cinnamic acid into capsaicin and norpluviine: lack of an apparent isotope effect following an NIH shift. J. Chem. Soc. Chem. Commun. 1075–1077 (1969)
Brochmann-Hanssen, E., Fu, C.-C., Leung, A.Y., Zanati, G.: Opium alkaloids X. Biosynthesis of 1-benzylisoquinolines. J. Pharm. Sci. *60*, 1672–1676 (1971a)
Brochmann-Hanssen, E., Fu, C.-C., Misconi, L.Y.: Opium alkaloids XI. Biosynthesis of aporphines in *Papaver somniferum*. J. Pharm. Sci. *12*, 1880–1883 (1971b)
Brochmann-Hanssen, E., Chen, C.-C., Chiang, H.C., McCurtrey, K.: Biosynthesis of magnoflorine from reticuline by direct oxidation coupling. J. Chem. Soc. Chem. Commun. 1269 (1972)
Brochmann-Hanssen, E., Chen, C.C., Chen, C.R., Chiang, H., Leung, A.Y., McMurtrey, K. J.: Opium alkaloids, Part XVI. The biosynthesis of 1-benzylisoquinolines in *Papaver somniferum* preferred and secondary pathways; stereochemical aspects. Chem. Soc. Perkin I. 1531–1537 (1975)
Bruce, I.T., Kirby, G.W.: Stereochemistry of protonation and hydroxylation in the biosynthesis of norpluviine and lycorine. J. Chem. Soc. Chem. Commun. 207–208 (1968a)

Bruce, I.T., Kirby, G.W.: Stereochemistry of protonation and hydroxylation in the biosynthesis of norpluviine and lycorine. Chimia *22*, 314–315 (1968b)
Bu'Lock, J.D., Ryles, A.P.: The biosynthesis of the fungal toxin gliotoxin–the origin of the "extra" hydrogens as established by heavy-isotope labelling and mass spectrometry. J. Chem. Soc. Chem. Commun. 1404–1406 (1970)
Bu'Lock, J.D., Ryles, A.P., Johns, N., Kirby, G.W.. A stereospecific exchange reaction of the β-CH_2 of phenylalanine, competing with gliotoxin biosynthesis. J. Chem. Soc. Chem. Commun. 100–101 (1972)
Chapman, O., Fitton, P.: A general synthesis of the troponoid system based on solvolysis of 1,4-dihydrobenzyl tosylates. J. Am. Chem. Soc. *85*, 41–47 (1963)
Daly, J., Jerina, D.: Migration of deuterium during aryl hydroxylation III. Effect of otho- and meta substituents. Arch. Biochem. Biophys. *134*, 266–268 (1969)
Dobberstein, R.H., Edwards, J.M., Schwarting, A.: Sequential appearance and metabolism of alkaloids in *Heimia salicifolia*. Phytochemistry *14*, 1769–1775 (1975)
Dobbins, M.: *Trichocereus pachanoi*, a mescaline cactus used in folk healing in Peru. Econ. Bot. *22*, 191–194 (1968)
Erdtmann, H., Wachtmeister, C.A.: Phenol dehydrogenation as a biosynthetic reaction. Festschrift A. Stoll. Basel: Birkhäuser (1957)
Fales, H.M., Wildman, W.C.: Biological interconversions in the amaryllidaceae. I. The haemanthamine-haemanthidine-tazettine sequence. J. Am. Chem. Soc. *86*, 294–295 (1964)
Frank, A.W., Massicot, J., Marion, L.: The biogenesis of alkaloids. Hordenine metabolism in barley. Can. J. Chem. *34*, 1641–1646 (1956)
Fuganti, C.: Evidence for the intermediacy of 11-hydroxyvittatine in the biosynthesis of narciclasine. Gazz. Chim. Ital. *103*, 1255–1258 (1973)
Fuganti, C.: Amaryllidaceae alkaloids. In: Alkaloids. Manske, R.H.F., Vol. 15, pp. 145–164 (1975)
Fuganti, C., Mazza, M., Relative stereochemistry of protonation and hydroxylation in the biosynthesis of lycorenine and haemanthidine from protocatechualdehyde. J. Chem. Soc. Chem. Commun. 1196–1197 (1971a)
Fuganti, S., Mazza, M.: Late intermediates in the biosynthesis of narciclasine. J. Chem. Soc. Chem. Commun. 1388–1389 (1971b)
Fuganti, S., Mazza, M.: The absolute configuration of narciclasine. A biosynthetic approach. J. Chem. Soc. Chem. Commun. 239 (1972)
Fuganti, C., Mazza, M.: Biosynthesis of amaryllidaceae alkaloids. Stereochemistry of hydrogen removal to the tertiary nitrogen atom in the biological conversion of norpluviine into lycorenine. J. Chem. Soc. Perkin I. 954–956 (1973)
Fuganti, C., Staunton, J., Battersby, A.R.: The biosynthesis of narciclasine. J. Chem. Soc. Chem. Commun. 1154–1155 (1971)
Gear, J.R., Spenser, I.D. The biosynthesis of hydrastine and berberine. Can. J. Chem. *41*, 783–803 (1963)
Golebiewski, W.M., Horsewood, P., Spenser, I.D.: Biosynthesis of securinine: the mode of incorporation of lysine. J. Chem. Soc. Chem. Commun. 217–218 (1976)
Gottlieb, D., Carter, H.E., Robbins, P.W., Burg, R.W.: Biosynthesis of chloramphenicol IV. Incorporation of carbon-14-labelled precursors. J. Bacteriol. *84*, 888–895 (1962)
Graf, W., Robert, J.-L., Vederas, J.C., Tamm, O., Solomon, P.H., Miura, I., Nakanishi, K.: Biosynthesis of Cytochalasans III. The building blocks of Cytochalasin D. Helv. Chim. Acta *57*, 1801–1815 (1974)
Hahn, G., Bärwald, L., Schales, O., Werner, H.: The synthesis of tetrahydroharman (4-carbolin) derivatives under physiological conditions. Annalen *520*, 107–123 (1935)
Harrison, J.W., Scrowston, R.M., Lythgoe, B.: The constitution of taxine-1. J. Chem. Soc. (*C*), 1933–1945 (1966)
Holland, H.L., Castillo, M., MacLean, D.B., Spenser, I.: Biosynthesis of corydaline and of ochotensimine. Can. J. Chem. *52*, 2818–2821 (1974)
Hurley, L.H., Zrujewski, M.J.: Biosynthesis of the antitumor antibotic anthramycin by streptomyces refuineus. J. Chem. Soc., Chem. Commun. 337–338 (1974)
Hurley, H.L., Gairola, C., Zrujewski, M.J.: Biosynthesis of the antiviral antibotic 11-demethyltomaymycin. J. Chem. Soc. Chem. Commun. 120 (1975)
Jeffs, P.W., Archie, W.C., Farrier, D.S.: Biosynthesis of mesembrine. The incorporation of one-carbon units and the origin of the C_6 unit. J. Am. Chem. Soc. *89*, 2509–2510 (1967)

Jeffs, P.W., Archie, W.C., Hawks, R.L., Farrier, D.S.: Biosynthesis of mesembrine and related alkaloids. The amino acid precursors. J. Am. Chem. Soc. *93*, 3752–3758 (1971)
Jeffs, P.W., Campbell, H.F., Farrier, D.S., Ganguli, G., Martin, N.H., Molina, G.: Incorporation of phenylalanine and examination of norbelladines as precursors of the mesembrine alkaloids. Phytochemistry *13*, 933–945 (1974)
Jeffs, P.W., Johnson, D.B., Martin, N.H., Rauckman, B.S.: Post-tyrosine intermediates in the biosynthesis of mesembrine alkaloids and stereochemistry of protonation at C-7 in the formation of the octahydroindole skeleton. J. Chem. Soc. Chem. Commun. 82–83 (1976)
Johns, N., Kirby, G.W., Bu'Lock, J.D., Ryles, A.: Stereospecific exchange of a β-methylene proton in phenylalanine proceeding biosynthetic incorporation into gliotoxin. J. Chem. Soc. Perkin I. 383–386 (1975)
Kapadia, G.J., Rao, G.S., Leete, E., Fayez, M.B.E., Vaishnav, Y.N., Fales, H.M.: On the origin of carbon 1 in tetrahydroisoquinoline alkaloids. J. Am. Chem. Soc. *92*, 6943–6951 (1970)
Kennard, O., Roberts, P.J., Isaacs, N.W., Motherwall, W.D.S., Gibson, K.H., Battersby, A.R.: X-ray determination of the structure of o,o-demethylipecoside. J. Chem. Soc. Chem. Commun. 899–900 (1971)
Kiechel, J.R., Binder, M., Tamm, C.: On the biogenesis of the antibiotic, Phomin. Helv. Chim. Acta *53*, 1797–1812 (1970)
Kirby, G.W.: Biosynthesis of the morphine alkaloids. Sience *155*, 170–173 (1967)
Kirby, G.W., Michael, J.: Labelling of aromatic amino acids stereoselectively with tritium in the β-methylene group: the stereochemistry of hydroxylation in the biosynthesis of haemanthamine. J. Chem. Soc. Chem. Commun. *187*, 415–416 (1971)
Kirby, G.W., Massy, S.R., Steinreich, P.: Biosynthesis of unnatural morphine derivatives in *Papaver somniferum*. J. Chem. Soc. Perkin I. 1642–1647 (1972)
Koo, S.H., Comer, F., Spenser, I.D.: Biosynthesis of the lythraceae alkaloids: mode of incorporation of phenylalanine. J. Chem. Soc. Chem. Commun. 897–898 (1970a)
Koo, S.H., Gupta, R.N., Wrobel, J.T.: Biosynthesis of the lythraceae alkaloids: incorporation of lysine. J. Chem. Soc. Chem. Commun. 396–397 (1970b)
Lebet, C.-R., Tamm, C.: Biosynthesis of cytochalasans III. ^{13}C-NMR of cytochalasin B (Phomin) and cytochalasin D. Incorporation of [1-^{13}C]- and [2-^{13}C]-sodium acetate. Helv. Chim. Acta *57*, 1785–1801 (1974)
Leete, E.: Biogenesis of *d*-norpseudoephedrine in *Catha edulis*. Chem. Ind. 1088–1089 (1958)
Leete, E.: Biogenesis of mescaline. Chem. Ind. 604 (1959)
Leete, E.: Biosynthesis of the alkaloids of *Chelidonium majus*. The incorporation of tyrosine into chelidonine. J. Am. Chem. Soc. *85*, 473–475 (1963a)
Leete, E.: The biosynthesis of the alkaloids of colchicum. III. Incorporation of phenylalanine-2-^{14}C into colchicine and demecolcine. J. Am. Chem. Soc. *85*, 3666–3669 (1963b)
Leete, E.: Biosynthesis of the tropolone ring of colchicine. Tetrahedron Lett. 333–336 (1965)
Leete, E.: Biosynthesis of the peyote alkaloids. The incorporation of tyrosine-2-^{14}C into mescaline and anhalonidine. J. Am. Chem. Soc. *88*, 4218–4221 (1966)
Leete, E., Bodem, G.G.: The biosynthesis of 3-dimethylamino-3-phenyl-propanoic acid in yew. Tetrahedron Lett. 3925–3927 (1966)
Leete, E., Braunstein, J.D.: Biosynthesis of anhalonidine: origin of the two carbon unit. Tetrahedron Lett. 451–452 (1969)
Leete, E., Marion, L.: The biogenesis of alkaloids. The formation of hordenine and N-methyltyramine from tyrosine in barley. Can. J. Chem. *31*, 126–128 (1953)
Leete, E., Nemeth, P.E.: The biogenesis of the alkaloids of *Colchicum*. I. The incorporation of phenylalanine into colchicine. J. Am. Chem. Soc. *82*, 6055–6057 (1960)
Leete, E., Marion, L., Kirkwood, S.: The biogenesis of alkaloids. The formation of hordenine and N-methyl-tyramine from tyramine in barley. Can. J. Chem. *30*, 749–760 (1952)
Leete, E., Kowanko, N., Newmark, R.A., Vining, L.C., McInnes, A.G., Wright, J.L.C.: The use of carbon-13 nuclear magnetic resonance to establish that the biosynthesis of tenellin involves an intramolecular rearrangement of phenylalanine. Tetrahedron Lett. 4103–4106 (1975)
Liss, I.: The occurrence of 3,4-dihydroxyphenylalanine in the latex of *Euphorbia lathyrus*. Flora (Jena) *157*, 351–367 (1961)

Lundström, J.: Biosynthetic studies on mescaline and related cactus alkaloids. Acta Pharm. Suec. *8*, 275–302 (1971)

Lundström, J., Agurell, S.: Biosynthesis of mescaline and tetrahydroisoquinoline alkaloids in *Lopophora Williamsii* (Lem.) coult. Acta. Pharm. Suec. *8*, 261–274 (1971)

MacDonald, J.C., Slater, G.P.: Biosynthesis of gliotoxin and mycelianamide. Can. J. Biochem. *53*, 475–478 (1975)

McFarlane, I.J., Slaytor, M.: Alkaloid biosynthesis in *Echinocereus Merkeri*. Phytochemistry *11*, 235–238 (1972)

McInnes, A.G., Smith, D.G., Wat, (C-K), Vining, L.C., Wright, J.L.C.: Tenellin and bassianin, metabolites of *Beauveria* species. Structure elucidation with ^{15}N- and doubly ^{13}C-enriched compounds using ^{13}C nuclear magnetic resonance spectroscopy. J. Chem. Soc. Chem. Commun. 281–282 (1974a)

McInnes, A.G., Smith, D.G., Walter, J.A., Vining, L.C., Wright, J.L.C.: New techniques in biosynthetic studies using ^{13}C nuclear magnetic resonance spectroscopy. The biosynthesis of tenellin enriched from singly and doubly labelled precursors. J. Chem. Soc. Chem. Commun. 282–284 (1974b)

Meister, A.: Biochemistry of amino acids, 2nd ed. New York: Academic Press 1965

Minale, L., Piatelli, M., Nicolaus, R.A.: Pigments of centrospermae. V. Betaxanthins from *Mirabilis Jalapa*. Phytochemistry *4*, 817–823 (1965)

Monkovic, I., Spenser, I.D.: Biosynthesis of berberastine. J. Am. Chem. Soc. *87*, 1137–1138 (1965a)

Monkovic, I., Spenser, I.D.: Biosynthesis of berberine and berberastine. Incorporation of catecholamines. Can. J. Chem. *43*, 2017–2026 (1965b)

Neubauer, D.: Biosynthesis of thebaine from tyrosine in *Papaver Bracteatum*. Arch. Pharm. *298*, 737–741 (1965)

Parry, R.J.: Biosynthesis of the securinega alkaloids. I. Specific incorporation of tyrosine. Tetrahedron Lett. 307–310 (1974)

Parry, R.J.: Biosynthesis of the securinega alkaloids. I. Stereospecificity of hydrogen loss from C-3 of tyrosine. J. Chem. Soc. Chem. Comm. 144–145 (1975)

Parry, R.J., Schwab, J.M.: Biosynthesis of cephalotaxus alkaloids. I. Novel mode of tyrosine incorporation into cephalotaxine. J. Am. Chem. Soc. *97*, 2555 (1975)

Paul, A.G., Khauna, K.L., Rosenberg, H., Takido, M.: Biosynthesis of peyote alkaloids. J. Chem. Soc. Chem. Commun. 838 (1969)

Poisson, M.J.: The presence of mescaline in a peruvian cactus. Ann. Pharm. Fr. *19*, 764–765 (1960)

Poupat, C., Kunesch, G.: Study of the biogenesis of the major alkaloids of dollar plant, *Lunaria biennis* Moench (Cruciferae). C. R. *273*, 433–436 (1971)

Robinson, R.: A theory of the mechanism of the phytochemical synthesis of certain alkaloids. J. Chem. Soc. *111*, 876–899 (1917)

Rosenberg, H., McLaughlin, J.L., Paul, A.G.: The cactus alkaloids. III. Phenylalanine, dopa and dopamine as precursors to mescaline in *Lophophora Williamsii*. Lloydia *30*, 100–105 (1967)

Rother, A., Schwarting, A.E.: Phenylalanine as a precursor for cryogenine biosynthesis in *Heimia salicifolia*. Phytochemistry *11*, 2475–2480 (1972)

Rother, A., Schwarting, A.E.: The phenylquinolizidines of the seedling of *Heimia Salicifolia*. Lloydia *38*, 477–488 (1975)

Schütte, H.R.: In: Biosynthese der Alkaloide. Mothes, K., Schütte, H.R. (eds.), p. 420. Berlin: VEB Deutscher Verlag der Wissenschaften 1969

Shibata, S., Imaseki, I.: Cultivation of medicinal plants X. Alkaloid biogenesis in *Ephedra* (2). Pharm. Bull. Tokyo *4*, 277–280 (1956)

Shibata, S., Imaseki, I., Yamazaki, M.: Biogenesis of ephedrine. Chem. Ind. 1625 (1958)

Smith, W.J., Kirshner, N.: Mechanism of 3,4-dihydroxyphenylethylamine-β-hydroxylase. J. Biol. Chem. *237*, 1890–1891 (1962)

Soloway, A.H.: Aromatic hydroxylation a new biochemical mechanism. J. Theoret. Biol. *13*, 100–105 (1966)

Spenser, I.D.: Alkaloids and other nitrogenous secondary metabolites. In: Comprehensive Biochemistry. M. Florkin, E.H. Stotz (eds.), pp. 231–413. Amsterdam: Elsevier 1968

Stermitz, R.F., Rapoport, F.: O-demethylation as a biosynthetic pattern in the formation of opium alkaloids. Nature (London) *189*, 310–311 (1961)

Stickings, C.E., Townsend, R.J.: Biochemistry of microorganisms CVIII. Metabolites of *Alternaria Tennis*. The biosynthesis of tenuazonic acid. Biochem. J. *78*, 412–418 (1961)

Stuart, K.L., Graham, K.: Alkaloids from croton species XIII. Alkaloid biosynthesis in *Croton Flavens*. Phytochemistry *13*, 1967–1972 (1973)

Stuart, K.L., Teetz, V., Franck, B.: Alkaloid biosynthesis in *Croton Flavens*. J. Chem. Soc. Chem. Commun. 333 (1969)

Suhadolnik, R.J., Zulalian, J.: Biosynthesis of the amaryllidaceae alkaloids IV. Incorporation of cinnamic, p-coumaric and caffeic acid into hemanthamine and lycorine. Proc. Chem. Soc. 217–218 (1963)

Suhadolnik, R.J., Fischer, A.G., Zulalian, J.: The biogenetic origin of the C_6-C_1 unit of lycorine. J. Am. Chem. Soc. *84*, 4348–4349 (1962)

Suhadolnik, R.J., Fischer, A.G., Zulalian, J.: Biogenesis of the amaryllidaceae alkaloids. II. Studies with whole plants, floral primordia and cell-free extracts. Biochem. Biophys. Res. Comm. *11*, 208–212 (1963)

Tani, C., Tagahara, K.: Berberine derivatives and related alkaloids VII. Biosynthesis of protopine. Chem. Pharm. Bull. *22*, 2457–2459 (1974)

Tewari, S., Bhakuni, D.S., Kapil, R.S.: Biosynthesis of the alkaloid boldine. J. Chem. Soc. Chem. Commun. 940–941 (1974)

Uprety, H., Bhakuni, D.S., Kapil, R.S.: Biosynthesis of papaverine. Phytochemistry *14*, 1535–1537 (1975)

Vederas, J.C., Graf, W., David, L., Tamm, C.: Biosynthesis of cytochalasans, Part 4. The mode of incorporation of common naturally occurring carboxylic acids into cytochalasin D. Helv. Chim. Acta *58*, 1886–1898 (1975)

Vining, L.C., McInnes, A.G., Wright, J.L.C.: The use of carbon-13 nuclear magnetic resonance to establish that the biosynthesis of tenellin involves an intramolecular rearrangement of phenylalanine. Tetrahedron Lett. 4103–4106 (1975)

Vollmer, K.O., Grisebach, H.: Biosynthesis of benzoic acids in *Gauetheria procumbens*. II. Z. Naturforsch. *19B*, 781–783 (1964)

Vollmer, K.O., Grisebach, H.: Biosynthesis of benzoic acids in *Gauetheria procumbens*. III. Z. Naturforsch. *21B*, 435–439 (1966)

Wildman, W.C., Battersby, A.R., Breuer, S.W.: Biosynthesis in the amaryllidaceae. Incorporation of 3-^{13}C-tyrosine and phenylalanine in *Nerine Bowdenii* W. Wats. J. Am. Chem. Soc. *84*, 4599–4600 (1962a)

Wildman, W.C., Fales, H.M., Highet, R.J., Breuer, S.W., Battersby, A.R.: Biosynthesis in the amaryllidaceae: evidence for intact incorporation of norbelladine into lycorine, crinamine, and belladine. Proc. Chem. Soc. 180–181 (1962b)

Wilson, M.L., Coscia, C.J.: Studies on the early stages of papaver alkaloid biosynthesis. J. Am. Chem. Soc. *97*, 431–432 (1975)

Wintersteiner, E., Trier, G.: Die Alkaloide. Berlin: Bornträger 1910

Wightman, R.H., Staunton, J., Battersby, A.R., Hanson, K.R.: Studies of enzyme-mediated reactions, Part I. Synthesis of deuterium or tritium-labelled (3S)- and (3R)-phenylalanines: stereochemical course of the elimination catalysed by L-phenylalanine ammonialyase. J. Chem. Soc. Perkin I. 2355–2364 (1972)

Wyler, H., Mabry, T.J., Dreiding, A.S.: Zur Struktur des Betanidins. Über die Konstitution des Rindenfarbstoffes Betanin. Helv. Chim. Acta *46*, 1745–1748 (1963)

Yamasaki, K., Sankawa, U., Shibata, S.: Biosynthesis of ephedrine in ephedra. Participation of C_6-C_1 unit. Tetrahedron Lett. 4099–4102 (1969)

Yamasaki, K., Tara, T., Uzawa, S., Shibata, S.: Biosynthesis of natural products VIII. Participation of C_6-C_1 unit in the biosynthesis of ephedrine in Ephedra. Phytochemistry *12*, 287–288 (1973)

Zenk, H., Müller, G.: Biosynthesis of p-hydroxybenzoic acid and other benzoic acids in higher plants. Z. Naturforsch. *19B*, 398–405 (1964)

Zulalian, J., Suhadolnik, R.J.: Biogenesis of the Amaryllidaceae alkaloids V. Caffeic acid and protocatechuic aldehyde as C_6-C_1 precursors of hemanthamine and lycorine. Proc. Chem. Soc. 422–423 (1974)

4.3 Alkaloids Derived from Tryptophan and Anthranilic Acid

D. GRÖGER

1 Indole Alkaloids

1.1 Introduction

Our knowledge of indole alkaloids has increased dramatically in the last two decades. The development in this field has been markedly influenced by the availability of modern isolation techniques (e.g., chromatography) and advanced instrumentation (e.g., mass and NMR spectroscopy). A classical paper in this respect is the mass spectroscopic investigation of the alkaloids of *Aspidosperma quebracho-blanco* Schlecht (BIEMANN et al., 1961), which paved the way for the application of this extremely useful method in alkaloid chemistry. A second major stimulus was the isolation of reserpine, one of the medically important principles of *Rauwolfia serpentina*, in the early fifties (SCHLITTLER and TAYLOR, 1960). Enormous efforts were made all over the world in searching for other indole alkaloids with valuable therapeutic properties. Now about 1200 alkaloids derived from tryptophan have been isolated, representing the largest group – about one quarter – of all known alkaloids. A comprehensive treatment of this subject here is therefore impossible. The reader is referred to the standard work in alkaloid chemistry: *The Alkaloids* (MANSKE, ed.) especially volumes VIII, XI, XIV, XV and the excellent compilation by HESSE (1964, 1968).

1.2 Structural Types of Indole Alkaloids

A generally accepted system of classification and nomenclature does not apparently exist. These problems have been discussed by HESSE (1968). Very often the stem of the alkaloid name is deduced from the name of the original plant source. According to the chromophore one may divide the indole alkaloids into seven groups (HESSE, 1968). Approximately 80% of all indole alkaloids possess an indole or indoline chromophore (Fig. 1). Most of these alkaloids (~85%) are "monomer" (Fig. 2) and contain a tertiary N-atom (N_b of the tryptamine residue). For a large group of indole alkaloids the numbering system of LE MEN and TAYLOR (1965) based on biogenetic ideas is now widely used. They distinguish the *Corynanthe/Yohimbe* class e.g., (VII), the *Aspidosperma* class e.g., (VIII) and the *Iboga* class e.g., (IX). TROJÁNEK and BLAHA (1966) divide the indole alkaloids into ten fundamental "parent" structures, and derivatives of them are designated according to the IUPAC rules. A more sophisticated system for the naming of tryptamine-

Fig. 1. Some indole chromophores

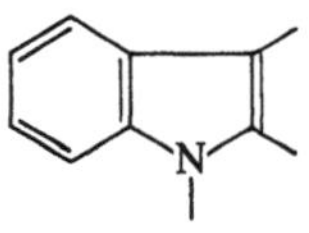

Indole
(I)

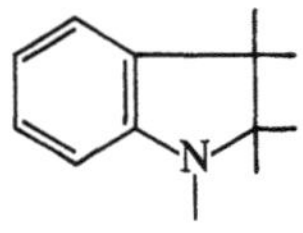

Indoline
(II)

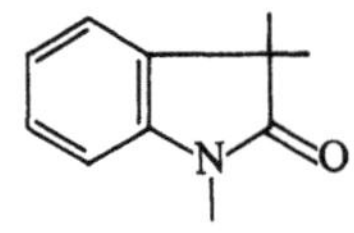

Oxindole
(III)

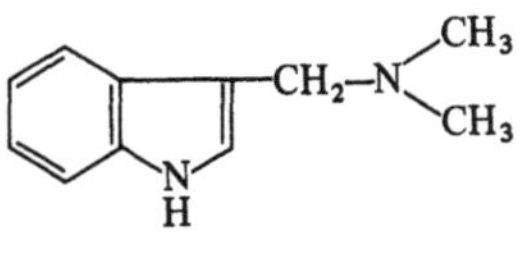

Gramine
(IV)

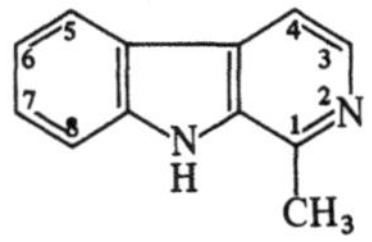

Harman
(V)

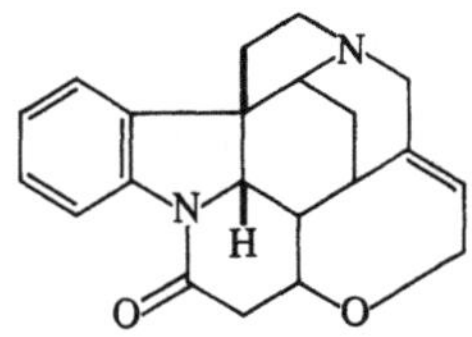

Strychnine
(VI)

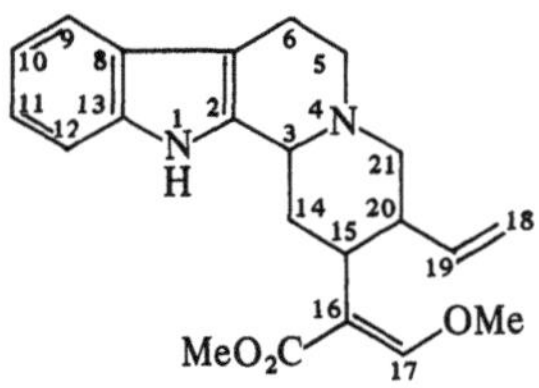

Corynantheine
(VII)

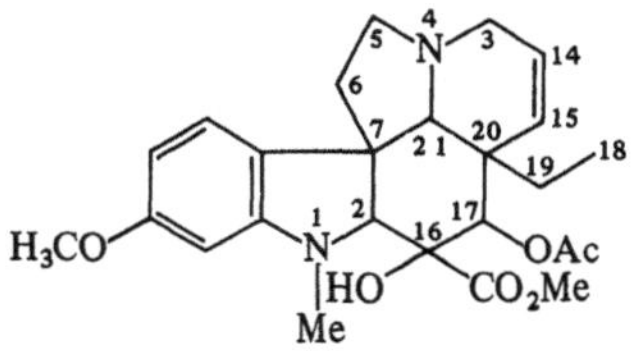

Vindoline
(VIII)

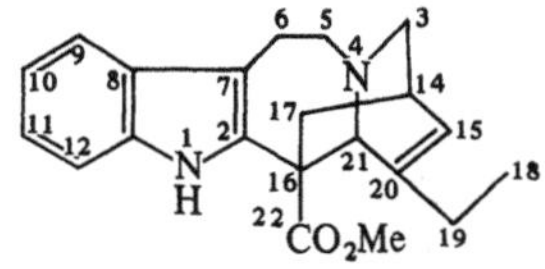

Catharanthine
(IX)

Fig. 2. Various types of monomeric indole alkaloids

monoterpene-derived indole alkaloids based on biogenetic experiments, in vitro (chemical) transformations and biogenetic hypotheses was proposed by KOMPIŠ et al. (1971). According to these authors indole alkaloids may be divided into five major classes and about 35 subgroups.

A smaller group of indole alkaloids are dimeric (Fig. 3). The so-called bis-indole alkaloids are formed from two monomeric indole units (X) which need not be identical (XI) (for review see: GARNIER et al., 1969). In some cases an indole moiety is connected with another structural type. About 100 monomeric bases serve as building blocks for 50 dimeric indole alkaloids (HESSE, 1968). The dimeric alkaloids are mainly found in the Apocynaceae and Loganiaceae.

One may roughly divide the tryptophan-derived alkaloids on a biogenetic basis: indolealkylamines, simple β-carboline bases, tryptophan-C_5-unit-derived alkaloids (e.g., ergolines) and alkaloids built up from tryptamine and a monoterpene moiety. By far the largest class is the latter one. Interestingly there are a number of tryptophan-derived bases which do not contain an indole nucleus (e.g., quinine alkaloids, calycanthine).

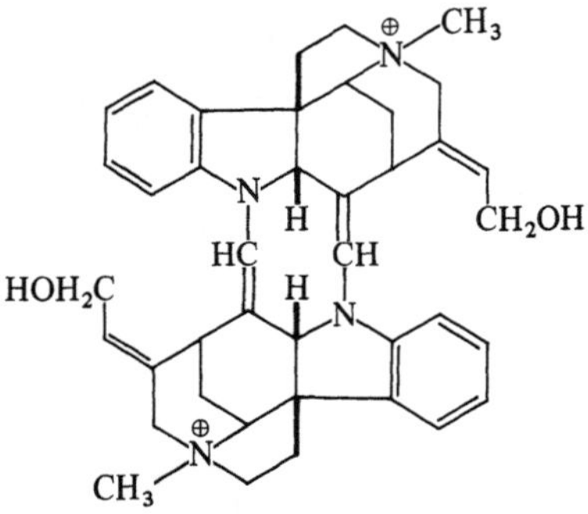

C–Toxiferine I
(X)

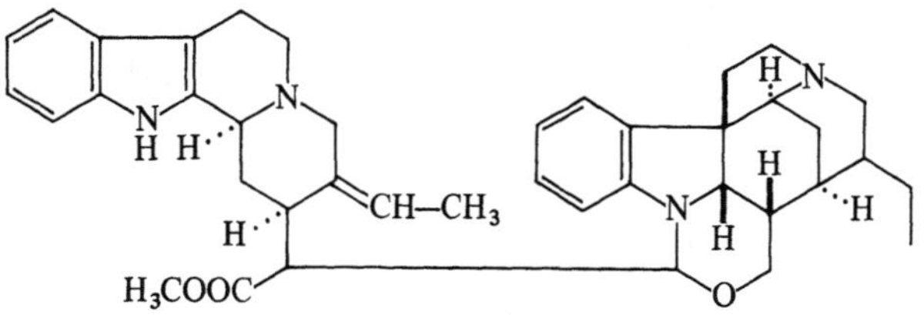

Geissospermine
(XI)

Fig. 3. Dimeric indole alkaloids

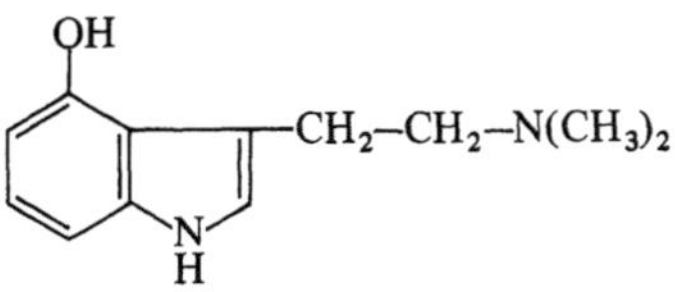

Psilocine
(XII)

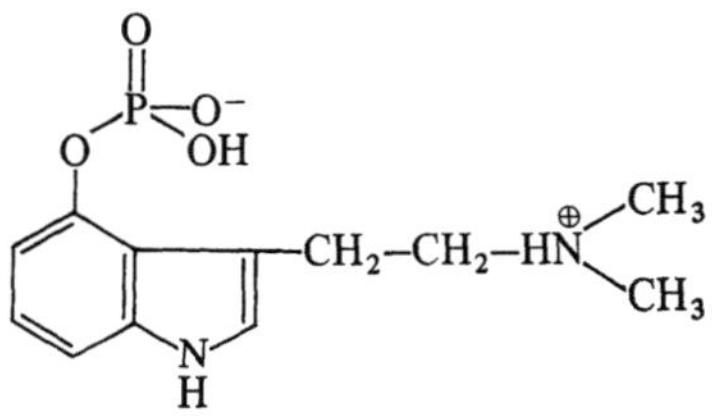

Psilocybine
(XIII)

1.3 Biochemistry of Indole Alkaloids

Indole alkaloids have been found in some groups of fungi and in flowering plants. They are not evenly distributed among higher plants. Most of the tryptamine-monoterpene-derived alkaloids are restricted to three families. Some genera are especially rich in alkaloids and a few species contain thirty or more alkaloids which belong to different structural types. Reviews on the distribution of indole alkaloids in the plant kingdom are available (WILLAMAN and SCHUBERT, 1961; SNIECKUS, 1968; WILLAMAN and LI, 1970).

A number of indole alkaloids are pharmacologically very active and are widely used in medicine, e.g., physostigmine, ergot alkaloids, *Curare-*, *Rauwolfia-*, *Vinca-*, *Catharanthus* alkaloids.

1.3.1 Simple Bases

According to the definition of HEGNAUER (1966), all the substances in this chapter are so-called protoalkaloids. These compounds contain no heterocyclic system except the indole chromophore. The chemistry of this class has been reviewed by SAXTON (1965). A compilation of the occurrence and biosynthesis of simple indole bases was given by GROSS (1969) and on a taxonomic basis recently by SMITH (1977). 5-Hydroxytryptophan is a constituent of *Panaeolus* species. N_b-Methyl-L-tryptophan (abrine) has been obtained so far only from the seeds of jequirity (*Abrus precatorius* L.). The corresponding dimethyl derivative occurs in the above mentioned plant and various *Erythrina* species. A variety of tryptamine and 5-hydroxytryptamine (serotonin) derivatives has been found in plants. In *Acacia* species tryptamine, its N_b-methyl derivative and N_b-dimethyl derivative are present. Especially interesting is the detection of the neurohormone serotonin in some edible fruits (e.g., banana, tomato, and pineapple). 5-Hydroxy-N,N-dimethyltryptamine (bufotenine) occurs in the secretions of the parotid gland of certain *Bufo* species, in some fungi (*Amanita*) and in higher plants. Gramine (3-dimethylaminomethylindole) (IV) is a constituent of sprouting barley but occurs also outside Gramineae. A "trimeric" compound di-β-indolylmethyleneindolenine was recently isolated from a strain of *Saccharomyces cerevisiae* (BUDZIKIEWICZ et al., 1972). Psilocine (XII) and psilocybine (XIII) are constituents of some mushrooms: *Psilocybe, Stropharia, Conocybe,* and *Panaeolus* species. They are 4-hydroxy-indoles which are rare in the plant kingdom. XII and XIII are the active principles of the hallucinogenic mushrooms used by Mexican Indians and were isolated for the first time by HOFMANN et al., 1959. South American hallucinogens contain either β-carbolines or various simple tryptamine derivatives (protoalkaloids). The fascinating ethnobotanical aspects of psychoactive drugs containing an indole chromophore were reviewed by SCHULTES (1976). According to feeding experiments with various labeled precursors and comparison of the specific incorporation rates the following pathway seems reasonable: tryptophan → tryptamine → N-methyltryptamine → N,N-dimethyltryptamine → XII → XIII (BRACK et al., 1961; AGURELL, 1966; AGURELL and NILSSON, 1968). In a study of the biosynthesis of another simple tryptophan-derived protoalkaloid, 5-methoxy-N,N-dimethyl-tryptamine in *Phalaris arundinacea* (BAXTER and SLAYTOR, 1972a) it was demonstrated that five of six possible pathways could lead to the endproduct. To find out the pathway which is operative in nature apparently requires studies at the enzymatic level. From the biogenetic point of view gramine (IV) is an interesting alkaloid because the side-chain nitrogen atom is separated only by one carbon atom from the indole ring compared with tryptamine (Fig. 4). It has been unambiguously established by multiple labeling that tryptophan is the precursor of (IV). Especially conclusive was the feeding of [3-^{14}C, 3-^{3}H]-tryptophan (O'DONOVAN and LEETE, 1963). The isotopes were confined with the same ratio as in the precursor in the 3-methylene group of the side-chain of the isolated gramine. The immediate precursors of gramine are (XVI) and (XVII). Both are present as "minor" alkaloids in barley and were incorporated with high efficiency by tracer experiments (GOWER and LEETE, 1963). An attractive hypothesis for the conversion of the tryptophan side-chain into gramine involving 2,3 cleavage of the amino acid side chain of an enzyme-bound tryptophan-pyridoxal phosphate intermediate (WENKERT, 1962).

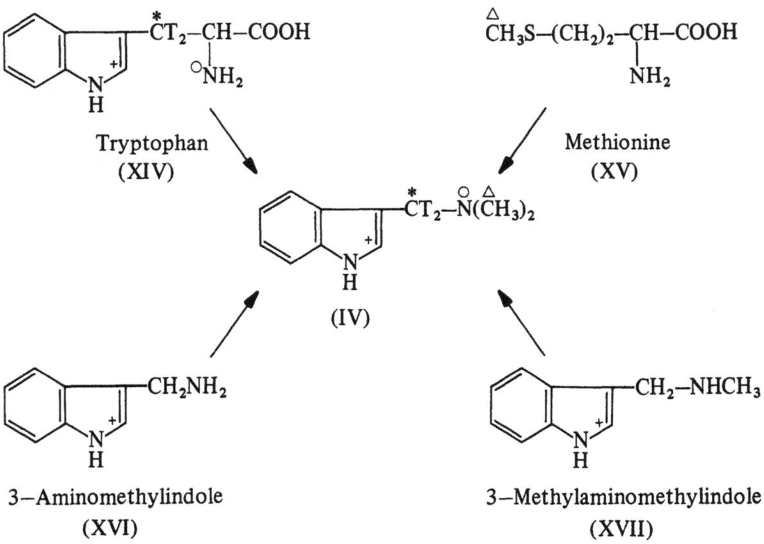

Fig. 4. Precursors of gramine

This idea is not entirely compatible with experiments of GROSS et al. (1967, 1974) who demonstrated using [3-^{14}C, 2-^{15}N]-tryptophan that the amino nitrogen of IV is derived from amino nitrogen of the precursor and could exclude glycine as intermediate in gramine biosynthesis.

1.3.2 Simple Derivatives of β-Carboline

Derivatives of harman (V) were first recognized in *Peganum harmala* L. – The so-called Syrian rue – which is found mainly in arid zones. About 15 plant families are now known to contain simple β-carboline alkaloids: e.g., Apocynaceae, Chenopodiaceae, Cyperaceae, Eleagnaceae, Gramineae, Malpighiaceae, Palmae, Passifloraceae, Rubiaceae and Zygophyllaceae. Most of the families are unrelated taxonomically. Many of the harman-alkaloids possess a methyl group in the C-1 position (V). The β-carbolines differ in the state of oxidation of their pyridine rings. Thus there are aromatic, dihydro- and tetrahydro-β-carbolines. Apparently it is unusual for all three types to occur in one plant species, an exception, however, is *Banisteria caapi* containing harmine, harmaline, and d-tetrahydroharmine. The latter compound is the only tetrahydro base in which optical activity has been noted. Some genera contain only tetrahydro-β-carbolines (*Eleagnus, Phalaris*). The red canary grass (*Phalaris arundinacea* L.) contains various 2-methyl-1,2,3,4-tetrahydrocarbolines. Besides β-carbolines, simple tryptamine derivatives and gramine were also isolated from red canary grass which is toxic to ruminant animals. Clones of this grass which possess gramine are not likely to contain N,N-dimethyltryptamine and/or β-carbolines (GANDER et al., 1976). 2-Methyl- and 1,2-dimethyl 6-methoxy-tetrahydro-β-carboline were detected for the first time in nature together with a number of tryptamine derivatives in various *Virola* species (Myristicaceae) (AGURELL et al., 1969). *Peganum harmala* which is rich in β-carbolines and quinazolines, also contains ruine which is 8-glucosyloxy-7-methoxy-1-methyl-β-carboline (NETTLESHIP and SLAYTOR, 1971). Certain more complex β-carbolines containing a fourth ring should be mentioned here on biogenetic grounds because they do not belong to the tryptophan-monoterpene-indole alkaloid group. Brevicolline, whose absolute configuration has been established (BLAHA et al., 1971), and perlolyrine, from the perennial grass *Lolium perenne* (JEFFREYS, 1970), are β-carbolines substituted in the pyridine ring. The psychoactive properties of some South American hallucinogens are based on the content of β-carbolines. The bark of *Banisteriopsis caapi* or *Banisteria inebrians* (Malpighiaceae) is used for preparing the famous drink ayahuasca, caapi, or yajé. Several *Virola* species (Myristicaceae) are the source of the hallucinogenic snuff epena or yakee of the natives and contain besides simple tryptamine derivatives tetrahydro-β-carbolines which act as monoamine oxidase inhibitors (SCHULTES, 1976).

A biogenetic relationship between tryptophan and β-carboline alkaloids is now well established. [3-^{14}C]-Tryptophan administered to root cultures of *Peganum harmala* gave radioactive harmine and harmalol (GRÖGER and SIMON, 1963). [1-^{14}C]-Tryptophan was specifically incorporated into eleagnine in *Eleagnus angustifolia* L. The label was localized at C-3 of this particular alkaloid (O'DONOVAN and KENNEALLY, 1967). STOLLE and GRÖGER (1968) have shown that the alanine side chain of tryptophan except the carboxyl group is incorporated into harmine after feeding of [3-^{14}C, 3-^{15}N]-tryptophan. Similar results were obtained after administration of [3-^{14}C, 3-^{15}N]-tryptamine. The remaining C_2 unit to complete the β-carboline skeletion is either derived from acetate or from pyruvate (O'DONOVAN and KENNEALLY, 1967; STOLLE and GRÖGER, 1968). The methyl group of the -OCH_3 group is derived by transmethylation from methionine (XV). 1-Methyl-1, 2,3,4-tetrahydro-β-carboline-3-carboxylic acid is not an intermediate in harmine biosynthesis (STOLLE and GRÖGER, 1968). This is compatible with the specific incorporation of [2-^{14}C]-acetyltryptamine into harman in *Passiflora edulis* and the nonutilization of N-acetyltryptophan (SLAYTOR and MCFARLANE, 1968). Apparently the decarboxylation of tryptophan and the acetylation of tryptamine are the first steps in β-carboline alkaloid biosynthesis.

1.3.3 Ergot Alkaloids

The ergot alkaloids are a therapeutically important group of indole alkaloids with many fascinating features. About 50 ergoline derivatives are known mostly from species of the genus *Claviceps*. Ergot fungi are mostly parasitic on rye and other wild grasses. The sclerotia of *Claviceps purpurea* (Fr.) Tulasne are designated as *Secale cornutum* in pharmacopoeias. *Claviceps purpurea* was one of the first recorded producers of myocotoxins, which were responsible for ergot poisoning (ergotism) which was known in the Middle Ages as Holy Fire.

Ergot alkaloids are 3,4-disubstituted indole derivatives and contain as a structural characteristic the tetracyclic ergoline (XVIII) ring system, or in a few cases a slight modification of it. They can be divided in two major groups: (a) amide-type derivatives of lysergic acid (6-methyl-8-carboxy-$\Delta^{9,10}$-ergolene) (XIX) and the stereoisomeric isolysergic acid (b) the so-called clavine alkaloids or clavines which are hydroxy- and dehydro derivatives of 6,8-dimethylergoline (XX) (Fig. 5).

The "classical" peptide ergot alkaloids possess a cyclic structure derived from proline, a 2-hydroxy-2-amino acid and some third amino acid (XXI). Further information on the physiology of ergot fungi and the chemical and biochemical

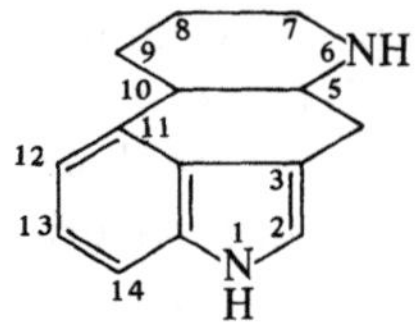

Ergoline
(XVIII)

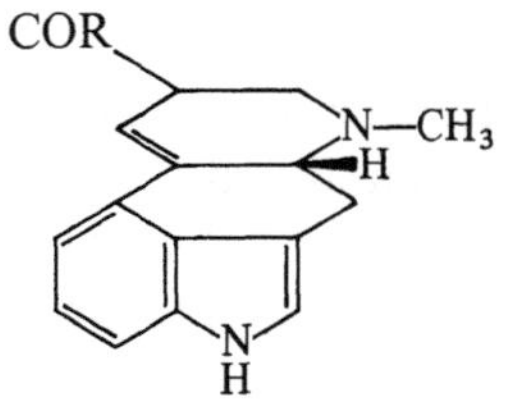

R = Tripeptide or
smaller units
Lysergic acid derivatives
(XIX)

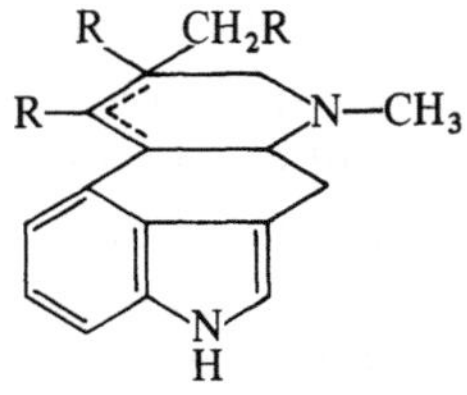

R = H = OH
Clavine alkaloids
(XX)

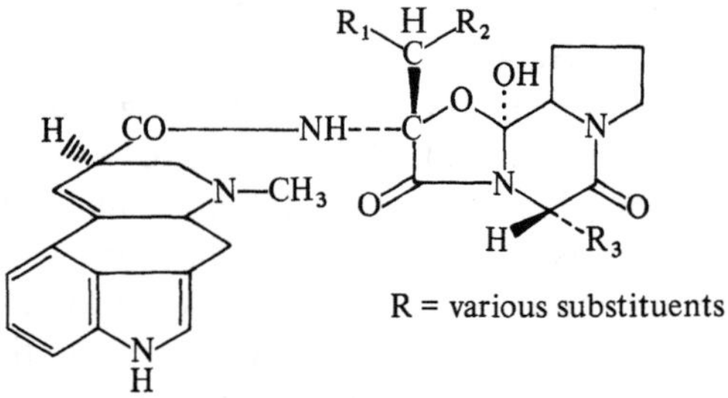

General formula of peptide
ergot alkaloids
(XXI)

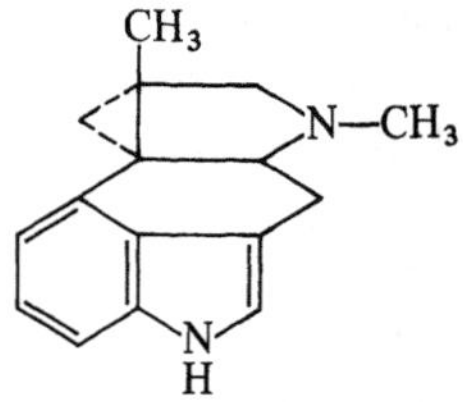

Cycloclavine
(XXII)

Fig. 5. Ergot alkaloids

features of ergolines is available in other reviews: (GRÖGER, 1972; STADLER and STÜTZ, 1975; FLOSS, 1976).

1.3.3.1 The Distribution of Ergoline Derivatives. Chemotaxonomical Problems

Ergolines have been regarded for many years as specific secondary metabolites of the Ascomycetes genus *Claviceps*, but now it is well established that ergot alkaloids are also constituents of other fungi and even higher plants. SPILSBURY and WILKINSON (1961) found festuclavine and fumigaclavine A and B in *Aspergillus fumigatus* Fres. The latter two have not yet been detected in *Claviceps*. YAMANO et al. (1962) confirmed this result and also found in different strains of *Aspergillus fumigatus* agroclavine, chanoclavine, elymoclavine, and an uncharacterized alkaloid fumigaclavine C. ABE et al. (1967) detected by chromatographic methods various clavine alkaloids in a number of *Aspergillus* and *Penicillium* species. AGURELL (1964) isolated costaclavine from *Penicillium chermesinum* Biourge. ABE et al. (1969) found the diastereoisomeric indole derivatives rugulovasine A and B and also chanoclavine-I in *Penicillium concavorugulosum*. The same group (OHMOMO et al., 1975) isolated from emerged cultures of *Penicillium roquefortii* three new ergoline derivatives roquefortine A, B, C, and festuclavine. SCOTT et al. (1976) investigated the same *Penicillium* species and isolated isofumigaclavine and roquefortine, which was identified as a diketopiperazine arising from tryptophan and dehydrohistidine.

A somewhat surprising finding was that of ergolines in some genera of the family Convolvulaceae. The active principles of the Mexican hallucinogen Ololiuqui include ergine, isoergine, and chanoclavine (HOFMANN and TSCHERTER, 1960), and to these may now be added elymoclavine, lysergol, and ergometrine (HOFMANN, 1963). Ololiuqui comprises seeds of *Rivea corymbosa* (L.) Hall. F. and *Ipomoea violacea* L. Other authors have screened many species of the Morning Glory family. The occurrence of ergolines in 28 species of the genera *Rivea, Ipomoea, Argyreia,* and *Strictocardia* was recorded (For review: CHAO and DER MARDEROSIAN, 1973). Surprisingly peptide alkaloids (ergosine and ergosinine) were isolated along with agroclavine from *Ipomoea argyrophylla* Vatke (STAUFFACHER et al., 1965). Cycloclavine (XXII) represents a new type of clavine alkaloid and was detected in *Ipomoea hildebrandtii* Vatke (STAUFFACHER et al., 1970) but has not yet been found in fungi.

The genus *Claviceps* contains about 30 species but not all have been intensively investigated. The most important for pharmaceutical purposes are *C. fusiformis* Loveless, *C. paspali* STEVENS and HALL, *C. purpurea* (Fr.) Tul., *C. fusiformis* is parasitic on wild grasses in Africa and the Far East as is *C. paspali* on *Paspalum* grasses.

The principal host for *C. purpurea* is the rye plant. *C. fusiformis* synthesizes only one type of ergoline, namely clavine alkaloids. *C. paspali* forms simple lysergic acid derivatives but only *C. purpurea* is able to synthesize all structural types of ergot alkaloids, including the peptide type. Four different chemical races have been found within the species *C. paspali* (KOBEL et al., 1964; GRÖGER, 1972). In some laboratories *C. purpurea* strains have been selected which are able to produce considerable amounts of peptide ergot alkaloids. At the present time six chemovarieties can be distinguished (GRÖGER, 1975a).

1.3.3.2 Biosynthesis of Ergot Alkaloids

In the past 20 years many attempts have been made by the use of tracer techniques and appropriate saprophytic cultures of *Claviceps* to elucidate the pathway of ergot alkaloid formation. The basic building blocks and some intermediates are now known and a first insight has been gained into the reaction mechanisms and enzymology involved. The results will be summarized here and some leading references mentioned. For a complete overview the reader is referred to the excellent compilation of FLOSS (1976). The ergoline ring system is constructed from tryptophan and mevalonic acid. L-Tryptophan (XIV) is incorporated into the ergolines intact except for the loss of the carboxyl carbon atom. An inversion of configuration takes place at the 2-carbon atom of the side chain of (XIV) (FLOSS et al., 1964). The N-methyl group of the ergot alkaloids is derived from methionine, but the exact stage at which methylation occurs is still unknown. For the first time in alkaloid biochemistry it has been shown (in the ergolines) that mevalonic acid (XXIII) can participate in the formation of alkaloids (GRÖGER et al., 1960; BIRCH et al., 1960). The isoprenylation of tryptophan (XIV → XXVI) takes place at C-4 forming 4-dimethylallyltryptophan. This is the first pathway-specific reaction. (XXVI) is specifically incorporated into clavines (PLIENINGER et al., 1962, 1967) and lysergic acid derivatives (AGURELL, 1966). The next step is the formation of the C ring of the ergoline system, this means synthesis of the obligate intermediate

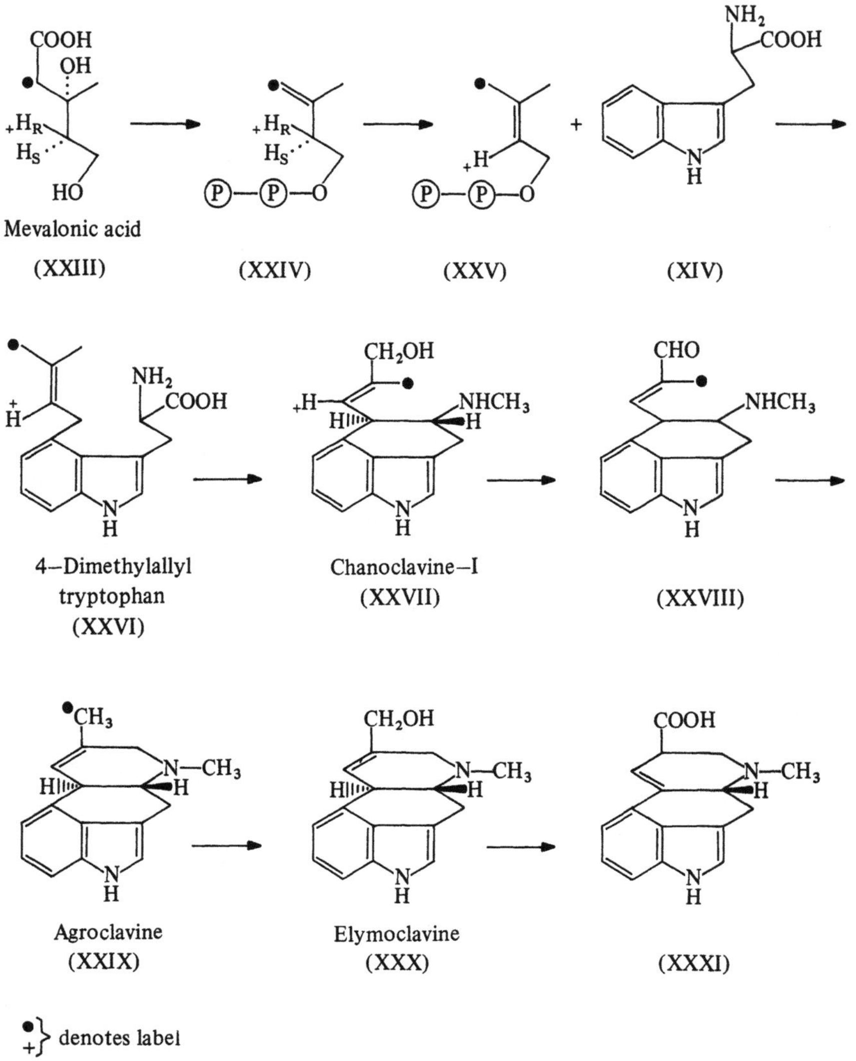

Fig. 6. Biosynthesis of the ergoline ring system

chanoclavine-I (XXVII). This alkaloid is oxidized to the corresponding aldehyde (XXVIII), which in turn is converted to the first tetracyclic ergoline alkaloid agroclavine (XXIX). The main pathway leads from agroclavine via elymoclavine (XXX) to lysergic acid derivatives (XXXI). The mechanism of the formation of rings C and D was intensively studied mainly in the laboratories of ARIGONI and FLOSS (ARIGONI, 1965; FEHR et al., 1966; FLOSS et al., 1967; FLOSS et al., 1968; SEILER et al., 1970; ABOU-CHAAR et al., 1972; FLOSS et al., 1974b; PACHLATKO et al., 1975). This was done using multiple labeled mevalonic acids, intricate precursors and sophisticated instrumentation. One surprising result was the apparently "normal" labeling of tetracyclic ergot alkaloids starting from [2-^{14}C]-mevalonic acid in the trans-carbon of the isoprene moiety. This is indeed the sequel of two isomerizations at the allylic double bond. One cis-trans isomerization takes place between

(XXVII)→(XXIX). Another cis-trans isomerization occurs between 4-dimethylallyl-tryptophan and chanoclavine-I. There is ample evidence that one of the methyl groups of the dimethylallyl residue of (XXVI) must be hydroxylated before ring C is closed; but neither E-OH-(XXVI) nor the promising Z-OH-(XXVI) (Fig. 6) are intermediates in the biosynthesis. Hence, PACHLATKO et al., 1975 postulated C-ring closure through an epoxide formed via a hydroperoxide of dimethylallyl-tryptophan (XXVI).

The biosynthesis of the peptide moiety of cyclol ergot alkaloids (XXI) is currently under investigation. (For review: GRÖGER, 1975b). Lysergic acid and the amino acids alanine, leucine, valine, phenylalanine, and proline are specifically incorporated into the appropriate constituents of the respective peptide alkaloids. The 2-hydroxy-2-amino acid residues of ergotamine and ergotoxine alkaloids originate from alanine (or a close relative) and valine, respectively. The question of whether free intermediates are involved is still unsolved. Lysergylamino acids and peptides, as well as various dipeptides and tripeptides and the corresponding diketopiperazines, are not precursors. These compounds are hydrolyzed prior to incorporation of the label. Based on these negative results and on inhibition experiments it was speculated that the synthesis of peptide ergot alkaloids is a nonribosomal process and takes place on the surface of a multienzyme complex.

1.3.4 Tryptophan-Monoterpene-Derived Alkaloids

1.3.4.1 Distribution in the Plant Kingdom

The majority of the tryptophan-derived alkaloids belong to the class of complex indole alkaloids, which are built up from tryptamine and a C_9-C_{10} unit. They are found in about eight families of vascular plants primarily in the Apocynaceae, Loganiaceae, and Rubiaceae, but also in some genera of the families Alangiaceae, Anonaceae, Euphorbiaceae, Icacinaceae and Sapotaceae. Especially rich in complex indole bases are the following genera: (a) Rubiaceae: *Adina, Corynanthe, Mitragyna, Pausinystalia, Uncaria* as well as *Cinchona* (b) Loganiaceae: *Strychnos* (c) Apocynaceae: *Aspidosperma, Catharanthus, Hunteria, Pleiocarpa, Tabernaemontana, Rauwolfia,* and *Vinca.* Alkaloids of the *Corynanthe* type occur in the Apocynaceae, Loganiaceae, and Rubiaceae, the closely related *Yohimbe*-type in the Apocynaceae, Euphorbiaceae, and Rubiaceae. The *Aspidospermine* type (*Plumerane* type) with about 250 representatives occurs in the Apocynaceae subfamily Plumerioideae. This topic has been reviewed recently by GANZINGER and HESSE (1976).

An interesting type is the tubulosine group of dimeric alkaloids which occur in the families Alangiaceae, Icacinaceae, and Rubiaceae. A tryptamine unit is here combined with a skeleton typical of *Ipecacuanha* bases.

Despite the enormous wealth of information about the constituents of some genera and species, e.g., *Catharanthus, Vinca, Rauwolfia,* our knowledge of the physiology and biochemistry of these alkaloids and their chemotaxonomic implication is still fragmentary. Apparently chemically distinct populations within a given species ("chemical races") have not been found. One reason is that suitable plant material from tropical and subtropical areas is difficult to obtain. This is especially true with regard to the ethnobotanically and medically important genus *Strychnos* (LEEUWENBERG, 1969; KRUKOFF, 1972; BISSET, 1974; BISSET and PHILLIPSON, 1976).

In South and Central America about 70 *Strychnos* species occur only half of which have been investigated. Typical constituents are monomeric and dimeric quarternary alkaloids. *Strychnos panamensis* seems to be the only species in the Western hemisphere which contains in addition to quarternary bases, strychnine, brucine, and diaboline. In Asia 44 species are recognized. In about 20 of them tertiary alkaloids were detected. *Strychnos* species characteristically contain bases like strychnine and brucine and lack dimeric indole alkaloids. In S. *nux-vomica* the main sites of location of the "normal" alkaloids are the roots and the seeds, whereas the N-methyl-sec-pseudobases are found only in the leaves. In Africa about 75 *Strychnos* species are indigenous, of which ten are more or less chemically characterized. There is considerable chemical variation. Within the genus, for example *S. camptoneura* Gilg et Busse contains serpentine and alstonine, typical Apocynaceae alkaloids, while dimeric quarternary alkaloids occur in *S. usambarensis* Gilg and *S. dolichothyrsa* Gilg ex Onichie et Hepper (ANGENOT, 1975; VERPOORTE and BAERHEIM-SVENDSEN, 1976).

1.3.4.2 Biosynthesis of Complex Indole Alkaloids

The biosynthesis of indole alkaloids has been intensively studied during the past 15 years. Remarkable progress has been achieved in our understanding of the pathways leading to the bewildering variety of different structures found in complex indole bases. The main intermediates are well known. A number of problems concerning the detailed mechanism and the relationship between the various groups of alkaloids need further investigation. The following reviews are available: BATTERSBY (1970); SCOTT (1970); ARIGONI (1972); CORDELL (1974), and should be consulted for original references.

Most of the experimental work has been done with *Catharanthus roseus* plants and seedlings because this species has many advantages for biosynthetic studies.

The incorporation of tryptophan (XIV) into all three major classes of indole alkaloids was shown by LEETE's group (LEETE, 1961; YAMAZAKI and LEETE, 1965; LEETE et al., 1965). [2-^{14}C]-Tryptophan specifically labeled reserpine and serpentine in *Rauwolfia serpentina* and vindoline (VIII) in *Catharanthus roseus*. The label from [3-^{14}C]-tryptophan was located at the predicted site in ibogaine as revealed by an unambiguous degradation.

Beside the indole nucleus, a nontryptamine portion is present in the three main types of indole alkaloids. The origin of the remaining nine or ten skeletal carbon atoms remained an intriguing question for a long time. It turned out that this "C_9–C_{10}"-unit is derived by fission of the five-membered ring of a cyclo-pentanoid monoterpene as suggested by THOMAS (1961) and WENKERT (1962). The sequence starts with mevalonic acid (XXIII) and leads via geraniol (XXXII) to the cyclopentane system (XXXIII). The fission of this ring system with rotation about the single bond gives the *Corynanthe* type (XXXIV) and further rearrangements should generate the *Aspidosperma* (XXXV) and *Iboga* skeletons (XXXVI) (Fig. 7). The ingenious experimental work to test this hypothesis was essentially done by the groups of ARIGONI, BATTERSBY, and SCOTT. First [2-^{14}C]-mevalonate was used as precursor (McCAPRA et al., 1965; GOEGGEL and ARIGONI, 1965; BATTERSBY et al., 1966a). Later on at various positions labeled "mevalonic acids" were administered. Unambiguous degradation procedures revealed that the sites

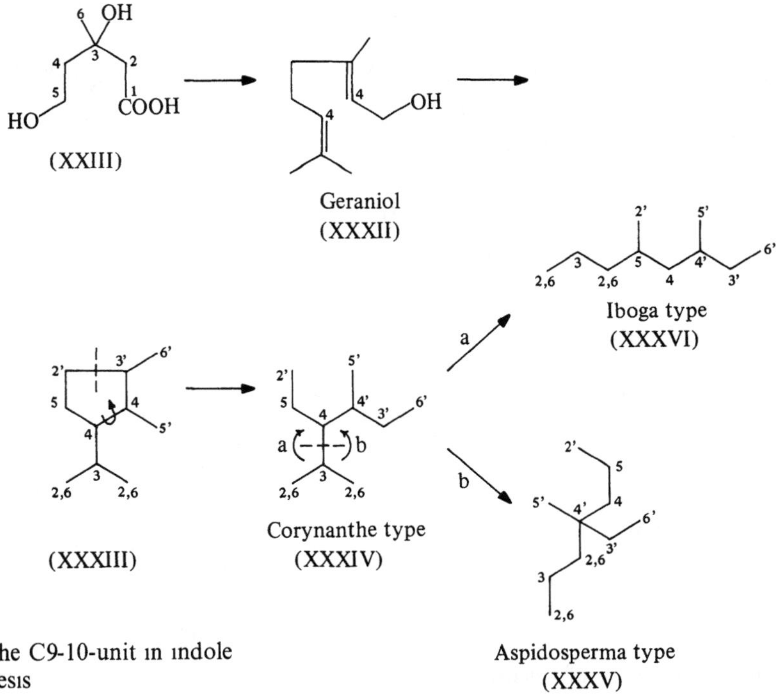

Fig. 7. Origin of the C9-10-unit in indole alkaloid biosynthesis

of labeling in alkaloids of the three main types are as outlined in Figure 7. For example, 25% of the radioactivity was found at C-22 of reserpinine, vindoline, and catharanthine. It should be noticed that also in alkaloid biosynthesis C-2 and C-6 of one mevalonate unit became equivalent as was found earlier for the monoterpene plumieride (YEOWELL and SCHMID, 1964). Also geraniol, an acyclic monoterpene, gave the expected labeling pattern.

The search for the appropriate cyclopentane system (XXXIII) involved in indole alkaloid biosynthesis was stimulated by the isolation of ipecoside (BATTERSBY et al., 1967). Actually four monoterpenes were tested and it was shown that only [O-methyl-^{3}H]-loganin gave rise to labeled ajmalicine (XLII), serpentine, catharanthine (IX), and vindoline (VIII) in *Catharanthus roseus* and labeling was solely at the ester methyl group (BATTERSBY et al., 1966b). Later on the absolute configuration of this monoterpene was established (BRECHBÜHLER-BADER et al., 1968) and [4-^{14}C]-loganin gave specifically also labeled ajmalicine (LOEW and ARIGONI, 1968). The next intermediate secologanin (XXXIX) was recognized, and found to be incorporated also into alkaloids of the *Corynanthe-*, *Aspidosperma-* and *Iboga* type as anticipated (BATTERSBY et al., 1968a). Loganin (XXXVIII) and secologanin are both found in minor quantities in *Catharanthus roseus* plants. In a crucial step the first indole alkaloid pathway specific intermediate is formed. In vitro condensation of tryptamine and secologanin afforded a mixture of two compounds, vincoside (XL) and isovincoside (Fig. 8). Both are epimers which differ in the configuration at the carbon atom 3. A mixture of [aryl-^{3}H; O-methyl-^{3}H)-vincoside and [aryl-^{3}H; 0-methyl-^{3}H]-isovincoside specifically labeled members of all three types of indole alkaloids. Furthermore vincoside (XL) bearing a hydrogen at C-3 in β-position was found to be the only precursor (BATTERSBY et al., 1968b, c). According to

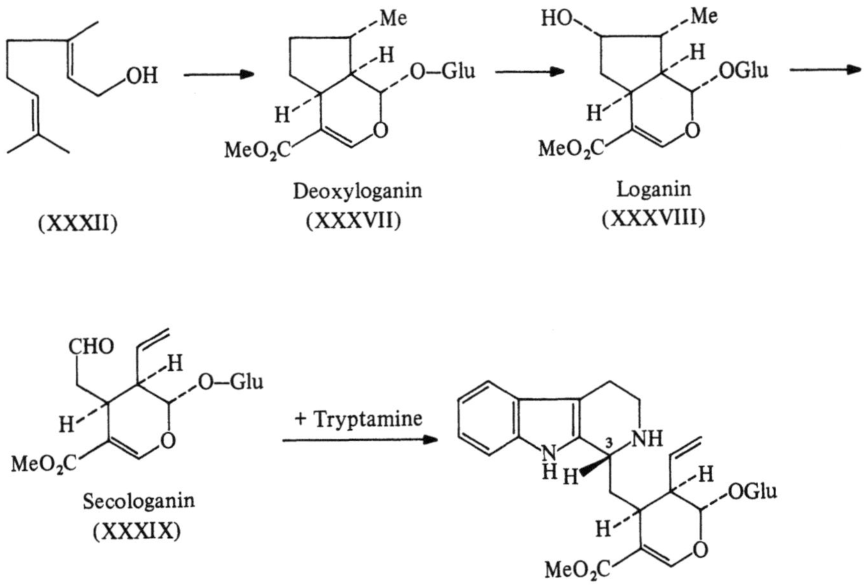

(XXXII) Deoxyloganin (XXXVII) Loganin (XXXVIII)

Secologanin (XXXIX)

Vincoside (XL)

Fig. 8. Formation of vincoside

these results an inversion must occur at that particular C-atom later on in the pathway leading to those indole bases which possess at C-3 the hydrogen in α-configuration. Opposite results were obtained recently by STÖCKIGT and ZENK (1977a). These authors did not observe an incorporation of vincoside (XL) into the three families of indole alkaloids. Using 30-day-old *Catharanthus roseus* shoots they obtained a good incorporation of isovincoside/strictosidine (XLa) into representatives of the three types of indole alkaloids. Also double label experiments revealed that isovincoside with 3α(S)-configuration is transformed into different indole alkaloids without appreciable change in the ratio of labels: it is surprising if XLa is fed in low concentrations no significant incorporation into alkaloids is observed (STÖCKIGT and ZENK, 1977a). Obviously the precursor is degraded by unspecific glucosidases. The reason for the conflicting results regarding the position of vincoside/isovincoside in indole alkaloid biosynthesis in differentiated plants is not yet clarified. Vincoside has been found also in nature (BATTERSBY et al., 1968b; SCOTT, 1970; SMITH, 1968).

Many efforts were made to clarify the steps beyond "vincoside" to the rearranged skeletons. (For review: SCOTT, 1970; SCOTT et al., 1973; PARRY, 1975). Geissoschizine (XLI) is an important intermediate and is incorporated into other *Corynanthe* type alkaloids (XL) and bases of the *Aspidosperma* and *Iboga* type (BATTERSBY and HALL, 1969; SCOTT et al., 1969). In this respect the labeling of akuammicine (XLIV) was of special interest. XLI was also found in *C. roseus.*

According to these results, available data from other tracer experiments and investigation of the sequential formation of alkaloids in germinating seeds of *C. roseus* (SCOTT, 1970) the following pathway seems to be well established: *Corynanthe*→*Corynanthe*/*Strychnos*→*Aspidosperma*→*Iboga* type alkaloids (Fig. 9a).

It is also interesting to note that also *Cinchona* alkaloids of the quinoline type possessing a further quinuclidine part, e.g., quinine are derived from tryptophan and the monoterpene loganin (LEETE and WEMPLE, 1969; BATTERSBY and HALL, 1970; BATTERSBY and PARRY, 1971). Firstly the *Corynanthe* sekeleton is formed and corynantheal is rearranged into ketoquinoline bases via cinchonaminal.

1.3.5 Regulation and Enzymes Involved in Indole Alkaloid Biosynthesis

Our knowledge about the regulation of the pathways leading to indole alkaloids is still fragmentary. This holds true also for other alkaloid classes. FLOSS et al. (1974a) gave an overview on control mechanisms in alkaloid biosynthesis.

A pyridoxal phosphate-dependent tryptophan decarboxylase was detected in *Phalaris arundinacea* seedlings, which reached a sharp maximum after four days (BAXTER and SLAYTOR, 1972b). The decarboxylation was inhibited at concentrations of 0.1 M by a number of indole derivatives including N,N-dimethyltryptamine one of the two major alkaloids of this plant. Apparently this alkaloid inhibits its own synthesis through a feedback mechanism.

A barley shoot extract catalyzes the conversion of 3-aminomethylindole and 3-methylaminomethylindole in the presence of S-adenosylmethionine into gramine (MUDD, 1961). Furthermore a 104,000 g supernatant fraction of a homogenate of 7-day-old barley shoots catalyze the conversion of [3-^{14}C]-tryptophan into gramine (BRECCIA et al., 1966). Some results of GROSS et al. (1970) seem to indicate that gramine inhibits its own biosynthesis by a feedback mechanism.

The first pathway-specific enzyme of ergot alkaloid biosynthesis, the dimethylallylpyrophosphate: L-tryptophan dimethylallyltransferase (DMAT synthetase) was found and characterized in the laboratory of H.G. FLOSS (HEINSTEIN et al., 1971; LEE et al., 1976). The DMAT synthetase is inhibited by the endproducts agroclavine and elymocalvine in the *Claviceps* strain SD 58. An enzyme system which catalyzes the ring closure of chanoclavine-I and chanoclavine-I-aldehyde into agroclavine was investigated by ERGE et al. (1973). The activities of DMAT synthetase and chanoclavine-I-cyclase increased rapidly during the transition of the trophophase to the idiophase and precedes alkaloid formation by about one day. Both enzymes are absent in nonalkaloid-producing *Claviceps* strains. Interesting to note that in particular *Claviceps* strains and under special environmental conditions tryptophan and some analogs of this amino acid act as inducer of alkaloid formation (FLOSS et al., 1974a).

Some progress has been also made in the class of tryptophan-monoterpene-derived alkaloids. MEEHAN and COSCIA (1973) isolated from *Catharanthus roseus* seedlings a microsomal mixed function oxidase which converts the monoterpene alcohols geraniol and nerol into their corresponding 10-hydroxy derivatives. This membrane-bound cytochrome P-450-dependent monooxygenase is inhibited by an endproduct indole alkaloid catharanthine, suggesting feedback control of this pathway (MCFARLANE et al., 1975) because both 10-hydroxy monoterpenes are good precursors for loganin. An S-adenosyl-L-methinonine-loganic acid methyltransferase was also found in leaves and shoots from *C. roseus*. Loganic acid and secologanic acid are equally good substrates (MADYASTHA et al., 1973). Vindoline exerts significant inhibition on an 18-fold purified preparation of this methyl transferase.

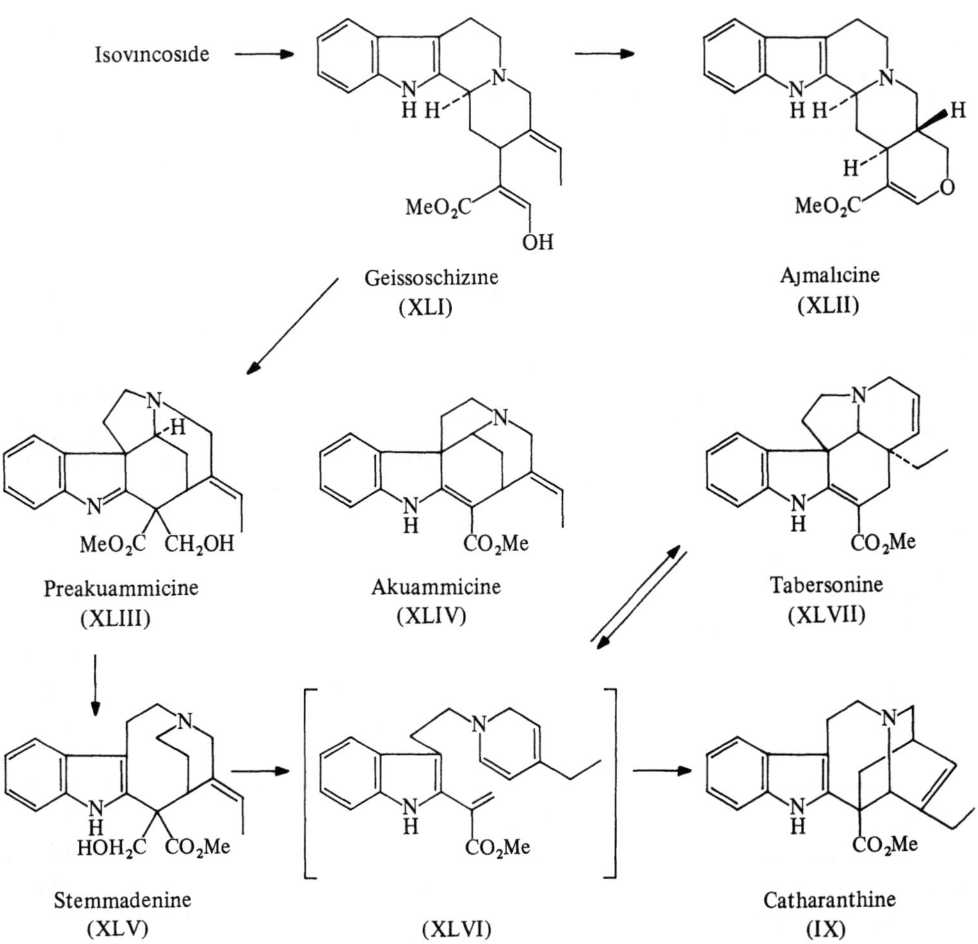

Fig. 9a. Formation of *Corynanthe-*, *Aspidosperma-*, *Iboga*-type alkaloids

Crude enzyme preparations of *C. roseus* seedlings or tissue cultures (callus) were able to convert a mixture of tryptamine and secologanin into ajmalicine (XLII) and geissoschizine (XLI). Geissoschizine as substrate was converted into ajmalicine. The enzymes are present in the 37,000 g supernatant fraction (SCOTT and LEE, 1975). STÖCKIGT et al. (1976) prepared an enzyme extract from cell suspension cultures of *C. roseus* and could show the conversion of tryptamine plus secologanin in the presence of NADPH into ajmalicine, 19-epi-ajmalicine and tetrahydroalstonine. If the incubation was performed in the absence of pyridine nucleotides a new alkaloid 20, 21-dihydroajmalicine (cathenamine) is accumulated. Cathenamine is apparently an obligate intermediate in the biosynthesis of ajmalicine and related indole alkaloids (STÖCKIGT et al., 1977). For a long time it was assumed that vincoside with 3β(R)configuration is the key intermediate in *Corynanthe* type alkaloid biosynthesis. Recently it was found (STÖCKIGT and ZENK, 1977b) that cell-free extracts of *Catharanthus* and also an enzyme preparation from *Rhazia stricta* cells catalyze the formation of isovincoside [3α(S)] = strictosidine after incubation of a mixture of tryptamine and secologanin (XXXIX). The addition of D-δ-glucono-

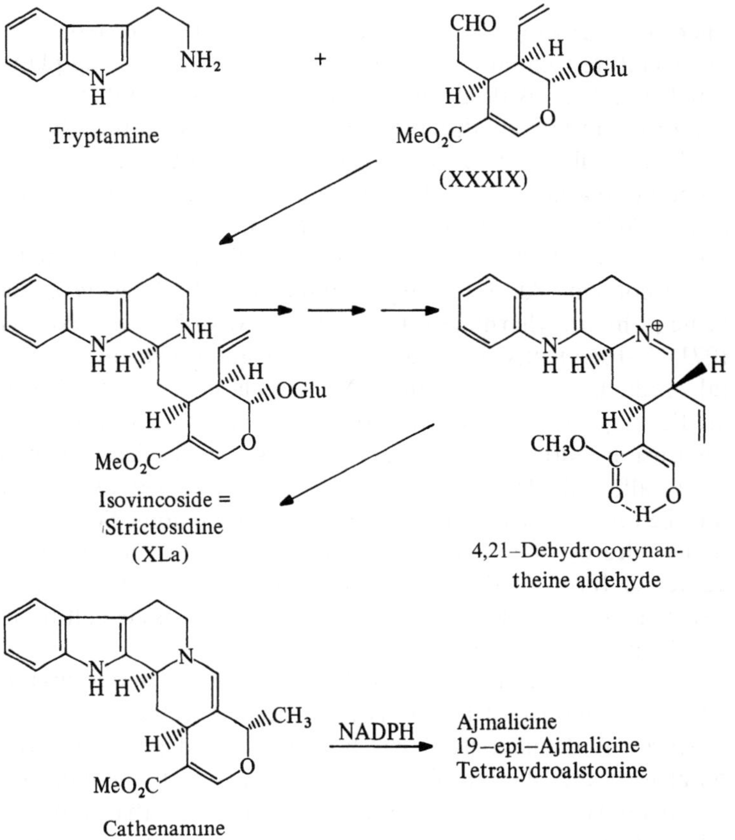

Fig. 9b. Biogenetic pathway of *Corynanthe*-type alkaloids

lactone, an inhibitor of glucosidases, to the enzyme preparation seems to be essential. Isovincoside (XLa) was then used as substrate and converted by a crude enzyme preparation from *C. roseus* to cathenamine and in the presence of reduced nucleotides in a yield of 14% into all three ajmalicine isomers. These results show unambiguously that isovincoside rather than vincoside is the true intermediate in indole alkaloid biosynthesis. For the widely distributed enzyme which catalyzes the condensation of tryptamine and (XXXIX) to yield isovincoside the name strictosidine synthetase was proposed (Stöckigt and Zenk, 1977a). As an intermediate between isovincoside and cathenamine 4,21-dehydrocorynantheine aldehyde might be involved (Stöckigt et al., 1978). The biosynthetic pathway of *Corynanthe* alkaloids based on in vitro studies with cell free systems is depicted in Figure 9b. Further attempts were made to purify the "ajmalicine synthetase" from *Catharanthus roseus* seedlings and callus extracts by Scott et al. (1977). A protein peak at 55,000 daltons was capable of synthesizing ajmalicine. This "ajmalicine synthetase" rapidly lost its ability to synthesize the alkaloid, but retained β-glucosidase activity. These promising results will allow the study of indole alkaloid biosynthesis on the enzymatic level in greater detail.

1.3.6 Indole Alkaloid Formation in Plant Tissue Cultures

There is a growing interest in the production of secondary metabolites by plant tissue culture techniques. Some indole alkaloid-producing plants have been investigated and the alkaloid spectra found in tissue cultures are often qualitatively and/or quantitatively different from those found in the whole plant.

Tissue cultures (callus) of *Peganum harmala* were able to form harmine but no other β-carboline nor the main alkaloid of the intact plant, vasicine (REINHARD et al., 1968). According to NETTLESHIP and SLAYTOR (1974) *P. harmala* callus cultures produced traces of harman-type alkaloids on a medium with a reduced 2,4-D content. After feeding harmine to a *P. harmala* callus the formation of ruine (8-hydroxyharmine-β-D-glucoside) was observed (NETTLESHIP and SLAYTOR, 1971). Subsequently the alkaloids harmol and dihydro-ruine were detected as natural products in callus cultures (MCKENZIE et al., 1975). Harman and norharman are the major products of the biotransformation of tryptophan in *Phaseolus vulgaris* suspension cultures (VELIKY and BARBER, 1975). The in vitro culture of root and stem callus of *Rauwolfia serpentina* was described by MITRA and KAUL (1964) but the reported detection of reserpine is ambiguous. Several cell lines of *Rauwolfia* tissue cultures, obtained after treatment with nitrogen mustard, showed an enhanced ability to form alkaloids compared with the untreated material (SHAMINA et al., 1974). Traces of reserpine appeared also in the tissue culture of *Alstonia constricta* F. Muell: (CAREW, 1965).

The tissue culture work with *Catharanthus roseus* was reviewed by CAREW (1975). A number of alkaloids have been detected by several groups but mostly in traces. A detailed study has been performed by PATTERSON and CAREW (1969) on growth and alkaloid formation in *C. roseus* tissue cultures. They found about 15 alkaloids, but surprisingly no vindoline or catharanthine which are the main alkaloids of the intact plant. Recently DÖLLER et al. (1976) reported the occurrence of serpentine as the main alkaloid and traces of raubasine in five different callus cell lines of *C. roseus*. On the dry weight basis 0.5% of alkaloid was present under favorable conditions.

For the first time the production of substantial amounts of serpentine and ajmalicine by cell suspension cultures of *Catharanthus roseus* was achieved by ZENK et al. (1977). Their strategy to develop a suitable cell line was the following: (a) callus was initiated from a high-yielding differentiated plant (b) a selection of variant cell strains capable of high alkaloid synthesis was performed (c) a special medium for alkaloid production was developed. The production medium consists of a basal medium with the addition of IAA, benzyladenine, 5% sucrose and 0.05% L-tryptophan. Based on this procedure, strains with a remarkably high yield of serpentine and ajmalicine equal to 1.3% of the total cell dry weight were selected, the highest alkaloid content which was ever observed in a given cell suspension culture. Most interestingly the alkaloid content was stimulated threefold by the addition of L-tryptophan. Variant strains were obtained: S-strain, yielding solely serpentine (162 mg/L) and the A-strain, yielding 77 mg/L serpentine plus 264 mg/L ajmalicine. The latter seems to be less stable than the S-strain. The growth and alkaloid production pattern during the course of fermentation in shake flasks and in an airlift fermenter were measured. Whilst in shake flasks growth and production parallel each other closely, in the batch culture a clear distinction between a growth phase and an idiophase was recorded. Various physiological aspects of alkaloid production by *Catharanthus roseus* as well as all general problems involved in secondary metabolite formation by cell suspension cultures were discussed recently (ALFERMANN and REINHARD, 1978).

2 Alkaloids Derived from Anthranilic Acid

2.1 Introduction

Beside some protein amino acids, nicotinic acid and anthranilic acid (XLVIII) may serve as nitrogen-containing precursors for alkaloid biosynthesis. Usually the amino acids are decarboxylated in the formation of N-heterocycles but this is not true with alkaloids derived from anthranilic acid. In general the carboxyl group is involved in the cyclization step or the carboxyl group is retained during the course of alkaloid formation. Anthranilic acid is a compound of the shikimate-chorismate metabolism and the first intermediate in the tryptophan branch of the aromatic amino acid pathway. Furthermore, (XLVIII) is a degradation product of tryptophan and is formed via formylkynurenine and kynurenine. During the course of tryptophan formation anthranilic acid is decarboxylated. Apparently, anthranilic acid derived directly from chorismic acid (with the intact carboxy group) acts as the alkaloid precursor rather than anthranilic acid formed from the degradation of tryptophan. Various ring systems in higher plants, microorganisms, and even animals are synthesized with direct participation of anthranilic acid. Some aspects of this topic will be discussed. For a more detailed description the reader is referred to a review by GRÖGER (1969).

2.2 Protoalkaloids

Numerous volatile oils contain traces of anthranilic acid as well as N-methylanthranilic acid methyl ester. A trimethylated derivative of 3-hydroxyanthranilic acid (IL) is damascenine (L), hitherto found only in the genus *Nigella* and the only alkaloid detectable in *Nigella damascena* L. The papillate cells of the seed epidermis are the specific storage sites of damascenine. In many respects "damascenine cells" resemble cells in which volatile oil is stored (MUNSCHE, 1964). [$^{14}CH_3$]-Methionine is an efficient precursor for all three variously bound methyl groups. The N-methyl group was always several times more heavily labeled than the O-methyl or the ester CH_3 group. The aromatic ring stems from anthranilic acid as was shown by tracer experiments especially with [U-^{14}C]-shikimic acid and [carboxy-^{14}C, amino-^{15}N]-anthranilic acid (MUNSCHE, 1966, Fig. 10). Apparently the anthranilic acid is directly hydroxylated in position 3 and in turn methylated. An anthranilic acid oxidase catalyzing the formation of 3-hydroxyanthranilic acid (IL) was detected in the leaves of *Tecoma stans* (NAIR and VAIDYANATHAN, 1964).

2.3 Quinoline Alkaloids

Quinoline derivatives are widely distributed in nature. It has been found that this particular ring system can be elaborated in vivo by various metabolic pathways.

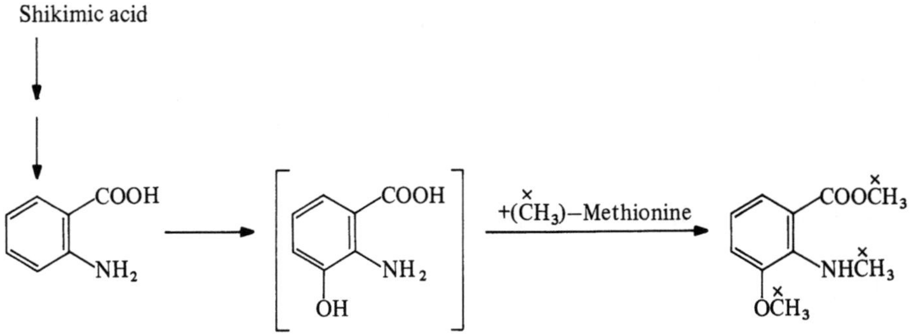

Anthranilic acid (XLVIII) — 3-Hydroxy-anthranilic acid (IL) — Damascenine (L)

Fig. 10. Biosynthesis of damascenine

From a biogenetic point of view the quinolines comprise a heterogenous group of natural products. Kynurenic acid and its derivatives arise as a result of tryptophan degradation in mammals and *Pseudomonas* species. The *Cinchona* alkaloids and some other alkaloids like perloline originate from tryptophan. Another route leading to the quinoline starts from anthranilic acid. This subject has been reviewed by LUCKNER (1969). Especially rich in anthranilic acid-derived quinolines are the Rutaceae. In this family these compounds are often associated with acridines, benzyltetrahydroisoquinolines, imidazole derivatives, oxazoles, and quinazolines. Structurally the quinolines from Rutaceae plants show great variability. They are found in many genera of the sub-families Rutoideae, Toddalioideae, and Aurantioideae. An excellent review on alkaloid distribution in the Rutaceae and their systematic significance was given by WATERMAN (1975). Ten 2-quinolones were found in seven genera mostly in the Asian taxa of *Haplophyllum.* About twenty 3-dimethylallyl-2-quinolones are known especially in the genera *Lunasia* and *Ptelea.* The largest group are the furoquinolines, especially well represented are skimmianine (LIV) (detected in 88 taxa from 37 genera); dictamnine (LIII) (40/23); kogusaginine (33/20) and γ-fagarine (22/14). Furoquinolones and linear and angular pyranoquinolones were also recorded, as were 2-alkyl- and 2-aryl quinolines and the corresponding 4-quinolones. The biosynthesis of furoquinolines has been extensively investigated, especially in *Choisya, Dictamnus, Fagara,* and *Ruta* species. The quinoline moiety is derived from anthranilic acid (carbon atoms 4–9 and 12, and nitrogen) and acetate. The carbon atom 10 stems from the carboxylic group of acetate (MONKOVIC and SPENSER, 1966; MATSUO et al., 1966). Label from [4-^{14}C]- and [5-^{14}C]-mevalonic acid appears specifically at C-2 and C-3 respectively of the furan ring. Interestingly, [^{14}C]-acetate was incorporated solely into the quinoline part of dictamnine in *Fagara japonica* but not into those carbon atoms which are derived from mevalonic acid (COLONNA and GROS, 1971). Experiments with potential intermediates to clarify the biosynthetic sequence starting from 4-hydroxy-2-quinolone via 4-methoxy-3-prenylquinoline (LI) to various furoquinoline alkaloids were performed by COBET and LUCKNER (1971) and GRUNDON et al.,1974, 1975; Fig. 11). The production of alkaloids with cell cultures of *Ruta graveolens* was recorded by STECK et al. (1971). Increased yields of alkaloids through precursor transformations were later described (STECK et al., 1973): 4-Hydroxy-2-quinoline

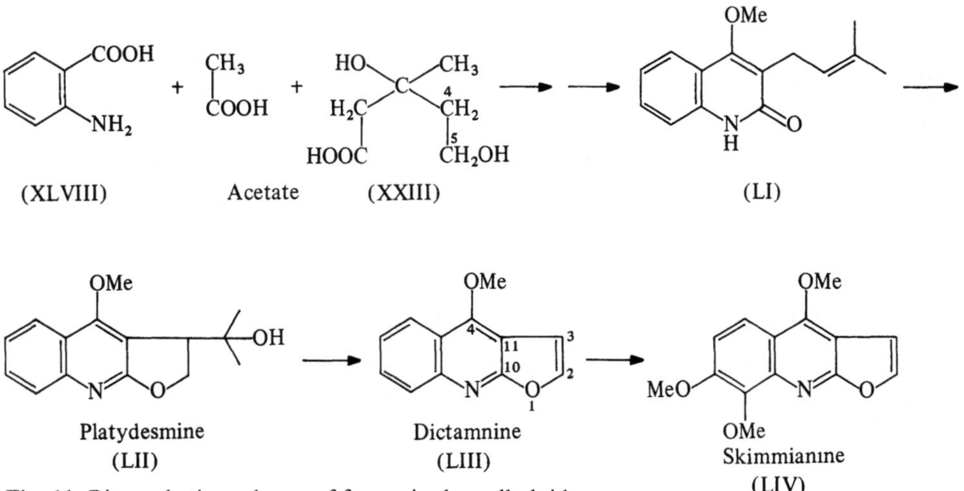

Fig. 11. Biosynthetic pathway of furoquinoline alkaloids

was efficiently converted to dictamnine and γ-fagarine which is not normally detectable in the cells. The alkaloid edulinine originated from administered 4-hydroxy-N-methyl-2-quinolone.

Some *Penicillium* species, especially *P. cyclopium* Westling and *P. viridicatum* Westling synthesize the quinoline alkaloids viridicatin (LIX) and viridicatol (LX). In a first step anthranilic acid, S-phenylalanine and the CH_3-group of methionine form cyclopeptine (LV) which is converted by a dehydrogenase to dehydrocyclopeptine (LVI). This compound serves as substrate for an epoxidase forming cyclopenin (LVII) which in turn might be hydroxylated by an m-hydroxylase to cyclopenol (LVIII). These benzodiazepine compounds are converted to the quinoline derivatives LIX and LX (Fig. 12). From *Penicillium* strains which predominantly produce the quinoline alkaloids, an enzyme cyclopenase was obtained which catalyzed the described rearrangement. The cyclopenase is localized in the conidiospores. Most work on biosynthesis and enzyme characterization of *Penicillium* alkaloids was done in the laboratory of LUCKNER. Also various problems of regulation processes and the organization of differentiation programs with regard to secondary metabolism were successfully elaborated (LUCKNER et al., 1977).

Certain strains of *Pseudomonas aeruginosa* form 2-n-alkyl-4-hydroxy-quinolines (pseudans) which differ in the length of the aliphatic side chain in position 2. Pseudans have also been found in some genera of the Rutaceae. Recently (REISCH et al., 1975) isolated 2-(n-undecyl)quinolone-4 from flowers of *Ptelea trifoliata* and roots of *Ruta graveolens*. Biosynthetically the pseudans are formed from anthranilic acid and acetate units. Apparently a β-ketoacid-CoA-ester corresponding to a given pseudan reacts with "activated" anthranilic acid. According to experiments with [^{14}C]-malonate the side-chain is build up from the terminal C-atoms (RITTER and LUCKNER, 1971).

2.4 Quinazolines

About 60 natural compounds with a quinazoline skeleton have been found in higher plants, microorganisms, and even animals. Several structural types may

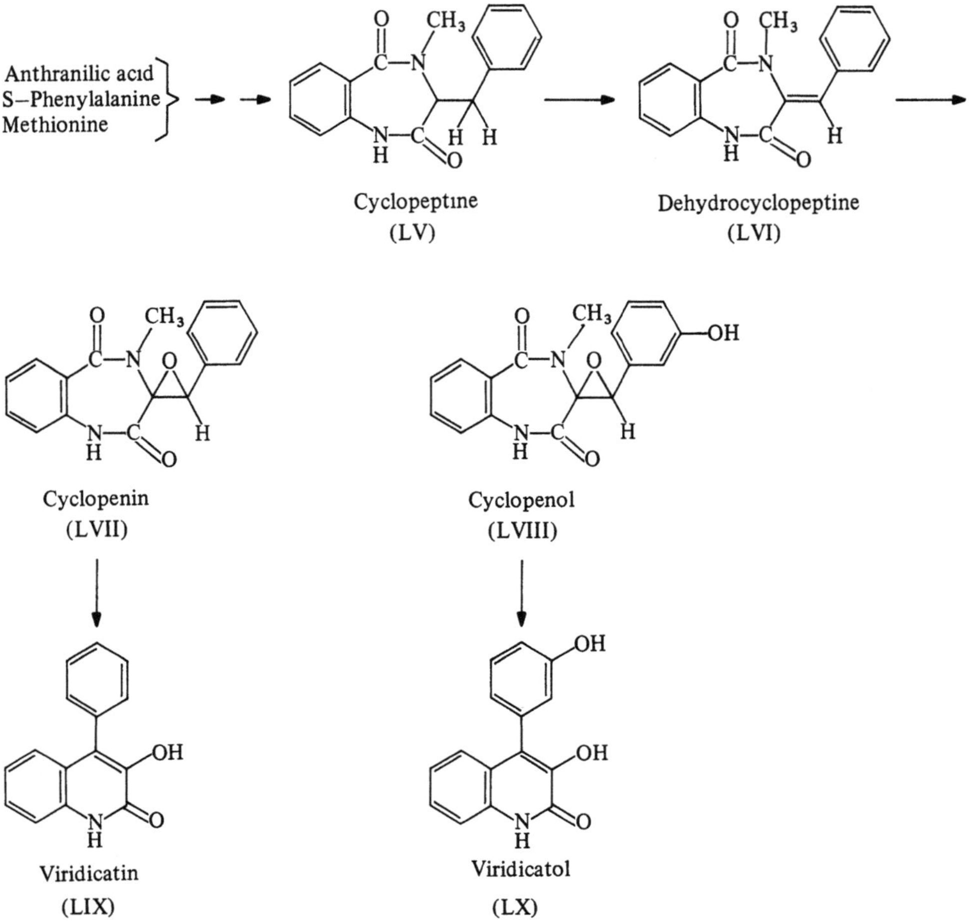

Fig. 12. Formation of *Penicillium* alkaloids

be distinguished. A number of quinazoline alkaloids occur in taxonomically unrelated plant families: Acanthaceae (*Adhatoda, Anisotes*), Araliaceae, (*Mackinlaya*); Leguminosae-Papilionatae (*Galega*); Malvaceae (*Sida*), Palmae (*Daemonorops*); Rutaceae (*Euxylophora, Evodia, Glycosmis, Hortia, Zanthoxylum*), Saxifragaceae (*Dichroa, Hydrangea*), Scrophulariaceae (*Linaria*) and Zygophyllaceae (*Peganum*). A review on naturally occurring quinazolines has been given by JOHNE and GRÖGER (1970).

A number of simply substituted derivatives of quinazol-4-one have been isolated from the genera *Dichroa, Hydrangea* and *Glycosmis*. Glycorine (1,4-dihydro-1-methyl-4-oxo-quinazoline) (LXI), arborine (LXII) and some related alkaloids are constituents of *Glycosmis arborea*. Febrifugine (LXIII) and isofebrifugine have been isolated from *Dichroa* species of Chinese origin. Much work has been done on these alkaloids because they exhibit strong anti-malarial activity, but hitherto they have not been used in medical practice. Glomerine (LXIV) and homoglomerine are the defensive agents of the millipede *Glomeris marginata*.

The pyrrolidino-quinazolines have been detected in six plant families: Acanthaceae, Leguminosae-Papilionatae, Malvaceae, Palmae, Scrophulariaceae, and Zygophyllaceae. The most common occurring alkaloid of this group is peganine-vasicine (LXV) which is found as the principal alkaloid in *Adhatoda vasica*, *Anisotes sessiliflorus*, *Peganum harmala* and a number of *Linaria* species. In *Anisotes* (ARNDT et al., 1967) and in young plants of *Adhatoda vasica* (JOHNE et al., 1971) minor alkaloids were found which possess an aromatic substituent, in position 3 of the pyrrolidino-quinazoline system e.g., anisotine (LXVI), which may be biosynthetically derived from anthranilic acid.

The β-carbolino-quinazolones are restricted to a few genera of the Rutaceae. Evodiamine (LXVII) and rutaecarpine are present in the fruits of *Evodia rutaecarpa*. The alkaloids of *Euxylophora paraënsis* Hub. have been investigated by DANIELLI et al. (1972) who isolated paraensine, the first indolopyridoquinazoline to be found which contains an isoprenoid moiety. From species of *Mackinlaya* (Araliaceae) 6,7,8,9-tetrahydropyrido(2,1-b)-quinazolines, e.g., (LXVIII) have been isolated (JOHNS and LAMBERTON, 1965). Quinazoline alkaloids have been found together with acridine and furoquinoline bases in the genus *Glycosmis*. In *Peganum* species peganine and harman-type alkaloids both occur and in *Sida* (GHOSAL et al., 1975) peganine, hypaphorine and ephedrine are present. Several 4-methylquinazolines were found in the culture broth of *Pseudomonas aeruginosa* when tryptophan was present as the nitrogen source (MANN, 1967) (Fig. 13).

The biosynthesis of pyrrolidino-quinazolines has been intensively investigated (JOHNE and GRÖGER, 1970). Anthranilic acid including the nitrogen atom is specifically incorporated into (LXV). There is still some controversy as to the origin of the "nonanthranilic acid moiety" of peganine. In the author's laboratory it was found that in *Adhatoda vasica* C-1, C-2, and N-11 of peganine may derive from aspartic acid and carbon atoms C-3 and C-10 of (LXV) from a C_2-unit (WAIBLINGER et al., 1972; JOHNE et al., 1973). Members of the glutamic acid family were unspecifically incorporated. LILJEGREN (1968, 1971) on the other hand found that in *Peganum harmala* labeled ornithine, proline, putrescine, and related compounds are more or less specifically incorporated into the pyrrolidino-ring system. The existence of two different pathways leading to peganine is still an open question. Arborine (LXII), the main alkaloid of *Glycosmis arborea*, is elaborated from phenylalanine and anthranilic acid (GRÖGER and JOHNE, 1968). The millipede alkaloid glomerine (LXIV) is also derived from [carboxy-^{14}C]-anthranilic acid (SCHILDKNECHT and WENNEIS, 1967). Japanese authors (YAMAZAKI and IKUTA, 1966; YAMAZAKI et al., 1967) clarified the biosynthesis of *Evodia* alkaloids. Tryptophan was found to be specifically incorporated into the tryptamine moiety of (LXVII) and rutaecarpine. The carbon atom C-3 and the N-CH_3 group derive from (CH_3)-methionine and the other aromatic portion from anthranilic acid. Experimental evidence is still lacking for the possible involvement of a β-carboline in *Evodia* alkaloid biosynthesis.

2.5 Acridine Alkaloids

The occurrence of acridine alkaloids seems to be limited to some genera of the Rutaceae. About 30 acridones have been found, notably in the following genera: *Acronychia*, *Atalantia*, *Balfourodendron*, *Evodia*, *Fagara*, *Glycosmis*, *Melicope*, *Ruta*, and *Teclea*. Other natural products with the skeleton of acridine (LXIX) have

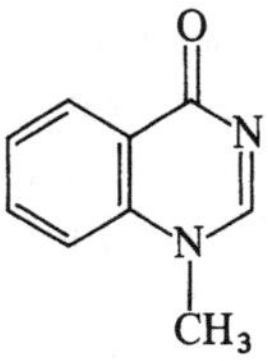

Glycorine
(LXI)

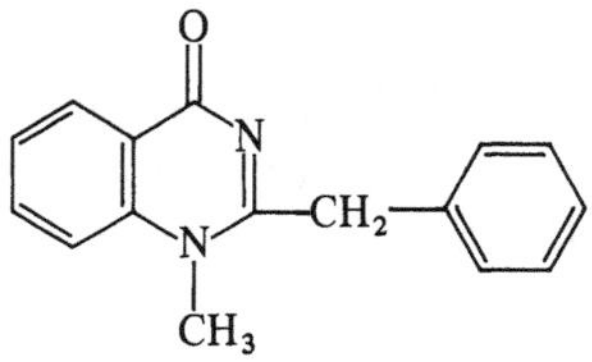

Arborine
(LXII)

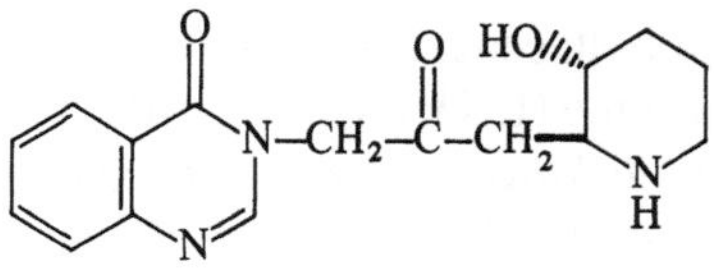

Febrifugine
(LXIII)

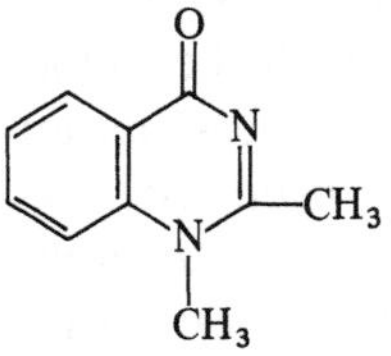

Glomerine
(LXIV)

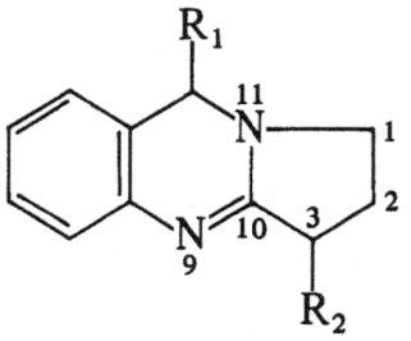

$R_1 = H_2$, $R_2 = OH$; Peganine
(LXV)

$R_1 = O$; $R_2 =$ 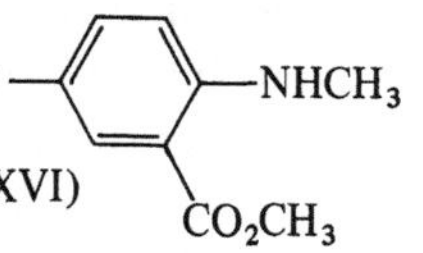

(LXVI)

Anisotine

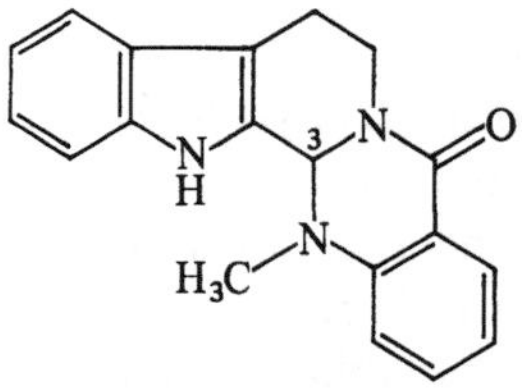

Evodiamine
(LXVII)

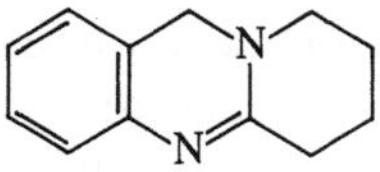

(LXVIII)

Fig. 13. Some structural types of quinazoline alkaloids

apparently not been isolated either from microorganisms or from other higher plants. The acridones are very weak bases which are yellow in colour. Some representatives have been listed in Figure 14. The methyl group in the 1 position can easily be eliminated by treatment with mineral acid. It has been a matter of some debate whether the nor-compounds are genuinely present in the plants or whether they only arise during isolation, but it seems quite certain that some e.g., arborinine (LXX) and evoprenine are true natural products. The remarkable anti-tumor activity of acronycine (LXXI) should be mentioned. Chlorine-containing alkaloids (LXXII) were isolated from the roots of *Ruta graveolens* (REISCH et al., 1972). The first two dimeric acridone alkaloids atalanine (LXXIII) and ataline were recently obtained from *Atalantia ceylanica* (FRASER and LEWIS, 1973). Reviews should be consulted for further information (JOHNE and GRÖGER, 1972; SAXTON, 1973). Rutacridone, 1-hydroxy-N-methyl acridone and the corresponding 3-methoxy derivative were isolated from callus cultures which had been obtained from meristematic cells of *Ruta graveolens*. The alkaloid pattern of the callus was different from that of the stem cells of intact plants (SCHARLEMANN, 1972). Remarkable

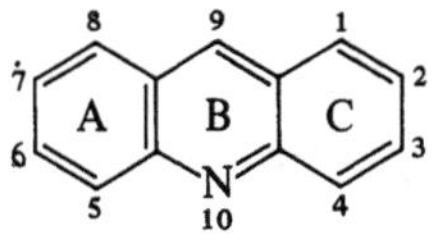

Acridine
(LIXX)

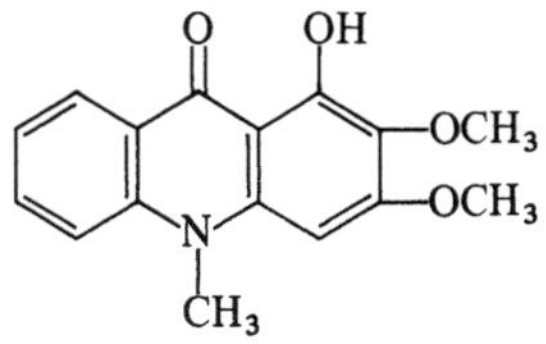

Arborinine
(LXX)

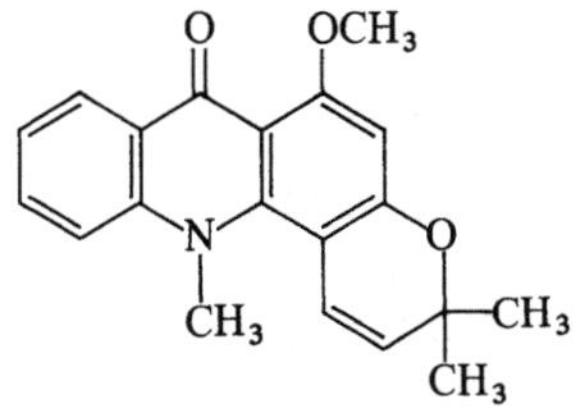

Acronycine
(LXXI)

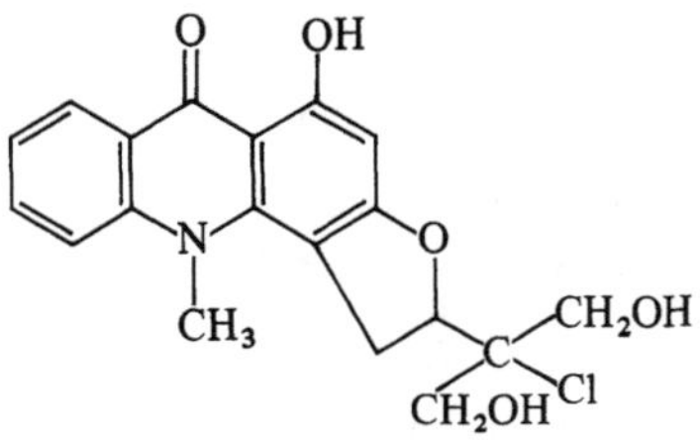

Gravacridonolchlorine
(LXXII)

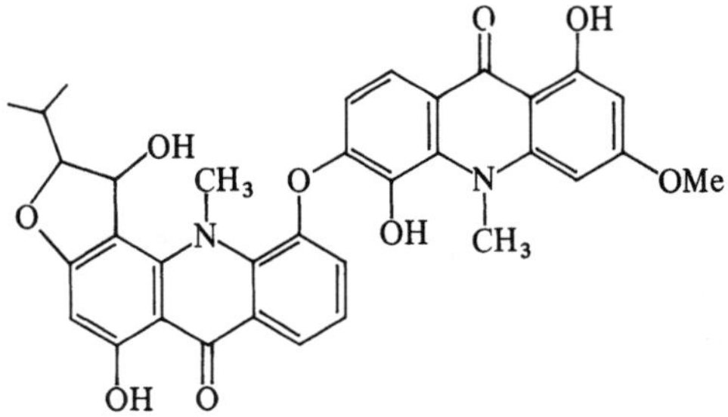

Atalanine
(LXXIII)

Fig. 14. Various types of acridone alkaloids

differences exist with respect to the localization of acridone alkaloids in different organs of *Ruta graveolens*. Especially rich were the roots of young and the xylem of old plants. The acridones are attached to the cell wall in aggregations or they are distributed more or less evenly throughout the cytoplasm. Idioblasts also contain large amounts of acridone alkaloids (VERZAR-PETRI et al., 1976).

The exact mechanism of acridone alkaloid biosynthesis is still obscure. On the basis of synthetic work under so-called physiological conditions HUGHES and RITCHIE (1951) postulated that in acridone alkaloids alkoxy groups are present only in ring C and that the 1 and the 3 position are always substituted. But some recently isolated alkaloids do not fit this scheme, e.g., atalaphylline and tecleanthine are substituted in position 5 of ring A and may derive from 3-hydroxy-anthranilic acid. Rutacridone and N-methylacridone do not show the above-mentioned substitution pattern in ring C. One may speculate that the prenyl side chains present in some acridones, e.g., acronycine and evoprenine, may be derived from mevalonate, but experimental evidence is still lacking. Anthranilic acid proved to be the precursor of ring A and C-9 and N-10 in the acridone (GRÖGER and JOHNE, 1968; PRAGER and THREDGOLD, 1969). The origin of ring B is uncertain; it may be formed from acetate via a polyketo acid or directly from phloroglucinol. Certain results (JOHNE et al., 1970) favor the acetate pathway, however. Presumably a benzophenone intermediate may be involved according to KHANSHAFIG et al. (1975), who were able to isolate tecleanone, a 2-methylaminobenzophenone from *Teclea grandifolia*.

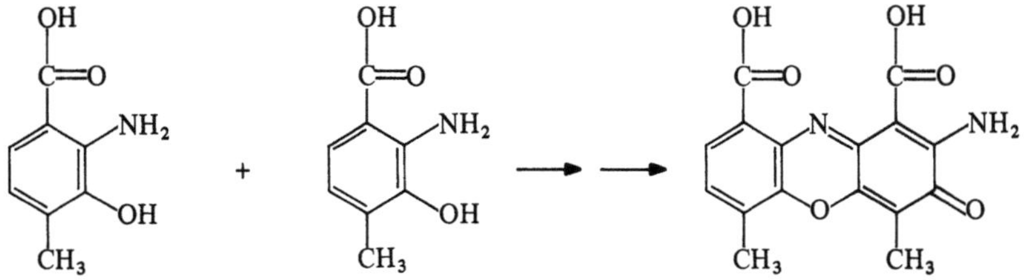

3–Hydroxy–4–methylanthranilic acid (LXXIV) + (LXXIV) → → Actinocin (LXXV)

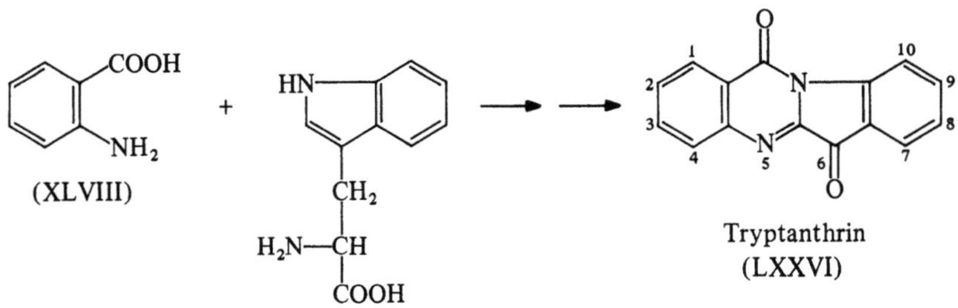

Fig. 15. Anthranilic acid-derived antibioties

2.6 Miscellaneous

A number of microbial metabolites are derived from anthranilic acid. Well-known examples are the actinomycins possessing a phenoxazinone skeleton. The actinomycin chromophore (LXXV) is derived from two molecules of tryptophan via 3-hydroxy-4-methylanthranilic acid (LXXIV) (KATZ and WEISSBACH, 1962). Apparently the following pathway is operative: tryptophan→kynurenine→3-hydroxykynurenine →3-hydroxy-4-methyl-kynurenine→(LXXIV). Methionine serves as the source of the aromatic methyl groups (HERBERT, 1974).

Tryptanthrin [indolo(2,1-b)quinazoline-6,12-dione, (LXXVI)] is a metabolite of the yeast *Candida lipolytica* which is formed in tryptophan-supplemented culture media (SCHINDLER and ZÄHNER, 1971). Biosynthetic studies (FIEDLER et al., 1976) revealed that one molecule of anthranilic acid and one molecule of tryptophan, after elimination of the side-chain, are combined (Fig. 15). When feeding anthranilic acid or tryptophan derivatives, the corresponding new "tryptanthrins" were isolated. After feeding tryptophan plus a substituted anthranilic acid two compounds were isolated: a substituted and an unsubstituted tryptanthrin. A monosubstituted and a disubstituted tryptanthrin were obtained using a substituted tryptophan plus anthranilic acid as substrates. The disubstituted tryptanthrin is formed by the condensation of the substituted tryptophan with the anthranilic acid derivative derived from the substituted tryptophan. For example, D,L-6-methyl tryptophan gives rise to 9-methyl tryptanthrin and 3,9-dimethyltryptanthrin.
During the biosynthesis of both antibiotic types (actinomycins and tryptanthrins) the anthranilic acid moiety results from tryptophan degradation.

References

Abe, M., Yamatodani, S., Yamano, T., Kozu, Y., Yamada, S.: Production of alkaloids and related substances by fungi. Part I. Examination of filamentous fungi for their ability of producing ergot alkaloids. J. Agric. Chem. Soc. Jpn. *41*, 68–71 (1967)

Abe, M., Ohmomo, S., Ohashi, T., Tabuchi, T.: Isolation of chanoclavine-I and two new interconvertible alkaloids, rugulo-vasine A and B, from the culture of *Penicillium concavo-rugulosum*. Agric. Biol. Chem. (Tokyo) *33*, 469–471 (1969)

Abou-Chaar, C.I., Günther, H.F., Manuel, M.F., Robbers, J.E., Floss, H.G.: Biosynthesis of ergot alkaloids. Incorporation of (5R)- and (5S)-mevalonate-5-T into chanoclavines and tetracyclic ergolines. Lloydia *35*, 272–279 (1972)

Agurell, S.: Costaclavine from *Penicillium chermesinum*. Experientia *20*, 25–26 (1964)

Agurell, S.: Biosynthetic studies on ergot alkaloids and related indoles. Acta Pharm. Suec. *3*, 71–100 (1966)

Agurell, S., Nilsson, J.G.L.: A biosynthetic sequence from tryptophan to psilocybin. Tetrahedron Lett. 1063–1064 (1968)

Agurell, S., Holmstedt, B., Lindgreen, J.-E., Schultes, R.E.: Alkaloids in certain species of *Virola* and other South American plants of ethnopharmacological interest. Acta Chem. Scand. *23*, 903–916 (1969)

Alfermann, A.W., Reinhard, E. (eds.): Production of natural compounds by cell culture methods. Proceedings of an international symposium on plant cell culture. Gesellschaft für Strahlen- und Umweltforschung. München: BPT-Report 1978

Angenot, L.: De nouveaux alcaloïdes quaternaires du *Strychnos usambarensis*. Planta med. *27*, 24–30 (1975)

Arndt, R.R., Eggers, S.H., Jordaan, A.: The alkaloids of *Anisotes sessiliflorus* C.B.C. (Acanthaceae) – Five new 4-quinazolone alkaloids. Tetrahedron *23*, 3521–3532 (1967)

Arigoni, D.: Some aspects of mevalonoid biosynthesis, at: Symposium on Organic Chemical approaches to Biosynthesis. London 1965

Arigoni, D.: Mechanismus der Biosynthese von Indolalkaloiden. Abh. Dtsch. Akad. Wiss. Berlin, 4. Int. Symp. Biochem. Physiol. Alkaloide, Band b. pp. 317–335 (1972)

Battersby, A.R.: Biosynthesis of terpenoid alkaloids. In: Natural substances formed biologically from mevalonic acid. Goodwin, T.W. (ed.), pp. 157–168. London: Academic Press 1970

Battersby, A.R., Hall, E.S.: The intermediacy of geissoschizine in indole alkaloid biosynthesis: rearrangement to the *Strychnos* skeleton. J. Chem. Soc. Chem. Commun. 793–794 (1969)

Battersby, A.R., Hall, A.R.: Biosynthesis of quinine from loganin. J. Chem. Soc. Chem. Commun. 194–195 (1970)

Battersby, A.R., Parry, R.J.: Biosynthesis of the *Cinchona* alkaloids: late stages of the pathway. J. Chem. Soc. Chem. Commun. 31–32 (1971)

Battersby, A.R., Brown, R.T., Kapil, R.S., Plunkett, A.O., Taylor, J.B.: Biosynthesis of the indole alkaloids. J. Chem. Soc. Chem. Commun. 46–47 (1966a)

Battersby, A.R., Brown, R.T., Kapil, R.S., Martin, A.J., Plunkett, A.O.: Role of loganin in the biosynthesis of indole alkaloids. J. Chem. Soc. Chem. Commun. 890–891 (1966b)

Battersby, A.R., Gregory, B., Spencer, H., Turner, J.C., Janot, M.-M., Potier, P., Francois, P., Levisalles, J.: Constitution of ipecoside; a monoterpenoid isoquinoline. J. Chem. Soc. Chem. Commun. 219–221 (1967)

Battersby, A.R., Burnett, A.R., Parsons, P.G.: Preparation of secologanin: its conversion into ipecoside and its role in indole alkaloid biosynthesis. J. Chem. Soc. Chem. Commun. 1280–1281 (1968a)

Battersby, A.R., Burnett, A.R., Parsons, P.G.: Partial synthesis and isolation of vincoside and isovincoside: biosynthesis of the three major classes of indole alkaloids from the β-carboline system. J. Chem. Soc. Chem. Commun. 1282–1284 (1968b)

Battersby, A.R., Burnett, A.R., Hall, E.S., Parsons, P.G.: The rearrangement process in indole alkaloid biosynthesis. J. Chem. Soc. Chem. Commun. 1582–1583 (1968c)

Baxter, C., Slaytor, M.: Biosynthesis and turnover of N,N-dimethyltryptamine and 5-methoxy-N,N-dimethyltryptamine in *Phalaris tuberosa*. Phytochemistry *11*, 2767–2773 (1972a)

Baxter, C., Slaytor, M.: Partial purification and some properties of tryptophan decarboxylase from *Phalaris tuberosa*. Phytochemistry *11*, 2763–2766 (1972b)

Biemann, K., Friedmann-Spiteller, M., Spiteller, G.: An investigation by mass spectrometry of the alkaloids of *Aspidosperma quebracho-blancho*. Tetrahedron Lett. 485–492 (1961)
Birch, A.J., McLoughlin, B.J., Smith, H.: The biosynthesis of the ergot alkaloids. Tetrahedron Lett. 1–3 (1960)
Bisset, N.G.: The asian species of *Strychnos*. Part III. The Ethnobotany. Lloydia *37*, 62–107 (1974)
Bisset, N.G., Phillipson, J.D.: The asian species of *Strychnos*. Part IV. The Alkaloids. Lloydia *39*, 263–325 (1976)
Blaha, K., Koblicová, Z., Pospišek, J., Trojánek, J.: Absolute configuration of the alkaloid brevicolline. Coll. Czech. Chem. Commun. *36*, 3448–3454 (1971)
Brack, A., Hofmann, A., Kalberer, F., Kobel, H., Rutschmann, J.: Tryptophan als biogenetische Vorstufe des Psilocybins. Arch. Pharm. *294*, 230–234 (1961)
Breccia, A., Crospi, A.M., Rampi, M.A.: The subcellular fractions in the biosynthesis of natural compounds I. Gramine alkaloids. Z. Naturforsch. *21b*, 1243–1245 (1966)
Brechbühler-Bader, S., Coscia, C.J., Loew, P., von Szepanski, Ch., Arigoni, D.: The chemistry and biosynthesis of loganin. J. Chem. Soc. Chem. Commun. 136–137 (1968)
Budzikiewicz, H., Eckau, H., Ehrenberg, M.: Bis (3-indolyl-)-3H-indolyliden-methan, ein von *Saccharomyces cerevisiae* gebildeter Farbstoff. Tetrahedron Lett. 3807–3810 (1972)
Carew, D.P.: Reserpine in a tissue culture of *Alstonia constricta* F. Muell. Nature (London) *207*, 89 (1965)
Carew, D.P.: Tissue culture studies of *Catharanthus*. In: The *Catharanthus* alkaloids. Taylor, W.I., Farnsworth, N.R. (eds.), pp. 193–208. New York: Marcel Dekker 1975
Chao, J.-M., Der Marderosian, A.H.: Identification of ergoline alkaloids in the genus *Argyreia* and related genera and their chemotaxonomic implications in the Convolvulaceae. Phytochemistry *12*, 2435–2440 (1973)
Cobet, M., Luckner, M.: 2,4-Dihydroxychinolin, ein direktes Zwischenprodukt bei der Biosynthese des Furochinolinalkaloids Skimmianin bei *Ruta graveolens*. Phytochemistry *10*, 1031–1036 (1971)
Colonna, A.O., Gros, E.G.: Biosynthesis of skimmianine in *Fagara coco*. Phytochemistry *10*, 1515–1521 (1971)
Cordell, G.A.: The biosynthesis of indole alkaloids. Lloydia *37*, 219–298 (1974)
Danielli, B., Manitto, P., Ronchetti, F., Russo, G., Ferrari, G.: Paraensine, a new indolopyridoquinazoline alkaloid from *Euxylophora paraënsis* Hub. Experientia *28*, 249–250 (1972)
Döller, G., Alfermann, A.W., Reinhard, E.: Produktion von Indolalkaloiden in Calluskulturen von *Catharanthus roseus*. Planta Med. *30*, 14–19 (1976)
Erge, D., Maier, W., Gröger, D.: Untersuchungen über die enzymatische Umwandlung von Chanoclavin-I. Biochem. Physiol. Pflanz. *164*, 234–247 (1973)
Fehr, T., Acklin, W., Arigoni, D.. The role of the chanoclavines in the biosynthesis of ergot alkaloids. J. Chem. Soc. Chem. Commun. 801–802 (1966)
Fiedler, E., Fiedler, H.P., Gerhard, A., Keller-Schierlein, W., König, W.A., Zähner, H.: Stoffwechselprodukte von Mikroorganismen. 156. Mitteilung. Synthese und Biosynthese substituierter Tryptanthrine. Arch. Mikrobiol. *107*, 249–256 (1976)
Floss, H.G.: Biosynthesis of ergot alkaloids and related compounds. Tetrahedron *32*, 873–912 (1976)
Floss, H.G., Mothes, U., Günther, H.: Über den Einbau der Tryptophan-Seitenkette und den Mechanismus der Reaktion am α-C-Atom. Z. Naturforsch. *19b*, 784–788 (1964)
Floss, H.G., Hornemann, U., Schilling, N., Gröger, D., Erge, D.: Chanoclavines and the biosynthesis of ergot alkaloids. J. Chem. Soc. Chem. Commun. 105–106 (1967)
Floss, H.G., Hornemann, U., Schilling, N., Kelley, K., Gröger, D., Erge, D.: Biosynthesis of ergot alkaloids. Evidence for two isomerizations in the isoprenoid moiety during the formation of tetracyclic ergolines. J. Am. Chem. Soc. *90*, 6500–6507 (1968)
Floss, H.G., Robbers, E.J., Heinstein, P.F.: Regulatory control mechanisms in alkaloid biosynthesis. Recent Adv. Phytochem. Vol. 8, pp. 141–178 New York: Academic Press 1974a
Floss, H.G., Tscheng-Lin, M., Chang, C.J., Naidoo, B., Blair, G.E., Abou-Chaar, C., Cassady, J.M.: Biosynthesis of ergot alkaloids. Studies on the mechanism of the conversion of chanoclavine-I into tetracyclic ergolines. J. Am. Chem. Soc. *96*, 1898–1909 (1974b)
Fraser, A.W., Lewis, J.R.: Two novel acridone alkaloids from *Atalantia ceylanica* (Rutaceae). J. Chem. Soc. Chem. Commun. 615–616 (1973)

Gander, J.E., Marum, P., Martin, G.C., Hovin, A.W.: The occurrence of 2-methyl-1,2,3,4-tetrahydro-β-carboline and variation in alkaloids in *Phalaris arundinacea*. Phytochemistry *15*, 737–738 (1976)
Ganzinger, D., Hesse, M.: A chemotaxonomic study of the subfamily Plumerioideae of the Apocynaceae. Lloydia *39*, 326–349 (1976)
Garnier, J., Koch, M., Kunesch, N.: Les alkaloïdes "bis-indoliques", repartition – essai de classification. Phytochemistry *8*, 1241–1254 (1969)
Ghosal, S., Ballav, R., Chauhan, P.S., Mehta, R.: Alkaloids of *Sida cordifolia*. Phytochemistry *14*, 830–832 (1975)
Goeggel, H., Arigoni, D.: The mevalonoid nature of vindoline and reserpine. J. Chem. Soc. Chem. Commun. 538–539 (1965)
Gower, B.G., Leete, E.: Biosynthesis of gramine: The immediate precursors of the alkaloid. J. Am. Chem. Soc. *85*, 3683–3685 (1963)
Gröger, D.: Anthranilic acid as precursor of alkaloids. Lloydia *32*, 221–246 (1969)
Gröger, D.: Ergot. In: Microbial toxins. Kadis, S., Ciegler, A. (eds.), Vol. 8, pp. 321–373. New York: Academic Press 1972
Gröger, D.: Chemische Rassen bei Alkaloidpflanzen. Planta Med. *28*, 269–288 (1975a)
Gröger, D.: Biosynthese des Peptidteils der Mutterkornalkaloide vom Cyclol-Typ. Planta Med. *28*, 37–51 (1975b)
Gröger, D., Johne, S.: Zur Biosynthese einiger Alkaloide von *Glycosmis arborea* (Rutaceae). Z. Naturforsch. *23b*, 1072–1075 (1968)
Gröger, D., Simon, H.: Zur Biogeneses der Harman-Alkaloide. Abh. Dtsch. Akad. Wiss. Berlin Kl. Chem. Geol. Biol. *4*, 343–348 (1963)
Gröger, D., Mothes, K., Simon, H., Floss, H.G., Weygand, F.: Über den Einbau von Mevalonsäure in das Ergolinsystem der Clavin-Alkaloide. Z. Naturforsch. *15b*, 141–143 (1960)
Gross, D.: Indolylalkylamine. In: Biosynthese der Alkaloide. Mothes, K., Schütte, H.R. (eds.), pp. 439–458. Berlin: VEB Deutscher Verlag der Wissenschaften 1969
Gross, D., Nemeckova, A., Schütte, H.-R.: Zur Biosynthese des Gramins. Z. Pflanzenphysiol. *57*, 60–67 (1967)
Gross, D., Lehmann, H., Schütte, H.-R.: Zur Physiologie der Graminbildung. Z. Pflanzenphysiol. *63*, 1–9 (1970)
Gross, D., Lehmann, H., Schütte, H.-R.: Zur Biosynthese des Gramins. Biochem. Physiol. Pflanz. *166*, 281–287 (1974)
Grundon, M.F., Harrison, D.M., Spyropoulos, C.G.: Biosynthesis of furoquinoline alkaloids, skimmianine, evoxine and choisyine: mechanism of formation of the furan ring. The timing of aromatic hydroxylation and of methylation. J. Chem. Soc. Chem. Commun. 51–52 (1974)
Grundon, M.F., Harrison, D.M., Spyropoulos, C.G.: Biosynthesis of aromatic isoprenoids Part III. Mechanism of formation of the furan ring and origin of the 4-methoxy-group in the biosynthesis of furoquinoline alkaloids. J. Chem. Soc. Perkin I, 302–304 (1975)
Hegnauer, R.: Comparative phytochemistry of alkaloids. In: Comparative phytochemistry Swain, T. (ed.), pp. 211–230. London: Academic Press 1966
Heinstein, P.F., Lee, S.-L., Floss, H.G.: Isolation of dimethylallylpyrophosphate: tryptophan dimethylallyl transferase from the ergot fungus (*Claviceps spec.*). Biochem. Biophys. Res. Commun. *44*, 1244–1251 (1971)
Herbert, R.B.: Biosynthesis of actinomycin: Incorporation of (methyl-$^{2}H_{3}$) methionine, (^{3}H)-kynurenine, and (^{3}H)-3-hydroxykynurenine. Tetrahedron Lett. 4525–4528 (1974)
Hesse, M.: Indolalkaloide in Tabellen und Ergänzungswerk. Berlin: Springer 1964, 1968
Hofmann, A.: The active principles of the seeds of *Rivea corymbosa* and *Ipomoea violacea*. Bot. Mus. Leafl. Harv. Univ. *20*, 194–212 (1963)
Hofmann, A., Tscherter, A.: Isolierung von Lysergsäure-Alkaloiden aus der mexikanischen Zauberdroge Ololiuqui (*Rivea corymbosa* (L.) Hall. F.). Experientia *16*, 414 (1960)
Hofmann, A., Heim, R., Brack, A., Kobel, H., Frey, A., Ott, H., Petrzilka, T., Troxler, F.: Psilocybin und Psilocin. Helv. Chim. Acta *42*, 1557–1572 (1959)
Hughes, G.K., Ritchie, E.: Experiments on the synthesis of the acridone alkaloids. Aust. J. Sci. Res. *4A*, 423–431 (1951)
Jeffreys, J.A.D.: The alkaloids of perennial rye-grass (*Lolium perenne* L.). Part IV. Isolation of a new base, perlolyrine; the crystal structure of its hydrobromide dihydrate, and the synthesis of the base. J. Chem. Soc. (C), 1091–1103 (1970)

Johne, S., Gröger, D.: Natürlich vorkommende Chinazolin-Derivate. Pharmazie *25*, 22–44 (1970)

Johne, S., Gröger, D.: Natürlich vorkommende Acridin-Derivate. Pharmazie *27*, 195–208 (1972)

Johne, S. Bernasch, H., Gröger, D.: Beitrag zur Biosynthese des Acridin-Alkaloids Arborinin. Pharmazie *25*, 777–779 (1970)

Johne, S.,, Gröger, D., Hesse, M.: Neue Alkaloide aus *Adhatoda vasica* Nees. Helv. Chim. Acta *54*, 826–834 (1971)

Johne, S., Waiblinger, K., Gröger, D.: Untersuchungen zur Biosynthese des Chinazolinalkaloids Peganin in *Adhatoda vasica* Nees. Pharmazie *28*, 403–406 (1973)

Johns, S.R., Lamberton, J.A.: Alkaloids of *Mackinlaya* species (family Araliaceae). Chem. Commun. 267–268 (1965)

Katz, E., Weissbach, H.: Biosynthesis of the actinomycin chromophore; Enzymatic conversion of 4-methyl-3-hydroxyanthranilic acid to actinocin. J. Biol. Chem. *237*, 882–886 (1962)

Khan Shafig, M., Lewis, J.R., Watt, R.A.: Conversion of tecleanone into 1,3-dimethoxy-10-methylacridone: a possible biosynthetic route to acridone alkaloids. Chem. Ind. 744–745 (1975)

Kobel, H., Schreier, E., Rutschmann, J.: 6-Methyl-$\Delta^{8,9}$-ergolen-8-carbonsäure, ein neues Ergolinderivat aus Kulturen eines Stammes von *Claviceps paspali* Stevens et Hall. Helv. Chim. Acta *47*, 1052–1064 (1964)

Kompiš, J., Hesse, M., Schmid, H.: An approach to the biogenetic classification of indole alkaloids. Lloydia *34*, 269–291 (1971)

Krukoff, B.A.: American species of *Strychnos*. Lloydia *35*, 193–271 (1972)

Lee, S.-L., Floss, H.G., Heinstein, P.: Purification and properties of dimethylallyl pyrophosphate: tryptophan dimethylallyl transferase, the first enzyme of ergot alkaloid biosynthesis in *Claviceps* sp. SD 58. Arch. Biochem. Biophys. *177*, 84–94 (1976)

Leete, E.: Biogenesis of the *Rauwolfia* alkaloids-II. The incorporation of tryptophan into serpentine and reserpine. Tetrahedron *14*, 35–41 (1961)

Leete, E., Wemple, J.N.: Biosynthesis of the *Cinchona* alkaloids II. The incorporation of tryptophan-1-^{15}N,2-^{14}C and geraniol-3-^{14}C into quinine. J. Am. Chem. Soc. *91*, 2698–2702 (1969)

Leete, E., Ahmad, A., Kompiš, I.: Biosynthesis of the *Vinca* alkaloids. I Feeding experiments with tryptophan-2-^{14}C and acetate-1-^{14}C. J. Am. Chem. Soc. *87*, 4168–4174 (1965)

Leeuwenberg, A.J.M.: The Loganiaceae of Africa, VIII. *Strychnos* III. Meded. Landbouwhogesch. Wageningen *69*, 1–316 (1969)

Le Men, J., Taylor, W.I.: Uniform numbering system for indole alkaloids. Experientia *21*, 508–511 (1965)

Liljegren, D.R.: The biosynthesis of quinazoline alkaloids of *Peganum harmala* L. Phytochemistry *7*, 1299–1306 (1968)

Liljegren, D.R.: Biosynthesis of quinazoline alkaloids of *Peganum harmala*. Phytochemistry *10*, 2661–2669 (1971)

Loew, P., Arigoni, D.: The biological conversion of loganin into indole alkaloids. J. Chem. Soc. Chem. Commun. 137–138 (1968)

Luckner, M.: Chinoline. In: Biosynthese der Alkaloide. Mothes, K., Schütte, H.R. (eds.), pp. 510–550. Berlin: VEB Verlag der Wissenschaften 1969

Luckner, M., Nover, L., Böhm, H.: Secondary metabolism and cell differentiation. In: Molecular biology, biochemistry and biophysics 23. Berlin, Heidelberg, New York: Springer 1977

Madyastha, K.M., Guarnaccia, R., Baxter, C., Coscia, C.J.: S-Adenosyl-L-methionine: loganic acid methyltransferase. A carboxyl-alkylating enzyme from *Vinca rosea*. J. Biol. Chem. *248*, 2497–2501 (1973)

Mann, S.: Chinazolinderivate bei Pseudomonaden. Arch. Mikrobiol. *56*, 324–329 (1967)

Manske, R.F.H. (ed.): The alkaloids: chemistry and physiology. Vol. VIII, XI, XIV, XV. New York: Academic Press Vol. VIII 1965, XI 1968, XIV 1973, XV 1975

Matsuo, M., Yamazaki, M., Kasida, Y.: Biosynthesis of skimmiamine. Biochem. Biophys. Res. Commun. *23*, 679–682 (1966)

McCapra, F., Money, T., Scott, A.I., Wright, I.G.: Biosynthesis of the indole alkaloids: Vindoline. J. Chem. Soc. Chem. Commun. 537–538 (1965)

McFarlane, J., Madyastha, K.M., Coscia, C.J.: Regulation of secondary metabolism in higher plants. Effect of alkaloids on a cytochrome P-450 dependent monooxygenase. Biochem. Biophys. Res. Commun. *66*, 1263–1269 (1975)

McKenzie, E., Nettleship, L., Slaytor, M.: New natural products from *Peganum harmala*. Phytochemistry *14*, 273–275 (1975)

Meehan, T.D., Coscia, C.J.: Hydroxylation of geraniol and nerol by a monooxygenase. Biochem. Biophys. Res. Commun. *53*, 1043–1048 (1973)

Mitra, G.C., Kaul, K.N.: In vitro culture of root & stem callus of *Rauwolfia serpentina* Benth. for reserpine. Indian J. Exp. Biol. *2*, 49–51 (1964)

Monković, I., Spenser, I.D.: Biosynthesis of dictamnine. J. Chem. Soc. Chem. Commun. 205–207 (1966)

Mudd, S.H.: 3-aminomethylindole and 3-methylaminomethylindole: new constituents of barley. Nature (London) *189*, 489 (1961)

Munsche, D.: Zytologische Untersuchungen zur Biosynthese des Damascenins in *Nigella damascena* L. 5. Mitteilung über Damascenin. Flora (Jena) *154*, 317–324 (1964)

Munsche, D.: Biosynthese des Damascenins. Abhandl. Deut. Akad. Wiss. Berlin Kl. Chem. Geol. Biol. *3*, 611–615 (1966)

Nair, P.M., Vaidyanathan, C.S.: Anthranilic acid oxidase system of *Tecoma stans* – II. Studies on the properties of a purified anthranilic acid oxidase system and its separation into different enzymic components. Phytochemistry *3*, 513–523 (1964)

Nettleship, L., Slaytor, M.: Ruine: a glucosidic β-carboline from *Peganum harmala*. Phytochemistry *10*, 231–234 (1971)

Nettleship, L., Slaytor, M.: Adaptation of *Peganum harmala* callus to alkaloid production. J. Exp. Bot. *25*, 1114–1123 (1974)

O'Donovan, P.G., Kenneally, M.F.: The biosynthesis of eleagnine. J. Chem. Soc. 1109–1110 (1967)

O'Donovan, D., Leete, E.: Biosynthesis of gramine: Feeding experiments with tryptophan-β-(H^3, C^{14}). J. Am. Chem. Soc. *85*, 461–463 (1963)

Ohmomo, S., Sato, T., Utagawa, T., Abe, M.: Isolation of festuclavine and three new indole alkaloids, roquefortine A, B and C from the cultures of *Penicillium roqueforti*. Agric. Biol. Chem. *39*, 1333–1334 (1975)

Pachlatko, P., Tobacik, C., Acklin, W., Arigoni, D.: Natürliche und unnatürliche Vorläufer in der Biosynthese der Ergotalkaloide. Chimia *29*, 526–527 (1975)

Parry, R.J.: Biosynthesis of *Catharanthus* alkaloids. In: The *Catharanthus* alkaloids. Taylor, W.I., Farnsworth, N.R. (eds.), pp. 141–191. New York: Marcel Dekker 1975

Patterson, B.D., Carew, D.P.: Growth and alkaloid formation in *Catharanthus roseus* tissue cultures. Lloydia *32*, 131–140 (1969)

Plieninger, H., Fischer, R., Liede, V.: 4-Dimethylallyltryptophan als Vorstufe der Clavin-Alkaloide. Angew. Chem *74*, p. 430 (1962)

Plieninger, H., Immel, H., Völkl, A.: Synthese und Einbau eines ^{14}C- und ^{3}H-markierten Dimethylallyl-tryptophans und eines ^{14}C-markierten 4-Dimethyl-allyltryptamins sowie Einbau eines ^{14}C-markierten Dimethylallyl-pyrophosphats. Liebigs Ann. Chem. *706*, 223–229 (1967)

Prager, R.H., Thredgold, H.M.: Studies using radioisotopes. III. The biosynthesis of acridone alkaloids. Aust. J. Chem. *22*, 2627–2634 (1969)

Reinhard, E., Corduan, G., Volk, O.H.: Nachweis von Harmin in Gewebekulturen von *Peganum harmala*. Phytochemistry *7*, 503–504 (1968)

Reisch, J., Szendrei, K., Rosza, Zs., Novak, I., Minker, E.: Chlorhaltige Acridon-Alkaloide aus *Ruta graveolens*. Phytochemistry *11*, 2359–2360 (1972)

Reisch, J., Rosza, Zs., Szendrei, K., Körösi, J.: 2-(n-Undecyl)-chinolon-(4) aus den Blüten von *Ptelea trifoliata* und den Wurzeln von *Ruta graveolens*. Phytochemistry *14*, 840–841 (1975)

Ritter, Ch., Luckner, M.: Zur Biosynthese der 2-n-Alkyl-4-hydroxychinolinderivate (Pseudane) bei *Pseudomonas aeruginosa*. Eur. J. Biochem. *18*, 391–400 (1971)

Saxton, J.E.: The simple bases. In: The alkaloids. Manske, R.H.F. (ed.), Vol. VIII, pp. 1–25. New York: Academic Press 1965

Saxton, J.E.: The acridines. Acheson, R.M. (ed.) In: The chemistry of heterocyclic compounds. Weissberger, A., Taylor, E.C. (eds.), Vol. 9, p. 379. New York: Wiley-Interscience 1973

Scharlemann, W.: Acridin-Alkaloide aus Kalluskulturen von *Ruta graveolens*. Z. Naturforsch. *27b*, 806–809 (1972)

Schildknecht, H., Wenneis, W.F.: Über Arthropoden-Abwehrstoffe XXV. Anthranilsäure als Precursor der Arthropoden-Alkaloide Glomerin und Homoglomerin. Tetrahedron Lett. 1815–1818 (1967)
Schindler, F., Zähner, H.: Tryptanthrin, ein von Tryptophan abzuleitendes Antibioticum aus *Candida lipolytica*. Arch. Mikrobiol. *79*, 187–203 (1971)
Schlittler, E., Taylor, W.I.: A view of progress in the chemistry of indole alkaloids. Experientia *16*, 244–248 (1960)
Schultes, R.E.: Indole alkaloids in plant hallucinogens. Planta Med. *29*, 330–342 (1976)
Scott, A.I.: Biosynthesis of the indole alkaloids. Acc. Chem. Res. *3*, 151–157 (1970)
Scott, A.I., Lee, S.-L.: Biosynthesis of the indole alkaloids. A cell-free system from *Catharanthus roseus*. J. Am. Chem. Soc. *97*, 6906–6908 (1975)
Scott, A.I., Cherry, P.C., Quereshi, A.A.: Mechanism of indole alkaloid biosynthesis. The *Corynanthe-Strychnos* relationship. J. Am. Chem. Soc. *91*, 4932–4933 (1969)
Scott, A.I., Reichardt, B.P., Slaytor, M.B., Sweeny, J.E.: Mechanism of indole alkaloid biosynthesis. Recognition of intermediacy and sequence by short-term incubation. Recent Adv. Phytochem. *6*, 117–145 (1973)
Scott, A.I., Lee, S.-L., Wan, W.: Indole alkaloid biosynthesis: partial purification of "ajmalicine synthetase" from *Catharanthus roseus*. Biochem. Biophys. Res. Commun. *75*, 1004–1009 (1977)
Scott, P.M., Merrien, M.A., Polonsky, J.: Roquefortine and isofumigaclavine A, metabolites from *Penicillium roqueforti*. Experientia *32*, 140–141 (1976)
Seiler, M., Acklin, W., Arigoni, D.: Biosynthesis of ergot alkaloids from (5R)- and (5S)-[5-$^3H_1$]-mevalonolactones. J. Chem. Soc. Chem. Commun. 1394–1395 (1970)
Shamina, Z.B., Kovaleva, T.A., Butenko, R.G.: Effect of nitrogen mustard on *Rauwolfia* tissue culture. Phytomorphology *22*, 260–264 (1974)
Slaytor, M., McFarlane, I.J.: The biosynthesis and metabolism of harman in *Passiflora edulis* – I. The biosynthesis of harman. Phytochemistry *7*, 605–611 (1968)
Smith, G.N.: strictosidine: a key intermediate in the biogenesis of indole alkaloids. J. Chem. Soc. Chem. Commun. 912–914 (1968)
Smith, T.A.: Tryptamine and related compounds in plants. Phytochemistry *16*, 171–175 (1977)
Snieckus, V.: The distribution of indole alkaloids in plants. In: The alkaloids. Manske, R.H.F., (ed.), Vol. XII, pp. 1–40. New York: Academic Press 1968
Spilsbury, S.F., Wilkinson, S.: The isolation of festuclavine and two new clavine alkaloids from *Aspergillus fumigatus* Fries. J. Chem. Soc. 2085–2091 (1961)
Stadler, P.A., Stütz, P.: The ergot alkaloids. In: The alkaloids. Manske, R.H.F. (ed.), Vol. XV, pp. 1–40. New York: Academic Press 1975
Stauffacher, D., Tscherter, H., Hofmann, A.: Isolierung von Ergosin und Ergosinin neben Agroclavin aus den Samen von *Ipomoea argyrophylla* Vatke (Convolvulaceae). Helv. Chim. Acta *48*, 1379–1380 (1965)
Stauffacher, D., Niklaus, P., Tscherter, H., Werner, H.P., Hofmann, A.: Cycloclavin, ein neues Alkaloid aus *Ipomoea hildebrandtii* Vatke. Tetrahedron *25*, 5879–5887 (1970)
Steck, W., Bailey, B.K., Shyluk, J.P., Gamborg, O.L.: Coumarins and alkaloids from cell cultures of *Ruta graveolens*. Phytochemistry *10*, 191–194 (1971)
Steck, W., Gamborg, O.L., Bailey, B.K.: Increased yields of alkaloids through precursor biotransformations in cell suspension cultures of *Ruta graveolens*. Lloydia *36*, 93–95 (1973)
Stolle, K., Gröger, D.: Untersuchungen zur Biosynthese des Harmins. Arch. Pharm. *301*, 561–571 (1968)
Stöckigt, J., Zenk, M.H.: Strictosidine (isovincoside): the key intermediate in the biosynthesis of monoterpenoid indole alkaloids. Chem. Commun. *1977a*, 646–648
Stöckigt, J., Zenk, M.H.: Isovincoside (strictosidine), the key intermediate in the enzymatic formation of indole alkaloids. FEBS Lett. *79*, 233–237 (1977b)
Stöckigt, J., Treimer, J., Zenk, M.H.: Synthesis of ajmalicine and related indole alkaloids by cell free extracts of *Catharanthus roseus* cell suspension cultures. FEBS Lett. *70*, 267–270 (1976)
Stöckigt, J., Husson, H.P., Kan-Fan, C., Zenk, M.H.. Cathenamine, a central intermediate in the cell free biosynthesis of ajmalicine and related indole alkaloids. J. Chem. Soc. Chem. Commun. 164–166 (1977)
Stockigt, J., Rueffer, M., Zenk, M.H., Hoyer, G.-A.. Indirect identification of 4,21-dehydroco-

rynantheine aldehyde as an intermediate in the biosynthesis of ajmalicine and related alkaloids. Planta Med. *33*, 188–192 (1978)

Taylor, W.I., Farnsworth, N.R. (eds.): The *Catharanthus* alkaloids. New York: Marcel Dekker 1975

Thomas, R.: A possible biosynthetic relationship between the cyclopentanoid monoterpenes and the indole alkaloids. Tetrahedron Lett. 544–545 (1961)

Trojánek, J., Blaha, K.: A proposal for the nomenclature of indole alkaloids. Lloydia *29*, 149–155 (1966)

Veliky, I.A., Barber, K.M.: Biotransformation of tryptophan by *Phaseolus vulgaris* suspension culture. Lloydia *38*, 125–130 (1975)

Verpoorte, R., Baerheim Sevendsen, A.: The alkaloids of *Strychnos dolichothyrsa*. Lloydia *39*, 357–362 (1976)

Verzár-Petri, G., Csedö, K., Möllmann, H., Szendrei, K., Reisch, J., Fluoreszenzmikroskopische Untersuchungen über die Lokalisierung von Acridon-Alkaloiden in Geweben von *Ruta graveolens*. Planta Med. *29*, 370–375 (1976)

Waiblinger, K., Johne, S., Gröger, D.: Zur Biosynthese des Pyrrolidinringes in Peganin. Phytochemistry II, 2263–2265 (1972)

Waterman, P.G.: Alkaloids of the rutaceae: their distribution and systematic significance. Biochem. Syst. Ecol. *3*, 149–180 (1975)

Wenkert, E.: Biosynthesis of indole alkaloids. The *Aspidosperma* and *Iboga* bases. J. Am. Chem. Soc. *84*, 98–102 (1962)

Willaman, J.J., Li, H.-L.: Alkaloid bearing plants and their contained alkaloids. 1957–1968. Lloydia Suppl. 3A *33*, 1–268 1970)

Willaman, J.J., Schubert, B.G.: Alkaloid bearing plants and their contained alkaloids. Washington, D.C.: Agricultural Research Service, U.S.D.A. Tech. Bull. No. 1234 (1961)

Yamano, T., Kishino, K., Yamatodani, S., Abe, M.: Researches on ergot fungus. Part XLIX. Investigation on ergot alkaloids occurred in cultures of *Aspergillus fumigatus* Fres. Annu. Rep. Takeda Res. Lab. *21*, 95–101 (1962)

Yamazaki, M., Ikuta, A.: Biosynthesis of *Evodia* alkaloids. Tetrahedron Lett. 3221–3224 (1966)

Yamazaki, M., Leete, E.: Biosynthesis of the *Iboga* alkaloids: The incorporation of tryptophan-3-^{14}C into ibogaine. Tetrahedron Lett. 1499–1501 (1964)

Yamazaki, M., Ikuta, A., Mori, T., Kawana, T.: Biosynthesis of *Evodia* alkaloids II. The participation of C-1 unit to the formation of indoloquinazoline alkaloids. Tetrahedron Lett. 3317–3320 (1967)

Yeowell, D.A., Schmid, H.: Zur Biosynthese des Plumierids. Experientia *20*, 250–252 (1964)

Zenk, M.H., El-Shagi, H., Arens, H., Stöckigt, J., Weiler, E.W., Deus, B.: Formation of the indole alkaloids serpentine and ajmalicine in cell suspension cultures of *Catharanthus roseus*. In: Plant tissue culture and its biotechnological application. Barz, W., Reinhard, E., Zenk, M.H. (eds.) Berlin, Heidelberg, New York: Springer 1977

4.4 Alkaloids Derived from Histidine and Other Precursors

G.B. FODOR

1 Alkaloids Derived from Histidine

1.1 Dolicotheline

Histamine was incorporated into the skeleton of dolichotheline (I) of *Dolichothele sphaerica*, by condensation with isovaleric acid (ROSENBERG and PAUL, 1969). DL-[2-^{14}C]-Histamine and [2-^{14}C]-leucine (via [1-^{14}C]-isovalerate) are the obvious precursors for this unusual alkaloid (O'DONOVAN and HORAN, 1971).

N-Cinnamoylhistamine (II) (Fig. 1), an alkaloid which has been isolated from an Australian legume can also be produced in *D. sphaerica* by the simultaneous feeding of histamine and cinnamic acid (FITZGERALD, 1964; ROSENBERG and STOHS, 1974a, b).

Histidine decarboxylase inhibitors allow the acylation of "artificial" amines, e.g., 4-aminomethylimidazole at the expense of the histamine derivative.

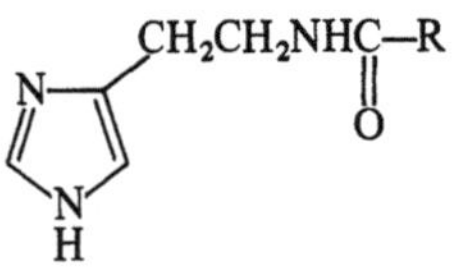

(I) R = Me_2CHCH_2–
(II) R = Ph–CH=CH–

Fig. 1

1.2 Pilocarpine

Surprisingly, feeding of Jaboranda plants (*Pilocarpus*) by the cut stem method with either labelled histidine or threonine and acetate did not show any appreciable incorporation into pilocarpine (V), whilst feeding with [S-^{14}C-Me]-methionine led to tracer in the N-methyl group of (V) (BROCHMANN-HANSSEN et al., 1975). The first of these results suggest that histidine and acetate are not the real precursors, however, an alternative explanation may be that the biosynthesis does not occur in the leaves but in the roots and the products are translocated to the leaves, as with the biosynthesis of nicotine (DAWSON, 1960).

The two possible pathways for the biosynthesis of pilocarpine are as in Scheme 1 (Fig. 2). Pathway A, suggested by ROBINSON (1955), assumes the combination of 4-imidazolyl acetol phosphate (III) and either two molecules of acetate or one molecule of acetoacetate. Another precursor, according to pathway B, could be the histidyl analogue of cinnamic acid (urocanic acid) (IV) and α-ketobutyrate, a metabolite

Fig. 2

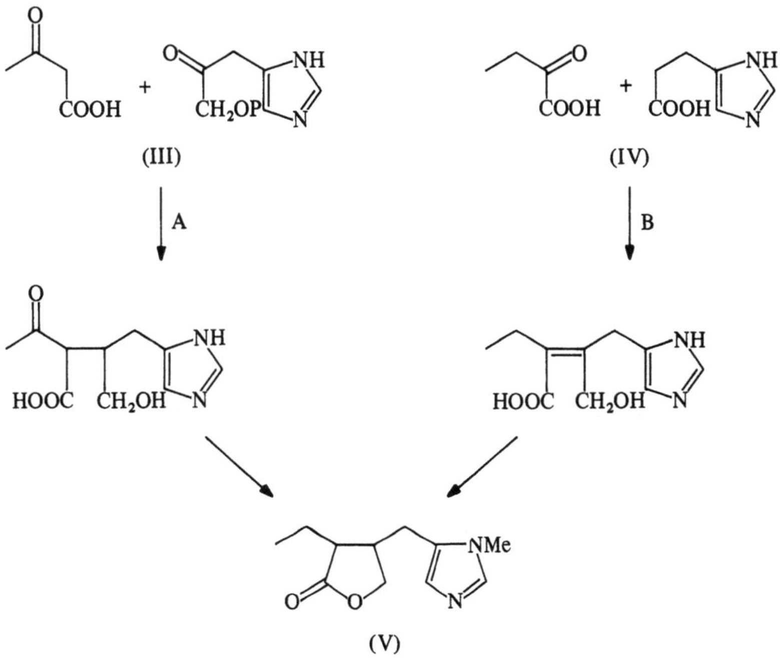

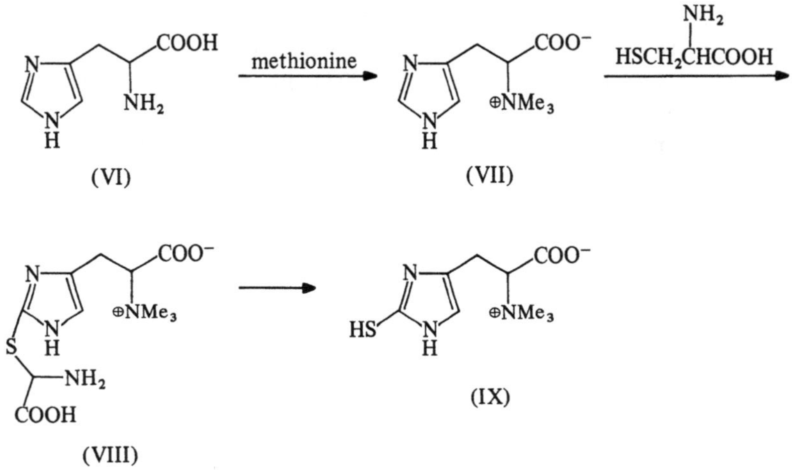

Fig. 3

of threonine. However, no definite answer to this question can be given at present. The methylation of norpilocarpine (pilocarpidine) certainly takes place in the leaves, methionine being the methyl group donor.

Histidine (VI) is a precursor for ergothioneine (IX) in *Neurospora crassa* (MELVILLE, 1959), in *Claviceps purpurea* (STOWELL, 1961) and in *Mycobacterium tuberculosis* (GENHOF and VAN DAMME, 1964) (Fig. 3).

[N-Me-^{14}C,2-^{14}C]-Hercynine (VII) gave ergothioneine (IX) labelled in the corresponding positions, indicating that methylation preceded thiolation (ASKARI and MELVILLE, 1962). The biosynthesis of histidine is closely related to that of the purine alkaloids (MEISTER, 1965).

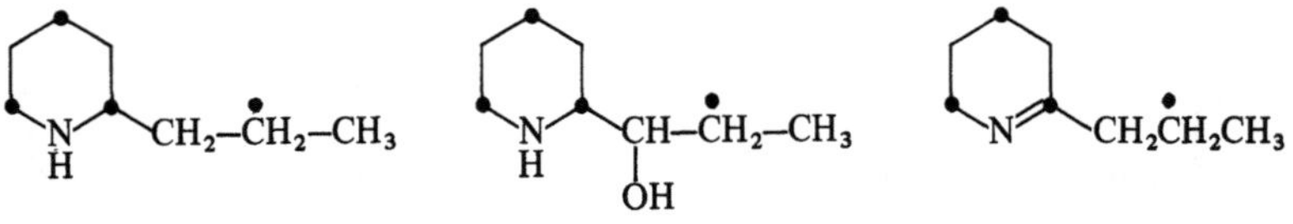

(X) (XI) (XII)

Fig. 4

2 Alkaloids Derived from Other Precursors

2.1 Alkaloids Derived by Linear Combination of Acetate

2.1.1 Coniine, Conhydrine, and γ-Coniceine

The hemlock bases, coniine (X), conhydrine (XI), and γ-coniceine (XII) (Fig. 4) are *not* formed from lysine or α-piperideine, as suggested by ROBINSON (1955), but from acetate. Carbons C-2′, C-2, C-4 and C-6 of these alkaloids carry $^{1}/_{4}$ of the specific activity after feeding with [1-^{14}C]-acetate (LEETE, 1963, 1964). Furthermore, [1′-^{14}C]-α-coniceine was converted into (X) and (XI) without scattering of the label (LEETE and ADITYACHAUDHURY, 1967). All this proves that the *Conium maculatum* alkaloids are formed from acetate units by linear attachment. Octanoic acid and its corresponding aldehyde are also precursors of coniine (LEETE, 1970; LEETE and OLSON, 1970; ROBERTS, 1971). Feeding experiments with L-[^{14}C-Me]-methionine and S-adenosylmethionine gave high incorporation into the methyl carbon (ROBERTS, 1974) of N-methylconiine. γ-Coniceine is enzymatically hydrogenated to coniine (ROBERTS, 1975) the hydrogen being derived from the β-H of NADPH (Fig. 5).

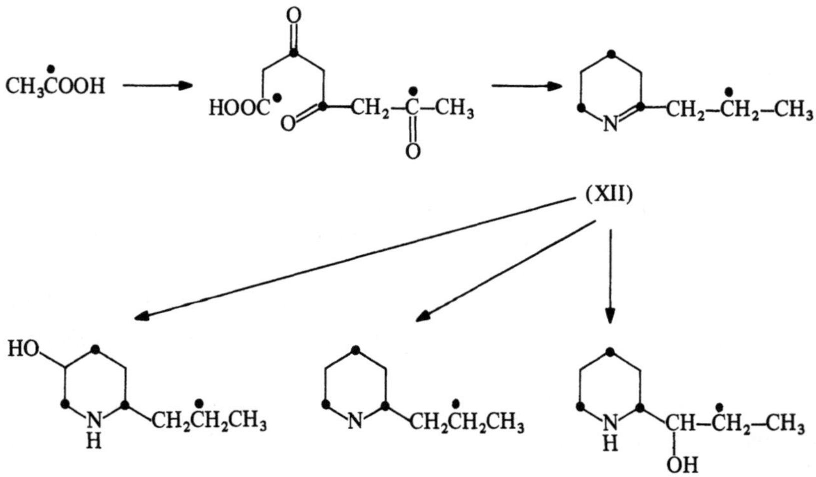

(XIII) (X) (XI)

Fig. 5

2.1.2 Pinidine

This piperidine alkaloid of *Pinus jeffreyi*, like the *Conium* alkaloids, is formed exclusively from acetate via a polyketide intermediate. Tracer (LEETE and OLSON, 1970) and enzyme studies (ROBERTS, 1971) point to the ketonic aldehyde as a possible intermediate for coniine formed from octanoic acid. Label from [1-^{14}C]-acetate appears in the alternate carbon atoms, C-2, C-4, C-6 and C-9 of pinidine (XIV) (LEETE and JUNEAU, 1969) (Fig. 6). However, neither [10-^{14}C]-5,9-dioxodecanoic acid (LEETE and CARVER, 1975) nor [1-^{14}C]-nonadione were precursors of pinidine (LEETE et al., 1975). On the other hand, when [1-^{14}C]-malonic acid was fed together with inactive sodium acetate (LEETE et al., 1975) the C-9 of (XIV) was more heavily labelled than C-2 (Fig. 6). Thus C-2 and C-7 represent the "starter" acetate unit and this would suggest the intermediacy of 5,9-dioxodecanoic acid in the biosynthetic route to pinidine (Fig. 7). The failure of the latter to serve as a precursor of pinidine may indicate that a double bond at C_3–C_4 is required (XV) for formation of the alkaloid.

2.1.3 Coccinelline

Coccinelline (XVI) is an alkaloid isolated from the autodefense secretion of Coccinellideae. Tracer from [1-^{14}C]-acetate was incorporated at C-2 and C-10 in (XVI) (KARLSSON and LOSMAN, 1972), possibly according to Scheme 2 (Fig. 8).

2.2 Alkaloids Derived from Isoleucine

2.2.1 Pyrrolizidine Alkaloids

The C10 necic acids of senecic acid type (XVII) form from two moles of L-isoleucine, while the alkamines (the necines) originate from ornithine and putrescine

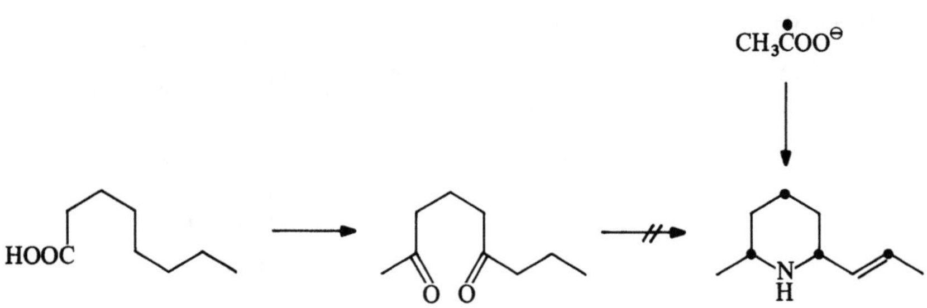

Fig. 6 (XIV)

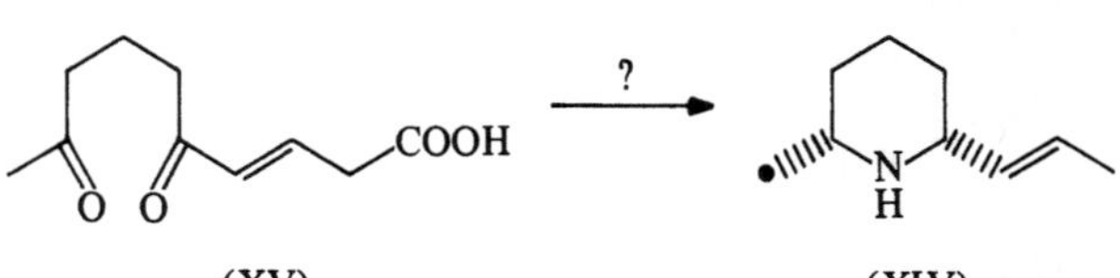

Fig. 7 (XV) (XIV)

(Fig. 9). Ornithine also gives the pyrrolidine ring in nicotine (LAMBERTS et al., 1959). Out of the four stereoisomers only L-isoleucine is incorporated (CROUT et al., 1972; DAVIES and CROUT, 1974) into senecionine (XVII) such specificity is rather unusual.

Evidence exists that monocrotalic acid (XVIII) (Fig. 10) is derived from acetate, (NOWACKI and BJERRUM, 1962), incorporation of L-[U-^{14}C]-isoleucine and L-[U-^{14}C]-threonine into the necic acid moiety of monocrotaline (XIX) has also been shown, however. Label from these two amino acids was found at carbons C-1, -2, -3, -6, and -7 of monocrotalic acid, with C-1, C-3, and C-7 being the most heavily labelled (ROBINS et al., 1974). Propionate is also a possible precursor; indeed, labelled propionate gave labelled (XIX) (Fig. 11) most probably via propionyl-coenzyme A (ROBINS et al., 1974). Valine proved to be the specific precursor of the echimidinic acid (XX) (Fig. 12) portion of heliosupine (CROUT, 1966).

Pulcherriminic acid (XXII), from *Candida pulcherrima*, is derived from two L-leucine units, via cyclodi-L-leucyl (XXI) (MACDONALD, 1965). Similar pathways are valid for a number of related metabolites, like flavacol, neoaspergillic, and neohydroxy aspergillic acids (MICETICH and MACDONALD, 1965) (Fig. 13).

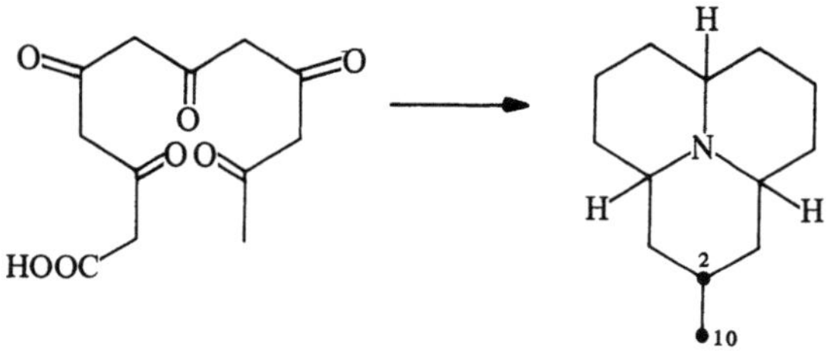

(XVI)

Fig. 8

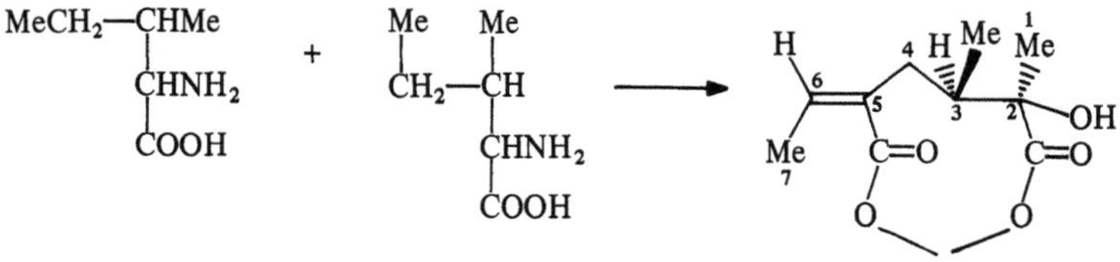

retronecine

(XVII)

Fig. 9

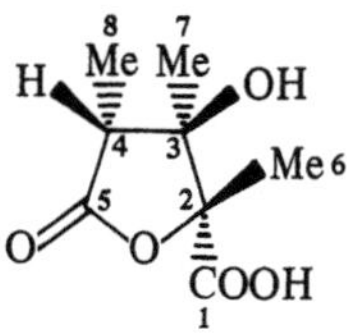

(XVIII)

Fig. 10

Fig. 11

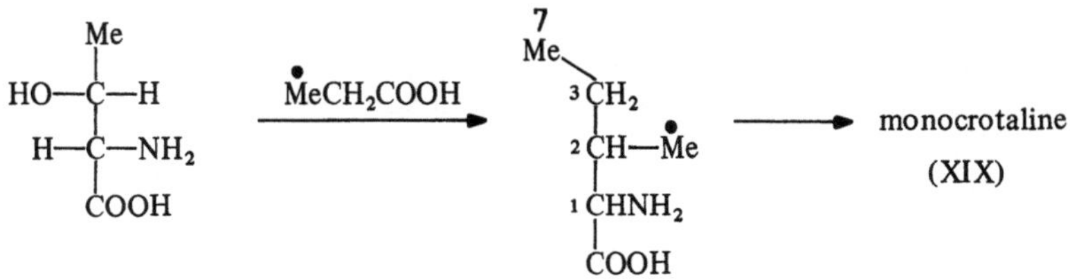

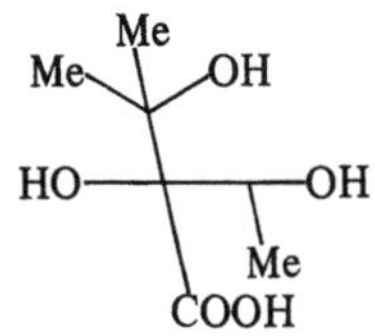

(XX)

Fig. 12

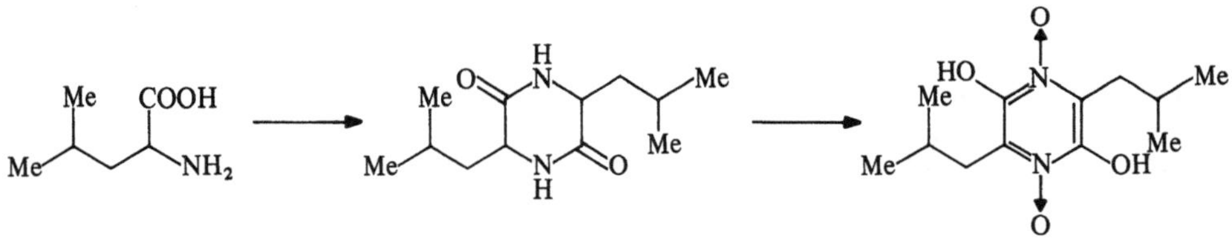

(XXI) (XXII)

Fig. 13

References

Askari, A., Melville, D.B.: J. Biol. Chem. Reaction sequency in ergothioneine biosynthesis herconine as an intermediate. *237*, 1615–1618 (1962)

Brochmann-Hanssen, E., Nunes, M.A., Olah, C.K.: On the biosynthesis of pilocarpine. Planta Med. *28*, 1–5 (1975)

Crout, D.H.G.: Pyrrolizidine alkaloids. The biosynthesis of echimidinic acid. J. Chem. Soc. (*C*), 1968–1972 (1966)

Crout, D.H.G., Davies, N.M., Smith, E.H., Whitehouse, D.: Pyrrolizidine alkaloids. The biosynthesis of senecic acid. J. Chem. Soc. Perkin I. 671–680 (1972)

Davies, N.M., Crout, D.H.G.: Pyrrolizidine alkaloid biosynthesis relative rates of incorporation of the isomers of isoleucine into the necic acid component of senecinine. J. Chem. Soc. Perkin I. 2079–2082 (1974)

Dawson, R.F.: Biosynthesis of the nicotiana alkaloids. Am. Sci. *48*, 321–364 (1960)

Fitzgerald, J.S.: Cinnamoylhistamine, the alkaloid of *Acacia argentea* and *A. polystacha*. Australian J. Chem. *17*, 375–378 (1964)

Genghof, D.S., van Damme, I.: Biosynthesis of ergothioneine and hercynine. J. Bacteriol. *87*, 852–862 (1964)

Karlsson, R., Losman, D.: The crystal structure of the hemihydrochloride of coccinellin, the defensive N-oxide alkaloid of the beetle *Coccinella septumpunctata*, a case of symmetrical hydrogen bonding. J. Chem. Soc. Chem. Commun. 626–627 (1972)

Lamberts, B.L., Dewey, L.J., Byerrum, R.U.: Ornithine as a precursor for the pyrrolidine ring in nicotine. Biophys. Acta *33*, 22–26 (1959)
Leete, E.: The biosynthesis of coniine from four acetate units. J. Am. Chem. Soc. *85*, 3523–3574 (1963)
Leete, E.: Biosynthesis of hemlock alkaloids. The incorporation of acetate-1-^{14}C into coniine and conhydrine. J. Am. Chem. Soc. *86*, 2509–2513 (1964)
Leete, E.: Biosynthesis of coniine from octanoic acid in hemlock plants. J. Am. Chem. Soc. *92*, 3835 (1970)
Leete, E., Adityachaudhury, N.: Biosynthesis of the hemlock alkaloids. II. The conversion of γ-coniceine to coniine and ψ-conhydrine. Phytochemistry *6*, 219–223 (1967)
Leete, E., Carver, R.A.: Synthesis of (−) pinidine and a putative biosynthetic precursor: 5,9-dioxodecanoic acid. J. Org. Chem. *40*, 2151–2153 (1975)
Leete, E., Juneau, K.N.: Biosynthesis of pinidine. J. Am. Chem. Soc. *91*, 5614–5618 (1969)
Leete, E., Lechleiter, J.C., Carver, R.A.: Determination of the "starter" acetate unit in the biosynthesis of pinidine. Tetrahedron Lett. 3779–3782 (1975)
Leete, E., Olson, J.O.: 5-Oxo-octanoic acid and 5-oxo-octanal precursors of coniine. J. Chem. Soc. Chem. Commun. 1651–1652 (1970)
MacDonald, J.C.: Biosynthesis of pulcherriminic acid. Biochem. J. *96*, 533–538 (1965)
Meister, A.: Biochemistry of amino acids, 2nd ed. New York: Academic Press 1965
Melville, D.B.: Ergothioneine. Vitam. Horm. *17*, 155–204 (1959)
Micetich, R.G., MacDonald, J.C.: Biosynthesis of neoaspergillic and neohydroxyaspergillic acid. J. Biol. Chem. *240*, 1692–1695 (1965).
Nowacki, E.K., Bjerrum, R.V.: Biosynthesis of the crotalaria alkaloids. Life Sci. *1*, 157–161 (1962)
O'Donovan, D.G., Horan, H.: Biosynthesis of dolicotheline in *Dolicothele sphaerica*. J. Chem. Soc., C 2083–2085 (1971)
Roberts, M.F.: Formation of γ-coniceine from 5-ketooctanal by transaminase of *Conium maculatum*. Phytochemistry *10*, 3057–3060 (1971)
Roberts, M.F.: Origin of the methyl carbon in methylconiine in *Conium maculatum*. Phytochemistry *13*, 1841–1845 (1974)
Roberts, M.F.: γ-Coniceine reductase in *Conium maculatum*. Phytochemistry *14*, 2393–2397 (1975)
Robins, D.J., Bale, N.M., Crout, D.H.G.: Pyrrolizidine alkaloids. Biosynthesis of monocrotalic acid, the necic acid component of monocrotaline. J. Chem. Soc. Perkin I. (1974)
Robinson, R.: The structural relations of natural products, Oxford: Clarendon Press 1955
Rosenberg, H., Paul, A.G.: Dolichothelin, an imidazole alkaloid from *Dolichothele sphaerica*. Tetrahedron Letters 1039–1042 (1969)
Rosenberg, H., Stohs, S.J.: Directed biosynthesis of unnatural alkaloids in *Dolichothele sphaerica*. Phytochemistry *13*, 823–827 (1974a)
Rosenberg, H., Stohs, S.J.: Aberrant formation of N-cinnamoyl histamine in *Dolichothele sphaerica*. Lloydia *34*, 313–314 (1974b)
Stowell, E.C.: In: Organic sulfur compounds. Kharasch, N. (ed.), Vol. 1, p. 462. Oxford: Pergamon Press 1961

4.5 Isoprenoid Alkaloids

J.G. RODDICK

1 General Introduction

Although the majority of plant alkaloids are of amine or amino acid origin (and where the nitrogen atom is part of a ring system, are referred to as "true alkaloids"), a considerable number of plants elaborate physiologically active nitrogenous bases which are wholly derived from isoprenoid precursors into which nitrogen is incorporated at a later stage in the biosynthetic sequence. Strictly, such compounds are nitrogenous isoprenoids, but since they satisfy most of the criteria which constitute the rather loose definition of an alkaloid they are commonly referred to as such. However, some qualification is desirable and, following the suggestion of HEGNAUER (1964), the term "pseudoalkaloid" has been adopted. Most classes of the naturally occurring lower isoprenoids are represented in this alkaloid group. Although, biogenetically, steroids are triterpenes, it is common practice to distinguish between the two groups, the former being characterised by possession of a cyclopentanoperhydrophenanthrene (sterane) ring system and less than 30 carbon atoms. Similarly, a distinction is usually made between terpenoid and steroidal alkaloids.

Certain alkaloids are partially isoprenoid in origin, part of the structure being derived from mevalonic acid (MVA) while the remainder (which contains the nitrogen) originates from some other pathway. Such alkaloids are usually classified according to their non-isoprenoid moiety (e.g., ajmalicine, reserpine, and vindoline which contain a monoterpenoid unit, and the ergot alkaloids which contain an isoprenoid C_5 unit are all classified as indole alkaloids) and will not be considered here.

Together, the terpenoid and steroidal alkaloids encompass a wide range of novel and interesting chemical forms and new structures are continually being discovered. However these two classes have not been widely utilised in plant chemotaxonomy (SMITH, 1976).

The literature on the terpenoid and steroidal alkaloids is quite extensive, but widely dispersed, and probably the most convenient single pool of information is Manske's celebrated series on the alkaloids (MANSKE and HOLMES, 1950–1954; MANSKE, 1955–1977). However, knowledge of the biochemistry and physiology of probably the majority of terpenoid and steroidal alkaloids is extremely limited and far outweighed by an abundance of chemical information. It is to be hoped that the recent upsurge of interest in what may be termed "natural product biology" will eventually rectify this imbalance.

2 Terpenoid Alkaloids

2.1 Introduction

The terpenoid alkaloids can be subdivided into the mono-, sesqui-, di-, and triterpenoid alkaloids. So far, no alkaloids of hemi-, sester-, tetra-, or polyterpenoid origin are known, but this does not rule out the possibility of the existence of at least some of these groups within the plant kingdom. This possibility is strengthened by the very recent confirmation of the existence of triterpenoid alkaloids (see Sect. 2.5). Of the large number of terpenoid alkaloids which have been categorised, only a few in each group have been shown conclusively to be derived from terpenoid precursors (Schütte, 1969); in the majority of cases, the terpenoid nature of an alkaloid has been inferred from its structure (Robinson, 1968).

2.2 Monoterpenoid Alkaloids

Monoterpenoid alkaloids have been isolated from plants belonging to the Actinidiaceae, Apocynaceae, Bignoniaceae, Dipsaceae, Gentianaceae, Loganiaceae, and Valerianaceae. Gentianine (I), the first monoterpenoid alkaloid to be identified, has been detected in a number of species, but the relatively high level reported in some plants has been shown (using $^{15}NH_3$) to arise from a reaction between ammonia in the extraction medium and non-nitrogenous precursors (Wildman et al., 1969). When the same plants were extracted with ammonia-free medium, gentianine was either not detectable or present only in trace quantities.

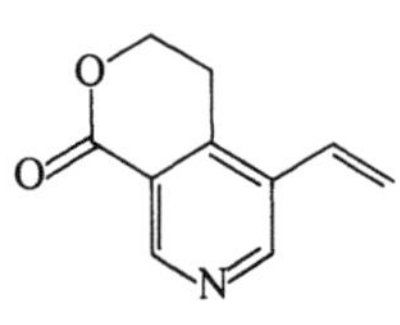

Gentianine
(I)

Gentianine and closely related alkaloids like gentianidine and gentioflavine are sometimes referred to as iridoid alkaloids, since they bear a close structural similarity to, and are probably derived from, monoterpenoid iridoid glycosides such as swertiamarin, gentiopicroside, etc. (Luckner, 1972). Other important monoterpenoid alkaloids, including actinidine (II) from *Actinidia polygama*, skytanthine (III) isomers (α, β- and δ-, of which β- is the most abundant) from *Skytanthus acutus*, tecostidine (IV), tecostanine, and tecomanine from *Tecoma stans* and the alkaloids of *Valeriana officinalis*, may also be biosynthesised from iridoid glycosides and here loganin would appear to be a possible precursor. This however remains purely speculative as Gross et al. (1972) did not observe incorporation of loganin or actinidine into the alkaloids of *T. stans*. Furthermore, although gentiopicroside can be converted to gentianine in vitro, it is not certain that a similar mechanism operates in vivo. Alternative proposals include synthesis from shikimic acid (Wen-

KERT and BRINGI, 1959) and from malonate (LEETE, 1965), and it is clear that more biosynthetic studies are required in this field.

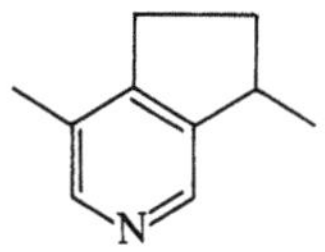

Actinidine
(II)

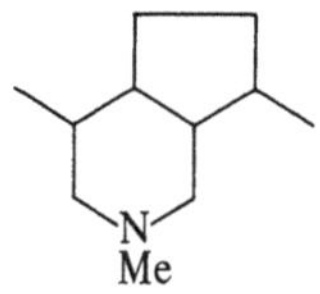

Skytanthine
(III)

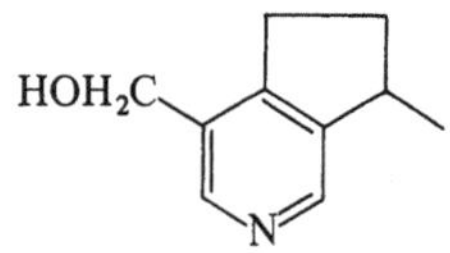

Tecostidine
(IV)

Although MVA is incorporated into skytanthine in mature *S. acutus* plants (AUDA et al., 1967), the route from this precursor is not clear and acetate has been reported to be a more efficient precursor of these alkaloids (EISENBRAUN personal communication, 1965). There are indications that the biosynthetic pathway may be variable and AUDA et al. (1967) found that the pattern of incorporation of tracer from [^{14}C]-MVA into β-skytanthine varied with the age of the plant. The alkaloid-synthesising ability of *T. fulva* has also been shown to vary widely with location (HILZ et al., 1973). Methionine may be involved in the methylation steps to β-skytanthine since the *N*-methyl group of the latter became labelled following incorporation of [methyl-^{14}C] methionine to *S. acutus* plants: on the other hand, radioactivity from [2-^{14}C]-lysine was not incorporated into the alkaloid (AUDA et al., 1967). Little is known about the distribution of skytanthine in the plant, but the root is probably a major organ of accumulation. β-Skytanthine has been detected in callus tissue derived from roots of *S. acutus* (LUCHETTI, 1965).

Monoterpenoid alkaloids are relatively non-toxic to micro-organisms and to mammals, the LD_{50} value for gentianine when administered intravenously to mice being of the order of 250 mg kg^{-1}. Nevertheless, high concentrations may cause stimulation of the CNS and respiratory rates (KENG TAO LIU et al., 1959), and anti-inflammatory (CHEN-YU SUNG et al., 1959) and hypoglycaemic (HAMMOUDA and AMER, 1966) effects.

2.3 Sesquiterpenoid Alkaloids

Four main groups of sesquiterpenoid alkaloid exist and they are classified according to the genus of the plant from which they were first isolated. The groups are *Nuphar* (Nymphaceae), *Dendrobium* (Orchidaceae), *Pogostemon* (Labiatae) and *Fabiana* (Solanaceae). Within each genus very few species have been examined.

A number of sesquiterpenoid alkaloids have been isolated from rhizomes of the water lily *Nuphar luteum*, e.g., nupharidine, deoxynupharidine (V), nupharamine, but some species of *Nuphar*, including *N. luteum*, elaborate C_{30}-derivatives which are dimers of C_{15} alkaloids. An interesting feature of these dimers, among which are thiobinupharidine (two nupharidine molecules) and neothiobinupharidine (VI, two deoxynupharidine molecules), is that they contain sulphur. Further, the sulphur atom which constitutes part of the bridge between the two C_{15} monomers is incorporated into a ring system. The terpenoid nature of these C_{30} alkaloids has been confirmed by incorporation experiments utilising [3,4-^{14}C] MVA as the precursor (SCHÜTTE and LEHFELDT, 1965). Scent glands of the North American beaver, *Castor canadensis,* contain a sesquiterpenoid alkaloid, castoramine (VALENTA and KHALEQUE, 1959), which is thought to be derived from deoxynupharidine obtained by ingestion of water lilies.

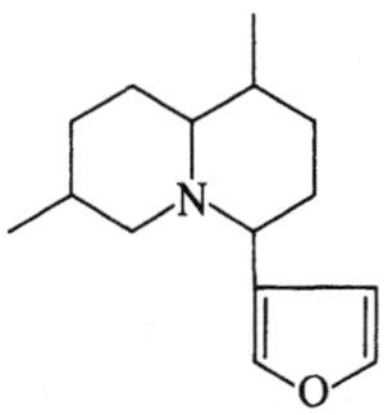

Deoxynupharidine
(V)

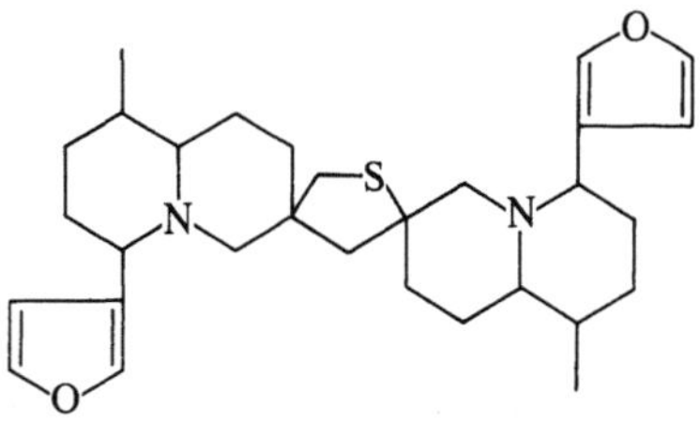

Neothiobinupharidine
(VI)

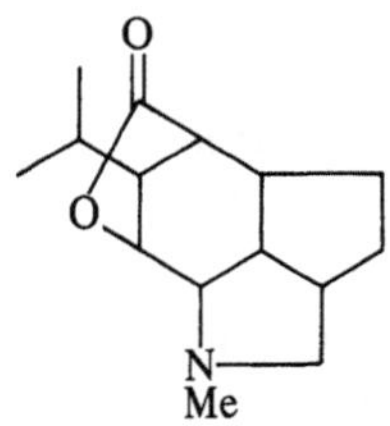

Dendrobine
(VII)

Two closely related major alkaloids viz. dendrobine (VII) and nobiline, as well as a number of minor alkaloids, have been isolated from the stem of the orchid *Dendrobium nobile* and other species of *Dendrobium*. Incorporation of [2-^{14}C]- and [4-^{14}C] MVA into dendrobine (YAMAZAKI et al., 1966; EDWARDS, 1969) and subsequent degradation indicated that this alkaloid is synthesised via the normal sesquiterpene biosynthetic pathway involving farnesol (YAMAZAKI et al., 1966). Dendrobine is one of the components of the Chinese drug Chin-Shih-Hu obtained from *D. nobile*, and causes contraction of isolated guinea pig uterus at low concentrations (ca. 0.003%) although the action is of brief duration (CHEN and CHEN, 1935).

Two C_{15} alkaloids, epiguaipyridine and patchoulipyridine, have been isolated from *Pogostemon patchouli* (BÜCHI et al., 1966). Although clearly similar in structure to known sesquiterpenes, their biogenetic relation to such remains to be confirmed. The biosynthesis of fabianine, a major alkaloid of *Fabiana imbricata*, is likewise uncertain. Debate centres on whether this C_{14} compound is synthesised from two isoprene units plus a C_4 unit or by degradation of a C_{15} skeleton such as cadinane (EDWARDS, 1969).

2.4 Diterpenoid Alkaloids

Although four different chemical classes exist viz. lycoctonine, heteratisine, veatchine, and atisine (PELLETIER and KEITH, 1970), the diterpenoid alkaloids are commonly, and more conveniently, classified into two broad groups based on structure and on pharmacological activity. These are: (1) the highly toxic and more heavily substituted ester bases known as the lycoctonines and aconitines – hydrolysis of compounds in this group yields a relatively non-toxic hexacyclic C_{19} alkamine: (2) the relatively simple and comparatively non-toxic C_{20} alkamines which show little substitution and are loosely referred to as the atisines. Of these two groups, the former is by far the most extensively documented.

The C_{19} alkaloids occur in a number of plants (e.g., *Inula royleana,* Compositae) but tend to be predominant in *Delphinium* sp. and *Aconitum* sp. (Ranunculaceae) where they occur in all the major plant organs. The basic hexacyclic skeleton consists of one 7-membered, three 6-membered and two 5-membered rings. All compounds contain a tertiary nitrogen substituted with either an ethyl or a methyl group and are generally heavily substituted with methoxyl or hydroxyl groups, one or two of the latter group usually being esterified with acetic acid and/or benzoic acid (or a derivative).

Although HERBERT and KIRBY (1963) could detect no radioactivity in delpheline when [2-^{14}C]-MVA was administered to the excised leaves of *D. elatum*, other workers have been able to demonstrate that the same precursor, MVA, was incorporated into brownine and lycoctonine in *D. brownii* (BENN and MAY, 1964) and delcosine in *D. ajacis* (FROST et al., 1967) when injected into the stem of the plant. These findings may have some bearing on the sites of synthesis in the plant. The relative synthetic capacity of the various organs is, however, not known, although the alkaloid-synthesising ability of roots is demonstrated by the detection of methyllycaconitine in cultured excised roots of *D. elatum* (STRZELECKA, 1966). The presence of methyllycaconitine in both roots and shoots of aseptically grown plants of the same age does not clarify the synthetic capacity of the shoot but the involvement of this organ is shown by the ability of detached leaves of *D. elatum* to incorporate label from [methyl-^{14}C] methionine into the *O*- and *N*-methyl groups of delpheline (HERBERT and KIRBY, 1963).

Levels of diterpenoid alkaloids appear to fluctuate from year to year and also with location (BRUTKO and MASSAGETOV, 1968). For example, during flowering, larkspur plants (*Delphinium* sp.) from the Illia region of the USSR had their highest alkaloid levels in the flowers, while those from the Dzhugar region showed greatest accumulation in the leaves (NIKISHCHENKO, 1966). Further support for an environmental influence on larkspur alkaloid levels comes from American work on the toxicity of these compounds to livestock (see below).

The C_{19} alkaloids of *Delphinium* sp. and *Aconitum* sp. are extremely toxic compounds, the lethal dose of aconitine (VIII) for man being ca. 2–5 mg, and aconite has a long history of criminal use. On the other hand, accidental poisoning in humans is rare in the USA and Europe, although slightly more frequent in India where roots of *A. chasmanthum* are widely used as a native medicine (CLARKE, 1970). Aconite preparations have been used medicinally for many centuries for the relief of such ailments as neuralgia, gout, and rheumatism. Topical application gives rise to a local anaesthetic effect whilst parenteral administration causes reductions in heart rate and blood pressure as well as curare-like neuromuscular effects.

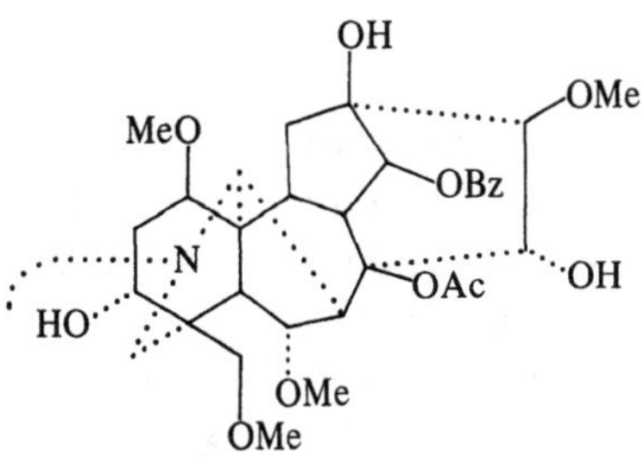

Aconitine
(VIII)

Poisoning by larkspur is the major cause of cattle deaths by poisonous plants in the USA (KINGSBURY, 1964). The most notorious plant is *D. barbeyi* which is widespread in the Western Ranges and is apparently attractive to livestock as they will consume it even when other forage is available (KEELER, 1975). This plant contains lycoctonine and a number of related compounds which cause death by paralysis of the respiratory centres. Horses and sheep are much less susceptible to larkspur poisoning, although this has been observed in sheep which have consumed *D. consolidum* present amongst garden refuse (CLARKE, 1970).

In contrast to the C_{19} alkaloids, the C_{20} atisines show little oxygen substitution and often contain no more than one methoxyl group. Usually they occur free although a few compounds have been identified as monoesters of acetic or benzoic acid. Prominent amongst this group are the *Garrya* alkaloids (Garryaceae) such as veatchine (IX) and garryfoline (X). These compounds are based on the veatchine skeleton which is derived from kaurene. Atisine (XI) and its derivatives, on the other hand, are found in a variety of genera including *Aconitum*, *Delphinium*, and *Spiraea* and thus often accompany one or more of the C_{19} alkaloids. Levels of atisines are normally low but roots of *A. heterophyllum* are reported to contain atisine at about 0.5%–1.0% (STERN, 1954).

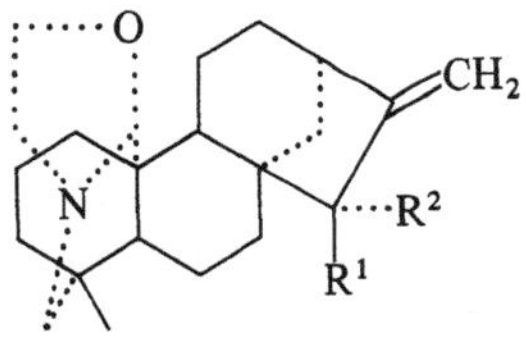

R^1 = H, R^2 = OH, Veatchine
(IX)
R^1 = OH, R^2 = H, Garryfoline
(X)

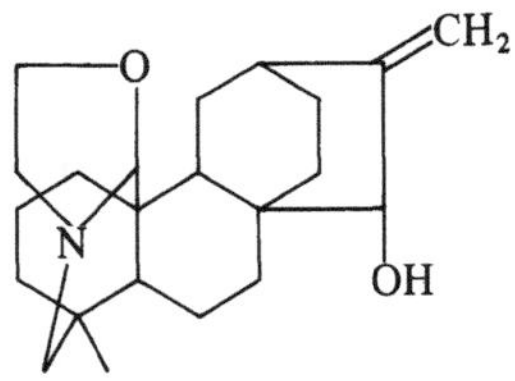

Atisine
(XI)

A scheme has been proposed (ROBINSON, 1968) for the biosynthesis of some of the C_{20} alkaloids (e.g., garryine and atisine) starting from the diterpenoid alcohol, nerolidol. The nitrogen atom and its attached C_2 unit are proposed to originate from ethanolamine or some other amino alcohol (ROBINSON, 1968; PELLETIER and KEITH, 1970) possibly by decarboxylation of an amino acid such as serine, but this still requires confirmation.

2.5 Triterpenoid Alkaloids

Of the alkaloids derived from a triterpenoid nucleus the steroidal alkaloids are by far the most thoroughly studied. Nevertheless, work conducted over the last decade points strongly to the existence of an alkaloid group based on the non-degraded triterpenoid skeleton (YAMAMURA and HIRATA, 1975).

A number of alkaloids have been isolated from *Daphniphyllum macropodum* and these may be divided into five groups according to the *N*-heterocyclic skeleton (YAMAMURA and HIRATA, 1975) although two types, [based on the C_{32} daphniphylline (XII) and the C_{27} yuzurimine (XIII)] predominate. The biosynthesis of these alkaloids is not yet fully understood but tracer studies with [^{14}C]-MVA support a terpenoid origin from 6 MVA units via a squalene-like intermediate (SUZUKI et al., 1973). This biosynthetic route has been further confirmed by degradation of labelled deacetyldaphniphylline and by incorporation of [^{14}C]-squalene into daphniphylline and the C_{30} compound codaphniphylline (SUZUKI et al., 1973). A scheme for the biosynthesis of daphniphylline from MVA has recently been proposed by YAMAMURA and HIRATA (1975).

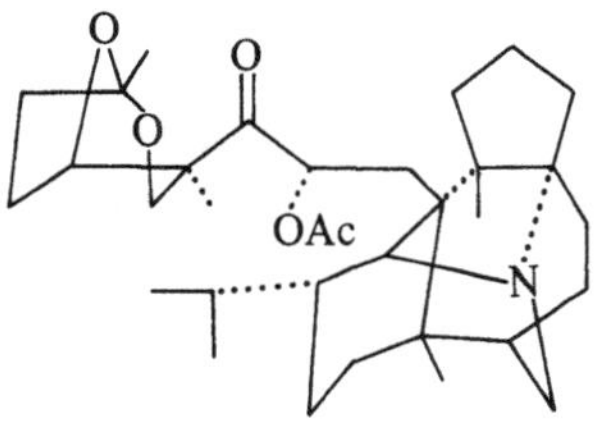

Daphniphylline
(XII)

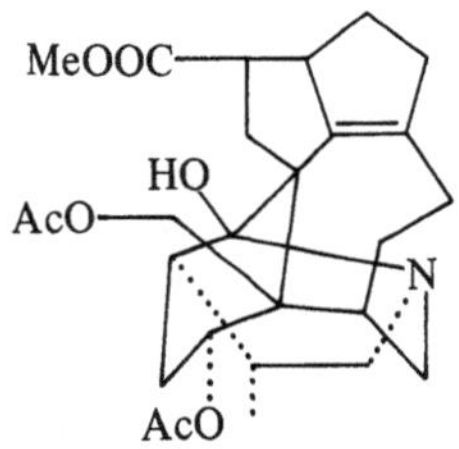

Yuzurimine
(XIII)

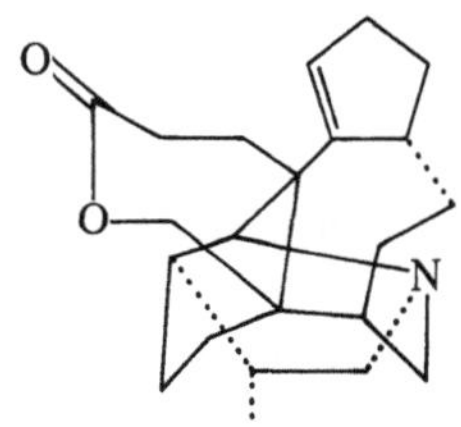

Daphnilactone B.
(XIV)

Related to the above alkaloids but with only 22 carbon atoms is daphnilactone B (XIV) which appears to be an intermediate between the two main types. Originally, it was thought to be derived from 4 MVA units together with one acetyl unit, but recent tracer studies by NIWA et al. (1973) support a route similar to that envisaged for the daphniphylline nucleus i.e., from 6 MVA via a squalene-like intermediate but with a later cleavage of a C_8 unit. Daphnilactone B has been isolated only from fruits of *Daphniphyllum* sp. (TODA et al., 1974) whereas daphniphylline and codaphniphylline are present also in leaves and bark (IRIKAWA et al., 1968). The types and amounts of these alkaloids apparently vary with the season (IRIKAWA et al., 1968), and SUZUKI et al. (1973) found that incorporation of [2-^{14}C] MVA into daphniphylline and codaphniphylline was maximal during the months of June and July.

A decoction of bark and leaves of *D. macropodum* has long been used in Japan as a vermicide and as a remedy for asthma. The *Daphniphyllum* alkaloids however are relatively toxic compounds and elicit a range of pharmacological effects from depression of the CNS to muscle relaxation and sedation (YAMAMURA and HIRATA, 1975).

3 Steroidal Alkaloids

3.1 Introduction

In contrast to the terpenoid alkaloids, the primary classification of which is based on biogenetic origin, the steroidal alkaloids tend to be categorised firstly according to plant taxonomic group and then according to chemical structure. Thus the four major groups of steroidal alkaloids are the Solanum, Veratrum, Apocynaceae, and Buxaceae alkaloids. This classification however is rather loose and may give rise to some confusion as the Solanum and Veratrum alkaloids are not restricted to these two genera respectively. (The words Solanum and Veratrum as applied to alkaloids are used in this review in an adjectival, rather than a generic, sense.) Steroidal alkaloids are also one of the very few groups of alkaloids found in the animal kingdom, although so far only in a few amphibians.

3.2 Solanum Alkaloids

The Solanum alkaloids occur as glycosides mainly in the Solanaceae and to a lesser extent in the Liliaceae and Asclepidiaceae. All alkamines are based on the C_{27} cholestane skeleton and can be divided into five groups according to structure, but most work has centred on the spirosolane (XV–XIX) and solanidane-type (XX–XXIII) compounds. The carbohydrate moiety may consist of up to four monosaccharides forming a branched structure linked to the steroid nucleus usually, but not always, at C-3. The leptines are monoacetylated glycoalkaloids found only in *Solanum chacoense* (SCHREIBER, 1968).

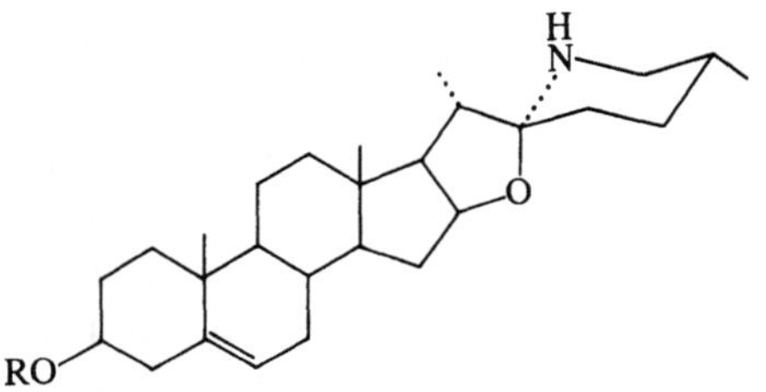

(XV) R = H, solasodine
(XVI) R = Rham–glu–gal–, α–solasonine

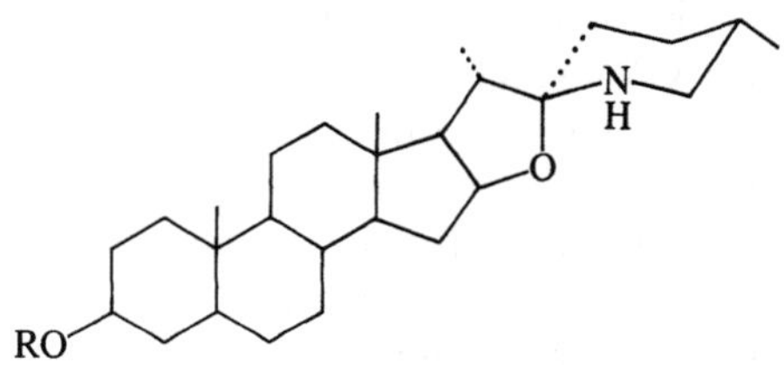

5αH { (XVII) R = H, tomatidine
5αH { (XVIII) R = Xyl–2glu–gal–, α–tomatine
Δ^5 (XIX) R = H, tomatidenol

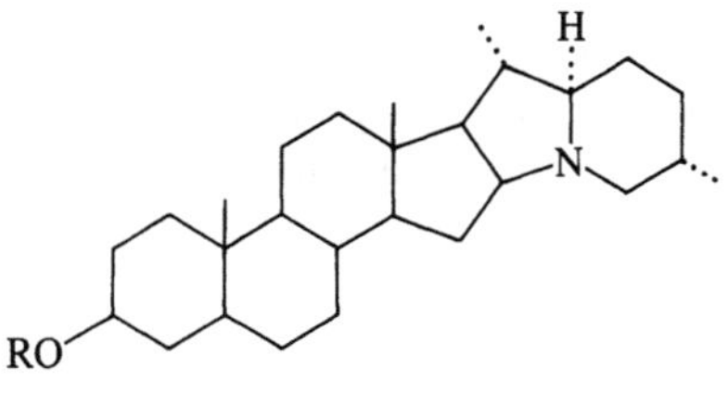

Δ^5 { (XX) R = H, solanidine
Δ^5 { (XXI) R = Rham–glu–gal–, α–solanine
Δ^5 { (XXII) R = 2Rham–glu–, α–chaconine
5αH (XXIII) R = Xyl–2glu–gal–, demissine

Most steroidal alkaloids are present throughout the plant, accumulating mainly in aerial organs, but some are more restricted in their distribution. Jurubine, for example, is present in roots of *S. paniculatum* but has not been detected in the leaves of this plant (SCHREIBER, 1974). In *S. laciniatum* and *Lycopersicon esculentum* the highest alkaloid levels are found in the flowers and young fruits but, as the fruits mature, alkaloid degradation occurs so that ripe fruits have very low or negligible amounts (SANDER, 1956; WILLUHN, 1967). As yet there is no convincing evidence for intercellular transport of steroidal alkaloids. Alkaloid levels vary not only with the species and variety but also with certain environmental factors. For example, the concentration of solasodine (XV) in *S. laciniatum* leaves varies with the maturity of the leaves (LANCASTER and MANN, 1975) whilst levels of potato alkaloids (XXI–XXII) are dependent upon location and cultural conditions (SINDEN and WEBB, 1974) and also subject to increase when the plants are stressed by damage or disease (LOCCI and KUĆ, 1967) or when the tubers are exposed to sunlight or UV-light (CONNER, 1937).

Experiments on grafting and the aseptic culture of excised roots of *L. esculentum* (SANDER, 1956) have established that synthesis of α-tomatine (XVIII) takes place independently in both root and shoot and is probably most active in meristematic regions (SANDER, 1956; RODDICK, 1974a). Nevertheless, unorganised cell cultures may still retain the capacity for alkaloid synthesis and α-solasonine (XVI) (KHANNA et al., 1976), solamargine (KADKADE and MADRID, 1977) and tomatine (RODDICK and BUTCHER, 1972) have been detected in callus tissue derived from a number of solanaceous plants, although levels were very low.

There is considerable evidence to suggest that the steroidal alkamines are synthesised via the normal steroid biosynthetic pathway operating in plants. Thus precursors such as [^{14}C]-acetate and/or [^{14}C]-MVA have been incorporated into the alkaloids of *S. laciniatum* (GUSEVA and PASESHNICHENKO, 1962), potato (GUSEVA et al., 1961) and tomato (SANDER and GRISEBACH, 1958), and [26, 27-^{14}C] cycloartenol (and to a lesser extent lanosterol) into solanidine (XX), tomatidine (XVII) and solanocapsine (RIPPERGER et al., 1971). The finding that [4-^{14}C]-cholesterol was incorporated into tomatidine (HEFTMANN et al., 1967) suggests that the route from cycloartenol to alkamine proceeds via lophenol and thence cholesterol or some other phytosterol. The co-occurrence of steroidal alkamines and sapogenins in many plants points to a biogenetic relationship between the two, and TSCHESCHE (1955) has proposed that both alkamines and sapogenins originate from a common pathway involving 16-dihydrokryptogenin. The nitrogen is thought to be incorporated into the alkamine prior to the final cyclizations (RONCHETTI et al., 1975) but the transamination process is not yet understood although evidence has recently been presented (KANEKO et al., 1976) that the nitrogen of solanidine in *Veratrum grandiflorum* is derived from L-arginine. Few studies have been made of the glycosylation reactions, but available evidence indicates that these are the final stages in glycoalkaloid biosynthesis involving UDP glycotransferase enzymes (LILJEGREN, 1971; LAVINTMAN et al., 1977). The intracellular sites of synthesis and glycosylation are uncertain but completed potato and tomato alkaloids accumulate mainly in the soluble phase of the cell, and probably in vacuoles (RODDICK, 1977).

Virtually nothing is known of the control of alkaloid biosynthesis. Thus the significance of reports that application of growth regulators increases (HEBLE et al., 1971; JEPPSEN et al., 1975) or decreases (EL-ANTABLY et al., 1975; MANN and LANCASTER, 1976) alkaloid levels cannot yet be fully assessed. The biochemical basis of UV-light enhancement of alkaloid synthesis in potato tubers is also obscure.

By using [4-^{14}C] tomatine, HEFTMANN and SCHWIMMER (1972) have shown that in developing tomato fruits tomatine is degraded to pregnenolone. No tomatine-degrading enzyme has yet been isolated from fruits although one (called tomatinase) has been identified in tomato leaves (PROKOSHEV et al., 1956). The report that trace quantities of pregnenolone have been detected in *L. pimpinellifolium* leaves (SCHREIBER and AURICH, 1966) may therefore be of some significance.

Steroidal alkaloids are toxic or inhibitory to a wide range of living organisms from bacteria to higher mammals (see NISHIE et al., 1971; RODDICK, 1974b; TUROVA et al., 1961). Of particular interest are the reputed tumour-inhibiting properties of β-solamarine from *S. dulcamara* (KUPCHAN et al., 1965). Amongst lower organisms inhibition is mostly a consequence of membrane disruption whilst in mammals toxicity invariably results from depression of the CNS and interference with respiratory and cardiovascular function. A few fatalities from glycoalkaloid poisoning have been recorded in man (WILLIMOTT, 1933) but poisoning of animals, especially livestock, is much more common (CLARKE, 1970). Of probably greater concern however is the reported teratogenicity of alkaloid-rich blighted potato tubers (POSWILLO et al., 1972) and embryotoxicity of α-solanine (XXI) (KLINE et al., 1961; BELL et al., 1976).

The mode of toxicity of the Solanum alkaloids is not well understood. Membrane lytic effects of some compounds may arise from surfactant properties but binding to membrane sterols is thought to be the principal mode of action (DÉFAGO, 1977). The claim that the aglycone is the active moiety (SCHÖNBECK and SCHLÖSSER, 1976), and the relative toxicity of glycoside and aglycone still require clarification. The pharmacological effects of solanine have been likened to those of the cardiac glycoside K-strophanthoside (NISHIE et al., 1971).

The contribution of Solanum alkaloids to the plant's defensive system against pathogens and predators is a matter of some debate. Some authors (e.g., SINDEN et al., 1973) hold that these compounds confer some protection against fungal pathogens, while others (e.g., FRANK et al., 1975) claim there is no correlation between alkaloid levels and resistance. Yet others (e.g., SCHREIBER, 1957; FRAENKEL, 1959) believe that these alkaloids may be more important as repellants of phytophagous insects and possibly other predators.

In view of the uncertainty of future supplies of the steroidal sapogenin diosgenin, certain spirosolane alkaloids e.g., solasodine (the N analogue of diosgenin) and tomatidenol (XIX), possess considerable potential as starting compounds for the industrial production of hormonal steroids and a number of chemical transformations to pregnane derivatives, as well as microbial hydroxylations, have been described (SCHREIBER, 1968). The possibility of large-scale cultivation of *S. laciniatum, S. aviculare*, and *S. khasianum* for solasodine production and *S. dulcamara* for tomatidenol production is currently being explored, but more knowledge is still needed of the factors which influence and govern the synthesis, accumulation, and degradation of these alkaloids.

3.3 Veratrum Alkaloids

The major alkamines of *Veratrum* sp. (Liliaceae) belong to the C-nor-D-homo-steroid series although, as indicated above, Solanum-type alkaloids based on the solanidane skeleton and also the 22, 26-epiminocholestane skeleton also occur in some plants. The C-nor-D-homo-steroidal alkaloids can be classified in various ways (e.g., TOMKO and VOTICKÝ, 1973; FIESER and FIESER, 1959), but the system of the latter authors which divides the alkaloids into jerveratrum (XXIV–XXV) and ceveratrum (XXVI) types based on oxygen substitution is adopted here. The jerveratrum alkamines contain 1–3 oxygen atoms and usually occur free or as glucosides, whereas the ceveratrum alkamines contain 7–9 oxygen atoms (mainly within hydroxyl groups) and normally occur as esters (of, for example, acetic acid, angelic acid, etc.) although occasionally they are found free (KUPCHAN and BY, 1968). Steroidal alkaloids of the C-nor-D-homo-series are also present in some species of *Fritillaria* (Liliaceae) and are usually of the jerveratrum type.

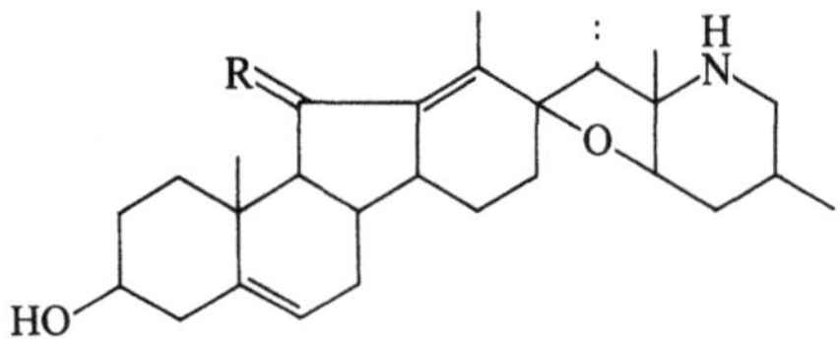

(XXIV) R = O, Jervine
(XXV) R = H_2, 11–deoxojervine (cyclopamine)

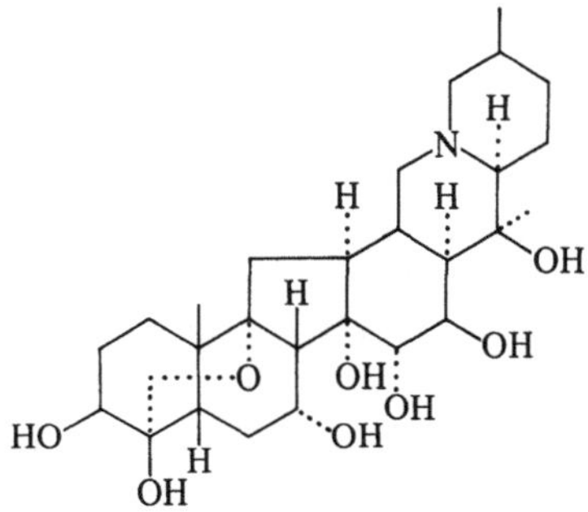

(XXVI) Germine

Not much work has been done on alkaloid production in the plant but Veratrum alkaloids are known to be present in both aerial and subterranean organs (particularly rhizomes) although the alkaloid complement may vary with the organ (KANEKO et al., 1976). Biosynthetic studies have established that ^{14}C-labelled acetate and cholesterol are incorporated into both jerveratrum and ceveratrum alkaloids in *V. grandiflorum* (KANEKO et al., 1970a). Sterol incorporation rates were low, probably due to solubility problems, but labelled acetate was incorporated into jerveratrum, ceveratrum, and solanidane-type alkaloids to approximately the same extent, suggesting that all three types originate from the normal steroid biosynthetic pathway. Some evidence for feedback control of alkaloid synthesis exists in that veratramine biosynthesis was influenced by the amount of jervine (XXIV) present (KANEKO et al., 1970a), and addition of 11-deoxojervine (XXV), a precursor of jervine, inhibited incorporation of [1-^{14}C]-acetate into jervine (KANEKO et al., 1970b). The finding that solanidyl glycoside accumulated in leaves of etiolated *V. grandiflorum* plants but was converted to jerveratrum alkaloids in the rhizome on illumination suggests that solanidine is an important precursor of alkaloids such as jervine and veratramine (KANEKO et al., 1972). A hypothetical scheme for solanidine biosynthesis in *Veratrum* sp. has recently been proposed (KANEKO et al., 1976).

Veratrum alkaloids are effective insecticides and also growth-inhibitory to some fungi and plants, and it has been proposed (OLNEY, 1968) that a specific effect on DNA stability is involved. *Zygadenus* sp. and *V. californicum* are particularly dangerous plants for sheep because of their alkaloid content, the former being highly toxic and causing considerable losses whilst the latter are teratogenic. A number of teratogenic symptoms are known, but the most striking is a cyclopian malformation in lambs caused mainly by cyclopamine (11-deoxojervine) (and possibly also by its 3-glucoside and jervine) ingested by the pregnant ewe (KEELER, 1975).

Some Veratrum alkaloids, particularly of the ceveratrum type, show effects on striated muscle contraction and cardiovascular function and may also induce respiratory depression (DALLEMAGNE, 1955). A number of alkaloids such as naturally occurring esters of germine (XXVI) and protoverine are very effective in reducing blood pressure and are used clinically as anti-hypertensive agents (KUPCHAN and FLACKE, 1967). The use of C-nor-D-homo-steroidal alkaloids such as jervine as starting compounds for the production of novel pharmaceutical steroids has been suggested (MASAMUNE and ORITO, 1972).

3.4 Alkaloids of the Apocynaceae

Plants belonging to a number of genera of this family e.g., *Holarrhena*, *Funtumia*, *Malouetia*, elaborate steroidal alkaloids which fall into one of two groups according to structure, viz. those based on a simple pregnane skeleton (XXVII–XXVIII) and those on a conanine skeleton (XXIX) (ČERNÝ and ŠORM, 1967). Most alkaloids found in this family exist in a free form but a few are found as esters or glycosides. Alkaloids appear in all plant organs, including bark and seeds, although the alkaloid complement varies between organs and also between species.

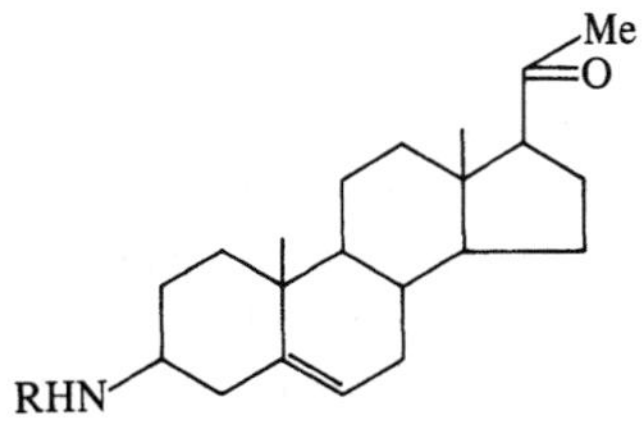

(XXVII) R = Me, Holaphylline
(XXVIII) R = H, Holaphyllamine

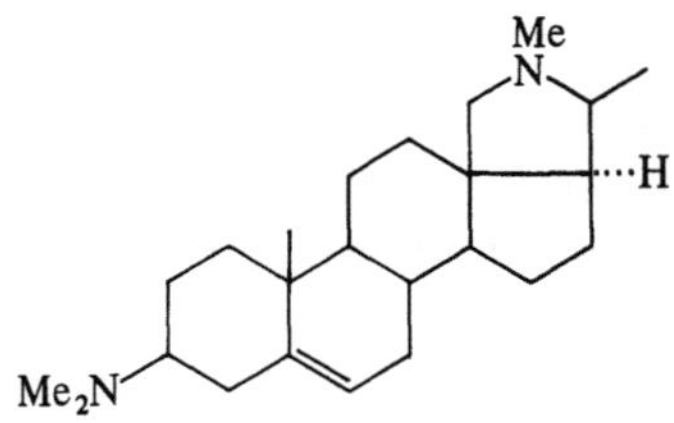

(XXIX) Conessine

The alkaloids of the Apocynaceae are thought to be derived from sterol or triterpenoid precursors via oxygenated intermediates, and neutral ketosteroids such as holadysone and progesterone have been detected in *Holarrhena* sp. Although progesterone is a likely intermediate in the biosynthesis of the alkaloids holaphylline (XXVII) and holaphyllamine (XXVIII), labelled progesterone was not incorporated into these compounds whereas [^{14}C]-pregnenolone and [^{14}C]-cholesterol were (BENNETT and HEFTMANN, 1965a, b). The exact biochemical sequence is not yet clear but synthesis from a C_{21} precursor seems likely (JANOT et al., 1963) and amination at C-20 is apparently preceded by introduction of oxygen at C-18 (ČERNÝ and ŠORM, 1967). Conessine (XXIX), a major alkaloid of *H. antidysenterica*, has been detected in trace quantities in hypocotyl callus (HEBLE et al., 1976a) and radioactive-feeding experiments have shown that it is synthesised by this tissue rather than being transported from the explant (HEBLE et al., 1976b).

The possibility of using certain alkaloids of this group for the preparation of pharmaceutical steroids has been raised, in particular C-18 substituted compounds like holarrhimine and conessine for the synthesis of aldosterone and related corticosteroids (GOUTAREL, 1971).

Whereas some alkaloids of this group e.g., holaphylline and holaphyllamine, are not particularly toxic, others are highly so, and malouetine from *Malouetia bequaertiana* has powerful curare-like properties. Both this compound and irehdiamine-A from *Funtumia* sp. are reported to inhibit DNA synthesis (MAHLER and GREEN, 1970) by interfering with helix formation and stability (WARING and CHISHOLM, 1972). However, the primary inhibitory effect is believed (SILVER and LEVINE, 1968) to be at the membrane level.

Alkaloids from *H. antidysenterica* (Kurchi bush) have long been used in India as antidysenterics and anthelmintics (MANSKE, 1955) but some doubts have been expressed as to their efficacy. Pharmacological studies of the *Holarrhena* alkaloid holamine have revealed a number of effects mainly associated with the functioning of the CNS (BLANPIN, 1960).

A new class of steroidal alkaloids, the aminoglycosteroids, have recently been detected in some *Holarrhena* species. These compounds, the first of which to be isolated and characterised was holacurtine (XXX) (JANOT et al., 1968), are composed of an amino sugar attached to a steroidal genin.

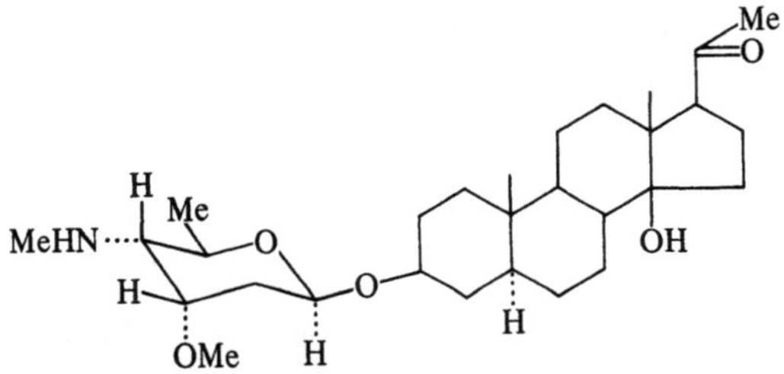

Holacurtine
(XXX)

3.5 Alkaloids of the Buxaceae

A number of methods of classifying *Buxus* alkaloids exists (TOMKO and VOTICKÝ, 1973). The majority of structures can be divided into two groups according to whether they have an additional cyclopropane ring [e.g., cyclobuxine-D (XXXI)] or an expanded ring B [e.g., buxamine-E (XXXII)]. Other important taxonomic features are the number of nitrogen atoms (at C-3 and/or C-20) and the substitution at these atoms. However, the number of possible structural variations based on these features is so large that both mono- and di-amino alkaloids are usually classified from tables which employ letter suffixes to represent substitution patterns (see GOUTAREL, 1971).

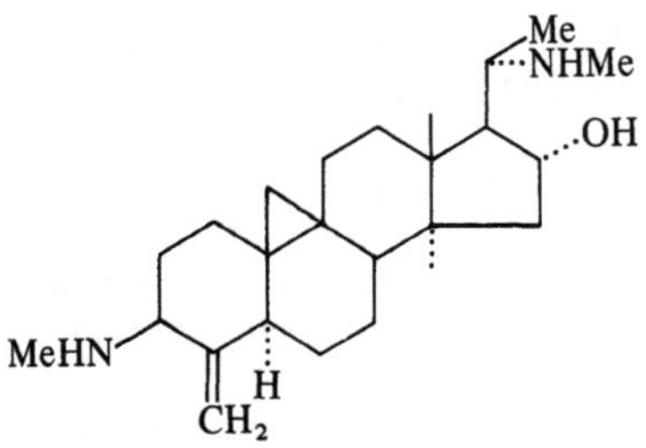

Cyclobuxine–D
(XXXI)

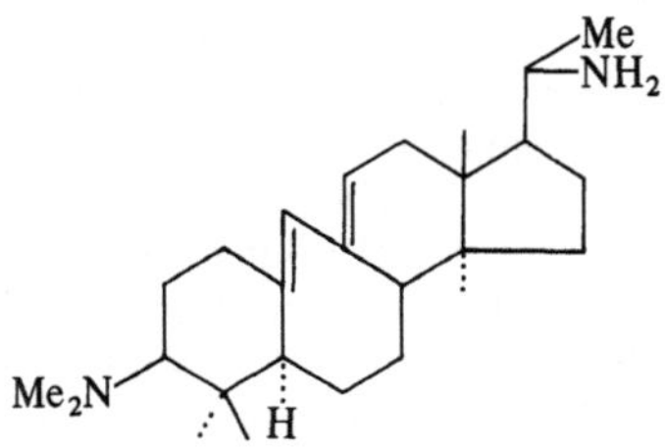

Buxamine–E
(XXXII)

Although no biosynthetic studies have yet been conducted, *Buxus* alkaloids appear to be constructed on a pregnane skeleton and their close similarity to cycloartenol and cycloeucalenol is strongly indicative of an origin in compounds of this type. Hypothetical biosynthetic schemes along such lines have in fact been proposed (KHUONG-HUU-LAINÉ et al., 1966). The presence of steroidal ketones in some *Buxus* sp. and *Paravallaris* sp. (POTIER et al., 1964) raises the further possibility of a biogenetic relationship between such compounds and the steroidal amines.

The common box (*Buxus sempervirens*) is a highly poisonous plant and fatalities have been observed in domesticated animals after consuming it (CLARKE, 1970). The principal alkaloid of this plant, cyclobuxine-D, is known to influence the stability of helical polynucleotides such as DNA and s-RNA (MAHLER and DURTON, 1964) but it is not certain if this is the basis of its in vivo toxicity.

More simple pregnane-type alkaloids, normally lacking both the cyclopropane ring or expanded ring B and substitution at C-4 and C-14, have been isolated from *Pachysandra* sp. and *Sarcococca* sp. No biosynthetic details are available but their structural similarity to some *Holarrhena* alkaloids suggests they may be synthesised by a similar route (ČERNÝ and ŠORM, 1967).

References

Auda, H., Juneja, H.R., Eisenbraun, E.J., Waller, G.R., Kays, W.R., Appel, H.H.: Biosynthesis of methylcyclopentane monoterpenoids I. *Skytanthus* alkaloids. J. Am. Chem. Soc. *89*, 2476–2482 (1967)

Bell, D.P., Gibson, J.G., McCarroll, A.M., McLean, G.A.: Embryotoxicity of solanine and aspirin in mice. J. Reprod. Fert. *46*, 257–259 (1976)

Benn, M.H., May, J.: The biosynthesis of diterpenoid alkaloids. Experientia *20*, 252–253 (1964)

Bennett, R.D., Heftmann, E.: Biosynthesis of *Holarrhena* alkaloids from pregnenolone and progesterone. Phytochemistry *4*, 873–879 (1965a)

Bennett, R.D., Heftmann, E.: Biosynthesis of *Holarrhena* alkaloids from cholesterol. Arch. Biochem. Biophys, *112*, 616–620 (1965b)

Blanpin, O.: Sur l'holamine, alcaloïde stéroïde des feuilles d'*Holarrhena floribunda* (G. Don) Dur. et Schinz. Apocynacées. I. Étude des propriétés pharmacodynamique. C.R. Séanc. Soc. Biol. *154*, 1587–1591 (1960)

Brutko, L.I., Massagetov, P.S.: Delphinium species as a source of raw materials for the production of drugs. I. The content of alkaloids in the aerial parts. Rastit. Resur. *4*, 46–49 (1968) [Chem. Abstr. *69*, 30133n (1968)]

Büchi, G., Goldman, I.M., Mayo, D.W.: The structures of two alkaloids from patchouli oil. J. Am. Chem. Soc. *88*, 3109–3113 (1966)

Černý, V., Šorm, F.: Steroid alkaloids: alkaloids of Apocynaceae and Buxaceae. In: The alkaloids. Chemistry and physiology, Ch. 8. Manske, R.H.F., (ed.), Vol. IX, pp. 305–426. London: Academic Press 1967

Chen, K.K., Chen, A.L.: The pharmacological action of dendrobine, the alkaloid of chin-shih-hu. J. Pharmacol. *55*, 319–325 (1935)

Chen-Yu Sung, Hsiu-Chuan Chi, Keng Tao Liu: Pharmacology of gentianine. I. Anti-inflammatory effect and action of pituitary-adrenal function of the rat. Sheng Li Hsueh Pao *22*, 201–205 (1958) [Chem. Abstr. *53*, 13415g (1959)]

Clarke, E.G.C.: The forensic chemistry of alkaloids. In: The alkaloids. Chemistry and physiology, Ch. 7. Manske, R.H.F. (ed.), Vol. XII, pp. 514–589. London: Academic Press 1970

Conner, H.W.: The effect of light on solanine synthesis in the potato tuber. Plant Physiol. *12*, 79–98 (1937)

Dallemagne, M.J.: Respiratory stimulants. In: The alkaloids. Chemistry and physiology, Ch. 40. Manske, R.H.F. (ed.), Vol. V, pp. 109–139. London: Academic Press 1955

Défago, G.: Rôle des saponines dans la resistance des plantes aux maladies fongiques. Ber. Schweiz. Bot. Ges. *87*, 79–132 (1977)

Edwards, O.E.: Sesquiterpene alkaloids. In: Cyclopentanoid terpene derivatives, Ch. 6. Taylor, W.I., Battersby, A.R. (eds.), pp. 357–408. New York: Marcel Dekker 1969

Eisenbraun, E.J.: Quoted in Wildman et al. (1969) p. 274

El-Antably, H.M.M., Ahmed, S.S., Eid, M.N.A.: The effect of gibberellic acid alone and when combined with [2-chloroethyl]-trimethyl ammonium chloride on the growth and alkaloid content of *Solanum laciniatum* Aiton. Pharmazie *30*, 397–399 (1975)

Fieser, L.F., Fieser, M.: Steroids. New York: Reinhold 1959

Fraenkel, G.: The *raison d'être* of secondary plant substances. Science *129*, 1466–1470 (1959)

Frank, J.A., Wilson, J.M., Webb, R.E.: The relationship between glycoalkaloids and disease resistance in potatoes. Phytopathology *65*, 1045–1049 (1975)

Frost, G.M., Hale, R.L., Waller, G.R., Zalkow, L.H., Girotra, N.N.: Biosynthesis of diterpenoid alkaloids in *Delphinium ajacis*. Chem. Ind. (London) 320–321 (1967)

Goutarel, R.: Steroidal alkaloids of the Apocynaceae and Buxaceae. In: The alkaloids, Ch. 17. Saxton, J.E. (ed.), Vol. 1, pp. 382–443. Specialist periodical reports. London: The Chemical Society 1971

Gross, D., Berg, W., Schütte, H.R.: Zur Biosynthese und zum Stoffwechsel der monoterpenoiden Alkaloide von *Tecoma stans*. Biochem. Physiol. Pflanz. *163*, 576–585 (1972)

Guseva, A.R., Paseshnichenko, V.A.: A study of solasodine biosynthesis by the method of oxidative breakdown. Biochemistry (USSR) *27*, 721–726 (1962)

Guseva, A.R., Paseshnichenko, V.A., Borikhina, M.G.: Synthesis of radioactive mevalonic acid and its use in the study of the biosynthesis of steroid glycoalkaloids from *Solanum*. Biochemistry (USSR) *26*, 631–635 (1961)

Hammouda, Y., Amer, S.: Antidiabetic effect of tecomine and tecostanine. J. Pharm. Sci. *55*, 1452–1454 (1966)

Heble, M.R., Narayanaswamy, S., Chadha, M.S.: Hormonal control of steroid synthesis in *Solanum xanthocarpum* tissue cultures. Phytochemistry *10*, 2393–2394 (1971)

Heble, M.R., Narayanaswamy, S., Chadha, M.S.: Studies on growth and steroid formation in tissue cultures of *Holarrhena antidysenterica*. Phytochemistry *15*, 681–682 (1976a)

Heble, M.R., Narayanaswamy, S., Chadha, M.S.: Metabolism of cholesterol by callus culture of *Holarrhena antidysenterica*. Phytochemistry *15*, 1911–1912 (1976b)

Heftmann, E., Schwimmer, S.: Degradation of tomatine to 3β-hydroxy-5α-pregn-16-en-20-one by ripe tomatoes. Phytochemistry *11*, 2783–2787 (1972)

Heftmann, E., Lieber, E.R., Bennett, R.D.: Biosynthesis of tomatidine from cholesterol in *Lycopersicon pimpinellifolium*. Phytochemistry *6*, 225–229 (1967)

Hegnauer, R.: Chemotaxonomie der Pflanzen. Vol. 3. Basel: Birkhäuser 1964

Herbert, E.J., Kirby, G.W.: Mevalonic acid and delpheline biosynthesis. Tetrahedron Lett. 1505–1506 (1963)

Hilz, W., Edelmann, G., Appel, A.H.: Chemical composition of *Tecoma fulva*. Rev. Latinoamer. Quim. *4*, 28–31 (1973). [Chem. Abstr. *79*, 123631 v (1973)]

Irikawa, H., Sakabe, N., Yamamura, S., Hirata, Y.: The structures of daphniphylline and codaphniphylline. Tetrahedron *24*, 5691–5700 (1968)

Janot, M.-M., Truong-Ho, M., Khuong-Huu, Q., Goutarel, R.: Alcaloïdes stéroïdiques XX: l'irehline, nouvel alcaloïde retiré des feuilles du *Funtumia elastica* (Preuss) Stapf. Bull. Soc. Chim. (Paris) 1977–1979 (1963)

Janot, M.-M., Devissaguet, P., Khuong-Huu, Q., Parello, J., Bisset, N.G., Goutarel, R.: Alcaloïdes stéroïdiques. Un nouveau type de glyco-alcaloïde stéroïdique: holacurtine et holacurténine, retirées des feuilles de l'*Holarrhena curtisii* King et Gamble (Apocynacées). C.R. Acad. Sci. Paris Ser. C *266*, 388–391 (1968)

Jeppsen, R.B., Salunkhe, D.K., Wu, M.T., Sharma, R.P.: Effects of N^6-benzyladenine on glycoalkaloids in potatoes as determined by thin layer chromatography. Experientia *31*, 345–346 (1975)

Kadkade, P.G., Madrid, T.R.: Glycoalkaloids in tissue cultures of *Solanum acculeatissimum*. Naturwissenschaften *64*, 147 (1977)
Kaneko, K., Mitsuhashi, H., Hirayama, K., Yoshida, N.: Biosynthesis of C-nor-D-homo-steroidal alkaloids from acetate-1-^{14}C, cholesterol-4-^{14}C and cholesterol-26-^{14}C in *Veratrum grandiflorum*. Phytochemistry *9*, 2489–2495 (1970a)
Kaneko, K., Mitsuhashi, H., Hirayama, K., Ohmori, S.: 11-deoxojervine as a precursor for jervine biosynthesis in *Veratrum grandiflorum*. Phytochemistry *9*, 2497–2501 (1970b)
Kaneko, K., Watanabe, M., Taira, S., Mitsuhashi, H.: Conversion of solanidine to jerveratrum alkaloids in *Veratrum grandiflorum*. Phytochemistry *11*, 3199–3202 (1972)
Kaneko, K., Tanaka, M.W., Mitsuhashi, H.: Origin of nitrogen in the biosynthesis of solanidine by *Veratrum grandiflorum*. Phytochemistry *15*, 1391–1393 (1976)
Keeler, R.F.: Toxins and teratogens of higher plants. Lloydia *38*, 56–86 (1975)
Keng Tao Liu, Hsiu-Chuan Chi, Chen-Yu Sung: The toxicity and pharmacology of gentianine isolated from *Gentiana macrophylla*. Sheng Li Hsueh Pao *23*, 203–10 (1959) [Chem. Abstr. *57*, 15764e (1962)]
Khanna, P., Uddin, A., Sharma, G.L., Manot, S.K., Rathore, A.K.: Isolation and characterization of sapogenin and solasodine from *in vitro* tissue cultures of some solanaceous plants. Indian J. Exp. Biol. *14*, 694–696 (1976)
Khuong-Huu-Lainé, F., Milliet, A., Bisset, N.G., Goutarel, R.: XLVII – Famille des Buxacées (5e communication) (1). Alkaloïdes du *Buxus rolfei* Vidal: cyclorolfeibuxine C, cyclobuxoxazine A, cyclométhoxazine B, cyclorolféine et cyclorolfoxazine. Bull. Soc. Chim. (Paris) 1216–1221 (1966)
Kingsbury, J.M.: Poisonous plants of the United States and Canada. New Jersey: Prentice-Hall – Englewood 1964
Kline, B.E., Von Elbe, H., Dahle, N.A., Kupchan, S.M.: Toxic effects of potato sprouts and of solanine fed to pregnant rats. Proc. Soc. Exp. Biol. *107*, 807–809 (1961)
Kupchan, S.M., By, A.W.: Steroid alkaloids: the *Veratrum* group. In: The alkaloids. Chemistry and physiology, Ch. 2. Manske, R.H.F. (ed.), Vol. X, pp. 193–285. London: Academic Press 1968
Kupchan, S.M., Flacke, W.E.: Hypotensive *Veratrum* alkaloids. In: Antihypertensive agents, Ch. 12. Schlittler, E. (ed.), pp. 429–458. London: Academic Press 1967
Kupchan, S.M., Barboutis, S.J., Knox, J.R., Lau Cam, C.A.: Beta-solamarine: tumor inhibitor isolated from *Solanum dulcamara*. Science *150*, 1827–1828 (1965)
Lancaster, J.E., Mann, J.D.: Changes in solasodine content during the development of *Solanum laciniatum* Ait. N. Z. J. Agric. Res. *18*, 139–144 (1975)
Lavintman, N., Tandecarz, J., Cardini, C.E.: Enzymatic glucosylation of steroid alkaloids in potato tuber. Plant Sci. Lett. *8*, 65–70 (1977)
Leete, E.: Biosynthesis of alkaloids. Science *147*, 1000–1006 (1965)
Liljegren, D.R.: Glucosylation of solasodine by extracts from *Solanum laciniatum*. Phytochemistry *10*, 3061–3064 (1971)
Locci, R., Kuć, J.: Steroid alkaloids as compounds produced by potato tubers under stress. Phytopathology *57*, 1272–1273 (1967)
Luchetti, M.A.: Biosintesi delle skitantine in vitro. Ann. Ist. Super. Sanita *1*, 563–565 (1965)
Luckner, M.: Secondary metabolism in plants and animals. London: Chapman and Hall 1972
Mahler, H.R., Durton, G.: Nucleic acid interactions. V. Effects of cyclobuxine. J. Mol. Biol. *10*, 157–175 (1964)
Mahler, H.R., Green, G.: Interaction of steroidal diamines with DNA and polynucleotides. Ann. N.Y. Acad. Sci. *171*, 783–800 (1970)
Mann, J.D., Lancaster, J.E.: Effects of herbicides and growth regulators on solasodine production by *Solanum aviculare* and *S. laciniatum* Ait. N. Z. J. Exp. Agric. *4*, 347–350 (1976)
Manske, R.H.F.: Minor alkaloids of unknown structure. In: The alkaloids. Chemistry and physiology, Ch. 48. Manske, R.H.F. (ed.), Vol. V, pp. 301–332. London: Academic Press 1955
Manske, R.H.F. (ed.): The alkaloids. Chemistry and physiology. Vols. V–XVII, London: Academic Press 1955–1977
Manske, R.H.F., Holmes, H.L. (eds.): The alkaloids. Chemistry and physiology, Vols. I–IV, London: Academic Press 1950–1954

Masumune, T., Orito, T.: The synthesis of an etiojervane analog of corticosterone. Bull. Chem. Soc. Jpn. *45*, 1888–1892 (1972)
Nikishchenko, T.K.: Phytochemical study of Illia larkspur and Dzhugar larkspur growing in Kazakhstan. Tr. Alma-At. Med. Inst. *23*, 524–526 (1966) [Chem. Abstr. *68*, 19585v (1968)]
Nishie, K., Gumbmann, M.R., Keyl, A.C.: Pharmacology of solanine. Toxicol. Appl. Pharmacol. *19*, 81–92 (1971)
Niwa, H., Hirata, Y., Suzuki, K.T., Yamamura, S.: Biosynthesis of daphnilactone B. Tetrahedron Lett. 2129–2132 (1973)
Olney, H.O.: Growth substances from *Veratrum tenuipetalum*. Plant Physiol. *43*, 293–302 (1968)
Pelletier, S.W., Keith, L.H.: The diterpene alkaloids: general introduction. In: The alkaloids. Chemistry and physiology. Manske, R.H.F. (ed.), Vol. XII, pp. xv–xvii. London: Academic Press 1970
Poswillo, D.E., Sopher, D., Mitchell, S.: Experimental induction of foetal malformation with "blighted" potato: a preliminary report. Nature (London) *239*, 462–464 (1972)
Potier, P., Kan, C., Le Men, J.: Présence de steroïdes cétoniques dans les feuilles du *Paravallaris microphylla* Pitard. (Apocynacées). Tetrahedron Lett. 1671–1674 (1964)
Prokoshev, S.M., Petrochenko, E.I., Paseshnichenko, V.A.: Tomatinase in tomato leaves. Dokl. Akad. Nauk USSR *106*, 313–316 (1956)
Ripperger, H., Moritz, W., Schreiber, K.: Zur Biosynthese von *Solanum* – Alkaloiden aus Cycloartenol oder Lanosterin. Phytochemistry *10*, 2669–2704 (1971)
Robinson, T.: The biochemistry of alkaloids. Berlin, Heidelberg, New York: Springer 1968
Roddick, J.G.: Biosynthesis and distribution of tomatine in cultured excised *Lycopersicon esculentum* roots. Phytochemistry *13*, 1459–1462 (1974a)
Roddick, J.G.: The steroidal glycoalkaloid α-tomatine. Phytochemistry *13*, 9–25 (1974b)
Roddick, J.G.: Subcellular localization of steroidal glycoalkaloids in vegetative organs of *Lycopersicon esculentum* and *Solanum tuberosum*. Phytochemistry *16*, 805–807 (1977)
Roddick, J.G., Butcher, D.N.: Isolation of tomatine from cultured exised roots and callus tissues of tomato. Phytochemistry *11*, 2019–2024 (1972)
Ronchetti, F., Russo, G., Ferrara, G., Vecchio, G.: The role of (25*S*)-5α-cholestan-3*β*, 26-diol and (25*S*)-5α-furostan-3*β*, 26-diol in the biosynthesis of tomatidine and neotigogenin. Phytochemistry *14*, 2423–2425 (1975)
Sander, H.: Studien über Bildung und Abbau von Tomatin in der Tomatenpflanze. Planta *47*, 374–400 (1956)
Sander, H., Grisebach, H.: Gewinnung von ^{14}C-markierten Tomatin. Z. Naturforsch. *13*, 755–756 (1958)
Schönbeck, F., Schlösser, E.: Preformed substances as potential protectants. In: Encyclopedia of plant physiology, New Series, Ch. 5.9. Pirson, A., Zimmermann, M. (eds.), Vol. 4, pp. 653–678. Berlin, Heidelberg, New York: Springer 1976
Schreiber, K.: Natürliche pflanzliche Resistenzstoffe gegen den Kartoffelkäfer und ihr möglicher Wirkungsmechanismus. Züchter *27*, 289–298 (1957)
Schreiber, K.: Steroid alkaloids: The Solanum group. In: The alkaloids. Chemistry and physiology, Ch. I. Manske, R.H.F. (ed.), Vol. X, pp. 1–192. London: Academic Press 1968
Schreiber, K.: Synthesis and biogenesis of steroidal *Solanum* alkaloids. Biochem. Soc. Trans. *2*, 1–25 (1974)
Schreiber, K., Aurich, O.: Isolierung von Nebenalkaloiden und 3*β*-hydroxy-5α-pregn-16-en-20-on aus *Lycopersicon pimpinellifolium* Mill. Phytochemistry *5*, 707–712 (1966)
Schütte, H.R., Isoprenoide Alkaloide. In: Biosynthese der Alkaloide, Ch. 24. Mothes, K., Schütte, H.R., (eds.), pp. 601–615. Berlin: VEB Deutscher Verlag der Wissenschaften 1969
Schütte, H.R., Lehfeldt, J.: Über die Biosynthese der Alkaloide in *Nuphar luteum*. Arch. Pharm. *298*, 461–465 (1965)
Silver, S., Levine, E.: Action of steroidal diamines on active transport and permeability properties of *Escherichia coli*. J. Bacteriol. *96*, 338–345 (1968)
Sinden, S.L., Webb, R.E.: Effect of environment on glycoalkaloid content of six potato varieties at 39 locations. USDA, ARS. Tech. Bull. *1472*, 1–30 (1974)
Sinden, S.L., Goth, R.W., O'Brien, M.J.: Effect of potato alkaloids on the growth of *Alternaria solani* and their possible role as resistance factors in potatoes. Phytopathology *63*, 303–307 (1973)

Smith, P.M.: The chemotaxonomy of plants. London: Edward Arnold 1976

Stern, E.S.: The aconitum and delphinium alkaloids. In: The alkaloids. Chemistry and physiology, Ch. 37. Manske, R.H.F., Holmes, H.L., (eds.), Vol. IV, pp. 275–333. London: Academic Press 1954

Strzelecka, H.: Untersuchungen an sterilen Wurzelkulturen von *Delphinium elatum* L. Abh. Dtsch. Akad. Wiss. Berlin Kl. Chem. Geol. Biol. *3*, 603–605 (1966)

Suzuki, K.T., Okuda, S., Niwa, H., Toda, M., Hirata, Y., Yamamura, S.: Biosynthesis of Daphniphyllum alkaloids. Tetrahedron Lett. 799–802 (1973)

Toda, M., Niwa, H., Irikawa, H., Hirata, Y., Yamamura, S.: The alkaloids from the fruits of the plant Daphniphyllaceae. Tetrahedron *30*, 2683–2688 (1974)

Tomko, J., Votický, Z.: Steroid alkaloids: the *Veratrum* and *Buxus* groups. In: The alkaloids. Chemistry and physiology, Ch. 1. Manske, R.H.F. (ed.), Vol. XIV, pp. 1–82. London: Academic Press 1973

Tschesche, R.: Neuere Vorstellungen auf dem Gebiet der Biosynthese der Steroide und verwandter Naturstoffe. Fortschr. Chem. Org. Naturst. *12*, 131–168 (1955)

Turova, A., Seifula, K.I., Belykh, M.S.: Pharmacological studies of solasodine. Farmakol. Toksikol. *24*, 469–474 (1961)

Valenta, Z., Khaleque, A.: The structure of castoramine. Tetrahedron Lett. 1–5 (1959)

Waring, M.J., Chisholm, J.W.: Uncoiling of bacteriophage PM2 DNA by binding of steroidal amines. Biochim. Biophys. Acta *262*, 18–23 (1972)

Wenkert, E., Bringi, N.V.: A stereochemical interpretation of the biosynthesis of indole alkaloids. J. Am. Chem. Soc. *81*, 1474–1481 (1959)

Wildman, W.C., Le Men, J., Wiesner, K.: Monoterpene alkaloids. In: Cyclopentanoid terpene derivatives, Ch. 4. Taylor, W.I., Battersby, A.R. (eds.), pp. 239–278. New York: Marcel Dekker 1969

Willimott, S.G.: An investigation of solanine poisoning. Analyst *LVIII*, 431–439 (1933)

Willuhn, G.: Untersuchungen zur chemischen Differenzierung bei *Solanum dulcamara* L. Planta Med. *15*, 58–73 (1967)

Yamamura, S., Hirata, Y.: The *Daphniphyllum* alkaloids. In: The alkaloids. Chemistry and physiology, Ch. 2. Manske, R.H.F. (ed.), Vol. XV, pp. 41–81. London: Academic Press 1975

Yamazaki, M., Matsuo, M., Arai, K.: Biosynthesis of dendrobine. Chem. Pharm. Bull. (Tokyo) *14*, 1058–1059 (1966)

5 The Isoprenoids

5.1 The Terpenoids

D.V. BANTHORPE and B.V. CHARLWOOD

1 Introduction

The terpenes (≡ terpenoids or isoprenoids) are a diverse class of secondary metabolites that may be formally considered to be constructed from isoprene (2-methylbutadiene) units (RUZICKA, 1953, 1959, 1973). Several thousand terpenes occur in many genera of higher plants and microorganisms (DEVON and SCOTT, 1972, DARNLEY-GIBBS, 1974) and although often the structures of the various classes seem unrelated, detailed biochemical studies have clearly revealed the pattern of biosynthesis. Much of this work has concerned the sub-class of steroids which is reviewed in Chapter 5.2, this volume.

Several hundred papers appear annually devoted to the biosynthesis of terpenes and, consequently, we here only sketch a brief outline and cite reviews and papers concerning important topics and containing leading references. Recent books and reviews have dealt with the chemistry (NEWMAN, 1972; NAKANISHI et al., 1974) and general aspects of biogenesis (LUCKNER, 1972; MILLER, 1973; MANN, 1978) of the class and of specific sub-classes (GOODWIN, 1970, 1971). Perusal of these, of the reviews cited at the head of the later sub-sections of this chapter, and of a series of annual reports (OVERTON, 1971; GEISSMAN, 1972, HANSON, 1977) enable an essentially complete literature coverage readily to be made.

Following the discovery that mevalonic acid (MVA; I) was an obligatory intermediate in steroid biosynthesis, its role together with that of the derived isopentenyl pyrophosphate (IPP; II) and 3,3-dimethylallyl pyrophosphate (DMAPP; III) in terpene biosynthesis in general was rapidly explored (CLAYTON, 1965) and the latter C5 pyrophosphates, taken in conjunction, were shown to be the biological equivalent of the isoprene building unit. Much recent work has concerned the irreversible formation of MVA from its immediate precursor: a step which is believed to be rate-limiting for the whole sequence (DEMPSEY, 1974; RODWELL et al., 1976). The present generally accepted view (largely due to RUZICKA , 1953)

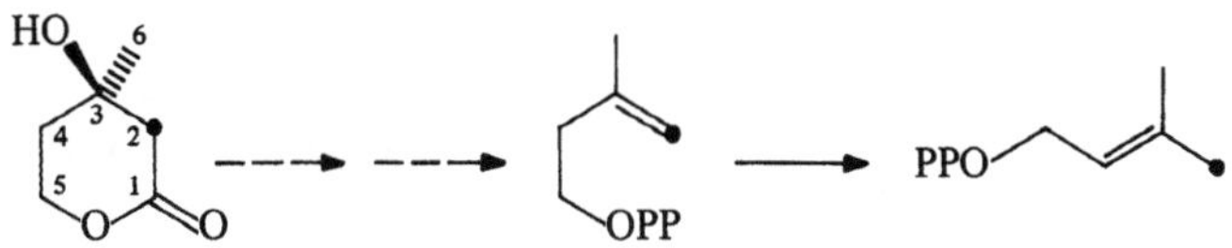

(I) (II) (III)

OPP = pyrophosphate
• = C–2 of MVA

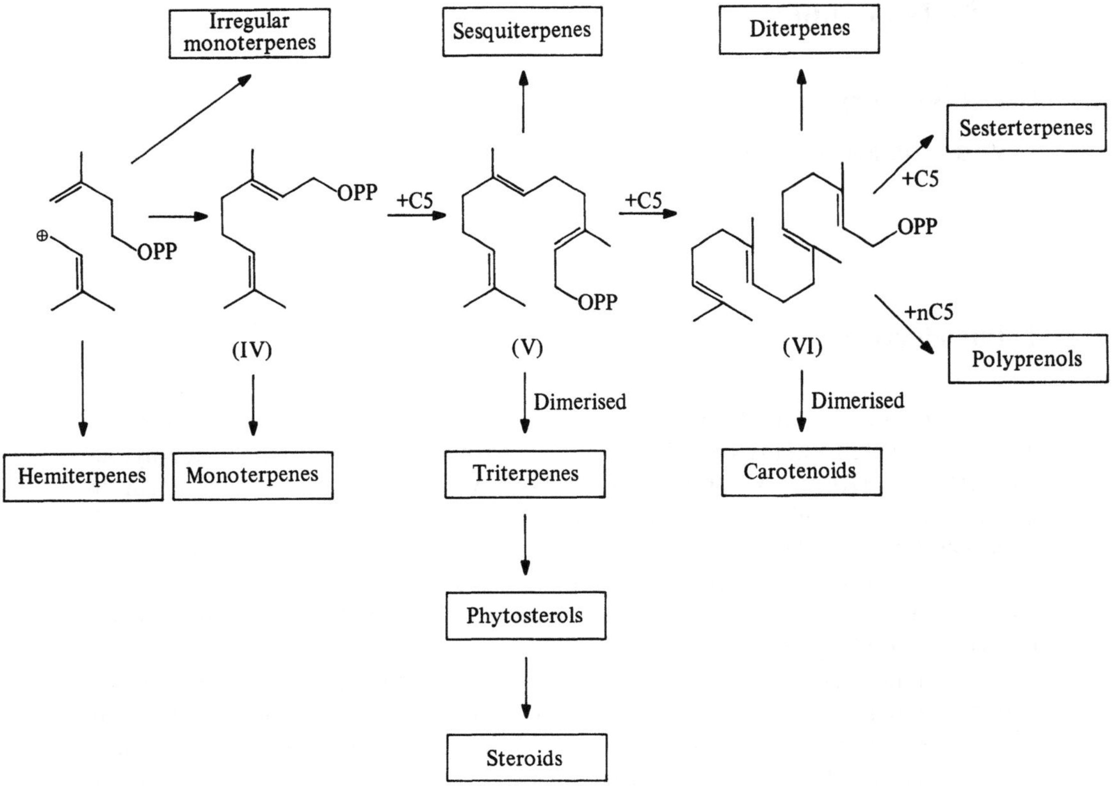

Fig. 1. Relationships of terpenes

is that IPP and DMAPP condense (probably in a carbonium ion-like process; POULTER et al., 1976) to form geranyl pyrophosphate (GPP; IV) and then yield in sequence farnesyl pyrophosphate (FPP; V) and geranylgeranyl pyrophosphate (GGPP; VI) that can all cyclise and/or dimerise to give the whole family of terpenes (see Fig. 1). The idea that these few simple acyclic compounds could act as the parents of all the sub-classes is called the Biogenetic Isoprene Rule (RUZICKA, 1953, 1959, 1973). In addition, several groups of compounds of partial mevalonoid origin (the meroterpenes) are known – e.g., the terpene alkaloids (cf. Chap. 4.5, this vol.).

Several detailed studies of the enzymic steps to the acyclic precursors have appeared, cf. an elegant investigation of MVA-kinase (HILL and ROGERS, 1974); studies on the feed-back inhibition of this by GPP and FPP (GRAY and KEKWICK, 1972), and determination of the variations in the activity of several enzymes of the terpenoid pathway during the germination of peas (GREEN and BAISTED, 1972). The stereochemical details of the MVA → IPP → DMAPP and of other conversions have been elucidated (CORNFORTH, 1973), and somewhat inconclusive investigations have sought to answer whether one or more prenyltransferases is involved in the extension of the carbon chain from C5 to C20 (GREEN and WEST, 1974; SHINKA et al., 1974; EBERHART and RILLING, 1975; REED and RILLING, 1975). Details of the conversions and functionalisations of the acyclic precursors to the individual members of the sub-classes have been largely explored by studies involving ^{14}C or double ($^{14}C/^{3}H$) labelling (HANSON, 1970). Recently the use of ^{13}C as a mechanistic probe has been introduced: the method of elucidation of the position of tracer (by Fourier Transform-^{13}C-NMR spectroscopy) is non-destructive and the use of, e.g., [1,2-$^{13}C_2$]-acetate as a biosynthetic precursor enables conclusions to be drawn concerning the direct incorporation of acetate units and rearrangements involving cleavage of C-C bonds in products derived from acetate (SCOTT, 1974; SIMPSON, 1975). However the method has severe limitations: it is expensive; high incorporations ($>5\%$) are necessary; and metabolic pools of precursors or of products must be absent: in practice, it has seldom been used for higher plants, but has found much application (see later) for studies on microorganisms.

2 Distribution of the Terpenoids

2.1 General

The isoprenoids in general are very widely distributed throughout the whole of the plant kingdom. All green plants have the ability to produce linear isoprenoids by the mevalonate pathway since it is a product of this pathway (viz. the phytyl side chain) that confers biological activity upon the chlorophyll molecule. However, whilst isoprenoids containing more than five isoprene units (that is to say the steroids and above) appear universal amongst plants (SWAIN, 1974) the less complex terpenoids (those between C5 and C25) are mainly restricted to the Tracheophyta, although sesquiterpenes have been found widely in the Bryophyta and the fungi and some species of the latter contain monoterpenes. It has been suggested (SWAIN, 1974) that the most primitive isoprenoid was probably a linear polyisoprenol of between C30 and C100, and that its function was associated with the transportation of polar metabolites. If this is correct then the formation of steroids and carotenoids would have been consequential on the development of enzyme systems able to bring about the tail-to-tail dimerisation of FPP and GGPP units respectively. On the other hand, the ability to accumulate lower terpenoids required the evolution of a mechanism for the removal of a suitable precursor from the particulate prenyl transferase(s) complex, coupled with enzyme systems capable of carrying out trans–cis isomerisation of these precursors.

2.2 Hemiterpenes

Although the biological "isoprene units" (DMAPP and IPP) are themselves hemiterpenes, they do not appear to be accumulated by plants, but 3,3-dimethylallyl alcohol and isopentenol, presumably derived from them, are found in a few oils (GILDEMEISTER and HOFFMAN, 1955–1960). Isoprene was long thought *not* to occur naturally, but recently it has been shown to be an emanation from leaves of certain trees (JONES and RASMUSSEN, 1975). Oxygenated C5 compounds (such as isovaleric acid, 3,3-dimethylacrylic acid, etc.) that have previously been found in plants are thought not to be products of the mevalonate pathway.

It may be that true hemiterpenes are only present in plants in very small amounts and hence have escaped detection. However, recent exhaustive searches for even minute pools of C5 terpenes (ALLEN et al., 1976) have proven negative.

2.3 Monoterpenes

The monoterpenes are very widely distributed in the plant kingdom and lists of their occurrence are available (GUENTHER, 1948; HAAGEN-SMIT, 1958; DE MAYO, 1959a; DARNLEY-GIBBS, 1974). Table 1 (see also diagram on page 560) lists the orders in which monoterpenes have been positively identified, but it must be pointed

Table 1. Occurrence of monoterpenes in flowering plants

Class Dicotyledones

Superorder I Magnolianae
1 Magnoliales (8); **4**
2 Laurales (II); **2**
3 Piperales (2); **1**
4 Aristolochiales (1); **1**

Superorder II Ranunculanae
7 Illiciales (2); **1**
9 Ranuculales (10); **1**[a]

Superorder III Hamamelidanae
16 Hamamelidales (3); **1**
18 Urticales (4); **2**
24 Myricales (1); **1**

Superorder IV Caryophyllanae
27 Caryophyllales (16); **1**

Superorder V Dillenianae
32 Paeonales (1); **1**
33 Theales (21); **2**
34 Violales (9); **1**[a]

Superorder VI Ericanae
41 Ericales (10); **1**
44 Primulales (3); **1**[a]

Superorder VIII Rosanae
48 Saxifragales (27); **2**
49 Rosales (3); **1**
50 Fabales (3); **1**[a]

Superorder IX Myrtanae
54 Myrtales (13); **1**

Superorder X Rutanae
56 Rutales (8); **4**
58 Geraniales (20); **1**

Superorder XI Aralianae
60 Cornales (10); **2**

Superorder XII Celastranae
63 Oleales (1); **1**
64 Santalales (13); **1**

Superorder XIV Lamianae
67 Dipsacales (4); **1**
71 Lamiales (4); **2**

Superorder XV Asteranae
74 Asterales (1); **1**

Class Monocotyledones

Superorder II Lilianae
79 Liliales (20); **2**
80 Iridales (4); **1**[a]
81 Zingiberales (8); **1**

Superorder IV Commelinanae
89 Poales (1); **1**

Superorder V Arecanae
92 Arales (2); **1**[a]
93 Pandanales (1); **1**[a]

The figure in parentheses after each order indicates the number of families in that order (according to TAKHTAJAN, 1969): the figure in bold type is the number of families in which monoterpenes have been reported.

[a] Rare occurrence of monoterpenes within this order.

out that less than 8% of the one half million species of plants have been investigated. It seems probable, however, that the orders derived from Violales do not produce monoterpenes, and this would be in accord with the view that these orders were derived from the primitive Violales (family Flacourtiaceae) in which monoterpenes have not been recorded. On the other hand it would seem surprising that the order Dilleniales is devoid of monoterpenes, since it is the direct progenitor of many orders that are abundant in terpenes. Further study of the families Dilleniaceae and Crossosomataceae might shed light on this situation.

The orders Rutales, Cornales, Lamiales, and Asterales are by far the most prolific in their production of monoterpenes, but the distribution of monoterpene content is not uniform throughout these orders: thus whilst four of the eight families of the order Rutales contain monoterpenes, only members of the Apiaceae

Table 2. Occurrence of sequiterpenes in flowering plants

Class Dicotyledones

Superorder I Magnolianae
1 Magnoliales (8); **4**
2 Laurales (11); **1**
3 Piperales (2); **1**
4 Aristolochiales (1); **1**

Superorder II Rununculanae
7 Illiciales (2); **1**

Superorder III Hamamelidanae
18 Urticales (4); **2**
24 Myricales (1); **1**

Superorder IV Caryophyllanae
27 Caryophyllales (16); **1**

Superorder V Dillenianae
33 Theales (21); **2**
34 Violales (9); **1**
40 Salicales (1); **1**

Superorder VI Ericanae
41 Ericales (10); **1**

Superorder VII Malvanae
45 Malvales (8); **1**

Superorder VIII Rosanae
50 Fabales (3); **1**

Superorder IX Myrtanae
54 Myrtales (13); **1**

Superorder X Rutanae
56 Rutales (8); **3**
58 Geraniales (20); **3**

Superorder XI Aralianae
60 Cornales (10); **2**

Superorder XII Celastranae
64 Santalales (13); **1**

Superorder XIV Lamianae
67 Dipsacales (4); **1**
68 Gentianales (11); **1**
70 Scrophulariales (18); **2**
71 Lamiales (4); **2**

Superorder XV Asteranae
74 Asterales (1); **1**

Class Monocotyledones

Superorder II Lilianae
80 Iridales (4); **1**
81 Zingiberales (8); **1**

Superorder III Juncanae
84 Cyperales (1); **1**

Superorder IV Commelinanae
89 Poales (1); **1**

Superorder V Arecanae
92 Arales (2); **1**

The figure in parentheses after each order indicates the number of families in that order (according to TAKHTAJAN, 1969): the figure in bold type is the number of families in which monoterpenes have been reported.

(Umbelliferae) of the ten families in the order Cornales are abundant in monoterpenes.

Generally a knowledge of the presence (or otherwise) of monoterpenes is of limited value in chemotaxonomy, although it has been pointed out (DARNLEY-GIBBS, 1974) that the separation of the Magnoliales from the Ranunculales is supported by the rare occurrence of monoterpenes in the latter compared with the former. At the levels of genus and species much useful taxonomic information can be obtained from a detailed study of specific monoterpene content.

2.4 Sesquiterpenes

The sesquiterpenes are also very widely distributed in the plant kingdom, as is indicated by Table 2 (p. 189), and lists of specific occurrences are available (HAAGEN-SMIT, 1958; DE MAYO, 1959a; DARNLEY-GIBBS, 1974). The orders Magnoliales, Rutales, Cornales and Asterales are most prolific in their production, and heterogeneous in the structures of sesquiterpenes, whilst further evidence in support of the separation of Magnoliales from Ranunculales is provided by the lack of this class of terpenoid in the latter. Many detailed studies using sesquiterpenes as taxonomic markers have been made at the genus and species level.

Table 3. Occurrence of diterpenes in flowering plants

Class Dicotyledones

Superorder I Magnolianae
2 Laurales (11); **1**
4 Aristolochiales (1); **1**

Superorder II Ranunculanae
9 Ranunculales (10); **1**

Superorder V Dillenianae
34 Violales (9); **1**

Superorder VI Ericanae
41 Ericales (10); **1**

Superorder VII Malvanae
46 Euphorbiales (7); **1**

Superorder VIII Rosanae
50 Fabales (3); **1**

Superorder X Rutanae
56 Rutales (8); **1**
57 Sapindales (9); **2**
58 Geraniales (20); **2**

Superorder XI Aralianae
60 Cornales (10); **1**

Superorder XII Celastranae
63 Oleales (1); **1**

Superorder XIV Lamianae
68 Gentianales (11); **1**
69 Polemoniales (8); **1**
70 Scrophulariales (18); **1**
71 Lamiales (4); **2**

Superorder XV Asteranae
72 Campanulales (7); **1**
74 Asterales (1); **1**

Class Monocotyledones

Superorder II Lilianae
80 Iridales (4); **1**

Superorder IV Commelinanae
89 Poales (1); **1**

The figure in parentheses after each order indicates the number of families in that order (according to TAKHTAJAN, 1969): the figure in bold type is the number of families in which monoterpenes have been reported.

2.5 Diterpenes

Apart from the biologically active gibberellins (see volume 9 of this series) which have a universal distribution, the cyclic diterpenes are more restricted than are the lower terpenoids (see Table 3, p. 190). Diterpenes are common in the orders derived

Table 4. Occurrence of triterpenes in flowering plants

Class Dicotyledones

Superorder I Magnolianae
2 Laurales (11); **1**

Superorder II Ranunculanae
9 Ranunculales (10); **3**
10 Papaverales (3); **1**

Superorder III Hamamelidanae
14 Eupteleales (1); **1**
16 Hamamelidales (3); **2**
18 Urticales (4); **1**
22 Betulales (1); **1**
23 Balanopales (1); **1**
24 Myricales (1); **1**

Superorder IV Caryophyllanae
27 Caryophyllales (16); **6**

Superorder V Dillenianae
33 Theales (21); **4**
34 Violales (9); **2**
36 Cucurbitales (1); **1**
38 Capparales (8); **2**
39 Tamaricales (3); **1**
40 Salicales (1); **1**

Superorder VI Ericanae
41 Ericales (10); **1**
43 Ebenales (5); **3**
44 Primulales (3); **3**

Superorder VII Malvanae
45 Malvales (8); **2**

Superorder VIII Rosanae
48 Saxifragales (27); **3**
49 Rosales (3); **1**
50 Fabales (3); **1**

Superorder IX Myrtanae
54 Myrtales (13); **3**

Superorder X Rutanae
56 Rutales (8); **5**
57 Sapindales (9); **2**
58 Geraniales (2); **4**
59 Polygalales (5); **1**

Superorder XI Aralianae
60 Cornales (10); **3**

Superorder XII Celastranae
61 Celastrales (11); **2**
62 Rhamnales (3); **2**
63 Oleales (1); **1**
64 Santalales (13); **2**

Superorder XIV Lamianae
67 Dipsacales (4); **2**
68 Gentianales (11); **6**
69 Polemoniales (8); **1**
70 Scrophulariales (18); **4**
71 Lamiales (4); **2**

Superorder XV Asteranae
72 Campanulales (7); **1**
74 Asterales (1); **1**

Class Monocotyledones

Superorder II Lilianae
79 Liliales (20), **1**

Superorder IV Commelinanae
89 Poales (1); **1**

Superorder V Arecanae
90 Arecales (1); **1**

The figure in parentheses after each order indicates the number of families in that order (according to Takhtajan, 1969): the figure in bold type is the number of families in which monoterpenes have been reported.

from the Gentianales, particularly prolific are the Lamiales and the Asterales, but the further distribution of this class is scattered with the Fabales and the Geraniales being the most important sources. Listings of occurrence of diterpenes in general (KARRER, 1958; SANDERMANN, 1962; DARNLEY-GIBBS, 1974) and of specific compounds (HANSON, 1968; DE MAYO, 1959b) are available.

2.6 Sesterterpenes

The sesterterpenoids constitute a very small group of secondary compounds found mainly in phytopathogenic fungi (TURNER, 1971). Although there are sporadic reports of the occurrence of these compounds (and of certain C20 compounds considered to be degraded sesterterpenes) in ferns, lichen, and a few flowering plants (CORDELL, 1974a) there is at present insufficient information to comment on the general distribution of this class.

2.7 Nonsteroidal Triterpenes

Although almost entirely restricted to the plant kingdom, the triterpenes are ubiquitous amongst the flowering plants (Table 4, p. 191: for listings see BASU and RASTOGI, 1967; KULSHRESHTHA et al., 1972). Triterpenes occur most frequently as O-glycosides (the saponins) and it is within this class that most taxonomic speculation has been made (DARNLEY-GIBBS, 1974).

3 Structural Diversity of Biosynthesis of C5 to C30 Compounds

3.1 Hemiterpenes (C5 Compounds)

The route of biosynthesis of isoprene is unknown, although dimethylvinylcarbinol derived by the allylic rearrangement of DMAPP is a logical precursor. The C5 building unit, presumably in the form of the 3,3,-dimethylallyl radical, is also a component of certain meroterpenes: thus IPP was incorporated into the furan ring and side chain of (VII) in cell-free extracts of *Eupatorium rugosum* (LIN and HEINSTEIN, 1974) and incorporation followed by degradation was revealed by tracer studies on the formation of (VIII) and (IX) in other plants (KUTNEY et al., 1973; LEISTNER, 1973).

3.2 Monoterpenes (C10 Compounds)

General reviews: LOOMIS, (1967); BANTHORPE et al., (1972a); CHARLWOOD and BANTHORPE, (1978).

The great majority of monoterpenes are "regular" in that they are formed by head-to-tail linkage of isoprene units (e.g., Fig. 1). The pattern of biosynthesis

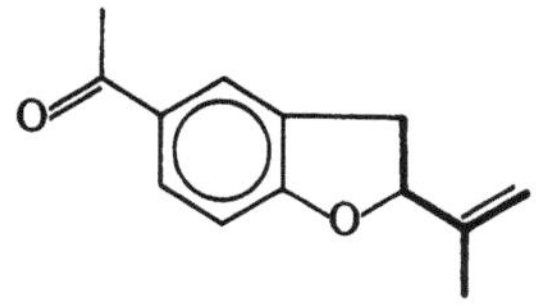

(VII)

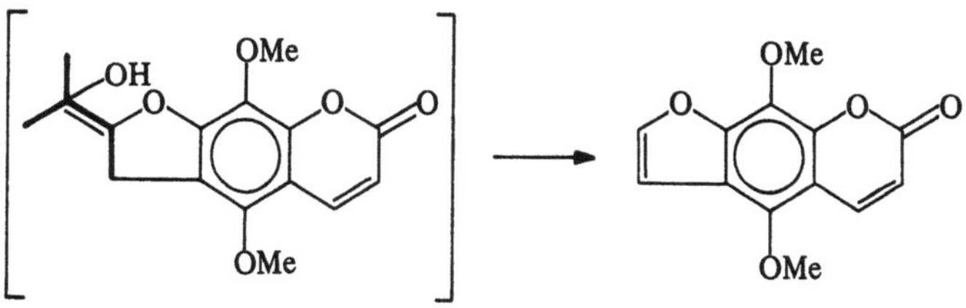

(VIII)

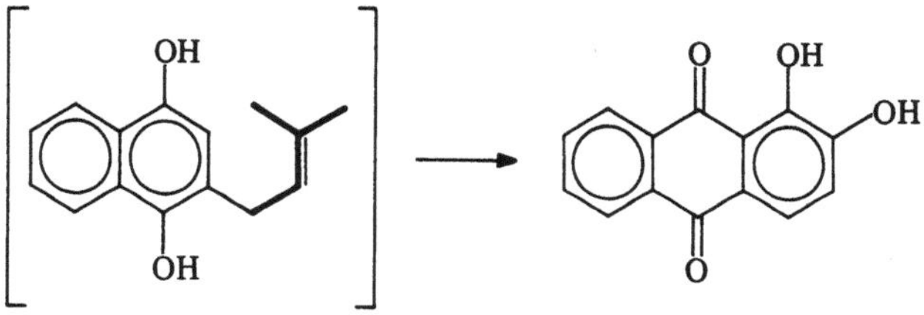

(IX)

of these compounds was hypothetically rationalised on the basis of the known chemical structures and reactions well before any biochemical studies were available (RUZICKA, 1953) and the proposed pathways have, almost without exception, been verified by later work. A modernised version of Ruzicka's scheme is in Figure 2: in this, and in later schemes, the representation of the intermediates as carbonium ions is purely formal, the actual reactants are not known in any particular case, but may be pyrophosphate esters, glycosides or protein-bonded species that are biogenetic equivalents of carbonium ions. The ion (X) derived from GPP cannot cyclise for stereochemical reasons but can form acyclic monoterpenes. In most cases it also isomerises to (XI), derived from neryl pyrophosphate (NPP), which can now cyclise to yield (XII) (derived from α-terpineol) and this by functionalisation can form menthane derivatives, by 1,3-elimination can give carane derivatives and by internal addition can yield pinane and bornane derivatives. Furthermore, a hydride shift to yield (XIII) enables thujane derivatives to be formed. The possibility that (X) and (XI) are derived from linalol (XIV) has also been proposed, but there is no evidence for the latter as an obligatory intermediate (BANTHORPE et al., 1978).

Elucidation of the route of formation of steroids led to the idea that MVA was an obligatory intermediate en route to GPP and all monoterpenes. This is universally accepted although there is actually very little direct evidence for the assumption. Thus feeding experiments on higher plants generally lead to very low (ca. 0.001%) incorporations of tracer from MVA into monoterpenes (BANTHORPE

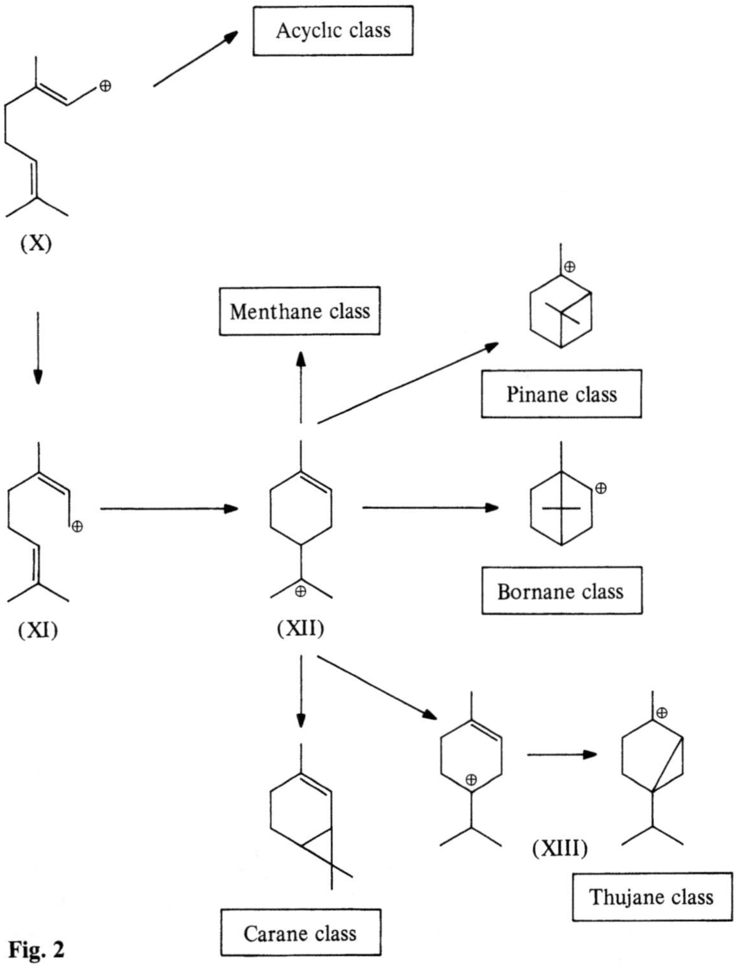

Fig. 2

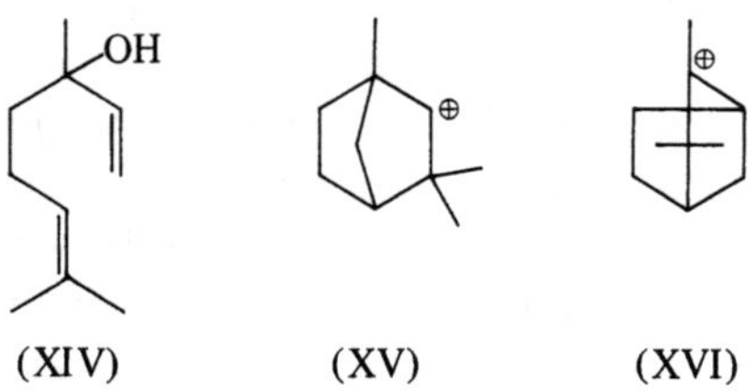

(XIV) (XV) (XVI)

et al., 1972a). Such incorporations would not be accepted as evidence for a direct precursor relationship if there were any other reasonable alternative route, and suggestions have been made that MVA is degraded to CO_2 and this is then incorporated into terpenes by some unspecified route presumably involving initial photosynthetic-like fixation. However, incorporation of tracer from MVA into a variety of monoterpenes is position-specific and the low incorporations are believed to be due to compartmentation effects whereby the exogenous precursor cannot penetrate into the biosynthetic sites (BANTHORPE et al., 1972a). The best pieces of evidence for the intermediacy of MVA are probably the high (ca. 22%) incorpora-

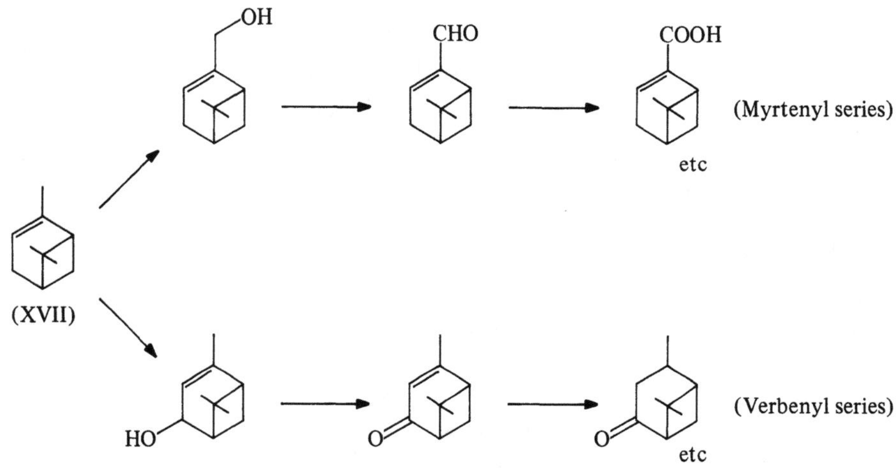

Fig. 3

tion achieved into geraniol and nerol and their glucosides in petals of *Rosa dilecta* (FRANCIS and ALLCOCK, 1969) and the incorporations (3% and 12%) of MVA and IPP into geraniol formed in cell-free extracts of *Tanacetum vulgare* (BANTHORPE et al., 1976); in each of these systems access to the biosynthetic sites should have been relatively unrestricted.

Formation of the skeletons in Figure 2 is enzymic and appears to be under fairly strict genetic control (HEFENDEHL and MURRAY, 1976). Other naturally occurring skeletons, e.g., those of fenchane (XV) and isocamphane (XVI) are probably formed from the pinyl and bornyl ions by Wagner-Meerwein rearrangements such as are well known in mechanistic organic chemistry. The primary and secondary skeletons may also be functionalised, probably by relatively unspecific enzymes acting in a metabolic grid. One example is the oxidation of α-pinene (XVII) to give a series of myrtenyl and verbenyl compounds which co-occur in many species of plants (Fig. 3: here, and later, details of stereochemistry and absolute configuration are ignored).

Another illustration of secondary transformations that have been extensively studied because of the commerical importance of the mixture of compounds produced is that elucidated by analytical and tracer studies on *Mentha* species (Fig. 4; BURBOTT and LOOMIS, 1967). Such examples could be multiplied and help to explain the extremely complex pattern of monoterpenes that occur in many oil-producing plants: as in all metabolism, the enzymes exploit the innate chemical reactivities of their substrates to promote, albeit very rapidly and sometimes extremely stereospecifically, the same reactions as are known or predicted from organic chemistry. However, it is worth noting that part of the complexity of the monoterpene pattern, and indeed of the entire essential oils of higher plants, may often be due to secondary chemical reactions during extraction, separation, storage, and work-up (PICKETT et al., 1975).

Tracer studies have been carried out whereby appropriately labelled precursors have been applied to plants in vivo or in vitro using cell-free extracts, and the monoterpenes (of all the main types) produced have been degraded to elucidate the position of label. Such work confirms the broad correctness of Ruzicka's hypothesis. One unexpected finding was that [2-^{14}C]-MVA was incorporated with

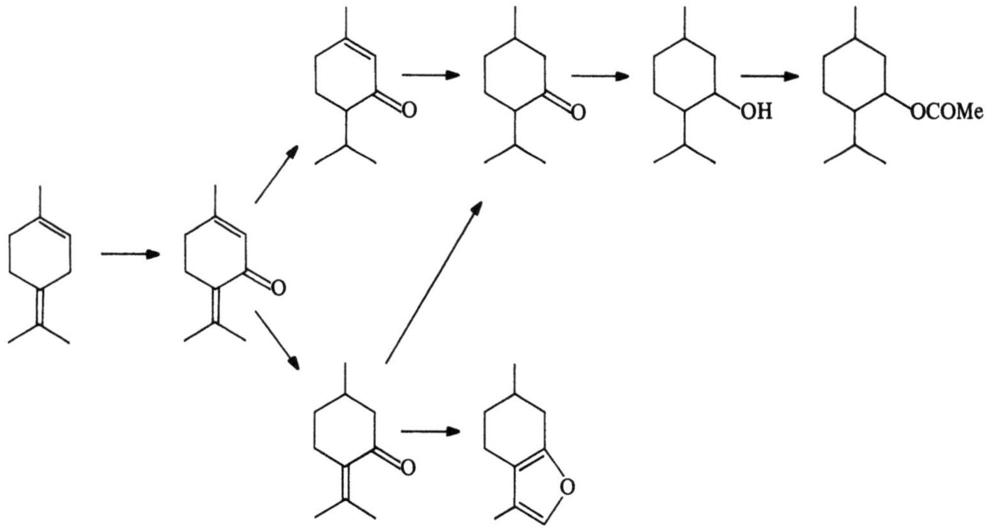

Fig. 4 (XVIII)

negligible degradation into a variety of types of monoterpenes but only that moiety derived from IPP was appreciably labelled (BANTHORPE et al., 1972a). Similar asymmetric labelling was later found when [^{14}C]-acetate and even $^{14}CO_2$ were used as precursors of monoterpenes. The latter observation is particularly pertinent as CO_2 is the only precursor that can be fed at physiological concentrations and so cannot perturb the usual patterns of metabolism. There are several possible explanations for these phenomena, but it transpired that the main factor was the existence (both in vivo and in cell-free extracts) of an endogenous pool of DMAPP or its biogenetic equivalent that was probably protein-bonded (ALLEN et al., 1976). At the low incorporations of precursor typically found (see above) this pool caused the observed anomaly: but when higher incorporations occurred (e.g., in *Rosa* species) the pool was swamped and position-specific labelling was found in both IPP and DMAPP-derived moieties (BANTHORPE et al., 1972c). One possible cause of the phenomenon of asymmetric labelling is that a C5 species other than DMAPP contributed to the construction of the monoterpene skeleton. A candidate for this moiety is 3,3-dimethylacrylic acid and claims were made that this was incorporated efficiently in toto into pulegone (XVIII) in *Mentha* species: however, such claims have been refuted (BANTHORPE et al., 1972b) and the role of this compound, and also of dimethylvinylcarbinol (3-methylbut-1-ene-3-ol, an isomer of 3,3-dimethylallyl alcohol), another possible candidate, is undecided.

Mevalonic acid is commercially available specifically tritiated at the 2R, 2S, 4R, 4S, and 5-carbon atoms and such labelled precursors have been used to determine stereochemical details of terpene biosynthesis. In particular, such compounds have recently been employed to elucidate two fundamental problems of monoterpene biosynthesis: viz (1) whether GPP is an obligatory intermediate or whether NPP (the parent of cyclic monoterpenes), rather than being formed from GPP, is produced directly by condensation of two C5 units; and (2) the possible role of linalol (XIV) or its biogenetic equivalent. In the event, studies on the

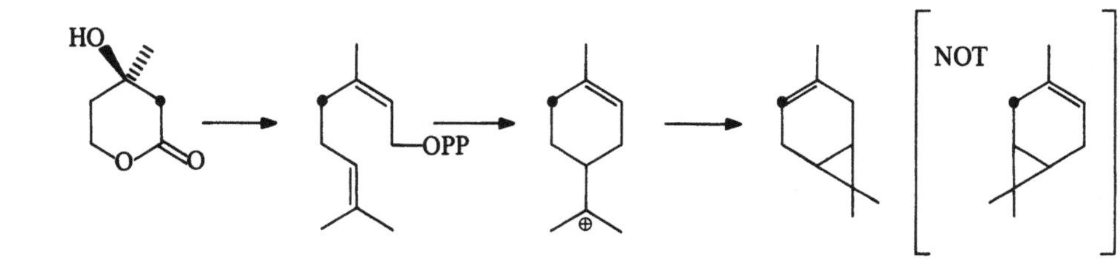

(I) (XII)

Fig. 5

(XIX) (XX)

formation of isothujone (trans-thujan-3-one; cf. Fig. 2), geraniol and nerol in *Tanacetum vulgare* have shown that the sequence GPP → NPP → cyclic terpenes almost certainly occurs, and linalol is not involved (BANTHORPE et al. 1978). In particular, the interconversion of geraniol and nerol appears to occur via a redox process involving dephosphorylation and stereospecific oxidation to the corresponding aldehydes which then rapidly equilibrate under physiological conditions.

The only point in which Ruzicka's hypothetical schemes appear in error, albeit in a minor way, is in the construction of the carane skeleton (cf. Fig. 2): feeding of [2-^{14}C]-MVA to *Pinus* species resulted in the labelling pattern as in Figure 5 whereby bicyclisation of the hypothetical intermediate (XII) occurred with migration of the double-bond (BANTHORPE and EKUNDAYO, 1976).

A class of regular monoterpenes that was not considered by Ruzicka is derived from the iridane skeleton (XIX) and they usually have a structure based on (XX). Compounds of this class were first discovered in species of ants but they occur fairly widely in higher plants as β-glucosides or as components of certain types of terpene-alkaloids. Reviews are available: TAYLOR and BATTERSBY (1970); PLOUVIER and FAVRE-BONVIN (1971); CORDELL (1974a); and SCOTT et al. (1973). The widespread distribution of the compounds in seed, root, bark and leaf tissue contrasts with the predominant storage in specialised oil glands of the other classes of monoterpenes and may be a consequence of water solubility endowed by the glucose residues, and thus of ease of transport. A function of the sugar group may also be to stabilise the monoterpene skeleton, as the aglycone is often unstable with respect to rearrangement and ring-opening. The parent of the class is loganin (XXI) and tracer studies on its formation (mainly in *Vinca rosea*) are consistent with the scheme in Figure 6. An additional deduction from these studies is that asymmetric labelling of the IPP-derived moiety, such as is common for the other monoterpene classes, was not widespread in the iridoids (GUARNACCIA et al., 1974). Several of the simpler types of iridoids and their derivatives have been investigated but little has been done beyond the demonstration of reasonable precursor relationships by means of feeding experiments (GUARNACCIA et al., 1972). Another sub-

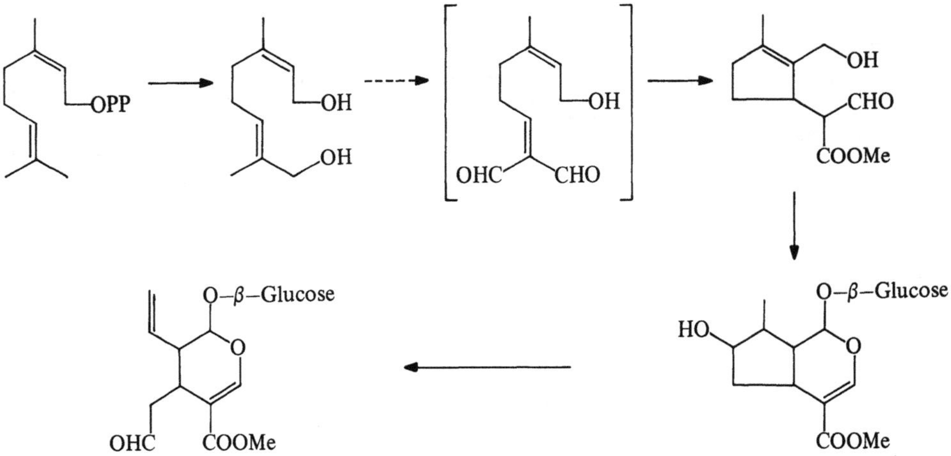

Fig. 6 (XXII) (XXI)

class, the secoiridoids, is a family of terpene glucosides that are believed to be formed from loganin by cleavage of the cyclopentane ring followed by functionalisation (TAKEDA and INOUYE, 1976): an example is (XXII). Few details of the mechanism or biochemistry of any of these conversions are known (MILES et al., 1976). Some studies on cell-free systems have been made (see Sect. 4.1) but these are mainly concerned with methylations, oxidations, etc. that are peripheral to the construction of the skeletons.

A small, but theoretically very interesting, class of monoterpenes are "irregular" in that they are *not* formed by head-to-tail linkage of isoprene units (EPSTEIN and POULTER, 1973). One group of these present no biogenetic problems as they are formed from regular monoterpenes by rearrangements and so obey the biogenetic isoprene rule in being ultimately derived from GPP: examples are fenchane compounds and secoiridoids. These are all believed to be formed enzymatically in vivo and although no biosynthetic studies have been carried out, the proposed (and usually obvious) routes are reasonable in having direct analogies in vitro. A second is far more interesting as it is believed to be composed of primary biosynthetic products which are not derived from GPP or any other regular precursor in any obvious manner. In this sense they break the biogenetic isoprene rule, and indeed appear to be the only class to do so, since the several types of higher terpenoids that appear irregular may always be referred back to regular parents as defined by Ruzicka (see Sect. 1) as precursors.

The most common irregular monoterpenes have the artemisyl (XXIV) and lavendulyl (XXV) skeletons and these, together with the much less frequently occurring santolinyl type (XXVI) have been proposed to be formed by cleavage of trans-chrysanthemyl alcohol (XXIII) (BATES and PAKNIKAR, 1965). The latter compound recently has been found to be naturally occurring and to have the correct stereochemistry to accommodate the proposed products (SHAW et al., 1975). Chrysanthemyl alcohol is an analogue of presqualene alcohol and of prephytoene, but the fissions postulated to lead to the irregular monoterpenes have no counterparts for the higher classes.

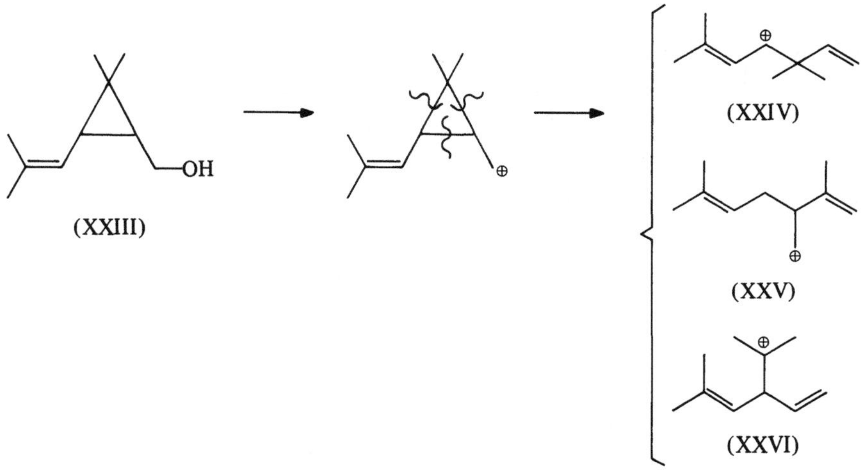

(XXIII) (XXIV) (XXV) (XXVI)

Several hypotheses for the biogenesis of chrysanthemyl alcohol and its presumed derivatives have been proposed and so-called "evidence" for some of these schemes has been presented on the basis of experiments with "model" (i.e., chemical) systems. The latter are of little value: it is altogether reasonable that the cleavages of (XXIII) to give (XXIV) to (XXVI) can occur, and so the demonstration that they take place under exotic conditions with carefully chosen and unusual leaving groups has no relevance whatsoever to the situation in vivo. As in all biosynthetic studies, one carefully planned phytochemical experiment is worth endless chemical speculation! Recently, the formation of artemisyl compounds in higher plants and in cell-free extracts has been studied in detail and this work suggests that chrysanthemyl alcohol is indeed the precursor of the three irregular skeletons (XXIV–VI). Furthermore, it was found that dimethylvinylcarbinol was a more effective precursor of chrysanthemyl alcohol than was either IPP or DMAPP, and it was demonstrated that GPP and NPP were not precursors of these classes of compounds (ALLEN et al., 1977; BANTHORPE et al., 1977b).

3.3 Sesquiterpenes (C15 Compounds)

General reviews: PARKER et al., (1967); RÜCKER, (1973); CORDELL, (1976). Sesquiterpene lactones: GEISSMAN (1973). Sesquiterpenes as stress metabolites: STOESSL et al. (1976).

Some fifty skeletons occur in this class and it is possible to derive all from farnesyl pyrophosphate (FPP) by chemically reasonable, but largely hypothetical mechanisms (RUZICKA, 1953). The characteristic of all these is an initial intramolecular attack by an electrophilic centre at C-1 on either the central or distal double bonds of the precursor: this addition being governed by either electronic or steric influences (HENDRICKSON, 1959). The basic scheme is in Figure 7; subsequent to the initial cyclisation, 1,2 and 1,3-hydride shifts, oxidations, cyclisations, and reductions (examples later) lead to the observed diversity of structures. Not surprisingly, this diversity is greater than for monoterpenes, but it is also much greater than for diterpenes and triterpenes. This situation presumably reflects the much

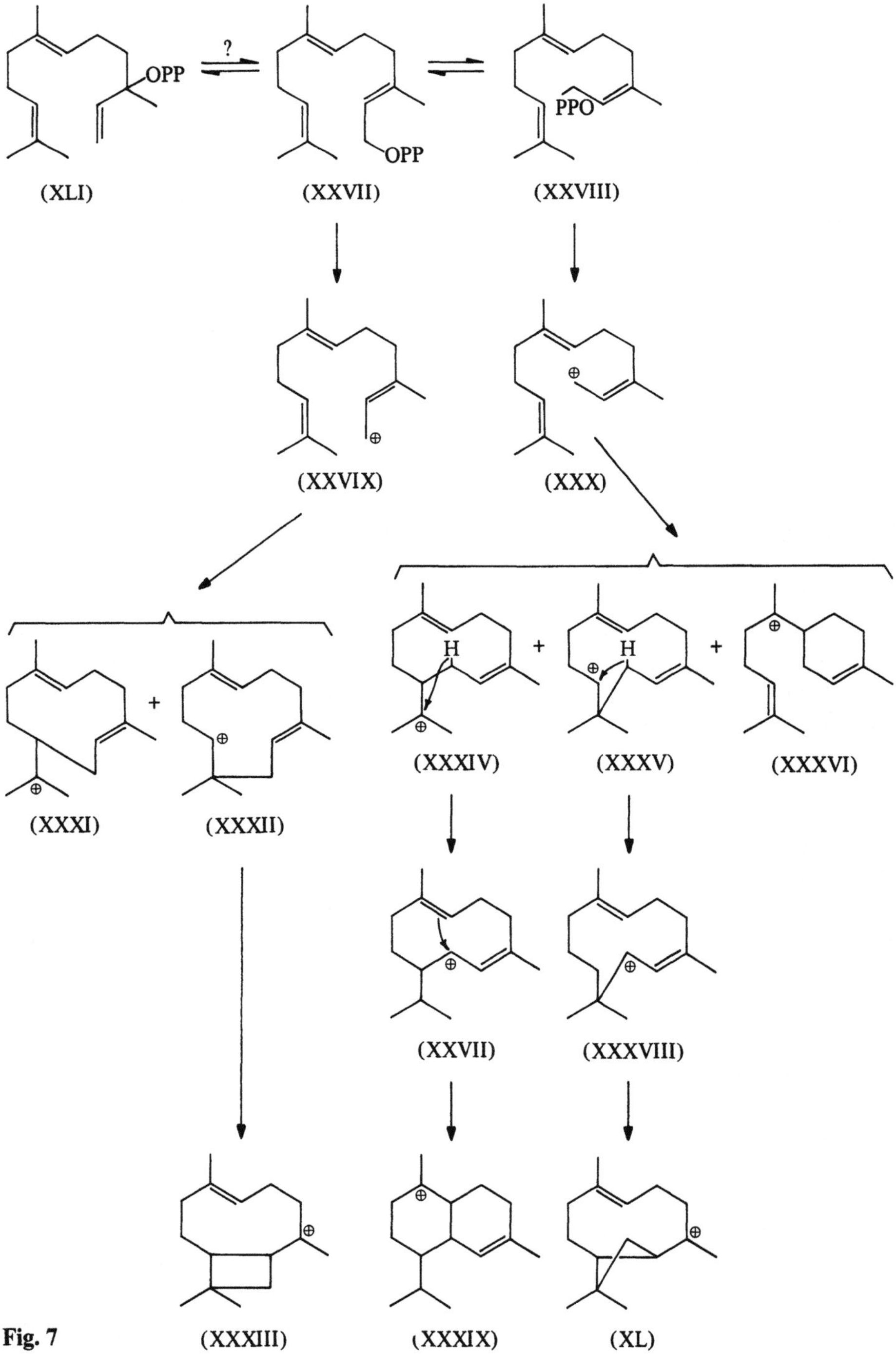
OPP
?
(XLI)
OPP
(XXVII)
PPO
(XXVIII)
⊕
(XXVIX)
⊕
(XXX)
⊕
+
⊕
(XXXI)
(XXXII)
H
⊕
+
H
⊕
+
⊕
(XXXIV)
(XXXV)
(XXXVI)
⊕
(XXVII)
⊕
(XXXVIII)
⊕
(XXXIII)
⊕
(XXXIX)
⊕
(XL)

Fig. 7

larger conformational flexibility of the C10 or C11 rings in the hypothetical intermediates (e.g., XXXI–XXXV) than that of the C6 rings that are usually involved in the intermediate species for the two latter classes.

Which isomer of FPP is involved in these biogenetic processes has been the subject of much debate and study. The 2-trans, 6-trans-isomer (XXVII) hypothetically gives rise to certain known skeletons (XXXI → XXXIII) but the 2-cis, 6-trans-isomer (XXVIII) has been thought necessary to form others for which (XXVII) seems stereochemically unsuitable (e.g., XXXIV → XL). Recent studies (MULLER and TAMM, 1975; EVANS and HANSON, 1976; MACKIE and OVERTON, 1977) have shown that in certain examples (XXVII) (or its biogenetic equivalent – known to be involved in steroid biosynthesis) is first formed by the prenyltransferase and is subsequently converted into (XXVIII) which in turn cyclises and is functionalised to form products based on the parent skeletons (XXXIV)→(XL). The conversion of (XXVII) into (XXVIII) is believed to involve dephosphorylation and redox interconversion of the appropriate aldehydes and alcohols (compare the situation for the monoterpenes, see previous sub-section) and studies using (5R)- and (5S)-[5-$^{3}H_{1}$]-MVA and (1R)- and (1S)-[1-$^{3}H_{1}$]-FPP as substrates have shown that the interconversion does indeed involve the loss of one prochiral hydrogen from C-1 of (XXVII). In addition, in certain cell-free extracts from *Citrus* species the presence of the necessary aldehydes has been demonstrated (CHAYET et al., 1973). In contrast, studies with other plants and fungi have provided evidence that (XXVII) may cyclise directly to form, e.g., (XXXIV), without the intermediacy of (XXVIII): i.e., both hydrogens from C-1 of (XXXIV) are retained in products (ARIGONI, 1975). At first sight such a cyclisation seems stereochemically impossible, but formation of a diene system could endow enough conformational flexibility in the C10 ring of (XXXI) (Fig. 8) to permit reaction, and to form, e.g., the fungal

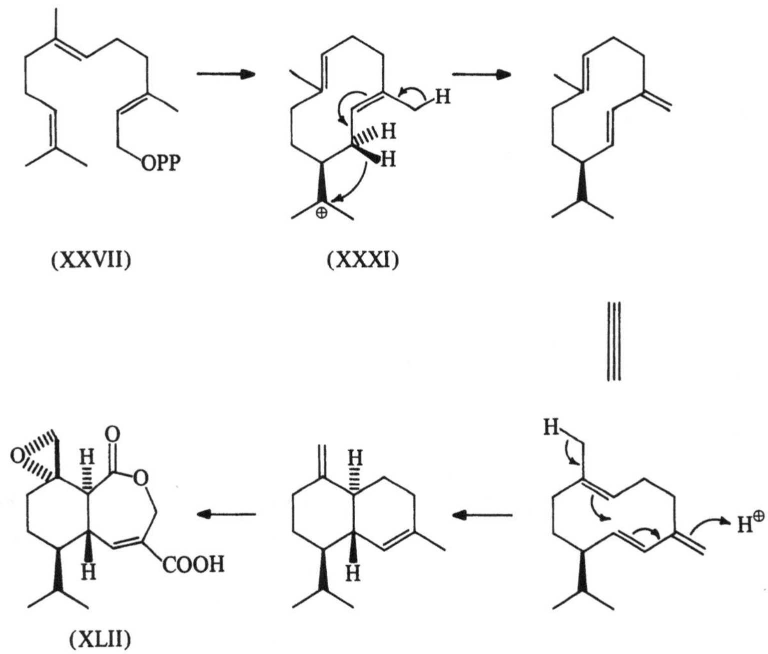

(XXVII) (XXXI)

Fig. 8 (XLII)

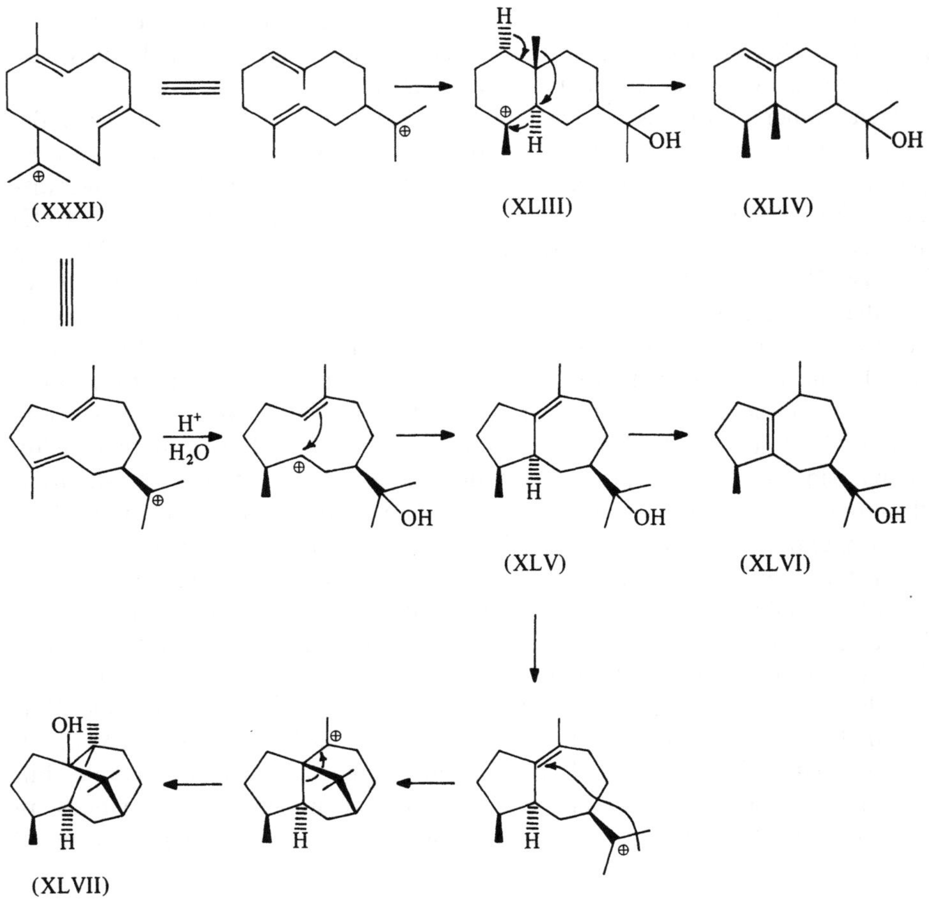

Fig. 9

metabolite avocettin (XLII). However, other explanations are possible. A redox process is definitely ruled out, but a direct trans → cis isomerisation of (XXVII) could be mediated by an appropriate ligase, or nerolidyl pyrophosphate (XLI; the C15 analogue of linalol) could be formed from (XXVII) and could subsequently cyclise.

Given the parent skeletons (XXXI) to (XL) (a few others are known, but the former account hypothetically for the vast majority of derived sesquiterpenes), easy routes can be envisaged to many common sesquiterpenes (PARKER et al., 1967). Thus (XXXI) can lead to the eudesmane (XLIII) and eremophilane (XLIV) series of compounds, or to bulnesol (XLV), guaiol (XLVI) and patchouli alcohol (XLVII) (Fig. 9). These hypothetically related classes of compounds are found in the same or related genera and species and some of the proposed conversions have been carried out in the laboratory. Similarly, it has been suggested (HENDRICKSON, 1959) that (XXX) would preferentially (for steric reasons) cyclise to give (XXXVI) rather than (XXXIV), and the former could be the parent of caryophyllene (XLVIII). Another reasonable biogenetic proposal is that a 1,3-hydride shift in (XXXVI) leads to the precursor of the longibornane series (e.g., longifolene,

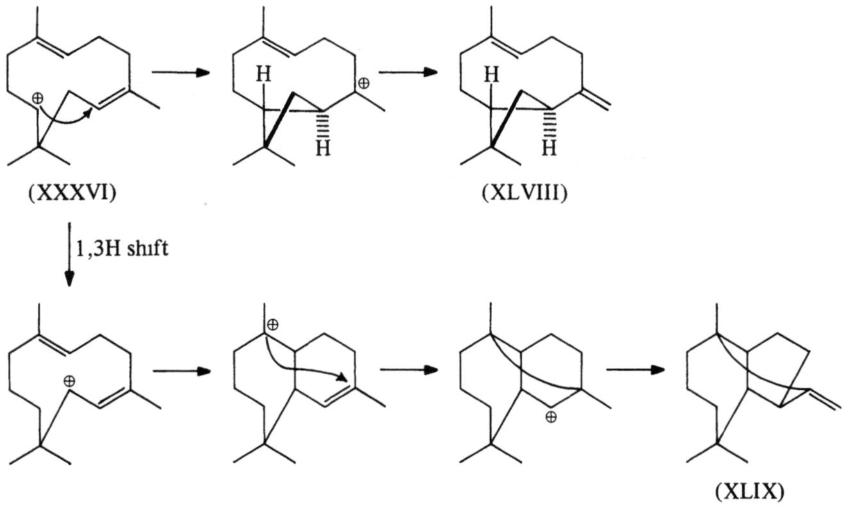

XLIX). Such examples could be multiplied and have been outlined in detail (PARKER et al., 1967; NEWMAN, 1972).

Recently a large amount of work has appeared that almost entirely confirms these hypotheses. ^{14}C or ^{3}H-labelled MVA have been fed to a variety of higher plants and fungi and the labelled products have been isolated, purified, and chemically or microbially degraded to determine the position of tracer, and in some cases to elucidate the stereochemistry of certain C-H bonds generated by hydride shifts. For example, the 1,3-hydride shift in the formation of (XLII) has been directly demonstrated, and ingenious stereochemical methods have been devised to show that the shift occurred before rotation about the C-C bond of the pendant $\overset{\oplus}{C}$-CMe_2 group, i.e., the stereochemical individualities of the methyl groups were retained (ARIGONI, 1975). Two interesting points, that may be general, were found. Firstly, the sites of monoterpene and sesquiterpene synthesis were probably quite distinct in *Mentha* species: although the latter type comprised only a few percent of the total extractable oil, it was labelled many-fold greater than were the monoterpenes when both were biosynthesised in vivo from [2-^{14}C]-MVA (CROTEAU and LOOMIS, 1972). Secondly, feeding of [2-^{14}C]-MVA to several higher plants resulted in asymmetric labelling of several different types of sesquiterpene (BIOLLAZ and ARIGONI, 1969; CROTEAU and LOOMIS, 1972): this situation was similar to that in monoterpenes in that the IPP-derived moieties were labelled to a greater extent than the DMAPP-derived portion. Doubtless the explanation is as for the monoterpenes (see previous sub-section). Similar feedings of fungi did not reveal this phenomenon, possibly because the percentage incorporation of tracer into the investigated products was now typically considerably greater than in the corresponding experiments with higher plants, and the effect of endogenous pools of C 5 intermediates was swamped.

We conclude this sub-section by outlining some recent tracer work relevant to the various types of hypothetical cyclisations. Much of this work involved fungal metabolites, but many important points emerged and it seems very likely that similar schemes are involved when sesquiterpenes of the appropriate classes are formed in higher plants.

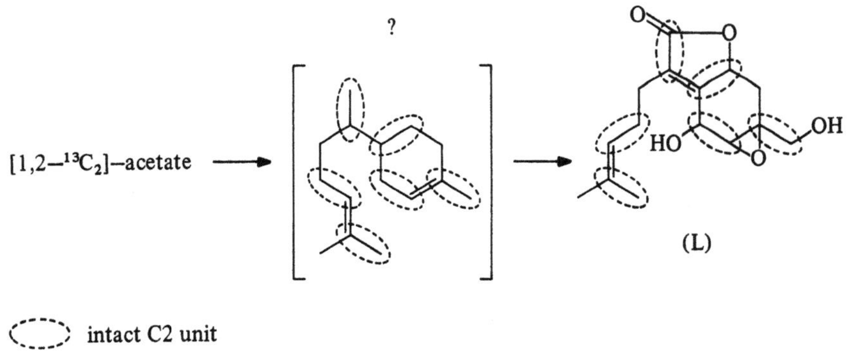

(L)

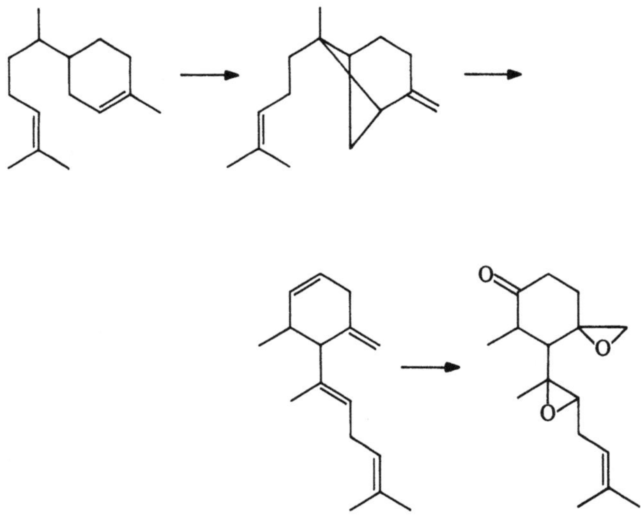

Fig. 10 (LI)

Derivatives of the bisabolane ion (XXXV) have been studied, mainly using ^{13}C-labelled precursors. Thus, [1,2-$^{13}C_2$]-acetate was incorporated into (L) as shown (OVERTON and PICKEN, 1976) and ^{13}C-labelled MVA was incorporated into ovalicin (LI) consistent with the route shown in Figure 10 (CANE and LEVIN, 1976). Use of ^{13}C-labelled precursors has also been made in extremely elegant studies that have elucidated the pathways to formannosin (LII) (CANE and NACHBAR, 1976) and the stress metabolite (LIII) (BIRNBAUM et al., 1976). More common are ^{14}C, ^{3}H-double labelling studies: these have elucidated the routes to the longibornanes (HANSON and NYFELER, 1975; ARIGONI, 1975) and enabled the biosynthesis of dendrobine (LIV) in the orchid *Dendrobium nobile* to be related to the longibornanes (Fig. 11) (ARIGONI, 1975). Another exotic transformation for which there is support from tracer studies is the formation of illudin M (LV) in certain fungi (HANSON et al., 1976). One of the best-studied routes is that to the mycotoxins of the trichothecane group, especially trichothecolone (LVI) (KNOLL and TAMM, 1975; EVANS et al., 1976): the pathway is supported not only by double-labelling and degradative studies but by the feeding of radioactively labelled compounds that were considered prospective intermediates (Fig. 12).

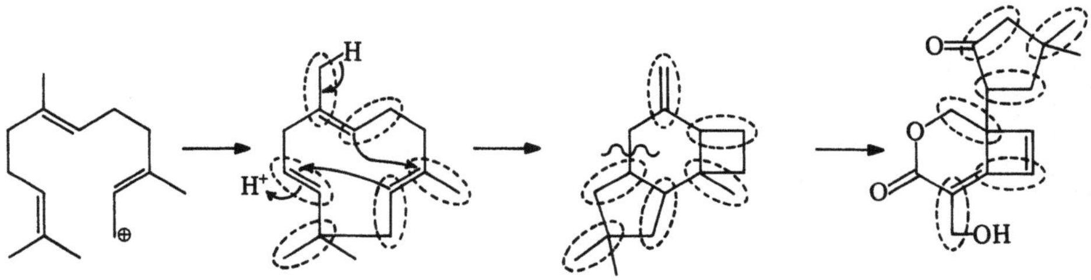

(LII)

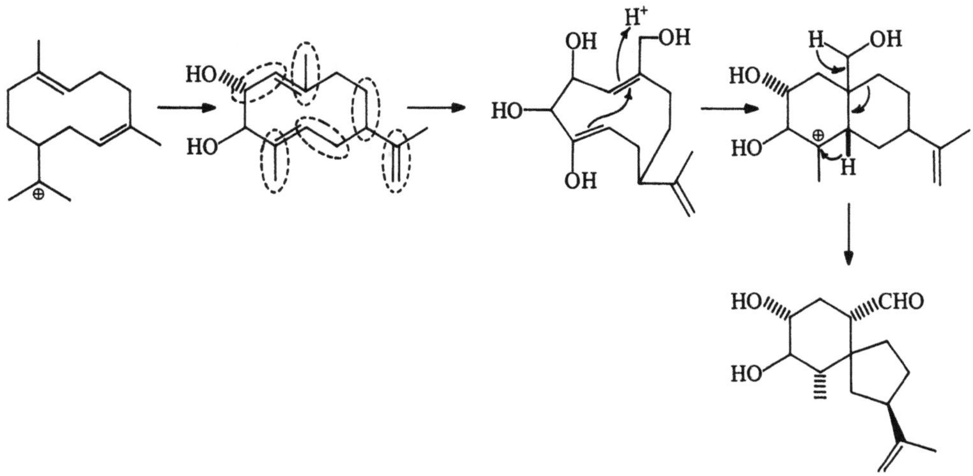

(LIII)

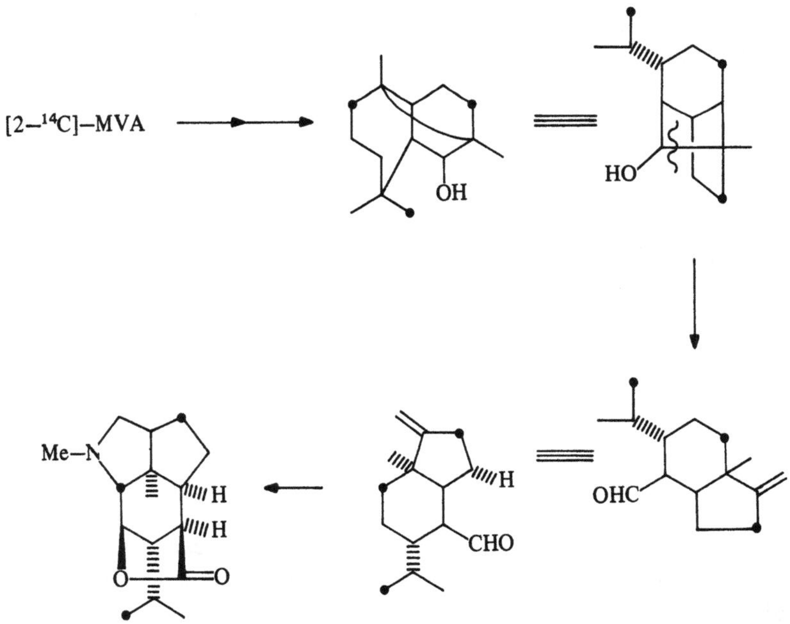

Fig. 11 (LIV)

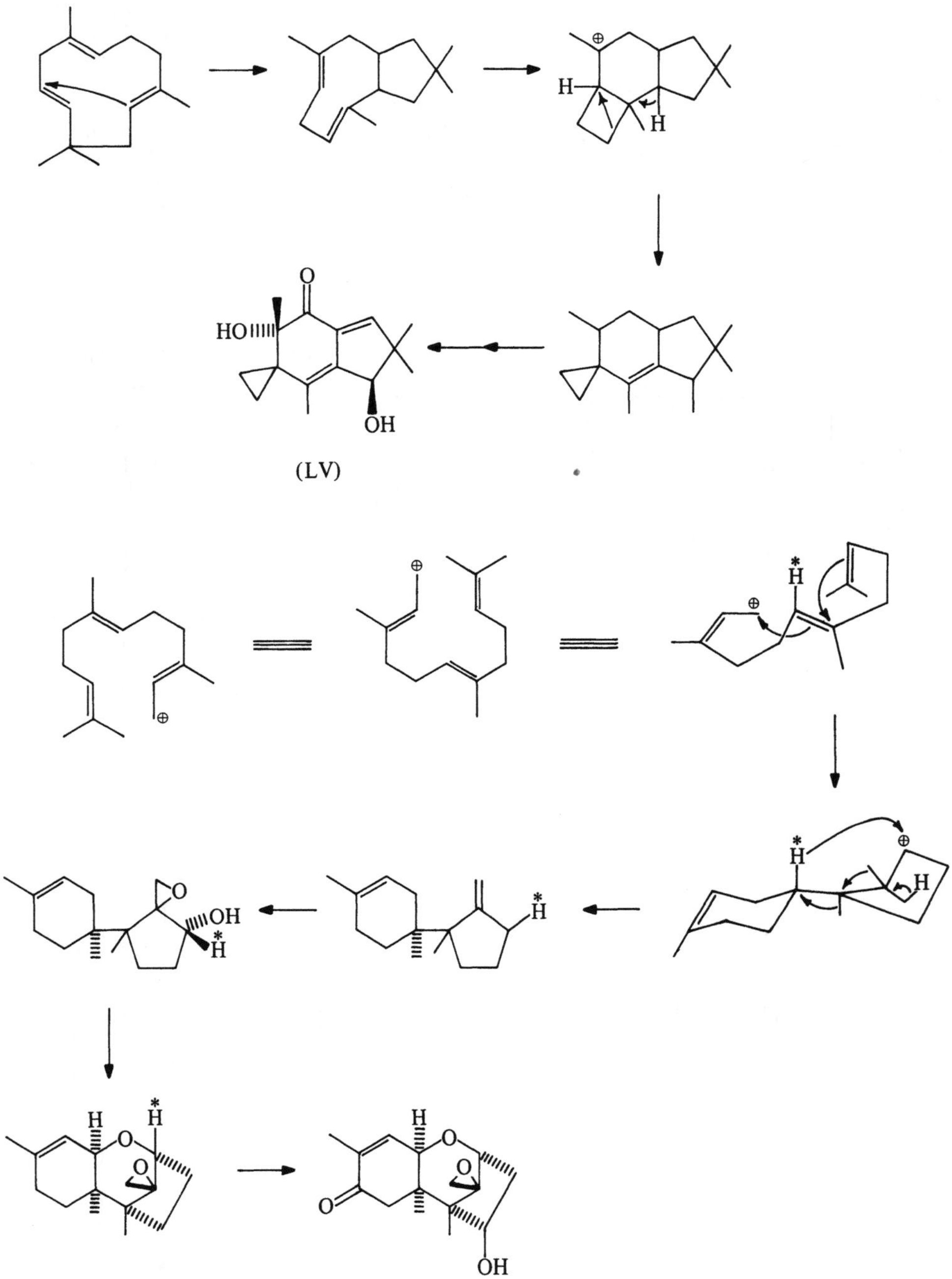

(LV)

(LVI)

Fig. 12.

A brief selection of the skeletal types has been illustrated in the previous section; possibly the most important and best-studied type of sesquiterpene that falls outside the above classifications is exemplified by abscisic acid (LVII) in which an unusual cyclisation to form a six-membered ring occurs (GROSS, 1972).

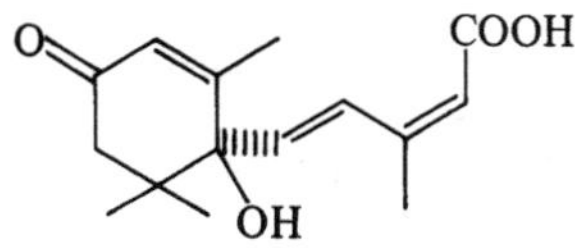

(LVII)

No more elegant unifying pattern for an apparently un-related group of compounds has been made in the whole of chemistry than that in the hypotheses and experiments here outlined for the class of sesquiterpenes. In essence, one can now predict with some confidence the detailed biogenetic route to any new compound of the class that is discovered.

3.4 Diterpenes (C20 Compounds)

General reviews: HANSON (1971, 1972); Gibberellins, HEDDEN et al. (1978); GRAEBE et al. (1978). Geranylgeranyl pyrophosphate (GGPP, VI) is invariably considered to be the parent of this class and a few members, e.g., phytol (LVIII), are directly formed from it. However, most diterpenes are cyclic with no OH group at C-3 (cf. sterols) and possess a trans-junction of the A and B rings that implies a stereospecific acid-catalysed (enzymic) cyclisation, cf. (LIX)→(LX). Hypothetical routes from (LX) to most of the class have been proposed (RUZICKA, 1953) and some of these have been demonstrated to occur in acid-catalysed rearrangements of the proposed precursors in purely chemical systems (OEHLSCHLAGER and OURISSON, 1967).

Formation from (LX) of the allylic ion (LXI) can reasonably lead to formation of sclareol (LXII) and manool (LXIII) and their derivatives, by cyclisation to, e.g., rimuene (LXIV) and abietic acid (LXV) and the former can be supposed to react further to yield compounds such as phyllocladene (LXVI). Stereochemical analysis suggests that the last compound is formed by the rearrangement shown

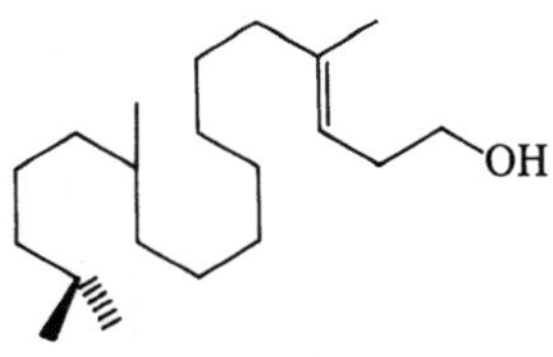

(LVIII)

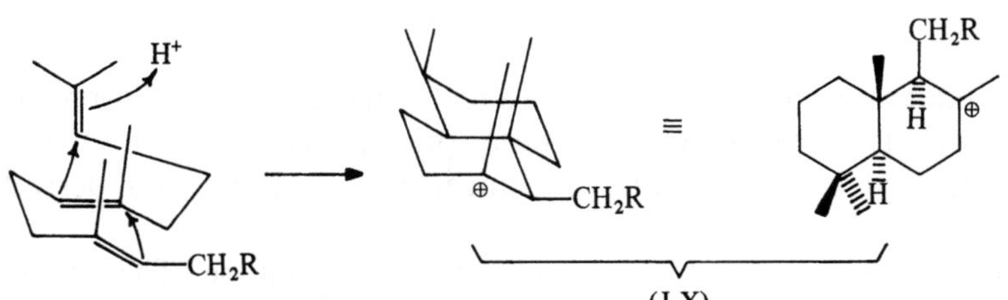

(LIX) (LX)

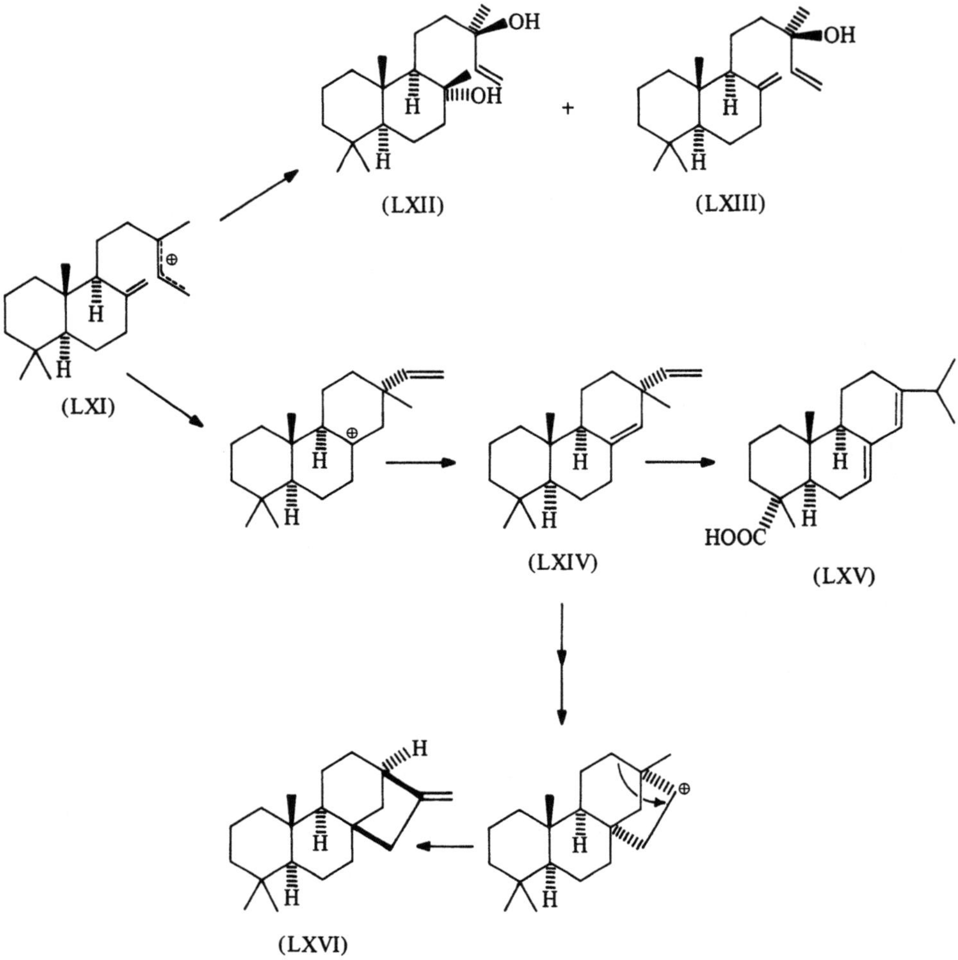

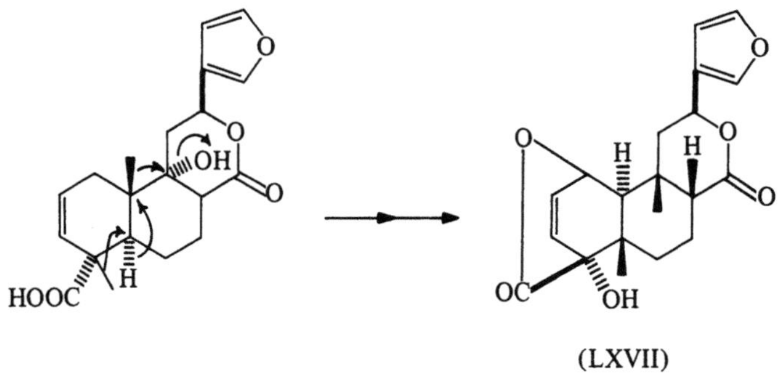

rather than by the more obvious shift of an angular methyl group (SCOTT et al., 1964). More exotic structures, such as the bitter principle columbin (LXVII), are believed to be formed by a backbone rearrangement of an obvious derivative of (LXV).

Several tracer studies have confirmed such routes in outline. The predicted route from GGPP to the mould metabolite rosenololactone (LXVIII) was confirmed

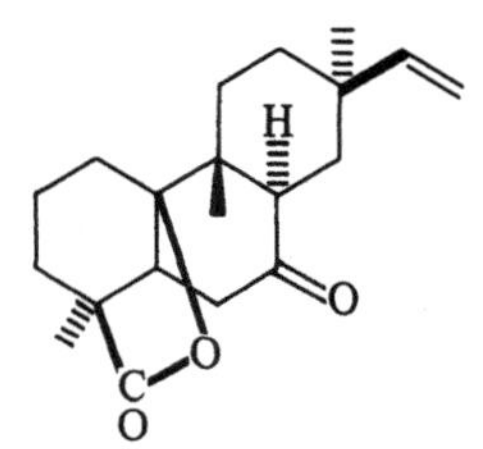

(LXVIII)

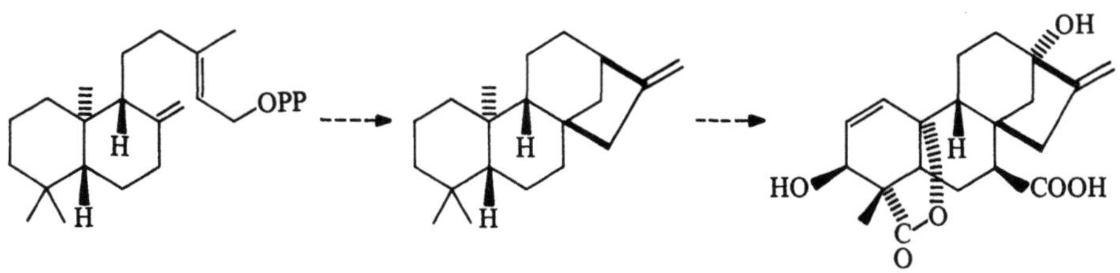

(LXIX) (LXX) (LXXI)

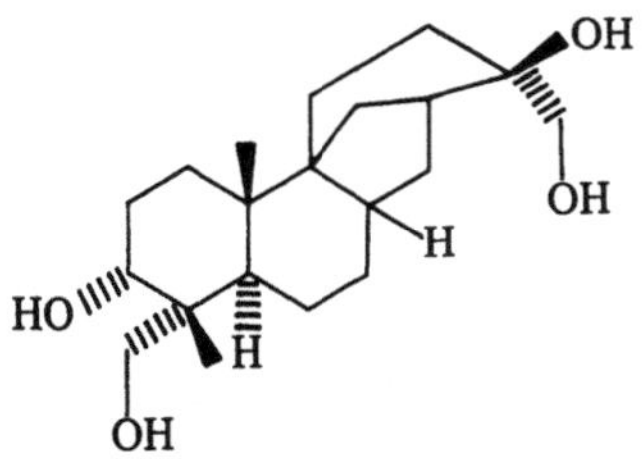

(LXXII)

(ACHILLADELIS and HANSON, 1969). The related gibberellins, e.g., gibberellic acid (LXXI), also gave a labelling pattern from [2-^{14}C]-MVA in accord with predictions (BIRCH et al., 1959) although the hypothetical precursor (LXIX) and the demonstrated intermediate ent-kaurene (LXX) were both of a series enantiomeric with that of the usual diterpenes that have been found in higher plants and fungi (cf. LXI). The theoretical proposals for the construction of the D-ring (see above) were confirmed by detailed tracer studies, and an interesting conclusion that follows from the demonstration that (LXX) is an obligatory intermediate of (LXXI) is that the OH at C-3 in the latter is not involved in the initiation of cyclisation. Most diterpenes have the basic skeletons described above. Some have slight additional modifications, e.g., aphidicolin (LXXII) the details of the formation of which have been elucidated by ^{13}C-studies (ADAMS and BU'LOCK, 1975), and a few classes have the kind of complexity found in sesquiterpenes, but quite lacking in the higher terpenoids. Thus pleuromutilins (LXXIII) biosynthesised from ^{14}C-labelled acetate or MVA gave a tracer pattern consistent with the route shown (ARIGONI, 1962) and ^{13}C-tracer studies on the formation of fusicoccin (LXXIV) showed that it was almost certainly formed by direct cyclisation of GGPP and

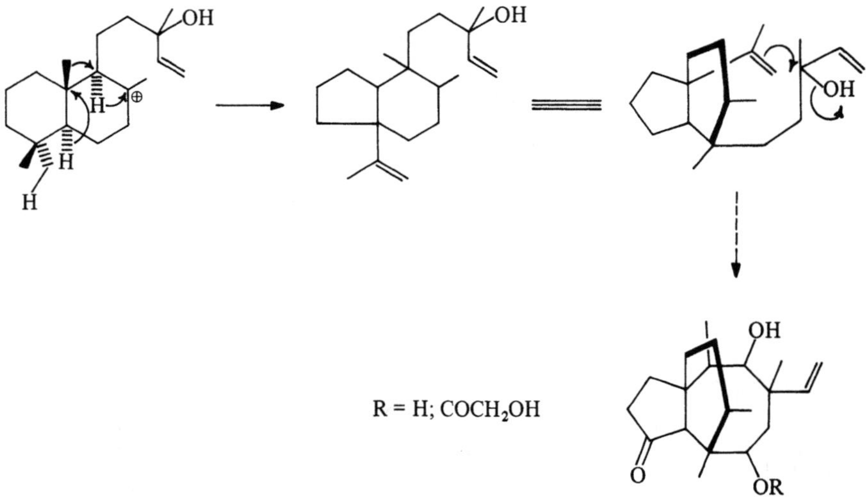

(LXXIII)

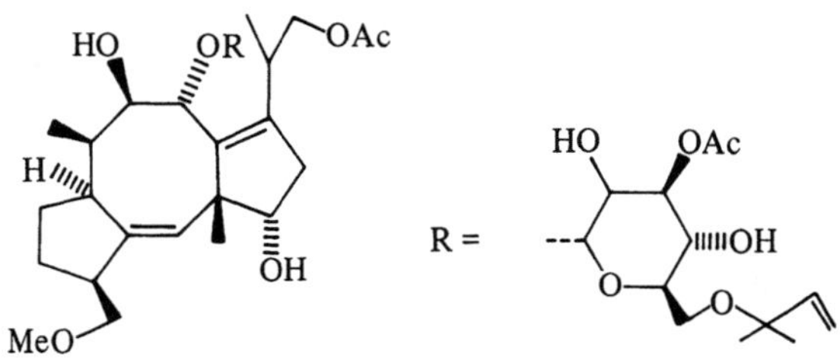

(LXXIV)

was not a degraded sesterterpene as had been thought possible (BARROW et al., 1975).

3.5 Sesterterpenes (C25 Compounds)

Review: CORDELL (1974a). For long it was thought that, with the exception of the higher acyclic polyprenols and the rubbers, the terpenoids with more than twenty carbons were formed by dimerisation of equal moieties. Thus the sequence was C5, C10, C15, C20, C30, and C40. Over the past decade, the existence of sesterterpenes (and also of a few C35 compounds) has been recognised, although relatively few structural types are known and these generally occur in obscure sources. Few tracer studies have appeared but the compounds appear to be derived from a C25 homologue of GGPP, all the C5 units being linked head to tail.

3.6 Nonsteroidal Triterpenes (C30 Compounds)

General reviews: CONNOLLY et al. (1970); KULSHRESHTHA et al. (1972).

The first proposals for the conversion of squalene (or as later recognised, 2,3-oxidosqualene) into lanosterol and thence into steroids was soon followed

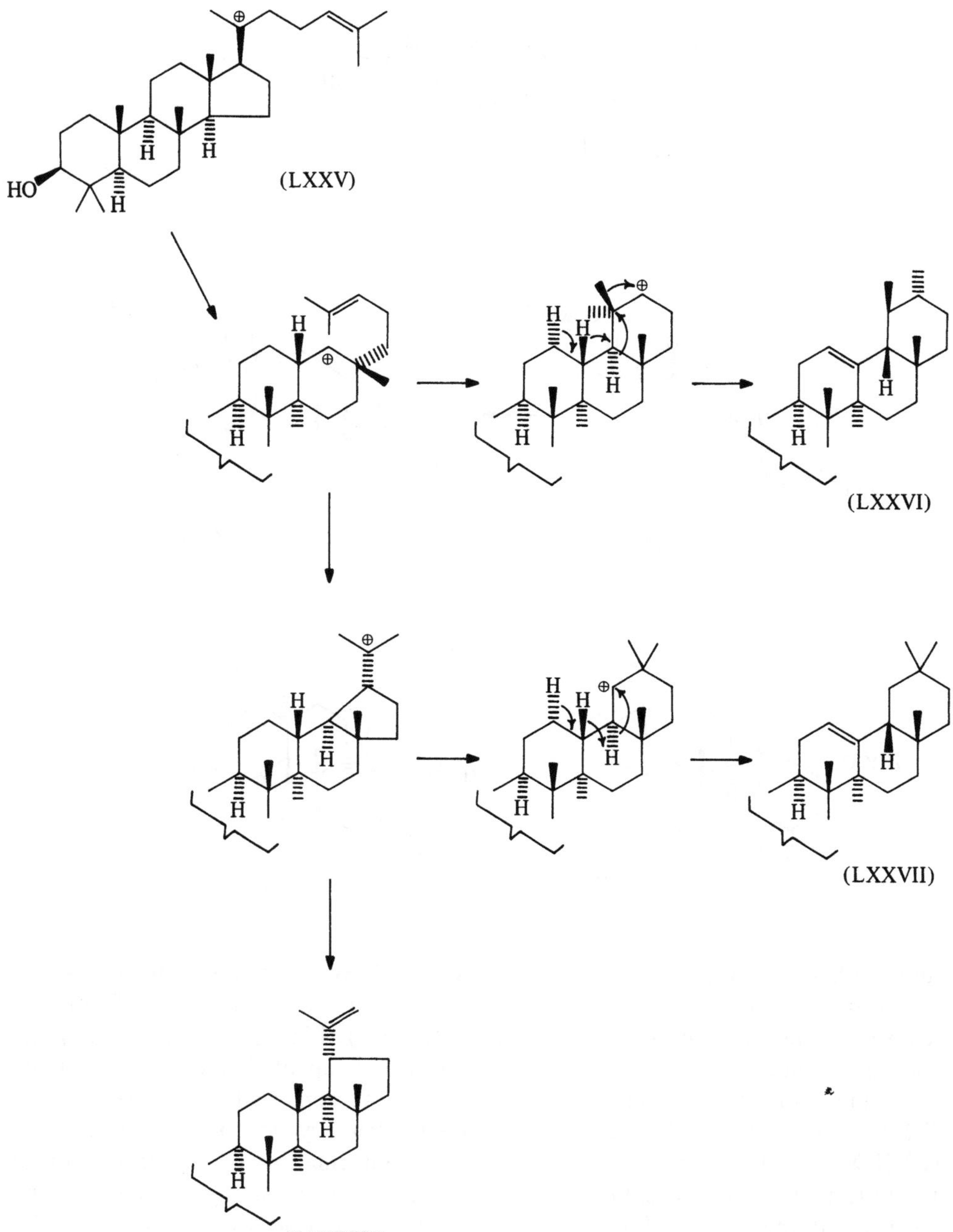

(LXXV)

(LXXVI)

(LXXVII)

(LXXVIII)

by suggestions for the derivation of the various types of nonsteroidal triterpenes from the same parent (ESCHENMOSER et al., 1955). There is little variation in the nature of the basic skeletons of the different types (cf. the situation in sesquiterpenes) and it was presumed that the various segments of squalene were arranged on enzyme surfaces in potential chair (*c*), boat (*b*), or in unfolded conformations (*u*), and that the usual stereoelectronic cyclisations then led to the observed structural types. The detailed proposals are impressive in their simplicity and generality

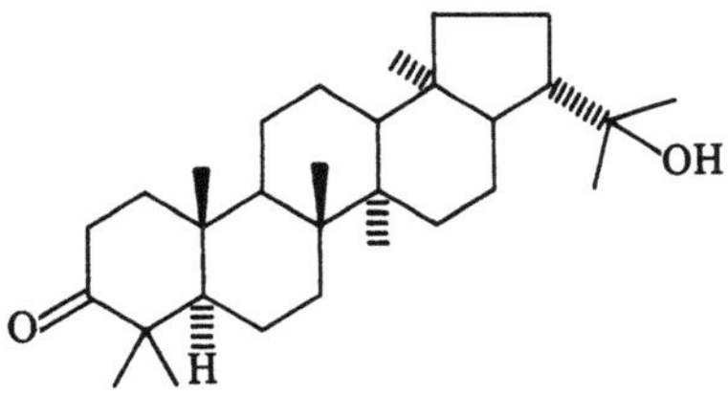

(LXXIX)

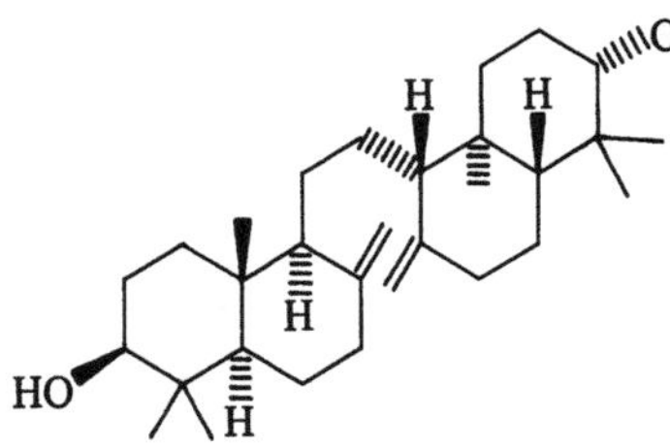

(LXXX)

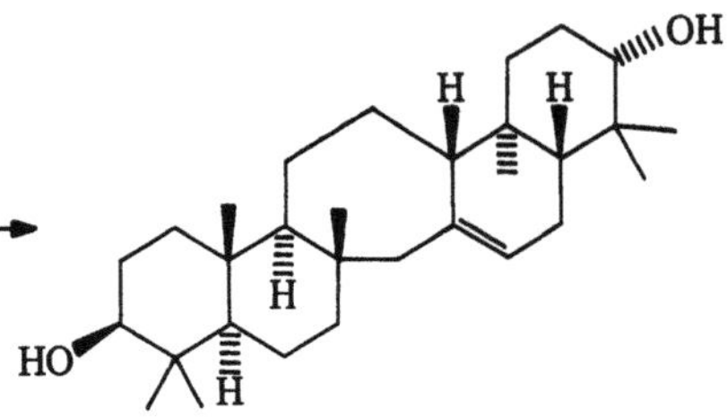

(LXXXI)

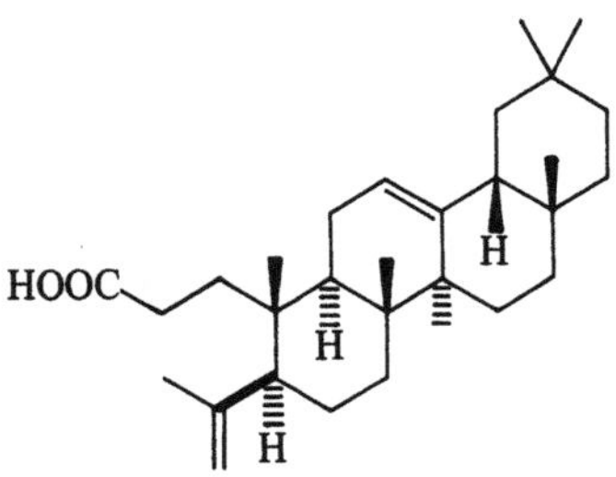

(LXXXII)

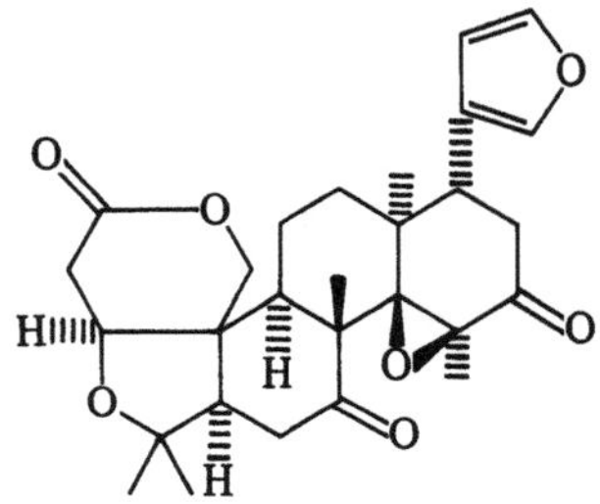

(LXXXIII)

and, with some modifications (cf. RICHARDS and HENDRICKSON; JOHNSON et al., 1974) are undoubtedly correct. Thus, alignment in the (*c-c-c-b-u*) conformation leads via a smooth cyclisation to the skeleton (LXXV) of lanosterol and pentacyclics such as α- and β-amyrin (LXXVI), (LXXVII) and lupeol (LXXVIII) arise from the same conformation but with a pause after the construction of the skeleton (LXXV) that permits further rearrangements. Pentacyclic compounds such as (LXXIX) can be derived from a (*c-c-c-c-c*) conformation of squalene without rearrangement, and cyclisation can start from each end leaving a potential ring enclosed at the centre: thus (*c-c-u-c-c*) squalene can lead to onocerin (LXXX) that can in turn yield serrantenediol (LXXXI). Formation of all the other skeletons may be hypothetically visualised along similar lines, and secondary transformations of these classes may lead to compounds such as nyctanthic acid (LXXXII) and limonin (LXXXIII).

As for the other classes of terpenoids, such tracer studies as have been carried out support these hypotheses (SLIWOWSKI and KASPRZYK, 1974; SUGA and SHISHIBORI, 1975). In particular, tracer studies using tissue cultures of appropriate plants have verified the hypothetical interconversions of the oleanane and ursane-type triterpenes similar to those in the scheme (LXXV)→(LXXVIII) (SEO et al., 1975).

4 Terpene Biosynthesis in Cell-Free Extracts and Tissue Culture

Brief reviews: NEWMAN (1972); BANTHORPE et al. (1972a); CHARLWOOD and BANTHORPE (1978). The use of tissue and cell cultures in the case of phenolic compounds discussed on page 386–391 of this volume may be mentioned for comparison.

4.1 Cell-Free Extracts

The study of the biosynthetic steps from acetate or MVA to squalene and steroids has relied heavily on enzymic studies using cell-free extracts, and several enzymes of the pathway in liver and yeast have been purified and partly characterised. However, extension of such work to secondary metabolism in general or to terpenoid biosynthesis in particular is much less common, and such studies on higher plants are least frequent of all. There are several reasons for the last situation; plant cells are extremely difficult to rupture without irretrievable damage to their enzymic contents; such rupture releases proteases and phosphatases which react with the contents and it also liberates phenols which (especially after oxidation to quinones) complex with and inactivate enzymes; and the success of the preparation of active extracts depends on the addition of cofactors, metal ions etc. that are often lost in the extraction process. In addition, these influences seem to vary markedly between species, and consequently a generally applicable methodology cannot be applied. Another problem is that the enzyme systems, especially for the synthesis of the higher terpenes may be spatially orientated (conveyor belt effect), and the system for a particular sequence may comprise partly extractable water-soluble proteins (those dealing with the substrates of lower molecular weight) and partly of particulate or membrane-bonded species (for the higher molecular weight compounds). Additionally it has recently been found that oxidative enzyme systems are often present in the extracts that can convert intermediates or final products into water-soluble epoxides and polyols (BANTHORPE et al., 1976).

Despite this catalogue of woe, attempts to develop techniques have been pursued energetically (EBEDO and MAGNUSSON, 1973; LOOMIS, 1974; RHODES and STEWART, 1974) as the rewards for success are great. The establishment of appropriate methods would enable specific tissues or organelles to be used for experimentation under defined optimum conditions, and would eliminate problems that have always bedevilled studies with whole plants, e.g., the occurrence of compartmentation that hinders access of exogenous precursor to the biosynthetic site; degradation reactions of added precursors; and climatic and environmental effects.

Much work, especially with the Compositae and on *Citrus, Pisum, Daucus*, and *Pinus* species has been reported, but extracts that convert less than 1% of added MVA into mono- or higher terpenes have been obtained in general. In addition, much of the work is inconclusive as rigorous purification of products to constant specific radioacitivity or even their proper characterisation was often missing. Few of the extracts have been further fractionated and no enzyme of terpene metabolism from a higher plant has been obtained in a remotely purified form; nor is anything known about the details of the mechanism of action of any enzyme. However, some meaningful studies have appeared inter alia concerned with geraniol synthetase (BANTHORPE et al., 1976), the enzymes responsible for formation of irregular monoterpenes (BANTHORPE et al., 1977b), and novel mechanisms for the formulation of borneol and camphor from NPP (CROTEAU and KARP, 1977). Fewer studies have concerned formation of sesqui- and higher terpenes. Extracts have been prepared from various species that convert MVA into FPP (DAVIES et al., 1975; MOORE and COOLBAUGH, 1976) and successful systems that sustain the formation of abscisic acid have been reported (MILBORROW, 1974). Amongst other studies those in the preparation of systems for the biosynthesis of ent-kaurene and gibberellins (GRAEBE et al., 1974, 1975) and of an oxidosqualene cyclase that generates α-onocerin (ROWAN and DEAN, 1972) are noteworthy, as is a detailed investigation of a cytochrome P-450-dependent hydroxylase which converts geraniol into its 10-hydroxy derivative – an intermediate in the formation of loganin (MADYASTHA et al., 1976).

Cell-free systems from moulds and fungi have proved more tractable and many successful systems have been achieved although most of these merely use extracts to demonstrate reactions from late intermediates to occur in the intact organisms. Noteworthy studies involve the biosynthesis and metabolism of gibberellins (BEARDER et al., 1976), of FPP (SUZUKI and NOZOE, 1974), of trichodiene (DOCKERILL and HANSON, 1977) and of oxidosqualene cyclase (KAWAGUCHI et al., 1973). One of the few more detailed biochemical studies involved the 170-fold purification of kaurene synthetase (FALL and WEST, 1971).

4.2 Tissue Cultures

The technique of growing explants of plant tissue as callus or suspensions under sterile, carefully defined conditions is well established (STREET, 1977). Most work has concerned morphological or physiological problems, but many attempts have been made to define conditions under which secondary metabolites are accumulated. Success would not only provide an invaluable biosynthetic tool for metabolic studies, but could provide a source for appropriate enzymes and could even lead to industrial applications for manufacture of essential oils, etc.

A few notable successes have been achieved using this technique as a method of investigating terpene biosynthesis. Most impressive are the stereochemical investigations concerning paniculides (sesquiterpenes) synthesised in cell-free extracts from callus cultures of an *Andrographis* species (OVERTON and PICKEN, 1976), demonstration of the formation of ent-kaurene in tomato cultures (YAFIN and SHECHTER, 1975), and studies concerning the details of triterpene formation (SEO et al., 1975). These investigations were notable for obtaining cultures with enzyme systems that sustained high levels of terpenoid synthesis. Other studies have reported the accumulation of mixtures of monoterpenes in callus obtained from various plant species (BANTHORPE and WIRZ-JUSTICE, 1972; DOWNING and MITCHELL, 1975; NAGEL and REINHARD, 1975; SUGISAWA and OHNISHI, 1976; KUZOVKINA et al., 1975) but generally the concentrations are much lower and the compositions differ greatly from the situation in the whole plants. Most other investigations have failed to demonstrate the occurrence of terpenoids or other secondary metabolites in their cultures, and consequently the technique has been widely considered as being of negligible application to biosynthetic studies of this class. The reasons for this apparent non-success are not clear: most cultures grow rapidly and sometimes even develop chloroplasts and so the pathways to steroids, and even carotenoids and phytol, must be present. It is not known whether the monoterpenes, etc., that accumulate in higher plants and fungi are formed as a direct offshoot of the route to steroids (via removal of GPP and FPP or the corresponding alcohols from the "conveyor belt") or whether they are formed by separate enzyme systems independent of the route to steroids: thus in the cultures that do not accumulate terpenoids there may be a lack of either a few "decoupling" enzymes or an entire pathway (or parts thereof). A possibility that has been completely ignored is that terpenoids may well be formed but can be oxidised or otherwise degraded so that accumulation does not take place. Recently, in another context, an oxidative degradation enzyme system of apparently low selectivity was demonstrated in certain callus cultures and this may be general. Studies to demonstrate the existence of the appropriate enzymes in cultures rather than the accumulation of products are almost non-existent.

A more fundamental reason for the biosynthetic incapabilities of cultures is that conditions may be quite inappropriate. Callus cultures that are colourless, in the stationary phase, in a medium which is not buffered, and being kept in a container of high CO_2-tension under constant illumination and temperature, are generally used. A good case could be made that suspension cultures, green, and in the log phase, maintained under physiological conditions as to gas atmosphere, light, and temperature should be investigated. In addition, the conditions thought by Loomis to be favourable to the formation of secondary metabolites, viz factors which lead to accumulation of an excess of acetyl-CoA, such as lack of oxygen, inhibition of formation or of operation of mitochondria, and excess of sugars in culture media, should be systematically examined. Nothing is known about the influence of such factors.

At present, the study of both cell-free systems[1] and tissue cultures as tools for biosynthetic studies has been disappointing in its results. However, the likely direction for future developments is clear, and one can be reasonably confident in stating that such development one day will revolutionise all studies of secondary metabolism.

References

Achilladelis, B., Hanson, J.R.: Biosynthesis of roseonolactone. J. Chem. Soc. (*C*), 2011–2014 (1969)

Adams, M.R., Bu'Lock, J.D.: Biosynthesis of the diterpene antiobiotic aphidicolin by radioisotope and ^{13}C-nuclear magnetic resonance methods. J. Chem. Soc. Chem. Commun. 389–390 (1975)

Allen, K.G., Banthorpe, D.V., Charlwood, B.V., Ekundayo, O., Mann, J.: Metabolic pools associated with monoterpene biosynthesis in higher plants. Phytochemistry *15*, 101–107 (1976)

Allen, K.G., Banthorpe, D.V., Charlwood, B.V., Voller, C.M.: Biosynthesis of artemisia ketone in higher plants. Phytochemistry *16*, 79–83 (1977)

Arigoni, D.: Structure of a new type of terpene. Gazz. Chim. Ital. *92*, 884–901 (1962)

Arigoni, D.: Stereochemical aspects of sesquiterpene biosynthesis. Pure Appl. Chem. *41*, 219–245 (1975)

1 Data on the use of cell free enzyme preparations for studies of gibberellin in biosynthesis will be presented in Vol. 9 of this Encyclopedia (MacMillan, J., ed.)

Banthorpe, D.V., Ekudayo, O.: Biosynthesis of (+)-Car-3-ene in *Pinus* species. Phytochemistry *15*, 109–112 (1976)

Banthorpe, D.V., Wirz-Justice, A.: Monoterpenes and carotenoids in tissue cultures of *Tanacetum vulgare* L. J. Chem. Soc. Perkin I. 1769–1772 (1972)

Banthorpe, D.V., Charlwood, B.V., Francis, M.J.O.: The biosynthesis of monoterpenes. Chem. Rev. *72*, 115–155 (1972a)

Banthorpe, D.V., Charlwood, B.V., Young, M.R.: Biosynthesis of (+)-pulegone in *Mentha pulegium* L. J. Chem. Soc. Perkin I. 1532–1534 (1972b)

Banthorpe, D.V., Le Patourel, G.N.J., Francis, M.J.O.: Biosynthesis of geraniol and nerol and their β-D-glucosides in *Pelargonium graveolens* and *Rosa dilecta*. Biochem. J. *130*, 1045–1054 (1972c)

Banthorpe, D.V., Bucknall, G.A., Doonan, H.J., Doonan, S., Rowan, M.G.: Biosynthesis of geraniol and nerol in cell-free extracts of *Tanacetum vulgare*. Phytochemistry *15*, 91–100 (1976)

Banthorpe, D.V., Bucknall, G.A., Gutowski, J.A., Rowan, M.G.: Epoxidation and hydration of prenyl pyrophosphates by plant extracts. Phytochemistry *16*, 355–358 (1977a)

Banthorpe, D.V., Doonan, S., Gutowski, J.: Biosynthesis of irregular monoterpenes in extracts from higher plants. Phytochemistry *16*, 85–92 (1977b)

Banthorpe, D.V., Ekundayo, O., Rowan, M.G.: Evidence for geraniol as an obligatory precursor of isothujone. Phytochemistry *17*, 1111–1114, 1978

Barrow, K.D., Jones, R.B., Pemberton, P. W., Phillips, L.: Biosynthesis of fusicoccin from [1-^{13}C]- and [2-^{13}C]-acetate. J. Chem. Soc. Perkin I. 1405–1410 (1975)

Basu, N., Rastogi, R.P.: Terpene saponins and sapogenins. Phytochemistry *6*, 1249–1270 (1967)

Bates, R.B., Paknikar, S.K.: Biogenesis of some Monoterpenoids not derived from a geranyl precursor. Tetrahedron Lett. 1453–1455 (1965)

Bearder, J.R., Frydman, V.M., Gaskin, P., Macmillan, J., Wels, C.M., Phinney, B.O.: Conversion of isosteviol and steviol acetate into gibberellin analogues by mutant B1-41a of *Gibberella fujikuroi* and the preparation of [^{3}H]-gibberellin A_{20}. J. Chem. Soc. Perkin I. 173–178 (1976)

Biollaz, M., Arigoni, D.: Biosynthesis of coriamyrtin and tutin. J. Chem. Soc. Chem. Commun. 633–634 (1969)

Birch, A.J., Richards, R.W., Smith, H., Harris, A., Whalley, W.B.: Studies in relation to biosynthesis: rosenonolactone and gibberellic acid. Tetrahedron *7*, 241–251 (1959)

Birnbaum, G.J., Huber, C.P., Post, M.L., Stothers, J.B., Robinson, J.R., Steossl, A., Ward, E.W.B.: Sesquiterpenoid stress compounds of *Datura stramonium*: biosynthesis of the three major metabolites from [1,2-^{14}C]-acetate. J. Chem. Soc. Chem. Commun. 330–331 (1976)

Burbott, A.J., Loomis, W.D.: Effects of light and temperature on the monoterpenes of peppermint. Plant Physiol. *42*, 20–28 (1967)

Cane, D.E., Levin, R.H.: Application of carbon-13 magnetic resonance to isoprenoid biosynthesis: ovalicin and the use of doubly-labelled mevalonate. J. Am. Chem. Soc. *98*, 1183–1188 (1976)

Cane, D.E., Nachbar, R.B.: Biosynthesis of fommanosin from [1,2-^{13}C]-acetate. Tetrahedron Lett. 2097–2100 (1976)

Charlwood, B.V., Banthorpe, D.V.: Biosynthesis of monoterpenes. Prog. Phytochem. *5*, 65–125 (1978)

Chayet, L., Pont-Lezicka, R., Georg-Nascimento, C., Cori, O.: Biosynthesis of sesquiterpene alcohols and aldehydes by cell-free extracts from orange flavedo. Phytochemistry *12*, 95–101 (1973)

Clayton, R.B.: Biosynthesis of sterol, steroids and terpenoids, Part I. Biogenesis of cholesterol and the fundamental steps in terpenoid biosynthesis. Q. Rev. *19*, 168–200 (1965)

Connolly, J.D., Overton, K.H., Polonsky, J.: The chemistry and biochemistry of the limonoids and quasinoids. Prog. Phytochem. *2*, 385–456 (1970)

Cordell, G.A.: The occurrence, structure, elucidation and biosynthesis of the sesterterpenes. Phytochemistry *13*, 2343–2364 (1974a)

Cordell, G.A.: Biosynthesis of indole alkaloids. Lloydia *37*, 219–298 (1974b)

Cordell, G.A.: Biosynthesis of sesquiterpenes. Chem. Rev. *76*, 425–460 (1976)

Cordell, G.A.: The sesterterpenes – a rare group of natural products. Prog. Phytochem. *4*, 209–256 (1977)
Cornforth, J.W.: The logic of working with enzymes. Chem. Soc. Rev. *2*, 1–20 (1973)
Croteau, R., Karp, F.: Demonstration of a cyclic pyrophosphate intermediate in the enzymatic conversion of neryl pyrophosphate to borneol. Arch. Biochem. Biophys. *184*, 77–86 (1977)
Croteau, R., Loomis, W.D.: Biosynthesis of mono- and sesquiterpenes in peppermint from [2-^{14}C]-mevalonate. Phytochemistry *11*, 1055–1066 (1972)
Darnley-Gibbs, R.: Chemotaxonomy of flowering plants. 4 Vols. Montreal: McGill – Queens University Press 1974
Davies, B.H., Rees, A.F., Taylor, R.F.: Preparation of labelled terpenyl pyrophosphates using extracts of barley seed embryos. Phytochemistry *14*, 717–722 (1975)
Dempsey, M.E.: Regulation of steroid biosynthesis. Ann. Rev. Biochem. *43*, 967–990 (1974)
De Mayo, P. In: Mono- and sesqui-terpenoids. New York: Interscience 1959a
De Mayo, P. In: The higher terpenoids. New York: Interscience 1959b
Devon, T.K., Scott, A.I.: Handbook of naturally-occurring compounds, Vol. II. New York: Academic Press 1972
Dockerill, B., Hanson, J.R.: Formation of pimara – 8(9),15-diene by *Trichothecium roseum*. J. Chem. Soc. Perkin I. 324–327 (1977)
Downing, M.R., Mitchell, E.D.: Mevalonate-activating enzymes in callus culture cells from *Nepeta cataria*. Phytochemistry *14*, 369–371 (1975)
Ebedo, L., Magnusson, K.E.: Disintegration of cells and protein recovery. Pure Appl. Chem. *36*, 325–338 (1973)
Eberhart, N.L., Rilling, H.C.: Prenyltransferase from *Saccharomyces cerevisiae*. J. Biol. Chem. *250*, 863–866 (1975)
Epstein, W.W., Poulter, C.D.: A survey of some irregular monoterpenes and their biogenetic analogies to presqualene alcohol. Phytochemistry *12*, 737–747 (1973)
Eschenmoser, A., Ruzicka, L., Jeger, O., Arigoni, D.: Eine stereochemische Interpretation der biogenetischen Isoprenregel bei den Triterpenen. Helv. Chim. Acta *38*, 1890–1904 (1955)
Evans, R., Hanson, J.R.: Formation of the sesquiterpene trichodiene. J. Chem. Soc. Perkin I. 326–329 (1976)
Evans, R., Hanson, J.R., Marten, T.: Formation of the sesquiterpenoid trichothecin. J. Chem. Soc. Perkin I. 1212–1214 (1976)
Fall, R.R., West, C.A.: Purification and properties of Kaurene synthetase from *Fusarium moniliforme*. J. Biol. Chem. *246*, 6913–6928 (1971)
Francis, M.J.O., Allcock, C.: Geraniol β-D-glucoside: occurrence and synthesis in rose flowers. Phytochemistry *8*, 1339–1347 (1969)
Francis, M.J.O , O'Connell, M.: The incorporation of mevalonic acid into rose petal monoterpenes. Phytochemistry *8*, 1705–1708 (1969)
Geissman, T.A. (ed.): Biosynthesis; specialist periodical report. Vol. I. London: The Chemical Society 1972
Geissman, T.A.: Biogenesis of sesquiterpene lactones of compositae. Recent Adv. Phytochem. *6*, 65–95 (1973)
Gildemeister, E., Hoffman, F.: Die ätherischen Öle. 8 Vols. Berlin: Akademie Verlag 1955–60
Goodwin, T.W. (ed.): Natural substances formed biologically from mevalonic acid. London: Academic Press 1970
Goodwin, T.W. (ed.): Aspects of terpenoid chemistry and biochemistry. London: Academic Press 1971
Graebe, J.E., Ropers, H.-J.: Gibberellins. In: Phytohormones and related compounds: A comprehensive treatise. Letham, D.S., Goodwin, P.B., Higgins, T.J.V. (eds.), Vol. I, pp. 107–204. Amsterdam: Elsevier, 1978
Graebe, J.E., Hedden, P., Gaskin, P., MacMillan, J.: Biosynthesis of gibberellins A_{12}, A_{15}, A_{24}, A_{36} and A_{37} by a cell-free system from *Cucurbita maxima*. Phytochemistry *13*, 1433–1440 (1974)
Graebe, J.E., Hedden, P., MacMillan, J.: The ring contraction step in gibberellin biosynthesis. J. Chem. Soc. Chem. Commun. 161–162 (1975)
Gray, J.C., Kekwick, R.G.O.: The inhibition of plant mevalonate kinase preparations by prenyl pyrophosphates. Biochim. Biophys. Acta *279*, 290–296 (1972)

Green, T.R., Baisted, D.J.: Development of the activities of enzymes of the isoprenoid pathway during early stages of pea-seed germination. Biochem. J. *130*, 983–995 (1972)
Green, T.R., West, C.A.: Purification and characterisation of two forms of geranyl transferase from *Ricinus communis*. Biochemistry *13*, 4720–4729 (1974)
Gross, D.: Chemistry and biochemistry of abscisic acid. Pharmazie *27*, 619–630 (1972)
Guarnaccia, R., Madyastha, K.M., Tegtmeyer, E., Coscia, C.J.: Geniposidic acid, an iridoid glucoside from *Genipa americana*. Tetrahedron Lett. 5125–5127 (1972)
Guarnaccia, R., Botta, L., Coscia, C.J.: Biosynthesis of acidic iridoid monoterpene glucosides in *Vinca rosea*. J. Am. Chem. Soc. *96*, 7079–7084 (1974)
Guenther, E.: The essential oils, 6 Vols. New York: Van Nostrand-Reinhold 1948
Haagen-Smit, A.J. In: Handbuch der Pflanzenphysiologie. Ruhland, W. (ed.), Vol. X, pp. 52–90. Berlin, Heidelberg, New York: Springer 1958
Hanson, J.R.: The tetracyclic diterpenes. Oxford: Pergamon Press 1968
Hanson, J.R.: The application of multiply-labelled mevalonate in terpene and steroid biosynthesis. Adv. Steroid Biochem. Pharmacol. *7*, 51–96 (1970)
Hanson, J.R.: The biosynthesis of diterpenes. Fortschr. Chem. Org. Naturst. *29*, 395–416 (1971)
Hanson, J.R.: The bicyclic diterpenes. Prog. Phytochem. *3*, 231–286 (1972)
Hanson, J.R. (ed.): Terpenoids, steroids: specialist periodical reports. Vol. VII. London: The Chemical Society 1977
Hanson, J.R., Nyfeler, R.: The mevalonoid origin of the hydrogen atoms of culmorin. J. Chem. Soc. Chem. Commun. 171–172 (1975)
Hanson, J.R., Marten, T., Nyfeler, R.. Biosynthesis of the sesquiterpenoid illudin M. J. Chem. Soc. Perkin I. 876–880 (1976)
Hedden, P., Macmillan, J., Phinney, B.O.: The metabolism of the gibberellins. Ann. Rev. Plant Physiol. *29*, 149–192 (1978)
Hefendehl, F.W., Murray, M.J.: Genetic aspects of the biosynthesis of natural odors. Lloydia *39*, 39–51 (1976)
Hendrickson, J.B.: Stereochemical implications in sesquiterpene biogenesis. Tetrahedron *7*, 82–89 (1959)
Hill, H.M., Rogers, L.J.: Mevalonate-activating enzymes and phosphatases in higher plants. Phytochemistry *13*, 763–777 (1974)
Johnson, W.S., Weidhaup, K., Brady, S.F., Olson, G.L.: Nonenzymic, biogenetic-like cyclistion of a tetraenic acetal to produce the D-homosteroid nucleus. J. Am. Chem. Soc. *96*, 3979–3984 (1974)
Jones, C.A., Rasmussen, R.A.: Production of isoprene by leaf tissue. Plant Physiol. *55*, 982–987 (1975)
Karrer, W : Konstitution und Vorkommen der organischen Pflanzenstoffe. Basel: Birkhäuser Verlag 1958
Kawaguchi, A., Kobayashi, H., Okuda, S.: Cyclisation of 2,3-oxidosqualene with microsomal fraction of *Cephalosporium caerulens*. Chem. Pharm. Bull. *21*, 577–583 (1973)
Knoll, W., Tamm, C.: Biosynthese der Verrucarine und Roridine. Helv. Chim. Acta. *58*, 1162–1171 (1975)
Kulshreshta, M.J., Kulshreshta, D.K., Rastogi, R.P.: The triterpenoids. Phytochemistry *11*, 2369–2381 (1972)
Kutney, J.P., Salisbury, P.J., Verma, A.K.: Studies on tissue cultures of *Thamnosma montana*. The role of mevalonate. Tetrahedron *29*, 2673–2681 (1973)
Kuzovkina, I.N., Kuznetsova, E.A., Smirnov, A.M.: Essential oils in the culture of isolated plant tissues. Izv. Akad. Nauk SSSR Ser. Biol. 377–381 (1975)
Leistner, E.: Mode of incorporation of precursors into alizarin (1,2-dihydroxy-9,10-anthraquinone). Phytochemistry *12*, 337–345 (1973)
Lin, T.J., Heinstein, P.: In vitro biosynthesis of isopentenylacetophenones in *Eupatorium rugosum*. Phytochemistry *13*, 1817–1823 (1974)
Loomis, W.D.: Biosynthesis and metabolism of monoterpenes. In: Terpenoids in plants. Pridham, J.B. (ed.), pp. 59–82. London: Academic Press 1967
Loomis, W.D.: Overcoming problems of phenolics and quinones in the isolation of plant enzymes and organelles. Methods Enzymol. *31A*, 528–544 (1974)

Luckner, M.: Secondary metabolism in plants and animals. London: Chapman and Hall 1972

Mackie, H., Overton, K.H.: Hydrolysis and isomerisation of *trans*, *trans*,- farnesyl diphosphate by *Andrographis* tissue culture enzymes. Eur. J. Biochem. *77*, 101–106 (1977)

Madyastha, K.M., Meehan, T.D., Coscia, C.J.: Characterisation of a cytochrome P-450 dependent monoterpene hydroxylase from the higher plant *Vinca rosea*. Biochemistry *15*, 1097–1102 (1976)

Mann, J.: Secondary metabolism. London: Oxford University Press 1978

Milborrow, B.V.: Biosynthesis of abscisic acid by a cell-free system. Phytochemistry *13*, 131–136 (1974)

Miles, D.H., Kokpol, U., Bhattacharyya, J., Atwood, J.A., Stone, K.E., Bryson, T.A., Wilson, C.: Structure of sarracenin. An unusual enol diacetal monoterpene from the insectivorous plant *Sarracenia flava*. J. Am. Chem. Soc. *98*, 1569–1573 (1976)

Miller, L.P. (ed): Phytochemistry. 3 Vols. New York: Van Nostrand-Reinhold 1973

Moore, T.C., Coolbaugh, R.C.: Conversion of geranylgeranyl pyrophosphate to *ent*-kaurene in enzyme extracts of sonicated chloroplasts. Phytochemistry *15*, 1241–1247 (1976)

Muller, B., Tamm, C.: The mode of incorporation of (3R)-[(5R)-5-^{3}H]-Mevalonate into verrucarol. Helv. Chim. Acta *58*, 483–488 (1975)

Nagel, M., Reinhard, E.: Volatile oil of tissue cultures of *Ruta irageolens*. Planta Med. *27*, 151–158 (1975)

Nakanishi, K., Goto, T., Ito, S., Natori, S., Nozue, S.: Natural product chemistry. New York: Academic Press 1974

Newman, A.A. (ed.): Chemistry of terpenes and terpenoids. London: Academic Press 1972

Oehlschlager, A.C., Ourisson, G.: Comparison of in vivo and in vitro skeletal transformations of diterpenes. In: Terpenoids in plants. Pridham, J.B. (ed.), pp. 83–110. London: Academic Press 1967

Overton, K.H. (ed.): Terpenoids and steroids, specialist periodical reports. Vols. I–VI. London: The Chemical Society 1971–1976

Overton, K.H., Picken, D.J.: Biosynthesis of bisabolene by callus cultures of *Andrographis paniculata*. J. Chem. Soc. Chem. Commun. 105–106 (1976)

Parker, W., Roberts, J.S., Ramage, R.: Sesquiterpene biogenesis. Q. Rev. *21*, 331–363 (1967)

Pickett, J.A., Coates, J., Sharpe, F.R.: Distortion of essential oil composition during isolation by steam distillation. Chem. Ind. 571–572 (1975)

Plouvier, V., Favre-Bonvin, J.: Les iridoides et seco-iridoides: repartition, structure, proprietes, biosynthese. Phytochemistry *10*, 1697–1722 (1971)

Poulter, C.D., Satterwhite, D.M., Rilling, H.C.: Prenyltransferase: the mechanism of the reaction. J. Am. Chem. Soc. *98*, 3376–3377 (1976)

Reed, B.C., Rilling, H.C.: Crystallisation and partial characterisation of prenyltransferase from avian liver. Biochemistry *14*, 50–54 (1975)

Rhodes, D., Stewart, G.R.: A procedure for the in vivo determination of enzyme activity in higher plant tissue. Planta *118*, 133–144 (1974)

Richards, J.H., Hendrickson, J.B.: The biosynthesis of steroids and acetogenins. New York: Benjamin 1964

Rodwell, V.W., Nordstrom, J.L., Mitschelen, J.J.: Regulation of HMG-CoA reductase. Adv. Lipid. Res. *14*, 2–74 (1976)

Ross, M.S.F., Lines, D.S., Stevens, R.G., Brain, K.R.: Epoxidase-epoxide hydrase activity in cell cultures of *Phaseolus vulgaris*. Phytochemistry *17*, 45–48 (1978)

Rowan, M.G., Dean, P.D.G.: Properties of squalene-2(3), 22(23)-diepoxide-α-onocerin cyclase from *Ononis spinosa* root. Phytochemistry *11*, 3111–3118 (1972)

Rücker, G.: Sesquiterpenes. Angew. Chem. Int. Ed. Engl. *12*, 793–806 (1973)

Ruzicka, L.. Isoprene rule and the biogenesis of terpenic compounds. Experientia *9*, 357–367 (1953)

Ruzicka, L.: History of the isoprene rule. Proc. Chem. Soc. 341–360 (1959)

Ruzicka, L.: In: The borderland between biorganic chemistry and biochemistry. Annu. Rev. Biochem. *42*, 1–20 (1973)

Sandermann, W.: In: Comparative biochemistry. Vol. 3, pp. 591–630. Florkin, M., Mason, H.S. (eds.). New York: Academic Press 1962

Scott, A.I.: Biosynthesis of natural products. Science *184*, 760–764 (1974)

Scott, A.I., McCarpra, F., Comer, F., Sutherland, S.A., Young, D.W., Sim, G.A., Ferguson, G.: Structure and stereochemistry of some polycyclic diterpenoids. Tetrahedron *20*, 1339–1358 (1964)

Scott, A.I., Reichardt, P.B., Slaytor, M.B., Sweeny, J.G.: Mechanisms of indole alkaloid biosynthesis: recognition of intermediacy and sequence by short term incubation. Recent Adv. Phytochem. *6*, 117–145 (1973)

Seo, S., Tomita, Y., Tori, K.: Biosynthesis of ursene-type triterpenes from sodium [1,2-^{13}C]-acetate in tissue cultures of *Isodon japonicus*. J. Chem. Soc. Commun. 954–955 (1975)

Shaw, J., Noble, T., Epstein, W.: Methyl (2R), (3S)-2,5-dimethy-3-vinylhex-4-enoate: a new irregular monoterpene constituent of *Artemisia tridentata*. J. Chem. Soc. Chem. Commun. 590–591 (1975)

Shinka, T., Ogura, K., Seto, S.: Farnesyl pyrophosphate synthetases during *Cucurbita pepo* germination. Phytochemistry *13*, 2130–2106 (1974)

Simpson, T.J.: Carbon-13 nuclear magnetic resonance in biosynthetic studies. Chem. Soc. Rev. *4*, 497–522 (1975)

Sliwowski, J., Kasprzyk, Z.: Stereospecificity of biosynthesis of triterpene alcohols in *Calendula officinalis* flowers. Phytochemistry *13*, 1441–1449 (1974)

Stoessl, A., Stotmers, J.B., Ward, E.W.B.: Sesquiterpenoid stress compounds of the solanaceae. Phytochemistry *15*, 855–872 (1976)

Street, H.E., (ed.): Plant tissue and cell culture, 2nd ed. Oxford: Blackwells 1977

Suga, T., Shishibori, T.: The stereospecificity of biosynthesis of squalene and *β*-amyrin in *Pisum sativum*. Phytochemistry *14*, 2414–2417 (1975)

Sugisawa, H., Ohnishi, Y.: Isolation and ıdentification of monoterpenes from cultured cells of perilla plant. Agric. Biol. Chem. *40*, 231–232 (1976)

Suzuki, K.T., Nozoe, S.: Biosynthesis of an antibiotic, siccanin. Bioorg. Chem. *3*, 72–80 (1974)

Swain, T.: In Comprehensive biochemistry. Florkin, M., Stotz, E.M. (eds.), Vol. 29A pp. 125–286 Amsterdam: Elsevier 1974

Takeda, Y., Inouye, H.: The fate of the C-8 proton of 7-deoxyloganic acid in the biosynthesis of secoiridoid glucosides. Chem. Pharm Bull *24*, 79–84 (1976)

Takhtajan, A.: Flowering Plants: origin and dispersal. Edinburgh: Oliver and Boyd 1969

Taylor, W.I., Battersby, A.R.: Cyclopentanoid terpene derivatives. New York: Dekker 1970

Turner, W.B.: Fungal metabolites. New York: Academic Press 1971

Yafin, Y., Shechter, I.: Comparison between biosynthesis of *ent*-kaurene in germinating tomato seeds and cell supervision cultures of tomato and tobacco. Plant Physiol. *56*, 671–675 (1975)

5.2 Steroids

C. GRUNWALD

1 Introduction

Steroids belong to a large group of compounds known as terpenoids or isoprenoids. Terpenes are formed by the polymerization of isoprene units (I) and steroids are triterpenes or triterpenoids. The term triterpene refers to a group of natural products containing 30 carbon atoms which are derived from six isoprene units. However, this definition is by no means rigid, and compounds with fewer than 30 carbon atoms, if derived from six isoprenes, are included in the steroids. Discussion in this section on isoprenoids will deal only with those triterpenes that have the steroidal ring structure of perhydro-1,2-cyclopentanophenanthrene (II). The numbering system of rings A, B, C, and D may be unusual, but this arrangement allows, with one exception, numbers assigned to an earlier incorrect formula to be transferred to the correct formula (FIESER and FIESER, 1967). The exception is the reversal of carbon positions 7 and 12.

The present discussion pertains mainly to the plant literature but a recent, more complete discussion of the biochemistry of steroids and other isopentenoids is available (NES and MCKEAN, 1977).

2 Sterols

2.1 Structure

In the literature dealing with plants, the sterols are the most often discussed steroids. Strictly speaking, sterols are secondary alcohols which are crystalline solids at room temperature. The name sterol is derived from the Greek *stereos* (solid) plus -ol (FIESER and FIESER, 1967). Plant sterols have the perhydro-1,2-cyclopentanophenanthrene ring system (II), with the three 6-membered rings in a nonlinear or

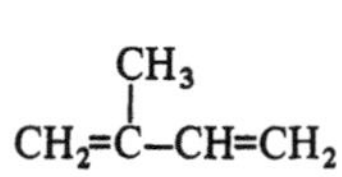

(I)

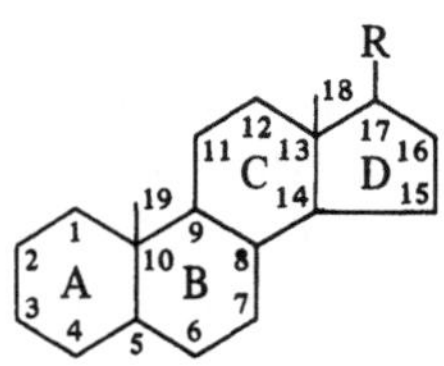

(II)

Table 1. Trivial and scientific names of selected steroids of the sterol biosynthetic pathway

Trivial name	Scientific name	sterol class
Cycloartenol	4,4,14α-trimethyl-9β,19β-cyclo-5α-lanost-24-en-3β-ol	4,4-dimethyl
24-Methylene cycloartanol	4,4,14α-trimethyl-9β,19β-cyclo-5α-ergost-24(28)-en-3β-ol	4,4-dimethyl
Cycloeucalenol	4α,14α-dimethyl-9β,19β-cyclo-5α-ergost-24(28)-en-3β-ol	4α-methyl
Obtusifoliol	4α,14α-dimethyl-5α-ergosta-8,24(28)-dien-3β-ol	4α-methyl
24-Methylene lophenol	4α-methyl-5α-ergosta-7,24(28)-dien-3β-ol	4α-methyl
24-Ethylidene lophenol *or* citrostadienol	4α-methyl-5α-stigmasta-7,Z-24(28)-dien-3β-ol	4α-methyl
Δ^7-Avenasterol	5α-stigmasta-7,Z-24(28)-dien-3β-ol	4-demethyl
Spinasterol	5α-stigmasta-7,22-dien-3β-ol	4-demethyl
Avenasterol *or* 28-Isofucosterol	stigmasta-5,Z-24(28)-dien-3β-ol	4-demethyl
Cholesterol	cholest-5-en-3β-ol	4-demethyl
Campesterol	(24R)-24-methylcholest-5-en-3β-ol	4-demethyl
Sitosterol	stigmasta-5-en-3β-ol; (24R)-24-ethylcholest-5-en-3β-ol	4-demethyl
Stigmasterol	stigmasta-5,22-diene-3β-ol; (24S)-24-ethylcholesta-5,22-dien-3β-ol	4-demethyl

phenanothrene arrangement fused to a 5-membered ring. Rings AB, BC, and CD are trans fused. The hydrocarbon skeleton has a secondary hydroxy group at C-3, with few exceptions angular methyl groups at C-10 and C-13, and generally an 8 to 10 carbon chain at C-17. The hydroxy, the two methyl groups, and the C-17 side chain all have the β-orientation.

The various sterols found in plants differ in the number of carbon atoms in the C-17 side chain, the degree and position of saturation in the ring system and in the side chain, substitutions at C-3 and C-4, and in the stereochemistry at asymmetric centers. The names of plant sterols are often perplexing because the most common sterols are referred to by their trivial names. The proper chemical name is derived from the hydrocarbon (stigmastane, ergostane, cholestane) or the parent alcohol (stigmastanol, ergostanol, cholestanol). The trivial and scientific names of a selected group of biosynthetically important higher plant sterols are given in Table 1.

2.2 Classification

The number of phytosterols encountered in nature is quite large and two classification schemes are often used. The first classification has three groups and takes advantage of the biosynthetic relationship, according to the number of methyl groups at the C-4 position. Sterols with two methyl groups are called 4,4-dimethyl sterols, and cycloartenol (III) is an important sterol of this group since it is the first cyclic product. Sterols with only one methyl group at C-4 are referred to as 4-methyl sterols, and obtusifoliol (IV) is a representation of this group. The single methyl group at C-4 has the α-orientation, and these sterols are important intermediates in the synthesis of compounds known as 4-demethyl sterols, or simply

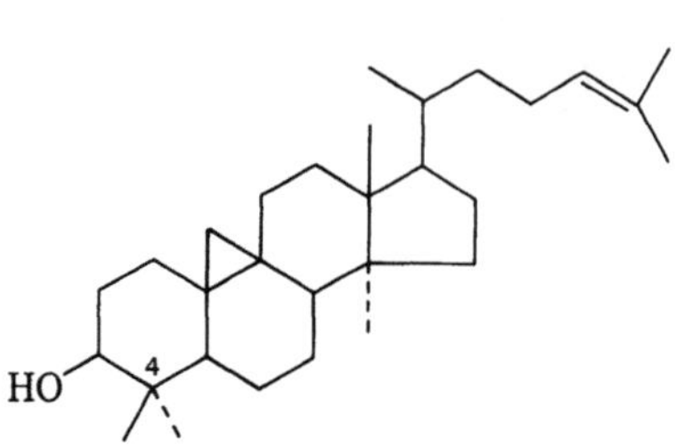

Cycloartenol
(III)

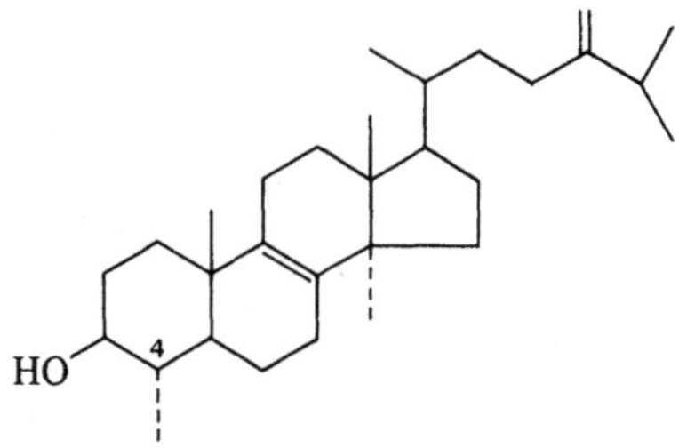

Obtusifoliol
(IV)

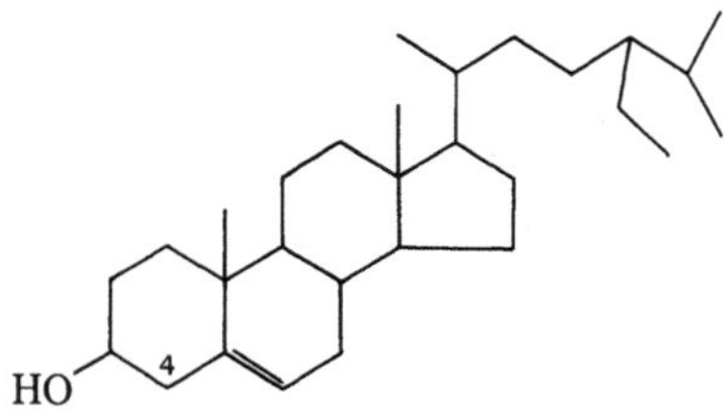

Sitosterol
(V)

sterols. Sitosterol (V) is the most common 4-demethyl sterol in higher plants (STOLL and JUCKER, 1955).

The second system of classifying sterols is according to the substitution at the C-3 position. This scheme is very convenient in discussing the physiological functions of sterols. At present, four groups are recognized. Sterols with a β-C-3-hydroxyl group are known as free sterols or simply sterols. Sitosterol represents this group. If a fatty acid forms an ester linkage with the sterol moiety at C-3, the compound is called a steryl ester, such as sitosteryl palmitate (VI). A large number of steryl esters are possible since the number depends on the number of fatty acids and sterols present. The most common fatty acids are palmitate, stearate, oleate, linolate, and linolenate (BUSH and GRUNWALD, 1972; KINTIA and WOJCIECHOWSKI, 1974; TORRES and GARCIA-OLMEDO, 1975). Sterols with a carbohydrate moiety at C-3 make up the third group, and these compounds are known as steryl glycosides. Sitosteryl glucoside (VII) is a steryl glycoside. The fourth group, the acylated steryl glycosides, are compounds in which a fatty acid moiety is linked, via an ester bond, to the C-6 position of the carbohydrate. 6-O-Palmitoyl-β-D-glucosyl sitosterol (VIII) is a member of this group. The number of steryl glycosides and acylsteryl glycosides possible is quite large, but the carbohydrate moiety most often isolated is glucose (LEPAGE, 1964; KIRIBUCHI et al., 1965) and mannose (EICHENBERGER and MENKE, 1966). The acylsteryl glycosides carry palmitic, stearic, oleic, linoleic, and linolenic acid as the main acyl moieties (KIRIBUCHI et al., 1966).

2.3 Distribution

Sterols have been isolated from a large number of species and probably occur in all angiosperms and gymnosperms. Less often studied are the sterols in liverworts,

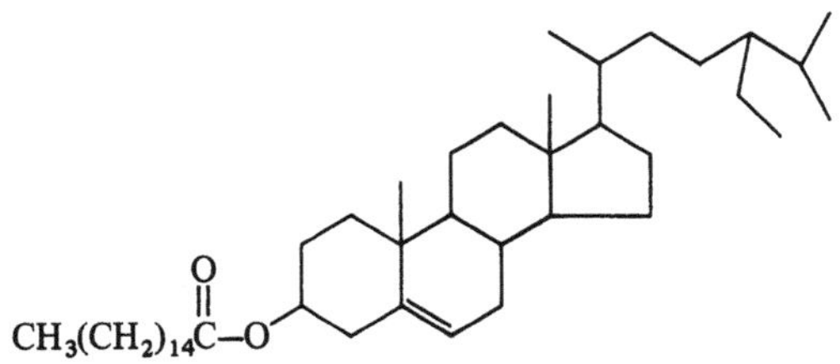

Sitosteryl palmitate
(VI)

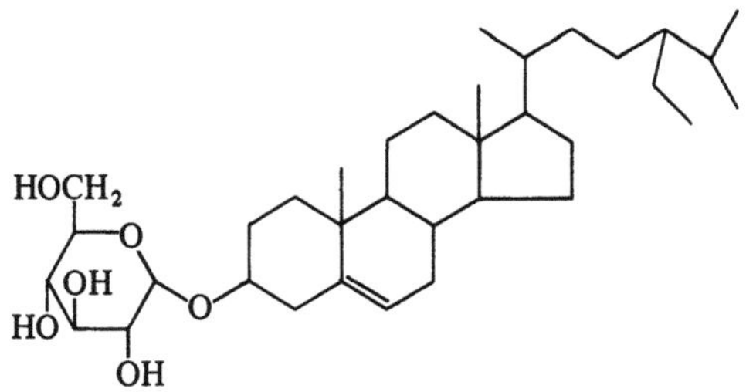

β–D–Sitosteryl glucose
(VII)

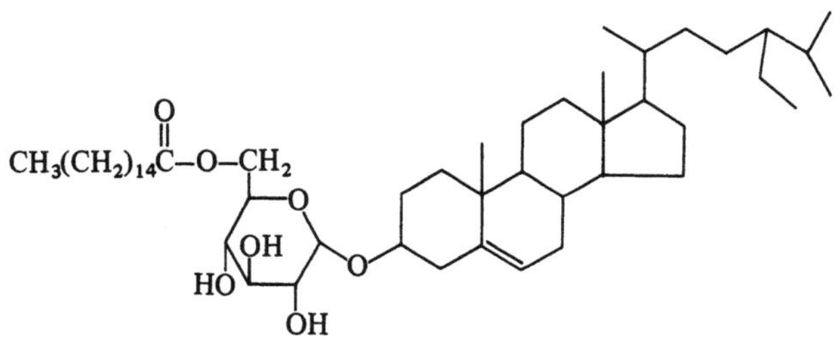

6–O–Palmitoyl–β–D–glucosyl–sitosterol
(VIII)

mosses, horsetail, and ferns. Sterols also occur in fungi, algae, lichens, and bacteria. A brief summary of some of the sterols isolated from various organisms is given in Table 2. For a more complete list see BEAN (1973).

The sterol most often isolated from higher plants is sitosterol (STOLL and JUCKER, 1955), but stigmasterol and campesterol are also quite common. Cholesterol, which for many years was considered to occur only in animals, has been found as a minor component in many plant tissues and it is a major component in *Solanum* (JOHNSON et al., 1963) and *Nicotiana* (KELLER et al., 1969). A large number of other sterols have been isolated, but a list of these compounds is beyond the scope of this summary. For a more complete review see GOAD and GOODWIN (1972).

Sterols have been isolated from all types of tissue such as leaves, roots, and stems (KEMP and MERCER, 1968a; WILLUHN and KÖSTENS, 1975); bark (ROWE, 1965); cotyledons (INGRAM et al., 1968); flowers (SLIWOWSKI and KASPRZYK, 1974); pollen (HÜGEL et al., 1964a, b); fruit (BENNETT et al., 1961); peel of fruit (WEIZMANN

Table 2. Various organisms and their selected sterols

Organism	Major selected sterols	Reference
Liverworts	Sitosterol, stigmasterol	MATSUO et al., 1973
Mosses	Sitosterol, stigmasterol, campesterol	MARSILI and MORELLI, 1968, 1970
Horsetail	Sitosterol	LUDWICZAK and STACHOWIAK, 1963
Ferns	Sitosterol	BERTI and BOTTARI, 1968
Fungi	Ergosterol	WEETE, 1973
Green algae	Depending on species, a variety of Δ^5, Δ^7, and $\Delta^{5,7}$ sterols, cholesterol, 24 methylene cholesterol, poriferasterol, clionasterol, 28-isofucosterol, ergosterol, 7-dehydroporiferasterol	PATTERSON, 1974
Brown algae	Fucosterol, 24-methylene cholesterol	PATTERSON, 1968
Red algae	Cholesterol, 22-dehydrocholesterol desmosterol	CHARDON-LORIAUX et al., 1976
Golden-brown algae and diatoms	Poriferasterol	RUBINSTEIN and GOAD, 1974
Blue-green algae	Cholesterol, sitosterol	REITZ and HAMILTON, 1968
Lichens	Ergosterol, lichesterol, poriferasterol	LENTON et al., 1973a, b
Amoeba	7-Dehydrostigmasterol, ergosterol	SMITH and KORN, 1968
Bacteria	Cholesterol, sitosterol	SCHUBERT et al., 1964, 1968

and MAZUR, 1958); pulp of fruit (KNAPP and NICHOLAS, 1969); seeds (KNIGHTS and LAURIE, 1967); seed coat, scutellum, and coleoptile (TORRES et al., 1976); and endosperm (KEMP et al., 1967). All plant tissues have the potential to synthesize sterols (EVANS, 1974; ATALLAH et al., 1975a) but the rate of synthesis depends upon the type and age of the tissue (GEUNS, 1973, 1975).

2.4 Biosynthesis

2.4.1 Squalene Biosynthesis

The formation of squalene is through the familiar pathway of acetate, acetoacetate, hydroxymethylglutaric acid, mevalonic acid, isopentenyl pyrophosphate, 3,3-dimethylallyl pyrophosphate, geranyl pyrophosphate, farnesyl pyrophosphate, and squalene. A number of recent reviews discuss this sequence (GOAD, 1967; CORNFORTH and CORNFORTH, 1970; JACOBSOHN, 1970; GOAD and GOODWIN, 1972; GRUNWALD, 1975b).

2.4.2 Squalene Cyclization

The cyclization of squalene to form the cyclopentanophenanthrene ring system is via squalene-2,3-oxide (Scheme 1). Cyclization was first demonstrated in liver

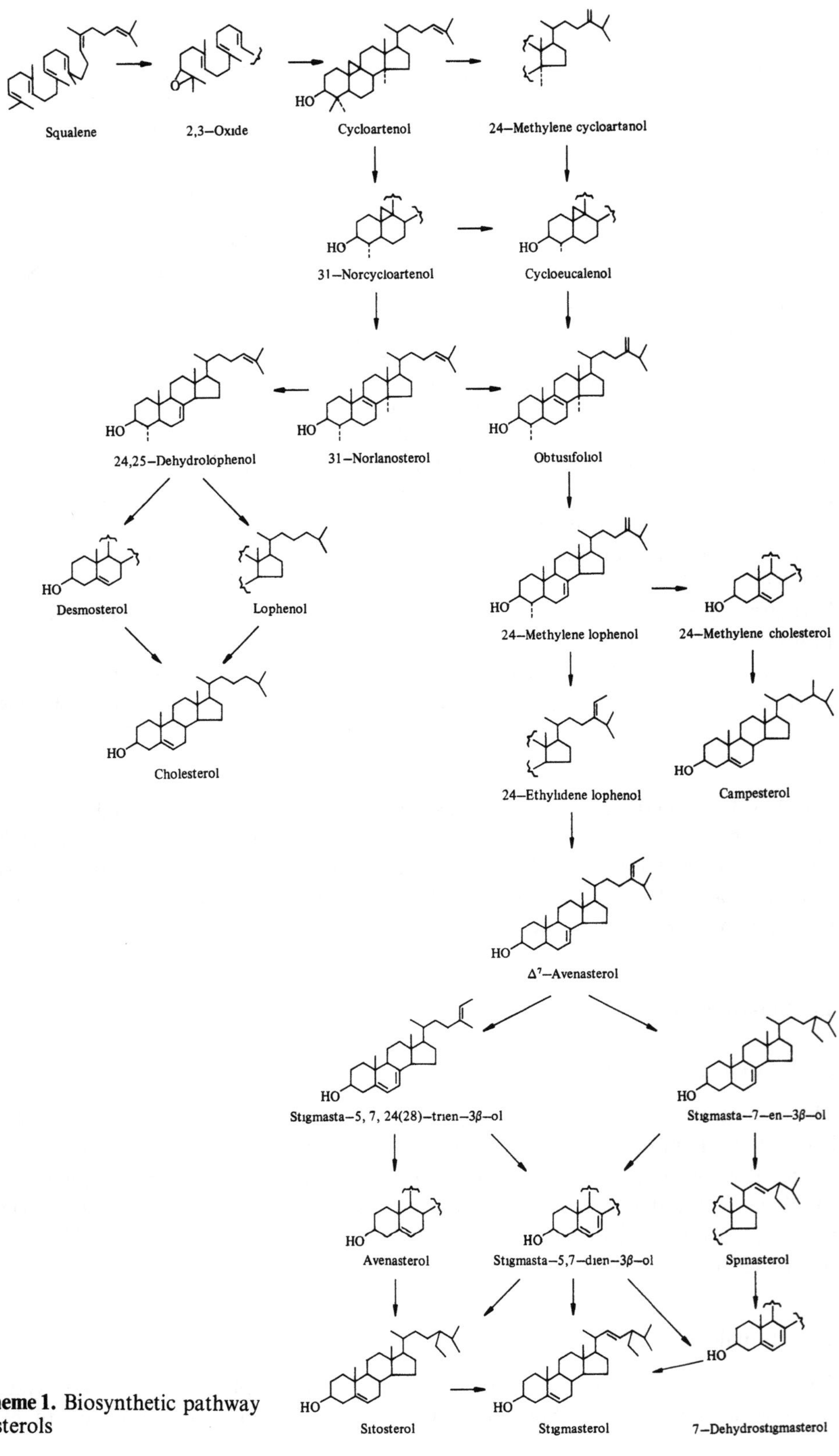

Scheme 1. Biosynthetic pathway of sterols

homogenates (COREY et al., 1966; VAN TAMELEN et al., 1966). Shortly thereafter, BENVENISTE and MASSY-WESTROPP (1967) and REID (1968) demonstrated that in plants squalene-2,3-oxide was also the intermediate during cyclization. However, the major difference is the formation of cycloartenol (REES et al., 1968a) instead of lanosterol as in animals (FRANTZ and SCHROEPFER, 1967) and fungi (WEETE, 1973). Squalene-2,3-oxide is also an intermediate in fungal sterol biosynthesis (MERCER and JOHNSON, 1969) and algae (REES et al., 1969). Cyclization is initiated by cation OH^+ attack at the squalene position which gives rise to C-3 of the sterol molecule. In plants, as in animals, the oxidation and the cyclization steps occur separately and not synchronously (EPPENBERGER et al., 1969). Formation of the tetracyclic steroid ring system is through molecular rearrangement, a migration of two hydrogen atoms and two 1,2-methyl shifts from C-8 to C-14 and from C-14 to C-13 (MAUDGAL et al., 1958; CORNFORTH et al., 1959; REES et al., 1968b). The 3-β-hydroxyl group is derived from atmospheric oxygen and not from water (TCHEN and BLOCH, 1957). The epoxidase, which converts squalene to the 2,3-oxide, is microsomal in nature and requires NADPH and molecular oxygen (YAMAMOTO and BLOCH, 1970), and addition of the sterol inhibitor, tris (2-diethylaminoethyl)phosphate, results in an accumulation of squalene-2,3-oxide (REID, 1968). The conversion of squalene-2,3-oxide to cycloartenol is anaerobic (EPPENBERGER et al., 1969) and the cyclase enzyme is also localized in the microsomal fraction (HEINTZ and BENVENISTE, 1970), but not very substrate specific (HEINTZ et al., 1970). Experiments with radioactive sterol precursors suggest that cycloartenol is a branch point, not only for cholesterol synthesis, but also for the synthesis of other major phytosterols. The most generally accepted pathway is through 24-methylene cycloartanol, but the sequence cycloartenol→31-norcycloartenol→31-norlanosterol→obtusifoliol has also been indicated (Scheme 1).

It is generally accepted that cycloartenol, and not lanosterol, is the first cyclic product in plants. Conclusive evidence was obtained with [2-^{14}C, (4R)-4-^{3}H]-mevalonic acid (BENVENISTE et al., 1966; EHRHARDT et al., 1967; REES et al., 1968b; KNAPP and NICHOLAS, 1971b).

2.4.3 Alkylation at C-24

To form the major phytosterols from cycloartenol, an alkylation at C-24 is probably the first step (Scheme 1), and this occurs through transmethylation involving S-adenosyl methionine (NICHOLAS and MORIARTY, 1963; CASTLE et al., 1963; KNIGHTS and LAURIE, 1967). The transmethylation reaction requires a $\Delta^{24(25)}$ intermediate (RUSSELL et al., 1967). During alkylation, a cationic site is created at C-24 which is stabilized by hydrogen migration from C-24 to C-25 (REES et al., 1966; RAAB et al., 1968). The product is 24-methylene cycloartanol, a 4,4-dimethyl sterol.

2.4.4 First C-4 Demethylation

Demethylation at C-4 (Scheme 1) is probably the next step, producing cycloeucalenol, the first 4-methyl sterol (BENVENISTE, 1968). The 4α-methyl of 24-methylene cycloartanol is oxidized, presumably by way of a primary alcohol, to carboxylic acid and removed by decarboxylation (OLSON et al., 1957; KNAPP and NICHOLAS, 1970). During demethylation of the 4α-methyl group, the 4β-methyl is epimerized to the 4α position (SHARPLESS et al., 1969). It is most likely that a C-3-ketone intermediate is involved (LINDBERG et al., 1963; KNAPP and NICHOLAS, 1970), but an alternate mechanism in which the ketone would be a product rather than an early intermediate has also been postulated (KNAPP and NICHOLAS, 1971b).

2.4.5 Opening of the Cyclopropane Ring

The C-4-β-methyl has to be absent before the 9β, 19β-cyclopropane ring can be opened most efficiently to form obtusifoliol (HEINTZ and BENVENISTE, 1974; RAHIER et al., 1977). Of course, this does not mean that a number of parallel mechanisms are not possible (BENVENISTE et al., 1969).

2.4.6 Removal of the C-14 Methyl

Enzyme attack at the C-14α-methyl position is hindered by the cyclopropane ring, therefore, the loss of the C-14 methyl group by oxidative decarboxylation occurs after the 9β, 19β-cyclopropane ring is opened (Scheme 1). Furthermore, for C-14 demethylation to occur, a $\Delta^{8(9)}$ bond is required and obtusifoliol and 31-norlanosterol have these requirements (BENVENISTE, 1968) to form 24-methylene lophenol, probably another important branch point in the biosynthesis of phytosterols. Formation of sitosterol and stigmasterol requires a second alkylation at C-28 to form ethylidene lophenol (SCHREIBER and OSSKE, 1964). The 28 carbon sterol, campesterol, does not require the second alkylation (Scheme 1).

2.4.7 Alkylation of C-28

As during the first alkylation, methionine is again the methyl donor for the second alkylation (RUSSELL et al., 1967). During this process, a cationic site at C-24 of the steroid molecule is created which is stabilized through the loss of a hydrogen atom from C-28 (GOAD et al., 1966). The formation of 24-ethylidene lophenol has been demonstrated with radioactive (CASTLE et al., 1963) and deuterium-labeled methionine (SMITH et al., 1967). The second alkylation does not appear to be substrate-specific since 24-methylene dihydrolanosterol, which is not found in higher plants, was converted to spinasterol and sitosterol when applied to the leaf (ALCAIDE et al., 1968; DEVYS et al., 1968). For a more detailed review concerning the alkylation process see GOAD and GOODWIN (1972).

2.4.8 Second Demethylation at C-4

The removal of the second C-4 methyl group from 24-ethylidene lophenol is also through oxidative decarboxylation (SWINDELL and GAYLOR, 1968). For demethylation to occur, the C-4 methyl group must be in the α configuration which, as described above, occurs through inversion of the 4β-methyl group during demethylation of the 4,4-dimethyl sterol. Demethylation of 4α-methyl sterols gives rise to the 4-demethyl sterols; and in the case of the sitosterol-stigmasterol pathway it is probably Δ^7-avenasterol. This sterol has been isolated from plants (KNIGHTS and LAURIE, 1967; SUCROW, 1968). In the formation of campesterol, which does not require the second methylation step, 24-methylene lophenol is demethylated at C-4 to give 24-methylene cholesterol (KNIGHTS and LAURIE, 1967; ALCAIDE et al., 1968).

2.4.9 Rearrangement of Double Bonds in the Ring System

At first glance, the conversion of Δ^7-avenasterol to the major phytosterols, sitosterol and stigmasterol, may appear to be straightforward, however, conclusive evidence for one specific pathway is difficult to find. At present, it appears that a number of pathways are possible and basically it involves a reduction of $\Delta^{24(28)}$ and the rearrangement of the double bond in ring B (Scheme 1).

One of the suggested pathways involves first the rearrangement of $\Delta^{7(8)} \rightarrow \Delta^{5,7} \rightarrow \Delta^{5(6)}$ to form avenasterol (GOAD et al., 1969). Avenasterol has been isolated from higher plants (IDLER et al., 1953; KNIGHTS, 1965). Additionally, experiments have demonstrated the in vivo conversion of radioactive mevalonic acid to avenasterol and sitosterol (VAN ALLER et al., 1968; BOLGER et al., 1969; BAISTED, 1969). Formation of sitosterol from avenasterol requires hydrogenation of $\Delta^{24(28)}$ and reduction of the 24-ethylidene, which must involve a $\Delta^{24(25)}$ intermediate since experiments with tritium-labeled mevalonic acid have shown the loss of the C-25 hydrogen atom (ARMAREGO et al., 1973). It is worth noting that the C-25 hydrogen atom which is lost, migrated from C-24 to C-25 during the first alkylation reaction. In this proposed pathway, formation of stigmasterol is assumed to occur through sitosterol by the enzyme 22,23-dehydrogenase (JOHNSON et al., 1964; ROWE, 1965), however, much of the evidence is indirect. Radioactive mevalonic acid is first detected in sitosterol, and with longer incubation the specific activity of sitosterol decreases, while that of stigmasterol increases (KNAPP and NICHOLAS, 1971a; BUSH and GRUNWALD, 1973). Direct conversion of sitosterol to stigmasterol, however, is low, and all attempts to demonstrate that this is the major pathway for stigmasterol biosynthesis are unconvincing (WATERS and JOHNSON, 1965; BENNETT and HEFTMANN, 1969). Formation of stigmasterol may also be via Δ^7-avenasterol → stigmasta-5,7,24(28)-trien-3β-ol → stigmasta-5,7-dien-3β-ol → stigmasterol (Scheme 1). This sequence would not involve avenasterol.

Another pathway for the biosynthesis of major higher plant sterols is first the reduction of $\Delta^{24(28)}$ of Δ^7-avenasterol to form stigmasta-7-en-3β-ol. This reaction must be through a $\Delta^{24(25)}$ intermediate since, as already discussed, the C-25 hydrogen atom is lost. Next, stigmasta-7-en-3β-ol goes through the $\Delta^{7(8)} \rightarrow \Delta^{5,7} \rightarrow \Delta^{5(6)}$ rearrangement to form sitosterol. Stigmasterol can be produced through dehydrogenation of sitosterol, or through branching at stigmasta-5,7-dien-3β-ol. The most likely route, however, for stigmasterol formation is through spinasterol (Scheme 1). Spinasterol has been isolated from plants (HART and HEYL, 1932), and it is highly labeled when radioactive precursors are supplied (ARMAREGO et al., 1973; SHARMA, 1970). The intermediate, 7-dehydrostigmasterol, has also been isolated from plants.

It is quite possible that all of the discussed pathways operate in plants, and depending upon species and environmental conditions, one pathway predominates over the others.

2.4.10 Cholesterol Biosynthesis

Cholesterol has been isolated from a number of higher plants and its possible importance as a precursor in the formation of other steroids is suggested (HEFT-

MANN, 1973). Cholesterol biosynthesis in higher plants has not been studied extensively. Formation of cholesterol does not require alkylation of C-24 and, therefore, oxidation decarboxylation at C-4 of cycloartenol to form 31-norcycloartenol is very likely the first step (Scheme 1). This step is probably followed by the opening of the 9β,19β-cyclopropane ring to give 31-norlanosterol (HEINTZ and BENVENISTE, 1974). It has been suggested that before the cyclopropane ring is opened, a second demethylation at C-4 might occur (DEVYS and BARBIER, 1967a), but tobacco leaves which contain high levels of cholesterol, and which are able to convert tritiated cycloartenol to cholesterol, did not reveal any pollinastanol (DEVYS et al., 1968) even though the leaves converted supplied pollinastanol to cholesterol (DEVYS et al., 1969a). Pollinastanol has been isolated from pollen (DEVYS and BARBIER, 1967b; DEVYS et al., 1969b) and it may be an intermediate in the cholesterol synthesis, especially in pollen.

In cholesterol synthesis, removal of the C-14 methyl group probably does not occur before the cyclopropane ring has been opened, since the ring hinders oxidative decarboxylation and 24(25)-dehydrolophenol is the suggested intermediate. The second demethylation at C-4, reduction of $\Delta^{24(25)}$, and rearrangement of $\Delta^{7(8)} \rightarrow \Delta^{5,7} \rightarrow \Delta^{5(6)}$ of the B ring are requirements, but the exact sequence of events has not yet been established. Intermediates 24(25)-dehydrolophenol (CHARLES et al., 1969), desmosterol (GIBBONS et al., 1967), and lophenol (DJERASSI et al., 1958) have been isolated.

2.5 Seed Germination and Plant Growth

2.5.1 Seed Germination

During seed germination, the sterol content increases and this increase is probably due to membrane biogenesis (DUPÉRON, 1971; BUSH and GRUNWALD, 1972). During seed imbibition, sterol synthesis does not occur since noninduced *Corylus avellana* seeds must be prechilled or treated with gibberellic acid to induce germination and sterol formation (SHEWRY and STOBART, 1974). During the first two to four days of germination, the accumulation of sterols is essentially nil (DUPÉRON, 1971; BUSH and GRUNWALD, 1972), and this is due to a low rate of biosynthesis (BAISTED, 1971) and the often observed increase in sterols during early germination is more a function of weight loss of the germinating seeds. Similarly, the steryl ester content does not change during germination, except for the scutellum, where a rapid increase in esters is observed (KEMP et al., 1967).

The vigor of the seedling also determines the rate of sterol biosynthesis, and seedlings grown with nutrient solution show higher sterol levels than those grown with distilled water (DUPÉRON, 1971).

2.5.2 Plant Development

Generally, the youngest growing tissue of plants contains the highest amount of sterol and it is also the most active area of sterol synthesis (GEUNS, 1973, 1975). As tissue ages, the rate of synthesis decreases, but the sterol level on a dry-weight basis continues to increase slightly (GRUNWALD et al., 1971) until the plant starts to senesce, when another increase in sterol content is registered (DAVIS, 1972; GRUNWALD, 1975a). The final increase in sterols is probably due to a loss in carbohydrates and protein without a corresponding loss in sterols (GRUNWALD, 1978).

The importance of metabolic activity on sterol biosynthesis is also observed (HARTMANN and BENVENISTE, 1974). Synthesis of sterols is very low in stored potato tuber slices, and as tissue slices are washed (aged) and the respiration rate increases, sterol biosynthesis also increases.

The individual sterol levels are also affected with plant development, especially the ratio of sitosterol to stigmasterol. During germination, seeds that are high in sitosterol show a large increase in stigmasterol without a corresponding increase in sitosterol (BUSH and GRUNWALD, 1972). Similarly, mung bean seedling tips are much higher in sitosterol than stigmasterol, and as one moves down the hypocotyl, the sterol ratio shifts in favor of stigmasterol (GEUNS, 1973). The same is true for the whole plant. The ratio of stigmasterol to sitosterol is higher in the older bottom leaves and lower in the younger top leaves (GRUNWALD, 1975a). Even a single leaf shows this difference in sterol ratio with age (DAVIS, 1972). Storage tissue is low in stigmasterol and high in sitosterol, and as the tissue is "aged" the ratio of stigmasterol to sitosterol increases (HARTMANN and BENVENISTE, 1974). Generally, meristematic tissue is high in sitosterol and low in stigmasterol, and stigmasterol appears to play a more important function at a later stage of cell development.

2.5.3 Growth Substances

The importance of growth substances on plant development has been studied extensively, but much less is known about their effect on sterol biosynthesis. The auxin, naphthalene acetic acid, produced a marked increase in the sterol content of mung bean hypocotyl sections (GEUNS and VENDRIG, 1974; GEUNS, 1975). Abscisic acid, which induces dormancy in a number of species, did not influence sterol biosynthesis (MERCER and PUGHE, 1969). Dwarf and normal pea seedlings treated with gibberellic acid did not show an influence on sterol metabolism, even though the growth of dwarf plants was stimulated (GRUNWALD, unpublished result). The report that gibberellic acid stimulated sterol synthesis of hazel seeds (*Corylus avellana*) is more of an induction requirement for germination than an induction in sterol metabolism.

2.5.4 Environmental Factors

Environmental factors most often considered in plant physiology are water, temperature, and light. Nothing is known about the effect of water stress on sterol metabolism, even though its importance on membrane structure is obvious.

Shoots of wheat plants grown under near-optimum temperatures had higher sterol levels than shoots from plants grown at below optimum temperatures, and most of the higher sterol level was due to sitosterol (DAVIS and FINKNER, 1972). The roots from the same plants had the reverse pattern, namely that lower temperatures produced roots with a higher sitosterol level. Transport of sterols from shoot to root is very unlikely.

Light influences a number of biochemical, physiological, and morphological characteristics of plants, including sterol metabolism. Generally, dark-grown plants contain more sterol on a dry-weight basis than light-grown plants (DUPÉRON, 1968; BUSH et al., 1971). The biosynthesis of sitosterol from mevalonic acid is higher in the light, while stigmasterol synthesis is not affected (BUSH and GRUNWALD, 1973). The light effect on sterol metabolism gives rise to a change in the sitosterol:stigmasterol ratio, very similar to that of senescence (GRUNWALD, 1978). It is well known that light is one environmental factor that not only acts on photosynthesis but also on the process of senescence (WOOLHOUSE, 1967). With aging, cell organelles show a loss in structure and eventually only the plasmalemma and some empty vesicles remain, and reduced light intensities enhance this process. The increased accumulation of stigmasterol during this phase of plant growth has been linked to the disorganization of intracellular organelles (DUPÉRON, 1971).

2.6 Physiological Function

Sterols have at least two functions, (1) as precursors in the formation of other steroids, and (2) as components of membranes. It is very unlikely that sterols themselves act as hormones, but they certainly are precursors to progestagens, estrogens, and androgens, which in turn may have hormonal properties.

The bulk of the sterols found in plants is localized in intracellular organelles (KEMP and MERCER, 1968b; GRUNWALD, 1970), and the plasmalemma (HARTMANN et al., 1975). It has been suggested that the interaction of sterols with phospholipids stabilizes membranes and, thereby, controls permeability (GRUNWALD, 1971). For effective phospholipid interaction, the C-3 hydroxyl group of the sterol molecule must be free and cannot be substituted. Furthermore, the cyclopentanophenanthrene ring system must have at least one double bond and the whole molecule must have a relatively flat molecular configuration. Branching in the steroidal C-17 side chain increases molecular bulkiness and decreases the effectiveness of the sterol molecule to stabilize membranes.

The importance of sterols in the function of membranes in several biological systems has been demonstrated. Filipin, a polyene antibiotic, is known to complex with sterols and causes cellular leakage which can be reversed by addition of sterol (MUDD and KLEINSCHMIDT, 1970). The amount of leakage induced with filipin in wheat endosperm of different sterol phenotypes is directly related to its free sterol concentration (CARBONERO et al., 1975). Sterols also restored the K^+ and NO_3^- uptake capability in pea stem sections treated with filipin (HENDRIX and HIGINBOTHAM, 1973). The polyene antibiotic also inhibited the photochemical activity in chloroplasts (BISHOP, 1973). The saponin, digitonin, forms equimolar insoluble complexes with free sterols, and when added to chloroplasts inhibits oxygen evolution (ROBINSON and WISKICH, 1975). Cholesterol, as well as $CaCl_2$, the classical membrane stabilizer, reversed phytochrome-induced anthocyanin production in red cabbage, probably by acting at the membrane level (BASSIM and PECKET, 1975).

Exposure of plants to ozone increases cell permeability and it is interesting to note that ozone-treated plants show a decrease in free sterols and an increase in steryl glycosides (TOMLINSON and RICH, 1971). Antisenescent compounds prevent or minimize ozone damage, and plants treated with benzimidazole, N-6-benzyladenine, and kinetin did not show the decrease in free sterols (TOMLINSON and RICH, 1973). Pretreatment with sterol inhibitors increased the susceptibility of plants to ozone (SPOTTS et al., 1975).

3 Steryl Esters

3.1 Occurrence and Distribution

The fatty acid esters of the 4,4-dimethyl (cycloartenol, 24-methylene cycloartanol), the 4α-methyl (cycloeucalenol, obtusifoliol, 24-methylene lophenol, 24-ethylidene lophenol), and the 4-demethyl (sitosterol, stigmasterol, campesterol, cholesterol) sterols have been isolated (WILLIAMS et al., 1967; KNAPP and NICHOLAS, 1969; BENVENISTE et al., 1969). Steryl esters probably occur in all higher plants and have been found in leaves, roots, tubers, and seeds (KEMP et al., 1967; KEMP and MERCER, 1968a), as well as in intracellular organelles such as nuclei, chloroplasts, mitochondria, and microsomes (KEMP and MERCER, 1968b; GRUNWALD, 1970; BRANDT and BENVENISTE, 1972). The most often observed fatty acid moieties of steryl esters are palmitate, oleate, linoleate, and linolenate (BUSH and GRUNWALD, 1972; NORDBY and NAGY, 1974; TORRES and GARCIA-OLMEDO, 1975).

3.2 Biosynthesis

Even though steryl esters are widespread and apparently occur in all higher plants, very little is known about their synthesis. If radioactive mevalonic acid is supplied to plant tissue, the first esters to be radioactive are the 4,4-dimethyl sterols (KEMP et al., 1968; KNAPP and NICHOLAS, 1971a). The 4α-methyl steryl esters are the second group to be labeled, and the 4-demethyl steryl esters are the last group. The esters of the 4,4-dimethyl and 4α-methyl sterol cannot be intermediates in the biosynthesis of 4-demethyl sterol, since demethylation at C-14 involves a C-3 ketone intermediate and the steroid esters do not fulfill this requirement.

It is postulated that esterification takes place after the various free sterols have been formed, and generally, the free sterols show mevalonic acid incorporation before the steryl esters (BUSH and GRUNWALD, 1973; ADLER and KASPRZYK, 1975a,

Table 3. In vivo formation of steryl esters from free sterol, and free sterol formation from steryl ester in six-day-old tobacco seedlings (unpublished)

Substrate	Incubation time (hrs)	DPM	Free sterol %	Steryl ester
[4-^{14}C]-Sitosterol	24	14,224	93.4	6.6
	48	12,584	88.4	11.6
[4-^{14}C]-Cholesterol	24	18,593	94.5	5.5
[4-^{14}C]-Cholesteryl palmitate	24	87,531	45.5	54.5

b). In vivo experiments showed, with sitosterol and cholesterol as substrate, esterification of the incorporated free sterols, but conversion of steryl esters to free sterol was generally much higher (Table 3). These observations agree with results obtained with mammalian tissue (FLINT and ARMSTRONG, 1973; STOKKE, 1972), and it appears that the steryl esterase enzyme functions mainly as a hydrolase.

3.3 Physiological Function

The physiological function of the various steryl esters is unknown, and it is very unlikely that the esters of the 4,4-dimethyl, 4α-methyl, and 4-demethyl sterols all have the same function. The hypotheses that have been proposed are based on steryl ester changes during plant development and differentiation, and steryl ester distribution in subcellular fractions. One suggestion is the involvement of 4-demethyl steryl esters in the sterol transport process analogous to their function in mammalian systems (KEMP et al., 1967). If these steroids are involved in this process, it is most likely that they function in the intracellular transport (GOODWIN, 1967) from site of synthesis (microsomes) to site of action in membranes. Intercellular steryl ester transport could not be demonstrated (ATALLAH et al., 1975b).

The esters of the 4,4-dimethyl and 4α-methyl sterols might play a role in the regulation of sterol biosynthesis (GRUNWALD, 1975b). Competition between ester formation and C-4 demethylation has been demonstrated (BRADY and GAYLOR, 1971). For example, aerobic conditions favor demethylation, but esterification would render the sterol intermediates inert to C-4 attack, thereby controlling 4-demethyl sterol formation.

Steryl esters are mainly localized in intracellular organelles (KEMP and MERCER, 1968b), but they do not appear to act as membrane stabilizers (GRUNWALD, 1971). NICHOLAS and his co-workers observed that many of the 4,4-dimethyl, 4α-methyl, and 4-demethyl sterols exhibit liquid crystalline states when esterified (ATALLAH and NICHOLAS, 1972, 1974). Liquid crystals possess characteristics reminiscent of both the crystalline and the liquid state, and in the liquid state the molecules still have a certain degree of order. Change in state can be brought about by temperature or addition of other chemical components. The question is, can changes in viscosity at critical steps in the biosynthetic pathways influence changes in membrane permeability? Experiments to prove this hypothesis have not been very successful.

4 Steryl Glycosides and Acylsteryl Glycosides

4.1 Occurrence and Distribution

The occurrence of steryl glycosides in plants (VII) has been known for over 60 years (POWER and SALWAY, 1913), and about 15 years ago the acylsteryl glycosides (VIII) were discovered (LEPAGE, 1964; MCKILLICAN, 1964). In the glycosides, the 3β-hydroxyl of the steryl moiety is linked to the hydroxyl of the C-1 position of the carbohydrate, and the acylated steryl glycosides are steryl-6-acyl-D-glycosides with palmitic, stearic, oleic, linoleic, and linolenic acid as the acyl moiety (LEPAGE, 1964; KIRIBUCHI et al., 1966). So far, only the 4-demethyl sterols (sitosterol, stigmasterol, campesterol, etc.) have been recovered as the sterol units from steryl glycosides and acylsteryl glycosides. The sugar most often found is sucrose (KIRIBUCHI et al., 1966), but mannose (LEPAGE, 1964; EICHENBERGER and MENKE, 1966) has also been reported. The steryl glycosides have been isolated from a number of tissues and probably occur in all tissues. At the subcellular level, chloroplasts generally contain the lowest quantity of steryl glycoside, while microsomes contain the highest (EICHENBERGER and GROB, 1970b; GRUNWALD, 1970). Acylsteryl glycosides are quantitatively highest in the mitochondria fraction (DUPÉRON et al., 1972).

4.2 Biosynthesis

Steryl glycoside formation was first demonstrated with cell particulate preparations from immature soybeans (HOU et al., 1968) and has since been confirmed with several other species (KAUSS, 1968; ONGUN and MUDD, 1970; PÉAUD-LENOËL and AXELOS, 1972). All 4-demethyl sterols are incorporated at about the same rate (BUSH and GRUNWALD, 1974) and the glycoside bond is mediated through nucleotide sugars (ONGUN and MUDD, 1970). Uridine diphosphate-glucose is the most active glycosyl donor. Steryl glycoside formation is stimulated by ATP (HOU et al., 1968; EICHENBERGER and GROB, 1969) and Ca^{2+} and Mg^{2+} (FANG and BAISTED, 1976). The optimum medium pH is critical and depends upon the tissue – generally it is near the neutral point (HOU et al., 1968; ONGUN and MUDD, 1970). If the pH is too low, acylsteryl glycoside formation is favored (BUSH and GRUNWALD, 1974).

Theoretically, acylsteryl glycosides can be synthesized via two pathways (Scheme 2). It is generally accepted that steryl glycosides are the precursors (Pathway I, Scheme 2), and in vitro experiments support this view (HOU et al., 1968; EICHENBERGER and GROB, 1970a; BUSH and GRUNWALD, 1974). In vivo experiments, however, suggest that a second pathway may also exist in which an acylglycosyl combines with the sterol moiety (Pathway II, Scheme 2) (BUSH and GRUNWALD, 1974; FRASCH and GRUNWALD, 1976).

Two acyl transfer reactions have been demonstrated – one involving phosphatidylethanolamine and a microsomal enzyme (EICHENBERGER and GROB, 1969; AXELOS and PÉAUD-LENOËL, 1971), and another involving galactolipids and a soluble enzyme (HEINZ et al., 1975). Acyltransferase activity has been found in leaves, roots, and flowers during all stages of vegetative growth (WOJCIECHOWSKI, 1974).

4.3 Physiological Function

The physiological function of steryl and acylsteryl glycosides is ill-defined, however, a number of suggestions have been made. SMITH (1969) proposed that sterols may serve as carriers in glucose transport. HOU et al. (1968) found that steryl glycosides accumulated during seed maturation and they suggested that sterols are stored as the glycoside. BUSH and GRUNWALD (1972) support this view since

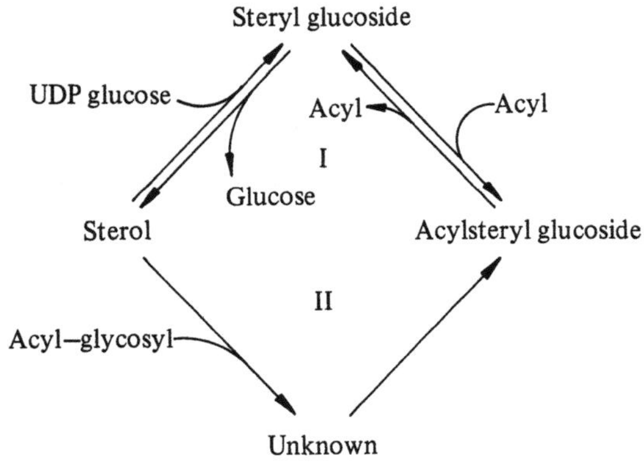

Scheme 2. Acylsteryl glycoside pathway

the steryl glycosides decreased during seed germination. It has also been reported that steryl glycosides have hormone-like activity in the *Avena* coleoptile test (KIMURA et al., 1975; TIETZ et al., 1977), however, whether or not this is a true hormonal effect needs further examination. The possibility does exist that steryl glycosides act at the membrane level, and thereby influence cell permeability, however, experiments to demonstrate permeability changes have been unsuccessful (GRUNWALD, 1971).

5 Ecdysteroids

5.1 Terminology, Chemistry, Isolation, and Distribution

The ecdysteroids are a group of terpenoids of relatively recent discovery by plant chemists. Most steroids are apolar, however, the ecdysteroids are polar and this may have contributed to the delay in their discovery (HEFTMANN, 1973).

The term "ecdysones" or "molting hormone" is also often used for this group of steroids because α-ecdysone (IX) and ecdysterone (X), first isolated from insects, induce ecdysis or molting in insects. The term "molting hormone" was carried over when these steroids were discovered in plants, but this is very unfortunate because this term does not describe the function of these steroids in plants. The term "ecdysones" for this group of steroids is also unsatisfactory since α-ecdysone is often called simply ecdysone. A more appropriate term for this group of terpenoids is "ecdysteroids."

Ecdysteroids are polyhydroxylated compounds with 27 to 29 carbon atoms. No single substitution is common to all ecdysteroids, however, most have four general features (1) a 6-keto group (2) 2β, 3β, 14α hydroxyl groups (3) 20(R), 22(R) hydroxyl groups, and (4) a $\Delta^{7(8)}$ bond (HIKINO and TAKEMOTO, 1974; HEFTMANN, 1975). The major difference in the various ecdysteroids is in the C-17 side chain, but some of these steroids have 2α, 3α hydroxyl groups, ponasterone B (XI), others may lack the 14β hydroxyl, podecdysone B (XII), or some other modification. One ecdysteroid-3β-glycoside, ponasteroside A (TAKEMOTO et al., 1968), as well as two cinnamate derivatives, crustecdysone 2-cinnamate and polypo-

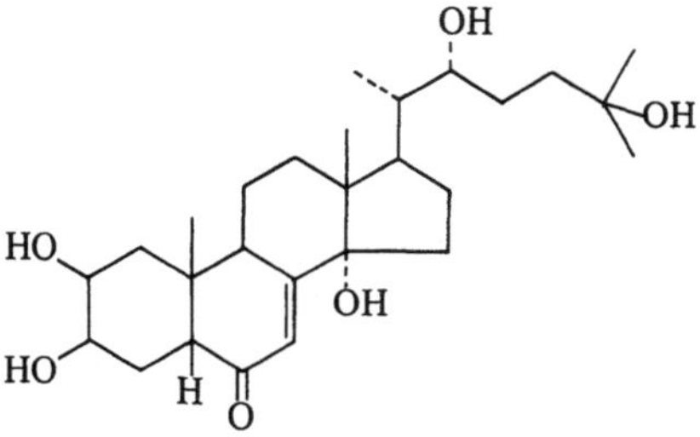

α–Ecdysone
Ecdysone
(IX)

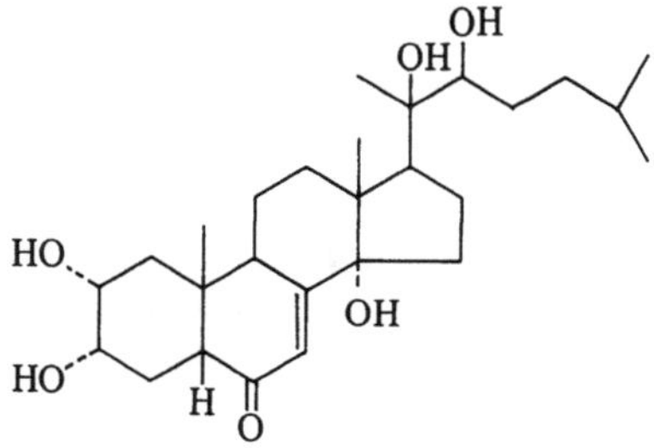

Ponasterone B
(XI)

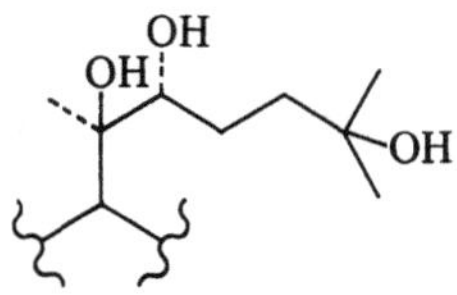

Ecdysterone
β–Ecdysone
Crustecdysone
Isoinokosterone
Polypodine A
Commisterone
20–Hydroxyecdysone
(X)

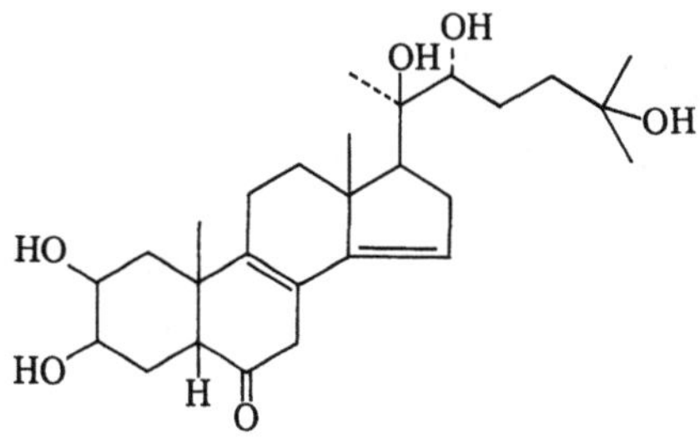

Podecdysone B
(XII)

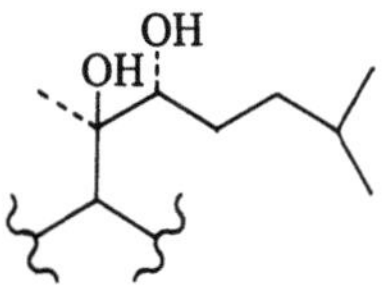

Ponasterone A
(XIII)

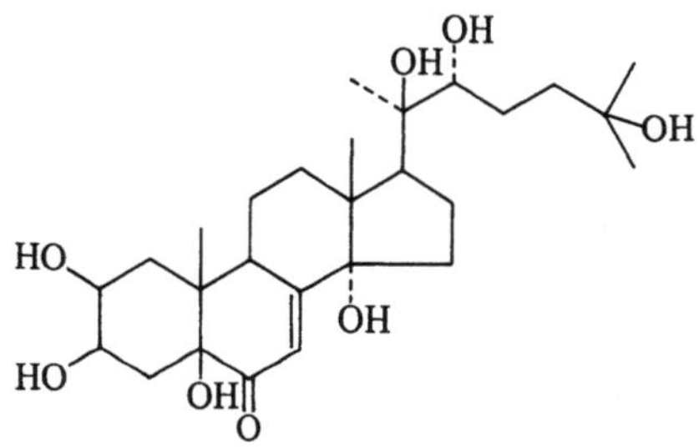

5β–Hydroxyecdysterone
(XV)

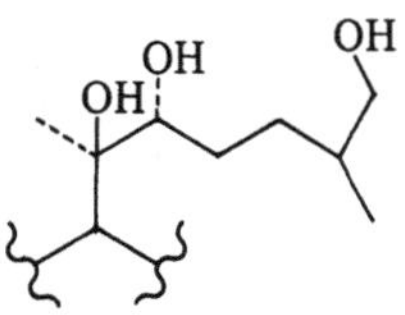

Inokosterone
(XIV)

dine B 2-cinnamate, have been isolated (RUSSELL et al., 1971). The most often isolated ecdysteroid from plants is, however, ecdysterone (X) (HIKINO and TAKEMOTO, 1974).

The first reports of isolation of phytoecdysteroids came in 1966 when two papers from two different laboratories appeared in the same issue of *Chemical Communications*. NAKANISHI et al. (1966) described the structure of ponasterone

A (XIII) which they isolated from leaves of a conifer (*Podocarpus nackaii*) of Formosan origin, and GALBRAITH and HORN (1966) found an ecdysteroid with the same chemical properties as crustecdysone (X) in the wood of an Australian coniferous timber tree (*Podocarpus elata*). Crustecdysone is also known as β-ecdysone, 20-hydroxyecdysone, isoinokosterone, polypodine A, commisterone, but it is probably most often referred to as ecdysterone. In 1967, in addition to other ecdysteroids, α-ecdysone (IX) was isolated from the bracken fern (*Pteridium aquilinum*) (KAPLANIS et al., 1967).

The first isolation of any ecdysteroid in relatively pure form occurred in 1954 by BUTENANDT and KARLSON (1954) with the crystallization of α-ecdysone from silk worm, however, it was not until 1965 that the chemical structure was elucidated (HUBER and HOPPE, 1965). Presently, the number of ecdysteroids of zoological origin are six, while the phytoecdysteroids number over 40, and only one of the zooecdysteroids, 20-dihydroxy-α-ecdysone, has not yet been isolated from plants. Generally, the ecdysteroid concentration in plants is much higher than that in insects and may reach a level of 0.1% of dry weight (HEFTMANN, 1970; OGAWA et al., 1974). Presently, ecdysteroids have been found in about 80 plant families, however, no taxonomic relationship can be established (HIKINO and TAKEMOTO, 1974).

5.2 Biosynthesis

At present, the pathway for the biosynthesis of ecdysteroids in plants is ill-defined and may differ from that in insects (LOCKLEY et al., 1975). The large number of ecdysteroids found in plants makes it unlikely that we are dealing with a single pathway. Most of the present knowledge relates to the biosynthesis of ecdysterone, even though mevalonic acid and cholesterol are incorporated into a number of other ecdysteroids.

The generalized Scheme 3 is based on a number of observations with various plant species. Administered cholesterol-4-^{14}C is recovered as ecdysterone from *Podocarpus elata* (SAUER et al., 1968), and from *Taxus baccata* seedlings (LLOYD-JONES et al., 1973). With *Podocarpus macrophyllus* seedlings, the supplied cholesterol is not only converted to ecdysterone but also to ponasterone A (HIKINO et al., 1969, 1970). With *Achyranthes fauriei* seedlings in addition to ecdysterone, labeled inokosterone (XIV) was also recovered (HIKINO et al., 1975). *Polypodium vulgare,* which has α-ecdysone (HEINRICH and HOFFMEISTER, 1967), converts cholesterol to ecdysterone, α-ecdysone, and 5β-hydroxyecdysterone (XV) (DE SOUZA et al., 1970). In *Sesuvium portulacastrum,* α-ecdysone was metabolized to ecdysterone but ecdysterone was not converted to α-ecdysone, therefore, the conversion of α-ecdysone to ecdysterone is probably an irreversible step (SIPAHIMALANI et al., 1972).

Generally, the formation of a 5β-hydrogen in plant steroids requires a Δ^4-3-ketone intermediate (CASPI and HORNBY, 1968), however, in the biosynthesis of ecdysteroids, no such intermediate apparently is needed (LLOYD-JONES et al., 1973; LOCKLEY et al., 1975). The introduction of the $\Delta^{7(8)}$ bond may occur while the $\Delta^{5(6)}$ unsaturation is still present and this rearrangement may not require a Δ^4-3-ketone. Furthermore, a 5,6-epoxide is not an obligatory intermediate in the

Scheme 3. Biosynthesis of ecdysteroids

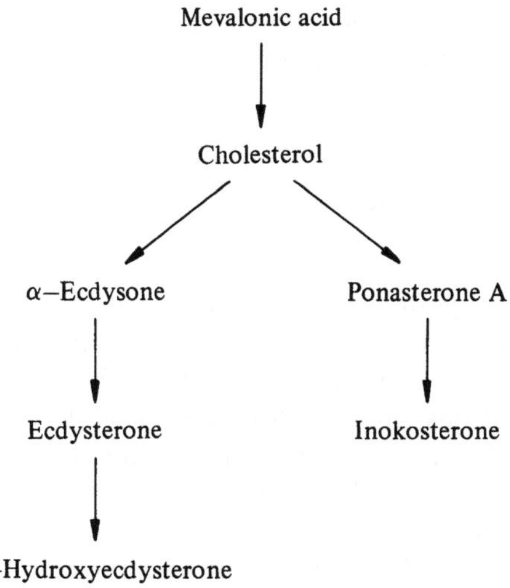

formation of the 6-keto group (Joly et al., 1969d), and the A/B-cis ring arrangement occurs before the C-17 side of ecdysone is hydroxylated (Tomita and Sakurai, 1974).

While it has been shown that cholesterol is a precursor for ecdysterone and α-ecdysone, it has not yet been established that all plants that contain 27 carbon ecdysteroids, especially ecdysterone, also contain cholesterol. Cholesterol may not be the obligatory intermediate. Analogous to the biosynthesis of 27 carbon ecdysteroids, 28 and 29 carbon ecdysteroids may be derived from 28 and 29 carbon sterols (Hikino and Takemoto, 1974). Experiments to prove such pathways with sitosterol and fucosterol have, so far, been unsuccessful (Boid et al., 1975).

5.3 Function

A number of suggestions have been put forward, but no definite function has been established for the presence of ecdysteroids in plants. The hypothesis that ecdysteroids act as potent deterrents and antifeeding agents against insects received probably the greatest support (Galbraith and Horn, 1966; Herout, 1970; Williams, 1970; Heftmann, 1970), although α-ecdysone and ecdysterone, when ingested by insects, have little or no effect (Robbins et al., 1968). However, when administered in the drinking water in concentrations as low as 1 ppm, ecdysterone has proved to be a powerful feeding deterrent (Williams, 1970). The analog, ajugalactone, which occurs in *Ajuga decumbens*, inhibits insect metamorphosis (Koreeda et al., 1970).

The mode of action of ecdysteroids, recently summarized by Karlson (1974), is probably at the gene level, by acting on specific sites and thereby regulating insect metamorphosis. Heftmann (1975) made the remark that it would not be surprising to find that ecdysteroids are also involved in the metamorphosis of plants. Ecdysteroids, however, did not produce qualitative or quantitative effects on flowering or vegetative development of shoot tips and intact plant cultures of *Xanthium pennsylvanicum*, irrespective of whether they were under inductive or noninductive photoperiodic conditions (JACOBS and SUTHERS, 1971).

6 Progestagens

Progestagens are steroids with 21 carbon atoms and a number of these steroids have been isolated from various plants (LEBOEUF et al., 1964; TSCHESCHE, 1966; GAWIENOWSKI and GIBBS, 1968). Generally, of all the steroids with 21 carbon atoms, pregnenolone and progesterone receive the greatest attention, partly because they are hormones in animals, but also because they are precursors to the cardenolides which are found in plants, and related compounds. The 21 carbon steroids are derived from sterols by oxidative cleavage of the C-17 side chain (PILGRIM, 1972). The conversion of cholesterol to pregnenolone Scheme 4 has been demonstrated with a number of plants (BENNETT and HEFTMANN, 1966; SAUER et al., 1967; CASPI et al., 1966; GAWIENOWSKI and GIBBS, 1969b). Intermediates in the degradation process are 22-hydroxycholesterol and 20,22-dihydroxycholesterol. The steroid 20α-hydroxycholesterol might also be an intermediate (BURSTEIN et al., 1970), but not all plants are capable of converting this compound (STOHS and EL-OLEMY, 1971). It has been suggested that the hydroxyl compounds might be by-products resulting from the true intermediate, since only a relatively small fraction of the pregnenolone formed from cholesterol can be accounted for by this sequence (HOCHBERG et al., 1974). Furthermore, not all plants contain cholesterol and, therefore, the question arises whether other sterols, especially the common

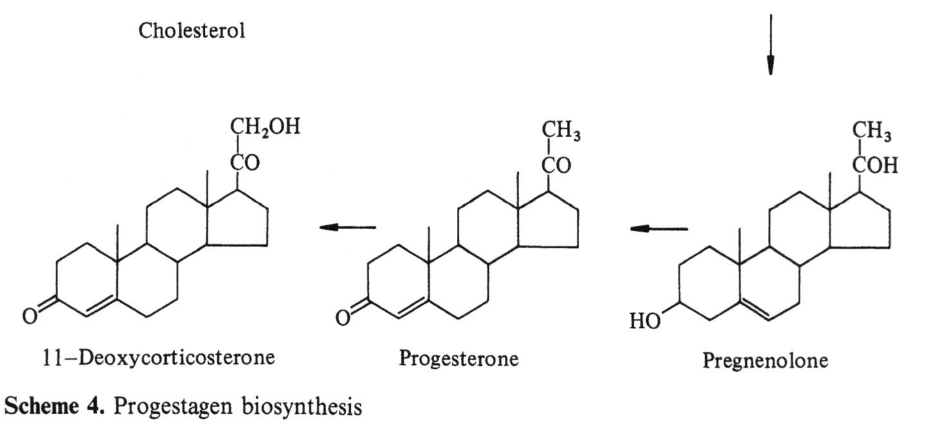

Scheme 4. Progestagen biosynthesis

29 carbon sterols, can be converted to pregnenolone. The answer was affirmative when experiments with [3-^{14}C]-sitosterol were carried out (BENNETT et al., 1969).

The conversion of pregnenolone to progesterone requires rearrangement of double bond $\Delta^{5(6)} \rightarrow \Delta^{4(5)}$ (Scheme 4). The mechanism proceeds through oxidation of the 3-hydroxyl to a 3-ketone, followed by isomerization of the olefinic bond from $\Delta^5 \rightarrow \Delta^4$ (CASPI and HORNBY, 1968). The reverse reaction, the conversion of progesterone to pregnenolone, has also been observed in plants (BENNETT and HEFTMANN, 1965b).

The function of progestagen is probably as an important intermediate in the biosynthesis of other steroids.

7 Corticosteroids

7.1 Occurrence

In animals the corticosteroids are very important 21 carbon steroids and are synthesized from progesterone (Scheme 4) in the cortex of the adrenal gland (GOWER, 1974). The possible importance of the corticosteroids in plants has also been discussed (GEUNS, 1978). In the mammalian system the corticosteroids function in two ways: (1) as glucocorticoids in controlling the carbohydrate metabolism, and (2) as mineralocorticoids in controlling the sodium-potassium ion concentration.

A definite occurrence of corticosteroids in plants has not yet been documented, however, one report of its isolation from plants does exist (BAHADUR and SRIVASTAVA, 1971). These workers isolated a component from *Oryza sativa* husk oil which corresponded in mobility by paper chromatography to 11-desoxycorticosterone and gave a brilliant yellow fluorescence under UV light. In an earlier paper it was reported that applied progesterone was converted in small amounts to 11-desoxycorticosterone by *Digitalis lanata* (CASPI et al., 1968), but any definite conclusion concerning their natural occurrence in plants must await further analysis and verification.

7.2 Physiological Function

The corticosteroids, cortisone and cortisol, stimulate plant growth (RATSIMAMANGA et al., 1958; RATSIMAMANGA and DIOT, 1959; DONNET et al., 1960) and root development (GEUNS, 1974), however, desoxycorticosterone, the only presently reported plant corticosteroid, was inactive (GEUNS, 1977). The corticosteroid, cortisone, also stimulates flower number (KOPCEWICZ, 1971b). Of the corticosteroids the most active are the 11-β-OH forms (corticosterone and cortisol). The 11-keto corticosteroids (cortisone) are less active and possibly for biological activity, a conversion to the 11-β-OH must occur. Cortisol, cortisone, and corticosterone are glucocorticoids, the inactive 11-desoxycorticosterone is a mineralocorticoid. The significance of these results is difficult to assess at the present time, since none of the active glucocorticoids has yet been isolated from plants.

8 Estrogens and Androgens

8.1 Occurrence

Estrogens and androgens are C_{18} and C_{19} steroids which have hormonal activity in mammalian systems. In has been debated for many years whether or not estrogens and androgens occur in plants, and even today no unanimous agreement has been reached.

In 1933, BUTENANDT and JACOBI (1933) and SKARZYNSKI (1933) reported that they had isolated, in crystalline form, compounds from plants which had estrogenic activity. The estrogen-like substance from the female willow flower agreed most closely, as its acetyl derivative, to estriol (XVI), however, its biological activity was only one-fourth that of estriol isolated from urine (SKARZYNSKI, 1933). Furthermore, mixed melting points with urine estriol gave a 1° C lowering. The isolate of estrogen from palm kernels was also physiologically active in female hormone tests (BUTENANDT and JACOBI, 1933). HASSAN et al. (1947) and later EL-RIDI et al. (1960), isolated an estrogenic compound from the pollen grains of palm trees which gave the estrus response in ovariectomized rats. All these reports used bioassay techniques, ultraviolet spectroscopy, and the Kober color test to establish the presence of estrogenic compounds.

In 1965, JACOBSOHN et al. (1965) challenged the conclusion of previous researchers that they were dealing with steroid estrogens. With the use of radioactive [16-^{14}C]-estrone (XVII) as a tag, they were unable to demonstrate the reported concentrations of up to 2 mg of estrogen/kg of tissue, and concluded that if any steroid estrogens were present, the concentration was not over 10 µg of estrogen/kg of tissue. They further proposed that estrogens, without the steroid nucleus, might

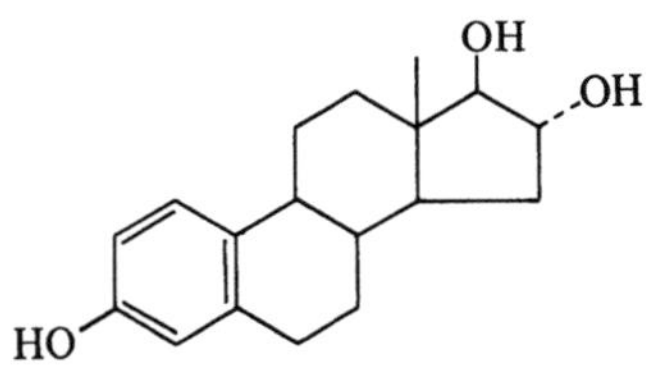

Estriol
(XVI)

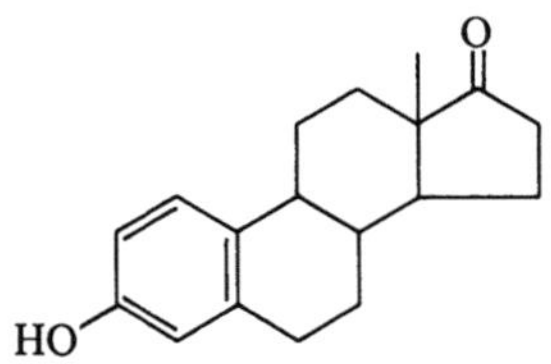

Estrone
(XVII)

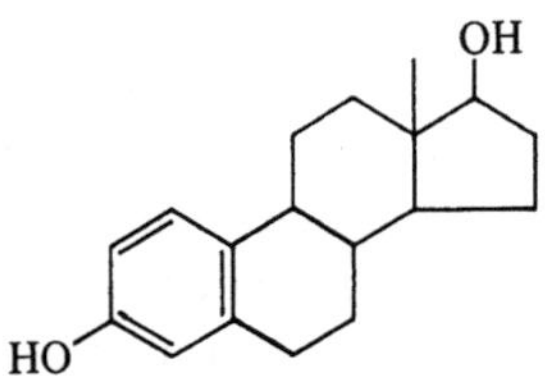

Estradiol
(XVIII)

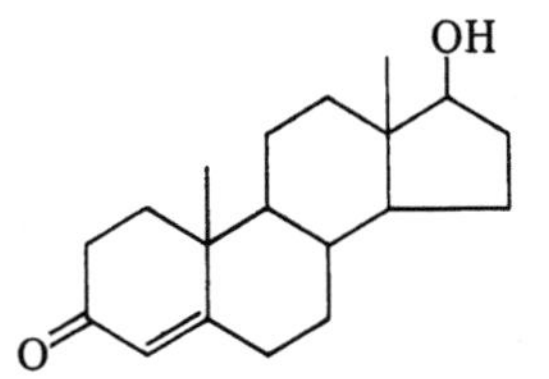

Testosterone
(XIX)

have been in some of the extracts used in bioassays. This is entirely possible since nonsteroid estrogens are known to occur in plants.

Since 1965, however, the number of reports that steroid estrogens are present in plant material has increased (HEFTMANN et al., 1965, 1966; AMIN et al., 1969; GAWIENOWSKI and GIBBS, 1969a; KOPCEWICZ, 1971a, 1972a, 1972b; AMIN and PALEOLOGOU, 1973; AWAD, 1974). The characterization of estrogen in these reports is generally based on color reactions, ultraviolet and/or infrared spectroscopy, mobility in chromatographic systems, melting points, and in some cases bioassays, and the reduction of estrogen to estradiol (XVIII). The quantity of estrogen reported varies with tissue and laboratory, but generally it is in the μg/kg range. DEAN et al. (1971) used the extremely sensitive competitive protein-binding technique to reexamine the estrogen question and they also found 3 to 9 μg of steroid/kg of tissue. With *Phaseolus vulgaris* plants, supplied ^{14}C labeled mevalonic acid was recovered as radioactive estradiol (YOUNG et al., 1977). Analysis was by radio-gas-liquid-chromatography and monitoring was by single-ion mass spectrometry.

Reports of androgen isolation from plants is rare, but testosterone (XIX) and androstenedione have been found in Scotch pine (SǍDEN-KREHULA et al., (1971).

8.2 Biosynthesis

The biosynthesis of estrogen and androgen has not yet been demonstrated in plants, but it has been postulated that it would be similar to that in microorganisms and animals (HEFTMANN, 1975). This process would require successive oxidation steps both at the C-17 and the C-19 position of the sterol molecule. Pregnenolone and progesterone, as well as the androgens, testosterone and androstenedione, would be intermediates, and all of these steroids have been isolated from plants.

8.3 Physiological Function

Interest in estrogens is stimulated by their possible involvement as plant hormones. In several well-established plant systems, exogenously applied estrogens and androgens influenced vegetative growth (HELMKAMP and BONNER, 1953; BISWAS et al., 1967; KOPCEWICZ, 1969) as well as flowering (CZYGAN, 1962; BISWAS et al., 1967; KOPCEWICZ, 1970), including sex expression (KOPCEWICZ, 1971b; GAWIENOWSKI et al., 1971). The magnitude of response depended upon the specific steroid applied, its concentration, and the age of the plant. In general, it appears that the exogenously applied estrogens had a greater biological activity than the androgens (HELMKAMP and BONNER, 1953; CZYGAN, 1962), possibly indicating a product precursor relationship, however, applied estrogens favored female flower production and androgens favored male flower production (KOPCEWICZ, 1971b). Endogenous estrogens by Kober color test appeared first at time of flower bud formation and reached their maximum at time of bud development and pod formation (KOPCEWICZ, 1971a).

9 Cardenolides

9.1 Chemistry and Distribution

Cardenolides are steroids with 23 carbon atoms and are characterized by a 14β-hydroxyl and an α,β-unsaturated γ-lactone (butenolide) ring. Cardenolides occur as glycosides in about 12 plant families (SINGH and RASTOGI, 1970). About 50 cardenolides have been isolated, and digitoxigenin (XX), digoxigenin (XXI), and gitoxigenin (XXII) are three important cardiac genins in *Digitalis*, a much-studied genera. Cardenolides are found in all tissues of the plant, but the relative quantity may vary with stage of development, ecological factors, time of collection, and sample handling. Cardenolides have poor stability in acidic and alkaline conditions.

9.2 Biosynthesis

Cardenolides are formed by condensation of a C_{21} pregnane derivative with acetate (TSCHESCHE and LILIENWEISS, 1964; EUW and REICHSTEIN, 1964; WICKRAMASINGHE et al., 1969). The sequence of formation is probably pregnenolone→progesterone →pregnane-3,20-dione→pregnane-3-ol-20-one→pregnane-3β,14β-diol-20-one→ cardenolides (CASPI and LEWIS, 1967; SAUER et al., 1967; TSCHESCHE et al., 1970).

Oxidation of the C-3 position of pregnenolone to form progesterone is an obligatory step in the biosynthesis of cardenolides (CASPI and HORNBY, 1968). Reduction of the progesterone Δ^4 bond gives rise to 5α-H-pregnane-3,20-dione and 5β-H-pregnane-3,20-dione (GRAVES and SMITH, 1967). In *Digitalis*, the 5α-isomer was 9× as radioactive as the 5β-isomer (BENNETT et al., 1968), and in other plant species only the 5α-isomer was recovered (STOHS and EL-OLEMY, 1972). Formation of pregnane-3β-ol-20-one is very likely the next step. Here again the 5α-isomer shows higher radioactivity when labeled precursors are supplied (BENNETT et al., 1968; STOHS and EL-OLEMY, 1972). These observations create some difficulty since the major cardenolides in *Digitalis* are 5β-(H)-derivatives. Supplied 5β-pregnane-3,20-dione and 5β-pregnane-3β-ol-20-one were readily converted to digitoxigenin, digoxigenin, and gitoxigenin (TSCHESCHE et al., 1970). The reaction is microsomal in nature (STOHS, 1969) and the involvement of a 5α-5β-isomerase is possible.

Introduction of the 14β-hydroxyl group precedes the butenolide ring formation (CASPI and LEWIS, 1968), and the most likely intermediate is pregnane-3β,14β-diol-20-one (TSCHESCHE et al., 1967). Introduction of the 14β-hydroxyl does not involve a $\Delta^{8(14)}$ (ABERHART et al., 1973) or $\Delta^{14(15)}$ (TSCHESCHE et al., 1970) intermediate.

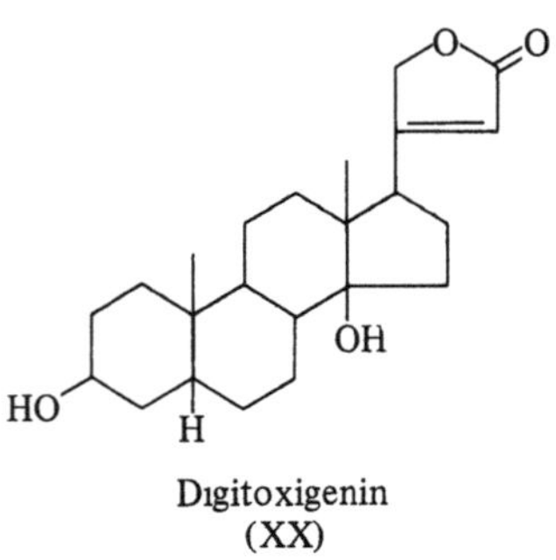

Digitoxigenin
(XX)

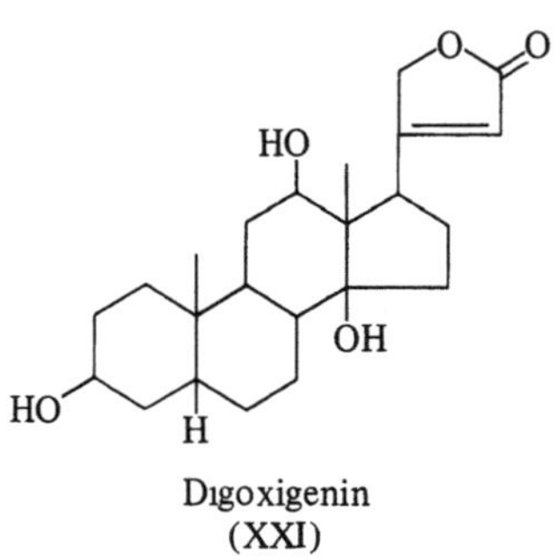

Digoxigenin
(XXI)

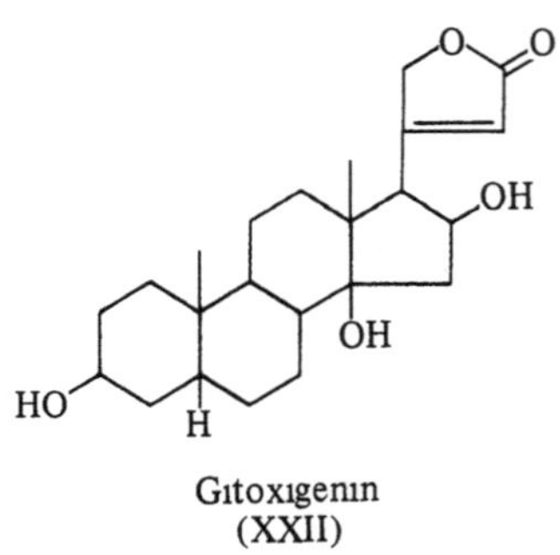

Gitoxigenin
(XXII)

The sugar moieties of cardenolides are normal sugars such as glucose or unbranched aldohexoses mostly of the 6-deoxy type and 2-deoxy type, but some novel sugar derivatives have also been reported (REICHSTEIN and WEISS, 1962). Up to five sugars may form a straight chain, which forms the glycoside linkage, always β with the C-3 of the aglycone.

9.3 Physiological Function

Many of the cardenolides are poisonous and have been used as arrow poison by primitive people. The importance of these steroids as a defense mechanism by the plant cannot be overestimated. Presently, cardenolides are used in the treatment of heart disease.

10 Sapogenins

Sapogenins are C_{27} steroids (Scheme 5), widely distributed in a number of monocot families, e.g., the Liliaceae, Amarylidaceae, and Dioscoreaceae, and in the dicots, e.g., the Scrophulariaceae and Solanaceae. In plants, the sapogenins are combined with sugars to form the saponins. Generally, the sugars are in a branched chain and are attached to the C-3 position of the steroid moiety. Saponins with the open side chains also exist (JOLY et al., 1969a, 1969b) by forming a glycoside (HEFTMANN, 1967). Sapogenins are used in the commercial preparation of steroidal hormones.

Cholesterol (BENNETT and HEFTMANN, 1965a; JOLY et al., 1969c) and sitosterol (STOHS et al., 1974) are precursors in the formation of saponins (Scheme 5). With sitosterol, however, removal of a two-carbon unit from C-24 must precede oxygenation. The sequence in which oxygen is introduced at positions 16, 22, and 26 is unanswered, but indirect evidence strongly suggests that oxygenation at C-26 is the first step (BENNETT et al., 1970), and is not dependent upon a Δ^{24} bond (JOLY et al., 1969c). Cholesterol, however, does not appear to be an obligatory step in the biosynthesis, since desmosterol is converted to saponins without going through cholesterol (TSCHESCHE et al., 1974).

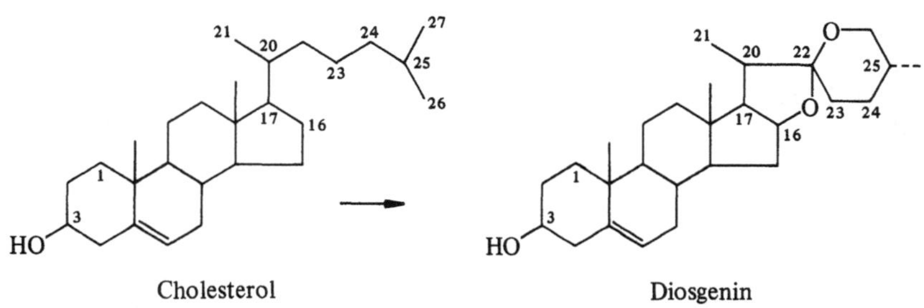

Scheme 5. Sapogenin formation from cholesterol

Acknowledgment. The author expresses his appreciation to (Mrs.) Betty Nelson for her technical assistance during the preparation of the manuscript.

References

Aberhart, D.J., Lloyd-Jones, J.G., Caspi, E.: Biosynthesis of cardenolides in *Digitalis lanata*. Phytochemistry *12*, 1065–1071 (1973)

Adler, G., Kasprzyk, Z.: Free sterols, steryl esters, glucosides, acylated glucosides and water-soluble complexes in *Calendula officinalis*. Phytochemistry *14*, 627–631 (1975a)

Adler, G., Kasprzyk, Z.: The incorporation of mevalonate-[2-^{14}C] into the sterol fractions of *Calendula officinalis*. Phytochemistry *14*, 723–726 (1975b)

Alcaide, A., Devys, M., Bottin, J., Fetizon, M., Barbier, M., Lederer, E.: Sur la biosynthèse des stérols du tabac *Nicotiana tabacum* à partir du méthylène-24 cholestérol et du méthylène-24 dihydrolanostérol. Phytochemistry *7*, 1773–1777 (1968)

Aller, R.T. van, Chikamatsu, H., De Souza, N.J., John, J.P., Nes, W.R.: The metabolic role of the 24-ethylidenecholesterols. Biochem. Biophys. Res. Commun. *31*, 842–844 (1968)

Amin, E.S., Paleologou, A.M.: Estrone in *Hyphaena thebaica* kernel and pollen grains. Phytochemistry *12*, 899–901 (1973)

Amin, E.S., Awad, O., Samad, El M.A., Iskander, M.N.: Isolation of esterone from moghat roots and from pollen grains of Egyptian date palm. Phytochemistry *8*, 295–297 (1969)

Armarego, W.L.F., Goad, L.J., Goodwin, T.W.: Biosynthesis of α-spinasterol from (2-^{14}C, (4R)-4-$^{3}H_1$) mevalonic acid by *Spinacea oleracea* and *Medicago sativa*. Phytochemistry *12*, 2181–2187 (1973)

Atallah, A.M., Nicholas, H.J.: What role have liquid crystals in nature? Chem. Tech. 486–589 (1972)

Atallah, A.M., Nicholas, H.J.: Function of steryl esters in plants: a hypothesis that liquid crystalline properties of some steryl esters may be significant in plant sterol metabolism. Lipids *9*, 613–622 (1974)

Atallah, A.M., Aexel, R.T., Ramsey, R.B., Nicholas, H.J.: Biosynthesis of sterols and triterpenes in *Pelargonium hortorum*. Phytochemistry *14*, 1529–1533 (1975a)

Atallah, A.M., Aexel, R.T., Ramsey, R.B., Threlkeld, S., Nicholas, H.J.: Translocation of sitosterol and related compounds in *Pelargonium hortorum* and *Helianthus annuus*. Phytochemistry *14*, 1927–1932 (1975b)

Awad, O.: Steroidal estrogens of *Prunus armeniaca* seeds. Phytochemistry *13*, 678–679 (1974)

Axelos, M., Péaud-Lenoël, M.C.: Identification de l'activité phosphatidylethanolamine: stérylglucoside transacylase extraite du Blé. C. R. Acad. Sci. Paris Ser. D *273*, 1434–1437 (1971)

Bahadur, K., Srivastava, R.L.: Study of the unsaponifiable matter of *Oryza sativa* husk oil. Plant Med. *20*, 340–343 (1971)

Baisted, D.J.: Sterols of *Euphorbia peplus*: the fate of 28-isofucosterol in phytosterol biosynthesis. Phytochemistry *8*, 1697–1703 (1969)

Baisted, D.J.: Sterol and triterpene synthesis in the developing and germinating pea seed. Biochem. J. *124*, 375–383 (1971)

Bassim, T.A.H., Pecket, R.C.: The effect of membrane stabilizers on phytochrome-controlled anthocyanin biosynthesis in *Brassica oleracea*. Phytochemistry *14*, 731–733 (1975)

Bean, G.A.: Phytosterols. In: Advances in lipid research. Paoletti, R., Kritchevsky, D. (eds.), Vol. II, pp. 193–218. New York: Academic Press 1973

Bennett, R.D., Heftmann, E.: Biosynthesis of *Dioscorea sapogenins* from cholesterol. Phytochemistry *4*, 577–586 (1965a)

Bennett, R.D., Heftmann, E.: Progesterone: biosynthesis from pregnenolone in *Holarrhena floribunda*. Science *149*, 652–653 (1965b)

Bennett, R.D., Heftmann, E.: Biosynthesis of pregnenolone from cholesterol in *Haplopappus heterophyllus*. Phytochemistry *5*, 747–754 (1966)

Bennett, R.D., Heftmann, E.: Biosynthesis of stigmasterol from sitosterol in *Digitalis lanata*. Steroids *14*, 403–407 (1969)

Bennett, R.D., Heftmann, E., Purcell, A.E., Bonner, J.: Biosynthesis of stigmasterol in tomato fruits. Science *134*, 671–673 (1961)

Bennett, R.D., Sauer, H.H., Heftmann, E.: Progesterone metabolism in *Digitalis lanata*. Phytochemistry *7*, 41–50 (1968)

Bennett, R.D., Heftmann, E., Winter, B.J.: Conversion of sitosterol to progesterone by *Digitalis lanata*. Naturwissenschaften *56*, 463 (1969)

Bennett, R.D., Heftmann, E., Joly, R.A.: Biosynthesis of diosgenin from 26-hydroxycholesterol in *Dioscorea floribunda*. Phytochemistry *9*, 349–353 (1970)

Benveniste, P.: La biosynthèse des stérols dans les tissus de tabac cultivés in vitro. Mise en évidence du cycloeucalènol et de l'obtusifoliol. Phytochemistry *7*, 951–953 (1968)

Benveniste, P., Massy-Westropp, R.A.: Mise en evidence de l'epoxyde-2,3 de squalene dans les tissus de tabac cultivés in vitro. Tetrahedron Lett. *37*, 3553–3556 (1967)

Benveniste, P., Hirth, L., Ourisson, G.: La biosynthèse des stérols dans les tissue de tabac cultivés in vitro. II. Particularités de la biosynthèse des phytostérols des tissus de tabac cultivés in vitro. Phytochemistry *5*, 45–58 (1966)

Benveniste, P., Hewlins, M.J.E., Fritig, B.: La biosynthèse des stérols dans les tissus de tabac cultivés in vitro. Cinétique de formation des stérols et de leurs précurseurs. Eur. J. Biochem. *9*, 526–533 (1969)

Berti, G., Bottari, F.: Constituents of ferns. In: Progress in phytochemistry. Reinhold, L., Liwschitz, Y. (eds.), Vol. I. pp. 589–685. London: Interscience Publishers 1968

Bishop, D.G.: Inhibition of photochemical activity in chloroplasts by polyene antibiotic, filipin. Arch. Biochem. Acta *283*, 469–482 (1973)

Biswas, P.K., Paul, K.B., Henderson, J.H.M.: Effects of steroids on chrysanthemum in relation to growth and flowering. Nature (London) *213*, 917–918 (1967)

Boid, R., Rees, H.H., Goodwin, T.W.: Studies in insect-moulting hormone biosynthesis. Biosynthesis of cyasterone in the plant, *Cyathula capitata*. Biochem. Physiol. Pflanz. *168*, 27–40 (1975)

Bolger, L.M., Rees, H.H., Goad, L.J., Goodwin, T.W.: Incorporation of [2-^{14}C]mevalonic acid into 28-isofucosterol by leaves of *Pisum sativum*. Biochem. J. *114*, 892–893 (1969)

Brady, D.R., Gaylor, J.L.: Enzymic formation of esters of methyl sterol precursors of cholesterol. J. Lipid Res. *12*, 270–276 (1971)

Brandt, R.D., Benveniste, P.: Isolation and identification of sterols from subcellular fractions of bean leaves (*Phaseolus vulgaris*). Biochim. Biophys. Acta *282*, 85–92 (1972)

Burstein, S., Kimball, H.L., Gut, M.: Transformation of labeled cholesterol, 20α-hydroxycholesterol, (22R)-22-hydroxycholesterol, and (22R)-20α,22-dihydroxycholesterol by adrenal acetone-dried preparations from guinea pigs, cattle and man: II. Kinetic studies. Steroids *15*, 809–848 (1970)

Bush, P.B., Grunwald, C.: Sterol changes during germination of *Nicotiana tabacum* seeds. Plant Physiol. *50*, 69–72 (1972)

Bush, P.B., Grunwald, C.: Effect of light on mevalonic acid incorporation into the phytosterols of *Nicotiana tabacum* L. seedlings. Plant Physiol. *51*, 110–114 (1973)

Bush, P.B., Grunwald, C.: Steryl glycoside formation in seedlings of *Nicotiana tabacum* L. Plant Physiol. *53*, 131–135 (1974)

Bush, P.B., Grunwald, C., Davis, D.L.: Changes in sterol composition during greening of etiolated barley shoots. Plant Physiol. *47*, 745–749 (1971)

Butenandt, A., Jacobi, H.: Über die Darstellung eines krystallisierten pflanzlichen Tokokinins (Thelykinins) und seine Identifizierung mit dem α-Follikelhormon. Untersuchungen über das weibliche Sexualhormon, 10. Mitteilung. Z. Physiol. Chem. *218*, 104–112 (1933)

Butenandt, A., Karlson, P.: Über die Isolierung eines Metamorphosehormons der Insekten in kristallisierter Form. Z. Naturforsch. *9b*, 389–391 (1954)

Carbonero, P., Torres, J.V., Garcia-Olmedo, F.: Effects of n-butanol and filipin on membrane permeability of developing wheat endosperms with different sterol phenotypes. FEBS Lett. *56*, 198–201 (1975)

Caspi, E., Hornby, G.M.: Biosynthesis of plant sterols. III. Mechanism of saturation of ring B in pregnenolone during its conversion to digitoxigenin in *Digitalis lanata*. Phytochemistry *7*, 423–427 (1968)

Caspi, E., Lewis, D.O.: Progesterone: its possible role in the biosynthesis of cardenolides by *Digitalis lanata*. Science *156*, 519–520 (1967)

Capsi, E., Lewis, D.O.: Biosynthesis of plant sterols. IV. An investigation of a possible mode of 14β-hydroxylation in digitoxigenin. Phytochemistry *7*, 683–691 (1968)
Caspi, E., Lewis, D.O., Piatak, D.M., Thimann, K.V., Winter, A.: Biosynthesis of plant sterols. Conversion of cholesterol to pregnenolone in *Digitalis purpurea*. Experientia *22*, 506–507 (1966)
Caspi, E., Wickramasinghe, J.A.F., Lewis, D.O.: The role of deoxycorticosterone in the biosynthesis of cardenolides in *Digitalis lanata*. Biochem. J. *108*, 503 (1968)
Castle, M., Blondin, G., Nes, W.R.: Evidence for the origin of the ethyl group of β-sitosterol. J. Am. Chem. Soc. *85*, 3306–3308 (1963)
Chardon-Loriaux, I., Morisaki, M., Ikekawa, N.: Sterol profiles of red algae. Phytochemistry *15*, 723–725 (1976)
Charles, G., Njimi, T., Ourisson, G., Ehrhardt, J., Conreur, C., Cavé, A., Goutarel, R.: Alcaloïdes stéroïdigues XCIV: Stéroïdes et alcaloïdes des graines de *Funtumia elastica* (Preuss) Stapf et de Funtumia latifolia Stapf: cycloarténol, nor-31 lanostérol, déhydro-24 lophénol, desmostérol et irehdiamines. C.R. Acad. Sci. Paris Ser. *C 268*, 2105–2108 (1969)
Corey, E.J., Russey, W.E., Ortiz de Montellano, P.R.: 2,3-Oxidosqualene, an intermediate in the biological synthesis of sterols from squalene. J. Am. Chem. Soc. *88*, 4750–4751 (1966)
Cornforth, J.W., Cornforth, R.H.: Chemistry of mevalonic acid. In: Natural substances formed biologically from mevalonic acid. Goodwin, T.W., (ed.), Vol. 29, pp. 5–15. Biochem. Soc. Symp. London: Academic Press 1970
Cornforth, J.W., Cornforth, R.H., Pelter, A., Horning, M.G., Popják, G.: Studies on the biosynthesis of cholesterol. 7. Rearrangement of methyl groups during enzymic cyclisation of squalene. Tetrahedron *5*, 311–339 (1959)
Czygan, F.: Blütenbildung bei *Lemna minor* nach Zusatz von Östrogenen. Naturwissenschaften *49*, 287–288 (1962)
Davis, D.L.: Sterol distribution within green and air cured tobacco. Phytochemistry *11*, 489–494 (1972)
Davis, D.L., Finkner, V.C.: Influence of temperature on sterol biosynthesis in *Triticum aestivum*. Plant Physiol. *52*, 324–326 (1972)
Dean, P.D., Exley, D., Goodwin, T.W.: Steroid oestrogens in plants: re-estimation of oestrone in pomegranate seeds. Phytochemistry *10*, 2215–2216 (1971)
De Souza, N.J., Ghisalberti, E.L., Rees, H.H., Goodwin, T.W.: Studies on insect moulting hormones: biosynthesis of ecdysone, ecdysterone and 5β-hydroxyecdysterone in *Polypodium vulgare*. Phytochemistry *9*, 1247–1252 (1970)
Devys, M., Barbier, M.: Isolement du pollinastanol et d'une série de Δ-7-stérols des pollens de deux composées. [Isolation of pollinastanol and a series of Δ^7 sterols from pollen from two compositae species.] Bull. Soc. Chim. Biol. *49*, 865–871 (1967a)
Devys, M., Barbier, M.: Sur les stérols des pollens. C. R. Acad. Sci. Paris Ser. D *264*, 504–506 (1967b)
Devys, M., Alcaide, A., Barbier, M., Lederer, E.: Biosynthèse de l'α-spinastérol à partir du méthylène-24 dihydrolanostérol par l'épinard *Spinacea oleracea*. Phytochemistry *7*, 613–617 (1968)
Devys, M., Alcaide, A., Barbier, M.: Biosynthesis of cholesterol from pollinastanol in the tobacco *Nicotiana tabacum*. Phytochemistry *8*, 1141–1444 (1969a)
Devys, M., Alcaide, A., Pinte, F., Barbier, M.: Pollinastanol dans la fougère *Polypodium vulgare* L. et la salsepareille *Smilax medica* Schlecht et Cham. C. R. Acad. Sci. Paris Ser. D *269*, 2033–2035 (1969b)
Djerassi, C., Krakower, G.W., Lemin, A.J., Liu, L.H., Mills, J.S., Villotti, R.: The neutral constituents of the cactus *Lophocereus schottii*. The structure of lophenol-4α-methyl-Δ^7-cholesten-3β-ol – a link in sterol biogenesis. J. Am. Chem. Soc. *80*, 6284–6292 (1958)
Donnet, V., Chevalier, J.M., Pruneyre, A.: Action de la cortisone sur la métamorphose et la croissance des têtards et la germination de certains végètaux. C. R. Soc. Biol. *154*, 1451–1453 (1960)
Dupéron, P.: Evolution des stérols au cours de la germination des semences de *Raphanus sativus*. Influence de la lumière et de la nutrition minérale. C. R. Acad. Sci. Paris Ser. D *266*, 1658–1661 (1968)

Dupéron, P.: Nature et comportement des stérols «libres» et estérifiés, au cours de la germination de divers types de semences. Hypothèses sur le rôle de ces substances chez les végétaux. Physiol. Vég. *9*, 373–399 (1971)
Dupéron, R., Brillard, M., Dupéron, P.: Localisation intracellulaire des différents composés stéroliques dans le tubercule de pomme de terre, avant sa germination. C. R. Acad. Sci. Paris Ser. D *274*, 2321–2324 (1972)
Ehrhardt, J.D., Hirth, L., Ourisson, G.: Études sur les triterpènes précurseurs des phytostérols. Recherche du cycloarténol et du lanostérol dans diverses espèces végétales. Phytochemistry *6*, 815–821 (1967)
Eichenberger, W., Grob, E.C.: Zur enzymatischen Bildung von Acyl-steringlycosid in Pflanzen. Chimia *23*, 368–369 (1969)
Eichenberger, W., Grob, E.C.: Enzymatische Acylierung von Sterin-^{14}C-glucosid durch ein lösliches Enzym aus Karotten. Chimia *24*, 394–396 (1970a)
Eichenberger, W., Grob, E.C.: Über die quantitative Bestimmung von Sterinderivaten in Pflanzen und die intrazellulare Verteilung der Steringlycoside in Blättern. FEBS Lett. *11*, 177–180 (1970b)
Eichenberger, W., Menke, W.: Sterole in Blättern und Chloroplasten. Z. Naturforsch. *21b*, 859–867 (1966)
El-Ridi, M.S., Mofty, El. A., Khalifa, K., Soliman, L.: Gonadotrophic hormones in pollen grains of the date palm. Z. Naturforsch. *156*, 45–49 (1960)
Eppenberger, U., Hirth, L., Ourisson, G.: Anaerobische Cyclisierung von Squalene-2,3-epoxyd zu Cycloartenol in Gewebekulturen von *Nicotiana tabacum* L. Eur. J. Biochem. *8*, 180–183 (1969)
Euw, J. v., Reichstein, T.: Die Biosynthese des Digitoxigenins, Herkunft des C-20. Helv. Chim. Acta *47*, 711–724 (1964)
Evans, F.J.: Uptake of [2-^{14}C]mevalonic acid by lipid and glycoside sterols. Planta *116*, 99–104 (1974)
Fang, T., Baisted, D.J.: UDPG:sterol glucosyltransferase in etiolated pea seedlings. Phytochemistry *15*, 273–278 (1976)
Fieser, L.F., Fieser, M.: Steroids. 2nd ed. New York: Reinhold Publ. Corp. 1967
Flint, A.P.F., Armstrong, D.T.: Activities of enzymes responsible for steroid biosynthesis and cholesterol ester metabolism in rabbit ovarian interstitial tissue and corpora lutea. A comparison of enzyme activities with flow rates. Biochem. J. *132*, 301–311 (1973)
Frantz, I.D., Jr., Schroepfer, G.J., Jr.: Sterol biosynthesis. Annu. Rev. Biochem. *36*, 691–726 (1967)
Frasch, W., Grunwald, C.: Acylated steryl glycoside synthesis in seedlings of *Nicotiana tabacum* L. Plant Physiol. *58*, 744–748 (1976)
Galbraith, M.N., Horn, D.H.S.: An insect-moulting hormone from a plant. J. Chem. Soc. Chem. Commun. 905–906 (1966)
Gawienowski, A.M., Gibbs, C.C.: Identification of cholesterol and progesterone in apple seeds. Steroids *12*, 545–550 (1968)
Gawienowski, A.M., Gibbs, C.C.: The isolation of estrone from apple seeds. Phytochemistry *8*, 685–686 (1969a)
Gawienowski, A.M., Gibbs, C.C.: Biosynthesis of pregnenolone from cholesterol in *Punica granatum*. Phytochemistry *8*, 2317–2319 (1969b)
Gawienowski, A.M., Cheney, R.W., Jr., Marsh, H.V.: Alteration of sex expression in the cucumber by testosterone and estradiol. Phytochemistry *10*, 2033–2034 (1971)
Geuns, J.M.C.: Variations in sterol composition in etiolated mung bean seedlings. Phytochemistry *12*, 103–106 (1973)
Geuns, J.M.C.: Physiological activity of corticosteroids in etiolated mung bean plants. Z. Pflanzenphysiol. 74, 42–51 (1974)
Geuns, J.M.C.: Regulation of sterol biosynthesis in etiolated mung bean hypocotyl sections. Phytochemistry *14*, 975–978 (1975)
Geuns, J.M.C.: Structure requirements of corticosteroids for physiological activity in etiolated mung bean seedlings. Z. Pflanzenphysiol. 81, 1–16 (1977)
Geuns, J.M.C.: Steroid hormones and plant growth and development. Phytochemistry 17, 1–44 (1978)
Geuns, J.M.C., Vendrıg, J.C.: Hormonal control of sterol biosynthesis in *Phaseolus aureus*. Phytochemistry *13*, 919–922 (1974)

Gibbons, G.F., Goad, L.J., Goodwin, T.W.: The sterols of some marine red algae. Phytochemistry *6*, 677–683 (1967)
Goad, L.J.: Aspects of phytosterol biosynthesis. In: Terpenoids in plants. Pridham, J.B. (ed.), pp. 159–190. London: Academic Press 1967
Goad, L.J., Goodwin, T.W.: The biosynthesis of plant sterols. In: Progress in phytochemistry, Reinhold, L., Liwschitz, Y. (eds.), Vol. 3, pp. 113–198. London: Interscience Publishers 1972
Goad, L.J., Hammam, A.S.A., Dennis, A., Goodwin, T.W.: Biosynthesis of the phytosterol side chain. Nature (London) *210*, 1322–1324 (1966)
Goad, L.J., Gibbons, G.F., Bolger, L.M., Rees, H.H., Goodwin, T.W.: Incorporation of [2-^{14}C, (5R)-5-$^{3}H_1$] mevalonic acid into cholesterol by a rat liver homogenate and into β-sitosterol and 28-isofucosterol by *Larix decidua* leaves. Biochem. J. *114*, 885–893 (1969)
Goodwin, T.W.: The biological significance of terpenes in plants. In: Terpenoids in plants. Pridham, J.B. (ed.), pp. 1–23. London: Academic Press 1967
Gower, D.B.: Modifiers of steroid hormone metabolism: a review of their chemistry, biochemistry and clinical applications. J. Steroid Biochem. *5*, 501–523 (1974)
Graves, J.M.H., Smith, W.K.: Transformation of pregnenolone and progesterone by cultured plant cells. Nature (London) *214*, 1248–1249 (1967)
Grunwald, C.: Sterol distribution in intracellular organelles isolated from tobacco leaves. Plant Physiol. *45*, 663–666 (1970)
Grunwald, C.: Effects of free sterols, steryl ester, and steryl glycoside in membrane permeability. Plant Physiol. *48*, 653–655 (1971)
Grunwald, C.: Phytosterols in tobacco leaves at various stages of physiological maturity. Phytochemistry *14*, 79–82 (1975a)
Grunwald, C.: Plant sterols. Annu. Rev. Plant Physiol. *26*, 209–236 (1975b)
Grunwald, C.: Shading influence on the sterol balance of *Nicotiana tabacum* L. Plant Physiol. *61*, 76–79 (1978)
Grunwald, C., Bush, L.P., Keller, C.J.: Variation in sterols, alkaloids, and polyphenols of two *Nicotiana* varieties under different nitrogen fertilization and drying processes. J. Agric. Food Chem. *19*, 216–221 (1971)
Hart, M.C., Heyl, F.W.: Spinasterol and some of its esters. J. Biol. Chem. *95*, 311–315 (1932)
Hartmann, M.A., Benveniste, P.: Effect of ageing on sterol metabolism in potato tuber slices. Phytochemistry *13*, 2667–2672 (1974)
Hartmann, M.A., Normand, G., Benveniste, P.: Sterol composition of plasma membrane enriched fractions from maize coleoptiles. Plant Sci. Lett. *5*, 287–292 (1975)
Hassan, A., Hassan, M., Wafa, El A.: An oestrogenic substance in pollen-grains of date palm tree *Phoenix dactylifera* L., Palmae. Nature (London) 159, 409–410 (1947)
Heftmann, E.: Biochemistry of steroidal saponins and glycoalkaloids. Lloydia *30*, 209–230 (1967)
Heftmann, E.: Insect molting hormones in plants. In: Recent advances in phytochemistry. Steelink, C., Runeckles, V.C. (eds.), Vol. 3, pp. 211–227. New York: Appleton-Century-Crofts 1970
Heftmann, E.: Steroids. In: Phytochemistry, organic metabolites. Miller, L.P. (ed.), Vol. II, pp. 171–226. New York: Van Nostrand Reinhold Company 1973
Heftmann, E.: Steroid hormones in plants. Lloydia *38*, 195–209 (1975)
Heftmann, E., Ko, S., Bennett, R.D.: Identification of estrone in date seeds by thin-layer chromatography. Naturwissenschaften *52*, 431–432 (1965)
Heftmann, E., Ko, S., Bennett, R.D.: Identification of estrone in pomegranate seeds. Phytochemistry *5*, 1337–1339 (1966)
Heinrich, G., Hoffmeister, H.: Ecdyson als Begleitsubstanz des Ecdysterons in *Polypodium vulgare* L. Experientia *23*, 995 (1967)
Heintz, R., Benveniste, P.: Cyclisation de l'epoxyde-2,3 de squalène par des microsomes extraits de tissues de tabac cultivés in vitro. Phytochemistry *9*, 1499–1503 (1970)
Heintz, R., Benveniste, P.: Plant sterol metabolism. Enzymatic cleavage of the 9β, 19β-cyclopropane ring of cyclopropyl sterols in bramble tissue cultures. J. Biol. Chem. *249*, 4267–4274 (1974)
Heintz, R., Schaefer, P.C., Benveniste, P.: Cyclisation of squalene 2,3; 22,23-diepoxide by microsomes from bramble (*Rubus fruticosa*) tissues grown in vitro. J. Chem. Soc. Chem. Commun. 946–947 (1970)

Heinz, E., Dieler, H.P., Rullkoeter, J.: Enzymatic acylation of steryl glycoside. Z. Pflanzenphysiol. *75*, 78–87 (1975)

Helmkamp, G., Bonner, J.: Some relationships of sterols to plant growth. Plant Physiol. *28*, 428–436 (1953)

Hendrix, D.L., Higinbotham, N.: Effects of filipin and cholesterol on K^+ movement in etiolated stem cells of *Pisum sativum* L. Plant Physiol. *52*, 93–97 (1973)

Herout, V.: Some relations between plants, insects and their isoprenoids. In: Progress in phytochemistry. Reinhold, L., Liwschitz, Y. (eds), Vol. 2, pp. 143–202. London: Interscience Publishers 1970

Hikino, H., Takemoto, T.: Ecdysones of plant origin. In: Invertebrate endocrinology and hormonal heterophylly. Burdette, W.J. (ed.), pp. 185–217. Berlin, Heidelberg, New York: Springer 1974

Hikino, H., Kohama, T., Takemoto, T.: Biosynthesis of ponasterone A, an insect-moulting substance from *Podocarpus macrophyllus*. Chem. Pharm. Bull. *17*, 415 (1969)

Hikino, H., Kohama, T., Takemoto, T.: Biosynthesis of ponasterone A and ecdysterone from cholesterol in *Podocarpus macrophyllus*. Phytochemistry *9*, 367–369 (1970)

Hikino, H., Jin, H., Takemoto, T.: Biosynthesis of inokosterone and ecdysterone from cholesterol and mevalonic acid in *Achyranthes*. Nihon Yakugakukai *95*, 590–595 (1975)

Hochberg, R.B., McDonald, P.D., Feldman, M., Lieberman, S.: Studies on the biosynthetic conversion of cholesterol into pregnenolone. Side chain cleavage and some 20-p-tolyl analogs of cholesterol and 20α-hydroxycholesterol. J. Biol. Chem. *249*, 1277–1285 (1974)

Hou, C.T., Umemura, Y., Nakamura, M., Funahashi, S.: Enzymatic synthesis of steryl glucoside by a particulate preparation from immature soybean seeds. J. Biochem. *63*, 351–360 (1968)

Huber, R., Hoppe, W.: Die Kristall- und Molekülstrukturanalyse des Insektenverpuppungshormons Ecdyson mit der automatisierten Faltmolekülmethode. Chem. Ber. *98*, 2403–2424 (1965)

Hügel, M., Barbier, M., Lederer, E.: Sur le pollinastanol, nouveau stérol du pollen. Bull. Soc. Chem. Fr. 2012–2013 (1964a)

Hügel, M., Vetter, W., Audier, H., Barbier, M., Lederer, E.: Analyse des stérols du pollen par spectrometric de masse. Phytochemistry *3*, 7–16 (1964b)

Idler, D.R., Nicksic, S.W., Johnson, D.R., Meloche, V.W., Schuette, H.A., Baumann, C.A.: Separation and tentative identification of two new sterols from oats. J. Am. Chem. Soc. *75*, 1712–1715 (1953)

Ingram, D.S., Knights, B.A., McEvoy, I.J., McKay, P.: Studies in the Cruciferae. Changes in the composition of the sterol fraction following germination. Phytochemistry *7*, 1241–1245 (1968)

Jacobs, W.P., Suthers, H.B.: The culture of apical buds of *Xanthium* and their use as a bioassay for flowering activity of ecdysterone. Am. J. Bot. *58*, 836–843 (1971)

Jacobsohn, G.M.: Sterol formation and transformation in *Digitalis*. In: Recent advances in phytochemistry. Steelink, C., Runeckles, V.C. (eds.), Vol. 3, pp. 229–247. New York: Appleton-Century-Crofts 1970

Jacobsohn, G.M., Frey, M.J., Hochberg, R.B.: The absence of steroid estrogens in plants. Steroids *6*, 93–99 (1965)

Johnson, D.F., Bennett, R.D., Heftmann, E.: Cholesterol in higher plants. Science *140*, 198–199 (1963)

Johnson, D.F., Heftmann, E., Houghland, G.V.C.: The biosynthesis of sterols in *Solanum tuberosum*. Arch. Biochem. Biophys. *104*, 102–105 (1964)

Joly, R.A., Bonner, J., Bennett, R.D., Heftmann, E.: Conversion of cholesterol to an open-chain saponin by *Dioscorea floribunda*. Phytochemistry *8*, 857–859 (1969a)

Joly, R.A., Bonner, J., Bennett, R.D., Heftmann, E.: Conversion of an open-chain saponin to dioscin by a *Dioscorea floribunda* homogenate. Phytochemistry *8*, 1445–1447 (1969b)

Joly, R.A., Bonner, J., Bennett, R.D., Heftmann, E.: The biosynthesis of steroidal sapogenins in *Dioscorea floribunda* from doubly labelled cholesterol. Phytochemistry *8*, 1709–1711 (1969c)

Joly, R.A., Svahn, C.M., Bennett, R.D., Heftmann, E.: Investigation of intermediate steps in the biosynthesis of ecdysterone from cholesterol in *Podocarpus elata*. Phytochemistry *8*, 1917–1920 (1969d)

Kaplanis, J.N., Thompson, M.J., Robbins, W.E., Bryce, B.M.: Insect hormones: alpha ecdysone and 20-hydroxyecdysone in bracken fern. Science *157*, 1436–1438 (1967)
Karlson, P.: Mode of action of ecdysones. In: Invertebrate endocrinology and hormonal heterophylly. Burdette, W. (ed.), pp. 43–54. Berlin, Heidelberg, New York: Springer 1974
Kauss, H.: Enzymatische Glucosylierung von pflanzlichen Sterinen. Z. Naturforsch. *23b*, 1522–1526 (1968)
Keller, C.J., Bush, L.P., Grunwald, C.: Changes in content of sterols, alkaloids, and phenols in flue-cured tobacco during conditions favoring infestation by molds. J. Agric. Food Chem. *17*, 331–334 (1969)
Kemp, R.J., Mercer, E.I.: The sterol esters of maize seedlings. Biochem. J. *110*, 111–118 (1968a)
Kemp, R.J., Mercer, E.I.: Studies on the sterols and sterol esters of the intracellular organelles of maize shoots. Biochem. J. *110*, 119–125 (1968b)
Kemp, R.J., Goad, L.J., Mercer, E.I.: Changes in the levels and composition of the esterified and unesterified sterols of maize seedlings during germination. Phytochemistry *6*, 1609–1615 (1967)
Kemp, R.J., Hammam, A.S.A., Goad, L.J., Goodwin, T.W.: Studies on phytosterol biosynthesis: observations on the esterified sterols of higher plants. Phytochemistry *7*, 447–450 (1968)
Kimura, Y., Tietz, A., Tamura, S.: Stigmasteryl-β-D-glucoside as an auxin synergist. Planta *126*, 289–292 (1975)
Kintia, P.K., Wojciechowski, Z.A.: Free and bound sterols in seedlings of *Cucumis sativus*. Phytochemistry *13*, 2235–2238 (1974)
Kiribuchi, T., Chen, C.S., Funahashi, S.: Systematic analysis of sterols in soybeans and other oil seeds. Agric. Biol. Chem. *29*, 265–267 (1965)
Kiribuchi, T., Mizunaga, T., Funahashi, S.: Separation of soybean sterols by florisil chromatography and characterization of acylated steryl glucoside. Agric. Biol. Chem. *30*, 770–778 (1966)
Knapp, F.F., Nicholas, H.J.: The distribution of sterols and steryl esters in the banana plant. Phytochemistry *8*, 2091–2093 (1969)
Knapp, F.F., Nicholas, H.J.: Phytosterol biosynthesis in banana peel. Initial removal of the 4α-methyl group of 24-methylenecycloartanol during its conversion into cycloeucalenol in *Musa sapientum*. J. Chem. Soc. Chem. Commun. 399–400 (1970)
Knapp, F.F., Nicholas, H.J.: The biosynthesis of phytosterols in *Musa sapientum*. Phytochemistry *10*, 85–95 (1971a)
Knapp, F.F., Nicholas, H.J.: The biosynthesis of 31-norcyclolaudenone in *Musa sapientum*. Phytochemistry *10*, 97–102 (1971b)
Knights, B.A.: Identification of the sterols of oat seed. Phytochemistry *4*, 857–862 (1965)
Knights, B.A., Laurie, W.: Application of combined gas-liquid chromatography-mass spectrometry to the identification of sterols in oat seed. Phytochemistry *6*, 407–416 (1967)
Kopcewicz, J.: Effect of estrone on the content of endogenous gibberellins in the dwarf pea. Naturwissenschaften *56*, 334 (1969)
Kopcewicz, J.: Influence of estrogens on flower formation in *Cichorium intybus* L. Naturwissenschaften *57*, 136 (1970)
Kopcewicz, J.: Estrogens in developing bean (*Phaseolus vulgaris*) plants. Phytochemistry *10*, 1423–1427 (1971a)
Kopcewicz, J.: Influence of steroidal hormones on flower sex expression in *Ecballium elaterium* (L.) A. Rich. Z. Pflanzenphysiol. *65*, 92–94 (1971b)
Kopcewicz, J.: Estrogens in the short-day plants *Perilla ocimoides* and *Chenopodium rubrum* grown under inductive and non-inductive light conditions. Z. Pflanzenphysiol. *67*, 373–376 (1972a)
Kopcewicz, J.: Oestrogens in the long-day plants *Hyoscyamus niger* and *Salvia splendens* grown under inductive and non-inductive light conditions. New Phytol. *71*, 129–134 (1972b)
Koreeda, M., Nakanishi, K., Goto, M.: Ajugalactone, an insect moulting inhibitor as tested by the Chilo dipping method. J. Am. Chem. Soc. *92*, 7512–7513 (1970)
Leboeuf, M., Cavé, A., Goutarel, R.: Alcaloïdes stéroïdiques. Présence de la progestérone dans les feuilles de l'*Holarrhena floribunda* (G. Don) Dür. et Schinz. C. R. Acad. Sci. Paris *259*, 3401–3403 (1964)

Lenton, J.R., Goad, L.J., Goodwin, T.W.: Sterols of *Xanthoria parietina*: evidence for two sterol 'pools' and the identification of a novel C_{28} triene, ergosta-5,8,22-trien-3β-ol. Phytochemistry *12*, 1135–1140 (1973a)
Lenton, J.R., Goad, L.J., Goodwin, T.W.: Sterols of the mycobiont and phycobiont isolated from the lichen *Xanthoria parietina*. Phytochemistry *12*, 2249–2253 (1973b)
Lepage, M.: Isolation and characterization of an esterified form of steryl glucoside. J. Lipid Res. *5*, 587–592 (1964)
Lindberg, M., Gautschi, F., Bloch, K.: Ketonic intermediates in the demethylation of lanosterol. J. Biol. Chem. *238*, 1661–1664 (1963)
Lloyd-Jones, J.G., Rees, H.H., Goodwin, T.W.: Biosynthesis of ecdysterone from cholesterol in *Taxus baccata*. Phytochemistry *12*, 569–572 (1973)
Lockley, W.J.S., Boid, R., Lloyd-Jones, G.J., Rees, H.H., Goodwin, T.W.: Fate of the C-4 hydrogen atoms of cholesterol during its transformation into ecdysones in insects and plants. J. Chem. Soc. Comm. 346–348 (1975)
Ludwiczak, R.S., Stachowiak, K.: Wyodrebnienie β-sitosterolu z ziela skrzypu polnego. Rocz. Chem. *37*, 575–579 (1963)
Marsili, A., Morelli, I.: Triterpenes from mosses. I. The occurrence of 22(29)-hopene in *Thamnium alopecurum* (L.) Br. Eur. ssp. *eu-alopecurum* Giac. Phytochemistry *7*, 1705–1706 (1968)
Marsili, A., Morelli, I.: Triterpenes from *Thuidium tamariscifolium*. Phytochemistry *9*, 651–653 (1970)
Matsuo, A., Nakayama, M., Maeda, M., Hayashi, S.: Distribution pattern of sterols in liverworts belonging to the Jungermanniineae. Phytochemistry *12*, 2413–2415 (1973)
Maudgal, R.K., Tchen, T.T., Bloch, K.: 1,2-Methyl shifts in the cyclization of squalene to lanosterol. J. Am. Chem. Soc. *80*, 2589–2590 (1958)
McKillican, M.E.: Studies on the phospholipids, glycolipids and sterols of wheat endosperm. J. Am. Oil Chem. Soc. *41*, 554–557 (1964)
Mercer, E.I., Johnson, M.W.: Cyclization of squalene-2,3-oxide to lanosterol in a cell-free system from *Phycomyces blakesleeanus*. Phytochemistry *8*, 2329–2331 (1969)
Mercer, E.I., Pughe, J.E.: The effect of abscisic acid on the biosynthesis of isoprenoid compounds in maize. Phytochemistry *8*, 115–122 (1969)
Mudd, J.B., Kleinschmidt, M.G.: Effect of filipin on the permeability of red beet and potato tuber discs. Plant Physiol. *45*, 517–518 (1970)
Nakanishi, K., Koreeda, M., Sasaki, S., Chang, M.L., Hsu, H.Y.: Insect hormones. The structure of ponasterone A, an insect-moulting hormone from the leaves of *Podocarpus nakaii* Hay. J. Chem. Soc. Chem. Commun. 915–917 (1966)
Nes, W.R., McKean, M.L.: Biochemistry of steroids and other isopentenoids. Baltimore: University Park Press 1977
Nicholas, H.J., Moriarty, S.: Biosynthesis of β-sitosterol. Fed. Proc. *22*, 529 (1963)
Nordby, H.E., Nagy, S.: Fatty acid composition of sterol esters from *Citrus sinensis*, *C. paradisi*, *C. limon aurantifolia* and *C. limettioides* sacs. Phytochemistry *13*, 443–452 (1974)
Ogawa, S., Nishimoto, N., Matsuda, H.: Source and availability of ecdysterones. In: Invertebrate endocrinology and hormonal heterophylly. Burdette, W.J. (ed.), pp. 218–232. Berlin, Heidelberg, New York: Springer 1974
Olson, J.A., Jr., Lindberg, M., Bloch, K.: On the demethylation of lanosterol to cholesterol. J. Biol. Chem. *226*, 941–956 (1957)
Ongun, A., Mudd, J.B.: The biosynthesis of steryl glucosides in plants. Plant Physiol. *45*, 255–262 (1970)
Patterson, G.W.: Sterols of *Laminaria*. Comp. Biochem. Physiol. *24*, 501–505 (1968)
Patterson, G.W.: Sterols of some green algae. Comp. Biochem. Physiol. *47[B]*, 453–457 (1974)
Péaud-Lenoël, C., Axelos, M.: D-glucosylation des phytostérols et acylation des stéryl-d-glucosides en présence d'enzymes de plantes. Carbohydr. Res. *24*, 247–262 (1972)
Pilgrim, H.: 'Cholesterol side-chain cleaving enzyme' Aktivität in Keimlingen und in vitro kultivierten Geweben von *Digitalis purpurea*. Phytochemistry *11*, 1725–1728 (1972)
Power, F.B., Salway, A.H.: The identification of ipuranol and some allied compounds as phytosterol glucosides. J. Chem. Soc. *103*, 399–406 (1913)
Raab, K.H., De Souza, N.J., Nes, W.R.: The H-migration in the alkylation of sterols at C-24. Biochim Biophys. Acta *152*, 742–748 (1968)

Rahier, A., Cattel, L., Benveniste, P.: Mechanism of the enzymatic cleavage of the 9, 19-cyclopropane ring of cycloeucalenol. Phytochemistry *16*, 1187–1192 (1977)
Ratsimamanga, A.R., Diot, J.: Action de l'hydrocortisone sur la croissance d'*Ervum lens*. C. R. Soc. Biol. *153*, 1978–1981 (1959)
Ratsimamanga, A.R., Nigeon-Dureuil, M., Boiteau, P.: Effets de la cortisone sur la croissance de *Ervum lens*. C. R. Soc. Biol. *152*, 960–963 (1958)
Rees, H.H., Mercer, E.I., Goodwin, T.W.: The stereospecific biosynthesis of plant sterols and α- and β-amyrin. Biochem. J. *99*, 726–734 (1966)
Rees, H.H., Goad, L.J., Goodwin, T.W.: Cyclization of 2,3-oxidosqualene to cycloartenol in a cell-free system from higher plants. Tetrahedron Lett. *6*, 723–725 (1968a)
Rees, H.H., Goad, L.J., Goodwin, T.W.: Studies in phytosterol biosynthesis. Mechanism of biosynthesis of cycloartenol. Biochem. J. *107*, 417–426 (1968b)
Rees, H.H., Goad, L.J., Goodwin, T.W.: 2,3-Oxidosqualene cycloartenol cyclase from *Ochromonas malhamensis*. Biochim. Biophys. Acta *176*, 892–894 (1969)
Reichstein, T., Weiss, E.: The sugars of the cardiac glycosides. Adv. Carbohydr. Chem. *17*, 65–120 (1962)
Reid, W.W.: Accumulation of squalene-2,3-oxide during inhibition of phytosterol biosynthesis in *Nicotiana tabacum*. Phytochemistry *7*, 451–452 (1968)
Reitz, R.C., Hamilton, J.G.: The isolation and identification of two sterols from two species of blue-green algae. Comp. Biochem. Physiol. *25*, 401–416 (1968)
Robbins, W.E., Kaplanis, J.N., Thompson, M.J., Shortino, T.J., Cohen, C.F., Joyner, S.C.: Ecdysones and analogs: effects on development and reproduction of insects. Science *161*, 1158–1160 (1968)
Robinson, S.P., Wiskich, J.T.: The effects of digitonin on photochemical activities of isolated chloroplasts. Plant Physiol. *55*, 163–167 (1975)
Rowe, J.W.: The sterols of pine bark. Phytochemistry *4*, 1–10 (1965)
Rubinstein, I., Goad, L.J.: Occurrence of (24S)-24-methylcholesta-5,22E-dien-3β-ol in the diatom *Phaeodactylum tricornutum*. Phytochemistry *13*, 485–487 (1974)
Russell, G.B., Horn, D.H.S., Middleton, E.J.: New phytoecdysones from *Dacrydium intermedium*. J. Chem. Soc. Chem. Commun. 11 (1971)
Russell, P.T., Van Aller, R.T., Nes, W.R.: The mechanism of introduction of alkyl groups at C-24 of sterols. II. The necessity of the Δ^{24} bond. J. Biol. Chem. *242*, 5802–5806 (1967)
Săden-Krehula, M., Tajić, M., Kolbah, D.: Testosterone, epitestosterone and androstenedione in the pollen of Scotch pine *P. silvestris* L. Experientia *27*, 108–109 (1971)
Sauer, H.H., Bennett, R.D., Heftmann, E.: Pregnenolone metabolism in *Digitalis lanata*. Phytochemistry *6*, 1521–1526 (1967)
Sauer, H.H., Bennett, R.D., Heftmann, E.: Ecdysterone biosynthesis in *Podocarpus elata*. Phytochemistry *7*, 2027–2030 (1968)
Schreiber, K., Osske, G.: Sterine und Triterpenoide. V. Über die 4α-Methyl-sterine der Kartoffelpflanze *Solanum tuberosum* L. Tetrahedron *20*, 2575–2584 (1964)
Schubert, K., Rose, G., Tümmler, R., Ikekawa, N.: Sterine in *Escherichia coli*. Hoppe-Seylers Z. Physiol. Chem. *339*, 293–296 (1964)
Schubert, K., Rose, G., Wachtel, H., Hörhold, C., Ikekawa, N.: Zum Vorkommen von Sterinen in Bakterien. Eur. J. Biochem. *5*, 246–251 (1968)
Sharma, R.K.: Biosynthesis of plant sterols and triterpenoids. The incorporation of (3RS)-[2-^{14}C,(4R)-4-$^{3}H_1$]mevalonate into α-spinasterol and β-amyrin in *Camellia sinensis*. Phytochemistry *9*, 565–568 (1970)
Sharpless, K.B., Snyder, T.E., Spencer, T.A., Maheshwari, K.K., Nelson, J.A., Clayton, R.B.: Biological demethylation of 4,4-dimethyl sterols. Evidence for enzymic epimerization of the 4β-methyl group prior to its oxidative removal. J. Am. Chem. Soc. *91*, 3394–3396 (1969)
Shewry, P.R., Stobart, A.K.: Effect of gibberellic acid on sterol production in *Corylus avellana* seeds. Phytochemistry *13*, 347–355 (1974)
Singh, B., Rastogi, R.P.: Cardenolides – glycosides and genins. Phytochemistry *9*, 315–331 (1970)
Sipahimalani, A.T., Banerji, A., Chadha, M.S.: Biosynthesis and interconversion of phytoecdysones in *Sesuvium portulacastrum* L. J. Chem. Soc. Chem. Commun. 692–693 (1972)
Skarzyński, B.: An oestrogenic substance from plant material. Nature (London) *131*, 766 (1933)

Śliwowski, J., Kasprzyk, Z.: Stereospecificity of sterol biosynthesis in *Calendula officinalis* flowers. Phytochemistry *13*, 1451–1457 (1974)
Smith, A.R.H., Goad, L.J., Goodwin, T.W., Lederer, E.: Phytosterol biosynthesis: evidence for a 24-ethylidene intermediate during sterol formation on *Ochromonas malhamensis*. Biochem. J. *104*, 56C–58C (1967)
Smith, F.R., Korn, E.D.: 7-Dehydrostigmasterol and ergosterol: the major sterols of an amoeba. J. Lipid Res. *9*, 405–408 (1968)
Smith, P.F.: The role of lipids in membrane transport in *Mycoplasma laidlawii*. Lipids *4*, 331–336 (1969)
Spotts, R.A., Lukezic, F.L., Lacasse, N.L.: The effect of benzimidazole, cholesterol, and a steroid inhibitor on leaf sterols and ozone resistance of bean. Phytopathology *65*, 45–49 (1975)
Stohs, S.J.: The metabolism of progesterone by plant microsomes. Phytochemistry *8*, 1215–1219 (1969)
Stohs, S.J., El-Olemy, M.M.: Pregnenolone and progesterone from 20α-hydroxycholesterol by *Cheiranthus cheiri* leaf homogenates. Phytochemistry *10*, 3053–3056 (1971)
Stohs, S.J., El-Olemy, M.M.: Pregnenolone and progesterone metabolism by cardenolide producing plants. Phytochemistry *11*, 2409–2413 (1972)
Stohs, S.J., Sabatka, J.J., Rosenberg, H.: Incorporation of 4-^{14}C-22,23-^{3}H-sitosterol into diosgenin by *Dioscorea deltoidea* tissue suspension cultures. Phytochemistry *13*, 2145–2148 (1974)
Stokke, K.T.: The existence of an acid cholesterol esterase in human liver. Biochim. Biophys. Acta *270*, 156–166 (1972)
Stoll, A., Jucker, E.: Phytosterine, Steroidsaponine und Herzglykoside. In: Modern methods of plant analysis. Paech, K., Tracey, M.V. (eds.) Vol. 3, pp. 141–271. Berlin: Springer 1955
Sucrow, W.: $\Delta^{7,24(28)}$-Stigmastadienol-(3β) aus Kürbis. Tetrahedron Lett. *20*, 2443–2445 (1968)
Swindell, A.C., Gaylor, J.L.: Investigation of the component reactions of oxidative sterol demethylation. J. Biol. Chem. *243*, 5546–5555 (1968)
Takemoto, T., Arihara, S., Hikino, H.: Structure of ponasteroside A, a novel glycoside of insect-moulting substance from *Pteridium aquilinum* var. *latiuschulum*. Tetrahedron Lett. *39*, 4199–4202 (1968)
Tamelen, E.E. van, Willett, J.C., Clayton, R.B., Lord, K.E.: Enzymic conversion of squalene 2,3-oxide to lanosterol and cholesterol. J. Am. Chem. Soc. *88*, 4752–4754 (1966)
Tchen, T.T., Bloch, K.: On the mechanism of enzymatic cyclization of squalene. J. Biol. Chem. *226*, 931–939 (1957)
Tietz, A., Kimura, Y., Tamura, S.: Sterin-glukoside: eine Gruppe pflanzlicher Wirkstoffe mit biphasischer Dosiswirkungskurve. Z. Pflanzenphysiol. *81*, 57–67 (1977)
Tomita, Y., Sakurai, E.: Biosynthesis of phytoecdysone: incorporation of 2β, 3β, 14α-trihydroxy-5β-cholest-7-en-6-one into β-ecdysone and inokosterone in *Achyranthes fauriei*. J. Chem. Soc. Chem. Commun. 434–435 (1974)
Tomlinson, H., Rich, S.: Effect of ozone on sterols and sterol derivatives in bean leaves. Phytopathology *61*, 1404–1405 (1971)
Tomlinson, H., Rich, S.: Anti-senescent compounds reduce injury and steroid changes in ozonated leaves and their chloroplasts. Phytopathology *63*, 903–906 (1973)
Torres, J.V., Garcia-Olmedo, F.: Genetic control of sterol esterification in developing wheat endosperm. Biochim. Biophys. Acta *409*, 367–375 (1975)
Torres, J.V., Carbonero, P., Garcia-Olmedo, F.: Endosperm sterol phenotype and germination in wheat. Phytochemistry *15*, 677–680 (1976)
Tschesche, R.: Pflanzliche Steroide mit 21 Kohlenstoffatomen. In: Progres dans la chimie des substances organiques naturelles. Zechmeister, L. (ed.), Vol. 24, pp. 99–148. New York: Springer 1966
Tschesche, R., Lilienweiss, G.: Cardenolid Biosynthese aus Pregnenolonglucosid. Z. Naturforsch. *19b*, 265–266 (1964)
Tschesche, R., Hulpke, H., Scholten, H.: Pregnan-3β, 14β-diol-20-on als Vorstufe der Cardenolide. Z. Naturforsch. *22b*, 677–678 (1967)
Tschesche, R., Hombach, R., Scholten, H., Peters, M.: Neue Beiträge zur Biogenese der Cardenolide in *Digitalis lanata*. Phytochemistry *9*, 1505–1515 (1970)

Tschesche, R., Güttel, H.W., Josst, G.: Desmosterol als biogenetische Vorstufe für Convallamarogenin. Phytochemistry *13*, 137–139 (1974)

Waters, J.A., Johnson, D.F.: Biosynthesis of sterols in the soybean plant. Arch. Biochem. Biophys. *112*, 387–391 (1965)

Weete, J.D.: Sterols of the fungi: distribution and biosynthesis. Phytochemistry *12*, 1843–1864 (1973)

Weizmann, A., Mazur, Y.: Steroids and triterpenoids of citrus fruit. II. Isolation of citrostadienol. J. Org. Chem. *23*, 832–834 (1958)

Wickramasinghe, J.A.F., Burrows, E.P., Sharma, R.K., Greig, J.B., Caspi, E.: The obligatory involment of a C_{21} intermediate in the biosynthesis of cardenolides from cholesterol. Phytochemistry *8*, 1433–1440 (1969)

Williams, B.L., Goad, L.J., Goodwin, T.W.: The sterols of grapefruit peel. Phytochemistry *6*, 1137–1145 (1967)

Williams, C.M.: Hormonal interaction between plants and insects. In: Chemical ecology. Sondheimer, E., Simeone, J.B. (eds.), pp. 103–132. New York: Academic Press 1970

Willuhn, G., Köstens, J.: Die quantitative Verteilung der Sterine und Sterinderivate in Organen von *Solanum dulcamara*. Phytochemistry *14*, 2055–2058 (1975)

Wojciechowski, Z.A.: Changes in UDPG: sterol glucosyltransferase activity in *Calendula officinalis*. Phytochemistry *13*, 2091–2094 (1974)

Woolhouse, H.W.: The nature of senescence in plants. In: Aspects of the biology of aging, Soc. of Exp. Biol. Symp. Vol. 21, pp. 179–213. New York: Cambridge University Press 1967

Yamamoto, S., Bloch, K.: Enzymatic studies on the oxidative cyclizations of squalene. In: Natural substances formed biologically from mevalonic acid. Goodwin, T.W. (ed.), Vol. 292, pp. 35–43. Biochem. Soc. Symp. London: Academic Press, New York 1970

Young, I.J., Knights, B.A., Hillman, J.R.: Oestradiol and its biosynthesis in *Phaseolus vulgaris* L. Nature (London) *267*, 429 (1977)

5.3 Carotenoids

T.W. GOODWIN

1 Introduction

Since the original edition of this Encyclopedia the amount of new information on carotenoids in plants has increased, but at the same time the space now allocated to the subject is less than previously. In order to give a balanced picture within the restraints imposed many general statements will have to be made and the reader will be referred to appropriate detailed reviews rather than to the primary literature.

All carotenoids belong to the large chemical group of terpenoids and most of them are tetraterpenes, C_{40} compounds made up of eight isoprenoid (*ip*) units joined so that the arrangements of the units is reversed at the centre of the molecule. The two central methyl groups are in a 1,6-position relative to each other, whilst the remaining non-terminal methyl groups are in a 1,5-relationship, thus:

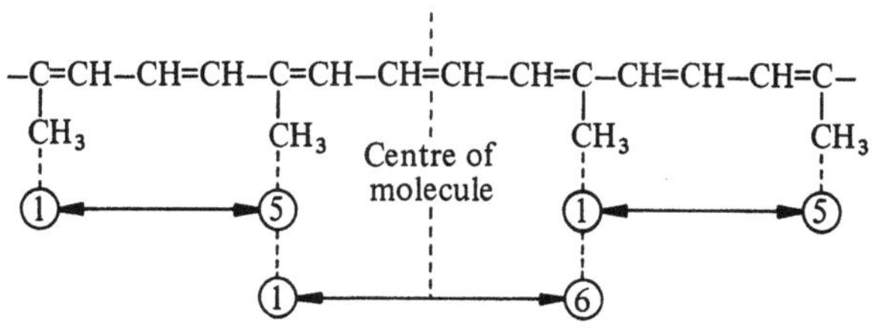

In addition to C_{40} carotenoids and their derivatives which contain fewer than forty carbon atoms (apocarotenoids), homocarotenoids (C_{45} and C_{50} terpenoids) and C_{30} carotenoids are found in some bacteria. They will not be discussed further here; see GOODWIN (1979) for further details.

All carotenoids can be formally derived from the acyclic $C_{40}H_{56}$ conjugated polyene lycopene (I) by reactions involving hydrogenation, dehydrogenation, cyclisation, double bond migration, methyl migration, chain elongation, chain contraction and by the insertion of oxygen in various forms. Carotenoid hydrocarbons are known as *carotenes* and oxygenated derivatives as *xanthophylls.* The structure and numbering of the parent lycopene are indicated in (I). Individual carotenes are named by the specific C_9 end groups. These groups are indicated in Figure 1. The nomenclature of carotenoids is complex (see Commission on Biochemistry Nomenclature, 1972), suffice it to say here that the great majority of carotenoids dealt with in this chapter are derived from lycopene (I) (ψ,ψ-carotene), α-carotene (II, β, ε-carotene) and β-carotene (III, β, β-carotene). If a carotenoid is unsymmetrical then the unprimed numbers [see (I)] are allotted to the end of the molecule

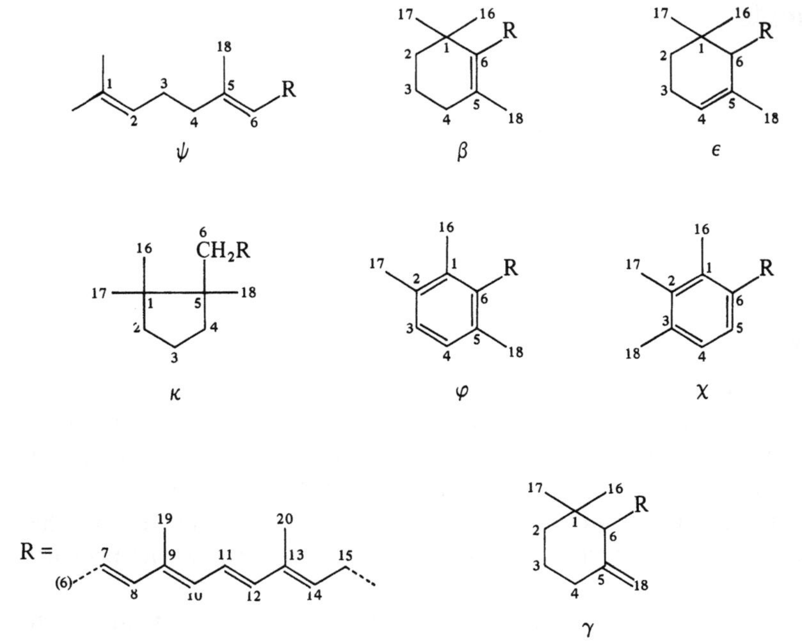

Fig. 1. Designation of carotenoid end groups

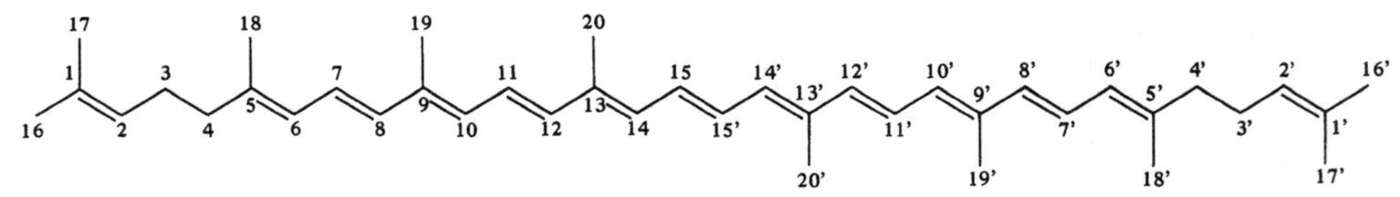

(I)

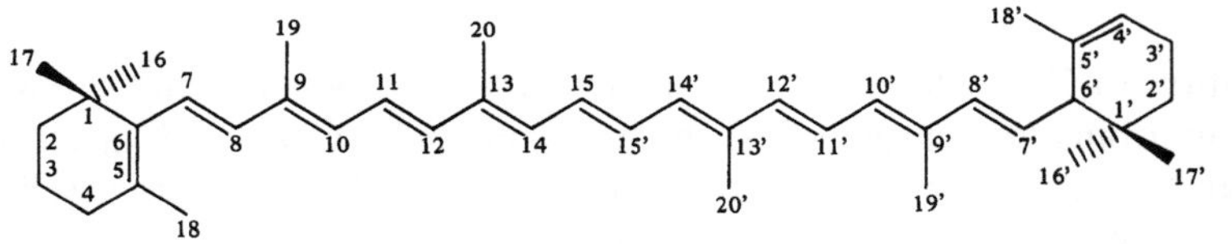

(II)

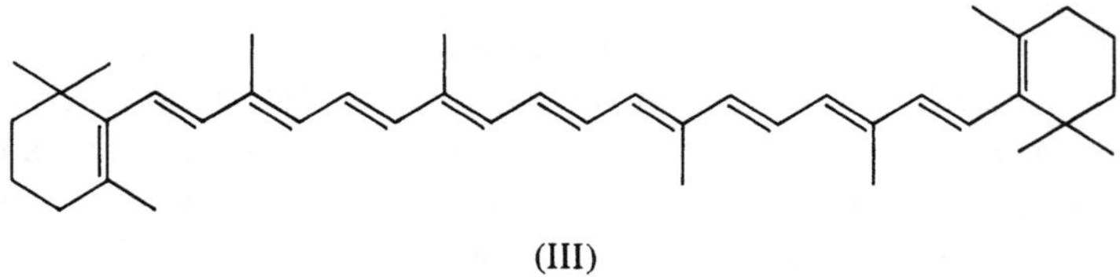

(III)

associated with the Greek letter prefix cited first in the name, as indicated in the case of α-carotene (II). Full details of the chemistry of the carotenoids, their general properties and methods of extraction and identification have recently been published (Moss and Weedon, 1976; Davies, 1976).

2 Biosynthesis

2.1 General Pathway

Carotenoids are formed from a C_5 building block, isopentenyl pyrophosphate (IPP) which is the common precursor of all terpenoids. IPP in turn is formed from the first specific terpenoid precursor, mevalonic acid (MVA), which itself arises from three molecules of acetyl-CoA (Fig. 2). Chain elongation takes place as indicated in Figure 3 until the C_{20} geranyl-geranyl pyrophosphate is produced, two molecules of which when then react tail-to-tail to form phytoene, the first C_{40} hydrocarbon in the biosynthetic sequence, via prephytoene pyrophosphate (Fig. 4). Phytoene is sequentially desaturated to yield phytofluene, ζ-carotene and its isomer 7,8,11,12-tetrahydrolycopene, neurosporene and lycopene (Fig. 5), the latter two compounds then cyclize to yield carotenes with β- and ε-rings. The exact steps in the cyclization reactions are not known, but each reaction indicated in Figure 6 has been observed in some organism. However, whatever the immediate precursors of the various cyclic carotenes, the mechanism involved in the formation of β- and ε-rings has recently been fully elucidated (Fig. 7); the β- and ε-rings are formed separately and not interconverted. A proton attack at C-2 is involved in cyclization (BRITTON et al., 1977a). The insertion of oxygen into a carotene normally occurs late in the biosynthetic pathway; for example in certain Flavobacteria zeaxanthin (3,3′-dihydroxy-β-carotene) is formed by hy-

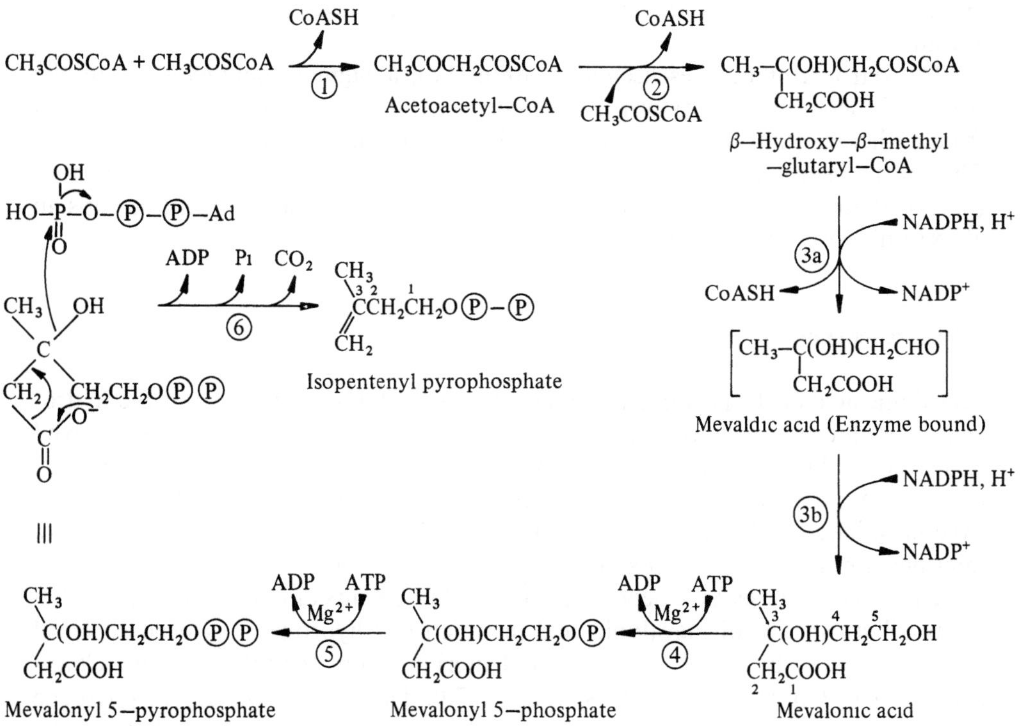

Fig. 2. Formation of isopentenyl pyrophosphate from acetyl-CoA via mevalonic acid

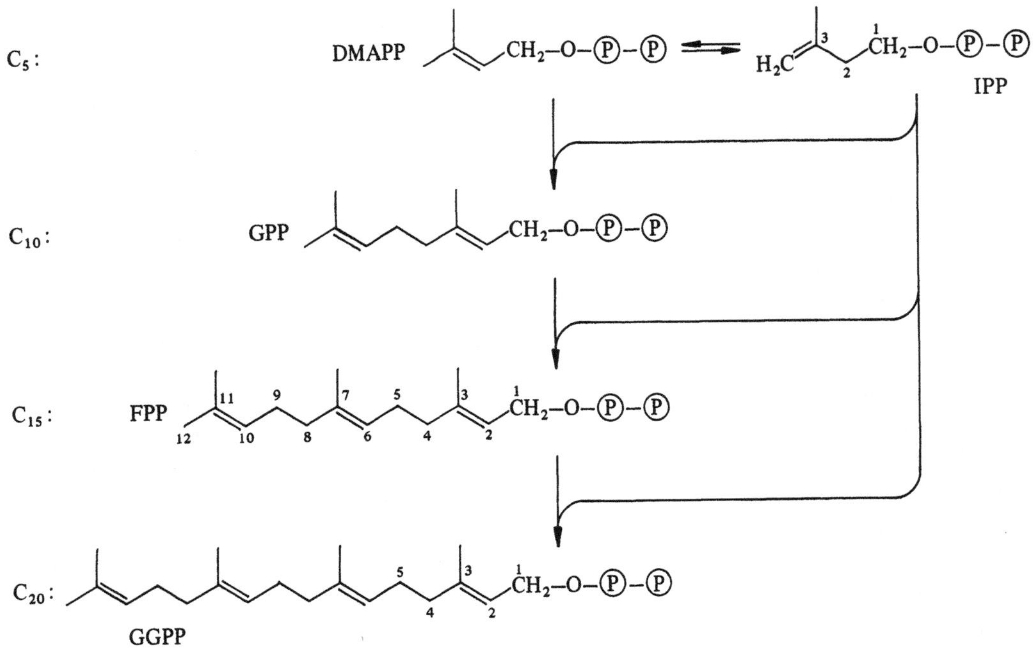

Fig. 3. Chain elongation in carotenoid biosynthesis

droxylation of β-carotene (see BRITTON 1976a, b for reviews) and in higher plants and algae antheraxanthin (5,6-epoxyzeaxanthin) is formed by epoxidation of zeaxanthin.

The retro-carotenoid eschscholtzxanthin (IV), a pigment characteristic of the petals of the californian poppy *Eschscholtzia californica*, is probably synthesized from antheraxanthin by a mechanism outlined in Figure 8 (WILLIAMS et al., 1966). The mechanisms involved in many other terminal metabolic reactions such as the formation of the allene group in neoxanthin (V) or the acetylene groups in alloxanthin (VI) are still unknown.

2.2 Stereochemistry

Mevalonic acid (VII) has three prochiral centres (C-2, C-4 and C-5) and the hydrogens at each centre are distinguished one from the other by the enzymes involved in terpenoid biosynthesis (see GOODWIN, 1973 for a review); for example H_{4S} of mevalonic acid (VII) is stereospecifically lost in converting IPP into dimethylallylpyrophosphate (DMAPP) and in condensing DMAPP with IPP to form the C_{10} intermediate geranyl pyrophosphate (Fig. 3). The developments in this aspect of terpenoid biosynthesis have depended on the achievements of CORNFORTH in synthesizing all six species of mevalonic acid in which one hydrogen has been replaced stereospecifically by either a tritium or deuterium atom (CORNFORTH and CORNFORTH, 1970; CORNFORTH and ROSS et al., 1970). The stereochemistry of phytoene desaturation to lycopene is given in Figure 9 (BRITTON, 1976a), and that for the formation of the ε-ring by loss of the pro-S hydrogen of C-2 of MVA from

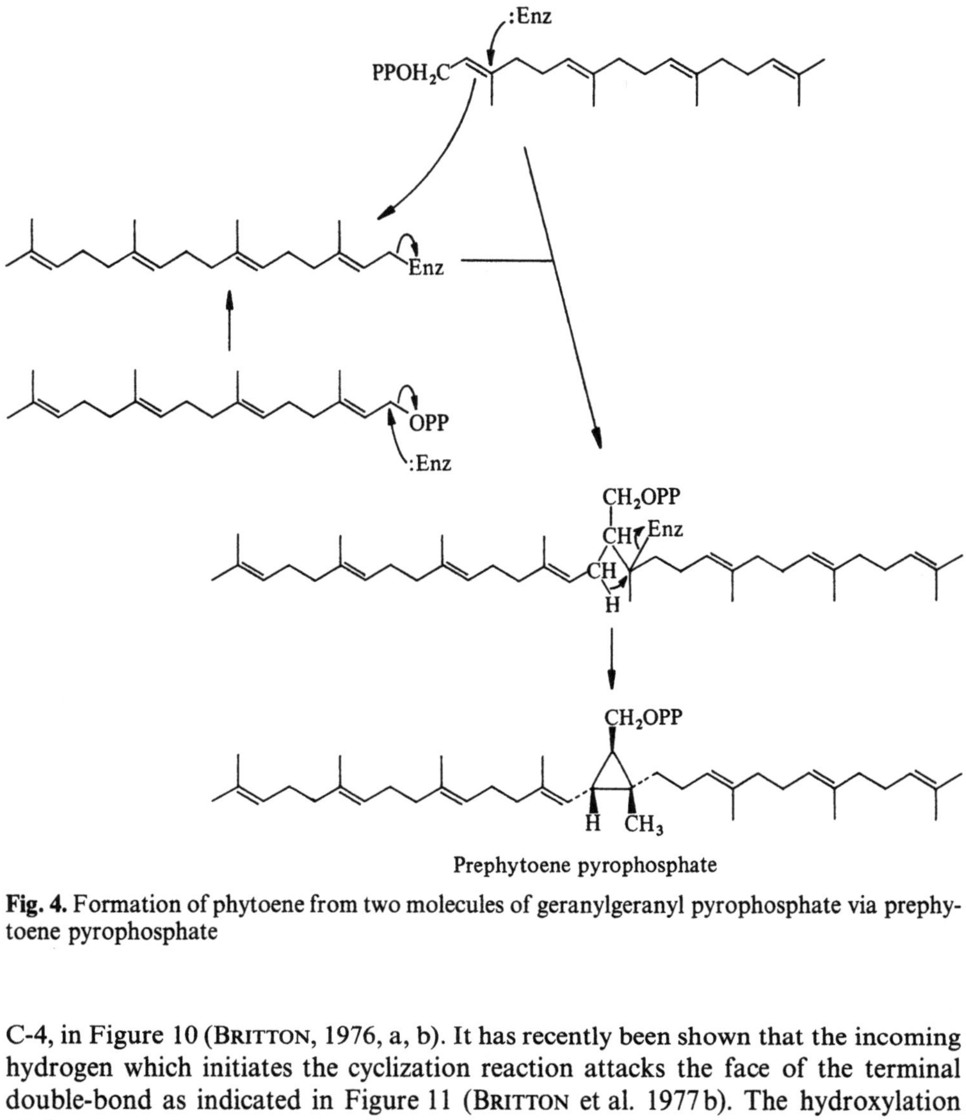

Fig. 4. Formation of phytoene from two molecules of geranylgeranyl pyrophosphate via prephytoene pyrophosphate

C-4, in Figure 10 (Britton, 1976, a, b). It has recently been shown that the incoming hydrogen which initiates the cyclization reaction attacks the face of the terminal double-bond as indicated in Figure 11 (Britton et al. 1977b). The hydroxylation of β-carotene to form zeaxanthin is also stereospecific with a direct insertion of OH without inversion of configuration (Fig. 12).

3 Higher Plants

3.1 Leaves

3.1.1 Nature and Distribution

The distribution of carotenoids in the leaves of higher plants is very similar in all species, the mixture consisting mainly of β-carotene (III), lutein (3,3′-dihydroxy α-carotene), violaxanthin (5,6,5′,6′-diepoxyzeaxanthin) and neoxanthin (V). α-Ca-

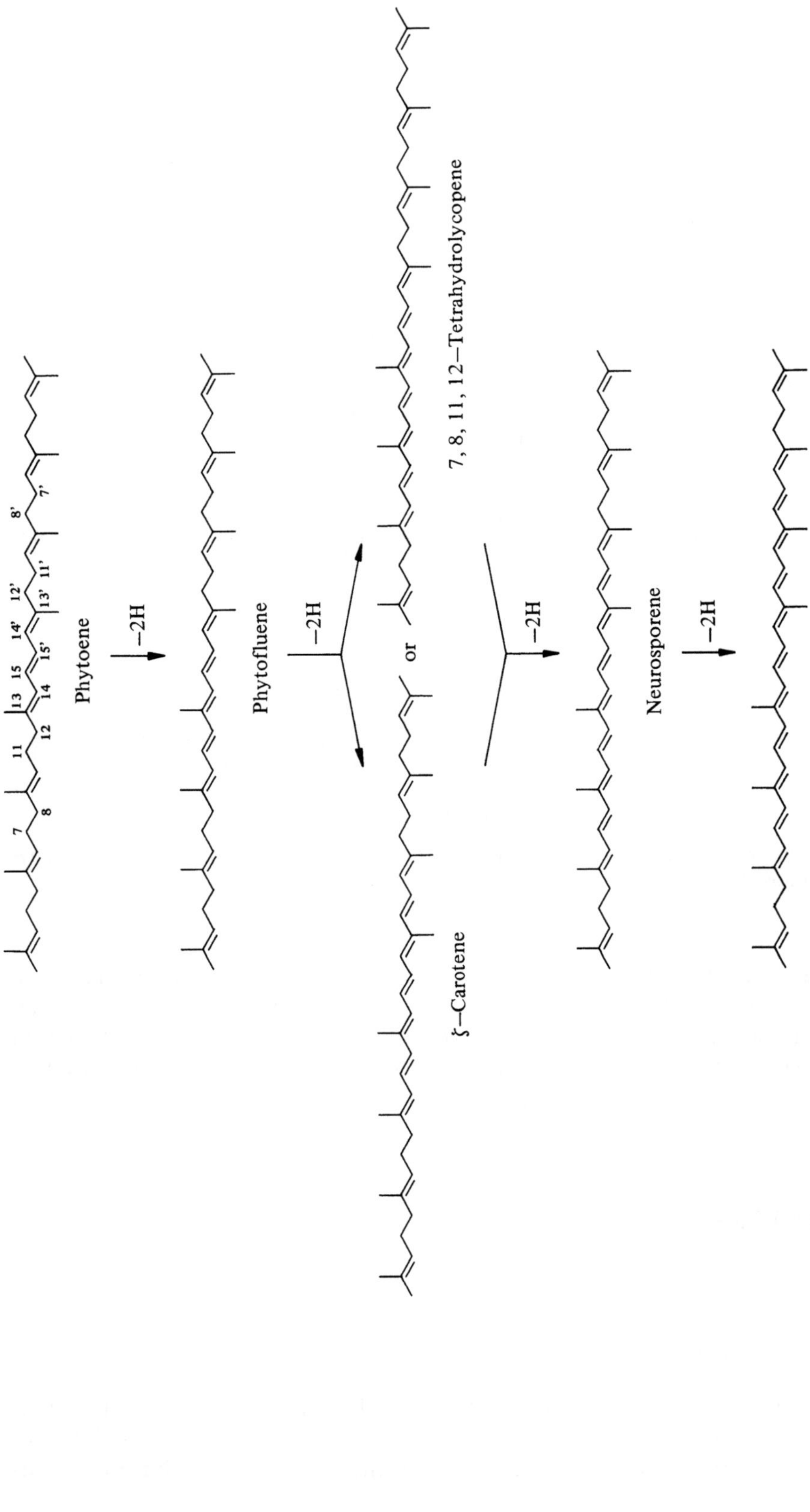

Fig. 5. Sequential desaturation of phytoene to yield lycopene

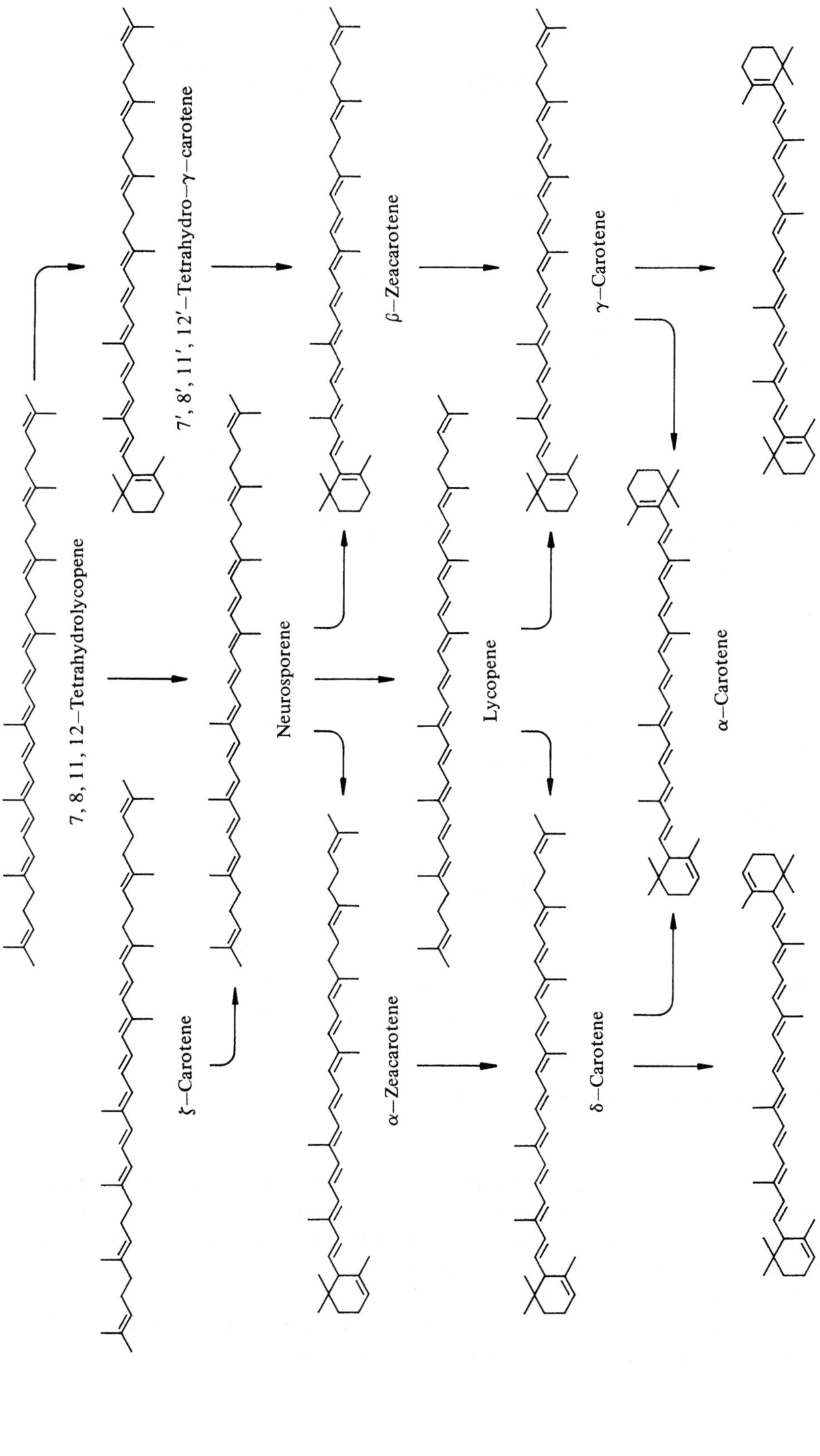

Fig. 6. Pathways in the formation of cyclic carotenes

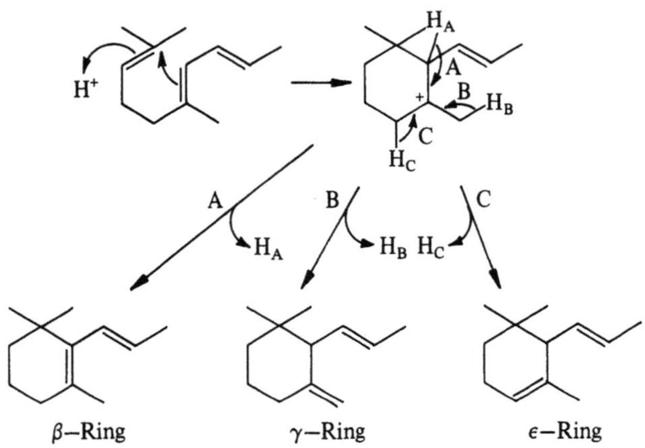

Fig. 7. The mechanism of cyclisation to form β-, γ- and ε-rings from acyclic carotenes

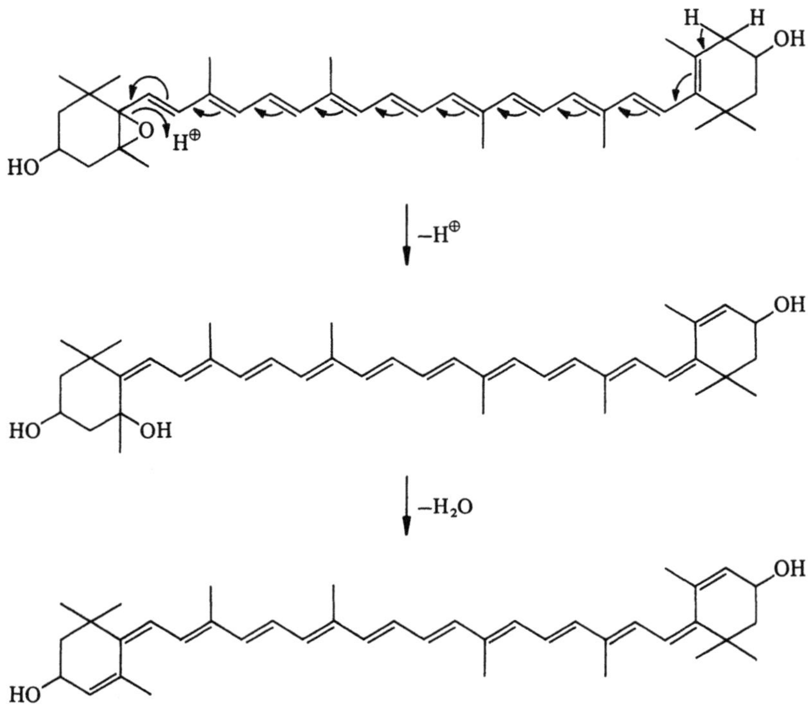

Fig. 8. Mechanism for formation of eschscholtzxanthin

rotene (II) can occur in amounts varying from 0% to 40% of the carotene fraction, and the carotene fraction can represent between 5% and 25% of the total carotenoids (WIERZCHOWSKI, 1965; GOODWIN, 1976). There are sporadic appearances of antheraxanthin (5,6-epoxyzeaxanthin), β-cryptoxanthin (3-hydroxy-β-carotene), zeaxanthin, and lutein-5,6-epoxide.

Although qualitative differences are slight amongst the most diverse plants grown under very different conditions (STRAIN, 1958, 1966), quantitative differences can

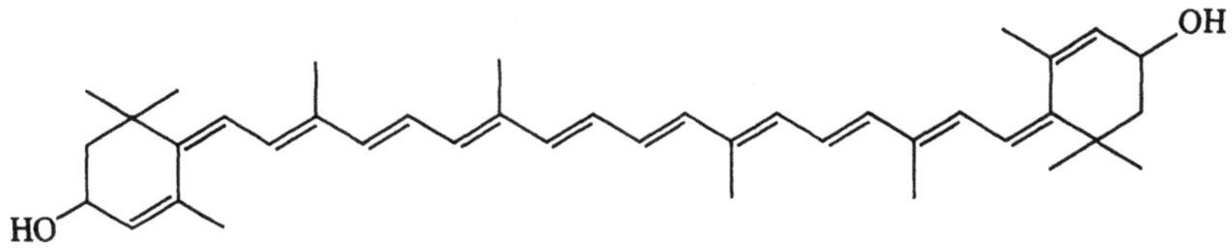

(IV)

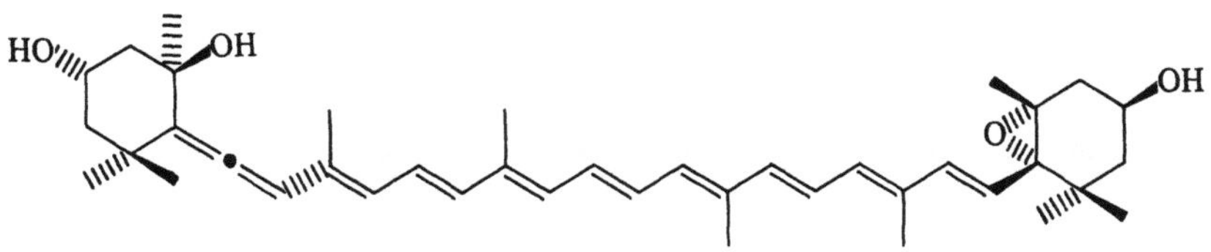

(V)

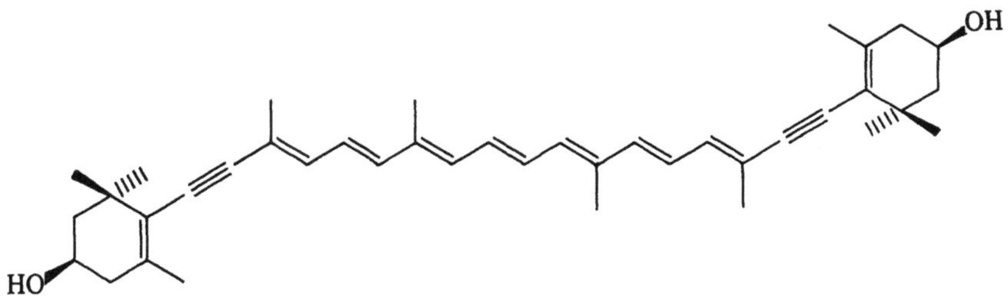

(VI)

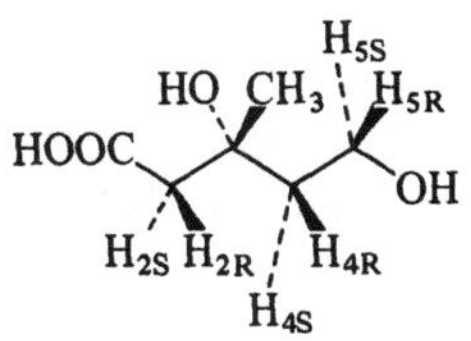

(VII)

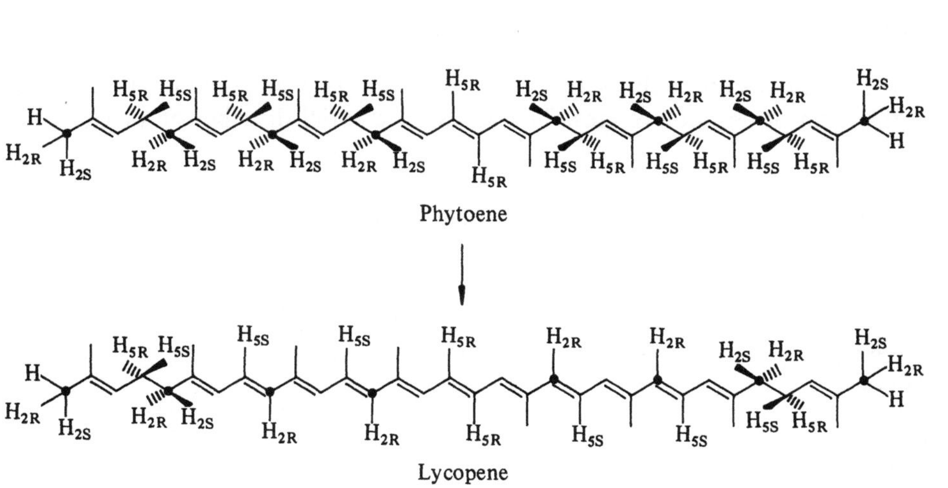

Fig. 9. The stereochemistry of the desaturation of phytoene to lycopene

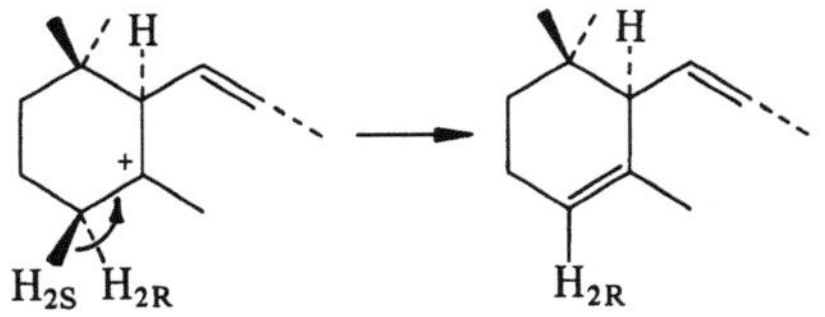

Fig. 10. Stereochemistry of the formation of ε-rings

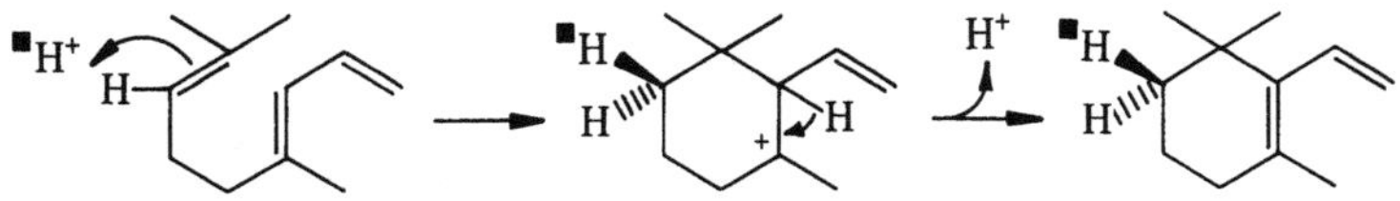

Fig. 11. Stereochemistry of protonation in the formation of β-rings

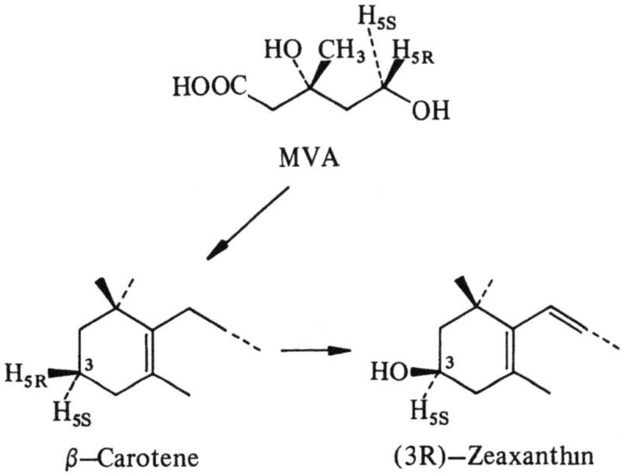

Fig. 12. Hydroxylation of β-carotene to zeaxanthin

be considerable. For example *Syngonium* (Araceae) contains much neoxanthin (V) and little violaxanthin, *Solidago* (Compositae) contains much violaxanthin and *Fremontia* (Sterculariaceae) contains as least as much zeaxanthin as lutein (STRAIN, 1958).

3.1.2 Localization

Leaf carotenoids are exclusively located in the grana of the chloroplasts, probably in the form of chromoproteins. There is a differential distribution of carotenoids

Table 1. Carotenoid distribution in sub-chloroplast particles obtained by detergent treatment (Triton X-100) of spinach chloroplasts (see GOODWIN, 1977)

Pigment	Light particle (photosystem I)	Heavy particle (photosystem II)	Intact thylakoids
β-carotene	16	6	6.3
Lutein	6	16	9.6
Violaxanthin	1	3	2.8
Neoxanthin (V)	1	2	2.5
Concentration mmol/100 mmol total chlorophyll			

between photosystem I and photosystem II of chloroplasts (see GOODWIN, 1978 for full details). β-Carotene is enriched in the light (photosystem I) fraction and the xanthophylls preponderate in the heavy (photosystem II) fraction (Table 1).

3.1.3 Mutants

Mutation which involves the deletion of the synthesis of coloured carotenoids in leaves is lethal (see Sect. 6.3), but such mutants will survive when grown heterotrophically in very dim light or in darkness. A number of such maize mutants have been described which either accumulate lycopene and ζ-carotene or δ-carotene (Fig. 6) (FALUDI-DANIÉL et al. 1967). Work on other mutants from maize and various other plants has recently been summarized (GOODWIN, 1978).

3.2 Anthers and Pollen

Not all anthers and pollen contain carotenoids, and antheraxanthin (5,6-epoxyzeaxanthin) first isolated from anthers of *Lilium tigrinum* (KARRER and JUCKER, 1950) is now known to be widespread and probably exists in small amounts in all green leaves as a component of the Xanthophyll cycle (see Sect. 6.4): furthermore not all anthers contain it. Of particular interest is that some anthers and pollen contain capsanthin (VIII) and related compounds previously considered unique to peppers (see Table 2) (GOODWIN, 1976).

3.3 Flowers

Carotenoids are not ubiquitous in flower petals, but if they do occur they accumulate in developing chromoplasts. Usually chlorophylls disappear during the development of chromoplasts, but in some cases, e.g., *Caltha palustris* and *Forsythia intermedia*, the mature flowers still contain chlorophylls.

Carotenogenic flowers can be separated into three main groups in which the major carotenoids are (1) highly oxidised xanthophylls, especially 5,8-epoxides such as flavoxanthin (5,8-epoxylutein) in yellow *Calendula officinalis*; (2) hydrocarbons such as lycopene (I) in orange *Calendula officinalis* and (3) highly species-specific pigments such as eschscholtzxanthin (IV) in *Eschscholtzia californica* (see GOODWIN, 1976, 1978 for reviews).

Xanthophylls of flower petals, unlike those in green leaves, are usually esterified. In *Helianthus annuus*, for example, not only does the dipalmitate of lutein (helenien) occur, but also esters of stearic (C_{18}), myristic (C_{14}) and lauric (C_{12}) acids (EGGER, 1968).

A unique situation is the presence of crocin, a digentiobiosyl ester of the diapocarotenoic acid crocetin (IX), in the petals of saffron (*Crocus sativus*).

3.4 Seeds

Of seeds which contain carotenoids there is no doubt that maize has been examined in most detail; indeed zeaxanthin and β-cryptoxanthin were first isolated from this source (Karrer and Jucker, 1950). More recently a new carotene β-zeacarotene (Fig. 6) (Petzold and Quackenbush, 1960) and a new xanthophyll α-cryptoxanthin (zeinoxanthin, 3-hydroxy-α-carotene) (Petzold et al. 1959) have been reported.

3.5 Fruit

3.5.1 Nature and Distribution

When the distribution of carotenoids in fruit is considered, a pattern emerges in which eight main groups can be detected. These are outlined in Table 2 but it must be emphasised that there are many examples in which the fruit could sensibly be placed in two groups. For full descriptions of carotenoids in fruit see Goodwin and Goad (1970); Goodwin (1979).

Pigment distribution between mesocarp (pulp) and exocarp (peel) varies both quantitatively and qualitatively and the exocarp is generally the region of highest concentration; this is particularly marked in citrus fruit where 50%–75% of the total pigment of an orange is in the peel.

3.5.2 Localization

As in flower petals, carotenoids generally appear in the chloroplasts as they develop from chloroplasts. This change is accompanied by profound changes in ultrastructure. For example, thylakoids (grana) disintegrate and are replaced by either large numbers of osmiophilic globules (plastoglobuli) and tubular filaments as in *Capsicum annuum*. Tomato fruit produce two types of chromoplast, as revealed by freeze etching; one is localized in the jelly part of the pericarp and contains osmiophilic globules rich in β-carotene, the other is found mainly in the outer part of the fruit and contains numerous sheets of lycopene in addition to osmiophilic globules (Harris and Spurr, 1969; Ben-Shaul and Naftali, 1969; Laval-Martin, 1974).

3.5.3 Mutants

Many mutant studies have been made on fruit but only tomatoes, which have been thoroughly examined, will be discussed here (see also Goodwin, 1978). Normal tomatoes possess the dominant allele r^+, whilst yellow tomatoes are homozygous for the recessive allele *r*; the r^+r gene controls the amount of pigment formed, the *rr* genotype synthesising only some 5% of the amounts formed by the r^+r^+ genotype. The gene hp^+hp also controls the amount of pigment synthesised, the recessive allele *hp* increasing the total amount by 100% (Tomes et al., 1958). *Apricot* tomatoes, homzygous for the recessive allele *at*, have reduced synthesis of lycopene and an unaltered synthesis of β-carotene (Jenkins and Mackinney, 1955). Red tomatoes carry the dominant allele t^+ while the recessive allele *t* expresses itself as *tangerine* tomatoes in which poly*cis* isomers, e.g. prolycopene (XI) and pro-γ-carotene, are formed at the expense of the all *trans*-isomers (Porter and Lincoln, 1950). The effects of *r* and *t* are reinforced by *at* in the formation of yellow-apricot

Table 2. Carotenoid distribution in representative groups of fruit

Group no.	Main pigments	References
I Traces		
Pyracantha rogersiana	–	GOODWIN (1956)
II Chloroplast carotenoids		
Sambucus nigra	1, 2, 3, 4	GOODWIN (1956)
III Lycopene series		
Diospyros kaki[a]	1, 2, 3, 4, 5, 6, 7, 8, 9, 10, 11, 12, 13, 14	BROSSARD and MACKINNEY (1963)
IV β-Carotene derivatives		
Mangifera indica	1, 2, 6, 7, 10, 12, 13, 14, 15	JUNGALWALA and CAMA (1963)
V Epoxides		
Crataegus pratensis	1, 12, 15	GOODWIN (1965)
VI Species-specific pigments[b]		
Capsicum annuum (red)	16, 17	(see GOODWIN and GOAD, 1970)
VII Poly-cis-derivatives		
Arum maculatum	18	GOODWIN (1956)
VIII Seco- and apocarotenoids		
Shepherdia canadensis	19[c]	KJØSEN and LIAAEN-JENSEN (1969)

Pigments

1. β-carotene (III)
2. Lutein
3. Flavoxanthin
4. Neoxanthin (V)
5. α-Carotene (II)
6. Zeaxanthin
7. Phytofluene } (see Fig. 5)
8. ζ-Carotene } (see Fig. 5)
9. Lycopene (I)
10. γ-Carotene (Fig. 6)
11. α-Cryptoxanthin (3-hydroxy-α-carotene)
12. β-Cryptoxanthin
13. Antheraxanthin
14. Violaxanthin
15. Mutatochrome (5,8-epoxy-β-carotene)
16. Capsanthin (VIII)
17. Capsorubin (X)
19. Apo-6′-lycopenoic acid (XII)

[a] Note the appearance also of type IV pattern.
[b] Only the specific pigments quoted.
[c] Reported as the methyl ester, but this may be an artefact.

and tangerine-apricot phenotypes, respectively (JENKINS and MACKINNEY, 1955). A cross in which lycopene is replaced by β-carotene (high β) is either homozygous or heterozygous for the dominant allele b^+, whilst normal fruit are homozygous for the recessive b (LINCOLN and PORTER, 1950). Expression of bb^+ is partly regulated by an independently inherited modifier $mo_B mo_B^+$; mo_B^+ in either the homozygous or heterozygous form reduces β-carotene by 50% and increases lycopene by the same amount (TOMES et al., 1954).

The delta strain of tomato which accumulates large amounts of δ-carotene (XIII) at the expense of lycopene carries the allele del^+ which is either incompletely

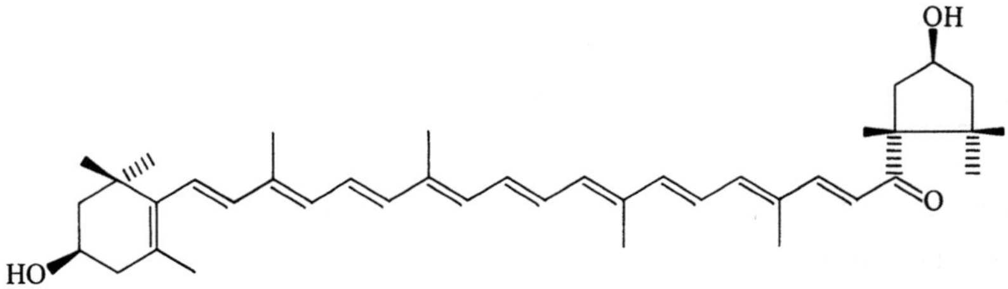

(VIII)

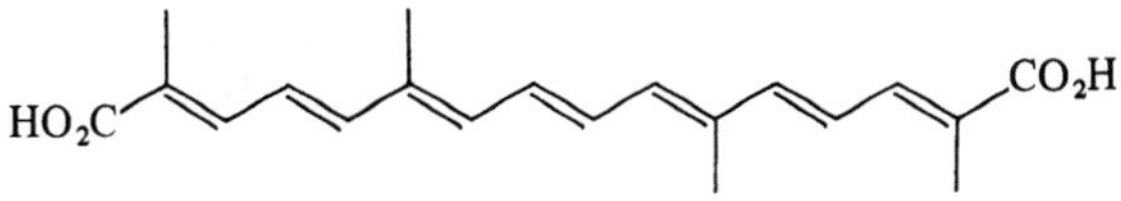

(IX)

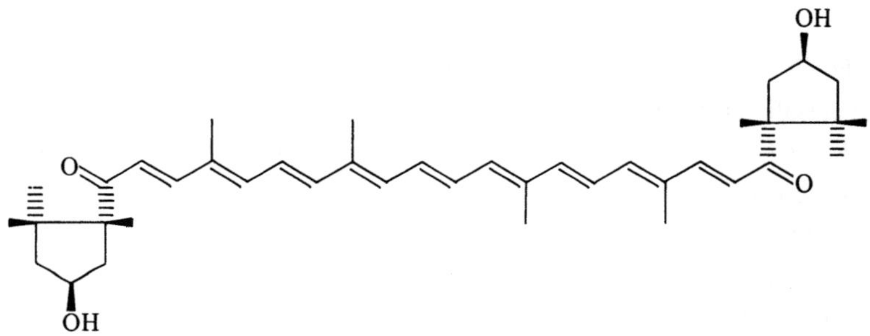

(X)

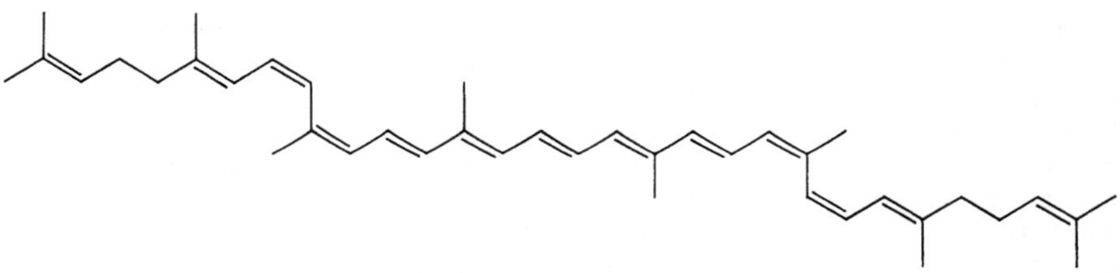

(XI)

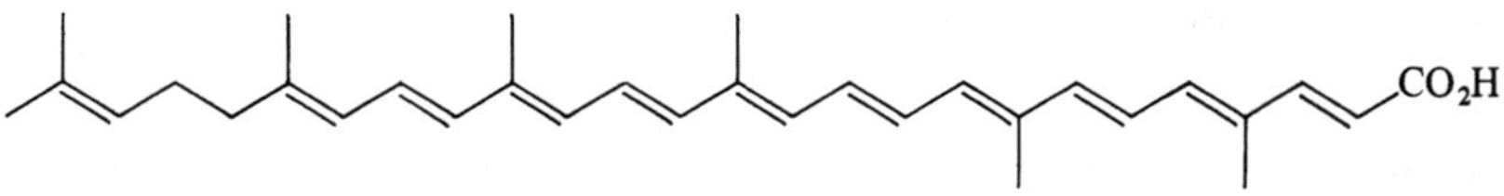

(XII)

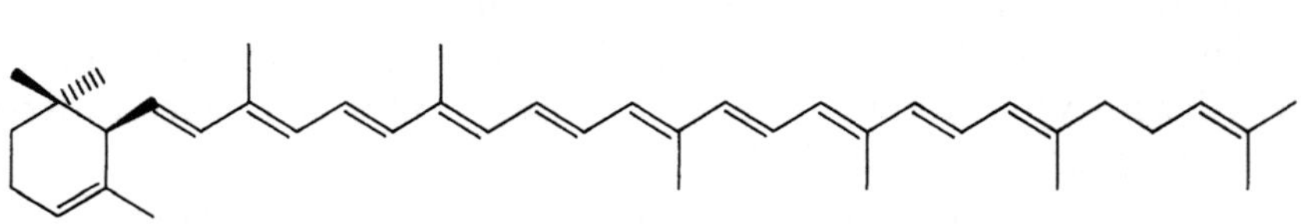

(XIII)

dominant or affected by a modifier gene (KARGL et al., 1960). The ghost phenotype appears spontaneously in lines carrying red fruit. Seedlings for the recessive allele *gh* germinate with green cotyledons which rapidly lose chlorophyll as they grow. With normal scions grafted onto ghost stock the plants produce milky white unripe and yellowish ripe fruit. The yellow colour is not due to carotenoid, for the fruit produce only the colourless polyene phytoene (Fig. 5). The $gh^{+}gh$ gene

Table 3. Carotenoid distribution in various tomato mutants (GOODWIN, 1976)

Genotypes	Pigments and colourless polyenes	Total concentration (µg/g fr. wt)	
		Total polyenes	Pigments only
Red (Normal)	1, 2, 3, 4, 5	87	50
High pigment (*hp*)	1, 2, 3, 4, 5	88	69
Tangerine ($r^{+}r^{+}tt$)	2, 3, 4, 5, 6	158	90
Yellow (*rr*)	1, 2, 4, 5	4	2
Yellow-Tangerine (*rrtt*)	1, 2, 4, 5, 7	24	14
Apricot (*atat*)	1, 2, 4, 5	13	11
Yellow-Apricot (*ttrratat*)	2, 3, 5 (trace)	2	1
Tangerine-Apricot (*ttatat*)	1, 2, 4, 5, 6, 7	29	16
Yellow-Tangerine-Apricot (*rrttatat*)	4, 5	10	0
Ghost (*ghgh*)	1, 4, 5	295	1
Intermediate β (*BB*)	1, 2, 3, 4, 5	50	46
High β ($BBB^{mo+}B^{mo+}$)	1, 2, 3, 4, 5	80	70
Delta (*deldel*)	1, 2, 3, 4, 5, 7, 8, 9, 10	84	68
High δ (*deldelhphp*)	1, 2, 3, 4, 5, 8, 9, 10	55	46

Key:

1. Lycopene (I)
2. β-Carotene (III)
3. γ-Carotene (Fig. 6)
4. Phytoene } (Fig. 5)
5. Phytofluene } (Fig. 5)
6. Prolycopene (poly-cis-lycopene) (XI)
7. ζ-Carotene (Fig. 5)
8. α-Carotene (II)
9. δ-Carotene (XIII)
10. Neurosporene (Fig. 5)

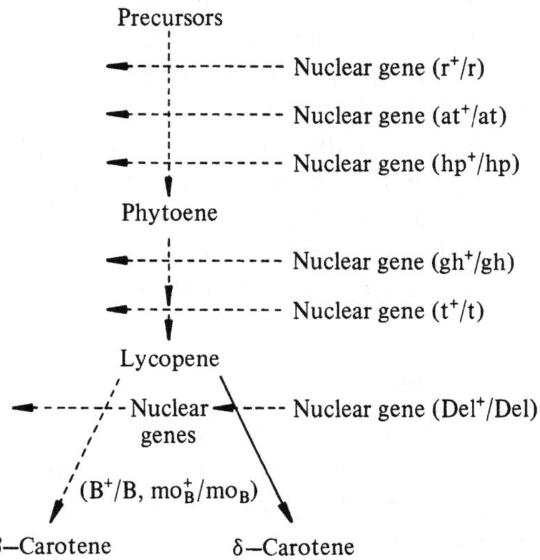

Fig. 13. The nuclear control of carotenogenesis in the tomato

also controls carotenoid synthesis in the leaves which accumulate only phytoene (MACKINNEY et al. 1956). The overall position in tomato crosses is summarised in Table 3.

The overall nuclear control of carotenoid synthesis in the tomato is summarised in Figure 13.

3.6 Roots

3.6.1 Nature and Distribution

Carrot roots are the classic source of "carotene" which was first isolated crystalline in 1831 by WACKENRODER, but was only separated into its isomers in the 1930's following the development of column chromatography. Carotenes (mainly β-carotene) represent over 95% of the total carotenoids in commercial carrots whilst in wild carrots xanthophylls represent 95% of the small amount of carotenoids present. Red varieties of carrots have lycopene and not β-carotene as their major pigment; it is accompanied by quite large amounts of other carotenes.

β-Carotene is also the main pigment in sweet potatoes (*Ipomea batatas edulis*) and only traces of xanthophylls are present. Roots of swede (*Brassica rutabaga*) produce small amounts of poly cis carotenes whilst the 10-apocarotenoid azafrin (XIV) has long been known to be present in the roots of *Escobedia* spp. (KARRER and JUCKER, 1950; GOODWIN, 1976, 1979).

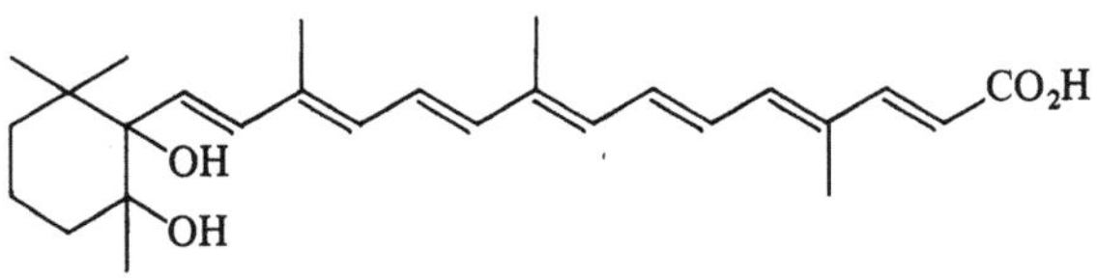

(XIV)

3.6.2 Localization

In carrots the pigments are located in chromoplasts, whilst the small amounts in potatoes occur in all sub-cellular fractions but are concentrated in membranes and mitochondria (COSTES et al., 1976).

In carrots the first particles to be observed are colourless vesicles (proplastids) about 1–2 μ diameter which begin to accumulate carotenoids only after they have reached 3–5 μ diameter when they appear as yellow vesicles. About 16–28 days after germination orange-red carotene crystals appear but in older cells, more than 60 days after germination, two types of plastid are seen, those which contain a starch-carotenoid complex and those which contain carotenoid crystals which may be as long as 40 μ (BEN-SHAUL and KLEIN, 1965).

3.6.3 Genetic Studies

Early inheritance studies in carrots concentrated on colour rather than carotenoid content. Only fairly recently have detailed studies on the inheritance of carotenoids in carrots been reported. Five colour phenotypes were examined: white, yellow, orange tinge, intermediate orange, and orange. It was found that white was dominant to all others and that the gene (*Y*) for the white phenotype almost completely inhibited carotene synthesis. A monogenic ratio was obtained for white and yellow crosses. Results with yellow and orange crosses revealed segregants for both dominance and non-dominance and the wide variation in carotene concentration within the orange class suggested the presence of modifier genes (IMAN and GABELMAN, 1968).

3.7 Formation and Metabolism

3.7.1 Developing Seedlings

Ungerminated seeds frequently contain traces of xanthophylls with little or no carotenes, but an important exception is maize (GOODWIN, 1978). New carotenoid synthesis generally begins within 3–5 days of germination and continues rapidly during the early period of active growth. Although the carotene concentration begins to fall quite early in the development of the plant, the amount continues to increase and it is only at the later stages of maturation that it begins to drop. Etiolated seedlings contain very little carotenoid but on illumination there is a rapid synthesis which parallels the conversion of etioplasts into chloroplasts. The signal triggered by light does not appear to be specific for carotenoid synthesis but for the formation of chloroplasts, in which carotenoid production is an essential part of an integrated process. The cotyledons of pine seedlings are probably the only regions of higher plants which can synthesize functional chloroplasts, and thus carotenoids, in the dark.

Whatever the triggering mechanism of light in illuminated, etiolated seedlings, a problem which has not yet been solved is whether or not the chloroplast can synthesize carotenoids without any outside help. There is no doubt that chloroplasts can synthesize enzymes concerned with carotenoid biosynthesis. There also seems little doubt that chloroplasts can make carotenoids from the first specific isoprenoid precursor mevalonic acid so the question narrows down to whether mevalonic acid can be formed from acetyl-CoA. The only evidence for this is the demonstration that the key enzyme HMG-CoA reductase is present in chloroplasts (BROOKER and RUSSELL, 1975). The most important question which then emerges is whether or not the chloroplast can make acetyl-CoA. This key intermediate is usually synthesized either from pyruvate by the action of pyruvate dehydrogenase, the pyruvate being formed by glycolysis or directly by the β-oxidation of fatty acids. Neither glycolysis nor β-oxidation has yet been demonstrated in chloroplasts and pyruvate dehydrogenase, even if it is located in chloroplasts, seems to be at a much lower level of activity than required for the massive synthesis of acetyl-CoA needed during chloroplast development. Thus one is led to the view that acetyl-CoA is formed in the chloroplast by an unusual route, or is formed in leaf mitochondria and transported to the chloroplast. A problem is that acetyl-CoA cannot penetrate the chloroplast membrane and thus a number of ways this can be overcome are indicated in Figure 14 (see GOODWIN, 1977, for fuller discussion). There is also the possibility that MVA formation in the chloroplast by-passes the acetyl-CoA pathway and that it formed from the carbon skeleton of an amino acid. The amino acid which could be involved is leucine (Fig. 14) which is certainly very carotenogenic in *Phycomyces blakesleeanus* (GOODWIN, 1979). This would remove

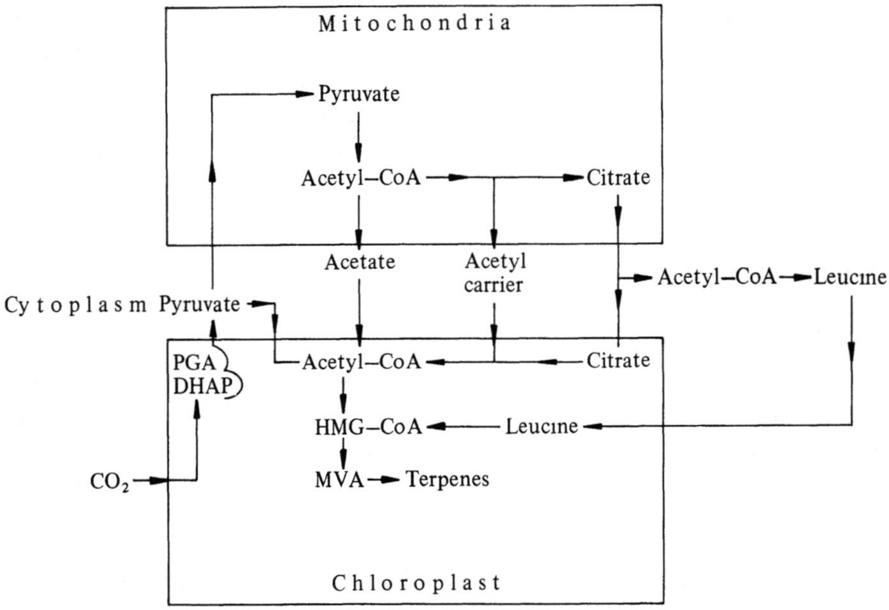

Fig. 14. Possible sources of precursors for chloroplastic synthesis of carotenoids

the problem of penetration into chloroplasts, which are freely permeable to amino acids, but it must be emphasized that it does not eliminate the basic requirement for acetyl-CoA. The carbon skeleton of leucine arises from three molecules of acetyl-CoA (GOODWIN, 1964). Synthesis of leucine occurs in the cytoplasm and the acetyl-CoA necessary for this could come from mitochondrial citrate (Fig. 14).

3.7.2 Senescence

In early classical experiments TSWETT found that the major proportions of carotenoids in senescing (autumn) leaves of deciduous trees behaved as carotenes in phase tests but were separable from carotenes by chromatography. He coined the name "autumn carotenes" for these pigments, but it is now known that these are esterified xanthophylls, the xanthophylls surviving longer than carotenes during the degenerative period. During senescence the internal structure of the chloroplast disappears and the organelles fill with plastoglobuli and eventually disintegrate. The surviving xanthophylls are then esterified in the cell-sap. The pattern of senescence in tobacco plants is different from that in deciduous trees, in that β-carotene is retained relative to xanthophylls and no esterified xanthophylls are formed (GOODWIN, 1979).

3.7.3 Flowers

There is a rapid synthesis of carotenoids as flowers develop and the pigments are localized in the chromoplasts where they can reach very high concentrations. For example in the bright red edges of the corona of *Narcissus* sp. the astounding value for β-carotene of 16.5% dry weight is reached with a rate of synthesis of 70 μg/mg dry wt/day which is more than 10,000 times faster than in carrot roots (BOOTH, 1957).

3.7.4 Fruit

In many fruits, ripening is accompanied by the conversion of chloroplasts into chromoplasts with the attendant disappearance of chlorophyll and appearance of large amounts of new carotenoids. Frequently one pigment dominates in ripe fruit, for example lycopene (I) in tomatoes and capsanthin (VI) in red peppers. This is clearly illustrated by comparing the carotenoids in unripe and ripe peppers (Table 4). The distribution in the green unripe fruit is very similar to that in green leaves, but on ripening there is a massive synthesis of the capsanthin group of pigments. Many fruits are autonomous from the point of view of carotenoid synthesis, for example a stored tomato can synthesise up to 1.2 mg lycopene per day (see GOODWIN, 1979 for details).

Table 4. The major carotenoids of leaves, unripe and ripe fruit of *Capsicum annuum* v. *lycopersiciforme rubrum* (CHOLNOKY et al., 1956)

Pigment	Leaves	Amount mg/100 g fr. wt.	
		Unripe fruit	Ripe fruit
β-Carotene (III)	7.9	0.095	2.35
β-Cryptoxanthin	0.45	0.027	1.10
Lutein	13.99	0.276	–
Zeaxanthin	–	–	1.75
Antheraxanthin	1.14	0.031	0.99
Violaxanthin	8.27	0.042	0.70
Neoxanthin (V)	5.66	0.058	–
Capsanthin (VIII)	–	–	9.60
Capsorubin (X)	–	–	1.46

3.7.5 Roots

Carotene concentration in carrot roots increases with growth and becomes maximal about 100 days after sowing. During maturation the ratio of pigments in the phloem and xylem changes for example from 1.5:1 to 1.2:1 in some strains. Both carrots and sweet potatoes can synthesise carotenoids in storage, the rate and extent of synthesis depending on variety (GOODWIN, 1979).

3.8 Contribution to Flower Colour

Many lemon and yellow flowers contain large amounts of xanthophylls, particularly epoxides, whilst deeper orange petals are often characterised by increased amounts of β-carotene as in the Red Emperor variety of *Mimulus cupreus* (GOODWIN and THOMAS, 1964) and the orange fringes of *Narcissus* spp. (BOOTH, 1957), or lycopene as in the very deep orange varieties of *Calendula officinalis* (GOODWIN, 1954). In some cases, as in yellow and orange roses, the deepening colour is caused by quantitative rather than qualitative changes in carotenoids (VALADON and MUMMERY, 1968).

3.9 Taxonomic Significance of Fruit Carotenoids

A full assessment of the chemotaxonomic significance of carotenoids in fruit is hampered by lack of data. It appeared that rubixanthin in relatively large amounts was characteristic of *Rosa* spp. (GOODWIN, 1956) but some cultivars contain little if any of this pigment (VALADON and MUMMERY, 1968). However, if a carotenoid extract contains a substantial amount of rubixanthin (3-hydroxy-γ-carotene) then it can be considered with confidence to have come from hips of a *Rosa* spp. This generalisation cannot be extended to all the Rosaceae family because other genera such as *Cotoneaster, Crataegus, Pyracantha,* and *Sorbus* do not synthesise rubixanthin (GOODWIN, 1956). Wide variations can also be seen amongst the Caprifoliaceae and even amongst different species of the same genus, for example in *Pyracantha* spp., *P. rogersiana* and *P. coccinea* synthesise "plastid" carotenoids, *P. flava* much β-carotene and *P. angustifolia* poly-cis-carotenes (GOODWIN, 1956).

4 Mosses, Liverworts and Spore-Bearing Vascular Plants

Rather little is known about carotenoids in these plants, but what is known is that they produce the same rather simple mixture found in the chloroplasts of higher plants. One interesting observation stands out in that the male gametophytes of *Equisetum* spp. contain rhodoxanthin (XV) (CYRONAK et al., 1977).

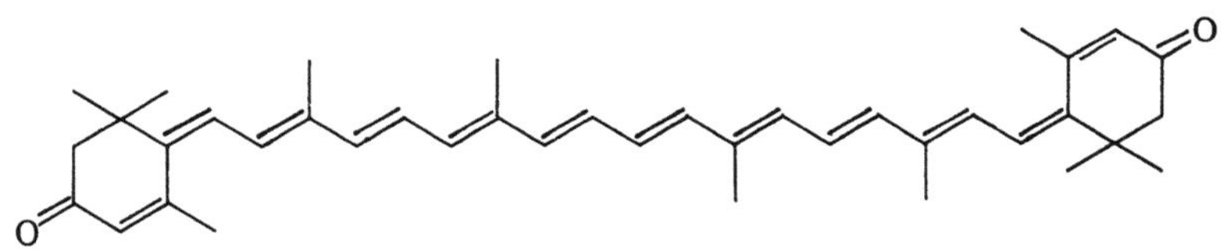

(XV)

5 Algae[1]

5.1 Distribution

The algae differ markedly from higher plants in that the different classes produce different carotenoids in their chloroplasts. Much recent work using definitive methods has clarified a complex situation and revealed the existence of a number of new pigments with unique structural features. It is not possible to detail these here, but the general pattern is summarised in Table 5. Full details have been reported elsewhere (GOODWIN, 1971, 1974, 1979).

1 As to carotenoids of fungi and bacteria, the reader is referred to GOODWIN, in press 1980

Table 5. Distribution of major carotenoids in algae (adapted from GOODWIN, 1976)

Pigment	Chlorophyta[a]	Rhodophyta	Xanthophyceae[c]	Chrysophyceae	Haptophyceae	Eustigmatophyceae	Bacillariophyceae[c]	Phaeophyta	Pyrrophyta	Euglenophyta	Cyanophyta	Cryptophyta	Chloromonadophyta
α-Carotene (II)	+	+										+	
β-Carotene (III)	+	+	+	+	+	+[e]	+	+	+	+	+	+	+
Echinenone (4-Keto-β-carotene)										+[g]	+[g]		
Lutein (3,3′-dihydroxy-α-carotene)	+	+											
Zeaxanthin (3,3′-dihydroxy-β-carotene)	+	+	+							+	+		
Antheraxanthin (5,6-epoxyzeaxanthin)		+[b]											
Violaxanthin (5,6,5′,6′-diepoxyzeaxanthin)	+	+[b]						+					
Neoxanthin (V)	+		+							+			
Fucoxanthin (XVIII)				+	+[d]		+	+	+[f]				
Diatoxanthin (XXII)			+	+			+			+			
Peridinin (XXI)									+				
Alloxanthin (VI)												+[h]	
Myxoxanthophyll (XXIII)											+		
Oscillaxanthin (XXIV)											+		

[a] Siphonaxanthin and siphonein are widely present in the Bryopsidophyceae. [b] Rarely encountered. [c] Also vaucheriaxanthin (XX) and heteroxanthin (XIX). [d] 19-Hexanoyloxyfucoxanthin. [e] Xanthophyll distribution rather varied. [f] Only rarely encountered; peridinin is usually the main pigment. [g] Other keto carotenoids also encountered. [h] Other acetylenic carotenoids also encountered.

5.2 Localization

As in higher plants, algal carotenoids are usually confined to the chloroplasts but extra-plastidic pigments are found in two situations. (1) Many Chlorophyta under adverse nutritional conditions, usually nitrogen deficiency, accumulate large amounts of carotenoids (secondary carotenoids) in intracytoplasmic deposits or in lipid vacuoles. The pigments which accumulate are generally β-carotene and/or its ketoderivatives such as echinenone (4-keto), canthaxanthin (4,4′-diketo) and astaxanthin (3,3′-dihydroxy-4,4′-diketo): (2) In the Phaeophyta, *Fucus vesiculosus*, *F. serratus* and *Ascophyllum nodosum*, there is differential distribution in male and female gametes. The bright orange-yellow of the male gametes is due almost entirely to β-carotene, whilst the olive-green of the ova is produced by a mixture of fucoxanthin and chlorophyll.

5.3 Carotenoids and Algal Taxonomy and Evolution

There is little doubt that the basic carotenogenic pathway from acetyl-CoA to lycopene (outlined in Sect. 2) is the pathway leading to the cyclic carotenoids in all algae. It is the variation of the metabolism of lycopene from phylum to phylum which can be of some interest to students of algal taxonomy and algal evolution. The changes relate the following enzymic activities: (1) β-cyclase which forms β-rings (2) ε-cyclase which forms ε-rings (3) hydroxylases which insert OH groups at C-3 (4) epoxidases which inserts O across C-3 and C-4 (5) 5,6-epoxide isomerases which form allenes such as neoxanthin (V). In addition important reactions are those involving hydroxylation of in-chain methyl groups as in loroxanthin (19-hydroxylutein), formation of acetylene linkages as in diadinoxanthin (XVI) and insertion of in-chain keto groups as well as in-chain oxidation of methyl groups as in siphonaxanthin (XVII).

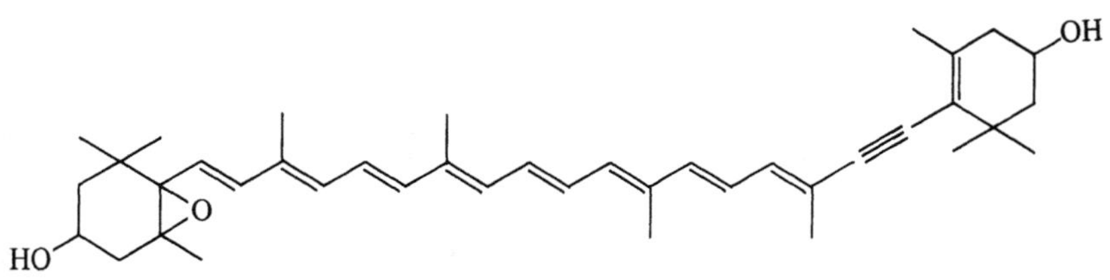

(XVI)

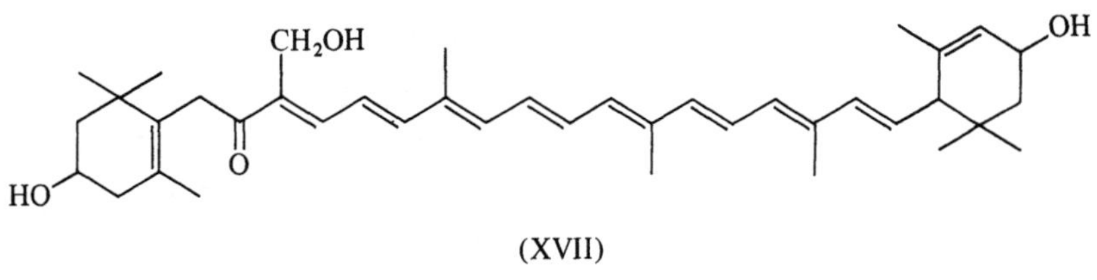

(XVII)

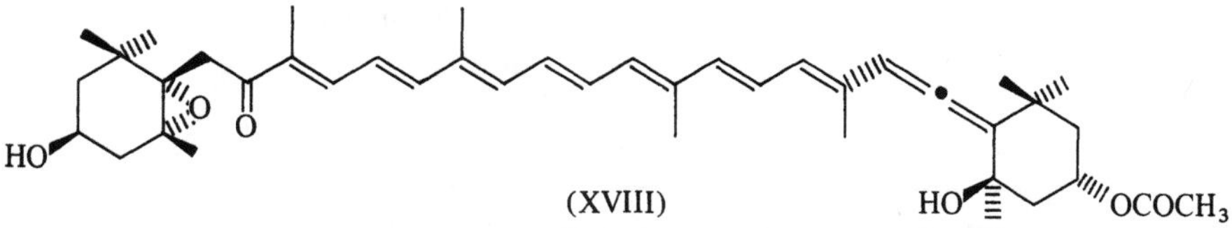

(XVIII)

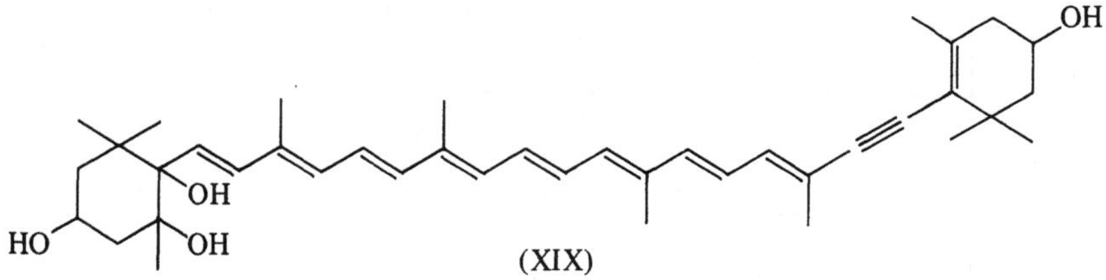

(XIX)

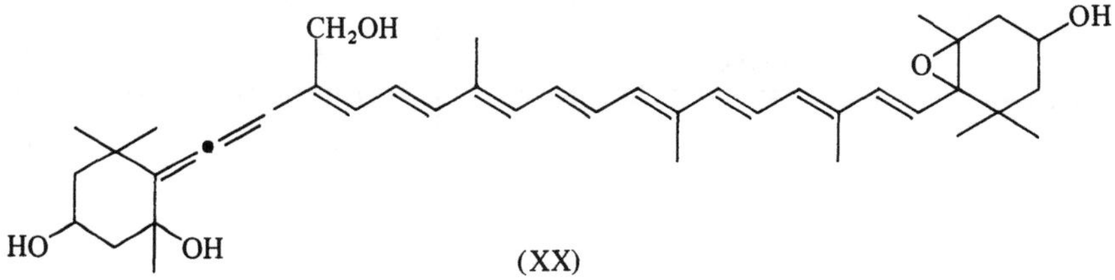

(XX)

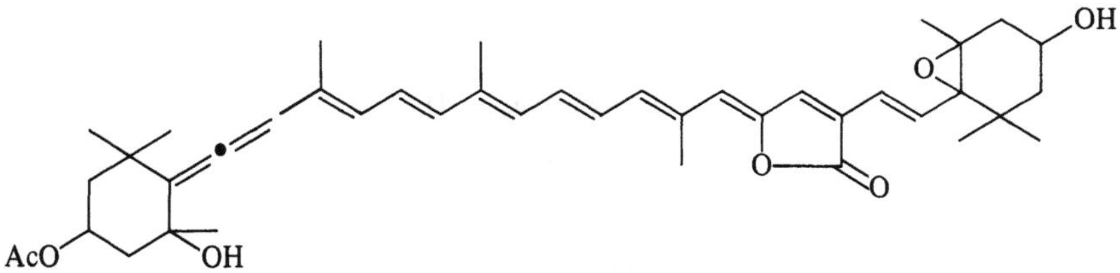

(XXI)

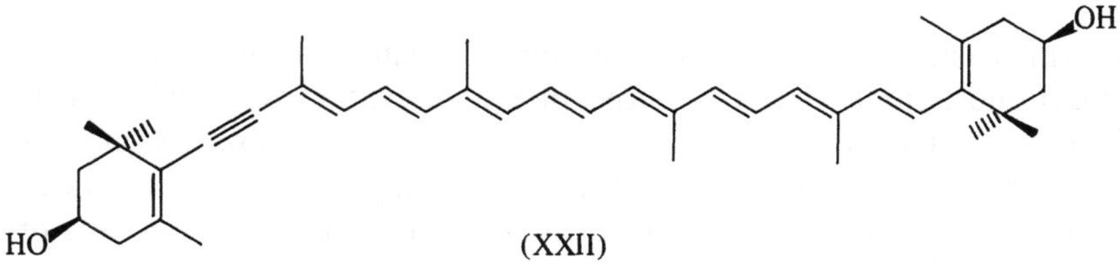

(XXII)

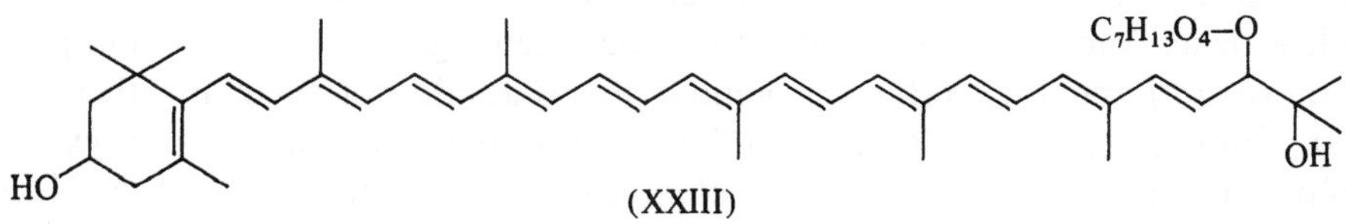

(XXIII)

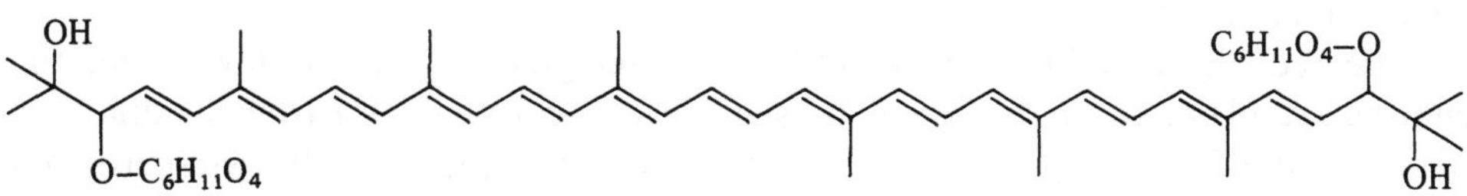

(XXIV)

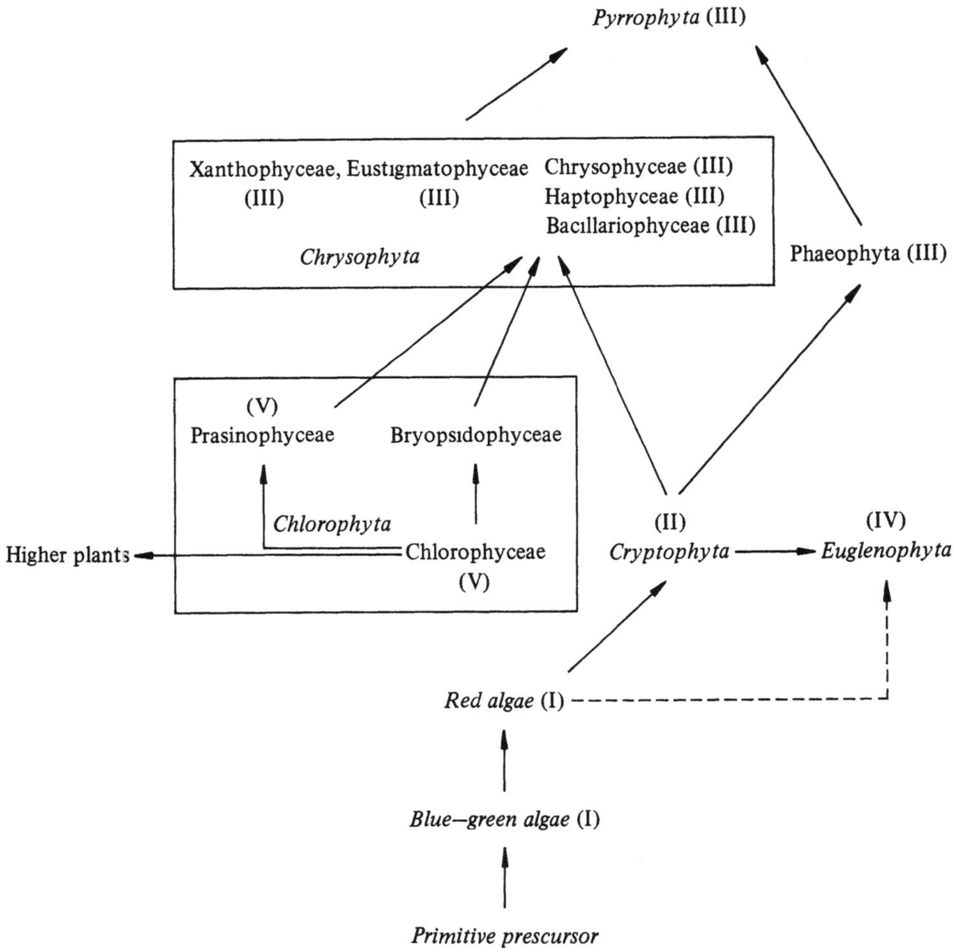

Fig. 15

With this basic information an evolutionary scheme for algae can be proposed (Fig. 15). If one assumes that a primitive precursor synthesizes acyclic carotenoids, then the first phylum to emerge is the blue-green algae. They possess only β-cyclase activity and thus produce only β-carotene (III) (2β-rings) and γ-carotene (see Fig. 6) (one β-ring and one acyclic end) and their derivatives. However one major β-carotene derivative in this phylum is echinenone in which oxygen has been inserted at C-4; this is a rather rare reaction and echinenone does not appear in chloroplasts of other algae. Another major pigment in blue-green algae is the carotenoid glycoside myxoxanthophyll (XXIII). Carotenoid glycosides do not occur in other algae but are found in bacteria (GOODWIN, 1979) and the structural features of myxoxanthophyll are similar to those of the carotenoids of photosynthetic bacteria (GOODWIN, 1979); this supports the current view that blue-green algae are prokaryotes and, in fact, are blue-green bacteria.

The red algae could have evolved from blue-green algae by developing the enzyme ε-cyclase, to enable them to synthesise α-carotene (II) (one β- and one ε-ring), and a 3-hydroxylase involved in forming lutein (3,3′-dihydroxy-α-carotene) and zeaxanthin (3,3′-dihydroxy β-carotene). At the same time they lost the ability to insert oxygen at C-4 and to glycosylate acyclic carotenoids; alternatively the blue-green algae might have evolved the enzymes concerned in these reactions after the red algae had branched off. The main development in the Chlorophyta is the appearance of 5,6-epoxidases and 5,6-epoxide isomerases which account for the formation of such compounds as antheraxanthin (5,6-epoxide of zeaxanthin) and neoxanthin (allene) (V) which are constant components of all the green algae examined. However, further developments within this phylum have occurred in the classes Bryopsidophyceae and Prasinophyceae. These developments involve additional oxidative steps; for example the Siphonales synthesise siphonaxanthin (XVII) in which the C-18 in-chain methyl group is oxidised to hydroxymethyl and a conjugated keto group appears at C-8. The next important evolutionary step would seem to be the appearance of acetylene bonds in carotenoids. These pigments, which could be formed by direct desaturation of compounds such as lutein and zeaxanthin, are characteristic of the Cryptophyta which can be considered as evolving directly from red algae. The Euglenophyta could, in turn, have developed from the Cryptophyta in a manner parallel to that by which the Chlorophyta developed from the Rhodophyta. A possibility is that the Euglenophyta evolved directly from blue-green algae without going through the red algae, because they cannot synthesise ε-rings and this restriction is also rather common in the Cryptophyta. On the other hand they could have evolved from those red algae which themselves never achieved the synthesis of ε-rings. Various suggestions that Euglenophyta have evolved from Chlorophyta have to explain the deletion of the ability to synthesise ε-rings.

There is only one clear picture relating the five classes in the phylum Chrysophyta. All except the Chrysophyceae synthesise acetylenic carotenoids and the appearance of these pigments suggests that the Chrysophyta arose from the Cryptophyta rather than from the Chlorophyta.

Other evolutionary developments have produced in the Xanthophyceae heteroxanthin (XIX), which involves in-chain methyl oxidation characteristic of loroxanthin, and formation of a conjugated keto group in fucoxanthin (XVIII), in the Chrysophyceae, Bacillarophyceae, and Haptophyceae, characteristic of siphonaxanthin. Vaucheriaxanthin (XX), which accompanies heteroxanthin, could have been formed by a loroxanthin-type oxidation of neoxanthin rather than of lutein. All these observations suggest the evolution of the Chrysophyta from the Chlorophyta via the Bryopsidophyceae. No ε-rings are observed in any Cryptophyta, which suggests that they have not evolved from the Chlorophyta. The Phaeophyta and the Pyrrophyta, in which the major pigments are fucoxanthin (XVIII) and peridinin (XXI) respectively, also do not make ε-rings. As peridinin is an oxidation product of fucoxanthin the Pyrrophyta might be considered as evolving from fucoxanthin-producing ancestors such as the Chrysophyta or Phaeophyta. The few Chloromonadophyta which have been examined bear strong pigment similarities to the Xanthophyceae. Thus a study of carotenoids in algae can make some contribution to the problem of algal evolution but they should be balanced by consideration of other parameters which have been discussed in detail by Klein and Cronquist (1967). However, it is interesting that the correlation of chloroplast ultrastructure, as exemplified by thylakoid-type, with carotenoid pigment distribution is marked (Bisalputra, 1974).

6 Function

6.1 Introduction

The well-established functions of carotenoids in plants are all related to their ability to absorb visible light. In the case of photosynthetic tissues there are two clearly defined functions: (1) in photosynthesis itself and (2) in protecting photosynthetic tissues against photosensitized oxidation. In non-photosynthetic tissues of higher plants and in fungi and non-photosynthetic bacteria carotenoids also take part in photoprotection, but the mechanism involved appears to be different from that in photosynthetic tissues.

6.2 Photosynthesis

Measurements of the action spectrum for photosynthesis indicate that light absorbed by carotenoids is photosynthetically active in both higher plants and algae, but that the efficiency of transfer is variable, being low in red and blue-green algae (when phycobilins are present) and very high (nearing 100%) in brown algae (Goodwin, 1978). The effective participation of carotenoids in photosynthesis can be demonstrated by measuring the enhancement of the fluorescence of chlorophyll on illumination of tissues with light of wavelength absorbed by carotenoids. As chlorophyll fluorescence is associated with photosystem II it is thought that carotenoid function is also associated with this system, although the pigments appear in both photosystem I and photosystem II particles (Goodwin, 1977). The exact mechanism of energy transfer has not yet been settled (see Krinsky, 1971 for a review) but it is probably associated with the *carotenoid band shift*. This is a decrease in absorbance around 478 nm accompanied by an increase around 515 nm which occurs on illumination of photosynthetic tissues and was first demonstrated in *Chlorella*: it is absent from mutants lacking β-carotene. A great deal of work on the carotenoid shift has been carried out on photosynthetic bacteria and the view is emerging that it represents an electrochromic effect and originates from a charge separation across the photosynthetic membrane. Properties and functions of carotenoids in context with photosynthesis are also considered in several chapters of Volumes 5 and 6 of this Encyclopedia (Photosynthesis I and II).

6.3 Photoprotection

6.3.1 Photosynthetic Tissues

Photoprotection was first demonstrated in photosynthetic bacteria when it was found that a mutant strain of *Rhodopseudomonas sphaeroides*, deficient in coloured carotenoids, was destroyed by a combination of light and oxygen but not by either alone, whilst the normal wild-type was not affected by such changes in cultural conditions (Stanier and Cohen-Bazire, 1957). The lack of protection has also been observed in mutants of algae and higher plants (Krinsky, 1971).

6.3.2 Non-Photosynthetic Tissues

Carotenoids in some non-photosynthetic bacteria and fungi play a role in protecting membranes against photodynamic killing (KRINSKY, 1971; GOODWIN, in press 1980) but no experiments on non-photosynthetic tissues of plants have been reported.

6.3.3 Mechanism of Lethal Oxidation

Experiments with the *albino* mutant of *Helianthus annuus* indicate that the lamellar system of the chloroplasts produced in plants germinated in dim light (20 lux) are completely disrupted on exposure to bright light (500 lux) with the loss of the chloroplast (90S) ribosomes which are the site of synthesis of some chloroplast proteins (WALLES, 1972). This has led to the view that destruction of the protein-synthesizing system is a major site of action for photodynamic killing. Nothing is known of the nature of the lethal killing in algae and photosynthetic bacteria. In non-photosynthetic bacteria evidence available suggests that the locus of action is the cell membrane.

6.3.4 Mechanism of Photoprotection

The mechanism of photoprotection in vivo is not yet clear but in vitro experiments suggest strongly that it involves the quenching of singlet oxygen [$^1O_2$]; for example low concentrations of β-carotene almost completely inhibit the methylene blue-sensitized oxidation of 2-methyl-2-pentene, a reaction involving [$^1O_2$]. It is suggested that triplet carotene [^{3}car] is formed as the singlet oxygen returns to its ground state thus:

$$^1O_2 + \text{car} \rightarrow {}^3O_2 + {}^3\text{car}$$

The ^{3}car then dissipates its energy to its surroundings and is thus available to continue the reaction in cyclic fashion (KRINSKY, 1971; GOODWIN, 1979).

6.4 The Xanthophyll Cycle

SAPOZHNIKOV et al. (1957) first showed changes in the epoxide content of leaves associated with illumination. Epoxide levels decrease on illumination in the absence of oxygen and are re-established in the dark in the presence of oxygen. It was originally thought that the cyclic process of epoxide formation was involved in oxygen evolution in photosynthesis but this is highly unlikely. Studies at the enzyme level (reviewed by HAGER (1975, 1980)) have revealed that the zeaxanthin de-epoxidase, M.W. 54,000, has a pH optimum of 5.2 and is essentially inactive at pH 7.0. The reductant for the reaction is ascorbic acid which is regenerated from dehydro-ascorbic acid by coupling with NADPH-dependent glutathione reductase and a glutathione-dependent dehydroascorbic acid reductase. The stimulation of activity on illumination is probably due to the drop of pH within the chloroplast to about 5.5 caused by the production of a proton gradient across the thylakoid mem-

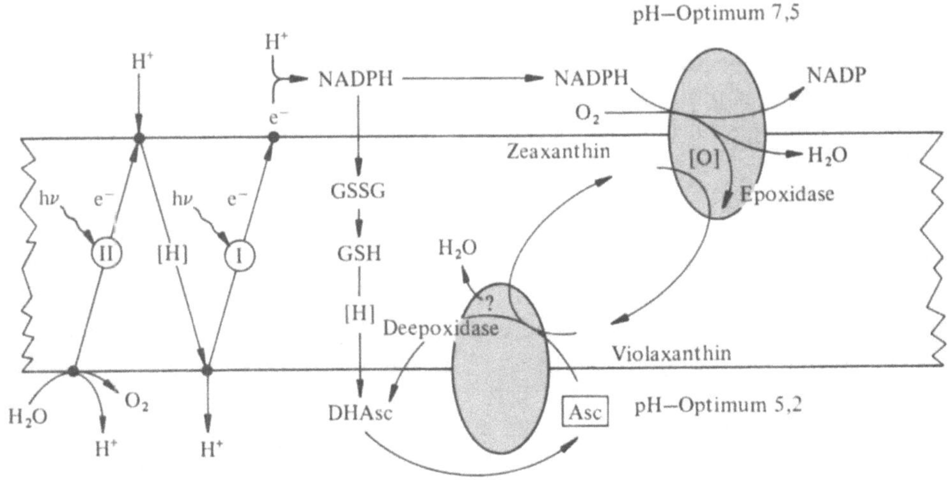

Fig. 16. Proposed mechanism for the xanthophyll cycle (adapted from HAGER, 1975)

brane which is associated with ATP production. The low optimum pH for the enzyme indicates that it is located on the inner surface of the thylakoid membrane. The zeaxanthin epoxidase, on the other hand, is a mixed function oxidase, requiring both oxygen and NADPH for activity, with an optimum pH at 7.3 and with no activity at pH 5.5; it would appear to be located on the outer surface of the thylakoid (SIEFERMANN and YAMAMOTO, 1975; see also WESTERHOFF, 1974).

The probable physical separation of the two enzymes has led to a proposal for the mechanism involved in the Xanthophyll cycle (HAGER, 1975) (Fig. 16). Because of the markedly different pH optima of the two enzymes the cycle depends on the light-induced gradient across the membrane and on the movement of zeaxanthin and violaxanthin across the membrane. There is as yet no evidence for such a movement, but the observation that the availability of endogenous violaxanthin for de-epoxidation varies with experimental conditions points either to a movement of the pigment across the membrane or to some conformational change which controls the exposure of the lipophilic pigment to the enzyme.

6.5 Phototropism

Phototropic responses can be elicited in all higher plants and in some fungi (e.g., Phycomycetes) and the action spectrum for the response suggests that carotenoids are involved. Although early difficulties in measuring absorption spectra in vivo have been largely overcome, the measurement of action spectra remains relatively inaccurate and studies of such spectra are not likely to resolve the argument as to whether a carotenoid or riboflavin is the photoreceptor for phototropism. A recent report of the isolation of a photoactive pigment complex from *Phycomyces* suggests an attractive opening into the problem (for further details see BURNETT 1976; GOODWIN, 1979).

For more information the reader is referred to Vol. 7 of this Encyclopedia (1979): Physiology of Movements (HAUPT, W., FEINLEIB, M.E., eds.), in particular to the chapter of DENNISON on phototropism (Chap. 5.1.2).

References

Ben-Shaul, Y., Klein, S.: Development and structure of carotene bodies in carrot roots. Bot. Gaz. *126*, 79–85 (1965)

Ben-Shaul, Y., Naftali, Y.: The development and ultra-structure of lycopene bodies in chromoplasts. Protoplasma *67*, 333–344 (1969)

Bisalputra, T.: Plastids. In: Algal physiology and biochemistry. Stewart, W.D.P. (ed.), pp. 124–160. Oxford: Blackwell 1974

Booth, V.H.: β-Carotene in the flowers of *Narcissus*. Biochem. J. *65*, 660–663 (1957)

Britton, G.: Biosynthesis of carotenoids. In: Chemistry and biochemistry of plant pigments. Goodwin, T.W. (ed.), Vol. I, pp. 262–327. New York: Academic Press 1976a

Britton, G.: Later reactions of carotenoid biosynthesis. Pure Appl. Chem. *47*, 223–236 (1976b)

Britton, G., Lockley, W.J.S., Patel, N.J., Goodwin, T.W.: The use of deuterium from deuterium oxide as a label in studies on biosynthetic pathways. FEBS Lett. *79*, 281 (1977)

Britton, G., Lockley, W.J.S., Powls, R., Goodwin, T.W., Heyes, L.M.: Carotenoid transformations during chloroplast development in *Scenedesmus obliquus* PG1 demonstrated by deuterium labelling. Nature (London) *268*, 81–82 (1977a)

Britton, G., Lockley, W.J.S., Patel, N.J., Goodwin, T.W., Englert, G.: J. Chem. Soc. Chem. Comm. 655–656 (1977b)

Brooker, J.D., Russell, D.W.: Properties of microsomal 3-hydroxy-3-methylglutaryl Coenzyme A reductase from *Pisum sativum* seedlings. Arch. Biochem. Biophys. *167*, 729–733 (1975)

Brossard, J., Mackinney, G.: The carotenoids of *Diospyros kaki* (Japanese Persimmons). Agric. Food Chem. *11*, 501–503 (1963)

Burnett, J.W.: Function of carotenoids other than in photosynthesis. In: Chemistry and biochemistry of plant pigments. Goodwin, T.W. (ed.), Vol. 1, pp. 655–678. New York: Academic Press 1976

Cholnoky, L., Gyorgfy, C., Nagy, E., Pancel, M.: Function of carotenoids in chlorophyll-containing organs. Nature (London) *178*, 410–411 (1956)

Commission on Biochemistry Nomenclature. Carotenoids. Biochem. J. *127*, 741–751 (1972)

Cornforth, J.W., Cornforth, R.H.: Chemistry of mevalonic acid: In: Natural substances formed biologically from mevalonic acid. Goodwin, T.W. (ed.). Biochem. Soc. Symp. *29*, 5–15 (1970)

Cornforth, J.W., Ross, F.P.: Synthesis of 5*S*-[$^3H_1$] mevalonic acid. J. Chem. Soc. Chem. Commun. 1395–1396 (1970)

Costes, C., Burghoffer, C., Carrayol, E., Ducet, G., Diano, M.: Occurrence of carotenoids in non-plastidial materials from potato tuber cells. Plant Sci. Lett. *6*, 253–259 (1976)

Cyronak, M.J., Britton, G., Simpson, K.L.: Rhodoxanthin, the red pigment of *Equisetum arvense* sporophytes. Phytochemistry *16*, 612–613 (1977)

Davies, B.H.: Carotenoids. In: Chemistry and biochemistry of plant pigments. Goodwin, T.W. (ed.), Vol. II, pp. 38–165. New York: Academic Press 1976

Egger, K.: Die Xanthophyllester der Sonnenblume. Z. Naturforsch. *23b*, 733–736 (1968)

Faludi-Daniél, A., Lang, F., Nagy, A., Faludi, B.: The inheritance of carotenoid types in maize. Acta Agron. Acad. Sci. Hung. *16*, 1–6 (1967)

Goodwin, T.W.: The carotenoids of the flower petals of *Calendula officinalis*. Biochem. J. *58*, 90–94 (1954)

Goodwin, T.W.: A survey of the polyenes in a number of ripe berries. Biochem. J. *62*, 346–352 (1956)

Goodwin, T.W.: The biosynthesis of vitamins and related compounds. New York: Academic Press 1964

Goodwin, T.W.: Algal carotenoids. In: Aspects of terpenoid chemistry and biochemistry. Goodwin, T.W. (ed.), pp. 315–356. New York: Academic Press 1971

Goodwin, T.W.: Prochirality in biochemistry. Essays Biochem. *9*, 104–160 (1973)

Goodwin, T.W.: Carotenoids and biliproteins. In: Algal physiology and biochemistry. Stewart, W.D.P. (ed.), pp. 161–205. Oxford: Blackwell, 1974

Goodwin, T.W.: Distribution of Carotenoids. In: Chemistry and biochemistry of plant pigments. Goodwin, T.W. (ed.), Vol. 1, pp. 225–261. New York: Academic Press 1976

Goodwin, T.W.: The prenyl lipids of the membranes of higher plants. In: Lipids and lipid polymers in higher plants. Tevini, M., Lichtenthaler, H.K. (eds.), pp. 29–47. Berlin, Heidelberg, New York: Springer 1977

Goodwin, T.W.: Biochemistry of the carotenoids, 2nd ed. Vol. I. London: Chapman & Hall in press 1980

Goodwin, T.W., Goad, L.J.: Carotenoids and triterpenoids. In: The biochemistry of fruits and their products. Hulme, A.C. (ed.), Vol. 1. pp. 305–368. New York: Academic Press 1970

Goodwin, T.W., Thomas, D.M.: The carotenoid pigments in the petals of *Mimulus cupreus* and *Mimulus tigrinus*. Phytochemistry *3*, 47–50 (1964)

Gross, J., Carmon, M., Lifshitz, A., Costes, C.: Carotenoids of banana pulp, peel and leaves. Lebensmittelwiss. Technol. *9*, 211–214 (1976)

Hager, A.: Die reversiblen, lichtabhängigen Xanthophyllumwandlungen im Chloroplasten. In: Farbstoffe in Pflanzen. Czygan, F.C. (ed.), pp. 27–44. Stuttgart: Fischer 1975; see also Ber. Deutsch. Bot. Gesellsch. *88*, 27–44 (1975)

Hager, A.: The reversible, light-induced conversions of xanthophylls in the chloroplast. In: F.-C. Czygan (ed.): Plant Pigments, Stuttgart: G. Fischer (1980)

Harris, W.M., Spurr, A.R.: Chromoplasts of tomato fruit. The red tomato. Am. J. Bot. *56*, 380–389 (1969)

Iman, M.K., Gabelman, W.H.: Inheritance of carotenoids in carrots. *Daucus carota*, L. Proc. Am. Soc. Hortic. Sci. *93*, 419–428 (1968)

Jenkins, J.A., Mackinney, G.: Carotenoids of the apricot tomato and its hybrids with yellow and tangerine. Genetics *40*, 715–720 (1955)

Jungalwala, F.B., Cama, H.R.: Carotenoids in mango (*Mangifera indica*) fruit. Indian J. Chem. *1*, 36–40 (1963)

Kargl, T.E., Quackenbush, F.W., Tomes, M.L.: The carotene polyene system in a strain of tomatoes high in delta-carotene and its comparison with eight other tomato strains Proc. Am. Soc. Hortic. Sci. *75*, 574–578 (1960)

Karrer, P., Jucker, E.: Carotenoids (E. A. Braude trs) London: Elsevier 1950

Kjøsen, H., Liaaen-Jensen, S.: Major carotenoids of *Shepherdia canadensis*. Isolation and synthesis of methyl apo-6-lycopenoate. Phytochemistry *8*, 483–491 (1969)

Klein, R.M., Cronquist, A.: A consideration of the evolutionary and taxonomic significance of some biochemical, micromorphological and physiological characters in the thallophytes. Q. Rev. Biol. *42*, 108–296 (1967)

Krinsky, N.I.: Function. In: Carotenoids Isler, O. (ed.), pp. 669–716. Basel: Birkhauser 1971

Laval-Martin, D.: La maturation du fruit de tomate cerise: Mise en evidence, par cryodécapage de l'evolution des chloroplastes en deux types de chromplastes. Protoplasma *82*, 33–59 (1974)

Lincoln, R.E., Porter, J.W.: Inheritance of β-carotene in tomatoes. Genetics *35*, 206–211 (1950)

Mackinney, G., Rick, C.M., Jenkins, J.A.: The phytoene content of tomatoes. Proc. Natl. Acad. Sci. USA *42*, 404–408 (1956)

Moss, G.P., Weedon, B.C.L.: Chemistry of carotenoids. In: Chemistry and biochemistry of plant pigments. Goodwin, T.W. (ed.), Vol. 1. pp. 149–224. New York: Academic Press 1976

Petzold, E.N., Quackenbush, F.W.: Zeinoxanthin, a crystalline carotenol from corn gluten. Arch. Biochem. Biophys. *86*, 163–165 (1960)

Petzold, E.N., Quackenbush, F.W., McQuistan, M.: Zeacarotenes, new provitamins A from corn. Arch. Biochem. Biophys. *82*, 117–124 (1959)

Porter, J.W., Lincoln, R.E.: *Lycopersicon* selections containing a high content of carotenes and colourless polyenes II. The mechanism of carotene biosynthesis. Arch. Biochem. *27*, 390–403 (1950)

Sapozhnikov, D.I., Krasowskaya, T.A., Mayevskaya, A.N.: Changes observed in the relation between the main carotenoids in the plastides of green leaves exposed to light. Dokl. Akad. Nauk. SSSR *113*, 465–467 (1957)

Siefermann, D., Yamamoto, H.: Properties of NADPH and oxygen-dependent zeaxanthin epoxidation in isolated chloroplasts. Arch. Biochem. Biophys. *171*, 70–77 (1975)

Stanier, R.Y., Cohen-Bazire, G.: The role of light in the microbial world: some facts & speculations. In: Microbial ecology. Williams, R.E.O., Spicer, C.C. (eds.), pp. 56–89. London: Cambridge U.P. 1957

Strain, H.H.: Chloroplast pigments and chromatographic analysis. 32nd Annual Priestley Lecture. University Park. Penn. State Univ. (1958)

Strain, H.H.: Chloroplast pigments and chromatographic analysis. pp. 180. Pennsylvania: University Park 1959

Strain, H.H.: Fat-soluble Chloroplast Pigments. In: Biochemistry of chloroplasts. Goodwin, T.W. (ed.), Vol. 1, pp. 387–406. New York: Academic Press 1966

Tomes, M.L., Quackenbush, F.W., McQuistan, M.: Modification and dominance of the gene governing formation of high concentrations of β-carotene in the tomato. Genetics *39*, 810–817 (1954)

Tomes, M.L., Quackenbush, F.W., Kargl, T.E.: Synthesis of β-carotene in tomato fruit. Bot. Gaz. *119*, 250–253 (1958)

Valadon, L.R.G., Mummery, R.: Carotenoids of some Rosaceae flowers. Phyton *25*, 151–157 (1968)

Walles, B.: An electron microscope study on photodestruction of plastid ribosomes in β-carotene deficient mutants of *Helianthus annuus*, L. Protoplasma 215–227 (1972)

Westerhoff, N.: Correlations between light dependent xanthophyll interconversions, photosynthetic electron transport and photophosphorylation. Ber. Deutsch. Bot. Ges. *87*, 545–551 (1974)

Wierzchowski, Z.: Karotenoidy roslin pastewnych i lisci drzew. Zesz. Probl. Postepow. Nauk. Roln. *53*, 305–310 (1965)

Williams, R.J.H., Britton, G., Goodwin, T.W.: A possible mechanism for the biosynthesis of eschscholtzxanthin. Biochim. Biophys. Acta *124*, 200–203 (1966)

5.4 Polyisoprenoids

5.4.1 Polyprenols and Terpenoid Quinones and Chromanols

D.R. THRELFALL

1 Introduction

These compounds are distributed throughout the higher plant and algal kingdoms and, in the cases of cis, trans-polyprenols and ubiquinones, the fungal kingdom as well. The function, if any, of the polyprenols per se is unknown, but the monophosphate esters of cis, trans-polyprenols are important intermediates in glycan biosynthesis in many organisms. The terpenoid quinone and chromanol group of compounds includes two quinones, plastoquinone-9 and ubiquinone, whose roles in photophosphorylation and respiratory chain phosphorylation are well established, and it is almost certain that in the very near future the functions of compounds such as phylloquinone and tocopherols will be elucidated.

In the limited amount of space allocated for this section it is impossible to give a full account of all the biochemical and physiological aspects of these compounds. The account is, therefore, based mainly on readily available review articles rather than on scientific papers. This means that many workers who either have made or are making important contributions to our knowledge of these compounds are not mentioned by name.

2 Polyprenols

The greater part of the information contained in this section was abstracted from the excellent reviews of HEMMING (1970, 1974). The sources of information not contained in these reviews are cited in the text.

2.1 Nature and Distribution

Polyprenols are acyclic primary monohydric alcohols, the carbon skeletons of which are made up of isoprene units joined head to tail. Two groups of polyprenols are found in plants: the all-trans-polyprenols and the cis, trans-polyprenols (Table 1).

All-trans-polyprenols are found in the tissues of higher plants (Table 1). It is of interest that solanesol (all-trans-nonaprenol) is found in tissues that are rich in plastoquinone-9 (a nonaprenyl substituted quinone), and that spadicol (all-trans-decaprenol) is found in a tissue with a high level of ubiquinone-10 (a decaprenyl substituted quinone).

Table 1. Nature and distribution of plant polyprenols

1. All-trans-polyprenols

General formula ω-T-T-[T]$_n$-T-OH*

Spadicol (n=6) (all-trans-decaprenol). Spadix of *Arum maculatum*

Solanesol (n=5) (all-trans-nonaprenol). Green leaves. Solanesol samples from flue-cured tobacco (*Nicotiana tabacum*) (the best source of this alcohol) contain small amounts of all-trans-octaprenol (n=4). Small amounts of acyl esters occur in the trichomes of tobacco leaves (CHAKRABORTY and WEYBREW, 1964).

Bombiprenone (VI). All-trans-C_{43} ketone found in mulberry leaves (*Morus bombycis*) (TOYODA et al., 1968) and flue-cured tobacco (IRVINE et al., 1972). Suggested that in tobacco leaves it is formed by aerobic photoxidation of plastoquinone-9 as the leaves die (IRVINE et al., 1972).

2. Cis, trans-polyprenols

ω-T-T-C-[C]$_n$-C-OH, n=6→9

Betulaprenols (di*trans*-, poly*cis*-prenols). Silver birch (*Betula verrucosa*) leaves. Acyl esters in silver birch wood.

ω-T-T-C-[C]$_n$-S-OH, n=9→13

Dolichols (dihydro-, di*trans*-, poly*cis* prenols). Yeast and *Phytophthora cactorum*. Also found in mammals and marine invertebrates.

ω-T-T-T-[C]$_n$-C-OH, n=4→8

Ficaprenols, castaprenols etc. (tri*trans*-, poly*cis*-penols). Green leaves. Acyl esters in latex of *Hevea brasiliensis*.

S-S-T-T-[C]$_n$-S-OH, n=14→18

Hexahydroprenols (hexahydro-, di*trans*-, poly*cis*-prenols). Occurs with acyl esters in *Aspergillus fumigatus*.

M-S-T-T-[C]$_n$-S-OH, n=14→18

Exo-methylene-hexahydroprenols (methylene-, hexahydro-, di*trans*-, poly*cis*-prenols). *Aspergillus niger*

ω, ω-isoprene residue (I); T, trans-isoprene residue (II); C, cis-isoprene residue (III); S, dihydro (saturated) isoprene residue (IV); M, exo-methylene-dihydroisoprene residue (V). HEMMING (1974) has proposed a nomenclature based on that recommended by IUPAC-IUB for the polyisoprenoid quinones. On this system the betulaprenols would be described as IV, V-ditrans-, polycis-prenols-6→9, where the roman numerals refer to the position of isoprene residues in the homologue having the shortest chain length and 6→9 gives the range of cis residues in the mixture. Their trivial name would then be ditrans-, polycis-prenols. The use of this type of trivial name has the merit that it conveys the order and type of the essential structural features of the molecules.

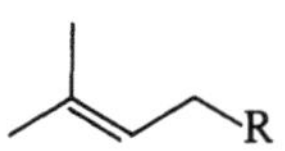

(I)

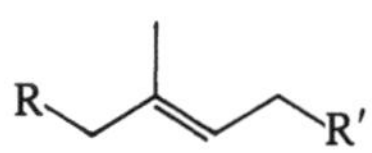

(II)

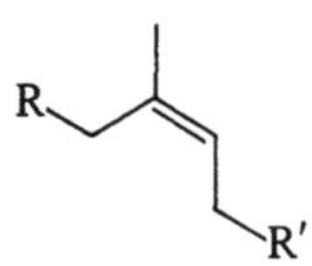

(III)

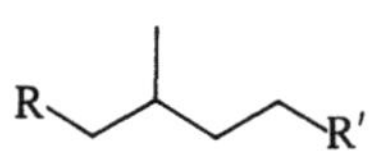

(IV)

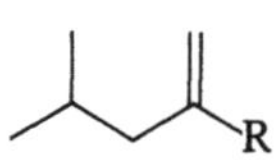

(V)

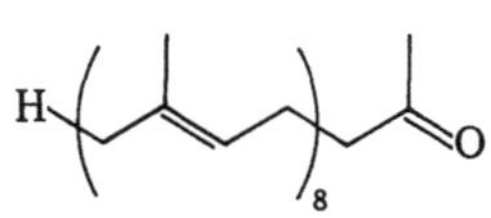

(VI)

Cis,trans-polyprenols occur in angiosperms, fungi, and a green alga (Table 1). They fall into a small number of well-defined families which differ from one another in the number of ω-trans-residues and cis-residues, the position and number of any saturated residues and, in the case of one family, the presence of an ω-exo-methylene residue (Table 1). Individual alcohols or families of alcohols are usually referred to by trivial names which indicate either some feature of their chemistry or the source from which they were first isolated (Table 1). Most analytical studies have given the impression that a given tissue or organism will contain a mixture of polyprenols from only one of the families, and that most of the mixture will be made up of those alcohols with an n value within two of the median value of the mixture. In some plants, however, members of two or even three families co-occur in the same tissue. Thus, leaves of *Magnolia campbellii* contain castaprenols and lesser amounts of di-trans and mono-trans analogues of castaprenols (SASAK et al., 1977) and cells of *Blakeslea trispora* contain α-dihydro-cis,trans-polyprenols with lesser amounts of unsaturated cis,trans-polyprenols (BU'LOCK and OSAGIE, 1973). An interesting plant in this respect is *Betula verrucosa*, since it contains esterified betulaprenols with lesser amounts of castaprenols in its wood and castaprenols with lesser amounts of betulaprenols in its leaves.

2.2 Seasonal Variation and Intracellular Distribution

Plant leaves contain unesterified unsaturated polyprenols, the concentrations of which increase with increasing age. In horsechestnut (*Aesculus hippocastanum*) leaves the levels of solanesol (all-trans nonaprenol) and castaprenols (cis, trans-polyprenols) rise from 6 pmol g wet wt.$^{-1}$ and 1.2 μg g wet wt.$^{-1}$ in April to 106 pmol g wet wt.$^{-1}$ and 84 μg g wet wt.$^{-1}$ in October, whilst in leaves of *Ficus elastica* the level of ficaprenols rises from 2.5 μg g wet wt.$^{-1}$ to 1.5 mg g wet wt. $^{-1}$. Solanesol, along with plastoquinone-9, is present in the chloroplasts. Most of the cis, trans-polyprenols are contained in the plastoglobuli, although some are associated with the thylakoid membrane and some with the cell wall.

The distribution of the acyl esters of betulaprenols in the wood of *B. verrucosa* and of ficaprenols in the latex of *Hevea brasiliensis* is unknown.

Moulds contain unesterified cis,trans-polyprenols and some of these may be present as acyl esters. In *Aspergillus fumigatus* the unesterified prenols are concentrated mainly in the mitochondria whilst the acyl esters are distributed more or less equally between the cell fractions. It is possible that both the unesterified and acyl-esterified alcohols are structural components of the various membranes. In *A. niger* some 3% of the total prenols are present as monophosphates at the stationary phase.

2.3 Biosynthesis

All-trans-polyprenols are believed to be formed as a result of the action of appropriate phosphatases on all-trans-polyprenylpyrophosphates which have been synthesised by polyprenylpyrophosphate synthetases that are capable of continuing the chain elongation reactions beyond the trans-trans-farnesylpyrophosphate and all-trans-geranylgeranylpyrophosphate stages described previously (Chap. 5.1). No cell-free preparation capable of synthesising either unesterified alcohols or polyprenylpyrophosphates has been isolated from plant tissues. Indirect evidence that mitochondrial membranes and chloroplasts are able to synthesise all-trans-polyprenylpyrophosphates is provided by the demonstrations that both yeast mitochondrial membranes and 10,000 g membrane preparations from broad bean seeds synthesise 3-polyprenyl-4-hydroxybenzoates with the same chain lengths as the constitutive ubiquinones (ubiquinone-6 in yeast, ubiquinone-10 in beans) from 4-hydroxybenzoate and IPP, and that isolated chloroplasts synthesise plastoquinone-9 from CO_2 (Sects. 3.3.1 and 3.3.2).

Cis,trans-polyprenols are also believed to be formed by the action of phosphatases, although in this case they could be formed either from their pyrophosphates or, if they have a functional role to play (Sect. 2.4), their monophosphates. Studies with stereochemically labelled MVA have provided evidence that the ω and chemically trans residues are biogenetically trans and that chemically cis-residues as well as any saturated residues at the α ends of the molecules are biogenetically cis. On the basis of these findings it has been proposed that the various families of cis, trans-polyprenylpyrophosphates are produced by synthetases which first of all synthesise either trans, trans-farnesyl-pyrophosphate or all-trans-geranylgeranylpyrophosphate and then continue the chain elongation reactions with IPP by cis-addition of isoprene residues. The saturation and substitution of residues presumably constitute the last steps in the biosynthetic sequences prior to removal of the phosphate group(s). There is one report of a cell-free system producing cis, trans-polyprenyl monophosphate and this is a post 25,000 *g* supernatant from pea hypocotyls that synthesises a dolichylmonophosphate from IPP (Daleo and Pont Lezica, 1977).

2.4 Function

Apart from a possible structural role in some fungal membranes (Sect. 2.2), the polyprenols and their acyl esters have no known functions in plant cells. There is, however, a substantial body of evidence that the monophosphate esters of dihydro-, ditrans-polycis-prenols (dolichols) (Table 1) are involved in the transfer of sugar residues (e.g., mannosyl-, N-acetylaminoglucosyl) from GDP-sugar to appropriate acceptor molecules in the formation of fungal glycans (Hemming, 1974): e.g.,

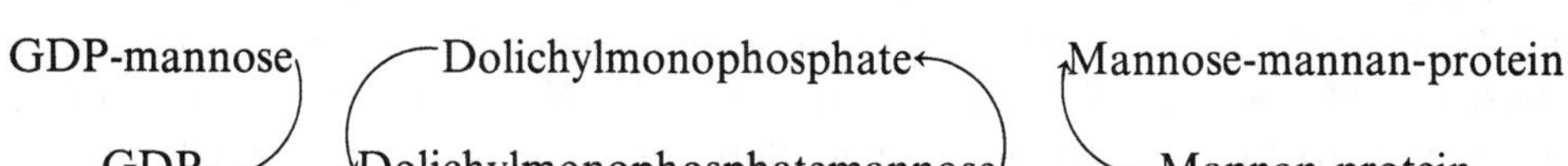

The involvement of cis,trans-polyprenylmonophosphates in the synthesis of glycans in photosynthetic plants remains to be demonstrated. There are, however, reports of the incorporation of mannose into lipid-phosphate-mannose in extracts of a variety of higher plants and the alga *Codium fragile*, and of the formation of lipid-phosphate-chitobiose and lipid-phosphate-oligosaccharides (which contain mannose and N-acetylglucosamine) by cotton fibre enzymes (see Brett and Leloir,

1977; DALEO and PONT LEZICA, 1977; HEMMING, 1974). In preparations from soya beans (*Glycine max*), pea (*Pisum sativum*) and wheat germ (*Triticum vulgare*), the lipid component of the lipid-phosphate-mannose has the properties expected of an α-dihydro-cis,trans-polyprenol (i.e., a dolichol) (BRETT and LEOIR, 1977; DALEO and PONT LEZICA, 1977) and not as might be expected from the nature of the polyprenols isolated from higher plants (Table 1), an unsaturated polyprenol of the ficaprenol-castaprenol type. It may be, however, that the eukaryotic type (α-dihydro) of cis, trans-polyprenylmonophosphate is used as a sugar acceptor outside the chloroplast, whereas the prokaryotic type (unsaturated) of cis, trans-polyprenylmonophosphate is used as a sugar acceptor inside the chloroplast.

3 Terpenoid Quinones and Chromanols

This section is based largely on the review articles by DICKS (1965), JANISZOWSKA and PENNOCK (1976), LICHTENTHALER (1973, 1977), THRELFALL (1971), THRELFALL and WHISTANCE (1971) and WALLWORK and CRANE (1970).

3.1 Nature and Distribution

The structural formulae and distribution patterns of the compounds under consideration in this section are given in Table 2. The first two groups of compounds, the *p*-benzoquinones and chromanols formed from homogentisate and the naphthaquinones formed from 2-succinylbenzoate, are synthesised only by photosynthetic plants, whereas the third group of compounds, the *p*-benzoquinones formed from 4-hydroxybenzoate, are synthesised by all aerobic eukaryotic cells and many Gram-negative bacteria.

3.2 Intracellular Distribution

3.2.1 Plastoquinones, Plastochromanols, Tocoquinones and Tocochromanols

There is general agreement that phylloquinone, plastoquinone-9, plastoquinol-9, plastoquinone-Bs and -Cs, α-tocopherolquinone, most of the α-tocopherol and when present plastoquinone-Zs, plastochromanol-8 and plastochromanol-Bs, Cs, and Zs are confined to the chloroplasts in the photosynthetic eukaryotic cell (JANISZOWSKA and PENNOCK, 1976; LICHTENTHALER, 1977; WALLWORK and CRANE, 1970). In addition to these compounds some chloroplasts may contain phylloquinone analogues, plastoquinone homologues or non-α-tocopherols and non-α-tocopherolquinones (WALLWORK and CRANE, 1970). Spinach chloroplasts, for example, have been reported to contain (μmol 100 μmol of chlorophyll^{-1}): demethylphylloquinone, 0.25; phylloquinone, 1; plastoquinone-9, 10; plastoquinone-Bs, 0.5; plastoquinone-Cs, 3.4; α-tocopherolquinone, 1; β-tocopherolquinone, 0.6; γ-tocopherolquinone, 0.3; δ-tocopherolquinone, 0.5 and α-tocopherol, 5 (WALLWORK and CRANE, 1970). It has been suggested that the concentrations of phylloquinone,

Table 2. Nature and distribution of plant terpenoid quinones and chromanols. The first abbreviation given after the name of a compound is either the one recommended by IUPAC-IUB or, in the case of a compound for which no recommendation has been made, the one in common usage

1. Quinones and chromanols formed from homogentisate

Plastoquinones and plastochromanols (THOMSON, 1971, THRELFALL, 1971, THRELFALL and WHISTANCE, 1971, WALLWORK and CRANE, 1970).

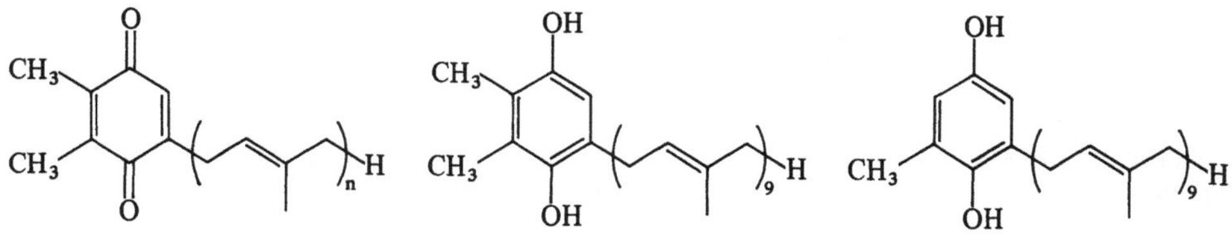

(VII) (VIII) (IX)

Plastoquinone-9 (VII, n=9) (PQ-9; Kofler's quinone; plastoquinone-A, PQA; plastoquinone 45, PQ-45). Most abundant member of this group of compounds. Found in all higher plants (most tissues), gymnosperms, bryophytes, mosses and algae. Highest levels in photosynthetic regions of plants.

Plastoquinol-9 (VIII) (PQH_2-9; plastohydroquinone). Reduced form of plastoquinone-9. Co-occurs with plastoquinone-9, to which it is probably oxidized in most extraction procedures.

Plastoquinone-8 (VII, n=8) (PQ-8) Leaves *A. hippocastanum*, shoots of *Zea mays* and cells of *Euglena gracilis.*

Plastoquinone-4 (VII, n=4) (PQ-4; plastoquinone-A20, PQ-A20). Leaves of *A. hippocastanum.*

Plastoquinone-3 (VII, n=3) PQ-3). Leaves of *Spinacia oleracea.*

2-Demethylplastoquinol-9 (IX). Bulbs of *Iris hollandica* (ETMAN-GERVAIS et al., 1977)

Plastoquinone-Cs (X) [PQ-Cs; Originally sub-divided into plastoquinone-Cs and -Ds; in analytical studies may be sub-divided into PQ-C_1, PQ-C_2..., and it has been proposed they should be

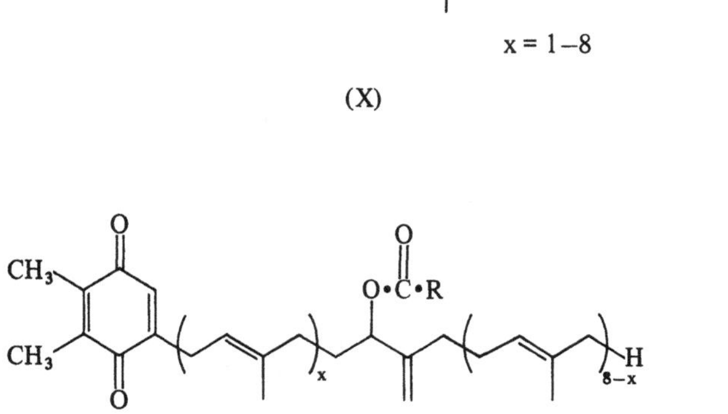

x = 1–8

(X)

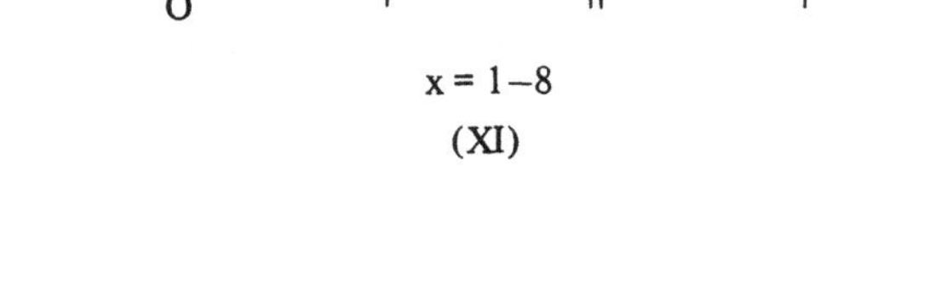

x = 1–8

(XI)

Table 2 (continued)

further subdivided into PQ-$C2_1$, PQ-$C3_2$..., where 2 and 3 indicate the presence of secondary and tertiary hydroxyl groups respectively (WALLWORK and CRANE, 1970)]. Allylic secondary and tertiary monohydroxy derivatives of plastoquinone-9. Allowing for the fact that first isoprene unit from the ring is apparently never modified and assuming that the hydroxylation reactions are stereospecific then potentially there are 16 isomeric forms, 10 of which have been isolated from plants. The distribution of these compounds is similar to that of plastoquinone-9.

Plastoquinone-Bs (XI) (PQ-Bs). Acyl esters of the allylic secondary monohydroxy forms of plastoquinone-Cs. There are 8 possible isomers, 6 of which have been isolated. Distribution as for plastoquinone-Cs.

Plastoquinone-Zs (PQ-Zs). Allylic secondary and tertiary monohydroxy derivatives of plastoquinone-Bs. Eight forms have been isolated. Distribution may be restricted to higher plants.

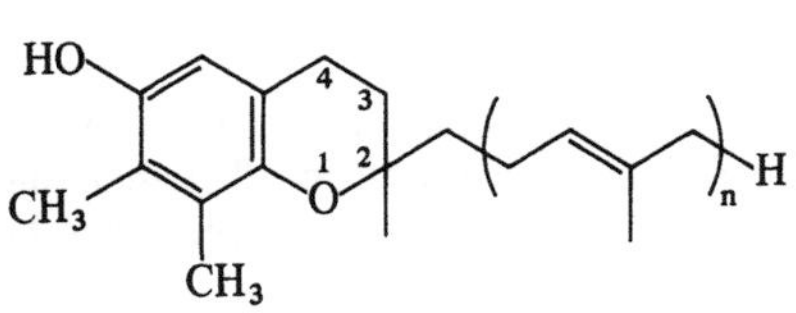

(XII)

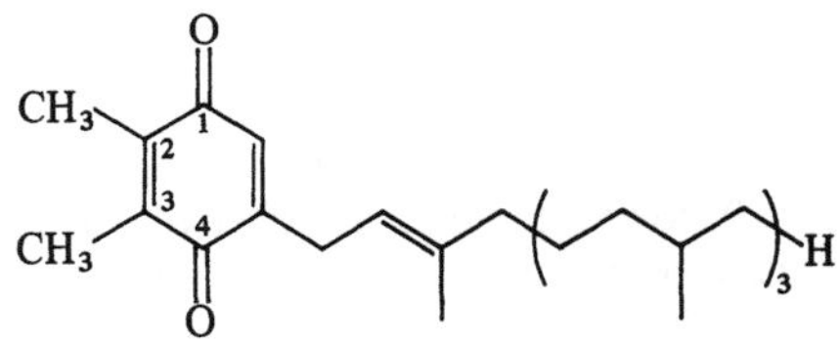

(XIII)

Plastochromanol-8 (XII, n=8) (PC-8). Cyclic derivative of plastoquinone-9. Green leaves of many plants.

Plastochromenol-8 (XII, n=8 with a Δ^3 double bond) (solanochromene). Artifact of plastoquinone-9 found in flue-cured tobacco.

Plastochromanol-Bs, -Cs and Zs (PC-Bs, PC-Cs and PC-Zs). Cyclic forms of plastoquinone-Bs, -Cs and -Zs. Green leaves of many plants.

Plastochromanol-3 (XII, n=3). This compound is γ-tocotrienol (see below).

Phytylplastoquinone (XIII) (PhytylPQ). Cells of *E. gracilis.*

2-Demethylphytylplastoquinone (2-methyl in (XIII) replaced by H). Cells of *E. gracilis.*

1-O-Methyl ethers of phytylplastoquinol and 2-demethyl-phytylplastoquinol. Cells of *E. gracilis.*

Chromanols of phytylplastoquinone and 2-demethylphytylplastoquinone. These are γ-tocopherol and δ-tocopherol (see below).

Tocochromanols (tocopherols and tocotrienols) and tocoquinones (DICKS, 1965, JANISZOWSKA and PENNOCK, 1976).

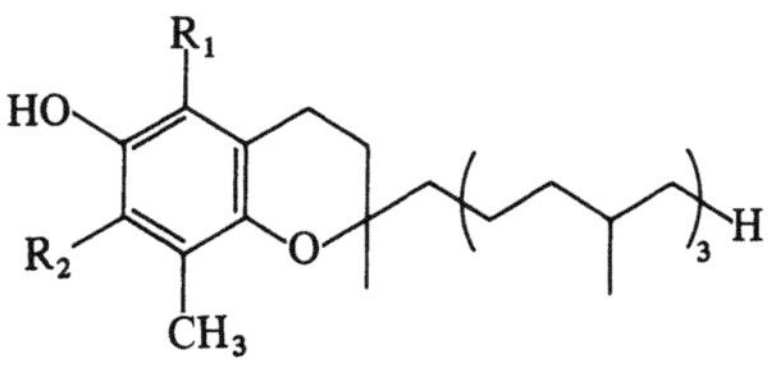

R_1	R_2	
H	H	δ—Tocopherol
CH_3	H	β—Tocopherol
H	CH_3	γ—Tocopherol
CH_3	CH_3	α—Tocopherol

(XIV)

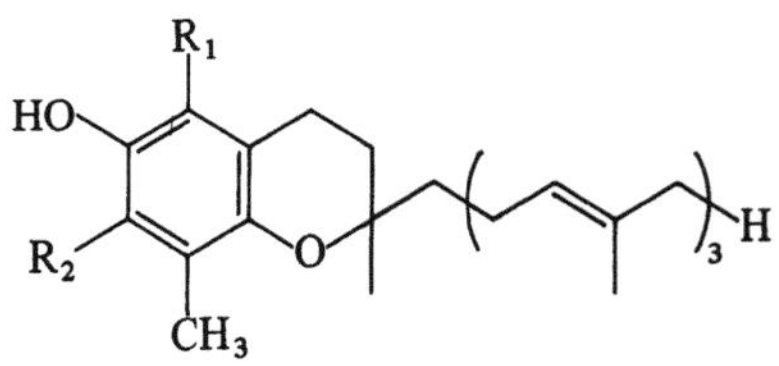

R_1	R_2	
H	H	δ—Tocotrienol
CH_3	H	β—Tocotrienol
H	CH_3	γ—Tocotrienol
CH_3	CH_3	α—Tocotrienol

(XV)

Table 2 (continued)

Tocopherols (XIV) (α-T, β-T, γ-T and δ-T; Vitamins E). Chromanol ring and side chain formed from a phytyl residue. Present in all phytosynthetic plants, with the exception of the blue-green alga *Anacystis nidulans*. α-Tocopherol most abundant form. In leaves of higher plants it is often accompanied by γ-tocopherol (β-tocopherol in some plants) and sometimes δ-tocopherol. Non-α-tocopherols often predominate in non-photosynthetic parts of plants. Compounds once described as η-, ε- and ζ-tocopherol are now known to be δ-, β- and α-tocotrienol.

Dehydrotocopherols (α-T-1, β-T-1...), α- and γ-dehydrotocopherols co-occur with the corresponding tocopherols in etiolated shoots of *Z. mays* and *H. vulgare* (THRELFALL and WHISTANCE, 1977).

Tocotrienols (XV) (α-T-3, β-T-3...; α-T_3, β-T_3...). Chromanol ring and side chain formed from geranylgeranyl residue. Restricted to some cereal seed oils and brans, palm oil and latex of *H. brasiliensis*. Acyl esters present in latex.

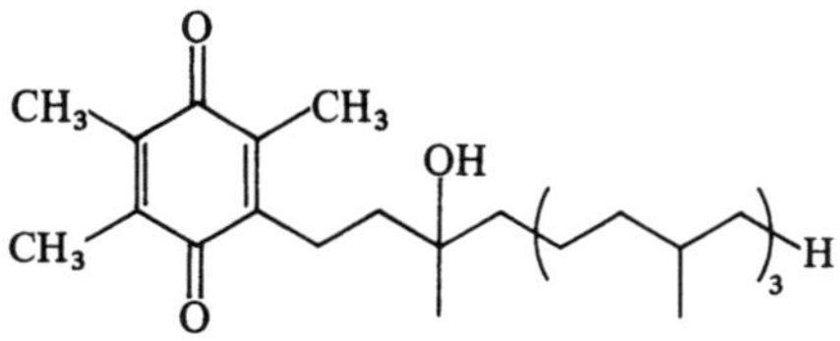

(XVI)

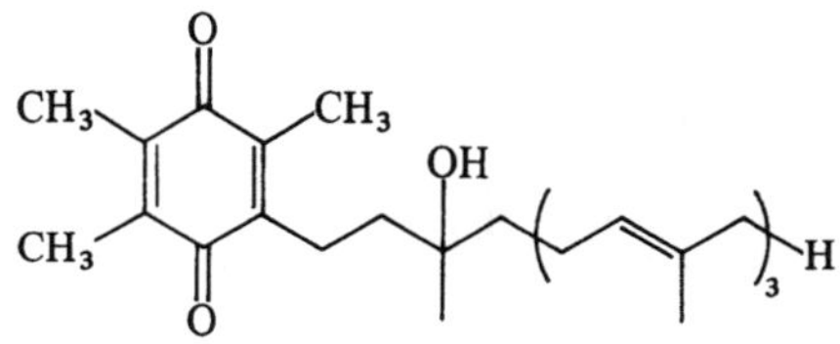

(XVII)

Tocopherolquinones (α-TQ(XVI), β-TQ....). Formed on mild oxidation of tocopherols. α-Tocopherolquinone restricted to photosynthetic tissues. δ-, γ- and β-Tocopherolquinones very limited distribution.

α-Dehydrotocopherolquinone (α-TQ-1). Etiolated shoots of *Z. mays* (THRELFALL and WHISTANCE, 1977).

Tocotrienolquinones (α-TQ-3 (XVII), β-TQ-3...; α-TQ_3, β-TQ_3...). α-Tocotrienolquinone isolated from latex of *H. brasiliensis*.

Dimers and trimers. Simple and mixed dimers and trimers of tocopherols and tocotrienols have been isolated from seed oils.

2. Quinones formed from 2-succinylbenzoate

Phylloquinones (THRELFALL, 1971)

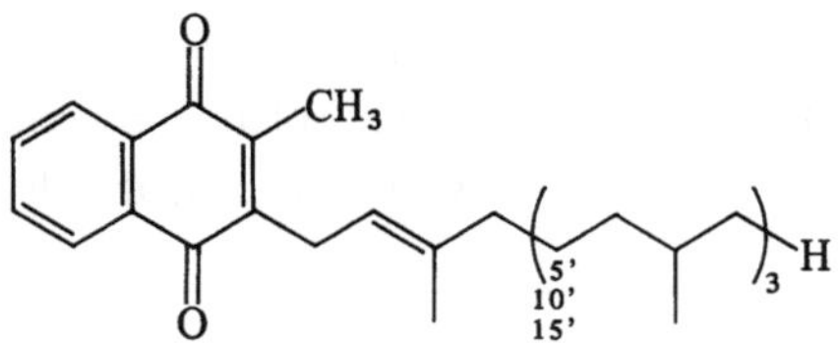

(XVIII)

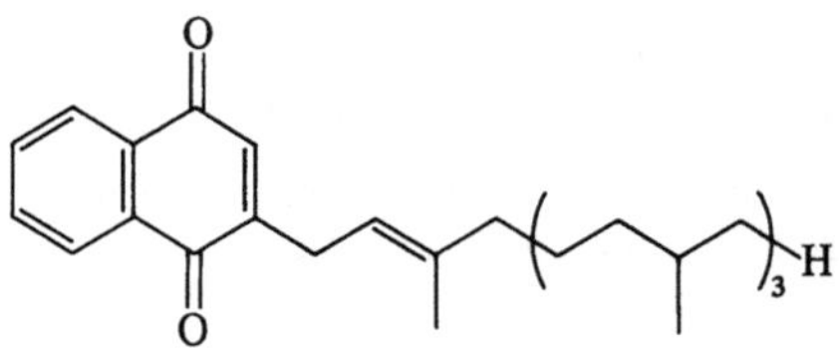

(XIX)

Phylloquinone (XVIII) (K; Vitamin K_1, K_1). Present in most photosynthetic cells.

Demethylphylloquinone (XIX) (DK). Spinach chloroplasts.

5′-Monohydroxyphylloquinone (5′-monohydroxy derivative of (XVIII)). Cells of *A. nidulans, Chlorella pyrenoidosa* and *E. gracilis*.

Dehydrophylloquinone (K-1). Etiolated shoots of *Z. mays* and *H. vulgare* (THRELFALL and WHISTANCE, 1977).

Table 2 (continued)

3. Quinones formed from 4-hydroxybenzoate

Ubiquinones and related compounds (THOMSON, 1971; THRELFALL and WHISTANCE, 1971).

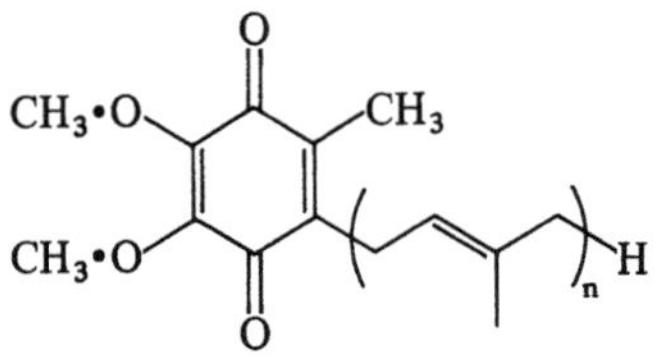

(XX)

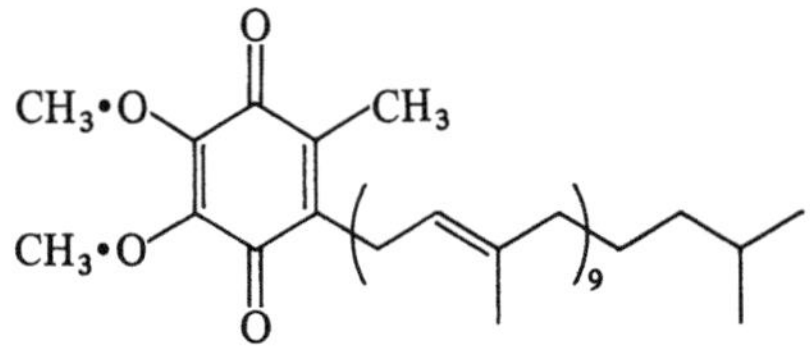

(XXI)

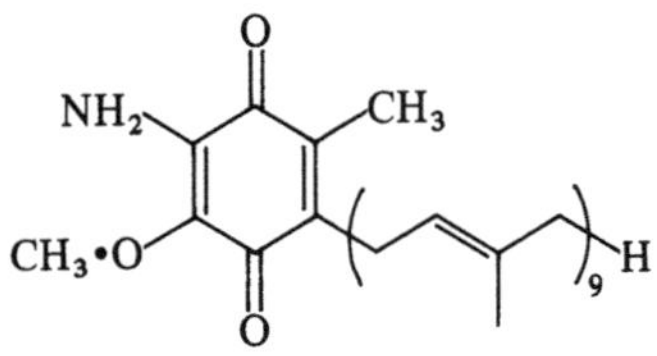

(XXII)

Ubiquinone-N (XX) (e.g., ubiquinone-10, Q-10, UQ-10, UQ_{10}, UQ-50, UQ_{50} (in the last two abbreviations the numbers refer to the number of carbon atoms in the polyprenyl side chain); Coenzyme Q-10, CoQ-10, CoQ_{10}). Homologous series of compounds of which the monoprenyl (XX, n=1) through to dodecaprenyl (XX, n=12) have been isolated from biological sources. All photosynthetic plants, with the exception of blue-green algae which do not contain ubiquinone, and many fungi contain either ubiquinone-10 or ubiquinone-9 and lesser amounts (<15%) of the immediately adjacent homologues e.g., fruit of *Capsicum annuum* contain ubiquinone-8 to ubiquinone-12 and cells of *Saccharomyces cerevisiae* contain ubiquinone-1 to ubiquinone-6 (major homologue).

Ubiquinol-N(QH_2-N; ubihydroquinone). Reduced form of ubiquinone (cf. VII and VIII). Co-occurs with ubiquinone, to which it is oxidized in most extraction procedures.

Dihydroubiquinone-N(Q-N(H_2)); when known the position of saturation is denoted by a roman numeral e.g., X-dihydroubiquinone-10 (XXI), Q-10 (X-H_2). Restricted to some moulds. Can co-occur with ubiquinones.

Rhodoquinone-9 (XXII) (RQ-9). Metabolite of ubiquinone-9 in cells of *E. gracilis*.

Monoepoxyubiquinones. Cells of *S. cerevisiae*. May be artifacts of the isolation procedure.

Ubichromenol-6 (UC-6). Isolated from dried cells of *Torula utilis*. Artefacts of the drying procedure (cf. solanochromene).

plastoquinone-9, plastoquinol-9, α-tocopherol, α-tocopherolquinone and carotenoids relative to chlorophyll concentration lie within certain standard concentration limits which are similar for plants of different systematic positions (LICHTENTHALER, 1968b):

chlorophyll a, 100; chlorophyll b, 33±5; plastoquinone-9, 10±5; plastoquinol-9, 15±10; α-tocopherol, 10±5; α-tocopherolquinone, 2.5±2; phylloquinone, 2±1.5.

Marked deviations from these values are found, however, in leaves of trees and shrubs which grow in strong sunlight, older leaves of perennial plants and leaves of aurea-variegated plants. In young chloroplasts of C_3 plants, as well as cells of blue-green algae, the quinones and chromanols are to be found in the thylakoid membranes, but as the chloroplasts age they are also laid down in the plastoglobuli. Thus, in horsechestnut chloroplasts the plastoquinone content (plastoquinone-9+plastoquinone-Cs) of the globules rises from a trace in May to 20% of the total accumulated plastoquinones in the chloroplast in October (BARR et al., 1967). When spinach plastoglobuli account for 40% of the total chloroplast quinones and α-tocopherol they have a mol% composition of: plastoquinone-9 (+plastoquinol-9), 50; α-tocopherol, 48: α-tocopherolquinone, 1.5 and phylloquinone, 0.5 (LICHTENTHALER and TEVINI, 1970). There has been no clear-cut demonstration of the distribution of the various quinones and chromanols between the two photosystems (Vol. 5 of this Encyclopedia), although some evidence has been presented for the association of plastoquinone-9 (and plastoquinol-9), α-tocopherol and α-tocopherolquinone with photosystem II, and of phylloquinone with photosystem I (LICHTENTHALER and TEVINI, 1970). It has been suggested that in the C_4 plant *Z. mays* plastoquinone-9, plastoquinol-9, α-tocopherol and α-tocopherolquinone are associated with the thylakoid membrane-rich, photosystem II-containing mesophyll chloroplasts rather than the bundle sheath chloroplasts (WILD et al., 1972).

The intracellular distribution of non-α-tocopherols and any extrachloroplastidic α-tocopherol in the eukaryotic photosynthetic cell has not yet been resolved. In the case of α-tocopherol, it has proved difficult to establish how much of this compound is actually present outside the chloroplast. Thus the values in the literature range from 0% for young broad bean leaves (BUCKE, 1968) to 40% for leaves of *Calendula officinalis* (JANISZOWSKA et al., 1976). Most workers, however, seem to accept that some α-tocopherol is found outside the chloroplast, and that it is probably located in the mitochondria (JANISZOWSKA and PENNOCK, 1976; WALLWORK and CRANE, 1970). In the cases of the non-α-tocopherols it has been reported that in the thallus of *Fucus spiralis* the β- and γ-tocopherol are in the 3,000 *g* and 100,000 *g* membrane fractions and δ-tocopherol is in the 200,000 *g* membrane and supernatant fractions (NEWTON and PENNOCK, 1971), whereas in leaves of *P. sativum* γ-tocopherol is present mainly in the chloroplast and microsomal fractions and δ-tocopherol is mainly in the chloroplast and supernatant fractions (HUGHES et al., 1971). Non-α-tocopherols are found in some prokaryotic photosynthetic cells, and it has been shown that in the unicellular *Gloeocapsa* the intracellular distribution of α-tocopherol follows that of chlorophyll, whereas no such relationship exists between the β- and δ-tocopherol and chlorophyll (NEWTON et al., 1977).

Non-photosynthetic tissues of higher plants can contain phylloquinone, plastoquinone-9, plastoquinol-9, plastoquinone-Bs and Cs, α-tocopherolquinone, α-tocopherol, and non-α-tocopherols (LICHTENTHALER, 1968a; WALLWORK and CRANE, 1970). The first six compounds and some of the α-tocopherol, depending on the type of tissue, are probably contained in etioplasts, leucoplasts, amyloplasts, or chromoplasts. In general the levels of these compounds in a tissue containing one or other of the first three types of plastid are much lower than the levels found in photosynthetic tissues. In chromoplast-containing tissue, however, the level of some of these compounds are higher than the levels found in the unripe green fruit (LICHTENTHALER, 1968a). The non-α-tocopherols and any extraplastidic α-tocopherol are presumably distributed within the cells of non-photosynthetic tissues in the same manner as they are in the cells of photosynthetic tissues.

3.2.2 Ubiquinones

The greater part of the ubiquinone or dihydroubiquinone content of aerobic eukaryotic cells is localised in the inner mitochondrial membrane. In the case of *Candida guilliermondi* cells grown on *n*-paraffins, however, only some 30% of the ubiquinone is contained in the mitochondria (TARASOVA et al., 1976).

3.3 Biosynthesis

As indicated already, it is possible to divide the terpenoid quinones and chromanols into three biogenetic groups (Sect. 3.1).

3.3.1 Photosynthetic Plants

The outlines and, as far as they are known, some of the details of the pathways are given in Fig. 1. Most of the information used to construct the pathways has come from feeding experiments with either whole plants or excised parts of higher plants. So far very little work has been done with cell-free systems.

The sequences of events leading from the specific distal precursors to the products of the various pathways have several features in common. Thus, the all-trans-polyprenyl-, all-trans-geranylgeranyl- and, if reduction of isoprene residues takes place at the allylic pyrophosphate stage, trans-phytyl-units are synthesised independently of the nuclei as pyrophosphates. The addition of the side chain to the nucleus is probably accompanied by the loss of the nuclear carboxyl group in the cases of the prenylation reactions involving 1,4-dihydroxynaphthoate (formed from 2-succinylbenzoate) and homogentisate, the resulting compounds then being converted by appropriate sequences of reactions into the products characteristic of the cell (organelle) in which the syntheses are taking place. In the case of the prenylation reaction involving 4-hydroxybenzoate decarboxylation will be the next step if the pathway to ubiquinone is similar to the one which operates in many Gram-negative bacteria (THRELFALL and WHISTANCE, 1971); however, it may be that methoxylation of the ring will precede decarboxylation (Sect. 3.2.2).

The feeding experiments which led to the establishment of the pathways also provided evidence that: the chloroplast is able to carry out the de novo synthesis of the chloroplast quinones and chromanols from CO_2 (THRELFALL and WHISTANCE, 1971); ubiquinone is synthesised at some site (mitochondrion) outside the chloroplast (THRELFALL and WHISTANCE, 1971); the extrachloroplastidic α-tocopherol and non-α-tocopherols are synthesized outside the chloroplast (JANISZOWSKA and PENNOCK, 1976). Direct support for the first proposal is provided by the fact that illuminated spinach chloroplasts are able to synthesise ^{14}C-labelled α-tocopherol and plastoquinone-9 from $^{14}CO_2$ or D-[1,6-^{14}C] shikimate or DL [3-^{14}C] tyrosine (BICKEL et al., 1978). Evidence is also available to support the second proposal

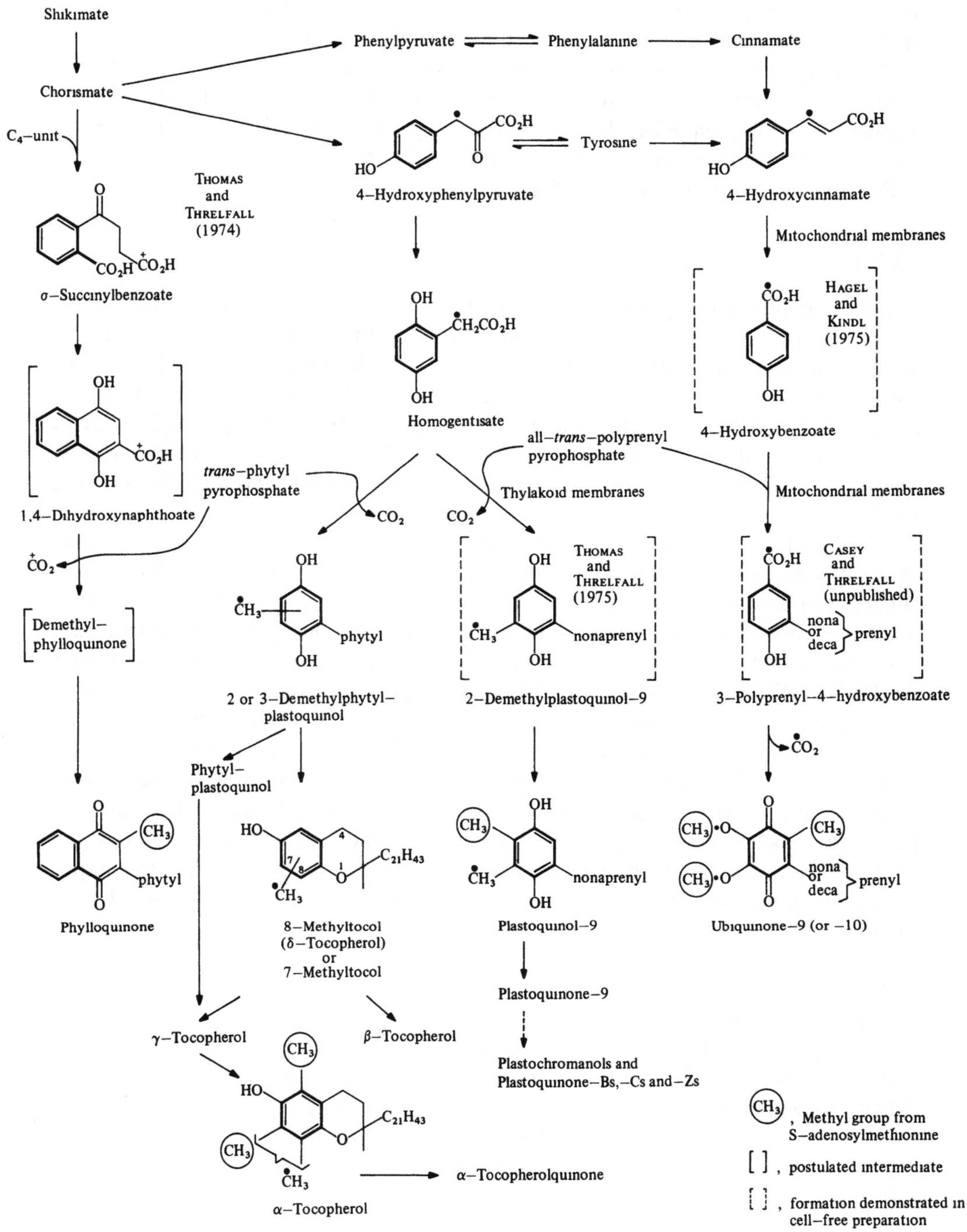

Fig. 1. Possible pathways for the biosynthesis of terpenoid quinones and chromanols [based on the reviews by JANISZOWSKA and PENNOCK (1976), THRELFALL (1971) and THRELFALL and WHISTANCE (1971)]. Recent (1979) experiments with isolated chloroplasts indicate that the terminal steps for the biosynthesis of α-tocopherol in the chloroplast are: 2-demethyl-phytylplastoquinol→phytylplastoquinol→γ-tocopherol→α-tocopherol (HUTSON and THRELFALL, unpublished work, SCHULTZ et al., personal communication)

that the synthesis of ubiquinone takes places in the mitochondrion. Thus, the inner mitochondrial membrane of cucumber roots is able to carry out the non-oxidative synthesis of 4-hydroxybenzoate from 4-hydroxycinnamate (HAGEL and KINDL, 1975) and mitochondrial membranes of broad bean seeds are able to synthesise 3-decaprenyl-4-hydroxybenzoate (Fig. 1), a probable intermediate on the pathway to ubiquinone-10, from 4-hydroxybenzoate and IPP (J. CASEY and D.R. THRELFALL, unpublished work).

3.3.2 Fungi

These organisms are capable of the synthesis of ubiquinones and dihydroubiquinones (Table 2). The specific distal precursors of the various carbon atoms, as in the case of photosynthetic plants (Fig. 1), are 4-hydroxybenzoate, MVA and the S-methyl group of *S*-adenosylmethionine (SAM). The origin of the 4-hydroxybenzoate is not known, although 4-hydroxycinnamate is incorporated (poorly) into ubiquinone-6 in *S. cerevisiae* (THRELFALL and WHISTANCE, 1971). It may be, however that it is formed directly from chorismate (THRELFALL and WHISTANCE, 1971).

Chemical analyses have shown that some fungi contain polyprenyl compounds of the type known to be late intermediates on the pathway leading from 4-hydroxybenzoate to ubiquinone in bacteria (THRELFALL and WHISTANCE, 1971). In addition it has been found that some yeast mutants accumulate 3-hexaprenyl-4,5-dihydroxybenzoate (GOEWERT et al., 1977) and that yeast mitochondria supplemented with NADPH, ATP, and succinate are able to synthesise 6-methoxy-2-hexaprenyl-phenol, 5-demethoxyubiquinone-6 and ubiquinone-6 from either 4-hydroxybenzoate, IPP and SAM or 3-hexaprenyl-4-hydroxybenzoate (CASEY and THRELFALL, 1978). Taken together the above results suggest that fungal mitochondria synthesise ubiquinones and dihydroubiquinones either by the bacterial pathway (a) or, as proposed for baker's yeast in particular and eukaryotic cells in general (GOEWERT et al., 1977), a deviant of the bacterial pathway (b). In the case of dihydroubiquinones the saturation of the side chain probably takes place either just before or just after the prenylation step:

4-Hydroxybenzoate
↓ Polyprenyl-pyrophosphate
3-Polyprenyl-4-hydroxybenzoate → (a) 2-Polyprenyl-phenol → 6-Methoxy-2-polyprenyl-phenol → Ubiquinone
3-Polyprenyl-4-hydroxybenzoate → (b) 3-Polyprenyl-4,5-dihydroxy-benzoate → 6-Methoxy-2-polyprenyl-phenol

3.4 Formation in Relation to Plant Development

3.4.1 Seed Germination

Seeds contain ubiquinone-9 (or-10), small amounts of plastoquinone, and a mixture of tocochromanols, which can consist of non-α-tocopherols with some α-tocopherol, predominantly tocopherols with some tocotrienols or predominantly tocotrienols with some tocopherols. They may also contain dimers and trimers of their constituent tocochromanols (Table 2). The types of tocochromanol found in a seed are species-specific, the levels and proportions being determined by the variety of the seed, maturity at harvest, location of growth, year of harvest, length of storage, and treatment (DICKS, 1965). The germination rate of grain is highly correlated with its tocopherol content. Thus, inland wheat with a germination rate of 75%–93% contains 4–5 mg% of tocopherol, whereas that with a germination rate of 3%–4% contains 3 mg% of tocopherol (DICKS, 1965). In general, it would appear that ubiquinone and tocopherols are present in all those parts of the seed capable of active metabolism, whereas tocotrienols are restricted to the starchy and oily endosperms of monocotyledenous seeds (GAUNT and STOWE, 1967; HALL and LAIDMAN, 1968a).

On germination the changes which occur in the quinone and chromanol content of wheat (*T. vulgare*) grain (contains ubiquinone-9, α- and β-tocopherol, and α- and β-tocotrienol) and pea (*P. sativum*) seeds (contain ubiquinone-10, plastoquinone-9 and γ-tocopherol) are very similar (GAUNT and STOWE, 1967; HALL and LAIDMAN, 1968b). In the first two to three days the most significant changes are a doubling in the ubiquinone contents of the aleurone layer of wheat and the cotyledons of pea. These increases are related to the increase in respiratory activities of these tissues, and in the case of the aleurone layer it is brought about by a gibberellin-independent synthesis of the quinone which occurs in step with the increase in size and development of the mitochondria. It is worth noting that it has been suggested that the α-tocopherol in the aleurone cells is associated with lipid-rich spherosomes, and that it is involved in the gibberellin-triggered development of the endoplasmic reticulum.

If the seeds are germinated in the dark (to eliminate effects due to photosynthesis) for periods of up to 7 (wheat) or 14 (pea) days, then the developing etiolated-shoots and roots contain ubiquinone-9 (wheat) or -10 (pea), plastoquinone-9, phylloquinone (pea) and α-tocopherol. Throughout these periods the tocochromanol content of the starchy endosperm of wheat and of the pea cotyledons undergo little change, and this means that contrary to earlier proposals the tocotrienols and non-α-tocopherols of seeds do not act as a source of precursors for α-tocopherol formation in the developing roots and shoots.

When seeds of wheat and pea as well as those of barley (*H. vulgare*), broad bean (*Vicia faba*), maize (*Z. mays*), oat (*Avena sativa*) and radish (*Raphanus sativum*), are allowed to germinate in the light the levels of plastoquinone-9 (in most analyses this probably represents plastoquinone-9 + plastoquinol, Table 2), plastoquinol-9 (barley and radish), phylloquinone and α-tocopherolquinone (absent from etiolated shoots) are higher than those for the corresponding parts of etiolated seedlings of the same age, whereas the levels of ubiquinone are unaffected (GAUNT and STOWE, 1967; HALL and LAIDMAN, 1968b; LICHTENTHALER, 1973; THRELFALL and GRIFFITHS, 1966; WALLWORK and CRANE, 1970). The α-tocopherol levels are either the same for both green and etiolated shoots (maize, wheat and pea) or higher in green shoots (barley and radish). If etiolated shoots are exposed to light for periods of up to 24 h, then concomitant with the formation of the thylakoid membranes there is an increase in the levels of plastoquinone-9 (plastoquinone-9 + plastoquinol-9), phylloquinone and, in barley and radish, α-tocopherol, and the appearance of α-tocopherolquinone, whereas the levels of ubiquinone (and, in maize, wheat, and pea, α-tocopherol) are unaffected. The actual magnitudes of the changes which take place depend on the nature of the compound and the age of the tissue. In broad bean, maize, oat and pea the levels of plastoquinone-Cs also increase on illumination whereas the plastoquinone-Bs disappear. It is of interest that the changes which take place on illumination of etiolated seedlings are similar to those which take place in bud opening of horse chestnut (WALLWORK and CRANE, 1970).

3.4.2 Higher Plant Development

There are many reports on the quinone and tocopherol contents of plant tissues of different types and at different stages of growth and maturity (DICKS, 1965: JANISZOWSKA and PENNOCK, 1976; WALLWORK and CRANE, 1970). Unfortunately only a few deal with the full quinone and chromanol complement of the tissue under investigation, and this makes it difficult to generalise as to the events which take place during the processes of growth and maturation.

In general terms, it would appear that the highest levels of chloroplast quinones and chromanols occur in leaves and the green and ripe parts of fruit. The levels are higher in green leaves than in pale green leaves and in large and mature leaves than in small and immature leaves, i.e., the levels are a reflection of the number, size, age, and plastoglobuli content of the chloroplasts within an individual leaf (Sect. 3.2.1).

WALLWORK and CRANE (1970), from an analysis of the literature, came to the conclusion that in the photosynthetic tissues of annual plants there is always a significant rise in the levels of plastoquinones and α-tocopherol in the latter part of the growing season and that in the case of herbaceous plants this rise is followed by a decline in levels as leaf fall is imminent. The level of α-tocopherolquinone remains fairly constant throughout the growth season, whereas that of phylloquinone either follows the same pattern as the plastoquinones or remains fairly constant. It also appears as if the accumulation of non-α-tocopherolquinones occurs late in the season as the number and size of plastoglobuli increase. In perennial plants the quinone and chromanol levels seem to undergo the same changes as those of annual plants before declining in late winter and early spring. In the succeeding year of growth the levels tend to be higher than at the corresponding stages of the previous year.

In the photosynthetic regions of green fruits the levels of chloroplast quinones and α-tocopherol rise as the fruit grows and then, with the exception of phylloquinone, continue to rise as the fruit ripens (LICHTENTHALER, 1968a).

The level of the mitochondrial quinone, ubiquinone, in green leaves either remains fairly constant or increases slowly throughout the growth season.

No studies have been carried out on the seasonal variations of non-α-tocopherols in leaves. In fruits and seeds, however, it is well established that they are accumulated, along with tocotrienols in some seeds, as the fruit or seed develops (DICKS, 1965; JANISZOWSKA and PENNOCK, 1976).

3.4.3 Regulation in Higher Plants

The regulation of quinone and α-tocopherol formation in photosynthetic tissues of higher plants has been studied by LICHTENTHALER and his colleagues (LICHTEN-

THALER, 1973, 1977). The experimental approaches used were (1) to follow the changes in the levels of plastoquinone-9, plastoquinol-9, phylloquinone, ubiquinone, α-tocopherolquinone and α-tocopherol in young etiolated seedlings of barley and radish exposed to white light, red light etc. and (2) to determine the effects of exogenous compounds on the production of the compounds listed under (1) in light-grown seedlings of barley and radish. The major findings of these studies were that active phytochrome, light quality (blue, red), light intensity, the phytohormones IAA and kinetin and the nitrogen status of the plant all exert effects on the formation of compounds by the developing chloroplast, but that with the exception of nitrogen status they have little effect on the formation of the mitochondrial quinone, ubiquinone (Fig. 2). Evidence was also obtained that protein synthesis in the cytoplasm and in the plastid are needed for the formation of chloroplast components.

In the fully developed chloroplast the production of plastoquinol-9 and α-tocopherol apparently proceeds in an unregulated manner and this leads to their deposition in plastoglobuli.

A further factor to be taken into consideration is that the quinones and chromanols undergo turnover. Thus, in the chloroplast the relative rates of turnover of these compounds are: plastoquinone-9 > phylloquinone > α-tocopherol and α-tocopherolquinone. In this connection, it is worth drawing attention to the diurnal variation of δ-tocopherol in *Xanthium strumarium* leaves (Sect. 3.5.3).

3.4.4 Algae

The levels of quinones and chromanols in unicellular green algae are regulated by many of the factors that control their levels in the photosynthetic parts of higher plants e.g., light intensity and quality, nitrogen status etc. (LICHTENTHALER, 1973, 1977; THRELFALL and GRIFFITHS, 1966; WALLWORK and CRANE, 1970). In addition, however, their formation is also governed by the presence or absence of metabolizable carbon sources in the growth medium. The highest levels of chloroplast quinones (phylloquinone or 5′-monohydroxyphylloquinone, plastoquinone-9 (this includes plastoquinol-9) and α-tocopherolquinone) and α-tocopherol are found in autotrophic cells. The levels tend to be highest at high light intensities and in old cells. The lowest levels are found in dark-grown heterotrophic cells, and intermediate values for the levels are found in photoheterotrophic cells. As in higher plants α-tocopherol is the least affected chloroplast component and this is possibly further support for the occurrence of some of this chromanol outside the chloroplast. The levels of chloroplast quinones and α-tocopherol in etiolated and photoheterotrophic cells increase on illumination and exhaustion of metabolisable carbon sources respectively.

The level of ubiquinone is usually higher in etiolated and photoheterotrophic cells than in autotrophic cells, and, as in leaves, it often increases slowly throughout the growth period of the alga.

It has been calculated that in autotrophic cells of *Chlorella pyrenoidosa* the approximate rates of turnover [$T^1/_2$ (h)] of phylloquinone, plastoquinone-9, plastoquinol-9, α-tocopherol and α-tocopherolquinone are 11, 120, 12, 113, and 13 respectively (LICHTENTHALER, 1977).

In brown sea weeds the levels and proportions of α-tocopherol and non-α-tocopherols vary between different parts of the plant (JENSEN, 1969). The rapidly

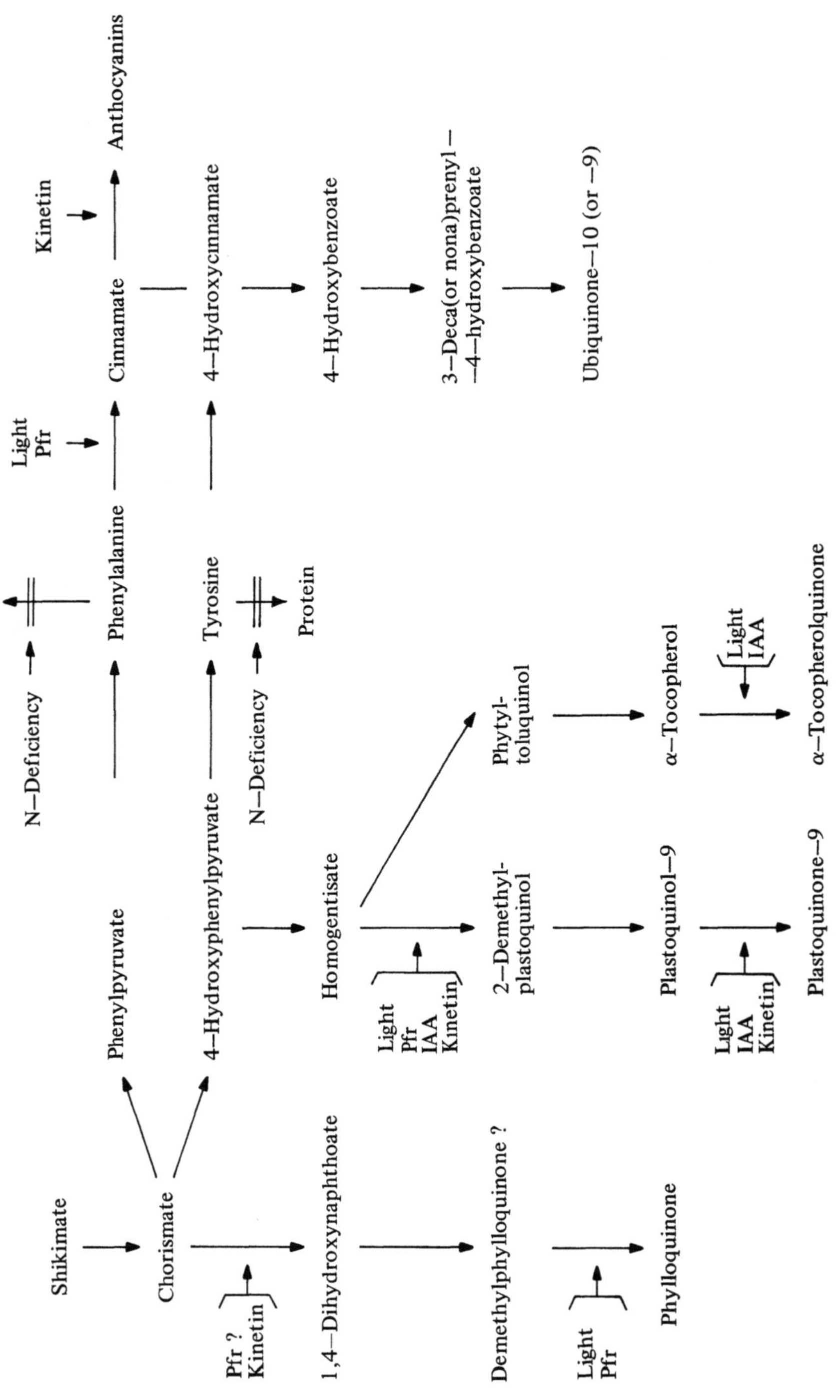

Fig. 2. Regulation of terpenoid quinone and chromanol formation in higher plants (LICHTENTHALER, 1973)

growing tips have lower levels of α-tocopherol than the oldest internodes whilst the highest levels of δ-tocopherol are found in non-pigmented parts of the plant. The highest levels of tocopherols are found in Autumn and the lowest in Spring, the effect being exaggerated by marked changes in the level of δ-tocopherol. The tocopherols also exhibit a diurnal variation, the α- and δ-tocopherols levels seeming to vary inversely to each other. The δ-tocopherol increasing in level towards the end of the periods of dessication and the first part of the submerged period, whereas the α-tocopherol decreases during the submerged period.

3.4.5 Tissue Cultures

Tissue cultures always contain ubiquinone, plastoquinone-9 and α-tocopherol, and in the cases of yellow-green and green tissue cultures they may also contain phylloquinone and α-tocopherolquinone (LICHTENTHALER, 1977; JANISZOWSKA and PENNOCK, 1976; THRELFALL and GRIFFITHS, 1966). The levels of the plastid components, as expected, are determined by the number, type, and degree of development of the plastids, and as in the case of algae they are repressed as the sugar content of the growth medium is increased (LICHTENTHALER, 1973; 1977).

3.4.6 Fungi

The synthesis of ubiquinones in fungi grown in liquid culture is affected by factors such as the oxygen tension of the medium and the nature and quantities of the carbon and nitrogen sources.

In the case of *S. cerevisiae* the highest levels of ubiquinone-6 are found in cells grown aerobically on a non-fermentable substrate. These levels are markedly reduced in catabolite repressed cells (this explains the low levels in petite mutants) and the quinone is barely detectable in anaerobic cells (GORDON and STEWART, 1969). On aeration of anaerobic cells or on exhaustion of fermentable substrates by repressed cells, the ubiquinone contents of the cells rise dramatically (GORDON and STEWART, 1969). The increase in level of ubiquinone in anaerobic cells undergoing respiratory adaptation is unaffected by chloramphenicol but is 50% inhibited by cycloheximide (GORDON and STEWART, 1971), and this shows that all of the enzymes involved in the biosynthesis of ubiquinone are constitutive to the organism and are present in the anaerobic cell in amounts sufficiently high to allow the synthesis of significant amounts of ubiquinone, even though the synthesis cannot proceed in the anaerobic cell due to the O_2-requirement of some of the terminal steps on the biosynthetic pathway.

As part of an investigation of the utilization of hydrocarbons by microorganisms, the effects of carbon source (alkanes, carbohydrates, TCA intermediates etc.), detergents, metal ions and vitamins on the synthesis of ubiquinone in *Candida tropicalis* pK 233 were studied (SHIMIZU et al., 1970). The results obtained showed that ubiquinone synthesis is repressed by readily fermentable substrates but not by *n*-alkanes and that there is a complicated interrelationship between 4-aminobenzoate and 4-hydroxybenzoate on ubiquinone production and yeast growth. The highest levels of ubiquinone (4.52 mg g dry cells^{-1}) are produced in cells grown in a vigorously aerated medium containing −5% of a hydrocarbon mixture rich in *n*-undecane, the detergent polyoxyethylene sorbitan oleyl ether and 4-hydroxybenzoate.

3.5 Function

3.5.1 Plastoquinones and Ubiquinones

Plastoquinone-9/plastoquinol-9 and ubiquinone/ubiquinol (or dihydroubiquinone/dihydroubiquinol) function as redox couples which, as well as transferring electrons, also act as vectorial translocators of protons in the chloroplast electron transport chain and mitochondrial electron transport chain respectively (Vols. 3 and 5 of this Encyclopedia). The functions of the other plastoquinones are not known.

3.5.2 Phylloquinone

There is no report of a function for this thylakoid naphthaquinone.

3.5.3 Tocopherols and Tocopherolquinones

The functions of these compounds in plant tissue have yet to be elucidated. In a recent review it was considered that there is some evidence for the function of tocopherols: as antioxidants; as structural components of chloroplast membranes and as factors in the initiation of flowering (JANISZOWSKA and PENNOCK, 1976). It was considered unlikely that α-tocopherol/α-tocopherolquinone could function as a redox couple in the chloroplast electron transport chain, but there was a possibility of an α-tocopherolquinone/α-tocopherolquinol redox couple. More recently there have been reports that the close correlation between δ-tocopherol concentrations in *Xanthium strumarium*, the phytochrome response and the process of flower induction (BATTLE et al., 1977) can be extended to include tocopherol oxidase (a tocopherol-destroying enzyme) activity (GAUNT and PLUMPTON, 1978), and that there may be a α-tocopherol/α-tocopherone redox couple functioning between the DCMU and DBMIB inhibition sites in the spinach chloroplast electron transport chain (BARR and CRANE, 1977).

References

Barr, R., Crane, F.L.: Evidence for α-tocopherol function in the electron transport chain of chloroplasts. Plant Physiol. *59*, 433–436 (1977)

Barr, R., Magree, L., Crane, F.L.: Quinone distribution in horse-chestnut chloroplasts. Am. J. Bot. *54*, 365–374 (1967)

Battle, R.W., Laidman, D.L., Gaunt, J.K.: The relationship between floral induction and δ-tocopherol concentrations in leaves of *Xanthium stumarium* L. Biochem. Soc. Trans. *5*, 322–324 (1977)

Bickel, H., Palme, L., Schultz, G.: Incorporation of shikimate and other precursors into aromatic amino acids and prenylquinones in isolated spinach chloroplasts. Phytochemistry *17*, 119–124 (1978)

Brett, C.T., Leloir, L.F.: Dolichyl monophosphate and its sugar derivatives in plants. Biochem. J. *161*, 93–101 (1977)

Bucke, C.: The distribution and stability of α-tocopherol in subcellular fractions of broad bean leaves. Phytochemistry *7*, 693–700 (1968)

Bu'Lock, J.D., Osagie, A.U.: Prenols and ubiquinones in single-strain and mated cultures of *Blakeslea trispora*. J. Gen. Microbiol. *76*, 77–83 (1973)

Casey, J., Threlfall, D.R.: Synthesis of 5-demethoxyubiquinone-6 and 3-hexaprenyl-4-hydroxybenzoate in yeast mitochondria. FEBS Lett. *85*, 249–253 (1978)

Chakraborty, M.K., Weybrew, J.A.: The chemistry of tobacco trichomes. Tob. Sci. *7*, 122–127 (1964)

Daleo, G.R., Pont Lezica, R.: Synthesis of dolichol phosphate by a cell-free extract from pea. FEBS Lett. *74*, 247–250 (1977)

Dicks, M.: Vitamin E content of foods and feeds for human consumption. Wyo. Agric. Exp. Stn. Bull. *435*, 1–193 (1965)

Etman-Gervais, C., Tendille, C., Polonsky, J.: 3-Déméthyl-plastoquinone-9 et 3-déméthylplastoquinone-8 isolées des bulbes d'*Iris hollandica*. Nouv. J. Chim. *1*, 323–325 (1977)

Gaunt, J., Plumpton, E.S.: Control in vitro of tocopherol oxidase by light in extracts from leaves of *Xanthium strumarium*. Biochem. Soc. Trans. *6*, 143–145 (1978)

Gaunt, J., Stowe, B.B.: Analysis and distribution of tocopherols and quinones in the pea plant. Plant Physiol. *42*, 851–858 (1967)

Goewert, R.R., Sippel, C.J., Olsen, R.E.: The isolation and identification of a novel intermediate in ubiquinone-6 biosynthesis by *Saccharomyces cerevisiae*. Biochem. Biophys. Res. Commun. *77*, 599–606 (1977)

Gordon, P.A., Stewart, P.R.: Ubiquinone formation in wild-type and petite yeast: The effect of catabolite repression. Biochem. Biophys. Acta *177*, 358–360 (1969)

Gordon, P.A., Stewart, P.R.: The effect of antibiotics on lipid synthesis during respiratory development in *Saccharomyces cerevisiae*. Microbios *4*, 115–132 (1971)

Hagel, P., Kindl, H.: *Para*-hydroxybenzoate synthetase: A complex associated with mitochondrial membranes of roots of *Cucumis sativus*. FEBS Lett. *59*, 120–124 (1975)

Hall, G.S., Laidman, D.L.: The determination of tocopherols and isoprenoid quinones in the grain and seedlings of wheat (*Triticum vulgare*). Biochem. J. *108*, 465–473 (1968a)

Hall, G.S., Laidman, D.L.: The pattern and control of isoprenoid quinone and tocopherol metabolism in the germinating grain of wheat (*Triticum vulgare*). Biochem. J. *108*, 475–482 (1968b)

Hemming, F.W.: Polyprenols. In: Natural substances formed biologically from mevalonic acid. Goodwin, T.W., (ed.), pp. 105–117. New York–London: Academic Press 1970

Hemming, F.W.: Lipids in glycan biosynthesis. In: MTP Int. Rev. Sci. Biochem. Lipids, Biochem. Ser. One. Goodwin, T.W. (ed.), Vol. 4, pp. 39–97. London: Butterworths 1974

Hughes, C.T., Gaunt, J.K., Laidman, D.L.: The subcellular distribution of tocopherols in green leaves of *Pisum sativum*. Biochem. J. *124*, 9P (1971)

Irvine, W.J., Woollen, B.H., Jones, D.H.: Bombiprenone from *Nicotiana tabacum*. Phytochemistry *11*, 467–469 (1972)

Janiszowska, W., Pennock, J.F.: The biochemistry of Vitamin E in plants. Vitam. Horm. *34*, 77–105 (1976)

Janiszowska, W., Michalsky, W., Kasprzyk, Z.: Polyprenyl quinones and α-tocopherol in *Calendula officinalis*. Phytochemistry *15*, 125–127 (1976)

Jensen, A.: Tocopherol content of seaweed and seaweed meal. II. Individual, diurnal and seasonal variations in some *Fucaceae*. J. Sci. Food Agric. *20*, 454–458 (1969)

Lichtenthaler, H.K.: Die Verbreitung der lipophilen Plastidenquinone in nicht-grünen Pflanzengeweben. Z. Pflanzenphysiol. *59*, 195–210 (1968a)

Lichtenthaler, H.K.: Verbreitung und relative Konzentration der lipophilen Plastidenquinone in grünen Pflanzen. Planta *81*, 140–152 (1968b)

Lichtenthaler, H.K.: Regulation der Lipochinonsynthese in Chloroplasten. Ber. Dtsch. Bot. Ges. *86*, 313–329 (1973)

Lichtenthaler, H.K.: Regulation of prenylquinone synthesis in higher plants. In: Lipids and lipid polymers in higher plants. Tevini, M., Lichtenthaler, H.K. (eds.). Berlin, Heidelberg, New York: Springer 1977

Lichtenthaler, H.K., Tevini, M.: Die Verteilung der Pigmente, Plastidenchinone und Plastoglobuli aus verschieden beschallten Spinatchloroplasten gewonnenen Partialfraktionen. Z. Pflanzenphysiol. *62*, 33–50 (1970)

Newton, R.P., Pennock, J.F.: The intracellular distribution of tocopherols in plants. Phytochemistry *10*, 2323–2328 (1971)

Newton, R.P., Walton, T.J., Moyse, C.D.: Non-α-tocopherols in the unicellular blue-green alga *Gloeocapsa*. Biochem. Soc. Trans. *5*, 1486–1489 (1977)

Sasak, W., Mankowski, T., Chojnacki, T.: Heterogeneity of C_{55}-polyprenol from leaves of *Magnolia campbellii*. Chem. Phys. Lipids *18*, 199–204 (1977)

Shimizu, S., Tanaka, A., Fukui, S.: Studies on the utilization of hydrocarbons by microorganisms. J. Ferment. Technol. *48*, 549–555 (1970)

Tarasova, N.V., Ivanonva, G.I., Gololobou, A.D.: Localisation of ubiquinone-9 in the electron transport chain of *Candida guilliermondii*. Mikrobiologiya *45*, 400–410 (1976)

Thomas, G., Threlfall, D.R.: Incorporation of shikimate and 4-(2-carboxyphenyl)-4-oxobutyrate into phylloquinone. Phytochemistry *13*, 807–813 (1974)

Thomas, G., Threlfall, D.R.: Synthesis of 3-polyprenyltoluquinols and 4-carboxy-2-polyprenylphenols by cell-free preparations of *Euglena gracilis*. Phytochemistry *14*, 2607–2615 (1975)

Thomson, R.H.: In: Naturally occurring quinones. London: Academic Press 1971

Threlfall, D.R.: The biosynthesis of Vitamins E and K and related compounds. Vitam. Horm. *29*, 153–200 (1971)

Threlfall, D.R., Griffiths, W.T.: Biosynthesis of terpenoid quinones. In: Biochemistry of chloroplasts. Goodwin, T.W. (ed.), Vol. 2, pp. 254–271. London: Academic Press 1966

Threlfall, D.R., Whistance, G.R.: Biosynthesis of isoprenoid quinones and chromanols. In: Aspects of terpenoid biochemistry. Goodwin, T.W. (ed.), pp. 357–404. London: Academic Press 1971

Threlfall, D.R., Whistance, G.R.: Dehydrophylloquinone, α-dehydrotocopherolquinone and dehydrotocopherols from etiolated maize and barley shoots. Phytochemistry *16*, 1903–1907 (1977)

Toyoda, M., Fukawa, H., Shimizu, T.: Isolation of new polyisoprenyl ketone from silkworm faeces. Tetrahedron Lett. *35*, 3837–3840 (1968)

Wallwork, J.C., Crane, F.L.: The nature, distribution and biosynthesis of prenyl phytoquinones and related compounds. Prog. Phytochem. *2*, 267–341 (1970)

Wild, A., Conrad, G., Zickler, H-O.: Der Gehalt an Plastidenchinonen und Pigmenten von *Zea mays* und *Atriplex rosea* in Abhängigkeit von der Lichtintensität und dem Alter der Pflanzen. Planta *103*, 181–187 (1972)

5.4.2 Polyisoprene

B.L. ARCHER

1 Introduction

Rubber is reputed to have been used as far back as the eleventh century by the natives of Honduras for their ball games. Chemically, pure rubber is a hydrocarbon, cis-1,4-polyisoprene $(C_5H_8)_n$ of high molecular weight. Gutta percha, obtained mainly from *Palaquium gutta* trees, is trans-1,4-polyisoprene of lower molecular weight than rubber whereas the similar but more resinous balata is produced from *Mimusops balata*. The term gutta refers to the trans hydrocarbon from either source. Chicle, a mixture of low molecular weight cis- and trans-polyisoprenes in the approximate relative proportions of 1:2, together with acetone soluble "resins", is obtained from *Achras sapota*.

2 Occurrence

2.1 Rubber, Gutta, and Chicle

MOYLE (1942) lists 1800 species of higher plants which contain polyisoprene but in most cases the amount present is small. STEWART et al. (1955) isolated from some unknown species of the fungal genera *Lactarius* and *Peziza* a low molecular weight polymer with an infra-red spectrum similar to that of rubber, but the identity of the material was not established further. KURIHARA et al. (1962, 1963, 1964) found cis-1,4-polyisoprene in a few species of *Lactarius, Russula*, and *Hygrophorus*, where the material accounted for up to 6% of the dry weight of the fungus.

ARCHER and AUDLEY (1973) list some of the more important rubber, gutta, and chicle-bearing species, most of which grow in the tropical regions of S.E. Asia, Africa and S. America. Usually the polyisoprene is rubber and only a relatively few species synthesise gutta. It appears that either the cis- or the trans-isomer, but not both, is in any species with the exception of *Achras sqpota*, or possibly *Dyera* spp. LEEPER and SCHLESINGER (1954) claimed to have detected small quantities of rubber in guttas from *Mimusops* and *Palaquium* spp. but no conclusive proof of its identity was obtained. The search for non-tropical species of rubber-bearing plants has recently been intensified, particularly in the U.S.A. (BUCHANAN et al., 1976; National Academy of Sciences, 1977), but apart from a pilot plant for the extraction of rubber from guayule (*Parthenium argentatum*) in Mexico

(CAMPOS-LOPEZ et al., 1976) and a very small amount from *Taraxacum kok-saghyz* in the U.S.S.R. there is no commercial production of rubber from any species other than *Hevea brasiliensis* at present anywhere in the world.

2.2 Mode of Occurrence of Polyisoprene in Plants

Polyisoprene occurs as minute particles in plants, often in the form of latex. In those plants which contain latex but no polyisoprene, the disperse phase is commonly some other isoprenoid material such as essential oils or resins (BONNER and GALSTON, 1947; ESAU, 1965; METCALFE, 1966). The latex is contained in specialised cells or vessels known as laticifers; these are often tubular and may or may not be interconnected. Laticifers can be of primary or secondary origin and may be either articulated or not (ESAU, 1965).

2.3 The Laticiferous System in *Hevea brasiliensis*

The formation and disposition of laticifers in plants has been described by METCALFE and CHALK (1957), ESAU (1965) and METCALFE (1966), but the cytology of the laticifers themselves has been studied mainly in *Hevea brasiliensis*. In this species rubber is largely formed and stored in the bark, in rings of latex vessels interspersed with the sieve tubes of the secondary phloem of the trunk, branches, and roots; laticifers are also present in the leaves, flowers, and fruits. In the cortex the latex vessels run at a slight angle to the longitudinal axis of the trunk or branches and originate from longitudinally contiguous rows of cells, from which the cross walls are resorbed. Anastomoses between adjacent vessels in each ring allow the latex to drain from a large area of the cortex on tapping, but there are few or no connections between the rings.

2.4 The Composition of Latex

Electron microscopy of laticifers from several plant families has shown that latex is a highly specialised cytoplasm containing typical structures such as nuclei and mitochondria, endoplasmic reticulum, and ribosomes, as well as polyisoprene and resin particles (DICKENSON 1963, 1965, 1969a; SÁRKÁNY et al., 1965; SASSEN, 1965; SCHULZE et al., 1967; GOMEZ, 1976). In addition, *H. brasiliensis* latex contains two other types of particle; one of these, the lutoid (L in Fig. 1) is a major component of latex (HOMANS and VAN GILS, 1948) and forms the bulk of the "bottom fraction" which is sedimented from fresh latex by centrifugation (MOIR, 1959; PUJARNISCLE, 1968). The lutoids are 1–3 μm in diameter and are osmotically sensitive (RUINEN, 1950), being bounded by a unit membrane characterised by a high content of phosphatidic acid and an excess of saturated over unsaturated fatty acids (DUPONT et al., 1976; D'AUZAC et al., 1977b). These latter authors claim that up to 54% of the RNA of the lutoids (MARIN and TROUSLOT, 1975) is associated with the membrane, probably on the inside (MARIN, 1976). In young vessels the lutoid particles contain bundles of helical microfibrils of diameter

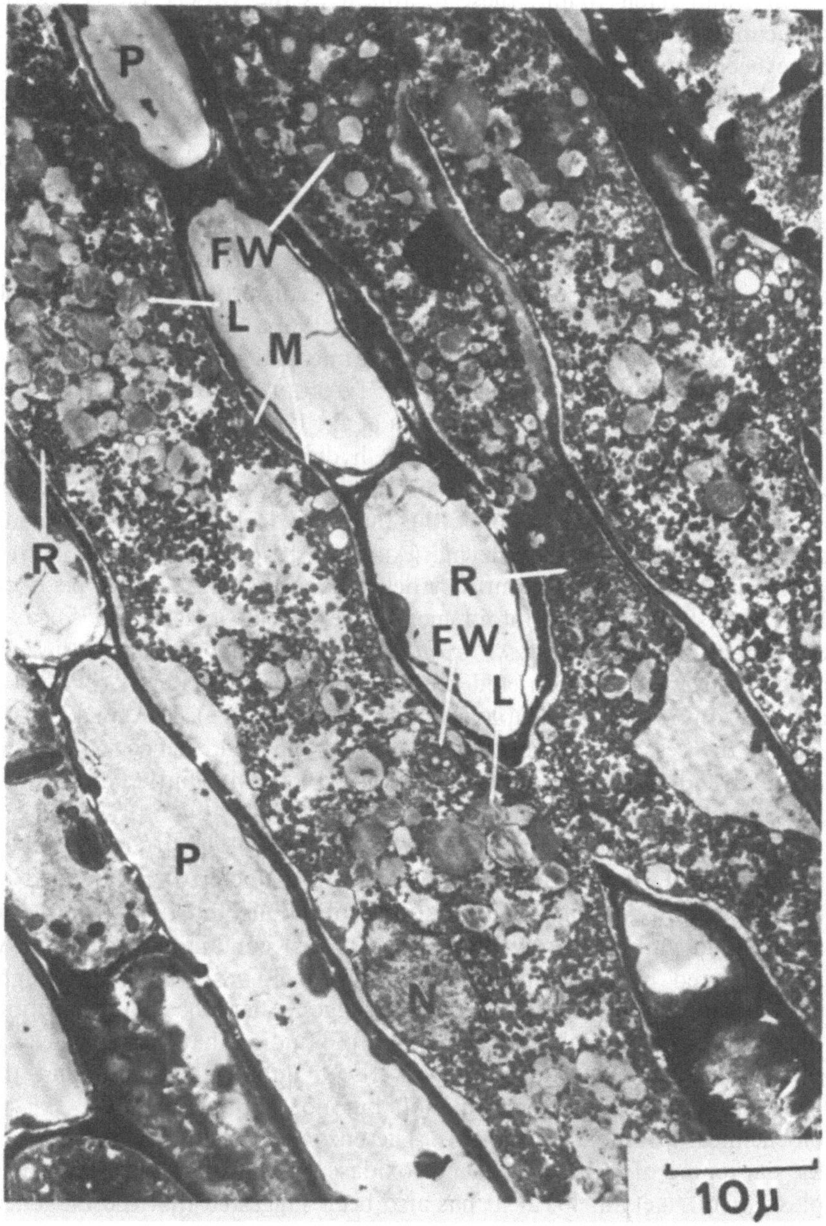

Fig. 1. Electron micrograph of ultra-thin longitudinal section of young latex vessels from the green branch region of a five-year-old *Hevea* tree; fixed in osmium tetroxide; embedded in epoxy resin. *FW* Frey-Wyssling complex; *M* mitochondrion; *P* parenchyma cell; *L* lutoid; *N* nucleus; *R* rubber (courtesy of P.B. Dickenson)

12.5 nm and pitch 10 nm; these disappear as the vessels age (DICKENSON, 1965) and consist almost entirely of acidic protein (AUDLEY, 1966); similar microfibrils have been found in *Ficus elastica* (DICKENSON, 1969a). As the lutoid matures the microfibrils disappear and sometimes they are replaced by alternative structures called microhelices. These have a diameter of 20–22 nm, a pitch of 30–35 nm and are composed partly of basic protein. The microhelices can be prepared in vitro by dialysis of lutoid extracts (GOMEZ and YIP, 1975). No physiological function has yet been ascribed to either the microfibrils or the microhelices.

WIERSUM (1957) suggested that the lutoids represent a polydisperse vacuome, and this view is supported by several of their properties including their mode of development and structure (DICKENSON, 1969a; GOMEZ, 1976) and their ability to concentrate substances such as inorganic and organic ions (RIBAILLIER et al., 1971; MONTARDY and LAMBERT, 1977; ARCHER et al., 1969; Rubber Research Institute of Malaysia, 1976; HANOWER et al., 1977; D'AUZAC et al., 1977a; MATILE, 1976). The lutoids contain a range of hydrolytic enzymes with acid pH optima and appear to be analogous to lysosomes (PUJARNISCLE, 1968; PUJARNISCLE et al., 1970; D'AUZAC, 1975). Two of the lutoid hydrolases, acid phosphatase (JACOB and SONTAG, 1974) and lysozyme (ARCHER, 1976; TATA et al., 1976), have been purified from lutoids. Recently cytochrome C reductase and type B cytochromes have been found in the lutoid membrane (MOREAU et al., 1975), as has a Mg-dependent ATPase (D'AUZAC et al., 1977b). Such vacuolar bodies as described above appear to be restricted to the articulated laticifers of Papaveraceae and certain Compositae and Euphorbiaceae (MATILE, 1975). Lutoids play an important role in the cessation of latex flow after tapping and hence in the determination of latex yield (BOATMAN, 1966; SOUTHORN, 1969; MILFORD et al., 1969; RIBALLIER, 1970) (see also Sect. 6).

The Frey-Wyssling complex (DICKENSON, 1965, 1969a; GOMEZ, 1976) is a minor component of *Hevea* latex and is 4–6 µm in diameter. It is bounded by a double membrane and contains a variety of membranous and tubular structures, as well as the lipid globules (Frey-Wyssling particles) containing β-carotene which were first detected by light microscopy (FREY-WYSSLING, 1929). The highly ordered structure of the Frey-Wyssling complex suggests that it has many functions although these are at present unknown. However, the finding that these particles can synthesise β-carotene (BARNARD and DICKENSON, unpublished) and that they contain plastoquinone and plastochromanol, but no ubiquinone (TAN, 1968) suggests a biochemical affinity with plastids. The enzyme long known to be responsible for the darkening of latex, polyphenol oxidase, is known to be located in these complexes (COUPÉ et al., 1972). It has also been suggested that the biogenesis of small rubber particles might originate within the complex (DICKENSON, 1969a).

The chemical analysis of *Hevea* latex has been summarised by ARCHER and AUDLEY (1973). More recently BRZOZOWSKA et al. (1974) have reported that the concentrations of amino acids in the sera of the cytoplasm and lutoids of *Hevea* latex are approximately equal, but there is a predominance of acidic amino acids in the cytosol and basic ones in the lutoids. NISHIMURA et al. (1977) using gas chromatography and mass spectroscopy have detected the presence in latex of 2-methylcyclobutanol together with smaller amounts of 2-methylcyclobutanone, and 2-methyl-2-hydroxycyclobutanone, but there is the possibility that these compounds were formed by degradation of other latex components during their isolation.

2.5 The Polyisoprene Particle in Plants

Hevea latex is a suspension in which the rubber particles are stabilised by an adsorbed film of phospholipids (Ho et al., 1976) and proteins, whilst triglycerides, together with β-sitosterol and related compounds are the principal constituents of the neutral lipids of the rubber phase. Bowler (1953) showed that the isoelectric point of the rubber particles was between 4.0 and 4.6, depending on the clone, but otherwise very little is known about the protein film surrounding the particles. In young trees the particles are spherical, ranging in diameter from 15 nm to 3000 nm with the mode at 100 nm, but in mature trees there is a higher proportion of larger particles, many of which may be pear-shaped (Schoon and Van der Bie, 1955; Dickenson, 1963). There is no evidence of internal structure in the rubber particles, [except possibly in some very small ones (Dickenson, 1969a)], indicating that the large particles are not formed by aggregation, as suggested by Lucas (1942). Stavely et al. (1961) found that the range of particle size of rubber from nine species was comparable with that of rubber from *Hevea*. There is no modern evidence to support the proposal of Hauser (1930), that the rubber particle in *Hevea* consists of a liquid core surrounded by a skin of rubber. On the other hand, the position might be different in those species which produce very resinous rubbers, since the location of the low molecular weight resins in the latex is unknown.

3 Chemical and Physical Properties of Polyisoprene

3.1 Purification

Pure rubber hydrocarbon can be produced from either dry rubber or latex. Lipids are removed from the former by acetone or alcohol extraction and the rubber dissolved in a mild solvent such as petroleum, without agitation; proteins etc. are left in the undissolved gel-phase (see below) (Bloomfield, 1951). Alternatively the rubber may be dissolved in a benzene-methanol mixture and fractionally precipitated by cooling (Bateman, 1947) or by the progressive addition of methanol (Burfield et al., 1976). Protein is precipitated with the first high molecular weight rubber fraction, the middle fractions contain pure rubber of decreasing molecular weight and any oxidised rubber remains to the last. With either method, rubber can be produced with a nitrogen content of 0.01%, but the hydrocarbon is not representative of that in the original sample. This drawback can be avoided by using latex as the starting material. The latex is treated with surfactant and/or proteolytic enzyme and then repeatedly diluted and reconcentrated by centrifugation. The rubber is finally coagulated with alcohol and extracted to remove surfactant. In this way rubber hydrocarbon containing less than 0.01% N and 0.05% ash was obtained by Allen and Bloomfield (1963). Gutta is readily purified by acetone extraction and fractional precipitation from petroleum (Morris, 1952). Chicle can be separated into gutta and rubber by precipitating the gutta from a benzene solution using ethyl acetate, followed by acetone to recover the rubber (SCHLESINGER and LEEPER, 1951).

3.2 Molecular Structure

Rubber and gutta are respectively cis- and trans-1,4-polyisoprenes (HARRIES, 1919; MEYER, 1942).

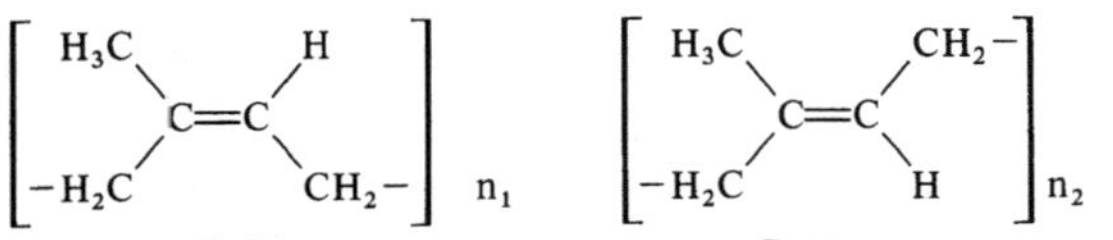

FRAGA (1959) showed by infrared spectroscopy that rubber, gutta, and balata all contained less than 0.2% of the 3:4 isomer, and CAMPOS-LOPEZ and PALACIOS (1976) have confirmed by nuclear magnetic resonance spectroscopy that the rubbers from *Guayule* and *Hevea* are identical in this respect also. GOLUB et al. (1962) used NMR to show that rubber and balata (gutta) were at least 99% cis and trans respectively, whilst TIDD (private communication) has recently reduced the minimum detection limit of the two isomers to 0.2% using Fourier transform NMR. Chicle is a mixture of 66% of trans and 34% of cis and there is no evidence of any molecules containing cis-trans or trans-cis linkages (TANAKA and SATO, 1976). However, it is not known whether the individual particles in chicle latex consist of one molecular species or are themselves mixtures.

3.3 Physical Properties

The most outstanding property of rubber is, of course, its elasticity, whilst gutta is a much harder and "hornier" material. The reason for this difference lies in differences in the crystallinity of the two substances. At ordinary temperatures, rubber is amorphous, with tangled, flexible, long molecules. When it is stretched, alignment of the molecules takes place, and crystallisation, with consequent hardening of the rubber, occurs. Gutta on the other hand is partly crystalline and therefore hard at room temperature. Gutta exists in three polymorphic forms (FISHER, 1953) depending on the degree of extension, and on the temperature, and can be made rubbery by warming: similarly rubber hardens on cooling but no distinct polymorphs of rubber have been detected.

3.4 The Gel-Phase in Natural Rubber

Although the hydrocarbon in the fresh latex from regularly tapped *H. brasiliensis* trees is almost entirely soluble in solvents such as benzene and chloroform, storage of the rubber in either the latex or dry form causes a fall in the solubility. The insoluble fraction, or gel, contains most of the nitrogenous impurities and consists of crosslinked rubber in a network of proteinaceous material from the surface of the original latex particles (FREEMAN, 1954; ALLEN and BRISTOW, 1963).

3.5 Molecular Weight

A wide range of molecular weights in any one sample of rubber from *H. brasiliensis* has been observed by WESTALL (1968) and by SUBRAMANIAM (1972, 1976). The distribution was bimodal with maxima at $1–2 \times 10^5$ and $1–2.5 \times 10^6$, whilst the weight average molecular weight (M_w) ranged from 1.66 to 2.26×10^6, and the number average (M_n) varied from $1.95–5.21 \times 10^5$ for the 19 clones examined. CAMPOS-LOPEZ and ANGULO-SANCHEZ (1976) report the M_w for guayule rubber to be 2.5×10^6 with $M_n = 8.8 \times 10^5$.

3.6 Chemical Properties

Rubber and gutta exhibit the typical reactions of olefins, for example the addition of bromine at the double bonds gives the bromide $(C_5H_8Br_2)_n$, a reaction which has been used for the estimation of rubber in plant tissue (MIDDLETON and WESTGARTH, 1963). Ozone initially yields an unstable molozonide, which then gives complex fragmentation products (BARNARD, 1970). Conditions have been found whereby high yields of laevulinic acid are obtained by oxidation of the molozonide (MCSWEENEY, 1968). This degradation was used to determine the chemical structure of rubber, and has since been invaluable in establishing the purity of rubber biosynthesized from labelled precursors (KEKWICK et al., 1959, ARCHER et al., 1961). The oxidation of rubber which proceeds via the formation of cylic- and hydroperoxides has been reviewed by CUNNEEN (1968), and the same author has shown that the isomerisation of rubber or gutta, which does not occur spontaneously, can be induced by chemical treatments (CUNNEEN and HIGGINS, 1963). Vulcanisation, which is the crosslinking of rubber hydrocarbon chains by sulphur or other reagents to give a strong elastic three-dimensional network, has recently been reviewed by PORTER (1977), whilst LEEFLANG (1963) and DICKENSON (1969b) have reported on the microbiological degradation of vulcanised rubber.

Although there is no doubt about the basic structure of the unvulcanised rubber molecule given above, evidence is accruing that there may be a small number of "abnormal" side or end-groups attached to the main hydrocarbon chain. Biogenetic considerations (see below) suggest that one end is terminated by an isopropylidene and the other by a pyrophosphate ester end-group. Hydrolysis of the pyrophosphate would lead to a hydroxyl group, which might migrate as a result of the isomerisation known to occur in allylic alcohols. No direct evidence for any hydroxyl structures has yet been obtained.

Other "abnormal" side groups which have been postulated for the hydrocarbon chain include aldehyde, carboxyl, δ-lactone, and epoxide. The hardening of commercial rubber which occurs on storage was explained by SEKHAR (1961) as due to cross-linking of the polyisoprene molecules by interaction of a small number of aldehyde groups with amines normally present in latex. Chemical reaction of rubber with aldehyde reagents such as hydroxylamine, dimedone or 2,4-dinitrophenylhydrazine inhibits storage hardening and indicates the presence of 1.6 to 5.4 mol of aldehyde per 10^6 g of rubber, dependent on the clone (SUBRAMANIAM, 1976). GREGG and MACEY (1973) from an investigation of the I.R.-spectra of natural and synthetic cis-polyisoprenes, concluded there were no aldehyde groups present in natural rubber and that a lactone formed from a hydroxyl group δ to a carboxylic acid group on the rubber chain was responsible for all the observed effects. BURFIELD et al. (1976) using a titration method claimed to be specific for epoxides, found 78 mol of these groups per 10^6 g of rubber. The origin of any of these putative groups has not yet been investigated. It is possible, but unlikely, that the imperfections are built into the rubber molecule during biosynthesis (see below), since the enzymes involved would probably not accept the necessary alternative substrates. It is more likely that oxidation of the methyl side-groups, or of a terminal hydroxyl group, mentioned above, takes place after the biosynthesis of the polyisoprene chain is complete. This latter oxidation might occur enzymically in the latex vessels, by a mechanism similar to that of the oxidation of side-chain methyl groups in steroids (AYRES et al., 1957).

4 Biosynthesis of Rubber

The biosynthesis of rubber has been the subject of numerous reviews, the latest of which are those by ARCHER and AUDLEY (1967, 1973) and LYNEN (1969). The survey by BEYTIA and PORTER (1976) included other isoprenoids besides rubber and, in addition, MOIR (1969) has reviewed the metabolism and general enzymology of *Hevea* latex. Rubber formation takes place in the latex vessels and all the intermediates of the carbon pathway from acetate shown in Figure 2, with the exception of acetoacetyl-CoA, have been shown to be utilised by *Hevea* latex in vitro and virtually all the enzymes involved have been individually demonstrated (LYNEN, 1967).

The formation of rubber from [^{14}C]-acetate was first demonstrated by ARREGUIN et al. (1951) using guayule plants, and proof that the incorporations of [1-^{14}C]- and [2-^{14}C]-acetate into rubber were in the ratio of 2:3, in accordance with the pathway of Figure 2, was obtained by RABINOWITZ and TEAS (1961) using *Hevea* latex in vitro. [^{14}C]-Acetoacetate is converted to rubber by *Taraxacum kok-saghyz* (VYVALKO et al., 1957) and by *Hevea* latex (BEALING, unpublished), but its in toto incorporation into the polyisoprene skeleton has not been proved. [^{14}C]-Acetyl-CoA, is incorporated into rubber on incubation with *Hevea* latex, but is, unexpectedly, a poorer precursor than acetate. This has been attributed by WEEKS (1965) to the formation of free CoA, which is inhibitory (KEKWICK, 1965). Citrate may be a source of acetyl-CoA in latex since radioactive hydroxymethylglutaric acid and rubber are apparently formed on incubating [^{14}C]-citrate with whole latex (D'AUZAC, 1965). HIGGINS and KEKWICK (1973) have shown that the malonyl-CoA pathway suggested for cholesterol biosynthesis by BRODIE et al. (1964) is not significant in polyisoprene biosynthesis.

Hevea latex contains all the glycolytic enzymes (D'AUZAC and JACOB, 1969; BEALING, 1969, 1976) and although it is generally assumed that sugars are the

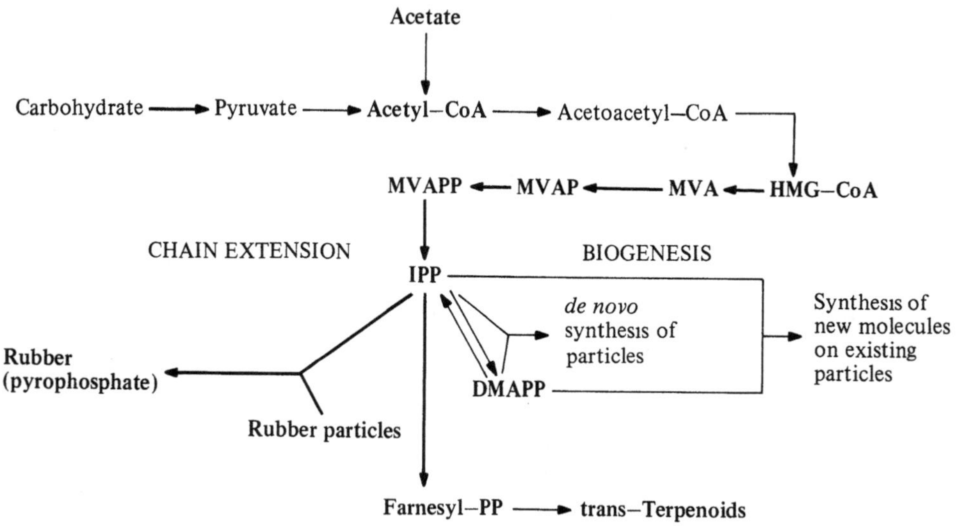

Fig. 2. Intermediates in the biosynthesis of rubber and other isoprenoids in Latex. *Heavy type,* established precursors and products; *light type,* presumed precursors and products; *heavy lines,* established reactions; *light lines,* presumed reactions

main source of carbon for rubber formation, there is no direct proof that this is true. Rubber isolated from latex incubated with [^{14}C]-glucose appears to be radioactive (D'AUZAC, 1965), but ozonolysis shows that the tracer is not present in the polyisoprene itself (BEALING, unpublished). There is also evidence that the rate of dissimilation of sugars in latex is insufficient to account for the rate of rubber synthesis in vivo (BEALING, 1976) and the same author has suggested that rubber may be formed from substances other than sugars which enter the latex vessels from the surrounding cells.

The enzymes converting pyruvate to acetyl-CoA in animal and plant cells are located in the mitochondria and this is almost certainly so in laticifers, there being no good evidence to the contrary. If this is the case, the observed poor in vitro incorporation of [^{14}C]-pyruvate and earlier precursors may be explained by the apparent scarcity of mitochondria in *tapped* latex (DICKENSON, 1969; 1969a; GOMEZ, 1976). It has recently been shown that the enzymes responsible for the conversion of acetyl-CoA to the hydroxymethylglutaryl-CoA (HMG-CoA) needed for cholesterol synthesis in animal cells are located in the cytosol (CLINKENBEARD et al., 1975) and it is probable that a similar situation exists in latex. The in vitro instability of latex HMG-CoA reductase (EC 1.1.1.34) (HEPPER and AUDLEY, 1969) may also contribute to the poor incorporation of earlier precursors.

In vitro synthesis of rubber by latex, from intermediates subsequent to acetoacetyl-CoA (Fig. 2) takes place more readily: HEPPER and AUDLEY (1969) showed that up to 50% of S-HMG-CoA is incorporated directly into the polyisoprene hydrocarbon on incubation with *Hevea* latex. The free acid is not incorporated, but it inhibits the formation of rubber from HMG-CoA. The reductase converting HMG-CoA to mevalonate (MVA) is sedimentable and probably present in the endoplasmic reticulum. NADPH is required as cofactor and it is assumed that the pentose phosphate pathway, the enzymes of which occur in latex, supplies hydrogen for the reduction. There is however, no good evidence for this and alternate sources of hydrogen are possible. The formation of MVA from HMG-CoA is an important rate-controlling step in cholesterol synthesis in animals, and has therefore been suggested as being involved in the regulation of rubber formation (LYNEN, 1967; HEPPER and AUDLEY, 1969).

PARK and BONNER (1958) and KEKWICK et al. (1959) showed that *Hevea* latex could convert MVA to high molecular weight rubber in substantial yield and the former workers showed by degradation studies that the positions of the individual carbon atoms in the repeating isoprene unit derived from MVA. The three enzymes which convert MVA to its 5′-phosphate (EC. 2.7.1.36), 5′-pyrophosphate (EC. 2.7.4.2) and then isopentenyl pyrophosphate (IPP) (EC. 4.1.1.33) in the latex serum have been isolated by WILLIAMSON and KEKWICK (1965); SKILLETER et al. (1966) and SKILLETER and KEKWICK (1968), respectively.

The biosynthesis of rubber from [^{14}C]-IPP on incubation with *Hevea* latex was first demonstrated by LYNEN (1960) and confirmed by ARCHER et al. (1961), who obtained a 97% incorporation of label and showed by ozonolytic degradation that the IPP had been converted to high molecular weight polyisoprene. In vitro, this stage in rubber biosynthesis occurs only on the surface of the rubber particles and according to ARCHER et al. (1963) the reaction taking place is the extension of pre-existing rubber molecules, which presumably carry allylic pyrophosphate end-groups. LYNEN (1969) however, considers that at least some of the incorporation of label represents the synthesis of completely new rubber molecules (see below).

By analogy with the mechanism of trans-isoprenoid biosynthesis, LYNEN et al. (1958) suggested that the first stage in the formation of rubber molecules is the isomerisation of IPP to dimethylallylpyrophosphate (DMAPP). The polyisoprene is then formed by repeated alkylation of successive IPP molecules by the growing allylic pyrophosphate chain; this however, has still to be formally proved. The addition of DMAPP or geranyl pyrophosphate (GPP; C_{10}-trans) to latex in the presence of [^{14}C]-IPP causes the enzymic production of farnesyl pyrophosphate (FPP; C_{15}-trans), which inhibits rubber formation (ARCHER et al., 1963; BARNARD, 1965). On the other hand, if DMAPP is added to a suspension of washed rubber particles, prepared by repeated dilution and centrifugal concentration of the latex, then a stimulation of [^{14}C]-IPP incorporation into the rubber is observed (ARCHER et al., 1963; LYNEN, 1967, 1969). This finding has been interpreted in two distinct ways; ARCHER et al. argue that the removal by washing of the serum, which contains the inhibitory FPP-synthesising system (EC. 2.5.1.1) allows the synthesis of new molecules of rubber (from DMAPP and IPP) to proceed in addition to chain extension. LYNEN (1969) interprets the evidence to mean that the isomerase catalysing the conversion of IPP to DMAPP is located on the surface of the rubber particles, and is preferentially removed by washing. The subsequent addition of DMAPP then stimulates the formation of new rubber molecules of high molecular weight.

Neryl pyrophosphate (C_{10}-cis), but not GPP (C_{10}-trans), accelerates rubber synthesis from IPP; i.e., the "polymerase" will not form rubber from a trans-starter molecule (BARNARD, 1965; ARCHER and COCKBAIN, 1969).

It is probable that both synthesis of new molecules and lengthening of pre-existing polyisoprene chains take place on the rubber particle surface, but there is as yet no direct evidence that isopentenyldiphosphate Δ-isomerase (EC. 5.3.3.2) is present in latex, or in fact that the rubber molecule contains an isopropylidene end-group.

By extensive washing it is possible to prepare a suspension of rubber particles which will not incorporate [^{14}C]-IPP into the rubber unless latex serum is added (McMULLEN and McSWEENEY, 1966). The enzyme(s) responsible for the activity of the serum has been partially purified and probably catalyses the alkylation of IPP by allylic pyrophosphate groups on the rubber particle (chain extension). The optimum pH is 7.0 and magnesium and a thiol are necessary cofactors. It is not known whether this enzyme can catalyse the de novo formation of rubber from DMAPP and IPP (ARCHER and COCKBAIN, 1969).

The biosynthesis of new rubber molecules at a site other than the surface of rubber particles, though it must logically occur in vivo has not been demonstrated in vitro. Candidates which have been suggested for such a site include Frey-Wyssling organelles or other latex particles with a specialised structure (DICKENSON, 1969a), and endoplasmic reticulum (ARCHER and AUDLEY, 1973). So far virtually no biochemical exploration of this problem has been carried out. Similarly no work has been done on the mechanisms controlling the molecular weight distribution of rubber (WESTALL, 1968; SUBRAMANIAM, 1972, 1976), or on the origin of the putative aldehyde or epoxide groups discussed above.

Hevea latex synthesises both high molecular weight cis-rubber and low molecular weight trans-isoprenoids such as sterols, from the same precursor. The enzymes involved are completely stereospecific; the one catalysing the formation of rubber eliminates a proton from C-2 of IPP, epimeric with that removed in the synthesis of FPP, the trans-precursor of the other isoprenoids. The possibility that rubber is first formed with a trans-configuration and subsequently isomerised is precluded by the experimental data (ARCHER et al., 1966). There is little doubt that there must be at least two stereospecific enzymes metabolising IPP in chicle latex, one producing rubber (cis-) and one or more yielding gutta and other trans-isoprenoids. It would be interesting to know not only what controls the relative yields of the two polyisoprenes, but also what determines the composition of a particular latex particle.

5 Role of Polyisoprene in Plants

Three hypotheses have been put forward so far to explain the part played by rubber in plants. The early suggestion that it protected the plant from animal attack has been discarded through lack of corroborative evidence (TOBLER, 1914), and the proposition that rubber acted as a reserve food supply is likewise not in accord with the fact that rubber, once it has been formed, is not further metabolised (PROKOFIEV, 1940; BONNER and GALSTON, 1947). More recently RITTER (1954) and BEALING (1965) suggested that the formation of rubber is a fermentative process analogous to the anaerobic production of ethanol. However the absence of stoichiometric coupling of the oxidative and reductive phases of the biosynthesis processes as discussed in detail by ARCHER and AUDLEY (1973) precludes this explanation. Although similar studies have not been made on the role of chicle and gutta, it seems likely that the end-products of polyisoprene synthesis have no function per se. However, in the case of laticiferous plants, other processes occurring in the laticifers may be of greater importance in the plants' physiology, than is polyisoprene formation itself.

6 The Artificial Stimulation of the Flow of Latex from *Hevea brasiliensis*

The kinetics of the flow of latex from the bark of a tree of *H. brasiliensis* after tapping, has recently been reviewed by LIORET et al. (1978). When the ends of the vessels are cut, latex is expelled initially at a high rate under the internal turgor pressure. Following the elastic, partial collapse of the severed vessels, there is a period of flow at a low rate under the residual pressure gradient, and during this latter period plugs of coagulated latex form at the ends of the vessels, thus terminating the flow (BOATMAN, 1966). MILFORD et al. (1969) have defined a "plugging index" *PI* as a measure of this effect, where

$$PI = \frac{100 \times \text{initial rate of flow}}{\text{total yield of latex}}$$

Further investigations by SOUTHORN (1968) and by PUJARNISCLE et al. (1970) implicated the lutoids in the clotting mechanism. Fifty years ago it was discovered that latex flow could be prolonged by the application of various mixtures to the bark of the tree and several hormonal compounds, of which 2,4-dichlorophenoxyacetic acid (CHAPMAN, 1951) is perhaps the best known, have since been used commercially (ABRAHAM et al., 1971).

The possibility of using ethylene to increase the yield of latex was suggested by BONNER (1968) and this gas was shown to be an effective yield stimulant by ABRAHAM et al. (1968). AUDLEY et al. (1977) confirmed the earlier suggestion of these latter authors that all non-gaseous stimulants of latex flow cause the evolution of ethylene from *Hevea* tissue, and it is therefore likely that this gas is the principal agent in yield stimulation, probably acting by increasing the stability of the lutoid particles and thus decreasing the plug formation. Yield stimulants bring about a number of other changes in the latex, including a fall in the total dry matter and in particular the magnesium content (RIBAILLIER, 1970). The copper (DE JONGE, 1955) and carbohydrate concentrations increase, as does the degree of polymerisation of the ribosomes (COUPÉ et al., 1977), and the pH of the latex (TUPY, 1973), but the mechanism of yield stimulation remains obscure. 2-Chloroethylphosphonic acid or ethephon, which decomposes to ethylene in neutral or alkaline solution, was found to stimulate the flow of latex from *Hevea* by ABRAHAM et al. (1968) and it is now widely used commercially. DICKENSON et al. (1976) have adsorbed ethylene on to molecular sieves (hydrated crystalline alumino-silicates) and dispersed them in a hydrocarbon grease for application to the rubber tree. Yields of latex of more than twice the normal have been reported with these materials with little apparent side effect. This method of applying ethylene to *Hevea* trees shows great promise for the future.

7 Breeding of *Hevea brasiliensis*

Practically the whole of the *H. brasiliensis* population of S.E. Asia is the progeny of the 1900 seedlings brought to Ceylon in 1876 and Malaya in 1877, but the selection of the first clones was not made until 1917 when bud-grafting was introduced (WYCHERLEY, 1968, 1969). Although the average yield of rubber per tree per tapping has increased from about 13 g in 1927 to some 65 g for the best commercial clones in 1975, the main impediment to the breeding of improved clones has been the long period of some 25 years from the cross-pollination of the parent flowers, through all the stages of propagation of sufficient mature trees of each clone, to their assessment for yield, disease resistance etc. over a number of years. In attempts to reduce this time and also the area of land needed, preliminary selections have usually been made largely on the seedlings' vigour, but it is questionable if this is an adequate basis for early selection (WYCHERLEY, 1969). Other genetic characteristics which have been found to be important in obtaining the maximum sustained yield per acre of plantation, are the girth of the trunk, the number of rings of latex vessels present in the bark (SANDERSON and SUTCLIFFE, 1930), and the thickness of the bark (ROSS and BROOKSON, 1966). Other factors of importance are the relative size and shape of the crown, which is related to the branching habit of the tree and to the incidence of damage by strong winds (Rubber Research Institute of Malaya, 1959). The resistance to disease (RAO, 1975) is of vital importance in the breeding of *Hevea*. The devastating leaf blight *Microcyclus ulei* (previously known as *Dothidella ulei*) is endemic to S. America but has not as yet been observed in S.E. Asia (CHEE, 1976). *Gleosporium* is more prevalent in southern Malaysia than in the north, whilst the climatic conditions in Ceylon and Vietnam favour attacks of *Oidium* (LANGLOIS, 1969).

Much of the variation of yield of latex can probably be explained by changes in the pattern of flow as expressed by the plugging index (see Sect. 6), which is a clonal characteristic. Other physiological factors include the partition of assimilates between tree growth and rubber production, and the onset of dryness associated with senescence of the latex vessels, known as brown bast (CHUA, 1967; PARANJOTHY et al., 1976).

8 Propagation and Tissue Culture of *Hevea brasiliensis*

Commercial rubber trees are propagated either from seed or vegetatively by budding of a selected clone on to seedling rootstocks. The striking of cuttings is difficult except with juvenile tissues, and has not been used commercially because of the absence of a tap-root on the resulting plants. Several attempts to tissue culture *Hevea* have been made: BOUYCHOU (1953) induced the formation of callus on pieces of young stem, and CHUA (1966) succeeded in forming roots on pieces of callus grown from seedling plumule segments. WILSON and STREET (1975) maintained subcultures of callus over long periods and detected differentiation into embryo cotyledons and possibly latex vessels, but attempts to promote the further development of embryos into plantlets were unsuccessful. PARANJOTHY and GHANDIMATHI (1975) obtained similar results by culturing *Hevea* anther tissue and the callus from epicotyl explants was frequently seen to exude latex when cut. These authors have also obtained rooted plantlets from seed embryos and from seedling shoot apices cultured in liquid media (PARANJOTHY and GHANDIMATHI, 1976). CHEN et al. (1979) claim to have produced five complete plants from the pollen grains of *Hevea brasiliensis*.

References

Abraham, P.D., Wycherley, P.R., Pakianathan, S.W.: Stimulation of latex flow in *Hevea brasiliensis* by 4-amino-3,5,6-trichloropicolinic acid and 2-chloroethanephosphonic acid. J. Rubber Res. Inst. Malaya *20*, 291–305 (1968)

Abraham, P.D., Blencowe, J.W., Chua, S.E., Gomez, J.B., Moir, G.F.J., Pakianathan, S.W., Sekhar, B.C., Southorn, W.A., Wycherley, P.R.: Novel stimulants and procedures in exploitation of *Hevea*. J. Rubber Res. Inst. Malaya *23*, 85–89 (1971)

Allen, P.W., Bloomfield, G.F.: Natural rubber hydrocarbon. In: The chemistry and physics of rubber-like substances. Bateman, L. (ed.), pp. 3–17. London: Maclaren 1963

Allen, P.W., Bristow, G.M.: The gel phase in natural rubber. J. Appl. Polym. Sci. *7*, 603–615 (1963)

Archer, B.L.: Hevamine: a crystalline basic protein from *Hevea brasiliensis* latex. Phytochemistry *15*, 297–300 (1976)

Archer, B.L., Audley, B.G.: Biosynthesis of rubber. In: Advances in enzymology. Nord, F.F. (ed.), Vol. 29, pp. 221–257. New York: Interscience 1967

Archer, B.L., Audley, B.G.: Rubber, gutta percha and chicle. In: Phytochemistry, Miller, L.M. (ed.), Vol. 2, pp. 310–343. New York: Van Nostrand Reinhold Company 1973

Archer, B.L., Cockbain, E.G.: Rubber transferase from *Hevea brasiliensis* latex. In: Methods in enzymology. Clayton, R.B. (ed.), Vol. XV, pp. 476–480. New York: Academic Press 1969

Archer, B.L., Ayrey, G., Cockbain, E.G., McSweeney, G.P.: Incorporation of (1-^{14}C) isopentenyl pyrophosphate into polyisoprene. Nature (London) *189*, 663–664 (1961)

Archer, B.L., Audley, B.G., Cockbain, E.G., McSweeney, G.P.: Biosynthesis of rubber. Biochem. J. *89*, 565–574 (1963)

Archer, B.L., Barnard, D., Cockbain, E.G., Cornforth, R.H., Cornforth, J.W., Popják, G.: The stereochemistry of rubber biosynthesis. Proc. R. Soc. London Ser. B *163*, 519–523 (1966)

Archer, B.L., Audley, B.G., McSweeney, G.P., Tan, C.H.: Studies on the composition of latex serum and 'bottom fraction' particles. J. Rubber Res. Inst. Malaya *21*, 560–569 (1969)

Arreguin, B., Bonner, J., Wood, B.J.: The mechanism of rubber formation in the guayule. Arch. Biochem. Biophys. *31*, 234–247 (1951)

Audley, B.G.: Isolation and composition of helical protein microfibrils from *Hevea brasiliensis* latex. Biochem. J. *98*, 335–341 (1966)

Audley, B.G., Archer, B.L., Runswick, M.J.: Ethylene production by *Hevea brasiliensis* tissues treated with latex yield-stimulatory compounds. Ann. Bot. (London) *41*, 63–71 (1977)

d'Auzac, J.: Relations entre la composition biochimique du latex, l'intensité de quelques réactions métaboliques et la productivité de l'*Hevea brasiliensis*. Rev. Gén. Caoutch. Plast. *42*, 1027–1036 (1965)

d'Auzac, J.: Caractérisation d'une ATP-ase membranaire en présence d'une phosphatase acide dans les lutoïds du latex d'*Hevea brasiliensis*. Phytochemistry *14*, 671–675 (1975)
d'Auzac, J., Jacob, J.-L.: Regulation of glycolysis in latex of *Hevea brasiliensis*. J. Rubber Res. Inst. Malaya *21*, 417–444 (1969)
d'Auzac, J., Brzozowska, J., Hanower, P., Lambert, C., Lioret, C., N'Goran, N.: Un modèle de structure vacuolaire isolée intacte: les lutoïds du latex d'*Hevea brasiliensis*. I. Accumulation et pénétration du citrate et de la L-lysine dans les lutoïdes. In: Colloque Int. C.N.R.S. No. 258, 1976 Rouen, 399–406 (1977a)
d'Auzac, J., Dupont, J., Jacob, J.-L., Lance, C., Marin, B., Moreau, F.: Un modèle de structure vacuolaire isolée intacte: les lutoïds du latex d'*Hevea brasiliensis*. II. Caractéristiques de la membrane lutoïdique. Exchanges ioniques transmembranaires chez les végétaux. Colloque Int. C.N.R.S. No. 258. 1976 Rouen, 391–398 (1977b)
Ayres, P.J., Hechter, O., Saba, N., Simpson, S.A., Tait, J.F.: Intermediates in the biosynthesis of aldosterone by capsule stripping of ox-adrenal gland. Biochem. J. *65*, 22P (1957)
Barnard, D.: Enzyme reactions in latex serum of relevance to rubber biosynthesis. In: Proc. Nat. Rubber Prod. Res. Assoc. Jubilee Conf. Cambridge 1964. Mullins, L. (ed.), pp. 89–100. London: Maclaren 1965
Barnard, D.: Natural rubber. In: Encyclopedia of polymer science and technology. Mark, H.F., Gaylord, N.E. (eds.), Vol. 12, pp. 227–229. New York: Interscience 1970
Bateman, L.: Photolysis of rubber. J. Polymer Sci. *2*, 1–9 (1947)
Bealing, F.J.: Role of rubber and other terpenoids in plant metabolism. In: Proc. Nat. Rubber Prod. Res. Assoc. Jubilee Conf. Cambridge 1964. Mullins, L. (ed.), pp. 113–122. London: Maclaren 1965
Bealing, F.J.: Carbohydrate metabolism in *Hevea* latex: availability and utilisation of carbohydrate substrates. J. Rubber Res. Inst. Malaya *21*, 445–455 (1969)
Bealing, F.J.: Quantitative aspects of latex metabolism: possible involvement of precursors other than sucrose in the biosynthesis of *Hevea* rubber. In: Proc. Int. Rubber Conf. 1975 Kuala Lumpur. *2*, 543–565 (1976)
Beytia, E.D., Porter, J.W.: Biochemistry of polyisoprenoid biosynthesis. Annu. Rev. Biochem. *45*, 113–142 (1976)
Bloomfield, G.F.: The rubber hydrocarbon in freshly tapped *Hevea* latex. Rubber Chem. Technol. *24*, 737–749 (1951)
Boatman, S.G.: Preliminary physiological studies on the promotion of latex flow by plant growth regulators. J. Rubber Res. Inst. Malaya *19*, 243–258 (1966)
Bonner, J.: Private communication (1968)
Bonner, J., Galston, A.: Rubber formation in plants – review. Bot. Rev. *13*, 543–596 (1947)
Bouychou, J.G.: La culture *in vitro* des tissus d'*Hevea*. Proc. Rubb. Conf. Bogor, 1952. Arch. Rubbercult. *30*, 50–53 (1953)
Bowler, W.W.: Electrophoretic mobility study of fresh *Hevea* latex. Ind. Eng. Chem. *45*, 1790–1794 (1953)
Brodie, J.D., Wasson, G.W., Porter, J.W.: Enzyme-bound intermediates in the biosynthesis of mevalonic and palmitic acids. J. Biol. Chem. *239*, 1346–1356 (1964)
Brzozowska, J., Hanower, P., Chezeau, R.: Free amino acids of *Hevea brasiliensis* latex. Experentia *30*, 894–895 (1974)
Buchanan, R.A., Cull, I.M., Otey, F.H., Russell, C.R.: Hydrocarbon and rubber producing crops: evaluation of U.S. plant species. A.C.S. Rubber Div. Fall Meet. 1976 San Francisco, No. 40 (1976)
Burfield, D.R., Chew, L.C., Gan, S.N.: Distribution of abnormal groups in natural rubber. Polymer *17*, 713–716 (1976)
Campos-Lopez, E., Angulo-Sanchez, J.L.: J. Polym. Sci. Lett. *14*, 649–652 (1976)
Campos-Lopez, E., Palacios, J.: Microstructure of natural Guayule rubber from 300 MHz NMR spectra. J. Polym. Sci. Chem. *14*, 1561–1563 (1976)
Campos-Lopez, E., Ponce, M.A., Neaves, V., Neaves, E., Canales, B.: Rubber from Guayule; research activities in Mexico. A.C.S. Rubber Div. Fall Meet. 1976, No. 35 (1976)
Chapman, G.W.: Plant hormones and yield in *Hevea brasiliensis*. J. Rubber Res. Inst. Malaya *13*, 167–176 (1951)
Chee, K.H.: South American leaf blight of *Hevea brasiliensis*: Spore behaviour and screening for disease resistance. Proc. Int. Rubber Conf. 1975 Kuala Lumpur *3*, 228–235 (1976)

Chen, C.-H., Chen, F.-T., Chien, C.-F., Wang, C.-H., Chang, S.-C., Hsu, H.-E., Ou, S.-H., Ho, Y.-T., Lu, T.-M.: A process of obtaining pollen plants of *Hevea brasiliensis*. Sci. Sinica *22*, 81–90 (1979)
Chua, S.E.: Studies on tissue culture of *Hevea brasiliensis*. J. Rubber Res. Inst. Malaya *19*, 272–276 (1966)
Chua, S.E.: Physiological changes in *Hevea* trees under intensive tapping. J. Rubber Res. Inst. Malaya *20*, 100–105 (1967)
Clinkenbeard, K.D., Reed, W.D., Mooney, R.A., Lane, M.D.: Intracellular localisation of the 3-hydroxy-3-methylglutaryl coenzyme A cycle enzymes in liver. J. Biol. Chem. *250*, 3108–3116 (1975)
Coupé, M., Pujarniscle, S., d'Auzac, J.: Compartimentation de diverses oxydoreductases (peroxydase, *O*-diphenol oxydase et malate déshydrogènase) dans le latex d'*Hevea brasiliensis*. Physiol. Vég. *10*, 459–480 (1972)
Coupé, M., Lambert, C., Primot, L., d'Auzac, J.: Cinétique d'action de l'acide 2-chloroethylphosphonique sur les polyribosomes du latex d'*Hevea brasiliensis*. Phytochemistry *16*, 1133–1136 (1977)
Cunneen, J.I.: Oxidative ageing of natural rubber. Rubber Chem. Technol. *41*, 182–208 (1968)
Cunneen, J.I., Higgins, G.M.C.: Cis-trans isomerism in natural polyisoprene. In: The chemistry and physics of rubber-like substances. Bateman, L. (ed.), pp. 19–40. London: Maclaren 1963
Dickenson, P.B.: The ultrastructure of latex and the cytology of latex vessels. In: The chemistry and physics of rubber-like substances. Bateman, L. (ed.), pp. 41–51. London: Maclaren 1963
Dickenson, P.B.: Ultrastructure of the latex vessel of *Hevea brasiliensis*. In: Proc. Natl. Rubber Prod. Res. Assoc. Jubilee Conf. Cambridge 1964. Mullins, L. (ed.), pp. 52–66. London: Maclaren 1965
Dickenson, P.B.: Electron microscopical studies of the latex vessel system of *Hevea brasiliensis*. J. Rubber Res. Inst. Malaya *21*, 543–559 (1969a)
Dickenson, P.B.: Natural rubber and its traditional use in underground pipe sealing rings. J. Rubber Res. Inst. Malaya *22*, 165–175 (1969b)
Dickenson, P.B., Sivakumaran, S., Abraham, P.D.: Ethad and other new stimulants for *Hevea brasiliensis*. Proc. Int. Rubber Conf. 1975, Kuala Lumpur *2*, 315–342 (1976)
Dupont, J., Moreau, F., Lance, C., Jacob, J.-L.: Phospholipid composition of the membrane of lutoids from *Hevea brasiliensis* latex. Phytochemistry *15*, 1215–1217 (1976)
Esau, K.: Distribution of latex in the plant kingdom. In: Plant anatomy, 2nd ed. pp. 318–319. New York: Wiley 1965
Fisher, D.: Crystal structure of gutta percha. Proc. Phys. Soc. B*66*, 7–16 (1953)
Fraga, D.W.: Absence of 3,4-structure in natural polyisoprenes. J. Polym. Sci. *41*, 522–524 (1959)
Freeman, R.: Microgel in latex and sheet rubber. In: Proc. 3rd Rubber Technol. Conf. Messenger, T.H. (ed.), pp. 3–12. Cambridge: Heffer 1954
Frey-Wyssling, A.: Microscopic investigations on the occurrence of resins in *Hevea* latex. Arch. Rubbercult. *13*, 392–412 (1929)
Golub, M.A., Fuqua, S.A., Bhacca, N.S.: High resolution nuclear magnetic resonance spectra of various polyisoprenes. J. Am. Chem. Soc. *84*, 4981–4982 (1962)
Gomez, J.B.: Comparative ultracytology of young and mature latex vessels in *Hevea brasiliensis*. Proc. Int. Rubber Conf. 1975, Kuala Lumpur *2*, 143–164 (1976)
Gomez, J.B., Yip, E.: Microhelices in *Hevea* latex. J. Ultrastruct. Res. *52*, 76–84 (1975)
Gregg, E.C., Macey, J.H.: The relationship of properties of synthetic polyisoprene and NR in the factory. The effect of non-rubber constituents of NR. Rubber Chem. Technol. *46*, 47–66 (1973)
Hanower, P., Brzozowska, J., N'Goran, M.: Absorption des acides amines par les lutoïdes du latex d'*Hevea brasiliensis*. Physiol. Plant. *39*, 299–304 (1977)
Harries, C.D.: Chemical composition of rubber hydrocarbon. In: Untersuchungen über die natürlichen und künstlichen Kautschukarten. Berlin: Springer 1919
Hauser, E.A.: Latex. (English translation by W.J. Kelly). pp. 62–68. New York: Chemical catalogue Co. 1930

Hepper, C.M., Audley, B.G.: The biosynthesis of rubber from β-hydroxy-β-methylglutaryl-coenzyme A in *Hevea brasiliensis* latex. Biochem. J. *114*, 379–386 (1969)
Higgins, M.J.P., Kekwick, R.G.O.: Investigation into the role of malonyl-coenzyme A in isoprenoid biosynthesis. Biochem. J. *134*, 295–310 (1973)
Ho, C.C., Subramaniam, A., Yong, W.M.: Lipids associated with the particles in *Hevea* latex. Proc. Int. Rubber Conf. 1975 Kuala Lumpur, *2*, 441–456 (1976)
Homans, L.N.S., Van Gils, G.E.: Fresh *Hevea* latex. A complex colloidal system. In: Proc. 2nd Rubber Technol. Conf. Dawson, T.R. (ed.), pp. 292–302. Cambridge: Heffer 1948
Jacob, J.-L., Sontag, N.: Purification et étude de la phosphatase acide lutoïdique du latex d'*Hevea brasiliensis*. Biochimie *56*, 1315–1322 (1974)
de Jonge, P.: Stimulation of yield in *Hevea brasiliensis*. J. Rubber Res. Inst. Malaya *14*, 383–406 (1955)
Kekwick, R.G.O.: Early stages of rubber biosynthesis. In: Proc. Nat. Rubber Prod. Res. Assoc. Jubilee Conf. Cambridge 1964. Mullins, L. (ed.), pp. 80–83 London: Maclaren 1965
Kekwick, R.G.O., Archer, B.L., Barnard, D., Higgins, G.M.C., McSweeney, G.P., Moore, C.G.: Incorporation of DL-(2-^{14}C)mevalonic acid lactone into polyisoprene. Nature (London) *184*, 268–270 (1959)
Kurihara, K., Kato, Y., Takei, Y., Shichiji, S.: Production of rubber by microorganisms. Rep. Ferment. Res. Ind. (Japan) *22*, 61–68 (1962)
Kurihara, K., Kato, Y., Takei, Y., Shichiji, S.: Studies on the rubber productive microorganisms. Rep. Ferment. Res. Ind. (Japan) *24*, 25–30 (1963)
Kurihara, K., Kato, Y., Takei, Y., Shichiji, S.: Studies on the rubber productive microorganisms. Rep. Ferment. Res. Ind. (Japan) *25*, 33–38 (1964)
Langois, S.: Results of vegetative selection in Cambodia. J. Rubber Res. Inst. Malaya *21*, 19–26 (1969)
Leeflang, K.W.H.: Microbiologic degradation of rubber. J. Am. Water Works Assoc. *55*, 1523–1535 (1963)
Leeper, H.M., Schlesinger, W.: Occurrence of both caoutchouc and gutta in additional plants. Science *120*, 185–186 (1954)
Lioret, C., Ribaillier, D., Combe, J.C.: Cinetique d'écoulement du latex d'*Hevea brasiliensis* après saignée. Rev. Gén. Caoutch. Plast. *578*, 91–98 (1978)
Lucas, F.F.: Motion picture study of balata and *Hevea* latices. Ind. Eng. Chem. *34*, 1371–1381 (1942)
Lynen, F.: Biological pathway to natural rubber. Angew. Chem. *72*, 454 (1960)
Lynen, F.: Biosynthetic pathways from acetate to natural rubber. J. Pure Appl. Chem. *14*, 137–167 (1967)
Lynen, F.: Biochemical problems of rubber synthesis. J. Rubber Res. Inst. Malaya *21*, 389–406 (1969)
Lynen, F., Eggerer, H., Henning, V., Kessel, I.: Farnesyl pyrophosphate and 3-methyl-Δ^3-butenyl-1-pyrophosphate, biological precursors of squalene. Angew. Chem. *70*, 738–742 (1958)
Marin, B.: Les problèmes posés par l'existence d'acides ribonucleiques dans les compartements lysosmales végétaux. Ann. Sci. Nat. Bot. Paris *17*, 361–374 (1976)
Marin, B., Trouslot, P.: The occurrence of ribonucleic acid in the lutoid fraction (lysosomal compartment) from *Hevea brasiliensia* latex. Planta *124*, 31–41 (1975)
Matile, Ph.: The lytic compartment of plant cells. p. 24. Berlin, Heidelberg, New York: Springer 1975
Matile, Ph.: Biochemistry and function of plant vacuoles. Proc. Int. Rubber Conf. 1975, Kuala Lumpur *1*, 123–141 (1978)
McMullen, A.I., McSweeney, G.P.: Biosynthesis of rubber. Incorporation of isopentenyl pyrophosphate into purified rubber particles by a soluble latex serum enzyme. Biochem. J. *101*, 42–47 (1966)
McSweeney, G.P.: Ozonolysis of natural rubber. J. Polym. Sci. *6*, 2678–2680 (1968)
Metcalfe, C.R.: Distribution of latex in the plant kingdom. Notes from the Jodrell Laboratory III, Kew, England (1966)
Metcalfe, C.R., Chalk, L.: Anatomy of the dicotyledons. Oxford: Clarendon Press 1957
Meyer, K.H.: Natural and synthetic high polymers. p. 121. New York: Interscience 1942

Middleton, K.R., Westgarth, D.R.: Determination of rubber in the vegetative parts of *Hevea brasiliensis*. Analyst *88*, 544–550 (1963)

Milford, G.F.J., Paardekooper, E.C., Ho, C.Y.: Latex vessel plugging, its importance to yield and clonal behaviour. J. Rubber Res. Inst. Malaya *21*, 274–282 (1969)

Moir, G.F.J.: Ultracentrifugation and staining of *Hevea* latex. Nature (London) *184*, 1626–1628 (1959)

Moir, G.F.J.: Latex metabolism. J. Rubber Res. Inst. Malaya *21*, 407–416 (1969)

Montardy, M.C., Lambert, C.: Diverses propriétés de l'absorption du citrate du malate et du succinate par les lutoïdes du latex d'*Hevea brasiliensis*. Phytochemistry *16*, 677–680 (1977)

Moreau, F., Jacob, J.-L., Dupont, J., Lance, C.: Electron transport in the membrane of lutoids from the latex of *Hevea brasiliensis*. Biochem. Biophys. Acta *396*, 116–124 (1975)

Morris, A.L.: The kinetics of oxidation of olefins with special reference to rubber. Ph.D. thesis. Univ. London (1952)

Moyle, A.E.: Bibliography and collected abstracts on rubber producing plants. Tex. Agric. Exp. Stn. Circ. 99 (1942)

National Academy of Sciences: Guayule: an alternative source of natural rubber. Washington: National Academy of Sciences 1977

Nishimura, H., Philp, R.P., Calvin, M.: Lipids of *Hevea brasiliensis* and *Euphorbia coerulescens*. Phytochemistry *16*, 1048–1049 (1977)

Paranjothy, K., Ghandimathi, H.: Morphogenesis in callus cultures of *Hevea brasiliensis*. Proc. Natl. Plant Tissue Cult. Symp. 1975, Kuala Lumpur. pp. 19–25 1975

Paranjothy, K., Ghandimathi, H.: Tissue and organ culture of *Hevea*. Proc. Int. Rubber Conf. 1975. Kuala Lumpur *2*, 59–84 (1976)

Paranjothy, K., Gomez, J.B., Yeang, H.Y.: Physiological aspects of brown bast development. Proc. Int. Rubber Conf. 1975, Kuala Lumpur *2*, 181–202 (1976)

Park, R.B., Bonner, J.: Enzymatic synthesis of rubber from mevalonic acid. J. Biol. Chem. *233*, 340–343 (1958)

Porter, M.: In: Organic chemistry of sulfur. Cae, S. (ed.) New York: Plenum Press 1977

Prokofiev, A.A.: Biological role of rubber. Bull. Acad. Sci. U.S.S.R., Ser. Biol. *4*, 606–607 (1940)

Pujarniscle, S.: Caractère lysosomal des lutoïdes du latex d'*Hevea brasiliensis*. Physiol. Vég. *6*, 27–46 (1968)

Pujarniscle, S., Ribaillier, D., d'Auzac, J.: Du rôle des lutoïdes dans l'écoulement du latex chez l'*Hevea brasiliensis*: II. Évolution des hydrolases lutoïdiques et de quelques propriétés du latex au cours de la saignée. Rev. Gén. Caoutch. Plast. *47*, 1317–1321 (1970)

Rabinowitz, J.L., Teas, H.J.: Biosynthesis of rubber from ^{14}C-acetate by stem slices and latex enzyme from Hevea. Proc. Int. Rubber Conf. 1960, Kuala Lumpur, Malaysia, 777–786 (1961)

Rao, B S.: Maladies of *Hevea* in Malaysia. Rubber Res. Inst. Malaysia, Kuala Lumpur 1975

Ribaillier, D.: Importance des lutoïdes dans l'écoulement du latex: action de la stimulation. Rev. Gén. Caoutch. Plast. *47*, 305–310 (1970)

Ribaillier, D., Jacob, J.-L., d'Auzac, J.: Sur certaines caractères vacuolaires des lutoïdes du latex d'*Hevea brasiliensis*. Physiol. Vég. *9*, 423–437 (1971)

Ritter, F.J.: Biosynthesis of rubber and other isoprenoid compounds. Rubber J. *126*, 55–71 (1954)

Ross, J.M., Brookson, C.W.: Progress of breeding investigations with *Hevea brasiliensis*. J. Rubber Res. Inst. Malaya *19*, 158–172 (1966)

Rubber Research Institute of Malaya.: Wind damage. Plrs'. Bull. Rubber Res. Inst. Malaya No. 43, 79–93 (1959)

Rubber Research Institute of Malaysia.: Rep. Rubber Res. Inst. Malaysia 1975, 47 (1976)

Ruinen, J.: Microscopy of the lutoids in *Hevea* latex. Arch. Rubbercult. *27*, 255–265 (1950)

Sanderson, A.R., Sutcliffe, H.: Vegetative characters and yield of *Hevea brasiliensis*. J. Rubber Res. Inst. Malaya *1*, 75–90, 151–200 (1930)

Sárkány, S., Fridvalsky, L., Lovas, B., Verzár-Petri, G.: Studien über die Feinstruktur der jungen Milchröhren, bzw. des Milchsaftes von *Papaver somniferum*. In: Proc. 3rd Eur. Reg. Conf. Electron Microsc. Prague, 1964. Titlbach, M. (ed.), pp. 161–162. London: Royal Microscopical Soc. 1965

Sassen, M.M.A.: Breakdown of the plant cell wall during the cell fusion process. Acta Bot. Neerl. *14*, 165–196 (1965)
Schlesinger, W., Leeper, H.M.: Chicle: cis and trans polyisoprenes from a single plant species. Ind. Eng. Chem. *43*, 398–403 (1951)
Schoon, T.G.F., Van der Bie, G.J.: Particle size distribution in brominated *Hevea* latices. J. Polym. Sci. *16*, 63–88 (1955)
Schulze, Ch., Schnepf, E., Mothes, K.: Über die Lokalisation der Kautschukpartikel in verschiedenen Typen von Milchröhren. Flora (Jena) *158*, 458–460 (1967)
Sekhar, B.C.: Inhibition of hardening in natural rubber. J. Rubber Res. Inst. Malaya *21*, 512–519 (1961)
Skilleter, D.N., Kekwick, R.G.O.: Some characteristics of pyrophosphomevalonate decarboxylase from *Hevea brasiliensis* latex. Biochem. J., *108*, 11P–12P (1968)
Skilleter, D.N., Williamson, I.P., Kekwick, R.G.O.: Phosphomevalonate kinase from *Hevea* latex. Biochem. J. *98*, 27P (1966)
Southorn, W.A.: Latex flow studies. J. Rubber Res. Inst. Malaya *20*, 226–235 (1968)
Southorn, W.A.: Physiology of *Hevea* (latex flow). J. Rubber Res. Inst. Malaya *21*, 494–512 (1969)
Stavely, F.W., Biddison, P.H., Forster, M.J., Dawson, H.G., Binder, J.L.: Structure of various natural rubbers. Rubber Chem. Technol. *34*, 423–432 (1961)
Stewart, W.D., Wachtel, W.L., Shipman, J.J., Yanko, J.A.: Synthesis of rubber by fungi. Science *122*, 1271–1272 (1955)
Subramaniam, A.: Gel-permeation chromatography of natural rubber. Rubber Chem. Technol. *45*, 346–348 (1972)
Subramaniam, A.: Molecular weight and other properties of NR: a study of clonal variations. Proc. Int. Rubber Conf. 1975, Kuala Lumpur *4*, 3–27 (1976)
Tan, C.H.: Non-rubber organic components of *Hevea brasiliensis* latex. Ph. D. thesis, Univ. Malaya (1968)
Tanaka, Y., Sato, H.: Sequence distribution of cis-1,4- and trans-1,4- units in polyisoprenes. Polymer *17*, 113–116 (1976)
Tata, S.J., Boyce, A.N., Archer, B.L., Audley, B.G.: Lysozymes: major components of the sedimentable phase of *Hevea* latex. J. Rubber Res. Inst. Malaysia *24*, 233–236 (1976)
Tobler, F.: Physiologische Milchsaft und Kautschuk Studien. Jahrb. Wiss. Bot. *54*, 265–308 (1914)
Tupy, J.: The regulation of invertase activity in the latex of *Hevea brasiliensis*. J. Exp. Bot. *24*, 516–524 (1973)
Vyvalko, I.G., Duschenchkin, A.I., Luschchevskaya, G.M., Matkovskii, K.I., Savivov, B.G., Shilov, E.A., Yasnikov, A.A.: Mechanism of biosynthesis of isoprenic compounds. Ukr. Khim. Zh *23*, 516–522 (1957)
Weeks, G.: Biosynthesis of isoprenoid compounds and fatty acids by latex of *Hevea brasiliensis*. Ph. D. thesis. Univ. Birmingham (1965)
Westall, B.: The molecular weight distribution of natural rubber latex. Polymer *9*, 243–248 (1968)
Wiersum, L.K.: Enkele Latexproblemen. Vakbl. Biol. *37*, 17–24 (1957)
Williamson, I.P., Kekwick, R.G.O.: Formation of 5-phosphomevalonate by MVA kinase in *Hevea brasiliensis* latex. Biochem. J. *96*, 862–871 (1965)
Wilson, H.M., Street, H.E.: The growth, anatomy and morphogenetic potential of callus and cell suspension cultures of *Hevea brasiliensis*. Ann. Bot. (London) *39*, 671–682 (1975)
Wycherley, P.R.: Introduction of *Hevea* to the Orient Planter, Kuala Lumpur *44*, 127–137 (1968)
Wycherley, P.R.: Breeding of *Hevea*. J. Rubber Res. Inst. Malaya *21*, 38–55 (1969)

6 Plant Phenolics

J.B. HARBORNE

1 Introduction

The expression "phenolic compounds" embraces a considerable range of substances which possess an aromatic ring bearing a hydroxyl substituent, and while a significant number of such compounds occur in animals, most are of plant origin. Indeed, the presence of a "phenolic fraction" is a characteristic feature of all plant tissues. Among the plant polyphenols of which several thousand have now been described, the flavonoids form the largest group. However, phenolic quinones, lignans, xanthones and coumarins and other groups exist in considerable numbers and there

Table 1. The major classes of phenolics in plants

Number of carbon atoms	Basic skeleton	Class	Examples
6	C_6	Simple phenols Benzoquinones	Catechol, hydroquinone 2,6-Dimethoxybenzoquinone
7	C_6-C_1	Phenolic acids	*p*-Hydroxybenzoic, salicylic
8	C_6-C_2	Acetophenones Phenylacetic acids	3-Acetyl-6-methoxybenzaldehyde *p*-Hydroxyphenylacetic
9	C_6-C_3	Hydroxycinnamic acids Phenylpropenes Coumarins Isocoumarins Chromones	Caffeic, ferulic Myristicin, eugenol Umbelliferone, aesculetin Bergenin Eugenin
10	C_6-C_4	Naphthoquinones	Juglone, plumbagin
13	C_6-C_1-C_6	Xanthones	Mangiferin
14	C_6-C_2-C_6	Stilbenes Anthraquinones	Lunularic acid Emodin
15	C_6-C_3-C_6	Flavonoids Isoflavonoids	Quercetin, cyanidin Genistein
18	$(C_6$-$C_3)_2$	Lignans Neolignans	Pinoresinol Eusiderin
30	$(C_6$-C_3-$C_6)_2$	Biflavonoids	Amentoflavone
n	$(C_6$-$C_3)_n$ $(C_6)_n$ $(C_6$-C_3-$C_6)_n$	Lignins Catechol melanins Flavolans (Condensed tannins)	– – –

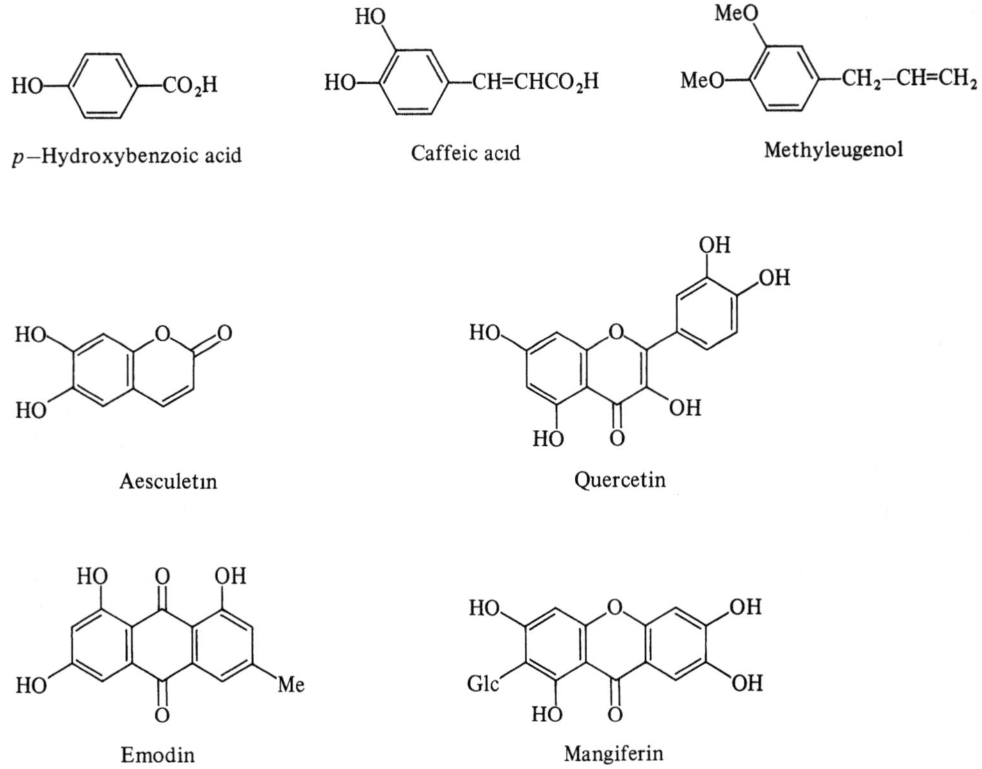

Fig. 1. Structures of some commonly occurring plant phenolics

are also many simple monocyclic phenols. In addition to monomeric and dimeric structures, there are also three important groups of phenolic polymer – the lignins, melanins, and tannins. Phenolic units are also encountered among nitrogen compounds (for example the aromatic amino acid tyrosine is phenolic) and among terpenoids.

Most of the major classes of plant polyphenol are listed in Table 1, according to the number of carbon atoms of the basic skeleton. Under the term "phenolic" are included compounds where this function may be masked by *O*-methylation or other substitution. Many of the compounds listed contain other functional groups besides phenolic hydroxyl groups and this clearly modifies their properties on occasion. Some illustrative structures of the more commonly occurring plant polyphenols are shown in Figure 1.

The seemingly bewildering array of natural phenols forms a more coherent picture when biogenetic considerations are taken into account. Practically all higher plant polyphenols are formed from shikimate, via the shikimic acid pathway and their production through the intermediacy of phenylalanine, the enzyme phenylalanine ammonia lyase, and cinnamic acid is illustrated in Figure 2. The role of acetate-malonate units in biosynthesis should not, however, be ignored, since these are needed, together with phenylpropanoid moieties, for the production of flavonoids. Many simple phenols and many plant quinones are also derived directly by the polyketide pathway from acetyl and malonyl coenzyme A.

Chemically, phenols are reactive substances and this must be borne in mind when considering their functions in plants. They are usually acidic and can often be separated from other plant constituents by their solubility in aqueous sodium carbonate. Unless sterically hindered, all phenols are capable of taking part in hydrogen bonding. This may be intramolecular, as between the 5-hydroxyl and 4-carbonyl in many flavonoids. More importantly, it may be intermolecular and bring about interactions between plant phenols and the peptide links of proteins and enzymes. Another important property of the many phenols with an *o*-dihydroxy (catechol) grouping is their ability to chelate metals and such metal chelates may be important biologically in plant systems. Finally, phenols are very susceptible to oxidation and there are special enzymes present in plants – the phenolases – which catalyse the oxidation of monophenols to diphenols and of diphenols to quinones and hence often to polymeric materials.

One of the problems in establishing a physiological role for plant polyphenols is the multitude of structures encountered in nature. It should, however, be emphasised that only a small number of compounds (cyanidin, quercetin, caffeic acid, etc.) are at all widespread; most of the others are of very limited occurrence. In terms of growth regulation, phenolics seem commonly to be inhibitors. Thus, the only established hormone with phenolic substitution – lunularic acid – is an inhibitor of growth and occurs throughout the liverworts, where it apparently replaces abscisic acid, the more universal dormancy hormone. The interaction between plant hormones and other classes of phenol has been under increasing study and while most of these experiments have been carried out in vitro, there are those who believe that phenols of higher plants may have an important controlling effect on plant growth in vivo.

In Volume X of the first *Encyclopedia of Plant Physiology*, there was very generous coverage of phenolics and chapters were included on anthocyanins, tannins, lignins, lichen phenols, and on biogenesis and metabolism. Since that time, progress in the study of plant phenols has been enormous and it is impossible in the space provided to cover the advances of the last twenty years at all comprehensively. Only a few topics can be selected for detailed treatment here. After a survey of the structural variation among the different classes of polyphenol, consideration will be given to the different ways that these compounds are conjugated in plant cells. A summary will then be given of their pathways of biosynthesis, their enzymology, and further metabolism. Their production in tissue culture will be compared with their formation in intact plants. Finally, an attempt will be made to summarise what is known of their functions in plants.

2 Phenolic Aglycones

2.1 Simple Phenols and Phenolic Acids

Although most of the more complex and widely occurring plant polyphenols contain catechol (I) or phloroglucinol (II) units as parts of their structures (see Fig. 1), these two simple phenols are relatively uncommon and of scattered occurrence. Catechol, in fact, has been reported with certainty in only a few plants, e.g., in leaves of three *Gaultheria* species as the glucoside (TOWERS et al., 1966). It may, however, be more frequently produced than this, since there is evidence (NICOLAUS and PIATELLI, 1965) that most plant melanins are catechol-based and are formed by oxidative polymerisation of catechol in the presence of the enzyme phenolase. Such black pigments, which yield catechol on degradation, have been found for example in seeds of sunflower and water melon and in the rust spores of the fungus *Ustilago maydis*. Finally, catechol derivatives with aliphatic substituents are known; urushiol (III), with a C_{15} hydrocarbon side chain, is the vesicant principle of poison ivy *Rhus toxicodendron* (Anacardiaceae).

Phloroglucinol (II) is also quite rare in plants, although it does occur as its glucoside, phlorin, in the peel of various *Citrus* fruits. Derivatives of phloroglucinol are also uncommon. Acylphloroglucinols occur in rhizomes of a few *Dryopteris* species (Filicopsida) and the hop resin compounds humulone and lupulone in *Humulus lupulus* are isoprenoid substituted phloroglucinols (SWAIN and COOPER-DRIVER, 1973).

Of other simple phenols in nature, the most widespread is undoubtedly hydroquinone (IV), which occurs in several families but is especially associated with *Pyrus* and *Docynia* in Rosaceae and with the tribe Arbuteae of Ericaceae (HARBORNE and WILLIAMS, 1973). Resorcinol itself does not appear to have been found in plants, but its 5-methyl derivative, orcinol (V) is present in *Erica arborea* and in a number of other Ericaceae (HARBORNE and WILLIAMS, 1969). A more complex natural derivative of resorcinol is Δ^1-tetrahydrocannabinol (VI) which is the hallucinogenic principle of leaves and flowers of *Cannabis sativa* (MECHOULAM, 1970).

By contrast with the simple phenols, phenolic acids seem to be universally distributed in plants; at least, acid or alkaline treatment of plant extracts normally yields a mixture of several such acids. The mode of binding is not entirely clear, but the acids are mainly present in the "alcohol-insoluble" fraction of leaf tissue and are probably at least partly linked to lignin by ester bonds. Four acids are apparently universal in the angiosperms: *p*-hydroxybenzoic (VII), protocatechuic (VIII), vanillic (IX) and syringic acids (X). The first three are also present widely in gymnosperms and ferns. The presence of bound syringic acid in leaves is correlated with the occurrence of syringyl residues specifically in angiosperm lignins (IBRAHIM et al., 1962).

Two other common acids in angiosperms are gentisic (XI) and salicylic acid (XII). Salicylic acid is occasionally present as its methyl ester in essential oils and this ester is the pungent-smelling oil of wintergreen of the British Pharmacopoeia. Salicylic acid is especially widespread in Ericaceae, where it occurs in association with *o*-pyrocatechuic acid (2,3-dihydroxybenzoic acid).

Gallic acid (XIII) is another well-known plant acid. It has recently been characterised as a specific flowering-inhibitor in leaves of *Kalanchoe blossfeldiana* and is also present in a bound form in the same plant after flowering (PRYCE 1972c). In its more usual occurrence in woody plants, gallic acid occurs in soluble form linked to glucose, as hydrolysable gallotannin. Its dimeric condensation product, hexahydroxydiphenic acid and the related dilactone, ellagic acid (XIV) are also common plant constituents, the former usually occurring bound as ellagitannin. Hydrolysable tannins are regularly present in woody dicotyledonous (but not monocotyledonous) angiosperms, occurring especially in a cluster of families with Rosalian affinities (BATE-SMITH, 1972).

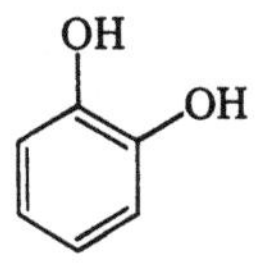

Catechol
(I)

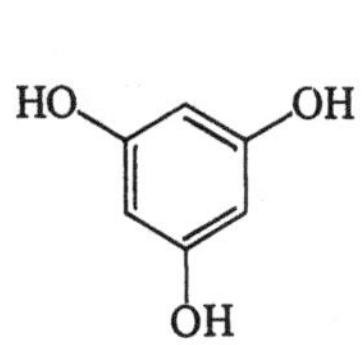

Phloroglucinol
(II)

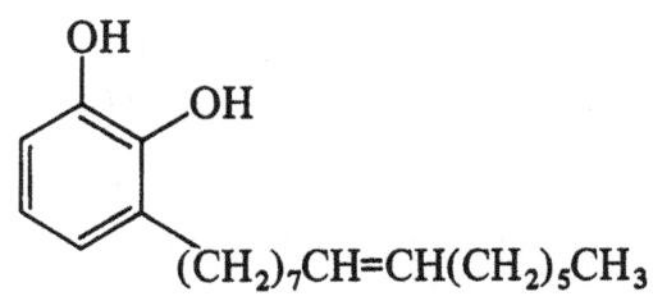

Urushiol
(III)

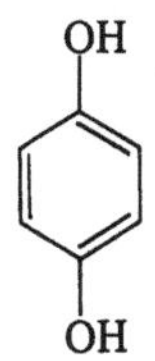

Hydroquinone
(IV)

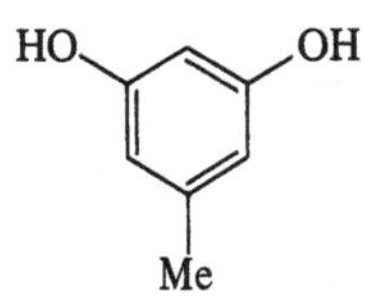

Orcinol
(V)

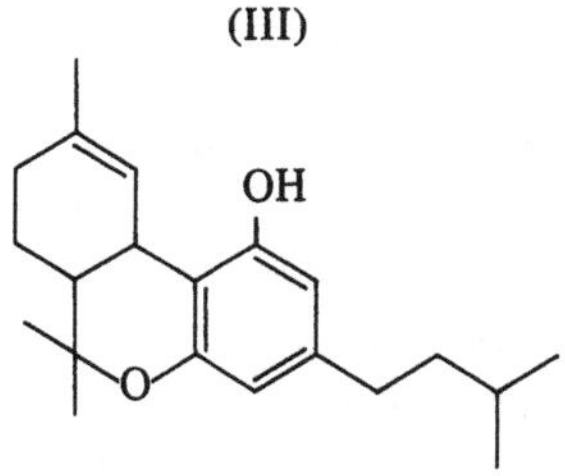

Δ^1–Tetrahydrocannabinol
(VI)

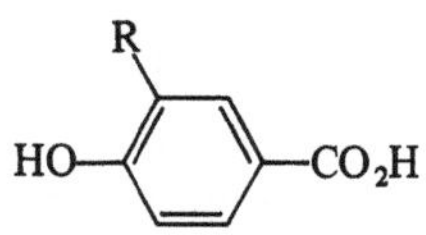

R = H, *p*–Hydroxybenzoic acid (VII)
R = OH, Protocatechuic acid (VIII)
R = OMe, Vanillic acid (IX)

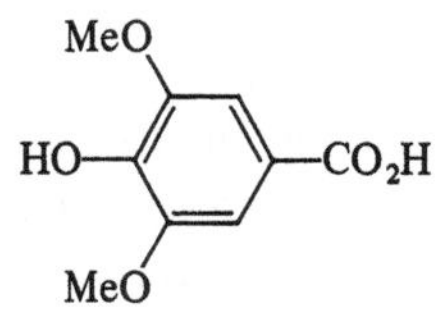

Syringic acid
(X)

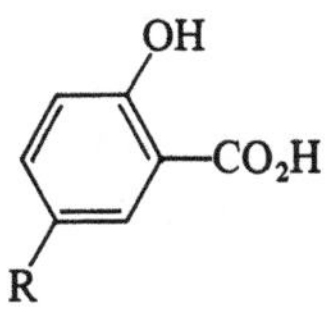

R = OH, Gentisic acid (XI)
R = H, Salicylic acid (XII)

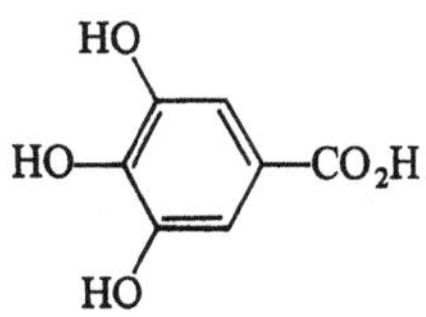

Gallic acid
(XIII)

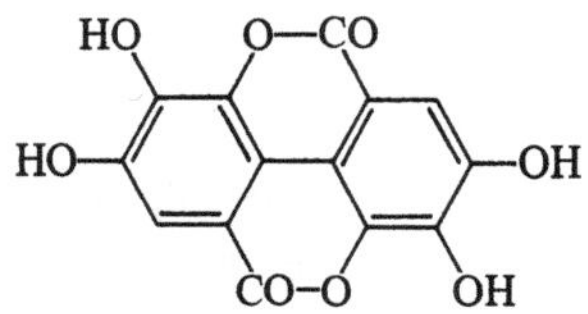

Ellagic acid
(XIV)

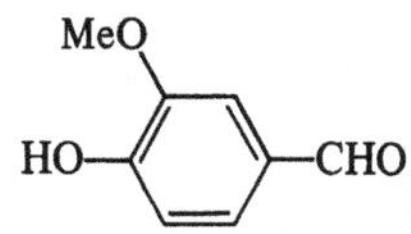

Vanillin
(XV)

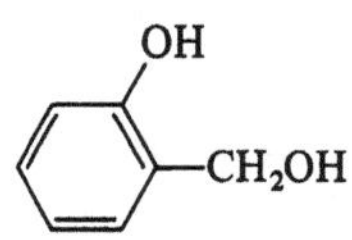

Salicylalcohol
(XVI)

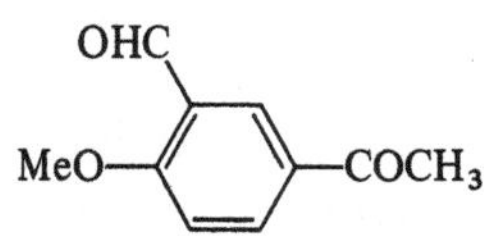

(XVII)

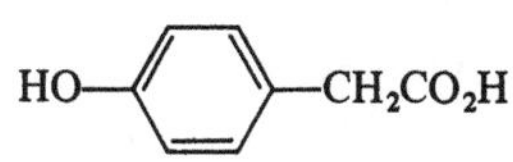

p–Hydroxyphenylacetic acid
(XVIII)

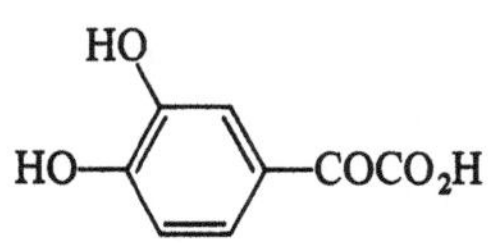

3,4–Dihydroxyphenyl–
glyoxylic acid
(XIX)

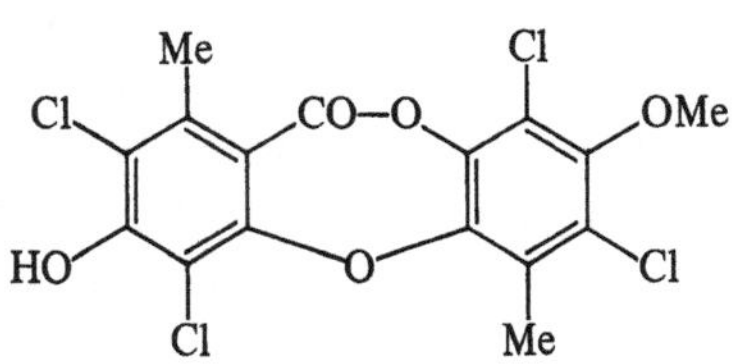

Diploicin
(XX)

Aldehydes related to the common C_6-C_1 acids are found in plants; for example, vanillin (XV) is the odour principle of vanilla pod *Vanilla planifolia* (Orchidaceae) and is present in a variety of plants. The corresponding alcohols are also known, an example being salicylalcohol (XVI) present in willow, *Salix* and in certain Rosaceae. Related phenolic ketones are also occasional plant products. One of particular ecological interest is 3-acetyl-6-methoxybenzaldehyde (XVII) which occurs as a growth inhibitor in leaves of the desert shrub *Encelia farinosa* and persists in the soil surrounding the plant causing the inhibition of germination in the seeds of annuals growing in the vicinity (GRAY and BONNER, 1948).

Related to the acetophenones, in being C_6-C_2 compounds, are the phenylacetic acids. Both the 2- and 4-hydroxyderivatives have been identified in plants, in *Astilbe* leaves and in roots of dandelion (*Taraxacum officinale*) respectively. The 4-hydroxy compound (XVIII) has recently been claimed as a natural growth regulator in certain higher plants (see GROSS, 1975) and also as an auxin-like substance in an alga (ABE et al., 1974).

A wide range of simple phenols and phenolic derivatives has also been found in fungi and lichens; fungal phenols have been listed by TURNER (1971) and lichen compounds by CULBERSON (1969). Two phenols which are unique to fungi are 3,4-dihydroxyphenylglyoxylic acid (XIX) of *Polyporus tumulosus* and 3,4,5-trihydroxybenzaldehyde of *Boletus scaber*. Illustrative of the more complex phenols of lichens is diploicin (XX), a depsidone from *Diploicia canescens,* which is biogenetically derived by esterification of a substituted 2,4-dihydroxybenzoic acid with a resorcinol derivative. Its structure, as with other lichen constituents, is complicated by nuclear substitution in the aromatic rings by methyl and chlorine groups.

2.2 Phenylpropanoids

As is apparent from Figure 2, a ubiquitous phenolic unit in plants is one with an aromatic ring attached to a C_3 aliphatic side-chain. Such phenylpropanoids may be exemplified by *p*-hydroxycinnamic acid (or *p*-coumaric acid) (XXI) which occurs universally in plants and is also the precursor of many other similar C_6-C_3 molecules. By *o*-hydroxylation and subsequent cyclisation, *p*-coumaric acid can give rise to hydroxycoumarins. On reduction, it yields *p*-coumaryl alcohol, which is one of the monomeric "building blocks" of the secondary cell wall polymer lignin. Dehydrogenation of this alcohol gives a phenylpropene, present as a group in volatile oils of plants. Dimerisation of *p*-coumaryl alcohol can give rise to a lignan. Finally, the *p*-coumaric acid moiety is an important unit in the structure of flavonoids, stilbenes and xanthones (Fig. 2). All the above classes of phenylpropanoid occur widely in plants.

The most common hydroxycinnamic acid of plants is caffeic acid (XXII) which is universal in higher plants; it is usually present as the depside chlorogenic acid, but many other combined forms are known. Besides *p*-coumaric and caffeic, two methylated acids are also very common: ferulic (XXIII) and sinapic (XXIV). Ferulic acid, in particular, can be found in many different contexts. It has been identified as an *N*-acyl terminal group in a protein in barley seed (VAN SUMERE et al., 1973), linked directly to glycine and phenylalanine. It also occurs, dimerised as diferulic acid (XXV), bound to the carbohydrate of cell walls in a number of grasses

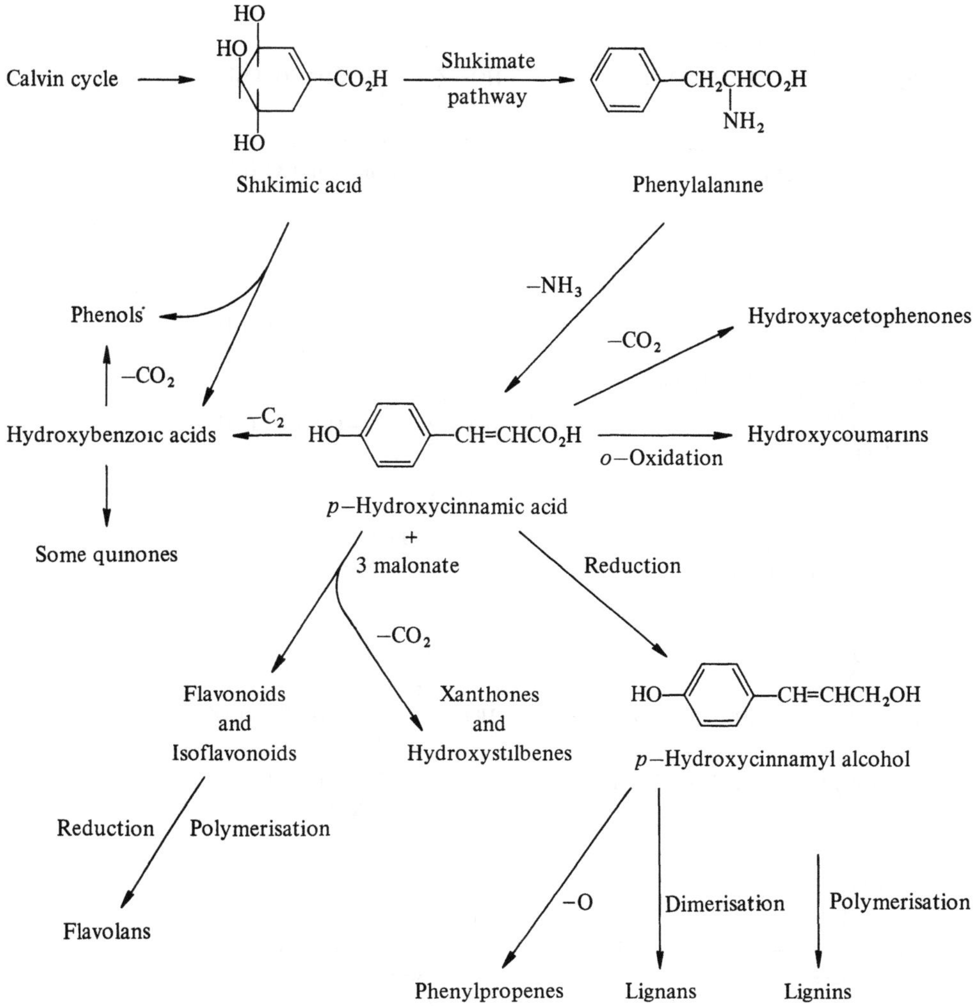

Fig. 2. Biosynthetic origin of plant phenolics from shikimate and phenylalanine

(HARTLEY and JONES, 1976). Again, ferulic acid is present free in sugar beet and cereal seeds and is reputedly a general germination inhibitor (VAN SUMERE et al., 1972). Hydroxycinnamic acids and their derivatives are capable of existing in cis and trans forms, and while there is evidence that the natural forms are all trans, isomerisation inevitably occurs during extraction, and mixtures of isomers are always isolated. Artefactual oxidation in the *o*-position can also occur and caffeic acid, in particular, may be partly converted to the coumarin, aesculetin (see below), during the isolation process.

Phenylpropionic acids, formed by reduction of the aliphatic double bonds of cinnamic acids, have been detected in nature. Melilotic acid (dihydro-*o*-coumaric acid) occurs in the Leguminosae, and dihydrocaffeic acid in beetroot (Chenopodiaceae). The related 3,4-dihydroxyphenyl-lactic acid (XXVI) occurs naturally as its caffeic acid ester (rosmarinic acid) which was first isolated from rosemary *Rosmarinus officinalis* (Labiatae) but is now known to be widespread in this and related families (HARBORNE, 1966).

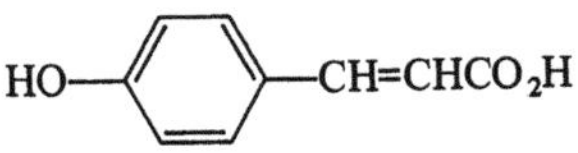

p–Coumaric acid
(XXI)

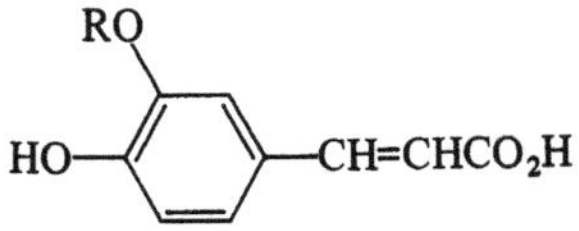

R = H, caffeic acid (XXII)
R = Me, ferulic acid (XXIII)

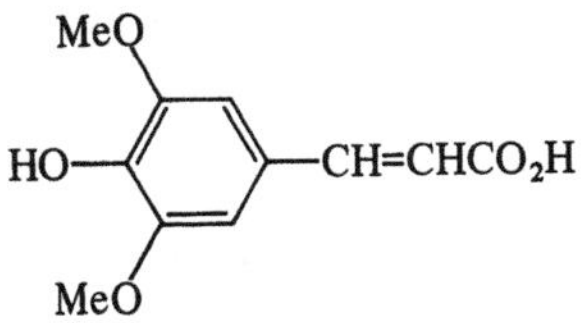

Sinapic acid
(XXIV)

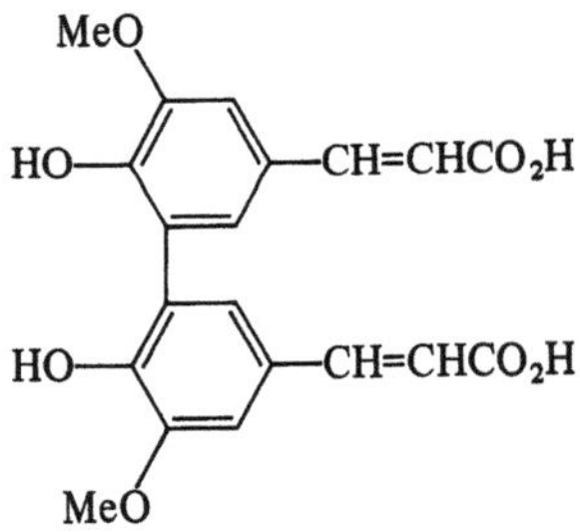

Diferulic acid
(XXV)

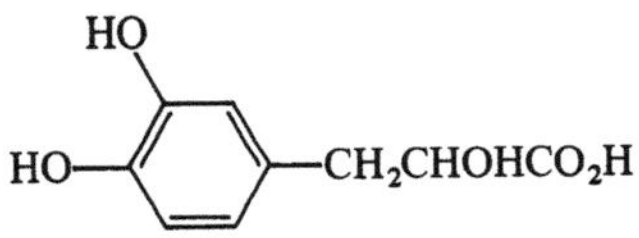

3,4–Dihydroxyphenyl–lactic acid
(XXVI)

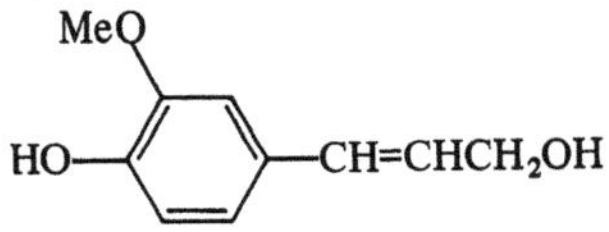

Coniferyl alcohol
(XXVII)

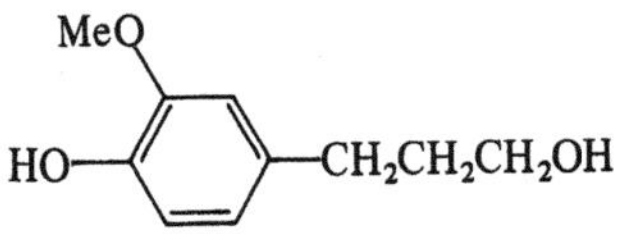

Dihydroconiferyl alcohol
(XXVIII)

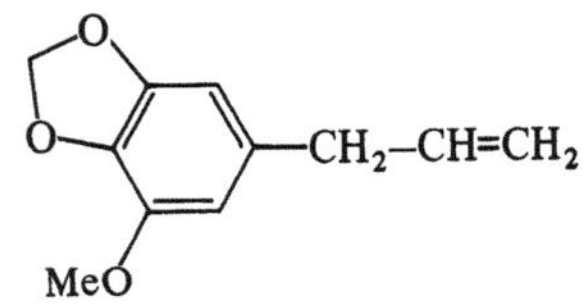

Myristicin
(XXIX)

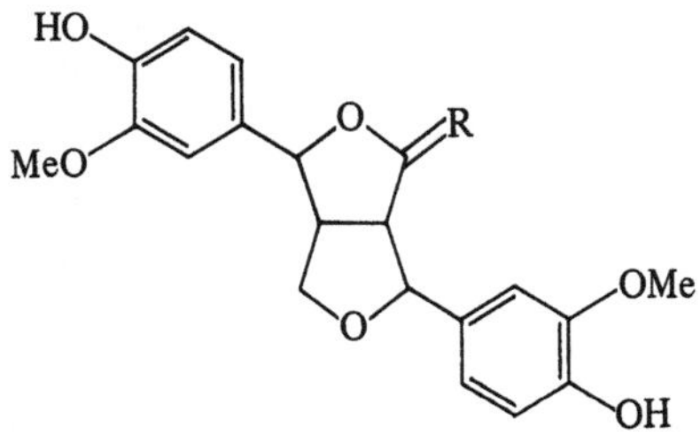

R = O, *Aegilops* lignan (XXXI)
R = H_2, pinoresinol (XXX)

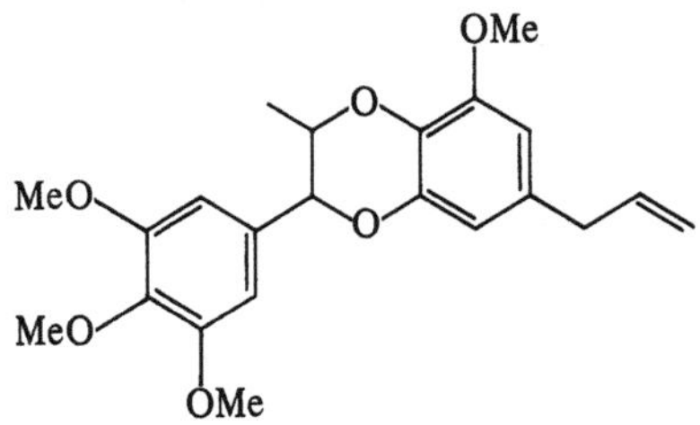

Eusiderin
(XXXII)

The three alcohols corresponding to *p*-coumaric, ferulic, and sinapic acids are undoubtedly widespread in plants, at least as trace constituents, since they are known to be the immediate precursors of plant lignins (Gross, 1978). In keeping with this role, coniferyl alcohol (XXVII), the ferulic acid analogue, has been found in cambial sap of both angiosperms and gymnosperms, while sinapyl alcohol occurs similarly only in angiosperm tissues. Dihydroconiferyl alcohol (XXVIII) has recently been reported as a plant product, since it has been identified as the "lettuce cotyledon factor" in *Lactuca*. It is a synergist of gibberellic acid in inducing elongation of hypocotyls and has this activity in lettuce and in several other plants (Sikurai et al., 1974).

The common plant hydroxycinnamyl alcohols are also precursors of two further classes of monomeric phenylpropanoid: allyl and propenylphenols. Thus coniferyl alcohol (XXVII) has been shown to be the immediate precursor of eugenol (side-chain-$CH_2-CH=CH_2$) and its methyl ether, methyleugenol, which occur together in leaves of the herb basil, *Ocimum basilicum* (Klischies et al., 1975). Allyl and propenyl phenols occur sporadically in higher plants, in essential oil fractions of leaves and fruits, but are not necessarily found together as isomeric pairs. One of the most widely occurring is myristicin (XXIX) which is reputedly hallucinogenic in man (Shulgin, 1966). While present in nutmeg *Myristica fragrans* (myristicaceae) together with isomyristicin, it occurs on its own in a range of umbellifers (Harborne et al., 1969).

Dimers, formed from the union of two coniferyl or sinapyl alcohol units, are known as lignans. Pinoresinol (XXX), for example, is derived by tail-to-tail linkage in the β-position of two coniferyl alcohol residues and occurs widely in *Pinus* and *Picea* (Hathway, 1962). The ketone (XXXI) derived from (XXX) has recently been characterised as one of the germination inhibitors in seeds of the cereal *Aegilops ovata* (Cooper et al., 1977). Related dimers, called neolignans, can be formed by other condensations between two C_6-C_3 units, e.g., joining head-to-tail instead of tail-to-tail. One such compound is eusiderin (XXXII) and this and related structures occur in heartwoods of Magnoliaceae, Piperaceae, and Lauraceae (Gottlieb, 1972).

Coumarins are lactones formally derived from *o*-hydroxycinnamic acids by cyclisation and ring closure between the *o*-hydroxy and carboxyl groups. Coumarin itself occurs in fresh plant tissues (e.g., in leaves of *Melilotus alba*) in bound form as the corresponding trans-*o*-glucosyloxycinnamic acid (XXXIII). On tissue damage, this bound form undergoes enzymic loss of sugar, trans$\rightarrow$cis isomerisation and ring closure, the volatile coumarin being released from the leaf surface as "the odour of new mown hay". Some methoxycoumarins occur naturally, bound in the same way. Thus 6,7-dimethoxycoumarin (XXXIV) is present in leaves of the orchid *Dendrobium densiflorum* as the corresponding 2-glucosyloxy-4,5-dimethoxycinnamic acid (Dahmen et al., 1975). By contrast, the common hydroxycoumarins occur in plants as such, with *O*-glycosidic attachment. Indeed, such glycosides (e.g., aesculin) have an intense blue fluorescence and their presence in plant tissues (e.g., bark of horse chestnut) can be readily detected by examining the material in UV light.

The three most common hydroxycoumarins of plants are umbelliferone (XXXV), aesculetin (XXXVI), and scopoletin (XXXVII) and correspond in structure to *p*-coumaric, caffeic, and ferulic acids. Curiously, the coumarin corresponding

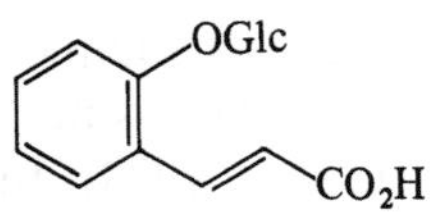

(XXXIII)

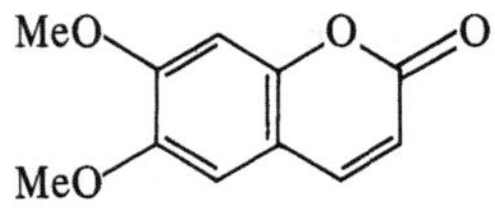

6,7–Dimethoxycoumarin
(XXXIV)

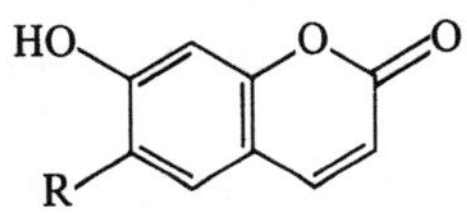

R = H, Umbelliferone (XXXV)
R = OH, Aesculetin (XXXVI)
R = OMe, Scopoletin (XXXVII)

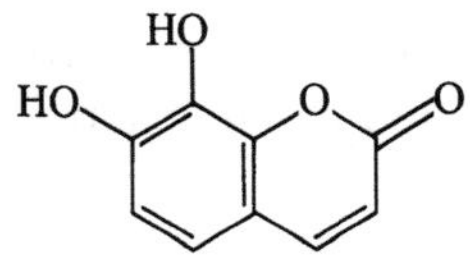

Daphnetin
(XXXVIII)

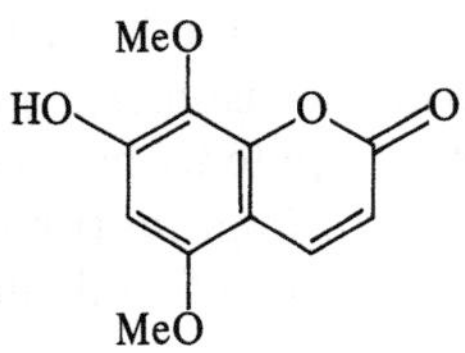

Leptodactylone
(XXXIX)

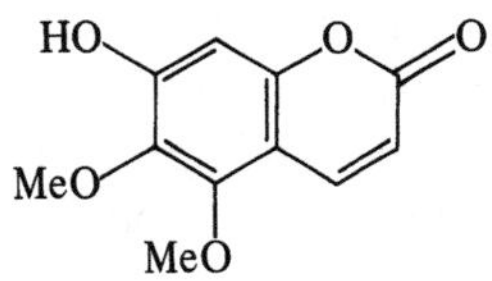

(XL)

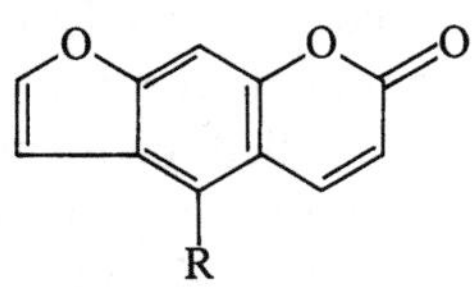

R = OMe, Bergapten (XLI)
R = H, Psoralen (XLII)

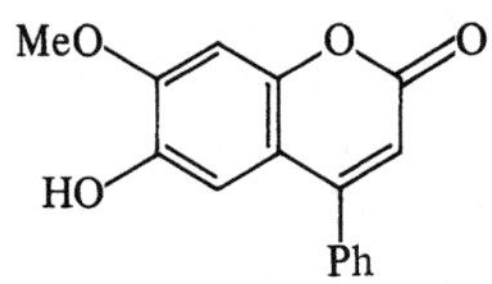

Dalbergin
(XLIII)

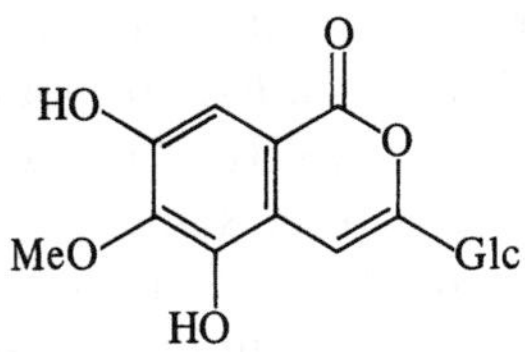

Bergenin
(XLIV)

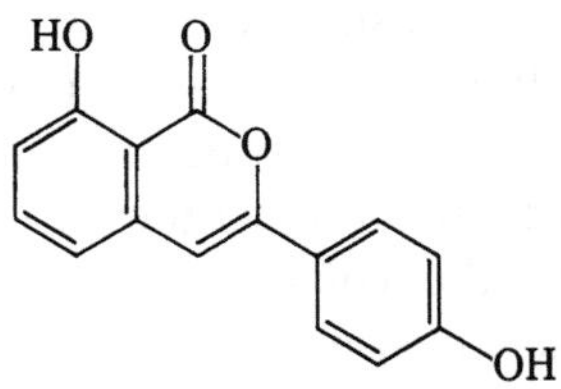

Hydrangenol
(XLV)

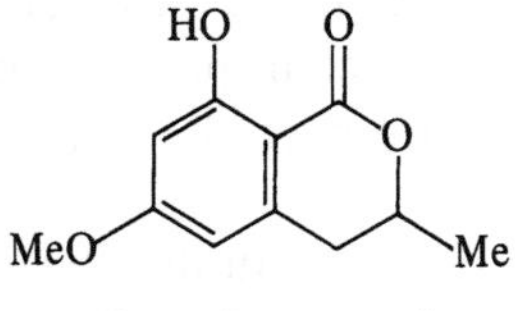

Carrot isocoumarin
(XLVI)

to sinapic acid, isofraxidin (6,8-dimethoxy-7-hydroxycoumarin), is quite rare, being restricted to bark of *Fraxinus* (Oleaceae). A fairly common coumarin, which has no obvious cinnamic acid counterpart, is daphnetin (XXXVIII), which was first isolated as the glucoside, daphnin, from bark of *Daphne mezereum* (Thymelaceae). It is also known from several other families and has recently been detected for the first time in Polemoniaceae (*Linanthus* and *Leptodactylon*) and in Juncaceae (*Juncus effusus*) (WILLIAMS and HARBORNE, 1975). Among other recently reported coumarins are leptodactylone (XXXIX), a yellow pigment in leaves of *Leptodactylon* (DEAN et al., 1978) and 5,6-dimethoxy-7-hydroxycoumarin (XL), present in roots of *Pelargonium reniforme* (Geraniaceae) (WAGNER and BLADT, 1975).

While structural variation among hydroxycoumarins is limited, there are more complex plant coumarins, known as furocoumarins, of which an almost infinite variety of structures has been described. These are mainly derived from umbelliferone (XXXV) by condensation with C_5 isoprenoid units and they are usually lipid-soluble, most or all phenolic hydroxyls being protected by *O*-methylation or isoprenylation. Bergapten (XLI) from bergamot oil is but one of several hundred such substances. Furocoumarins are of physiological interest because of their growth-inhibiting effects in relation to seed germination. Thus, psoralen (XLII) has been identified as an endogenous inhibitor in the seedcoat of *Psoralea subacaulis* (Leguminosae). It inhibits germination of the seed and has to be leached out of the coat before the seed will develop. On leaching into the soil, it may also prevent germination of the seed of other plant species in the vicinity (BASKIN et al., 1967). A chemical account of the furocoumarins is provided by DEAN (1963). More recent reviews on the structural variation and natural distribution of these compounds in Umbelliferae and Rutaceae can be found in HEYWOOD (1971) and GRAY and WATERMAN (1978).

One final class of phenolic coumarin should be mentioned: 4-phenylcoumarins or neoflavanoids. A typical member is dalbergin (XLIII) which occurs in heartwood of *Dalbergia* trees (Leguminosae). About 42 structures are known and they are found almost exclusively in plants of the Leguminosae or Guttiferae (DONNELLY, 1975).

Only a few isocoumarins are known, but all carry one or more phenolic groups. One is bergenin (XLIV), in roots of *Astilbe, Bergenia, Peltoboykinia*, and *Rodgersia* (Saxifragaceae). Another is hydrangenol (XLV), a 3-phenyl substituted isocoumarin present specifically in *Hydrangea macrophylla* (BILLEK and KINDL, 1962). A third, 3-methyl-6-methoxy-8-hydroxy-3,4-dihydroisocoumarin (XLVI), is of plant pathological interest, because although a fungal metabolite, it is formed in carrot, *Daucus carota*, tissue following infection by the pathogen *Ceratocystis fimbriata*. Its role as a carrot phytoalexin has been disputed, but it is now generally accepted that it is produced by the higher plant and not by the fungus in this interaction (cf. DEVERALL, 1976).

Chromones are isomeric with coumarins but fewer are known and most occur in only one or two plant sources. Eugenin (XLVII) is an example of a simple 2-methyl-chromone, present in *Eugenia aromatica* (Myrtaceae). Simple 5,7-dihydroxychromones unsubstituted in the 2-position e.g., (XLVIII), have been reported in several plants, including the peanut *Arachis hypogaea* (PENDSE et al., 1973), but recent evidence indicates that they are not actually plant products, but are formed as artefacts from flavanone glycosides present during the drying of the plant tissue

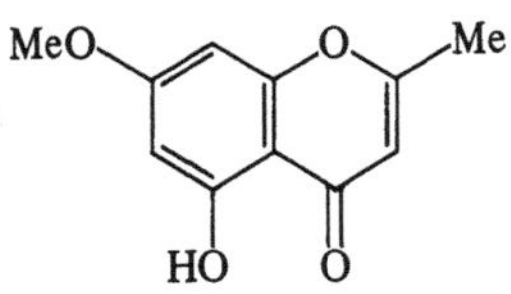

Eugenin
(XLVII)

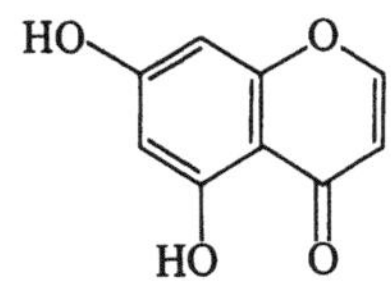

5,7-Dihydroxychromone
(XLVIII)

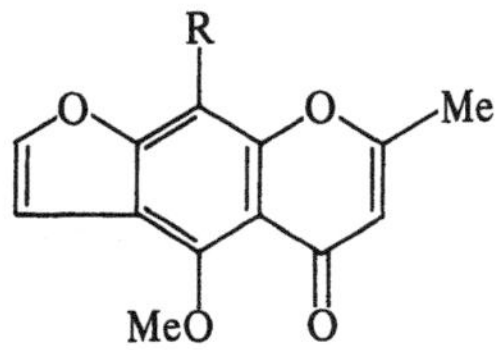

R = OMe, Khellin (XLIX)
R = H, Visnagin (L)

before extraction (STOCKER and POHL, 1976). The only plant chromone unsubstituted at position 2 seems to be an unusual 3-ethylchromone which has been identified as a phytoalexin of the sweet pea, *Lathyrus odoratus* and *L. hirsutus* produced in the seed or leaf in response to fungal infection (ROBESON, D. and HARBORNE, J.B., unpublished results). Finally, mention must be made of furochromones, such as khellin (XLIX) and visnagin (L), which occur in fruits of *Ammi visnaga* (Umbelliferae), because of their pharmaceutical importance in the treatment of heart disease.

2.3 Flavonoids

2.3.1 The Common Flavonoids

A large number of naturally occurring flavonoids have been described; according to a recent estimate (HARBORNE et al., 1975), there are about 2000 known structures. In spite of a bewildering array of structural variation, some simple patterns can be discerned. All are based on the same C_{15} skeleton of flavone (LI) and are formed by the same pathway from malonate units condensing with a phenylalanine-derived C_6-C_3 precursor.

Flavonoids are conveniently divided into some twelve classes, according to the oxidation level of the central pyran ring (Table 2). Most of these classes will be mentioned here but emphasis will be given to the three more important and widespread classes – the anthocyanins, flavones, and flavonols. It should be observed (Table 2) that the different classes vary in their biological properties; many (e.g., anthocyanins) are strongly pigmented but others (e.g., flavanones) are completely colourless. It is among the colourless flavonoids that compounds with significant physiological activity have been reported.

Of the several hundred flavonoid aglycones that have been isolated from plants only eight of these occur widely. One or other of these eight structures may be expected to be present in the hydrolysed extract of any higher plant studied.

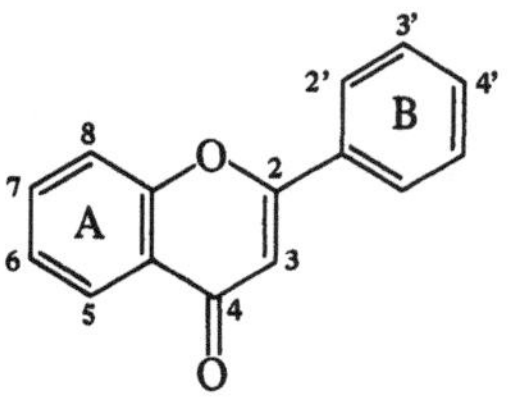

Flavone
(LI)

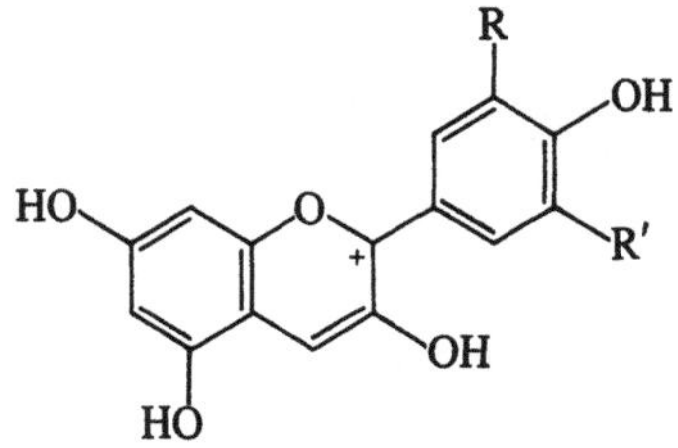

Pelargonidin, R = R' = H (LII)
Cyanidin, R = OH, R' = H (LIII)
Delphinidin, R = R' = OH (LIV)

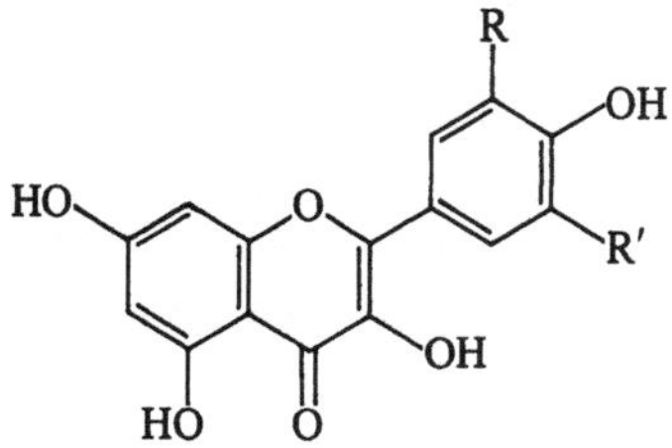

Kaempferol, R = R' = H (LV)
Quercetin, R = OH, R' = H (LVI)
Myricetin, R = R' = OH (LVII)

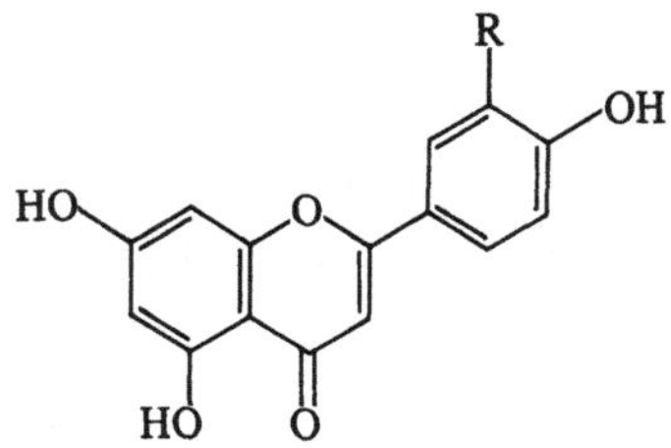

Apigenin, R = H (LVIII)
Luteolin, R = OH (LIX)

Table 2. The major known classes of flavonoid

Class	No. of known structures	Biological properties
Anthocyanins	250	Red to blue pigments
Chalcones	60	Yellow pigments
Aurones	20	
Flavones	350	Cream pigments in flowers; feeding repellents (?) in leaves
Flavonols	350	
Flavanones	150	Some have bitter tastes
Dihydrochalcones	10	
Proanthocyanidins	50	Astringent substances, some with tanning ability
Catechins	20	
Flavan-3,4-diols	20	
Biflavonoids	65	None known
Isoflavonoids	150	Oestrogenic and fungitoxic

All eight of these flavonoids [(LII) to (LIX)] have similar hydroxylation in the A-ring and differ mainly in the oxidation level of the central pyran nucleus and in the number of hydroxyl groups in the B-ring: one, two, or three. All these common structures occur naturally in water soluble form as *O*-glycosides (see Sect. 3.2). Three are anthocyanidins: the scarlet-coloured pelargonidin (LII), the crimson cyanidin (LIII), and the mauve delphinidin (LIV). These three pigments or their simple methyl ethers occur very widely in plants, being particularly abundant in coloured flowers and fruits. Within the angiosperms, they are almost ubiquitous, but are replaced in one group of families in the order Centrospermae (e.g., in beetroot, *Beta vulgaris*, Chenopodiaceae) by a different class of purple pigment, betacyanin, derived biosynthetically from the aromatic amino acid L-DOPA. Pelargonidin and delphinidin occur especially frequently in cyanic flowers, their synthesis being correlated with natural selection by animal pollinators for bright scarlet colours (hummingbirds) and deep blue colours (bees) respectively. By contrast, pigmented leaves normally contain cyanidin, the other two anthocyanidins being relatively rarely found in vegetative tissue.

The flavonols corresponding in structure to the three main anthocyanidins are kaempferol (LV), quercetin (LVI) and myricetin (LVII), in order of increasing B-ring hydroxylation. While these flavonols usually accompany the anthocyanidins in flowers where they have an important role as co-pigments, they occur even more frequently, indeed almost universally, in plant leaves. A leaf survey of over 1000 angiosperms showed, for example, that 48% of species had kaempferol, 56% quercetin and 10% myricetin (SWAIN and BATE-SMITH, 1962). In all, over 60% of the sample had one or other of the three flavonols. While kaempferol and quercetin occur throughout the angiosperms, myricetin is more restricted in its occurrence, being present mainly in leaves of woody plants, in association with tannins.

Flavones lack the 3-hydroxyl group present in flavonols and anthocyanidins. Only two structures are common: apigenin (LVIII) and luteolin (LIX). These two flavones, besides occurring with flavonols, are also found on their own in many herbaceous plants, there being a tendency for them to replace flavonols in more specialised angiosperm families. Flavones, unlike the other common flavonoids, are frequently found in *C*-glycosidic combinations and are thus often reported as their *C*-glycosidic derivatives, vitexin (8-*C*-glucosylapigenin) and orientin (8-*C*-glucosyl-luteolin). They also occur widely in the more usual *O*-glycosidic combinations.

Tricetin, the flavone corresponding in structure to myricetin or delphinidin, is known, but so far appears to be of rare occurrence. It has been isolated from only a few plants, such as *Lathyrus pratensis* (Leguminosae), *Thuja occidentalis* (Cupressaceae) and *Lachenalia uniflora* (Liliaceae) (WILLIAMS et al., 1976). A methylated derivative of tricetin, namely tricin, is, however, more regular in occurrence, being particularly common as a constituent of grasses. Another peculiarity of flavones is their ability to link together by carbon–carbon bonds to form dimers. Over sixty biflavones are known, one of the commonest being amentoflavone or 3′,8″-biapigenin. These substances are largely restricted to and are typical of gymnosperm tissues, but they do occasionally occur elsewhere in the plant kingdom (GEIGER and QUINN, 1975).

2.3.2 Variations in Hydroxylation Pattern

Many flavonoids differing in hydroxylation pattern from the eight common structures are known, but the great majority are of limited occurrence in nature. Some typical examples are compounds (LX)–(LXIX). The range extends from flavone itself and simple mono- and dihydroxyflavones (e.g., LX, LXI) which occur exclusively in farina on leaves of *Primula* species to compounds such as digicitrin

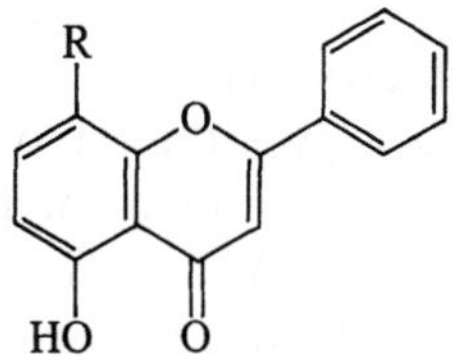

5–Hydroxyflavone, R = H (LX)
5,8–Dihydroxyflavone, R = OH (LXI)

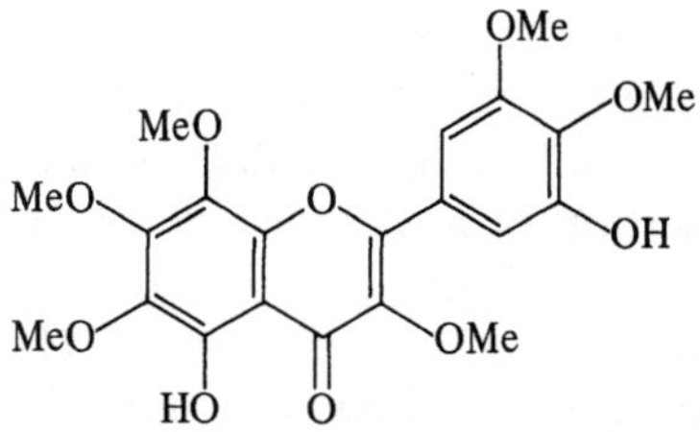

Digicitrin
(LXII)

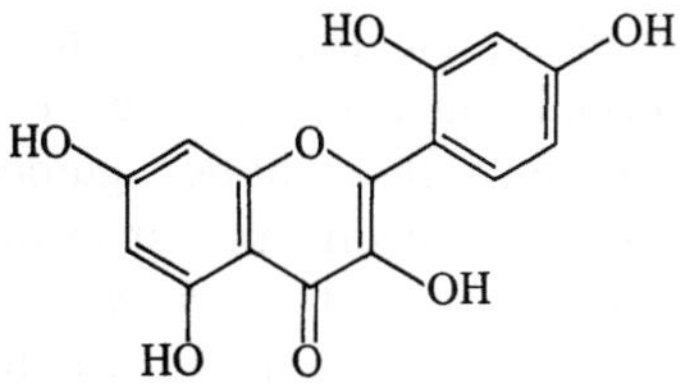

Morin
(LXIII)

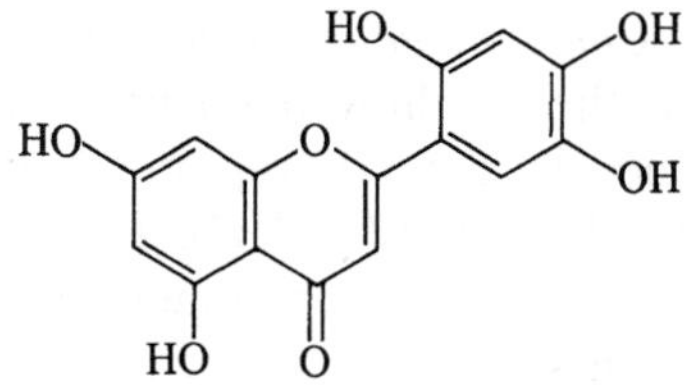

Isoetin
(LXIV)

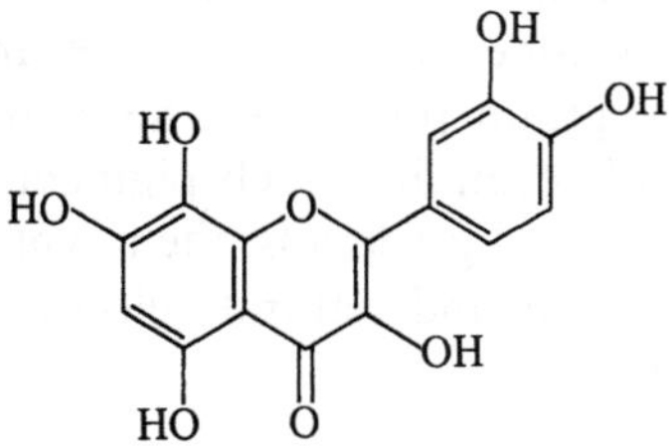

Gossypetin
(LXV)

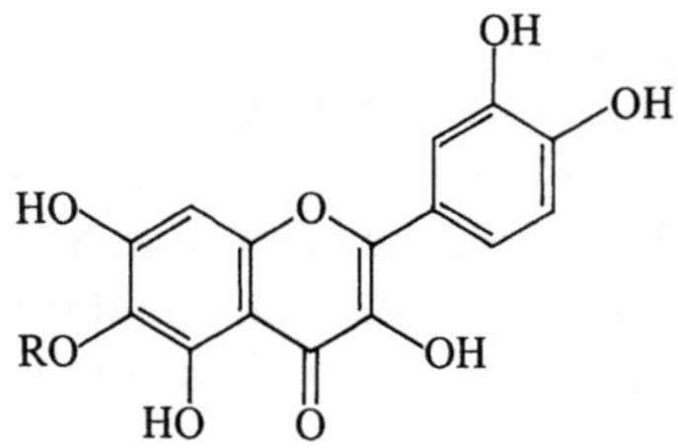

Quercetagetin, R = H (LXVI)
Patuletin, R = Me (LXVII)

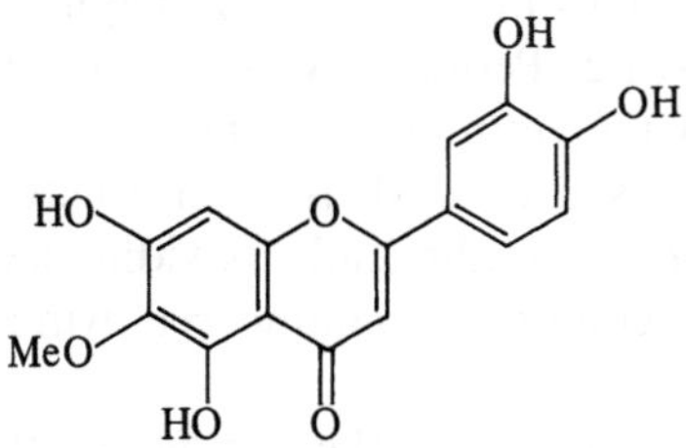

6–Methoxyluteolin
(LXVIII)

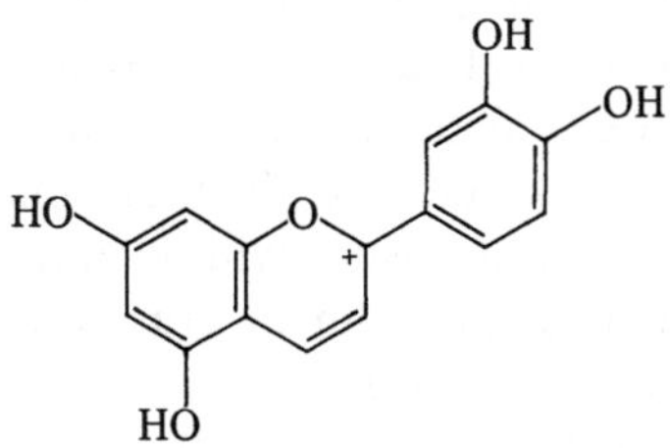

Luteolinidin
(LXIX)

(LXII) from the foxglove, which has eight hydroxyl groups, seven of which are protected by methylation. Most of the rarer flavonoids are closely related to one or other of the eight common structures, being formed from them by insertion of an extra hydroxyl function in an otherwise unoccupied site in the flavonoid nucleus. Thus, morin (LXIII) is derived from kaempferol by the insertion of an extra hydroxyl in the 2′-position. This substance occurs in *Morus alba* leaves and is one of two flavonoids, the other being quercetin 3-glucoside, which contributes to the selective feeding of silkworm larvae on mulberry plants.

Similar insertion of a 2′-hydroxyl into luteolin (LIX) gives the flavone isoetin (LXIV) which has recently been identified as a yellow flower pigment in several members of the Cichorieae (Compositae) (HARBORNE, 1978b). Yellow flower colour of a flavonoid nature is, however, more frequently based on the substitution of an extra hydroxyl group into the A-ring. Introduction of a hydroxyl at the 6- or 8-positions causes a significant shift in colour and such compounds, e.g., gossypetin (LXV) and quercetagetin (LXVI), are yellow instead of pale cream, as in quercetin (LVI). These two yellow pigments contribute colour to such well-known plants as the primrose, *Primula vulgaris*, cotton flower, *Gossypium hirsutum*, and corn marigold, *Chrysanthemum segetum*. In some plants, such as the composite *Rudbeckia hirta*, these yellow pigments have a special function because of their strong absorbance in the UV region, in that they can act as honey guides to insect pollinators. Patuletin (LXVII), a methyl ether of quercetagetin, for example, is differentially distributed in the inner part of the ray flower of *Rudbeckia*, while carotenoid pigments, which only reflect light in the UV, are evenly distributed over the whole ray (THOMPSON et al., 1972). Flavones [e.g., (LXVIII)] and flavonols with extra oxygen functions in the A-ring also occur regularly in leaves of angiosperms, particularly in herbaceous plants.

Other rarer flavonoids are those in which a particular hydroxyl function found in the common structures is removed rather than added to. One such example is luteolinidin (LXIX) which lacks the 3-hydroxyl group present in the common anthocyanidins and is referred to as a 3-desoxyanthocyanidin. Such pigments are rare in nature, but where they do occur, as in New World gesnerads, their synthesis seems to be correlated with hummingbird pollination and natural selection for scarlet or orange flower colour (HARBORNE, 1967b).

2.3.3 Methylated Flavonoids

The most common flavonoid *O*-methyl ethers are peonidin (LXX), petunidin (LXXI), and malvidin (LXXII), which are derived from cyanidin or delphinidin by methylation of the hydroxyl groups in the B-ring. Peonidin was first isolated from peony flowers, petunidin from *Petunia* and malvidin from *Malva*, but they are by no means restricted to their original plant sources. All three are actually widely present in plant tissues with cyanic colour. Petunidin and malvidin occur, for example, with delphinidin in grapes and the colour of genuine red wines is derived from these pigments.

Very rarely, methylation may affect the hydroxyl groups of the A-ring of anthocyanidins, and two series are known with 5-or 7-*O*-methylation. 5-Methyl ethers of delphinidin, petunidin, and malvidin have been found exclusively in Plumbaginaceae, e.g., in the light blue corollas of *Plumbago capensis*, while 5-*O*-methylcyanidin (LXXIII) has recently been reported in *Egeria* (Elodeaceae) (MOMOSE et al., 1977). The 7-methyl ethers of peonidin and malvidin are called rosinidin and hirsutidin (LXXIV), respectively, and are likewise very restricted in their distribution. They only occur to any extent in one plant family, the Primulaceae, and here only in *Primula* and *Dionysia* species.

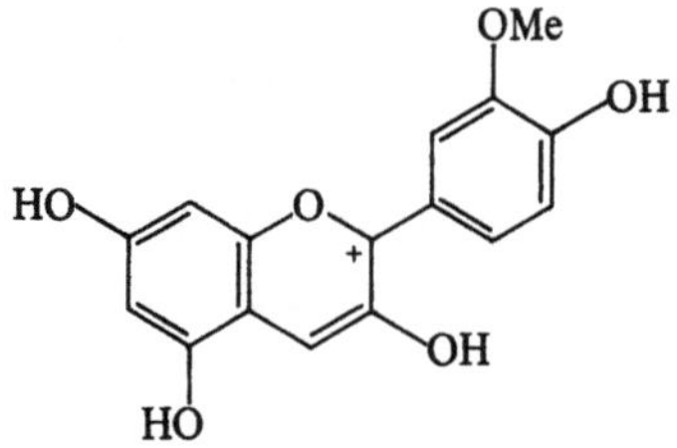

Peonidin
(LXX)

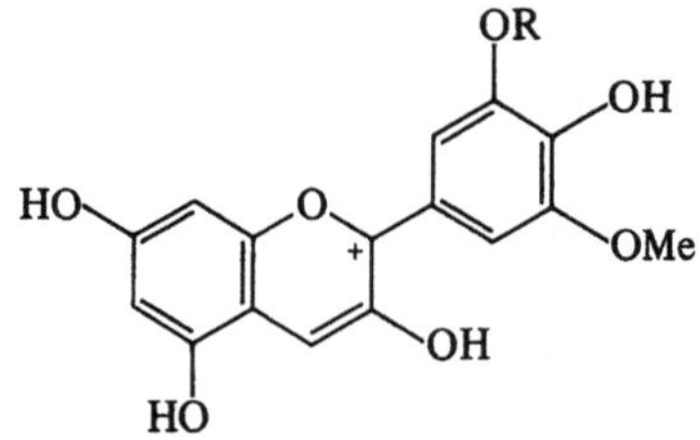

Petunidin, R = H (LXXI)
Malvidin, R = Me (LXXII)

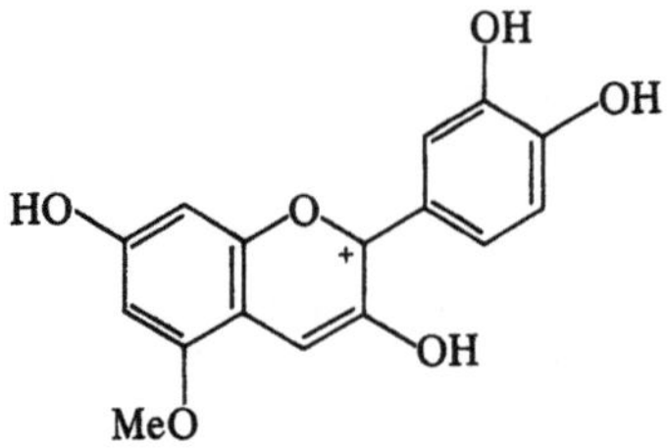

5–Methylcyanidin
(LXXIII)

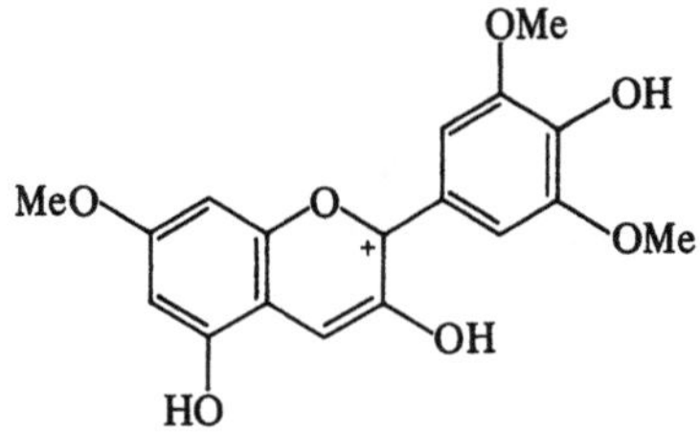

Hirsutidin
(LXXIV)

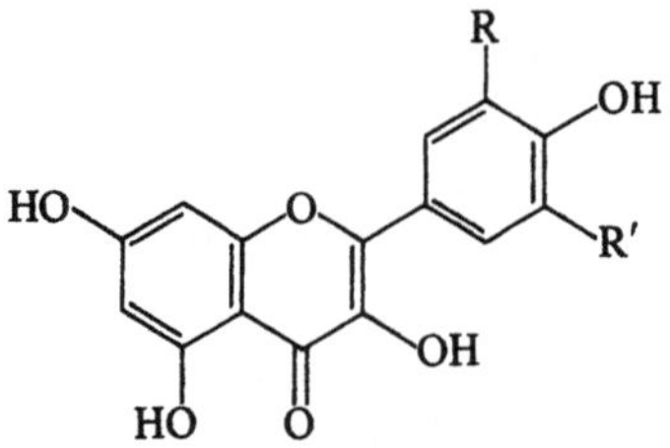

Isorhamnetin, R = OMe, R′ = H (LXXV)
Larycitrin, R = OMe, R′ = OH (LXXVI)
Syringetin, R = R′ = OMe (LXXVII)

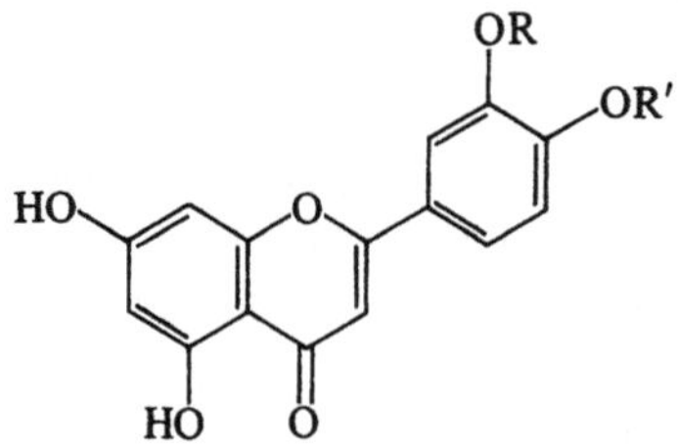

Chrysoeriol, R = Me, R′ = H (LXXVIII)
Diosmetin, R = H, R′ = Me (LXXIX)

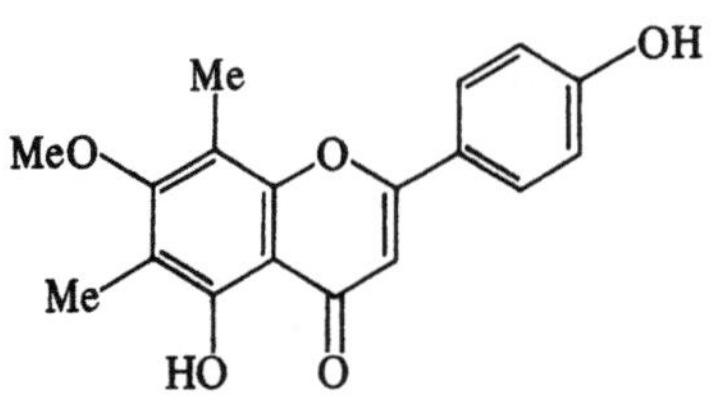

Sideroxylin
(LXXX)

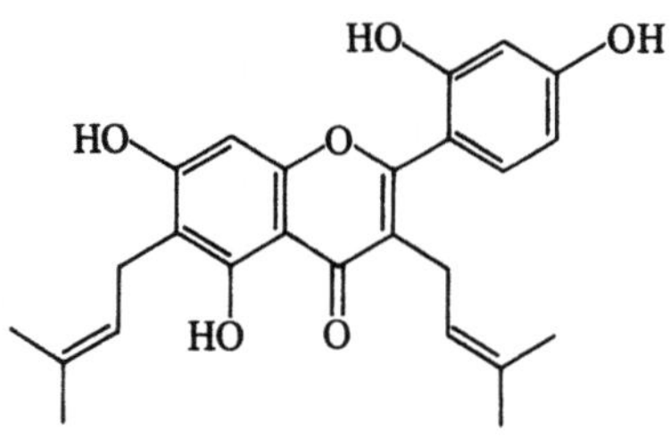

Mulberrin
(LXXXI)

The three flavonol methyl ethers corresponding in structure to peonidin, petunidin, and malvidin are isorhamnetin (LXXV), larycitrin (LXXVI), and syringetin (LXXVII). At one time, they were thought to be relatively rare, but recent evidence suggests that they are in fact quite common. Isorhamnetin (LXXV) is now known to occur in leaves and petals of many plants and is also rather common as a pollen constituent. On the other hand, syringetin was not known as a natural product until its identification in *Lathyrus pratensis* flowers twelve years ago (HARBORNE, 1965). However, it has subsequently been found in a range of other dicotyledons, e.g. in *Philydrum* (Philydraceae) (BOHM and COLLINS, 1975) and also in monocotyledons, in *Elegia* (Restionaceae) and *Hedychium* (Zingiberaceae) (WILLIAMS and HARBORNE, 1977).

There are many other flavonol methyl ethers of more restricted natural occurrence. In the case of quercetin (LVI), all possible five monomethyl ethers are known, while at least seven isomeric dimethyl, two trimethyl, and three tetramethyl ethers have been described (GOTTLIEB, 1975). The frequency of methylation at the different hydroxyl groups of flavonols varies with their chemical reactivity, and while the 3-, 7- and 3′-hydroxyls often carry methyl substituents, 4′-methylation is less common and 5-methylation quite rare. Furthermore, if extra hydroxyl groups are substituted in the nucleus, e.g., in the 6-position, to give quercetagetin (LXVI), this extra hydroxyl is often methylated at the same time, e.g., giving compounds such as patuletin (LXVII). In the case of flavones, the most common methyl ether is chrysoeriol (luteolin 3′-methyl ether) (LXXVIII) but diosmetin (luteolin 4′-methyl ether) (LXXIX) and acacetin (apigenin 4′-methyl ether) are not uncommon, since both are known from a range of plants.

Methylation in the flavonoid series masks the reactive phenolic groups, at the same time changing the physical properties significantly by introducing lipid solubility into molecules which are otherwise hydrophilic. While substances with one *O*-methyl group (e.g., peonidin, isorhamnetin, chrysoeriol) usually occur in glycosidic combination in the water-soluble vacuolar fraction of the leaf, those with two or more methyl groups [e.g., digicitrin (LXII)] are often present without sugar attachment in a lipid-soluble fraction. The location of such methylated flavonols and flavones within the leaf has not been fully explored, but lipid solubility would suggest that they are either in the cytoplasm or in the wax of the leaf surface. Indeed, highly methylated derivatives have been isolated from leaf waxes in some plants, e.g., in *Eucalyptus*, and from bud exudates of *Populus* and *Alnus*. One structure from *Eucalyptus* leaf wax is sideroxylin (LXXX), which is unusual in being *C*-methylated as well as *O*-methylated.

Finally, there is another structural feature in the flavonoid series which has the same effect as methylation on solubility: this is attachment of C_5 isoprenoid residues. One such compound is mulberrin (LXXXI), which is present in the bark of *Morus alba*. About 60 such structures have been listed, some having up to six C_5 residues attached (VENKATARAMAN, 1975).

2.3.4 Chalcones and Aurones

Chalcones and aurones are of principal biological interest because they contribute to yellow flower colour in a number of plants, especially members of the Compositae. Nevertheless, they do occur as colouring matters in other tissues, for example

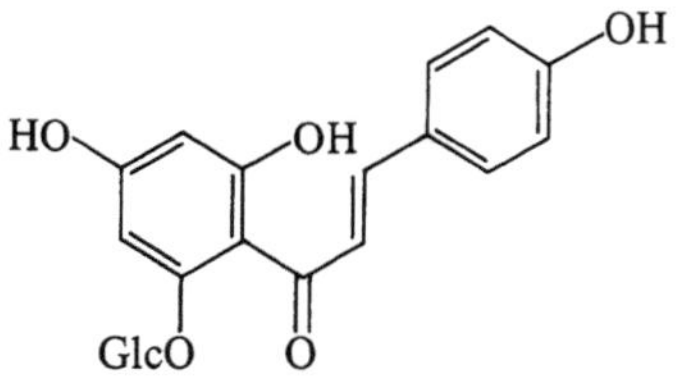

Isosalipurposide
(LXXXII)

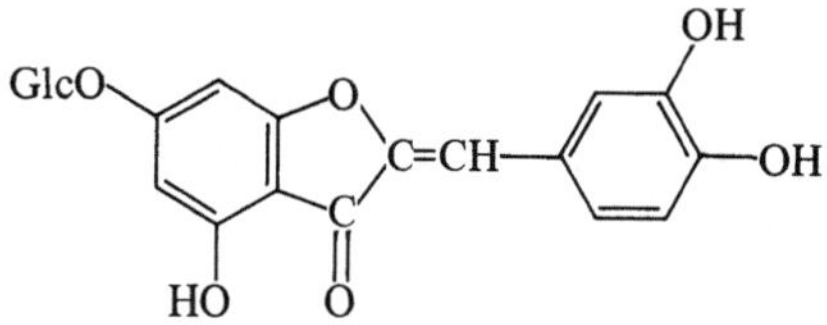

Aureusin
(LXXXIII)

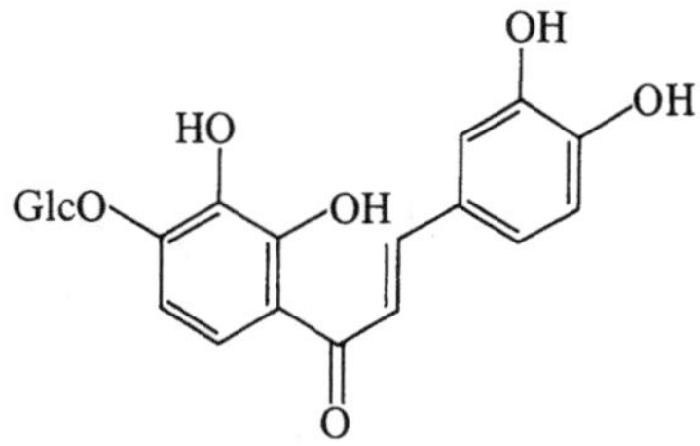

Marein
(LXXXIV)

in heartwoods and seeds of certain legumes. They are readily detected as flower pigments by the fact that when yellow petals are fumed with the alkaline vapour of a lighted cigar or of a bottle of ammonia, there is a dramatic colour change to orange or red. Because of this colour response, which is not shown by the more common yellow carotenoid pigments, these flavonoids are known as anthochlor pigments.

Such anthochlors may be the only class of yellow pigment in certain flowers. For example, the chalcone isosalipurposide (LXXXII) is the sole yellow colouring matter of yellow carnations, while the aurone aureusidin, occurring as the 6-glucoside aureusin (LXXXIII), is the major yellow pigment in the snapdragon, *Antirrhinum majus* (Scrophulariaceae). Anthochlors also occur regularly in yellow flowers accompanying lipid-soluble carotenoids. In such cases, the two classes of pigment may be differentially distributed in the flower tissue in such a way that the carotenoid is spread over the whole flower and acts as a general attractant to an insect visitor, while the yellow chalcone is present only in the inner ray and provides a UV absorbing honey guide to the centre of the flower. This happens in *Coreopsis bigelovii*, where the chalcone involved is marein (LXXXIV).

2.3.5 Flavanones and Dihydroflavonols

These two classes of flavonoid are simple reduction products of flavones and flavonols, respectively, and in terms of biosynthesis are formed first and then undergo oxidation in the 2,3-position to give rise to the more highly oxidised compounds. As biosynthetic intermediates, they are presumably universally present as trace constituents, but do not usually accumulate in quantity in leaf or flower tissue. Nevertheless, the two flavanones, naringenin (LXXXV) and eriodictyol (LXXXVI), corresponding in structure to apigenin and luteolin, have been recorded with some frequency in plants. They occur in glycosidic form and one particular

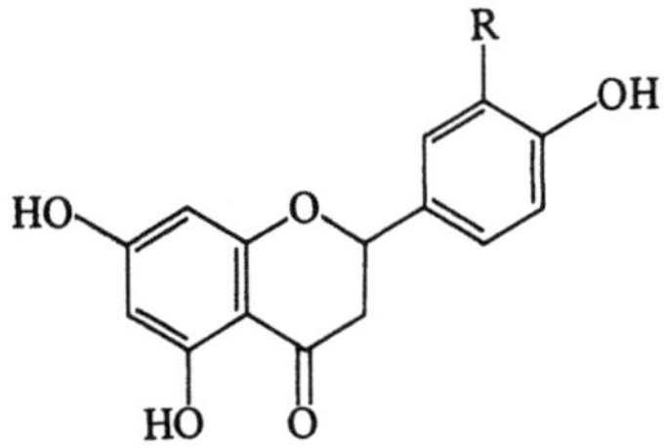

Naringenin, R = H (LXXXV)
Eriodictyol, R = OH (LXXXVI)

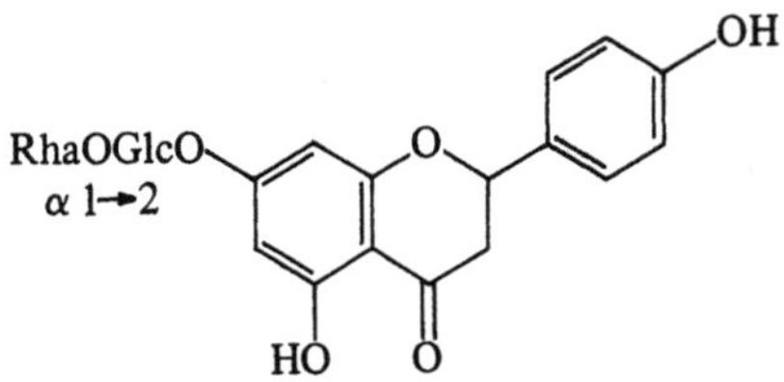

Naringin
(LXXXVII)

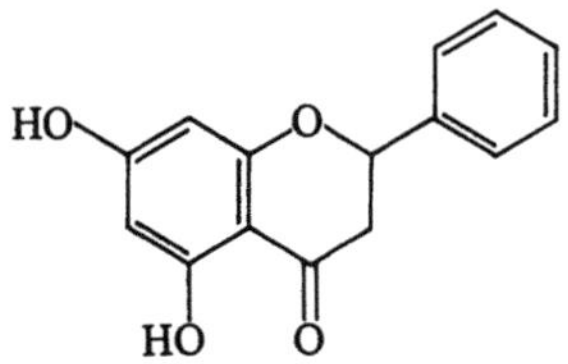

Pinocembrin
(LXXXVIII)

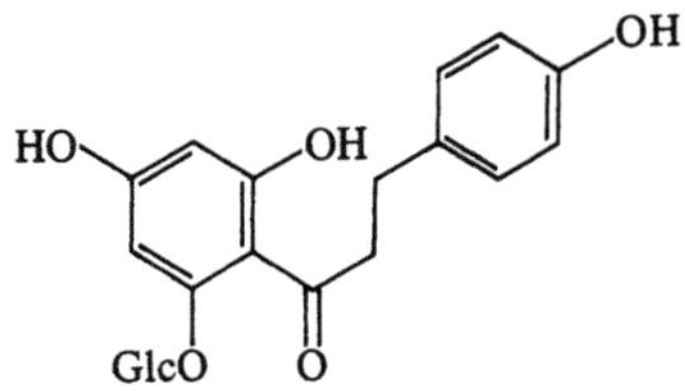

Phloridzin
(LXXXIX)

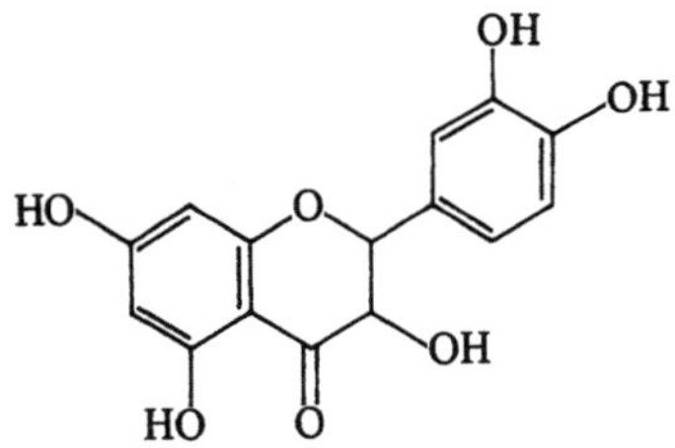

Dihydroquercetin
(XC)

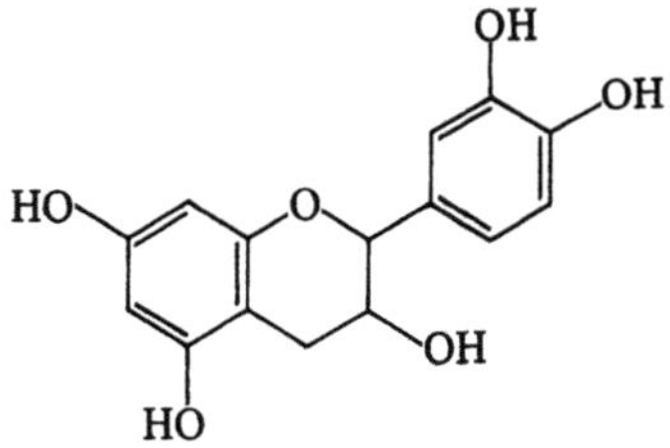

Catechin
(XCI)

series of flavanone glycosides occurring in *Citrus* fruit with the disaccharide neohesperidose attached to the 7-position as in naringin (LXXXVII) are of special interest because of their bitter taste to humans. Free naringenin has been reported to occur in dormant peach buds, where it acts as an antagonist to gibberellins (PHILLIPS, 1962). Flavanones which lack B-ring hydroxyls, e.g., pinocembrin (LXXXVIII), are also known.

On ring opening, flavanones give rise to the corresponding chalcones and the two classes of flavonoids are closely inter-related. Indeed, there is an enzyme chalcone-flavanone isomerase present in plants which catalyses this inter-conversion. One type may also be formed from the other during extraction and work-up of plant extracts, unless special care is taken to use very mild conditions. Reduction of the chalcone formed by ring opening of a flavanone gives rise to yet another flavonoid class, the dihydrochalcone. Such compounds are quite rare in nature. However, one member of this group, phloridzin (LXXXIX), must be mentioned, because of its great pharmacological activity in man in producing glucosuria. Phloridzin occurs universally in the apple genus *Malus* and could function in leaves and roots as a herbivore-feeding deterrent because, like naringin, it has a bitter taste. It is also a potent inhibitor of plant growth and may be responsible for the autotoxic effects of apple root extracts on the growth of apple seedlings (see RICE, 1974).

Dihydroflavonols corresponding in structure to the three common flavonols are all known, e.g., dihydroquercetin (XC), and are found most frequently in the free state in the heartwood of trees. They also occur occasionally in glycosidic form in leaf tissues. More highly reduced flavonoids, often co-occurring with dihydroflavonols, are the related flavan-3,4-diols and flavan-3-ols, where the 4-carbonyl is successively reduced to OH and H. Flavan-3-ols are also known as catechins and may have significant biological activity. While their main occurrence is in wood tissue, catechins also occur occasionally in leaves. For example, catechin (XCI) and its galloyl esters occur in high concentration in fresh leaves of *Camellia sinensis* and their oxidation products, formed during leaf fermentation, are responsible for much of the taste, flavour, and colour of tea. Catechins are also important as precursors of condensed tannins or proanthocyanidins (see Sect. 3.6).

2.3.6 Isoflavones and Isoflavonoids

Isoflavones such as genistein (XCII) are isomeric with the flavones (e.g., apigenin) and are derived biosynthetically from the same flavanone intermediate by aryl migration (see Sect. 4). They appear to be more restricted in their natural distribution than flavones, since they have only been found regularly in one subfamily of the Leguminosae, the Papilionoideae. There are, however, occasional records in Amaranthaceae (*Iresine*), Iridaceae (*Iris*), Myristicaceae (*Myristica*), Moraceae (*Maclura*), and Rosaceae (*Prunus*). This suggests they could be more widespread and, since there is no satisfactory method of screening plant tissues for their presence, our present knowledge of their natural occurrence is certainly imperfect.

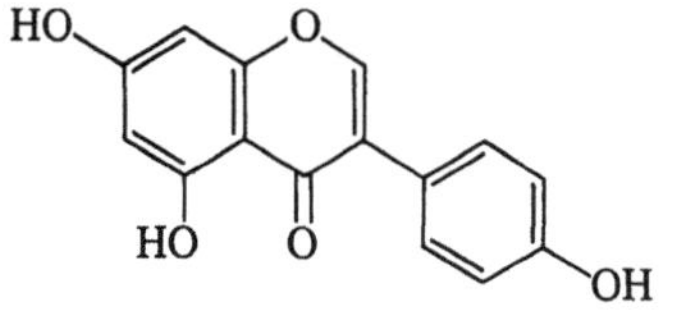

Genistein
(XCII)

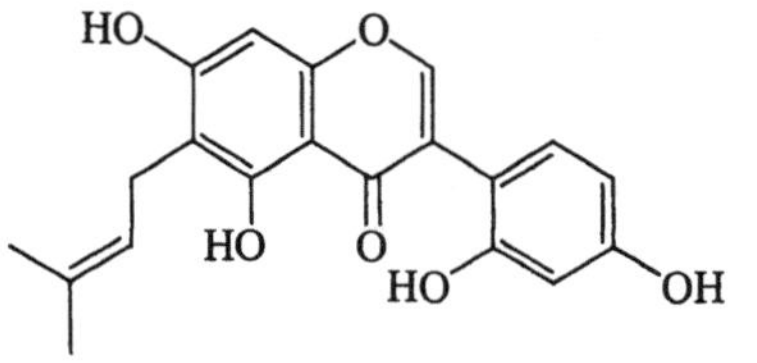

Luteone
(XCIII)

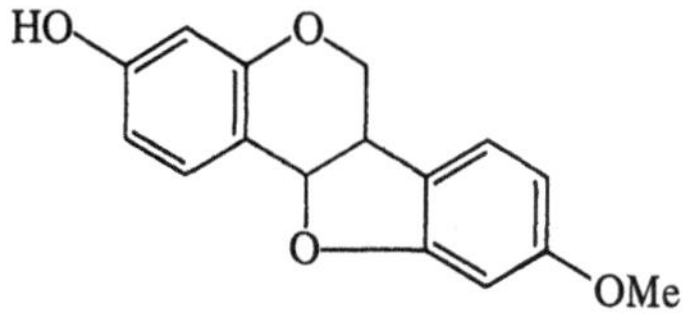

Medicarpin
(XCIV)

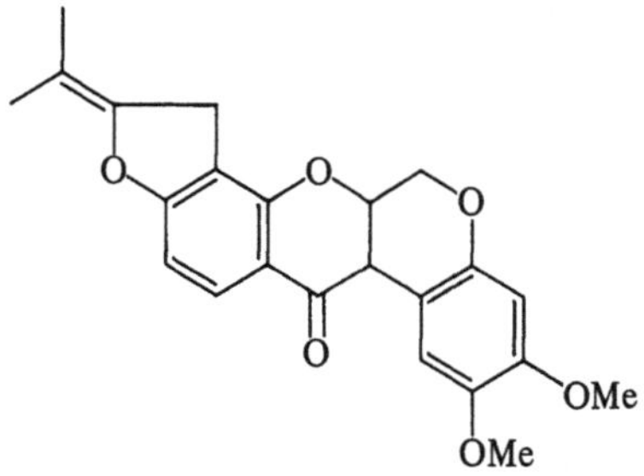

Rotenone
(XCV)

As a group, isoflavones and their derivatives are significantly more biologically active than the corresponding flavones. While simple isoflavones may not be highly active as such, they can become active after structural modification. 6-Isopentenylgenistein, for example, together with the 2′-hydroxy derivative, luteone (XCIII), is significantly antifungal in its occurrence on leaf surfaces of *Lupinus* species (HARBORNE et al., 1976). More regular fungitoxic activity is displayed by reduced derivatives, isoflavans and pterocarpans such as medicarpin (XCIV), which are formed as phytoalexins in many legumes. Reduction in vivo of isoflavones, during ingestion by mammals, also leads to biological activity of an oestrogenic nature (see SHUTT, 1976). Finally, one other series of complex isoflavonoids containing isopentenyl substitution, must be mentioned. These are the rotenoids, e.g., rotenone (XCV), which are well known for their insecticidal activity. Rotenone is in fact highly toxic to all forms of life, since it inhibits respiration at an essential step in the terminal electron transport pathway, and blocks the catalysis of an $NADH_2$-dependent dehydrogenase.

2.4 Xanthones and Stilbenes

2.4.1 Xanthones

Although structurally related to the flavonoids and similar to them in chromatographic behaviour, the xanthones are much more restricted in their natural distribution. Indeed, practically all the known hydroxyxanthones have been obtained from either the Guttiferae or the Gentianaceae. They occur especially in the wood, roots, or leaves of these plants. More than 70 structures are known (CARPENTER et al., 1969), ranging from compounds with one hydroxyl (e.g., 7-hydroxyxanthone (XCVI)) to those with five hydroxyl (or methoxyl) substituents (e.g., 4,7-dimethoxy-1,3,8-trihydroxyxanthone (XCVII)). Although originally most xanthones reported were found in the free state in heartwood or root tissues, an increasing number of *O*-glycosides have been described more recently from leaves of *Gentiana* and *Swertia* species (HOSTETTMANN and WAGNER, 1977).

Mangiferin (XCVIII) is unique among the natural xanthones in having a much wider natural occurrence than any of the others: this is the 2-*C*-glucoside of 1,3,6,7-tetrahydroxyxanthone and was first found in leaves of the mango tree *Mangifera indica*. It has since been found not only in ferns, e.g., in *Asplenium* (HARBORNE et al., 1973) but also sporadically in at least 15 angiosperm families (CARPENTER et al., 1969). More recent reports of its occurrence include *Heptaptera* (Umbelliferae) (HARBORNE, 1971), *Senecio* and *Dahlia* (Compositae) (GIANNASI, 1975) and *Polystachya* (Orchidaceae) (WILLIAMS, 1979).

2.4.2 Stilbenes

The most widespread natural stilbene is lunularic acid (XCIX), actually a dihydrostilbene, which occurs in all liverworts as a natural growth inhibitor, replacing abscisic acid, the dormancy hormone of most other plant groups. Lunularic acid is usually accompanied in liverworts by its decarboxylated derivative, lunularin (C), which is inactive as a hormone (GORHAM, 1977). The acid is assumed to have a

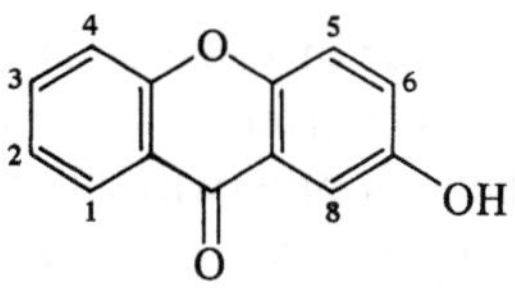

(XCVI)

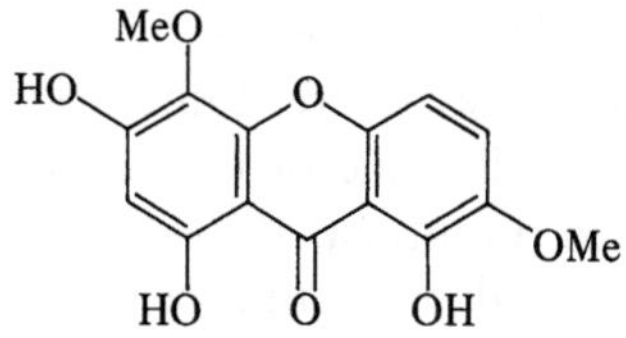

(XCVII)

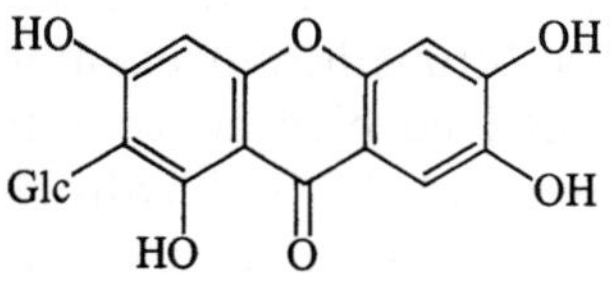

Mangiferin
(XCVIII)

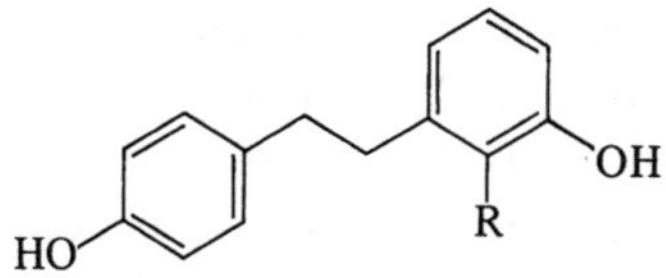

R = CO_2H, Lunularic acid (XCIX)
R = H, Lunularin (C)

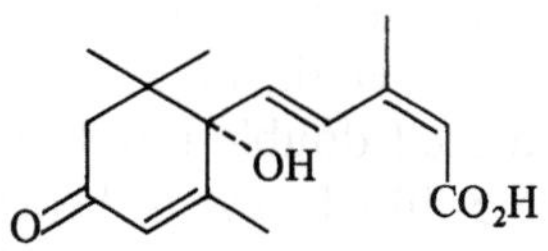

Abscisic acid
(CI)

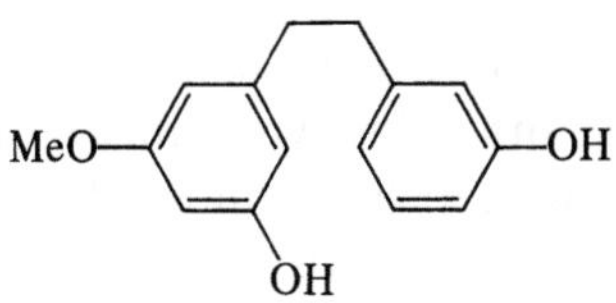

Batatasin III
(CII)

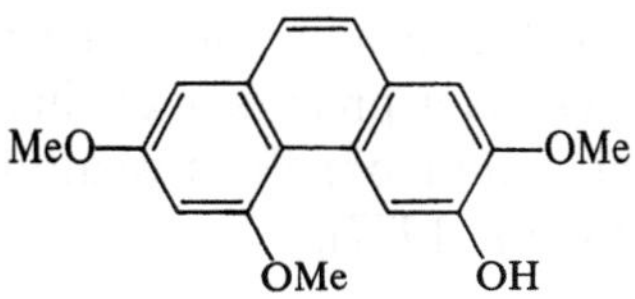

Batatasin I
(CIII)

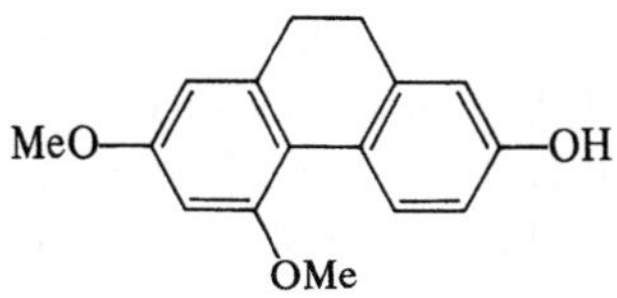

Orchinol
(CIV)

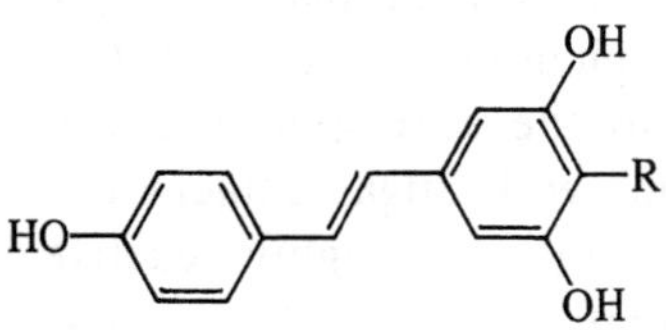

R = H, Resveratrol (CV)
R = Isopentenyl (CVI)

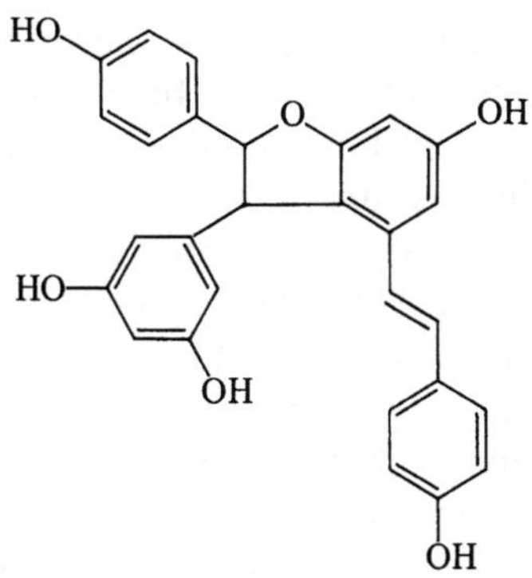

ϵ–Viniferin
(CVII)

similar hormonal role to abscisic acid (CI) and although not a sesquiterpene, it does have an analogous structure. Lunularic acid may also occur in some algae (PRYCE, 1972a, b), although some of the earlier reports of such occurrences have recently been questioned (GORHAM, 1977). This hormonal stilbene is otherwise absent from plants, although it does occur exceptionally in one angiosperm species, in *Hydrangea macrophylla*.

Another dihydrostilbene with physiological activity is batatasin III (CII), which is one of three phenolic growth inhibitors responsible for inducing dormancy in bulbils of the yam *Dioscorea batatis* (HASHIMOTO et al., 1974). A second compound, batatasin I, has been characterised as 6-hydroxy-2,4,7-trimethoxyphenanthrene (CIII) (LETCHER, 1973), a compound which could be derived biogenetically from a stilbene precursor. It is interesting (see below) that three related 9,10-dihydrophenanthrenes, e.g., orchinol (CIV), occur as antifungal agents in infected corms of *Orchis militaris* and *Loroglossum hircinum* (Orchidaceae) (FISCH et al., 1973).

Phenolics such as (CIII) with a phenanthrene skeleton may act as endogenous growth inhibitors in other plants besides the yam. It is significant that REISCH et al. (1973) have reported several similar phenanthrenes in rhizomes of *Tamus communis*, also Dioscoreaceae, and the same type of compound occurs in an unrelated dicot source in *Combretum* (Combretaceae) (LETCHER et al., 1972).

Most of the other known stilbenes have an unsaturated double bond between the benzene rings, typical members being pinosylvin (3,5-dihydroxy) in heartwood of *Pinus* and resveratrol (CV) in *Eucalyptus* trees (Myrtaceae). Hydroxystilbenes are of biological interest because of their antifungal properties and there is presumptive evidence that they protect the heartwoods of trees from fungal infections. This view of their antimicrobial activities has been confirmed by recent discoveries of resveratrol and its isopentenyl derivative (CVI) as phytoalexins produced in several legumes, e.g., *Arachis hypogaea* and *Trifolium*, in response to fungal attack (KEEN and INGHAM, 1976). Resveratrol has also been found as a phytoalexin in the unrelated *Vitis vinifera* (Vitaceae), where it also acts as a precursor of dimeric and trimeric derivatives, ε- and α-viniferin (CVII), which are more effective antifungal agents than the parent stilbene (LANGCAKE and PRYCE, 1977).

2.5 Quinones

Most of the naturally occurring quinone pigments contain phenolic or methoxyl substituents and can thus be classified under the heading of plant polyphenols. However, even those quinones which do not have phenolic substitution (e.g., the plastoquinones) are related simply by reduction to the corresponding *p*-quinols and may actually occur in vivo in reduced form, with sugar attached to one of the quinol hydroxyls. Indeed, simple benzoquinones are highly reactive substances of considerable toxicity and do not normally occur in living tissues in any quantity. The parent substance, *p*-benzoquinone (CVIII), is used by a variety of arthropods in their defense secretions; the bombardier beetle in particular employs it in an explosive discharge of vapour. It is not surprising therefore, that benzoquinone only occurs in plants in reduced form as hydroquinone (1,4-dihydroxybenzene), as its glucoside, arbutin (CIX). The regular presence of hydroquinone in Ericaceae and Rosaceae has already been mentioned (Sect. 2.1).

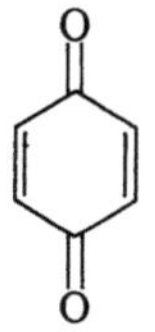

p–Benzoquinone
(CVIII)

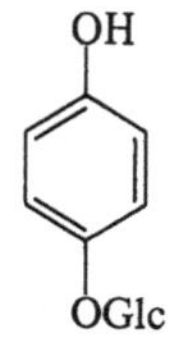

Arbutin
(CIX)

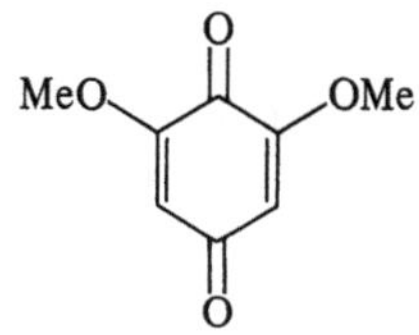

2,6–Dimethoxybenzoquinone
(CX)

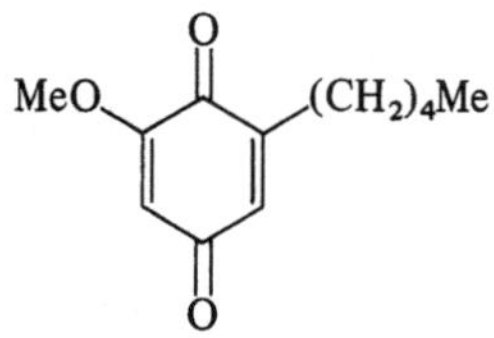

Primin
(CXI)

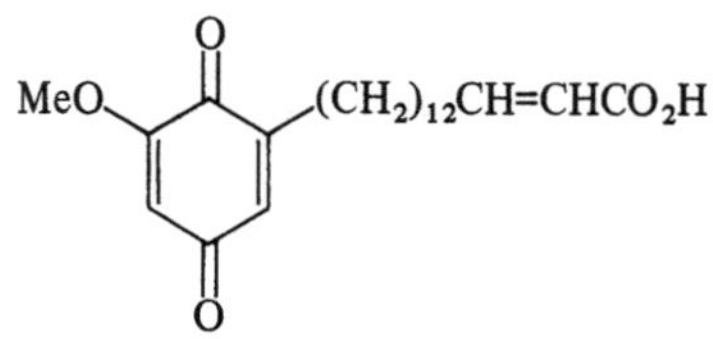

Sarcodontic acid
(CXII)

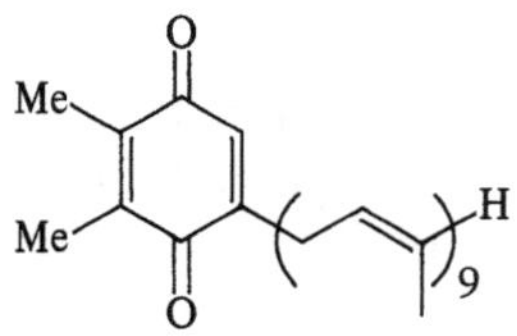

Plastoquinone–9
(CXIII)

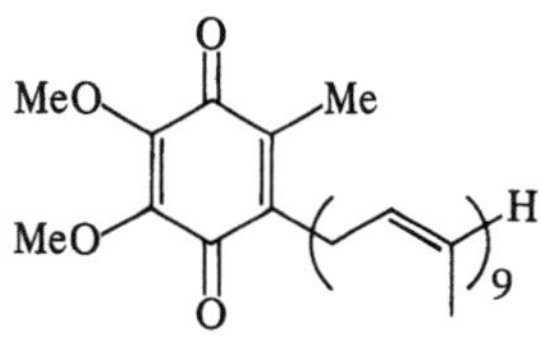

Ubiquinone–9
(CXIV)

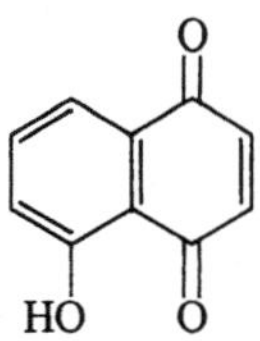

Juglone
(CXV)

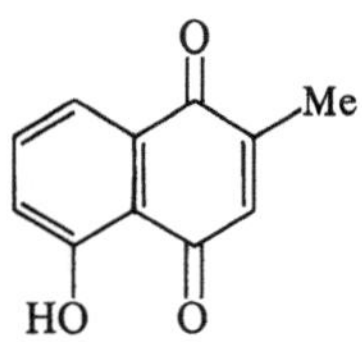

Plumbagin
(CXVI)

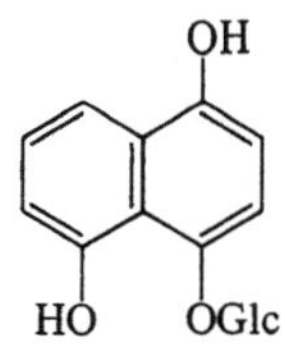

(CXVII)

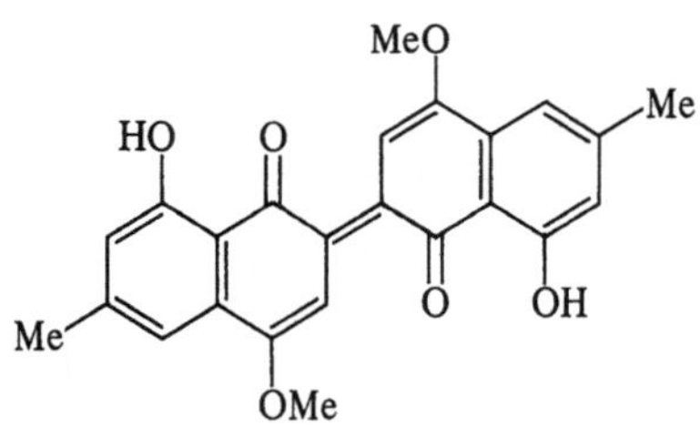

Diosindigo
(CXVIII)

The only simple substituted benzoquinone to occur widely in plants is the 2,6-dimethoxy derivative (CX), which occurs in wheat grains and also in a variety of other miscellaneous sources (see THOMSON, 1971). Again, it occurs in the reduced form as a monoglucoside and, in its occurrence in wheat germ, it appears to play a part in the germination process. The toxic effect of simple benzoquinones is well illustrated by the case of primin (6-methoxy-2-*n*-pentylbenzoquinone) (CXI), which is present in the glandular hairs of leaves of *Primula obconica* (Primulaceae), and is responsible for the allergic effects of this popular ornamental plant. Some people handling this *Primula* species develop skin dermatitis through the irritant properties of this quinone pigment.

Most of the other known simple benzoquinones (see THOMSON, 1971, 1976) occur in fungi, mainly in the Hyphomycetes and Basidiomycetes, where they may contribute to the colour of the fruiting bodies. Thus sarcodontic acid (CXII) is responsible for the intense sulphur yellow of the wood-rotting fungus *Sarcodontia setosa* found on old apple trees. More complex benzoquinones, with isoprenoid sidechains, also occur in plants but these are universally present, as carrier molecules in the electron transport chain. The two main classes are the plastoquinones, present in plant chloroplasts, and the ubiquinones, which are located in the mitochondria. In both groups, variation occurs in the length of the isoprenoid sidechain; two of the major components in plants are ubiquinone-9 (CXIII) and the corresponding plastoquinone-9 (CXIV).

A second class of quinone pigment are those based on naphthalene – the naphthoquinones. Most of these are of higher plant origin and although intensely coloured are not often apparent as plant-colouring matters; they are often present in heartwood or bark, where their presence is masked. When present in living tissues, in leaves and roots, they are usually in colourless form and colour is produced only after extracts have been treated with acid to bring about hydrolysis of sugar linkages and oxidation of quinol to quinone. This is true of both juglone (CXV) of walnut and plumbagin (CXVI) of *Plumbago capensis*. Juglone (5-hydroxynaphthoquinone) is widely present in leaves and roots of members of the Juglandaceae but occurs as the 4-glucoside of the corresponding 1,4,5-trihydroxynaphthalene (CXVII). The quinone is released in the walnut during ripening of the fruit and is responsible for the yellow-brown staining of the skin caused by handling these nuts. Juglone (CXV) is also released in the soil around walnut trees, either as a root exudate or as washing from the leaves, and exerts an allelopathic effect in limiting the seed germination of certain annual plants which might otherwise grow in the immediate vicinity (RICE, 1974).

Plumbagin (CXVI), an orange pigment, is also present like juglone in bound form in plants. It is a more widespread compound, having been detected regularly in Plumbaginaceae, Droseraceae, and Ebenaceae and also once in Euphorbiaceae (*Pera ferruginea*). Like juglone, it is physiologically active; it has an irritating odour, affects the mucous membrane, stains the skin and produces blisters. Its allelopathic potential, however, does not seem to have been investigated.

When present in heartwoods of trees, naphthoquinones occur free and a wide range of structures may be present together in any one given source. *Diospyros* (Ebenaceae) heartwoods are particularly rich, containing many derivatives of 7-methyljuglone in which molecules are linked together to form dimers, trimers, and tetramers. One notable *Diospyros* dimer, named diosindigo (CXVIII), is actually blue in colour (THOMSON, 1976).

Much the largest group of natural quinones is the tricyclic anthraquinones and a wealth of structures are known with varying numbers of hydroxyl and other substituents. Over 200 occur in flowering plants, where they are found especially in the families: Leguminosae (*Cassia*), Liliaceae (*Aloe*), Polygonaceae (*Rheum, Rumex*), Rhamnaceae (*Rhamnus*), Rubiaceae and Scrophulariaceae (*Digitalis*). Members of the Rubiaceae (e.g., the madder plant, *Rubia tinctoria*) are characterised by having anthraquinones with phenolic groups in only one of the two benzene rings, e.g., alizarin (2,3-dihydroxy), anthragallol (2,3,4-trihydroxy) and purpurin (1,2,4-trihydroxy) (CXIX); these pigments have been used in the past for dyeing textiles. Other more common plant anthraquinones have hydroxyl substituents in both rings. Such compounds as emodin (1,3,8-trihydroxy-6-methyl) (CXX) and

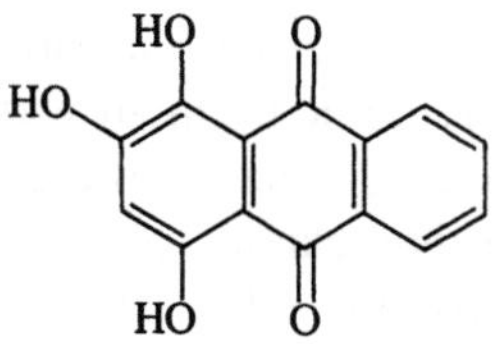

Purpurin
(CXIX)

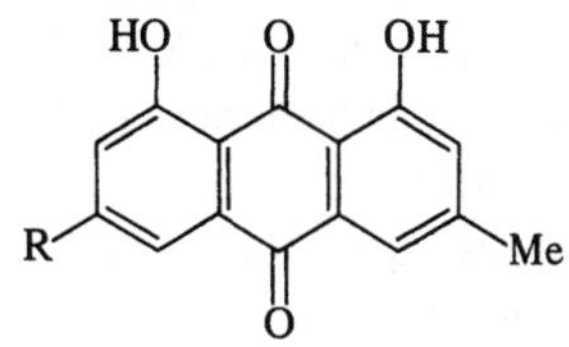

R = H, Chrysophanol (CXXI)
R = OH, Emodin (CXX)

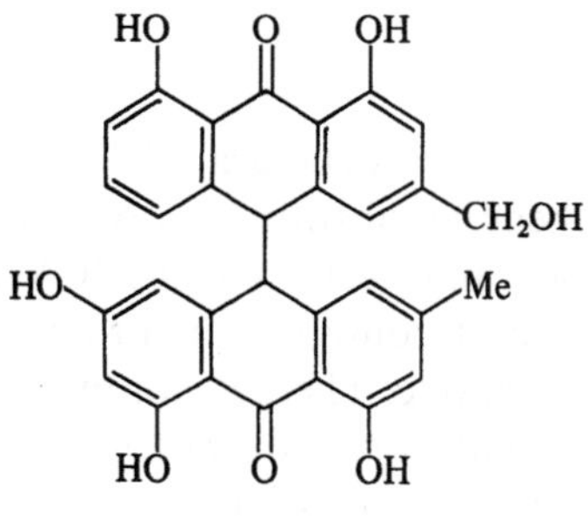

Palmidin A
(CXXII)

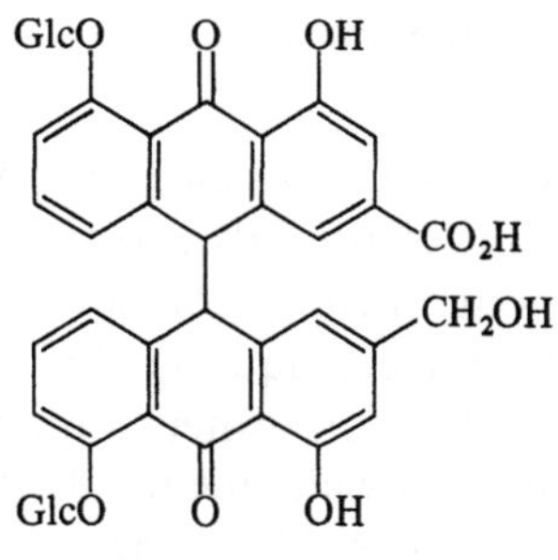

Sennoside B
(CXXIII)

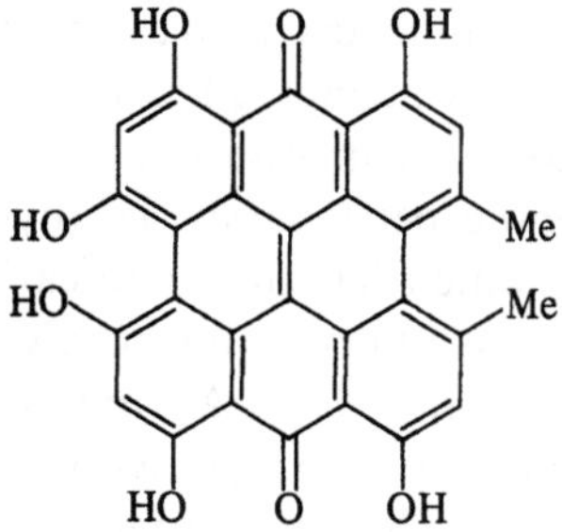

Hypericin
(CXXIV)

chrysophanol (1,8-dihydroxy-6-methyl) (CXXI) occur in several unrelated plant groups and also in fungi.

Emodin and chrysophanol, together with the related aloe-emodin (as emodin, but the 6-methyl is replaced by 6-CH_2OH) and rhein (1,8-dihydroxy-6-carboxy) occur characteristically in several plants used medicinally as a source of purgative drugs, i.e., the petioles of rhubarb, the rhizomes of *Rheum palmatum*, bark of buckthorn *Rhamnus cathartica*, and dried bark of *R. purshiana* (cascara). The free quinones, in fact, show little purgative activity and do not occur to any extent in these plants. They are present either in glycosidic combination (e.g., emodin as its 8-glucoside) or in reduced form (as anthrone) or dimeric reduced form (as bianthrone). Many of the naturally occurring bianthrones, which are the active purgative agents, are mixed dimers derived from more than one parent quinone. Rhubarb root, for example, contains palmidin A (CXXII), which is aloe-emodin: emodin bianthrone, while *Cassia* (senna pod) contains a bianthrone diglucoside, such as sennoside B (CXXIII). Among more highly condensed derivatives of emodin bianthrone is hypericin (CXXIV), a yellow photodynamic pigment widely present in leaves and stems of *Hypericum* species (Guttiferae). Finally, there are a variety of natural quinones which formally belong to other classes of natural product. Most of these are terpenoids but a few are flavonoids. The structural variation among such miscellaneous plant quinones has recently been reviewed by THOMSON (1979).

2.6 Miscellaneous Phenols

A number of plant substances belonging formally to other structural classes such as alkaloids and terpenoids also contain incidental phenolic substitutions. A familiar example among nitrogen compounds is the aromatic amino acid tyrosine (CXXV) which is of course ubiquitous in nature, both free and combined in protein. A number of related non-protein amino acids are also phenolic. There is, for example, 3,4-dihydroxyphenylalanine or DOPA (CXXVI), present in broad bean *Vicia faba* tissues and in seeds of a variety of other legumes (see BELL, Chap. 7 this vol.). Phenolic alkaloids are also regularly found in plants. Morphine, the pain-killing principle of the opium poppy, *Papaver somniferum*, has one phenolic hydroxyl, while betanidin (CXXVII), the purple pigment of beetroot and of other members of the Centrospermae, has two.

Among the various plant terpenoids, there are phenolic derivatives in most classes. Indeed, any alicyclic compound containing a six-membered carbocyclic ring with oxygen substituents is capable of undergoing aromatisation to yield a phenolic compound. A simple case in the monoterpene series is thymol (CXXVIII), a fairly common essential oil constituent, which has the same carbon skeleton as limonene. An example in the diterpene series is inuroyleanol (CXXIX), present in roots of *Inula royleana* (Compositae) (BHAT et al., 1975). An important group of pigments in the cotton plant is of the same type, the gossypols, the parent compound being gossypol (CXXX) itself. Another phenolic terpenoid pigment is 3-hydroxyisorenieratene (CXXXI), which is formally a carotenoid derived from β-carotene and has been isolated from *Streptomyces mediolani* (ARCAMONE

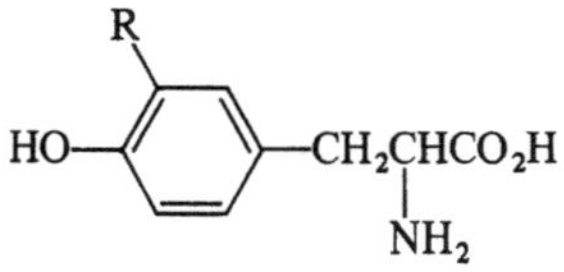

R = H, Tyrosine (CXXV)
R = OH, DOPA (CXXVI)

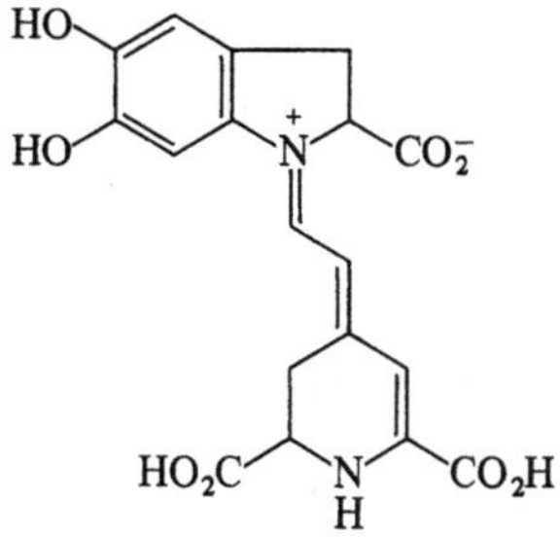

Betanidin
(CXXVII)

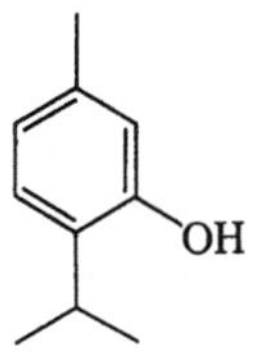

Thymol
(CXXVIII)

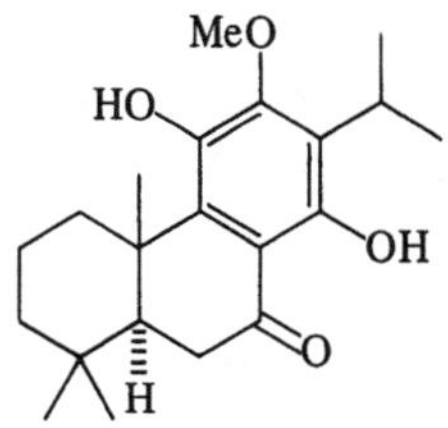

Inuroyleanol
(CXXIX)

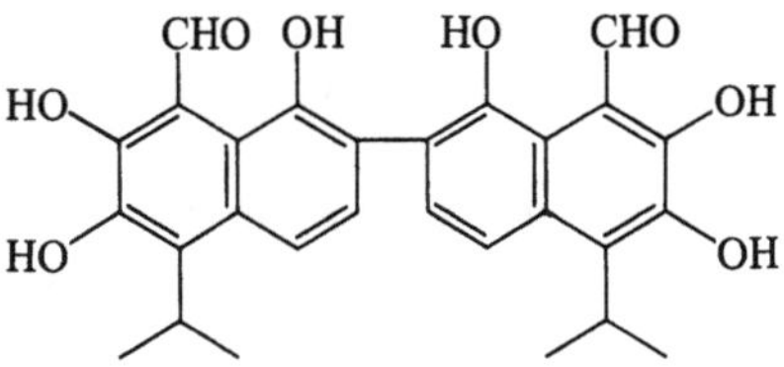

Gossypol
(CXXX)

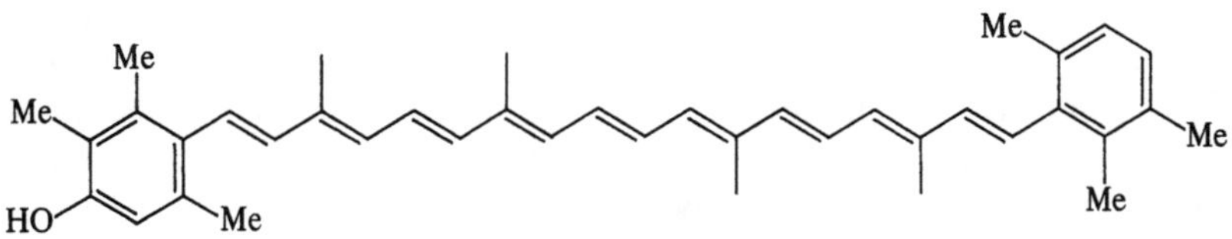

3–Hydroxyisorenieratene
(CXXXI)

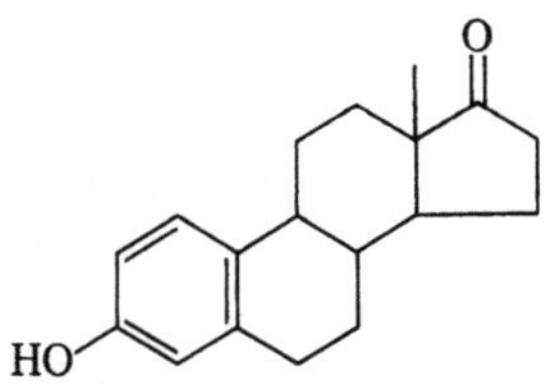

Oestrone
(CXXXII)

et al., 1970). Finally, it is worth noting that mammalian female sex hormones, which are usually phenolic (e.g., oestrone, (CXXXII)) occur occasionally in plant tissues, e.g., in date palm seed (see HARBORNE, 1977a).

3 Phenolic Conjugates

3.1 Functional Significance of Conjugation

Phenolic compounds rarely occur in the free state in living plant tissue; they are practically always present in conjugated form. In the simplest instance, they are bound to sugar as β-D-glucopyranosides but a wide array of other bound forms are known. Undoubtedly one of the reasons why diphenols such as hydroquinone, catechol, or protocatechuic acid are bound to sugar is because of their potential toxicity in the free form to many forms of life. Simple phenols are caustic substances and well known to be potent antimicrobial agents. In medicine, the first successful antiseptic surgery was achieved following the use of phenol itself. The corresponding quinones are also toxic to most forms of life (see Sect. 2.5). The relative infrequency in plants of simple phenolic derivatives may well be related to their significant phytotoxicity. Where they do occur, they may become involved in allelopathic reactions between plants. Compounds such as hydroquinone or salicylic acid occur bound in the plant; they may be released into the environment from leaf or root in free form to exert an inhibitory effect on seed germination or plant growth in the surrounding soil (see HARBORNE, 1977a).

In the case of polyphenols such as the flavonoids, a bewildering array of conjugated forms is known. Glycosides are particularly abundant, but other forms of binding including sulphation or acylation are also possible. Reasons for the conjugation of flavonoids are complex but a major consequence of binding with sugar is their sequestration in the plant vacuole in a form and site where they cannot interfere with the vital enzymic processes of plant metabolism. It is significant here to note that when flavonoids occur in chloroplasts, as they apparently do in small amounts in almost all plants, they are always glycosidically bound. Although flavonols have often been considered to be quite harmless, there is increasing evidence of toxicity associated with some of these structures, particularly in the free state. Quercetin has recently been reported to be mutagenic (BJELDANES and CHANG, 1977) and has also been found to be an active inhibitor of cell membrane transport by its interference with the calcium-dependent ATPases (RACKER, 1975). In fact, free quercetin is capable of inhibiting a whole series of enzyme activities (see WAGNER, 1979).

One particularly important active site in polyphenols is the catechol nucleus, which has the ability to chelate metals and which may be directly involved in a number of biological reactions. For example, flavonoids with a 3′,4′-dihydroxyphenyl B-ring (e.g., quercetin) have a significant sparing effect in vitro on the destruction of auxin by IAA oxidase and thus if present at the growing point could indirectly stimulate growth in plants (see Sect. 6.1). Similarly in mammals, it has been found that catechol oestrogens inhibit elicited accumulation of hypothalamic cyclic AMP and hence may have a role as endogenous anti-oestrogens (PAUL and SKOLNICK, 1977). In either case, masking of the catechol nucleus by *O*-glycosylation would undoubtedly have a profound effect on such biological activities.

Glycosylation of phenolic hydroxyl groups has a functional role in the case of those flavonoids which are coloured and occur in flower tissues. Also certain variations in the position of attachment of sugar to the flavonoid nucleus can produce significant shifts in visible colour. In the case of flavonoids of the leaf, it is more difficult to assign a role to conjugation but there are experiments which suggest the nature of the sugar present may determine whether a particular flavonoid is an effective feeding deterrent to insects and grazing animals. Glycosylation may thus be significant in relation to the ecological importance of flavonoids in plants (see HARBORNE, 1977a).

Finally, it may be observed that glycosylation also affects the physiological activities of flavonoids and other phenolics in plants. This is apparent from the experiments of STENLID (1970) and others; this subject is considered in more detail in Section 6.1.

3.2 Glycosidic Variation

Most phenolics occur in plant leaf, stem, or flower in water-soluble form as glycoside and every plant that has been examined has yielded a water-soluble fraction in which the bulk of the low molecular weight phenolic material is present. The most common form of conjugate is phenolic glucoside, but combination with rutinose, as in the quercetin glycoside rutin, is also widespread. Hydroxycinnamic acids differ from most other phenols in occurring most frequently as quinic acid esters (see Sect. 3.3). When hydroxycinnamic acids occur with sugar, this is usually attached to the carboxylic acid grouping rather than to a phenolic residue. However, a caffeic acid derivative with both the 3-hydroxyl and the carboxyl group carrying glucose moieties has been described recently (IMPERATO, 1976).

Glycosidic variation is most marked among the flavonoids and in particular among the anthocyanins, flavonols, and flavones. In the case of the anthocyanins (glycosides of anthocyanidins), a comprehensive listing of known pigments in 1967 revealed the presence in plants of some 120 sugar derivatives (HARBORNE, 1967a); the number known today must be at least double this (see e.g., TIMBERLAKE and BRIDLE, 1975). With the common anthocyanidin cyanidin, about forty different glycosides are at present known. Although cyanidin 3-glucoside (CXXXIII) remains the most widely distributed anthocyanin, cyanidin derivatives with branched (CXXXIV) or linear (CXXXV) trisaccharides are by no means uncommon. The most highly glycosylated anthocyanin to date, however, is an acylated delphinidin derivative, isolated from flowers of *Lobelia erinus* (Lobeliaceae) (YOSHITAMA, 1977). This has five monosaccharide units: glucose and rhamnose (as the disaccharide rutinose) attached to the 3-position and glucose moieties separately linked at the 5-, 3′- and 5′-positions (CXXXVI). This pigment is also remarkable in being one of the first anthocyanins to be found with sugar substitution on the B-ring hydroxyls; glycosylation is much more commonly associated with only the 3- or the 3- and 5-positions.

Quercetin is an extremely common plant polyphenol so it is not surprising that as many as 80 of its glycosides have been identified variously in plants (HAR-

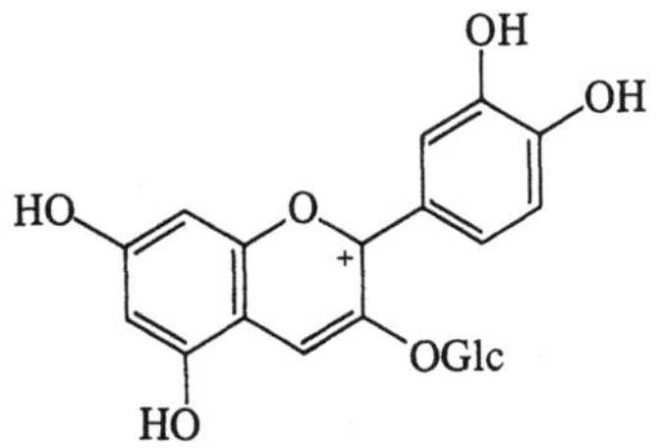

Cyanidin 3–glucoside
(CXXXIII)

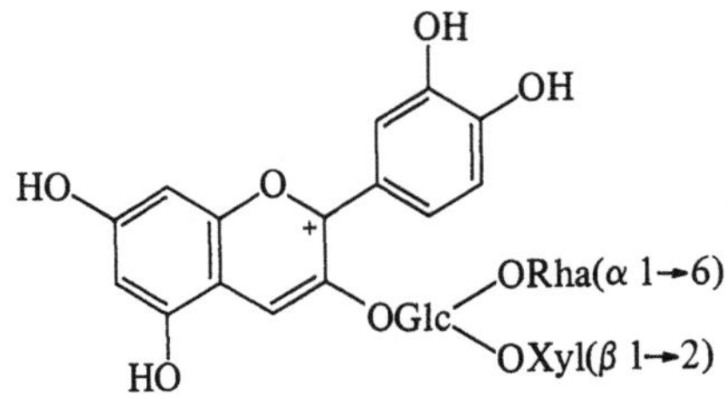

Cyanidin 3–(2G–xylosylrutinoside)
(CXXXIV)

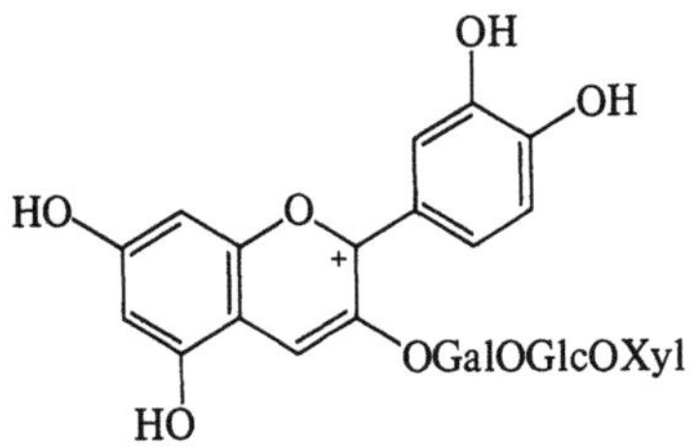

Cyanidin 3–(xylosylglucosylgalactoside)
(CXXXV)

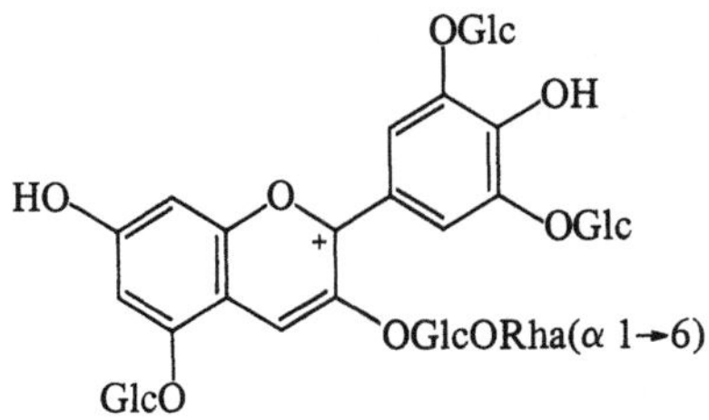

Delphinidin pentaglycoside
(CXXXVI)

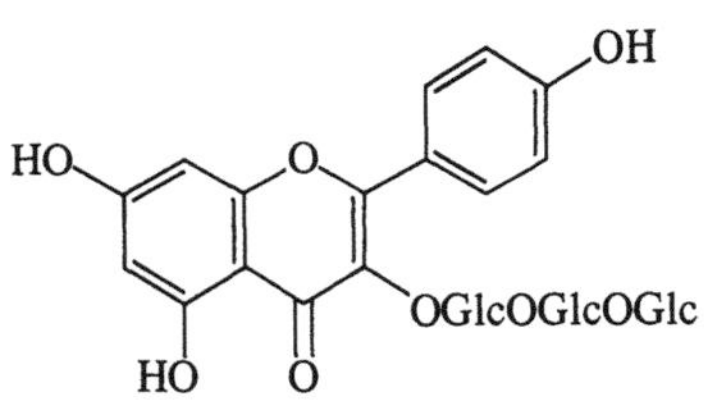

Kaempferol 3–triglucoside
(CXXXVII)

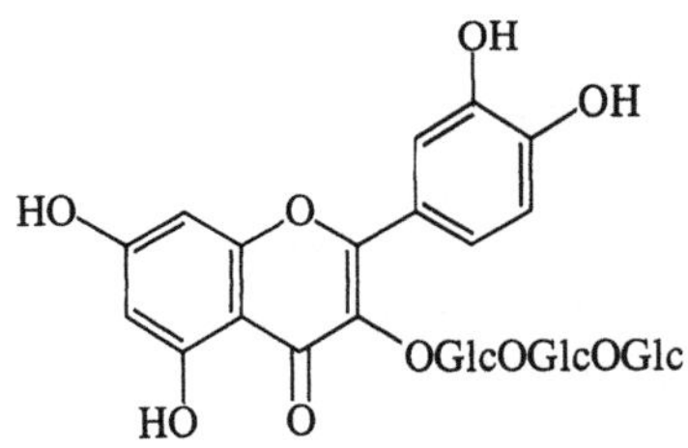

Quercetin 3–triglucoside
(CXXXVIII)

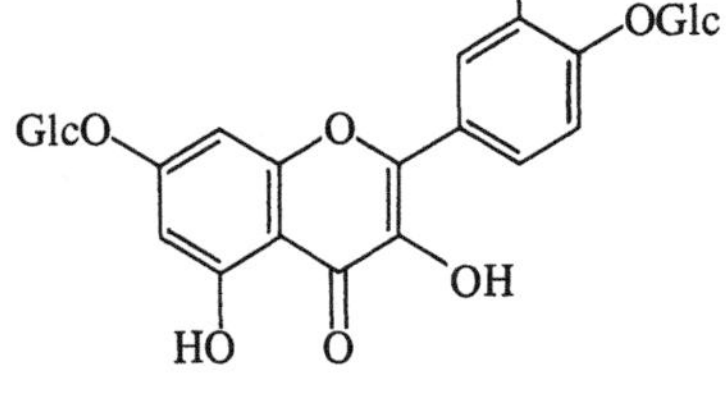

Quercetin 7,4′–diglucoside
(CXXXIX)

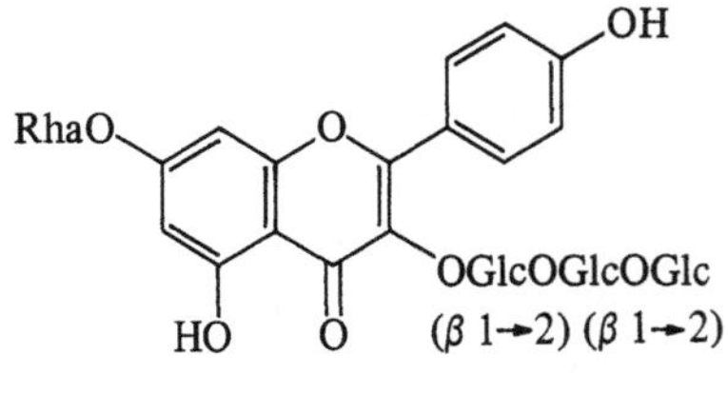

Kaempferol 3–sophorotrioside–7–
rhamnoside (CXL)

BORNE and WILLIAMS, 1975). Almost as many different glycosides of kaempferol and myricetin are also reported. An array of several glycosides is likely to be encountered in any given plant tissue. Pea plants, for example, contain at least four glycosides, the 3-triglucosides (CXXXVII), (CXXXVIII) of kaempferol and quercetin together with their *p*-coumaryl derivatives. These four substances have been implicated by GALSTON (1969) in growth responses in *Pisum sativum*, while the quercetin acylated 3-triglucoside has been reported to be involved in tendril coiling in the same plant.

While quercetin 3-glucoside, cf. cyanidin 3-glucoside, is the commonest flavonol glycoside, the 3-rhamnoside, 3-galactoside, 3-glucuronide and 3-arabinoside of quercetin are also present relatively frequently in plants; all these simple monoglycosides co-occur, remarkably enough, in the skin of the common apple, *Malus pumila*. Rarer glycosidic types include the 3,4′- and 7,4′-diglucosides (CXXXIX) present in onion bulbs and the 3,3′-diglucoside present in horsechestnut seed. The most highly glycosylated flavonol glycoside remains the kaempferol derivative with four sugars (CXL) present in potato seed (see HARBORNE and WILLIAMS, 1975).

With the flavones, glycosidic complexity is compounded by the fact that sugars can be bound to phenolic hydroxyl groups and to the carbon nucleus of the aromatic ring at the 6- and/or 8-positions. Such *C*-glycosides are called glycoflavones and are capable of having *O*-sugars attached as well, either to one of the carbon-bound sugars or to a phenolic group. In case of simple flavone *O*-glycosides, the commonest are apigenin and luteolin 7-glucosides. Two examples of more highly substituted derivatives are luteolin 7-neohesperidoside-4′-sophoroside (CXLI) which occurs in the moss *Hedwigia ciliata* (ÖSTERDAHL and LINDBERG, 1977) and the acacetin pentaglycoside (CXLII) present in leaves of *Coptis japonica* (Ranunculaceae) (FUJIWARA et al., 1976). This latter compound has a branched tetrasaccharide in the 7-position and its structure is further complicated by the presence of no less than three acetyl groups attached to various sugar hydroxyls.

Among simple glycoflavones of plants is saponarin (CXLIII), a major compound of barley leaves, which is present not only in the vacuole but also in chloroplast preparations of this plant (SAUNDERS and MCCLURE, 1972). A more complex glycoflavone is vitexin 2″-sophoroside (CXLIV), reported in leaves of *Polygonatum odoratum* (Liliaceae) (MORITA et al., 1976). Particularly complex mix-

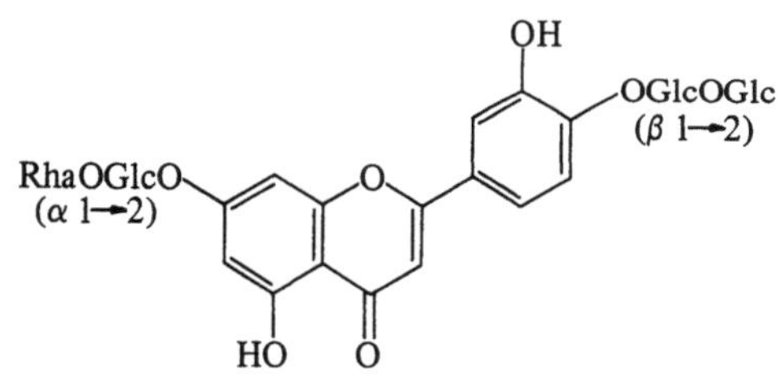

Luteolin 7–neohesperidoside–4′–sophoroside
(CXLI)

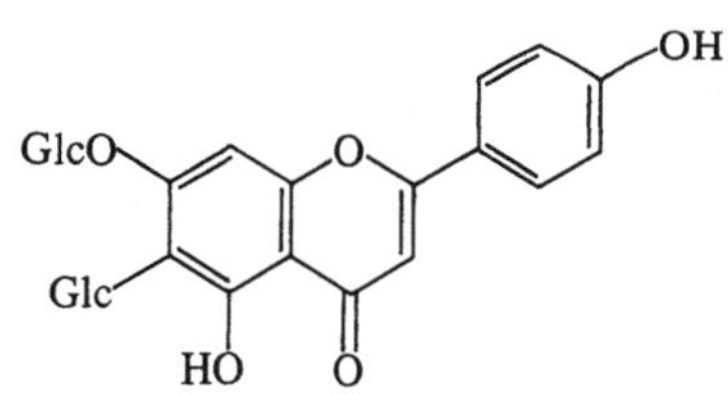

Saponarin
(CXLIII)

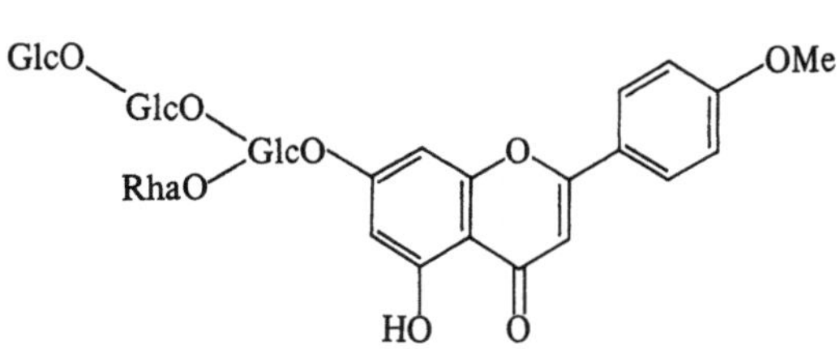

(CXLII)

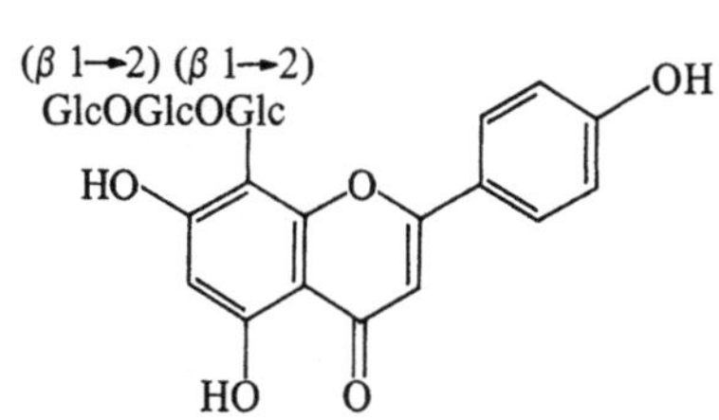

Vitexin 2″–sophoroside
(CXLIV)

tures of *O*- and *C*-glycosides occur in grasses: no less than 26 compounds are present in oat leaves (POPOVICI et al., 1977) and 27 in barley leaves (FRÖST et al., 1977). The ability of Gramineae to accumulate unusual glycoflavones is illustrated by the isolation of the first *C*-galactoside, 8-*C*-galactosylapigenin, from *Briza media* (CASTLEDINE and HARBORNE, 1976). A recent review of plant glycoflavones is that of CHOPIN and BOUILLANT (1975).

3.3 Bound Forms of Hydroxycinnamic Acids

Hydroxycinnamic acids occur naturally in a much wider range of combined forms than any other group of plant phenols. While most frequently present as quinic acid esters, they may nevertheless occur conjugated with organic acids, sugars, amino compounds, lipids and terpenoids (Table 3). They may also be linked to other phenolic compounds and hydroxycinnamic acids are regularly found linked through sugar to flavonoid glycosides (see Sect. 2.3). In addition, as already men-

Table 3. Some simple esters of hydroxycinnamic acids in plants

Class of compound	Ester group	Esterified acid and trivial name of conjugate
Cyclohexane carboxylic acid	Quinic acid	CAF: chlorogenic acid FER/COUM
	Shikimic acid	CAF/COUM
Cyclitol	*m*-Inositol	COUM
Sugar	Glucose	CAF/FER/COUM
	Rhamnose	CAF
	Fructose	COUM
	Gentiobiose	CAF/FER
	Rutinose	COUM
Organic acid	Tartaric acid	CAF: chicoric acid
	Malic acid	CAF: phaselic acid
Amino compounds	Dopaldehyde	CAF
	DOPA	CAF: clovamide
	Tryptamine	FER/COUM
	Putrescine	CAF: paucine FER/COUM
	Choline	SIN: sinapine FER
	Lupinine	COUM
Phenolics	Salicin	CAF: populoside
	3,4-Dihydroxyphenyl-lactic acid	CAF: rosmarinic acid
	Cyanidin 3-glucoside	COUM: hyacinthin
	Kaempferol 3-glucoside	COUM: tiliroside
Lipids	Hexacosylalcohol	FER
	Glycerol	CAF/COUM
Terpenoids	Borneol	COUM
	α-verbesinol	COUM
	Catalpol	COUM
	Queretaroic acid	CAF

Key: COUM = *p*-coumaric, CAF = caffeic, FER = ferulic, SIN = sinapic
In some cases, only a few of many known derivatives are listed

tioned (Sect. 2.2), ferulic acid can be associated in plants with protein and also with the cellulose and hemicellulose of plant cell walls. An account of the bound forms of hydroxycinnamic acids was given by HARBORNE (1964a). Since then, many more derivatives have been characterized, so that it is only possible to

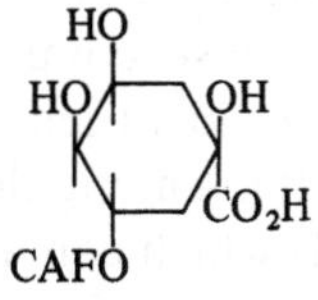

Chlorogenic acid
(CXLV)

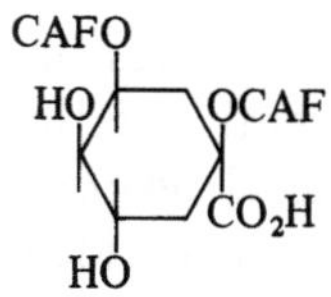

Cynarin
(CXLVI)

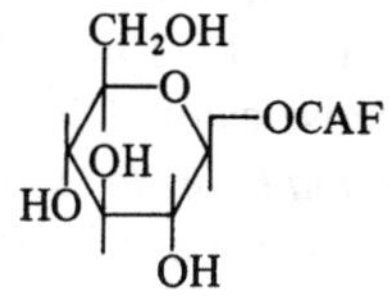

Caffeylglucose
(CXLVII)

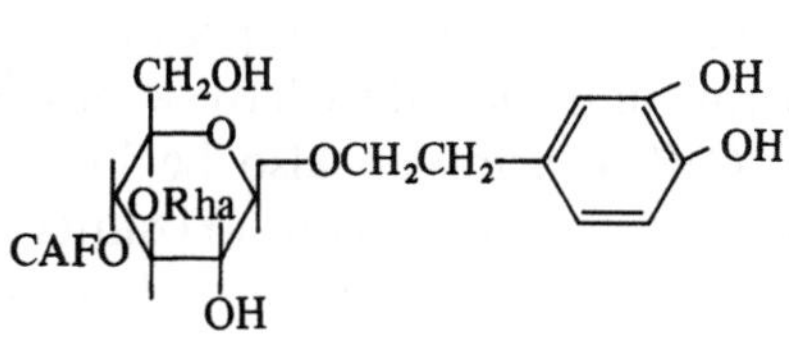

Acteoside
(CXLVIII)

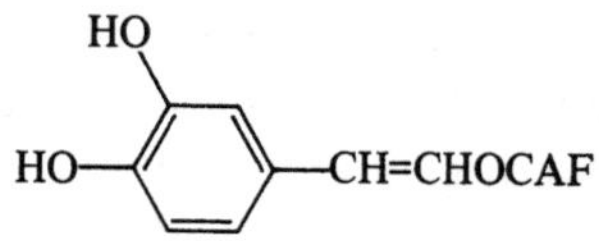

Caffeyldopaldehyde
(CXLIX)

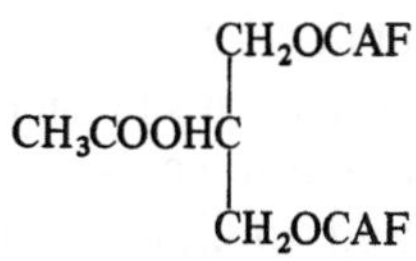

Lasiocarpin
(CL)

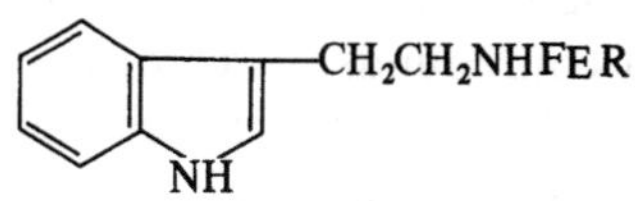

N–ferulyltryptamine
(CLI)

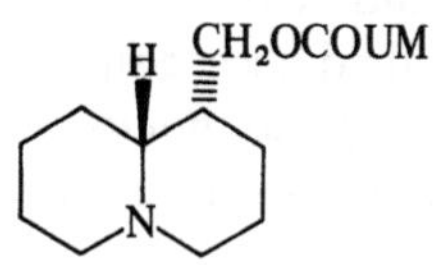

p–Coumaryl–lupinine
(CLII)

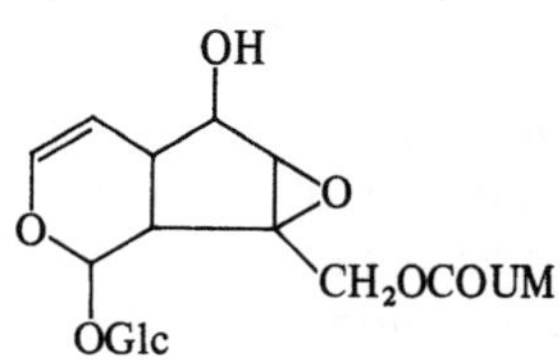

10–*p*–Coumarylcatalpol
(CLIII)

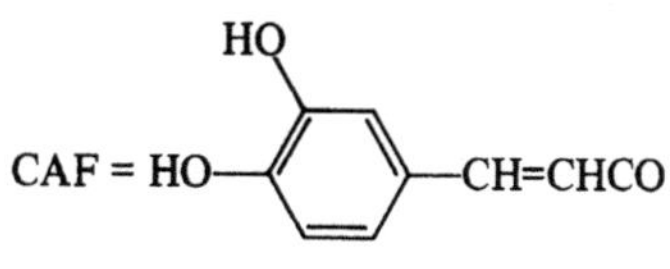

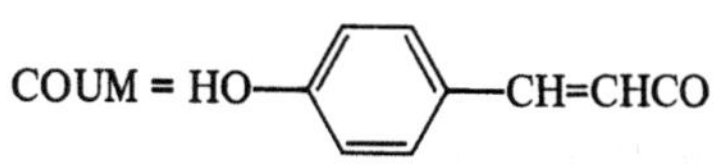

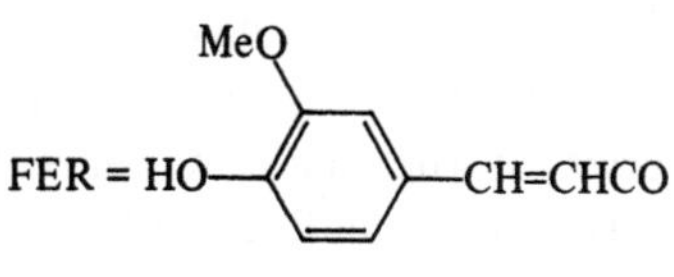

mention here a selection of recent structures involving the three most common acids, caffeic, ferulic, and *p*-coumaric.

Caffeic acid occurs most widely as the simple quinic acid ester, chlorogenic acid (CXLV), but dicaffeyl esters of quinic are also possible. Cynarin (CXLVI), for example, is the choleretic principle of the artichoke *Cynara scolymus* (PANIZZI and SCARPATI, 1965). Sugar derivatives of caffeic are also common, particularly the glucose ester (CXLVII); recently the rhamnose ester has been found in flowers of *Lantana hybrida* (Verbenaceae) (IMPERATO et al., 1975). Among the more complex sugar derivatives of caffeic acid is acteoside (CXLVIII) which contains both glucose and rhamnose and is found in flowers of *Syringa vulgaris* (Oleaceae) (BIRKOFER et al., 1968).

Among an increasing number of amino derivatives of caffeic acid to be reported in the last few years, the derivative with putrescine, $NH_2(CH_2)_4NH_2$, may be mentioned. This and related acylated amines appear to be especially associated with the reproductive tissues of angiosperms and may thus have a physiological role in flowering (TANGUY et al., 1978). Caffeic acid can also be linked to the 3-hydroxyl of the amino acid DOPA as in *Trifolium pratense* (YOSHIHARA et al., 1977) and to the aldehyde oxygen of dopaldahyde to give a yellow conjugate (CXLIX) recently detected in leaves of *Plectranthus caninus* (Labiatae) (ARIHARI et al., 1975). Several lipid-based derivatives of caffeic acid are also present in plants. The monoglyceride of caffeic acid is present in oat seeds and pineapple stems (TAKATA and SCHEUER, 1976) while a triglyceride, lasiocarpin C (CL), has been discovered in *Populus* bud extracts by ASAKAWI et al. (1977). In this and related compounds, glycerol is substituted in the 2-position by acetic acid, while the 1- and 3-positions are occupied by caffeic and/or *p*-coumaric acids.

Ferulic acid, like caffeic, has also been detected bound to lipid components and hexacosyl ferulate in needles of *Pinus roxburghii* (CHATTERJEE et al., 1977) is only one of a series of such hydrocarbon derivatives. Among other novel ferulyl conjugates is *N*-ferulyltryptamine (CLI) which is present in kernels of *Zea mays* (EHMANN, 1974). Conjugation with nitrogen compounds extends to the alkaloids and a variety of acylated alkaloids have been described (see SMITH, 1977). One recent example is *p*-coumaryl-lupinine (CLII) from *Lupinus luteus*, which occurs as such and also with rhamnose attached to the phenolic group of the *p*-coumaryl residue (MURAKOSHI et al., 1977).

Finally, mention should be made of the many hydroxycinnamyl residues which are linked to aliphatic hydroxyl groups among terpenoids (Table 3). *p*-Coumaric acid, for example, is acylating the monoterpene borneol in *Seseli mucronatum* (DUKHOVLINOVA et al., 1975), the monoterpene lactone catalpol (CLIII) in *Scutellaria altissima* (WEINGES et al., 1975) and the sesquiterpene alcohol, α-verbesinol in *Verbesina virginiana* (GARDNER et al., 1961).

3.4 Lipid-Soluble Derivatives

The phenolic compounds that occur in lipid-soluble plant extracts have been less well studied than those in water-soluble extracts, so that much less is known about them. One of the more unexpected discoveries of the last few years has been of water-soluble phenolics occurring frequently, to a small extent, in the characteristically lipid-based environment of the plant chloroplast (see McClure, 1975). It is conceivable that lipid-soluble phenolics may well be found to be regular constituents of this and other cell organelles, e.g., in the cytoplasm, but so far this is an area for future investigation. The only situation where lipid-soluble phenols have been found regularly is in the waxy hydrocarbon coating of plant leaves. Even here, few detailed studies have been made of the phenolic substances present.

Lipid-soluble phenolics have been more fully characterised only in the more exceptional situations of bud exudates from angiosperms and gymnosperm trees, of the crystalline pigmented deposits on the under-surfaces of fern fronds and of the waxy farina present on *Primula* plants. In such instances, the most significant chemical feature is that of a variable number of phenolic groups carrying *O*-methyl substitution. *O*-Methylation thus appears to be an important process, at least in polyphenols, which masks the reactivity of the phenolic groups and at the same time significantly increases the lipid-solubility and volatility. With simple phenols, such as the hydroxyphenylpropenes, *O*-methylation or methylenedioxy ring formation is the rule rather than the exception and this makes the substances so volatile that they are not only readily extracted into ether but are also steam-volatile and appear with monoterpenes in the "essential oil" fraction. Such masked phenols as methyleugenol or myristicin are readily detected by insects and play a significant role, when present in flowers or leaf, in attracting or deterring insects to feed or oviposit on plants containing them (see Harborne, 1977a).

The importance of methylation for masking phenolic groups is further emphasised by the fact that polyhydroxy compounds are rarely found in any concentration in any of these lipid-soluble fractions. In the lipid-soluble farina of *Primulas*, flavone itself is, in fact, the major component. Trace amounts of mono- and dihydroxyflavones may accompany it, but only one trihydroxyflavone (5,8,2′-) has ever been found and this only occurs in three of several hundred known species (Bouillant et al., 1971). Similarly, in the bud exudates of *Betula* and *Populus* studied mainly by Wollenweber (1976), the flavones and flavonols present rarely have more than one or two free hydroxyls. Typically present are kaempferol 3,7- and 3,4′-dimethyl ethers and various isomeric trimethyl ethers of quercetin, together with apigenin 7,4′-dimethyl ether and luteolin 7,3′,4′-trimethyl ether. More highly *O*-methylated derivatives may also be present, e.g. 3,7,3′,4′,5′-pentamethyl ether of myricetin. It is possible that the only situations where flavonols such as kaempferol, quercetin, and myricetin occur in the free state, without methylation or glycosylation, is in dying or dead plant tissue, especially in seed coats or in the heartwoods of trees.

An alternative or additional means of introducing lipid-solubility into phenolic nuclei is by the attachment of isoprene residues. These are most frequently attached directly to the aromatic ring, but they may also be attached to a phenolic group. One example of such a compound is luteone, 6-isopentenyl-5,7,2′,4′-tetrahydroxyisoflavone, which occurs in leaves of *Lupinus* (Leguminosae). While it occurs to some extent within the leaf, its main site is on the surface in the leaf wax. It is highly fungitoxic and occurs in sufficient amount to act as an important barrier to microbial infection in these plants. Isopentenyl substitution, besides conferring lipid solubility to luteone, considerably enhances its biological properties, since related isoflavones lacking this substituent have only one-tenth the fungitoxicity exhibited by luteone (HARBORNE et al., 1976). Luteone is only one of some 400 phenolic compounds with isoprenoid substituents, most of them being coumarins (see Sect. 2.2).

Finally with regard to lipid-soluble hydroxycinnamic acids, as mentioned in the previous section, these are present in many different tissues. Unlike the above phenolics, these derivatives actually have lipid components, including glycerol itself and long-chain hydrocarbon alcohols.

3.5 Sulphates

A recent discovery is that flavonoid sulphates are relatively widespread in the leaves of plants. Approximately 50 conjugates have been identified in over 200 species drawn from 20 families (HARBORNE, 1977b). Some of these flavone and flavonol sulphates contain sugar as well (CLIV)–(CLV), but the majority simply have one or more bisulphate groups directly attached to phenolic hydroxyls (CLVI)–

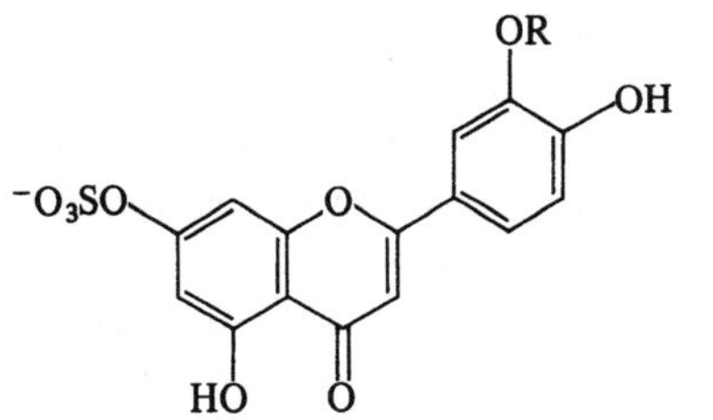

R = H, Luteolin 7–sulphate (CLVII)
R = Glc, Luteolin 7–sulphate 3′–glucoside (CLIV)
R = SO_3^-, Luteolin 7,3′disulphate (CLVIII)

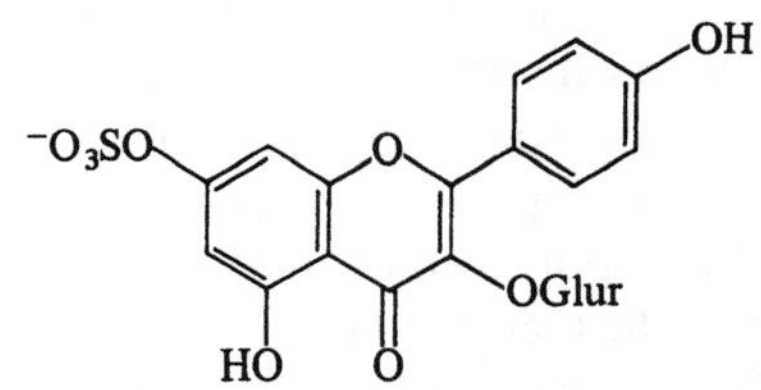

Kaempferol 3–glucuronide–7–sulphate (CLV)

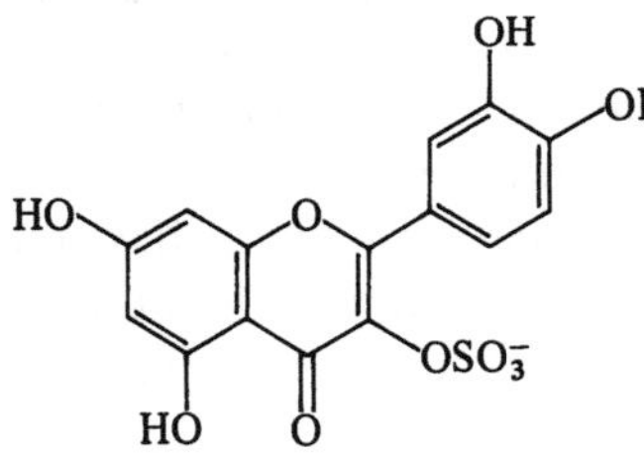

Quercetin 3–sulphate (CLVI)

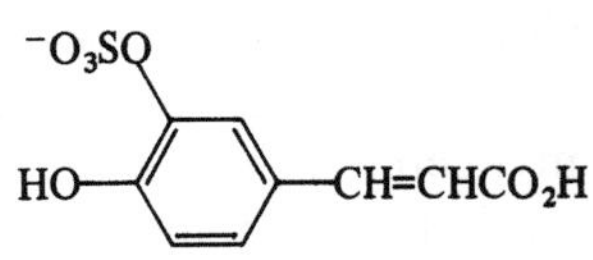

Caffeic acid 3–sulphate (CLX)

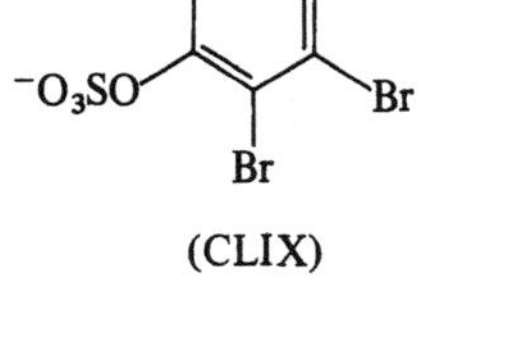

(CLIX)

(CLVIII). Such an attachment converts a neutral molecule to an anionic compound and has a dramatic effect in providing very considerable water and hence sap solubility. The direct effect of sulphation is thus similar to that of glycosylation and it is an interesting question why such an obviously alternative system of conjugation should have arisen in certain plants.

If sulphation is a simple alternative to glycosylation in plants, one might expect it to affect other classes of phenol. This is in fact so, and other phenolic conjugates of this type are being discovered quite regularly. One of the simplest is a sulphated catechol (CLIX) that has been found in the red alga *Polysiphonia* (see THOMSON, 1977). Other simple compounds include the 3-sulphates of caffeic acid (CLX) and gallic acid that have been detected, together with quercetin 3-sulphate (CLVI), in plants of the Polygonaceae. Yet another type has been found in the same family: the first reported anthraquinone sulphates. These occur in leaves of *Rumex pulcher* and include emodin 1(or 8)-glucosidesulphate and its dianthrone derivative (HARBORNE and MOHKTARI, 1977).

Another area where flavonoid sulphates occur regularly is in marine higher plants and especially in members of the order Fluviales. For example, eel grass, *Zostera marina*, contains all its flavone conjugated with sulphate, luteolin 7,3′-disulphate (CLVIII) being one of its characteristic constituents. In other Fluviales, e.g., turtle grass *Thalassia testudinum*, flavonoid sulphates appear to be replaced by simpler conjugates, since several isomers of chlorogenic acid containing sulphate occur in these plants (HARBORNE and WILLIAMS, 1976).

Finally, one other type of sulphated caffeic acid derivative has been noted, that based on the glucose ester, and such compounds have been noted in several ferns and in *Paspalum convexum* (Gramineae) (COOPER-DRIVER and SWAIN, 1975).

It is clear that conjugation of phenols with sulphate is a general phenomenon, not restricted to flavonoids, and may occur with a variety of phenolic substrates. Undoubtedly, more types will be found in future investigations. Their significance in plant metabolism is not yet clear, but their regular presence in water plants and also in halophytes (see HARBORNE, 1977a) suggests they may play a part in inorganic ion absorption, storage or metabolism. Apart from the covalent binding of sulphate, these phenolic derivatives also bind potassium ion and this may be as important as their anionic sequestration.

3.6 Polymers

One process which probably reduces the chance of phenolic compounds interfering with enzymic reactions in the cell is polymerisation. Undoubtedly, as the molecular weight of a phenolic compound increases, its transport within the cell is considerably diminished. The two main types of phenolic polymers in plants are the lignins and the tannins. Lignins are sequestered in the secondary layer of the cell wall in close association with a cellulose matrix, where the phenolic hydroxyl groups may be hydrogen bonded or even covalently linked to carbohydrate. The number of phenolic groups in lignin is low because many phenolic functions are lost as part of the polymerisation process and appear in lignin structure as ether linkages. It is interesting in this context that the three monomers of lignin (namely *p*-coumaryl, ferulyl, and sinapyl alcohols) only have a single *p*-hydroxyl group, i.e., lignins

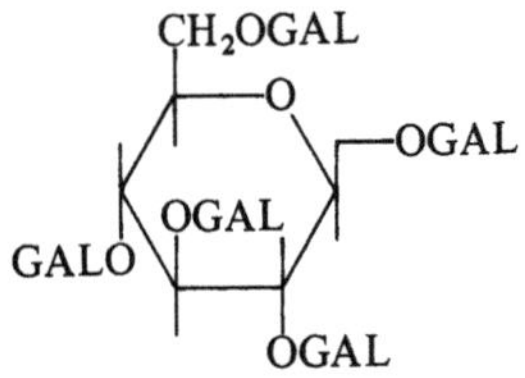

Pentagalloylglucose
(CLXI)

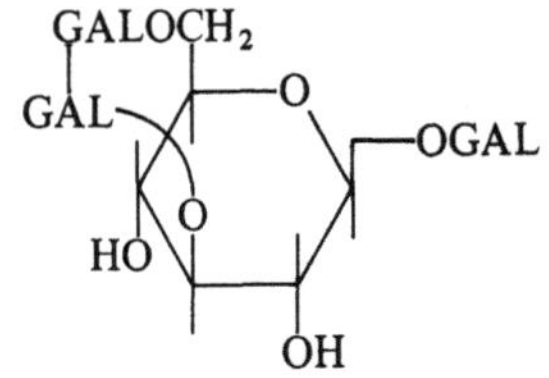

Corilagin
(CLXII)

GAL = galloyl

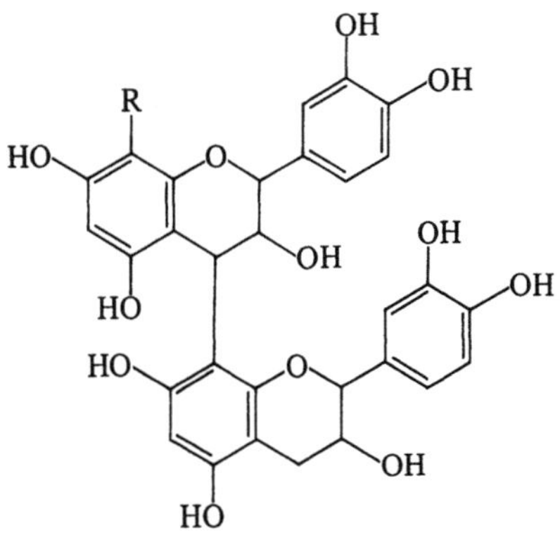

Procyanidin (CLXIII)
R = H, dimer
R = Flavan, trimer
R = further flavans, polymer

never incorporate the more reactive catechol-containing caffeyl residues. For a recent review of lignin biochemistry, see GROSS (1979).

In the case of the plant tannins, there is a big difference in molecular size between the hydrolysable and condensed tannins, the latter being of much higher molecular weight than the former. The two groups also differ in that hydrolysable tannins usually contain glucose as an integral part of their structures. Pentagalloylglucose (CLXI), the simplest hydrolysable tannin, is thus a conjugated form of gallic acid, even if the ratio of glucose to gallic acid is rather low. In other hydrolysable tannins (CLXII), glucose may be linked to phenolic groups to render the molecules more water-soluble and ensure their sequestration in the vacuole.

Condensed tannins (e.g., CLXIII) or proanthocyanidins (HASLAM, 1975) by contrast, appear never to be associated with sugar, although there have been occasional reports of glucose being present. The biological activity of condensed tannins is limited by their high molecular weight and their relative immobility. Nevertheless, the absence of any protective groups on the hydroxylated matrix of these polymers is probably the reason why they are so important in plant–animal interactions. When cell organisation is disrupted, e.g., when an insect bites into a leaf, these tannins readily complex with proteins, significantly reducing their enzymic functions or nutritive value in the process. It is interesting that no detoxification process is known in animals which is capable of dealing with these reactive polymers.

The ability of polymeric proanthocyanidins to interfere with growth processes in plants must be severely limited by their relative immobility. The degree of polymerisation is critical also for the ability to tan protein (see SWAIN, 1965) and the lower oligomers of flavolan have little tanning ability. This probably explains why FEUCHT and NACHIT (1977) have been able to observe growth-promoting effects in young shoots of *Prunus avium* by application of catechins and dimeric flavolans. These relatively low molecular weight substances presumably are able to have a sparing effect on IAA, without interfering at the same time with any of the vital processes of growth.

4 Biosynthesis and Metabolism of Phenolics

4.1 Carbon Pathway

4.1.1 Introduction

The biosynthetic origins of the carbon atoms in most classes of phenolic compounds are now well established, as a result of a variety of tracer studies applied to whole plants, organ cultures and tissue cultures. Some of the main pathways to the more common phenolic constituents are outlined in Figure 3. It should be emphasised that many classes may be formed by more than one route; this applies particularly to the quinone pigments, but is also becoming apparent for certain types of phenylpropanoid. The biosynthesis of phenolics has been extensively reviewed in recent years (see especially HAHLBROCK and GRISEBACH, 1975; HASLAM, 1974; HARBORNE, 1977c). Here only a summary of the main findings will be included.

4.1.2 Phenolic Acids

There are two possible routes to these benzoic acid derivatives. They may be formed by the β-oxidation of C_6-C_3 hydroxycinnamic acids, or more directly by aromatisation of shikimic acid or of a related cyclohexane derivative (*dotted line* in Fig. 3). The weight of evidence is in favour of the former route being the most important biosynthetically; thus extensive radiocarbon studies by TERASHIMA et al. (1975) of *p*-hydroxybenzoic acid synthesis in *Populus nigra* support the route via *p*-coumaric acid. The actual mechanism of the oxidation of the C_3 sidechain has been examined by FRENCH et al. (1976) in potato tuber tissues, and it does not appear to involve the coenzyme A ester, as might be expected in the normal β-oxidation pathway. Instead, the double bond is hydrated, there is cleavage of a two-carbon fragment and the aldehyde produced is then enzymically oxidized to carboxylic acid, the route being: $RCH{=}CHCO_2H \rightleftharpoons RCHOHCH_2CO_2H \rightleftharpoons R$-$CHO \rightleftharpoons RCO_2H$. The enzymes involved are heat-labile and unstable and thus have not yet been studied in detail.

4.1.3 Hydroxycinnamic Esters and Hydroxycoumarins

The CoA esters of hydroxycinnamic acids have a central role in phenolic biosynthesis (ZENK, 1979) and these substances are known to be the essential intermediates

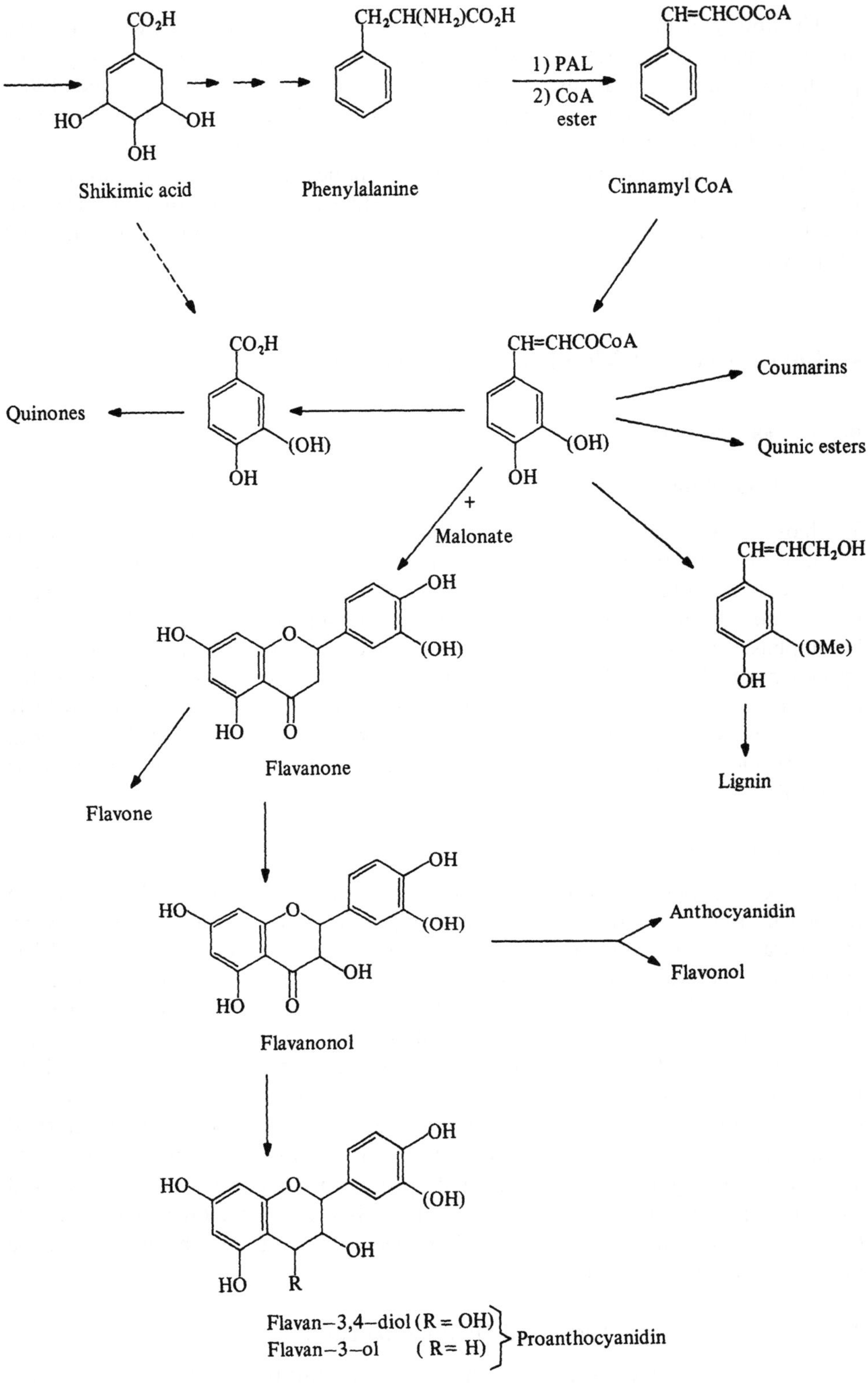

Fig. 3. Pathway of biosynthesis of flavonoids and related plant phenolics

Fig. 4. Alternative pathways of hydroxycinnamic acid ester and hydroxy coumarin biosynthesis

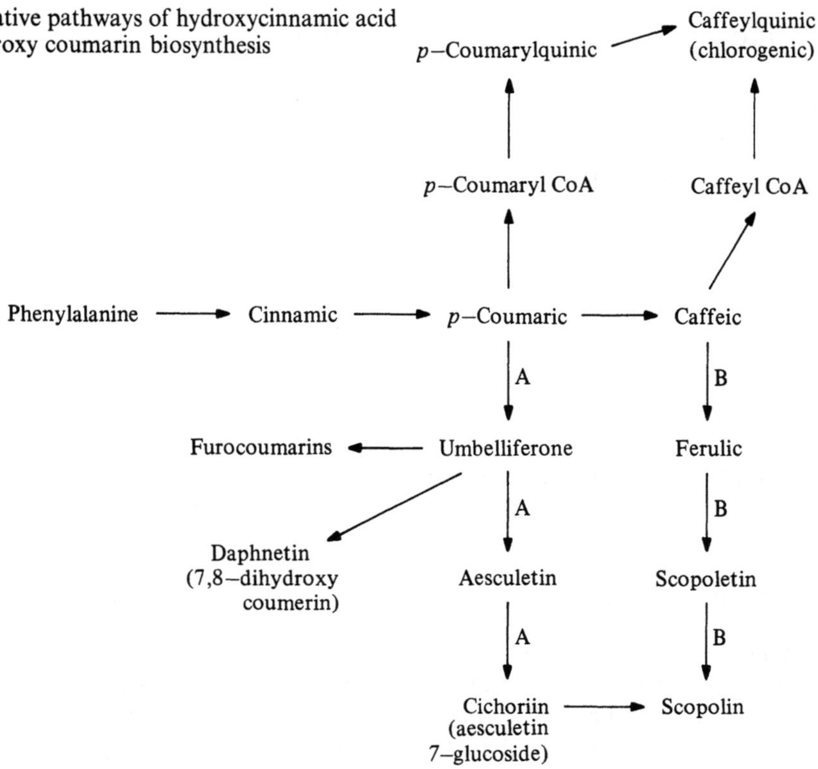

in the production of the many and varied bound forms of *p*-coumaric, caffeic, and ferulic acids present in nature. There are still, however, some uncertainties in the actual pathways of biosynthesis. Two alternative routes, differing in the stage at which a second hydroxyl group is introduced into the phenyl ring, have been proposed for the synthesis of the well-known quinic ester, chlorogenic acid (Fig. 4). It is of course possible that both routes are used, depending on the availability of substrates at a particular site of synthesis. More work is needed to establish the relative importance of these alternative pathways.

The actual route to hydroxycoumarins in plants may also vary depending again on the stage at which the second hydroxyl is introduced into the phenylpropanoid intermediate. Feeding experiments in *Daphne odorata* and *Cichorium intybus* have established route A (Fig. 4) for aesculetin and cichoriin synthesis in these plants, while experiments in tobacco tissue culture show that scopoletin and scopolin are manufactured via route B. In this case, the involvement of caffeic and ferulic acids as intermediates in coumarin synthesis, as in route B, might be attributed to the abnormal conditions of tissue culture growth except that the same pathway has been detected in tobacco leaves. Although route B is thus established, it may not be widely used, since there is evidence from other tracer studies that the conversion of *p*-coumaric to umbelliferone (route A) has a central place in the biosynthesis of most plant coumarins.

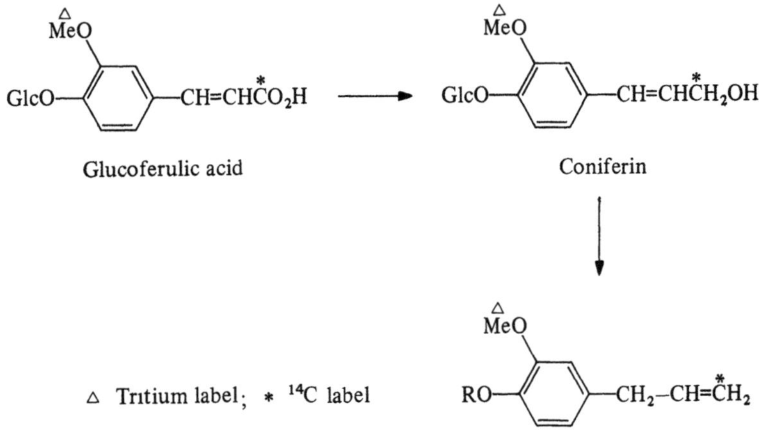

R = H, Eugenol
R = Me, Methyleugenol

Fig. 5. Biosynthesis of eugenol and methyleugenol in *Ocimum*

4.1.4 Allyl and Propenyl Phenols

Studies of the biosynthesis of allyl and propenylphenols in plants have given rise to some controversy. Labelling experiments in basil (*Ocimum basilicum*) suggested to CANONONICA et al. (1971) that the allylphenol, eugenol, was formed from ferulic acid by loss of a terminal sidechain carbon and its subsequent replacement by a C_1 unit derived from methionine. Loss of sidechain carbon did not occur, however, in the biosynthesis of the isomeric propenylphenol, isoeugenol. A pathway involving loss of C_1 seems improbable and indeed more critical feeding experiments in the same plant, using labelled ferulic acid instead of labelled phenylalanine as precursor, have shown that both eugenol and methyleugenol are formed from ferulic acid by the more likely route without loss of carbon (Fig. 5) (KLISCHIES et al., 1975).

These differing results indicate some of the problems of interpretation in plant-feeding experiments and in particular the possibility that fed precursors, especially those remote from the end product (e.g., phenylalanine), may undergo "abnormal" metabolism instead of being incorporated into a desired biosynthetic route.

4.1.5 Flavonoids

The biosynthetic origin of flavonoids was established in feeding experiments carried out by NEISH and others with buckwheat and red cabbage in the early 1950's, and has been extensively studied since then, largely by GRISEBACH and his co-workers. A vast body of information is thus available now (for summary see HAHLBROCK and GRISEBACH, 1975) so that the main outline of the pathway and the inter-relationships of the different classes of flavonoids are very clear (Fig. 3). The only major change in the sequence of events in synthesis leading from phenylalanine to the more highly oxidised end-products of synthesis is that it is no longer necessary to include chalcones as intermediates in the main pathway. Thus, recent enzymic studies establish that the flavanone naringenin is the primary product of the key condensation of *p*-hydroxycinnamyl CoA with three malonate

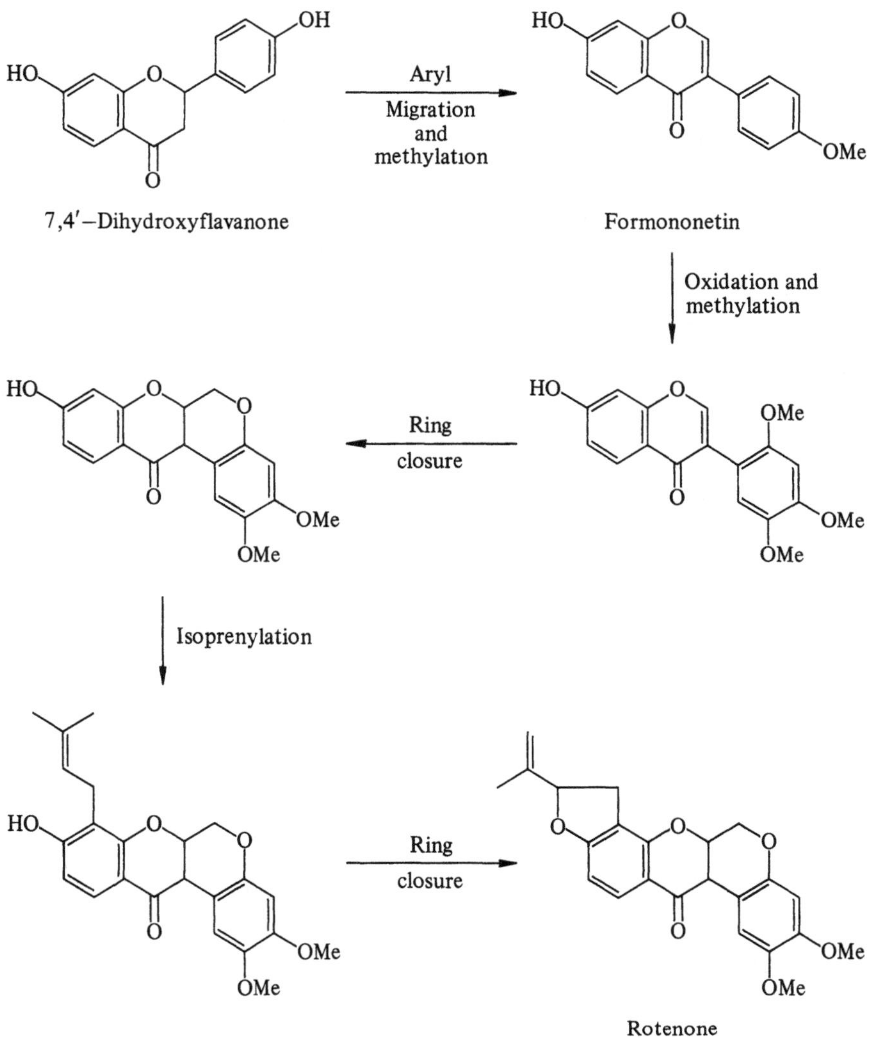

Fig. 6. The biosynthesis of the isoflavonoid rotenone

units (see Sect. 4.2) so that the obligatory intermediacy of 2′,4′,6′,4-tetrahydroxy-chalcone has been eliminated.

The precise mechanism of conversion of dihydroflavonol to anthocyanidin (e.g., dihydroquercetin → cyanidin, Fig. 3) has yet to be determined, but progress in studying anthocyanin biosynthesis has been aided by the realisation that most of the reactions take place partly or completely on membranes. There is still some doubt about the stage at which the second and third hydroxyl groups are introduced into the B-ring, but most evidence points to such hydroxylation occurring at the dihydroflavonol stage to give quercetin and myricetin respectively. Isoflavonoids are formed by a special aryl migration occurring very early on in synthesis (see Fig. 3).

The detailed pathway to an isoflavonoid such as rotenone is now known, the various steps in the synthesis having been established by feeding experiments in young seedlings of *Amorpha fruticosa* (Crombie et al., 1971; Fig. 6).

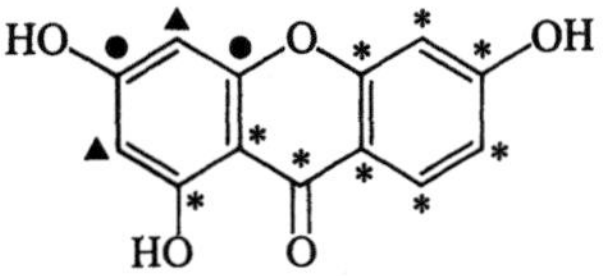

Xanthone: gentisein
Gentiana lutea
(GUPTA and LEWIS 1971)

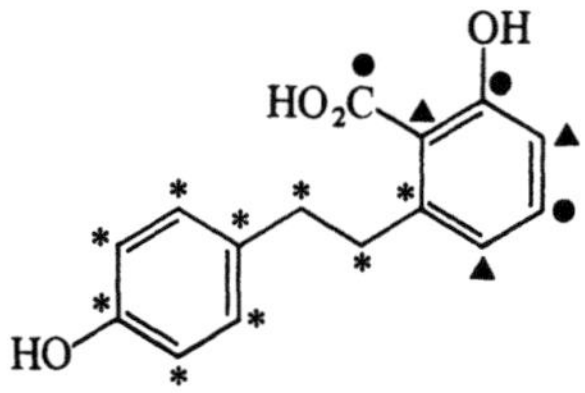

Stilbene: lunularic acid
Lunularia cruciata
(PRYCE 1972a)

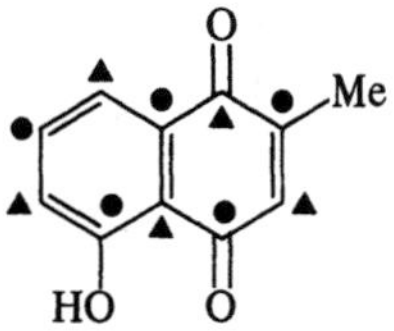

Naphthoquinone: plumbagin
Plumbago europeaus
(DURAND and ZENK 1971)

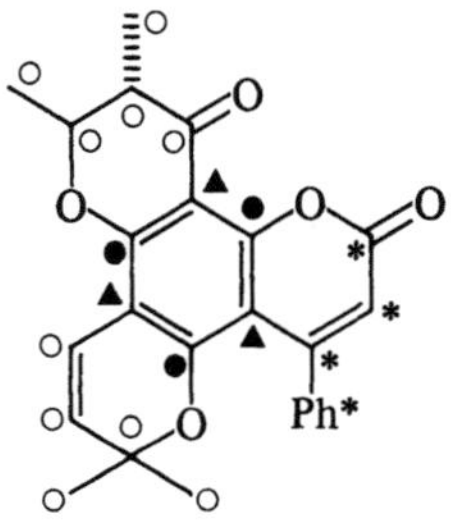

4–Phenylcoumarin: inophyllide
Calophyllum inophyllum
(GAUTIER et al. 1972)

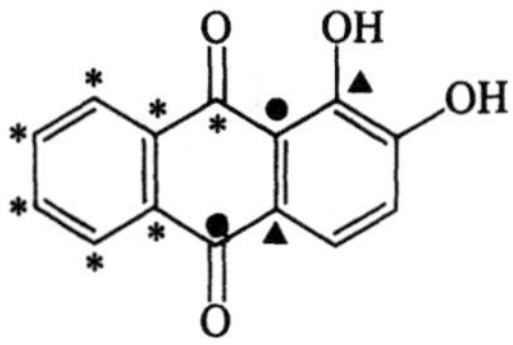

Anthraquinone: alizarin
Rubia tinctorum
(LEISTNER and ZENK 1968)

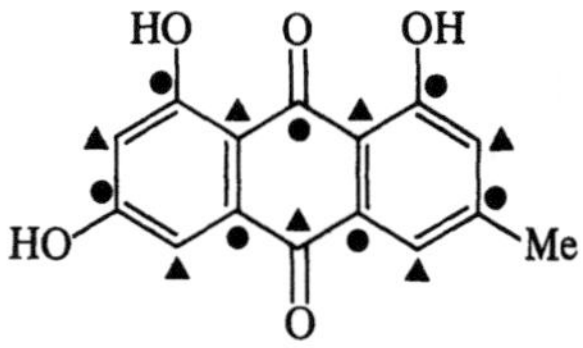

Anthraquinone: emodin
Rhamnus frangula
(LEISTNER 1971)

▲ Methyl carbon/acetate; ● Carboxyl carbon/acetate
* Shikimate carbons; ○ Isoprenyl carbons

Fig. 7. Biosynthetic origin of carbon skeletons of various quinones and other phenolics

4.1.6 Other Phenolics

The carbon pathways of many other groups of plant polyphenols have been elucidated by feeding experiments and some typical results are shown in Figure 7. In the case of stilbenes and xanthones, very few experiments have been carried out to date so that our present knowledge of the various steps along the pathway is distinctly limited. It is possible that for a given class of compound more than one pathway may operate. Indeed, there is evidence for two routes to xanthones, since feeding experiments on mangiferin biosynthesis in *Anemarrhena asphodeloides* indicate it is formed by a different route from that indicated for gentisein in Figure 7, one involving the intact incorporation of *p*-coumaryl CoA (FUJITA and INOUE, 1977).

In the case of the quinones, there are considerable problems in predicting the biosynthesis of any particular pigment because there are at least five known pathways to benzoquinones, five to naphthoquinones, and two to anthraquinones. One of the simplest routes to naphthoquinones is by polyketide synthesis via acetate and malonate, a pathway commonly followed in microorganisms. As indicated in Figure 7, plumbagin is formed in *Plumbago* roots in this way. Other closely similar naphthoquinones, e.g., lawsone (2-hydroxynaphthoquinone) from *Impatiens balsamina*, are, however, shikimate-derived. Only in the case of anthraquinones is the choice of pathway more limited. Typically, alizarin is formed in *Rubia* from shikimate, acetate, and mevalonate, while emodin is formed in *Rhamnus* purely from acetate (malonate) (see Fig. 7). Further details of the pathways to quinones are given in the recent comprehensive review of BENTLEY (1975).

4.2 Enzymology

4.2.1 Introduction

Spectacular progress has been made in advancing our knowledge of phenolic biosynthesis through enzymic investigations; this is particularly true in the case of phenylpropanoids, flavonoids, and lignin. The results of studying the specificities of the enzymes that catalyse the synthesis of phenolic compounds have been especially important in confirming and extending our knowledge of the intermediates in synthesis. There have been occasional conflicts in the results derived from tracer and enzymic studies. As already mentioned, the enzymic approach to flavonoid biosynthesis has ruled out the need to postulate chalcones as obligate intermediates.

In this section, a brief account will be given of the enzymology of biosynthesis, as far as it has been explored at the present time. Enzymic endeavours have concentrated on phenylpropanoid and flavonoid synthesis and almost nothing is known of enzymes concerned in xanthone or stilbene biosynthesis. The only other area of active investigation seems to be that of furocoumarin formation (see e.g., DHILLON and BROWN, 1976).

4.2.2 Phenylalanine Ammonia Lyase

Phenylalanine ammonia lyase (PAL), the key enzyme of phenolic biosynthesis, was first reported in *Hordeum vulgare* by KOUKOL and CONN (1961). It catalyzes the deamination of phenylalanine and by antiperiplanar elimination of the pro-*S*-proton from carbon 3 and of NH_3 from carbon 2 it produces trans-cinnamic acid (Fig. 8). Since 1961, its properties have been extensively studied and it has been detected in a very wide range of higher plants and also in many microorganisms (CAMM and TOWERS, 1977). The enzyme PAL has an average molecular weight of 330,000, a pH optimum of 8.8 and no cofactor requirements. Some

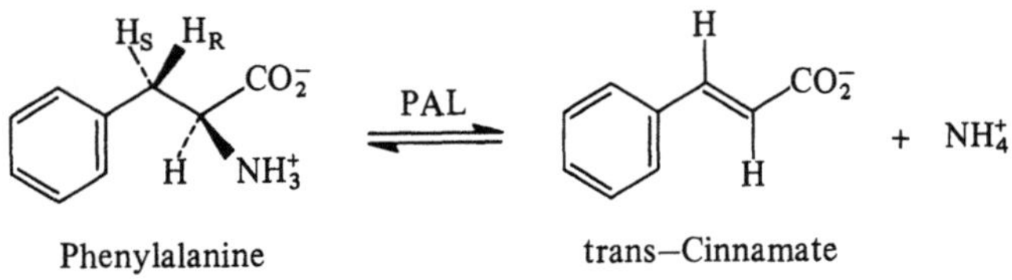

Fig. 8. The reaction catalyzed by PAL

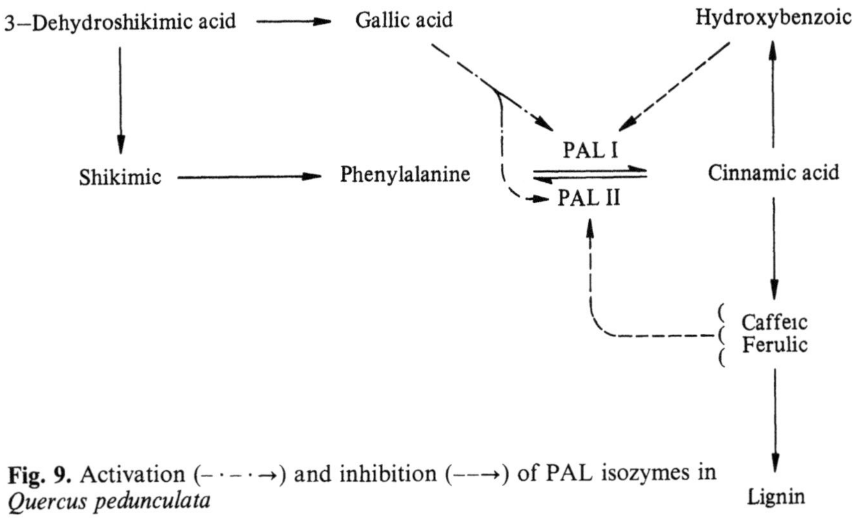

Fig. 9. Activation (–·–·→) and inhibition (––→) of PAL isozymes in *Quercus pedunculata*

preparations of PAL, especially those from grasses, show activity towards tyrosine, but it has not yet been established whether in such cases there exists a separate enzyme for the deamination of tyrosine. In general, PAL is otherwise highly specific towards its named substrate.

Most PAL preparations are inhibited by cinnamic acid and also by later products of the phenolic pathway (e.g., *p*-coumarate, kaempferol, etc.). BOUDET et al. (1971) have isolated two isozymes of PAL from *Quercus pedunculatus* leaves, which are differentially inhibited by phenolic products (Fig. 9). On the basis of these inhibition studies and of their separate compartmentation within the cell, BOUDET et al. suggest that PAL I is required for benzoic acid synthesis, while PAL II is specific for the lignin pathway (see Fig. 9).

Much controversy has centred on the question of how far PAL activity controls the flow of phenolic biosynthesis in plants. Evidence in favour of control through PAL is its rapid activation by light, through phytochrome, and its deactivation in the dark by complexing with a partly characterized high molecular weight inhibitor (see SMITH et al., 1977). If PAL is the control, it is, however, difficult to understand why its activities in plant tissues are generally much greater than those required for normal phenolic synthesis, and also why there is often a lack of correlation between changes in PAL levels and the rate of accumulation of phenylpropanoids in those tissues.

MARGNA (1977) has cogently argued a case that the major control on phenolic biosynthesis is through substrate supply. Thus, under critical conditions of rapid plant growth, phenylalanine will be preferentially incorporated into protein synthesis *before* it becomes available in quantity for incorporation into flavonoids and other phenolics via PAL. Presumably, under conditions when protein synthesis is minimal, more phenylalanine becomes available for conversion into phenolics. This suggestion of MARGNA, while very plausible, does of course only push the point of control one further step back along the pathway. It is still necessary to define the mechanism by which phenylalanine is preferentially channelled into protein and how the substrate is also allowed to flow along the pathway into phenolic production.

4.2.3 Hydroxylases and Phenolases

The enzymic conversion of cinnamic acid to 4-hydroxycinnamic acid was first reported by NAIR and VINING (1965) in extracts from spinach leaves and the enzyme was more fully characterised by RUSSELL and CONN (1967) from *Pisum sativum*. Cinnamate 4-hydroxylase is a mixed function oxidase, which requires molecular oxygen, NADPH and mercaptoethanol; there is good evidence that cytochrome P-450 is required in vivo for this oxidation to take place. The mechanism of oxidation (Fig. 10) involves the well-known NIH shift; the H atom at position 4 in the benzene ring, which is replaced by OH, is not lost but ends up attached to carbon 3.

The enzymic oxidation of cinnamic acid in the *o*-position to give *o*-coumaric acid (Fig. 10) is a key step in coumarin biosynthesis. The enzyme has only been partly characterised because it is membrane-bound. GESTETNER and CONN (1974) have obtained it from chloroplast preparations of *Melilotus alba* seedlings and record its optimum pH as being 7.0.

The introduction of a second hydroxyl group into *p*-coumaric to give caffeic (Fig. 10) is catalysed by a well-known group of plant enzymes – the phenolases or polyphenoloxidases. These enzymes normally oxidise the caffeic acid further to the corresponding *o*-quinone and this then polymerises to dark products. The

Fig. 10. Enzymic oxidation of cinnamic acids

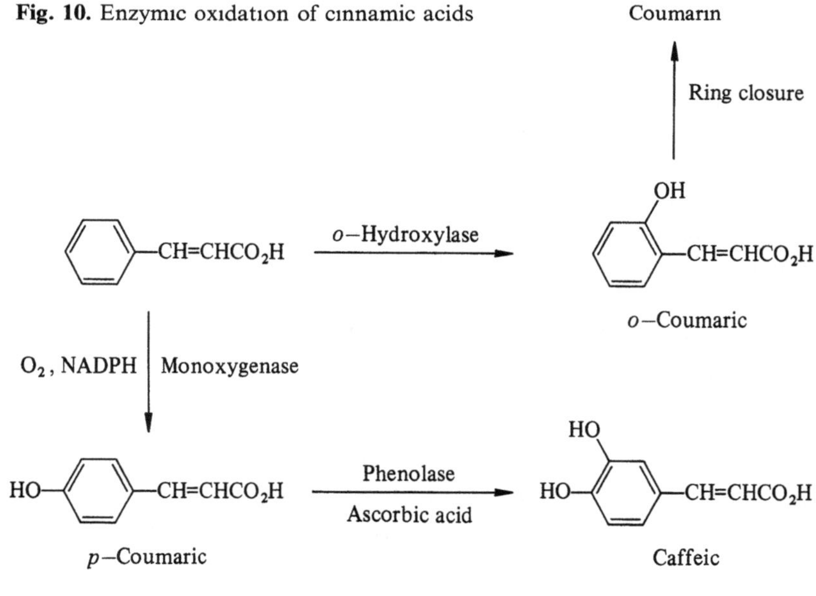

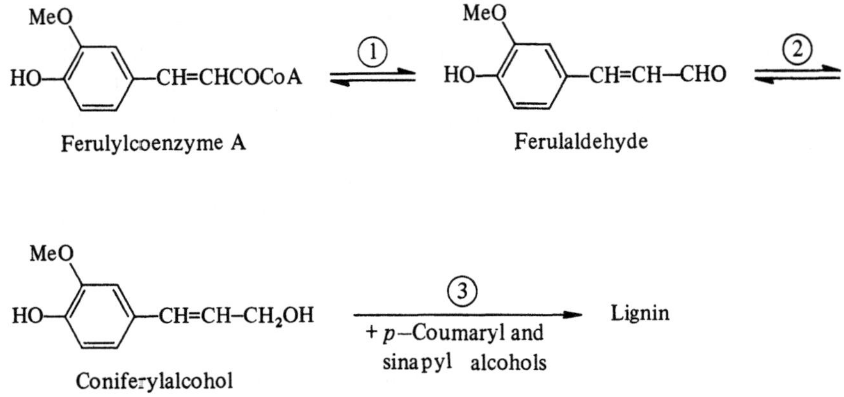

Fig. 11. Enzymes of lignin biosynthesis. *1* cinnamyl CoA NADPH reductase, *2* cinnamyl alcohol NADP oxidoreductase, *3* peroxidase

synthetic role of phenolases has been explored and, in the presence of ascorbic acid as reductant, the reaction can be limited to the first stage. Indeed, phenolase preparations from spinach beet (VAUGHAN et el., 1969) and parsley suspension cultures (SCHILL and GRISEBACH, 1973) will readily catalyse the conversion of *p*-coumaric to caffeic acid. Unlike the parsley enzyme, that from spinach beet also catalyses oxidation of naringenin, dihydrokaempferol and kaempferol in the 3′-position, so that it could be also important in flavonoid biosynthesis.

4.2.4 *O*-Methyltransferases

Enzymes capable of specifically catalysing the *O*-methylation of caffeic acid in the 3-position to give ferulic acid (3-methoxy-4-hydroxycinnamic acid) have been described from a variety of plant tissues. Like other *O*-methyltransferases, they require *S*-adenosylmethionine as methyl donor. Many of the enzyme preparations described will act on several catechol substrates, besides caffeic, but a few are more highly specific. Thus parsley suspension cultures have yielded an *O*-methyltransferase which is only really active with flavone substrates and will methylate luteolin to chrysoeriol (luteolin 3′-methyl ether) in good yield (EBEL et al., 1972). More recently, two methyltransferases have been separated from soya bean cultures, one acting on caffeic acid and being involved in lignification, the other acting only on flavonoid substrates (POULTON et al., 1976). These two enzymes are not isozymes, since they have clearly distinct substrate requirements, and presumably they have different roles in overall phenolic metabolism. Many other methylations of hydroxyl substituents occur in phenolic and particularly in flavonoid biosynthesis but the enzymes concerned have yet to be characterised.

4.2.5 Enzymes of Lignin Biosynthesis

The two enzymes required for the reduction of hydroxycinnamic acids, as their CoA esters, to the corresponding alcohols (Fig. 11) have been characterised and detected in a wide variety of plant tissues. The NADPH-dependent cinnamyl alcohol oxidoreductase has been purified 600-fold from *Forsythia suspensa* tissues and

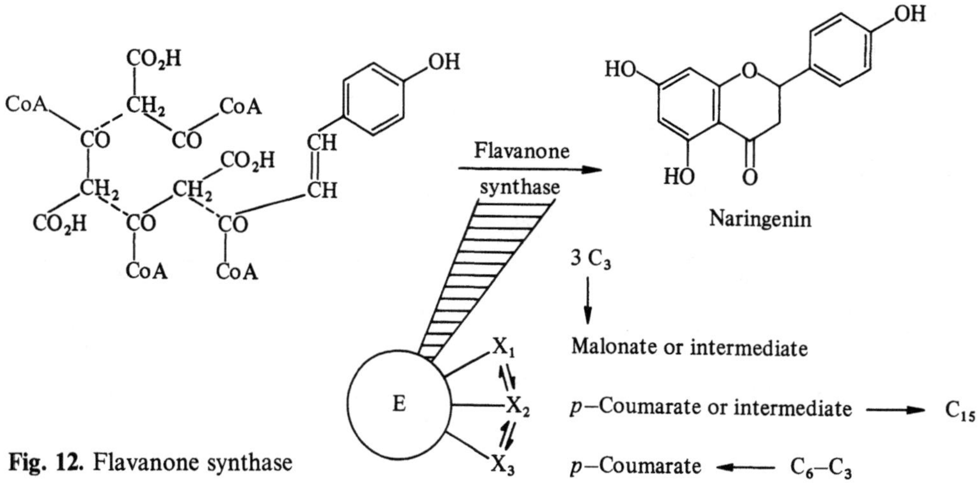

Fig. 12. Flavanone synthase

been extensively studied (Mansell et al., 1974). In its properties, it resembles other plant alcohol dehydrogenases but is absolutely specific towards cinnamyl substrates so that its role in lignin biosynthesis is assured. It has a MW of 80,000 and a pH optimum of 7.6. In keeping with its biosynthetic importance, measurements have shown that there is ten times more activity in gymnosperm cambial tissues than in herbaceous angiosperm non-lignifying tissues.

The last stage in lignin biosynthesis is assumed to involve peroxidase, cellulose and mixtures of hydroxycinnamyl alcohols (see e.g., Nakamura et al., 1974). Gymnosperm lignin is based on *p*-coumaryl and ferulyl alcohols, while angiosperm lignin has sinapyl alcohol units as well. Efforts to determine the critical biochemical factors controlling these differences in lignin make up have not so far been completely successful, but the enzymes of hydroxylation and *O*-methylation of *p*-coumarate and ferulate may be most important.

4.2.6 Enzymes of Flavonoid Synthesis

Flavanone synthase which catalyses the condensation of *p*-coumaryl CoA with three malonate units to give naringenin (Fig. 12) has been isolated from parsley and *Machaeranthera* cell cultures and from leaves, flowers, and anthers of tulip (Hrazdina et al., 1976, 1978; Saleh et al., 1978). It is highly specific for *p*-coumaryl CoA with a pH optimum of about 8.0 and is inactive with cinnamyl, *p*-methoxycinnamyl, and isoferulyl CoA esters. It is, however, very active with caffeyl CoA, giving eriodictyol, but this reaction has a different pH optimum, between 6.5 and 7.0. The fact that flavanones are formed directly in this reaction rules out earlier suggestions that chalcones were the first C_{15} compounds to be formed in synthesis. It also means that flavanone-chalcone isomerases, which are actually widespread in plants, do not have, as was supposed, a key biosynthetic role.

Flavanone synthase probably has three binding sites, as indicated in Figure 12, which catalyse the stepwise addition of malonate units to the activated *p*-coumaric or caffeic acid. Thus *p*-coumaryl CoA is first attached at X_3 and is shuttled to and fro from sites X_2 and X_1, receiving at X_1 three successive malonate units. The completed C_{15} molecule is then lost from site X_2.

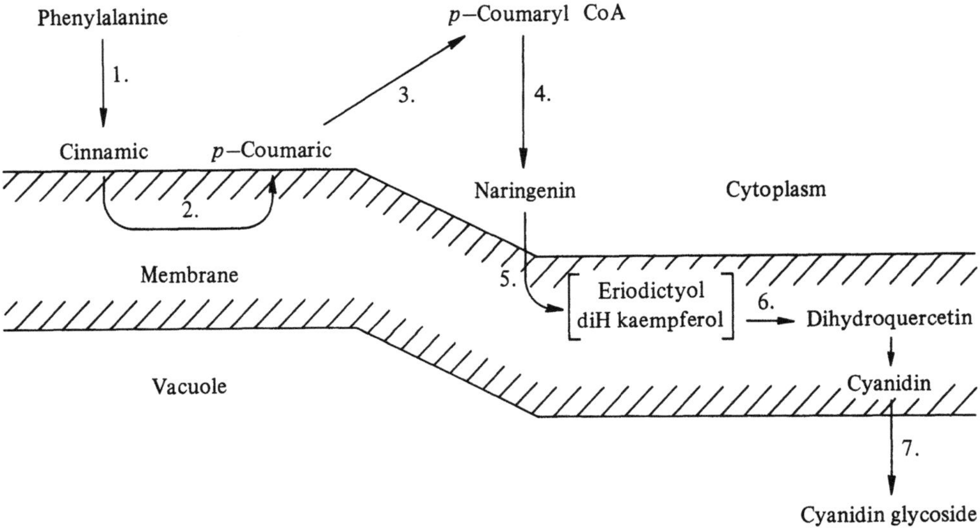

Fig. 13. Enzymology of anthocyanin biosynthesis. *1* PAL, *2* cinnamate 4-hydroxylase, *3* *p*-coumarate CoA ligase, *4* flavanone synthase, *5,6* oxidases, *7* UDP glucose:cyanidin 3-*O*-glucosyltransferase

The final step in the synthesis of flavones requires the oxidation of flavanones in the 2,3-position and a flavanone oxidase with just these properties had been detected in young primary leaves of parsley (SUTTER et al., 1975). It catalyses the oxidation of naringenin to apigenin and also of 7,4′-dihydroxyflavanone to 7,4′-dihydroxyflavone. It requires O_2, Fe^{2+} and a heat-stable, dialysable cofactor for its operation. Notably, it fails to oxidize chalcones to flavones and this again shows (see above) that flavanones and not chalcones are key intermediates in flavonoid synthesis.

The biosynthesis of anthocyanins has been examined enzymically (FRITSCH and GRISEBACH, 1975) especially in suspension cultures of *Machaeranthera*, and the pathway is probably as shown in Figure 13. Unfortunately, the enzyme catalyzing the step: dihydroquercetin → cyanidin has so far eluded detection. However, a major step forward in comprehending pigment synthesis has been the realisation that most of the enzymic reactions take place partly or completely on membrane surfaces. Recent experiments with leaves and petals of *Hippeastrum* and *Tulipa* on the subcellular localisation of these enzymes suggest that their highest activity is in the cytosol fraction of the protoplasts (HRAZDINA et al., 1978).

4.2.7 *O*-Glycosyltransferases

Enzymes capable of transferring glucose from UDP glucose to phenolic hydroxyl groups, as in hydroquinone, have been known for a long time and the enzymic synthesis of hydroxycinnamic acid glucose esters in the presence of UDP glucose has also been well described. Similar glycosyl transferases are also involved in the final stages of flavonoid synthesis and such catalytic activities have now been detected in over a dozen plants (Table 4). The four main glycosyl transferases detected mainly operate on only one class of flavonoid and they catalyse the

Table 4. Source of glycosyl transferase enzymes in flavonoid synthesis

UDP Glucose: quercetin 3-*O*-glucosyltransferase	
Brassica oleracea seedlings	*Phaseolus aureus* leaves
Impatiens balsamina petals	*Ph. vulgaris* leaves[a]
Leucaena glauca leaves	*Zea mays* pollen
Petroselinum crispum cells	
TDP Rhamnose: quercetin 3-*O*-glucoside *O*-rhamnosyltransferase	
Phaseolus aureus leaves	*Ph. vulgaris* leaves
UDP Glucose: cyanidin 3-*O*-glucosyltransferase	
Brassica oleracea seedlings[b]	*Machaeranthera* cells
Hippeastrum leaves	*Tulipa* leaves, petals
UDP Glucose: apigenin 7-*O*-glucosyltransferase[c]	
Petroselinum crispum cells	

[a] Also contains a glucuronic acid transferase enzyme
[b] Enzyme from this source will also glucosylate flavonols and other anthocyanidins, besides cyanidin
[c] This enzyme will also glucosylate flavonols in the 7-position; a second enzyme from parsley cells has been characterized which will transfer apiose to apigenin 7-glucoside to give the 7-apiosylglucoside (see HAHLBROCK and GRISEBACH, 1975)

addition of sugar specifically to a particular hydroxyl (either in the 3- or 7-positions). The glycosyl donor molecule is normally UDP glucose for glucosylation and TDP rhamnose for rhamnosylation, but at least in vitro the enzymes will transfer sugars from other donor molecules (see HAHLBROCK and GRISEBACH, 1975).

Occasionally, the sugars of flavonoid glycosides carry acyl components and the enzymology of such acylation has been explored in at least two cases. Thus, a malonyl transferase adding malonic acid to apigenin 7-apiosylglucoside has been described in parsley cultures (HAHLBROCK and GRISEBACH, 1975). A similar enzyme catalysing the transfer of *p*-coumaric acid to kaempferol 3-triglucoside has been reported in leaves of *Pisum sativum* (SAYLOR and MANSELL, 1977).

4.3 Physiology of Biosynthesis

Physiological aspects of flavonoid formation in plants were extensively reviewed by BLANK in Volume X of the Encyclopedia of Plant Physiology, first series. More recent work has been critically reviewed by MCCLURE (1975) and excellent accounts are available on the photocontrol of flavonoid biosynthesis via phytochrome (SMITH, 1972) and of the regulation of PAL synthesis in plants (SMITH et al., 1977). Less has been written about the physiology of formation of other classes of phenol, but presumably many of the parameters affecting flavonoid production apply to phenolic compounds in general, since they are almost all formed ultimately via the key enzyme system PAL. The present position regarding the major physiological variables in flavonoid synthesis is summarised in Table 5.

A major problem in interpreting the results of physiological studies in this area is that rarely, until recently, has the concentration of more than one type

Table 5. Physiological factors in phenolic synthesis

Light

1. low energy R$\rightleftharpoons$FR phytochrome system
   general requirement in all plants studied.
2. High Intensity Response (HIR), either FR or blue light
   often a requirement for anthocyanin synthesis.
3. Photosynthetic involvement for precursor or cofactor
   a presumed requirement for anthocyanin synthesis in tuber or root.
4. UV light
   occasionally induces synthesis or may augment it.
5. Ionizing radiation
   increases anthocyanin levels.

Temperature

Lower temps. generally favour increased anthocyanin and isoflavone levels in leaves. Exceptionally, proanthocyanidin synthesis is depressed by lowering temp.

Water

Waterlogged plants may show increased flavonoid production, while desiccated plants (20% water deficit) in general produce less phenolic

Carbohydrate

General, sometimes marked, increase in synthesis.

Growth Hormones

Response unpredictable, depending more on other factors.

Mineral Nutrition

N or P deficiency usually increase flavonoid levels.
B deficiency may increase hydroxycinnamic acid levels at the expense of flavonoids or lignin.

of flavonoid constituent been determined. In fact, attention has nearly always focused on anthocyanin, since it is so easy to measure it by spectrophotometry at 520 nm. Unfortunately, in determining the effect of any physiological variable on phenolic synthesis, it is necessary to examine all major classes, since they are all formed on a biosynthetic metabolic grid. An inhibition in one pathway may be correlated with an increase in synthesis along another. It is possible, for example, for some physiological factor to inhibit anthocyanin synthesis by perhaps 50%, but at the same time to increase flavonol glycoside synthesis by, perhaps, 300%. If levels of both classes are approximately the same in the control, then the change in the physiological parameter in reality has on overall beneficial effect on flavonoid production, even though if anthocyanin levels alone are measured it would appear to be harmful to synthesis.

Where the effect of light has been studied on more than one end product of synthesis, it has been found to have different effects on different pigments and the effects produced depend on the tissue used. For example, in the case of *Spirodela intermedia*, dark-grown plants only synthesise flavonoids with a monohydroxylated B-ring, i.e., vitexin and kaempferol 3-glucoside. Plants grown under low intensity white light start synthesising compounds with a dihydroxylated B-ring, namely orientin and quercetin 3-glucoside. Only high light intensity for several days will induce anthocyanin (cyanidin 3-glucoside) synthesis in addition (McCLURE, 1968). Again, with flavonol glycoside synthesis in *Pisum sativum* (GALSTON, 1969), brief red light causes the accumulation of quercetin glycoside in the plumule and kaempferol glycoside in the internode tissues, whereas both groups of flavonol glycoside are synthesised in the leaf.

Many observations have been made on the effect of adding organic precursors, plant hormones, enzyme inhibitors and inorganic salts on phenolic synthesis in plants. Rarely, however, do any clear-cut conclusions emerge from such studies (see Table 5). One common observation, viz. that adding sugar nearly always stimulates phenolic synthesis more than adding a known precursor, continues to pose a problem in terms of the accepted routes of biosynthesis. CREASY and SWAIN (1966) in their study of phenolic synthesis in strawberry leaf discs, however, found that high sugar concentrations could actually mask the usual phytochrome response in flavonoid synthesis. Clearly, in such cases, the results of artificial feeding experiments have to be interpreted with caution.

4.4 Turnover and Metabolism

It is now widely accepted by plant biochemists that phenolics are not end products that accumulate unchanged in plant cells. Instead, they are thought to be part of a dynamic equilibrium in which, even when produced in quantity, there is continual synthesis, turnover, and degradation. Turnover can be particularly rapid in floral tissues, since in many angiosperms, the requirement for floral pigmentation is quite short, i.e., only a day or a few days while the floral organs are ripe for pollination. This is true in *Petunia hybrida*, where the half life of anthocyanins based on delphinidin has been determined by tracer studies to be between 25 and 31 h (STEINER, 1971).

A study of anthocyanin synthesis in relationship to flower development in the pea flower has also demonstrated the presence of a very active system of synthesis and turnover. As the pea flower senesces, the supply of "synthetic" enzymes runs out and the pigments produced lack some of the methyl and sugar residues that are present in the pigments of the flower at maturity (STATHAM and CROWDEN, 1974). Similar studies in *Impatiens* (HAGEN, 1966) and other flowers have shown a progressive change with time in pigment patterns. The catabolism of these floral anthocyanins is not fully understood, but one of the first steps is the removal of sugar from the 3-hydroxyl and specific glycosidases, called anthocyanases, are known which carry out this step. Another catabolic reaction in the case of anthocyanidins with catechol nuclei (e.g., cyanidin) is oxidation and phenolase or peroxidase-type enzymes have been reported in plant tissues (e.g., cherry fruit) where cyanidin glycosides are decolorised.

Leaf flavonoids are probably turned over more slowly than floral pigments, but there is good evidence that metabolism occurs. Pulse-labelling measurements on kaempferol and quercetin glycosides in *Cicer arietinum* leaves have clearly indicated a half life of 7 to 12 days (BARZ and HÖSEL, 1971). The catabolism of flavonols involves at an early stage the addition of oxygen to the 2,3-double bond to give a 2-hydroxy derivative (Fig. 14). This 2,3-dihydroxyflavanone is then ring-cleaved to give a substituted benzoic acid and a phloroglucinol derivative,

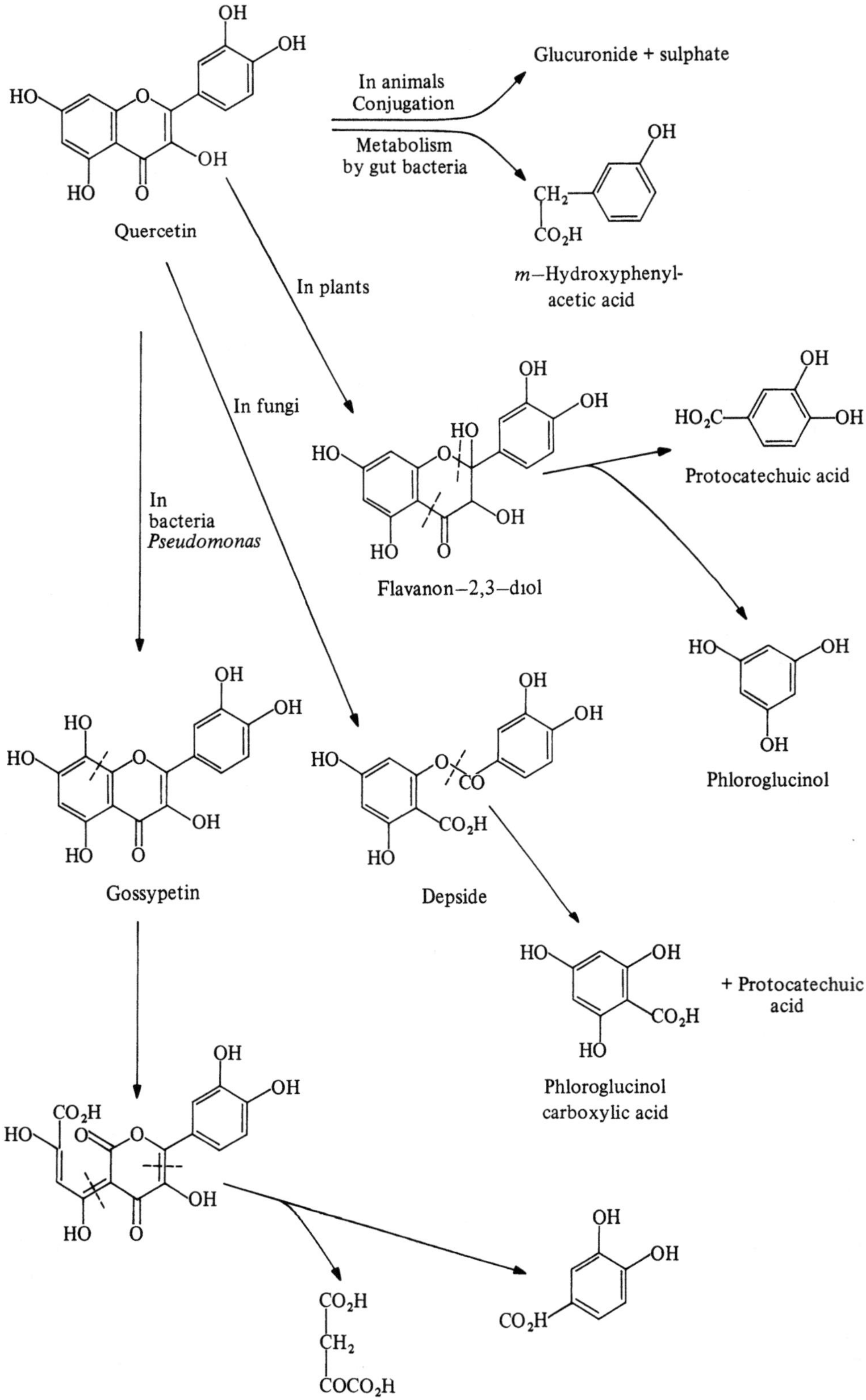

Fig. 14. Metabolic fate of the flavonol quercetin in animals, plants, and micro-organisms

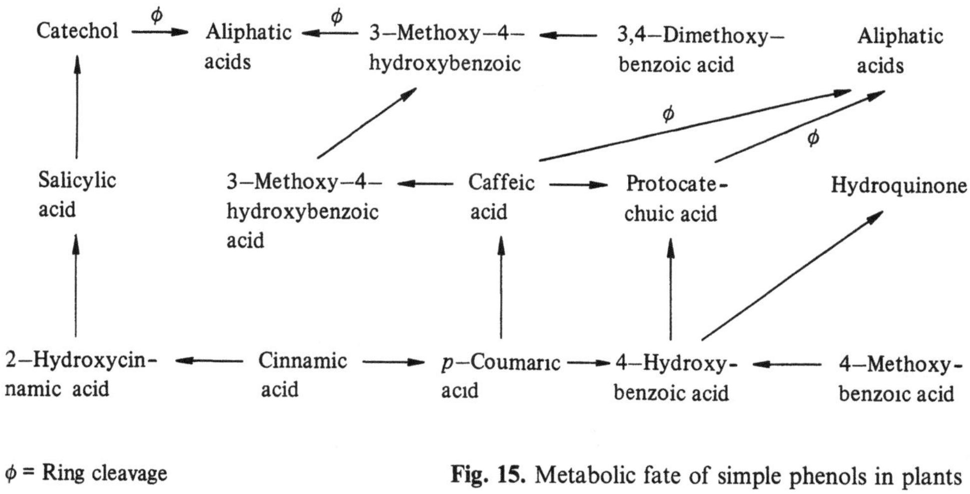

Fig. 15. Metabolic fate of simple phenols in plants

which may not be detectable because it so readily undergoes further oxidation. Undoubtedly the end products are eventually returned to the atmosphere as respired CO_2.

The catabolism of other leaf flavonoids has been less well studied, but it appears that flavanones may be oxidised by ring cleavage at the 2-position to furnish 5,7-dihydroxychromones and hydroxybenzenes (see Sect. 2.2), which are then further broken down in the same way to give CO_2 as their final fate. Alternatively, flavanones may undergo isomerisation to chalcones, which are then metabolised by the pathway: hydroxycinnamic acid → → hydroxybenzoic acid → → → CO_2.

The fate of simpler phenolic compounds in plant tissues has been extensively studied (ELLIS, 1974) and is summarised in part in Figure 15. One main reaction is the β-oxidation of hydroxycinnamic acids to hydroxybenzoic acids. Also methoxybenzoic acids may undergo demethylation and/or decarboxylation. In general, catechol derivatives are formed and these finally undergo ring cleavage to aliphatic acids and then to CO_2. Recent experiments have shown that higher plants possess active aromatic ring-cleaving enzyme systems, similar to those well known in microorganisms, and ring fission is a common feature of phenolic turnover in plants. Ring-cleaving enzymes have also been detected in plant tissue cultures and the catabolism of phenolics in cell cultures has been well studied (BARZ and HÖSEL, 1975; BARZ, 1977).

From the viewpoint of comparative biochemistry, it is instructive to relate the fate of a flavonol such as quercetin in plant tissues with its fate in animals and microbes (Fig. 14). In animals the major reaction that phenolics undergo on feeding is conjugation, especially with glucuronic acid but also with ethereal sulphate (WILLIAMS, 1964). Considerable degradation may also occur in vivo but this is essentially due to the activities of enzymes produced by the gut bacteria (GRIFFITHS and BARROW, 1972). Dehydroxylation of aromatic compounds is a special feature of such bacterial degradation and quercetin, on feeding to a range of animals, is often recovered as *m*-hydroxyphenylacetic acid. By contrast, in the fungal turnover of quercetin, the first reaction is cleavage at the heterocyclic ring by quercetinase, with the expulsion of the 3-carbon atom as carbon monoxide – the resulting depside (Fig. 14) then undergoes esterase attack with the production

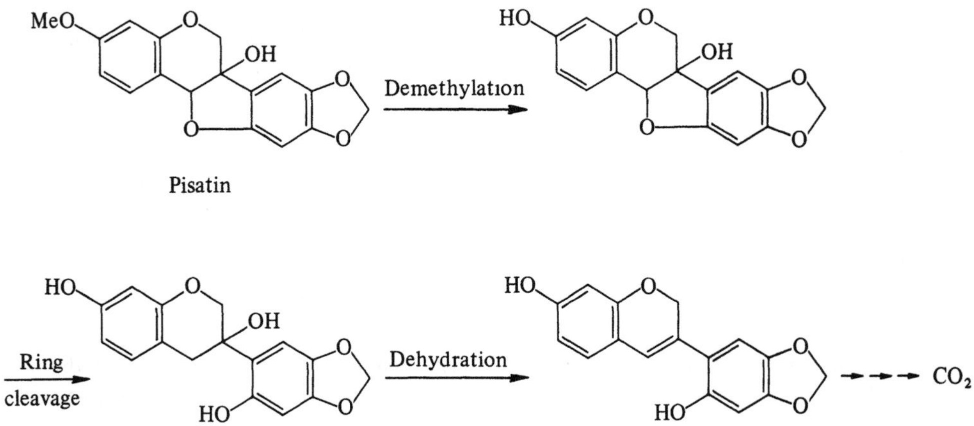

Fig. 16. Metabolism of the pterocarpan pisatin by the fungus *Fusarium oxysporum*

of phloroglucinol carboxylic and protocatechuic acids. Bacteria are also capable of catabolizing quercetin and employ yet another degradative route. Thus, in *Pseudomonas*, the first step is the introduction of oxygen at the 8-position to give gossypetin (8-hydroxyquercetin). This then undergoes specific ring cleavage at the C-C bond adjacent to the introduced oxygen, with the eventual production of protocatechuic and aliphatic acids (Fig. 14).

Less is known of the comparative catabolism of isoflavonoids in different organisms, but the in vivo reduction in mammals of isoflavones to isoflavans has been actively studied, because the isoflavans produced have oestrogenic effects in farm animals (see SHUTT, 1976). The fungal modification of isoflavonoids has also been of particular interest, because of the role of certain of these phenolics as antifungal agents produced in legume plants de novo in response to fungal infection (see HARBORNE and INGHAM, 1978). Fungal metabolism includes introduction of hydroxyl groups and demethylation of methoxyl substituents; the metabolites are usually significantly less fungitoxic than the original isoflavonoids. Only the first products of metabolism have been characterised. Such metabolites then apparently undergo ring-cleavage reactions and the small molecules so produced are rapidly dissimilated. One of the best known isoflavonoid phytoalexins is pisatin from *Pisum sativum* and its metabolism by the pea pathogen *Fusarium oxysporum* is illustrated in Figure 16.

5 Phenolic Production in Tissue and Cell Culture

5.1 Introduction

Phenolic synthesis is a characteristic and universal feature of plant metabolism and it is a property which extends to plants grown in organ culture, as undifferentiated callus tissue or in cell suspension culture. While the pattern of products may be dramatically altered under tissue culture conditions, nevertheless the same phenolic materials continue to be made in such cells. Initially, the phenolic concentrations may be very low and the compounds produced of the simplest type; on subculturing, however, strains may be obtained with enhanced synthetic ability. In general, the substances formed in tissue culture are the same as those in the intact plant and as indicated in Table 6, almost every class of phenolic encountered in plants can be found in tissue culture as well. Unfortunately, the phenolic repertoire of a cell culture has rarely been compared in detail with that of the corresponding intact plant, so that it is still not certain whether the few differences that have been noted in culture are real or apparent.

Tissue culture studies have been most important in providing information on the enzymology of phenolic biosynthesis. Indeed our present sound basis of knowledge of the pathways of flavonoid production (see Sect. 4.1) could not have been obtained without the extensive use of parsley and soya bean tissue cultures as convenient sources of enzyme activities. In many cases, extraction and characterisation of new enzymes along the pathway using tissue-cultured cells has been an essential preliminary to the often more daunting task of detecting the same activity in whole plants.

Several accounts of phenolic production in plant tissues can be found inter alia in general reviews such as those of BARZ et al. (1977), BUTCHER (1977) and STREET (1977). Here, attention will be focused on the synthetic capacity of tissue-cultured cells and on the regulation of such synthesis.

5.2 Qualitative Aspects

The fact that anthocyanins are produced in tissue culture was realised over two decades ago, when STEWARD (1958), one of the early pioneers of the technique, recorded that some callus tissues derived from carrot *Daucus carota* turned deep red following subculturing. Since then, a range of plants has been found which consistently produces anthocyanin pigments in callus or shake culture. One of the most striking examples is *Machaeranthera gracilis*, which produces relatively large amounts (usually 3% dry wt.) of the 3-glucoside and 3-rutinoside of cyanidin in cultured cells. It is remarkable in that the plant from which these cultures are derived shows no signs of anthocyanin colour in normal growth. However, the two cyanidin derivatives formed in culture are those characteristic of plants of the Compositae, to which family *Machaeranthera* belongs (see HARBORNE, 1978a), and no doubt this plant could be induced to produce the same pigments in leaves or stems by subjecting it to some kind of physiological stress. In fact, in almost all cases where a proper comparison has been made of anthocyanin pigments in culture and the whole plant (e.g., in *Dimorphotheca* and in *Solanum tuberosum*, see Table 6) the same pigments have been found. Thus, the capacity for anthocyanin synthesis normally reflects that found in the intact tissue from which the callus has been derived.

One possible anomaly is flax tissue culture, which is reported to be coloured by cyanidin 3,5-diglucoside (see Table 6). In this case, the pigments have been fully analysed in flowers and leaves of various flax genotypes (DUBOIS and HARBORNE, 1975) and although cyanidin is one of the aglycones present, it occurs

Table 6. Phenolics identified in plant callus or suspension cultures

Plant	Substances identified	References
Anthocyanins		
Daucus carota (carrot)	Cyanidin 3-glycosides[a]	SCHMITZ and SEITZ (1972)
Dimorphotheca sinuata	Cyanidin and delphinidin 3-glucosides	BALL et al. (1972)
Linum usitatissimum (flax)	Cyanidin 3,5-diglucoside (?)	IBRAHIM et al. (1971)
Machaeranthera gracilis[b]	Cyanidin 3-glucoside and 3-rutinoside (also quercetin glycosides and chlorogenic acids)	HARBORNE (1964b); STICKLAND and SUNDERLAND (1972)
Parthenocissus quinquefolia	Cyanidin 3,5-diglucoside (and quercetin 3-glycosides)	BLEICHERT and IBRAHIM (1974)
Solanum tuberosum (potato: cv. Congo)	Negretein (acylated malvidin 3-rutinoside-5-glucoside)	HARBORNE and SIMMONDS (1962)
Other flavonoids		
Camellia sinensis (tea)	Quercetin 3-rutinoside, catechins and procyanidins	FORREST (1969)
Cicer arietinum (chickpea)	Formononetin and biochanin A	BARZ (1977)
Citrus cvs.	Sinensetin, nobiletin and other methylated flavones	BRUNET and IBRAHIM (1973)
Glycine max (soya bean)	Apigenin and other flavones	BARZ (1977)
Petroselinum crispum[c] (parsley)	Apigenin 7-apiosylglucoside and 23 other flavones	KREUZALER and HAHLBROCK (1973)
Phaseolus aureus	Daidzein, 2′4′4-trihydroxychalcone, coumestrol and soyagol	BERLIN and BARZ (1971)
Ph. vulgaris (bean)	Phaseollin	INGRAM (1977)
Pisum sativum (pea)	Pisatin	BAILEY (1970)
Silybum marianum	Silybin	SCHRALL and BECKER (1977)
Theobroma cacao (cocoa)	(−)-Epicatechin, procyanidins, flavanones (?) (*p*-coumaric and caffeic acids)	JALAL and COLLIN (1977)
Phenylpropanoids		
Ammi visnaga	Visnagin (chromone)	KAUL and STABA (1967)
Coleus blumei	Rosmarinic acid	RAZZAQUE and ELLIS (1977)
Nicotiana tabacum	Scopolin, scopoletin and hydroxycinnamic acid esters	FRITIG et al. (1970)
Ruta graveolens	Umbelliferone, bergapten, psoralen and other coumarins	STECK et al. (1971)

[a] Pigments in carrot cultures have not been fully characterised but it is likely that they are identical to those reported in the intact carrot plant: cyanidin 3-lathyroside, cyanidin 3-xylosylglucosylgalactoside and its ferulyl derivative (see HARBORNE, 1976a)

[b] This is now the correct name for the plant referred to in the literature as *Happlopappus gracilis*

[c] This is now the correct name for the plant referred to in the literature as *Petroselinum hortense*

Table 6 (continued)

Plant	Substances identified	References
Quinones		
Cassia tora	Chrysophanol, emodin, physcion	TABATA et al. (1975)
Digitalis lanata	Digitolutein and five other anthraquinones	FURUYA et al. (1972)
Lithospermum erythrorhizon	Shikonin and related naphthoquinones	TABATA (1977)
Morinda citrifolia	Alizarin, damnacanthal, morindone	ZENK et al. (1975)
Plumbago zeylanica	Plumbagin	HEBLE et al. (1974)

in petals as the 3-glucosylrutinoside, 3-triglucoside, 3-rutinoside or 3-glucoside. The report by IBRAHIM et al. (1971) of the 3,5-diglucoside in cell culture is, therefore, unexpected. However, it could well be a misidentification, since the same authors (THAKUR and IBRAHIM, 1974) spuriously reported some unusual methylated pigments in the hypocotyl of flax seedlings, which on re-examination was found to contain only cyanidin 3-rutinoside (DUBOIS and HARBORNE, 1975). It should also be noted (a point not realised by BUTCHER (1977) in his review) that the earlier report by ARDENNE (1965) of cyanidin 3,5-diglucoside in *Machaeranthera* was erroneous. The second pigment in this culture is the 3-rutinoside (HARBORNE, 1964b), an identification that has been confirmed by later workers (STICKLAND and SUNDERLAND, 1972).

A range of other flavonoids and of hydroxycinnamic acid derivatives have been identified variously in tissue cultures (Table 6) and the compounds are usually the same as those produced in the parental plants although this point has not always been checked. The most elaborate investigation has been of parsley tissue cultures, which produce 24 flavones. While 14 of these have been identified, it is not clear how many of these constituents actually occur in intact parsley plants: only a few have been recorded positively in seed or leaf tissue.

One plant where a direct comparison has been conducted between phenolic patterns in cotyledon, leaf, stem, and callus derived from these tissues is *Theobroma cacao* (JALAL and COLLIN, 1977). While there are anthocyanins and flavonol glycosides in the cotyledons and glycoflavones in the leaves, none of these substances was detected in the respective callus tissues. All four callus clones had the same pattern, containing epicatechin and two procyanidins, three reduced flavonoids uniformly present throughout the cocoa plant. Several apparently new flavanone-like derivatives were also present in callus. Whether these unidentified constituents are really distinctive of tissue culture is not clear, since they may be formed in the plant in amounts below the normal level of detection. Alternatively they may represent biosynthetic intermediates of the more highly oxidised flavonoids of intact tissue, which accumulate because one or more of the enzymes for flavonoid oxidation are absent from the culture.

Hydroxycinnamic acids have been widely observed in cultured cells, occurring both free and combined. Apparently "new" conjugates of *p*-coumaric, caffeic, and ferulic with the diamine putrescine were found for the first time in tobacco callus by MIZUSAKI et al. (1971). Subsequent investigations, however, have shown that these conjugates are not peculiar to tissue culture, but occur in whole plants, in tobacco and in other species (TANGUY et al., 1978).

One situation where tissue cultures produce phenolics not present in healthy plants is that of phytoalexin synthesis in legumes such as *Pisum* and *Phaseolus*. The pterocarpans pisatin and phaseollin are not detectable in healthy leaves or stems of peas and beans, but are specifically produced in these tissues in response to fungal invasion. What is remarkable, therefore, in the tissue cultures of these two legumes is their ability to synthesise pterocarpans without the "trigger" mechanism necessary in intact tissue. In the case of *Pisum sativum*, the necessary stimulus is provided by the coconut milk in the medium (BAILEY, 1970). It is clear from other pathological studies in tissue culture that normal resistance mechanisms to disease may be modified by the special physiological conditions of cell culture. For example, potato tissue cultures derived from plants with R gene resistance against blight, on infection with *Phytophthora infestans* spores, produce a toxic mixture of salicylic, vanillic and *p*-hydroxybenzoic acids as a response. By contrast, in tubers with R gene resistance, the fungus induces synthesis of the sesquiterpenoid rishitin (see INGRAM, 1977).

In conclusion, then, there is at present little evidence for the synthesis of any phenolics in cell cultures which are not produced at some stage or other in the life cycle of the plant. The only suggestion of abnormal synthesis seems to be in lignification in tissue culture, since NIMZ et al. (1975) have found in soya bean cultures that there are simple cinnamyl residues in the lignin and also a specific reductase converting cinnamic acid to cinnamyl alcohol. Such residues have not so far been recorded as monomeric units in normal plant lignin. It may be noted, however, that lignification in tissue culture is somewhat aberrant generally, since whatever the plant used for producing callus, the lignin in the culture is essentially the same, based almost entirely on guaicyl residues (see BUTCHER, 1977).

5.3 Metabolic Aspects

One of the major features of plant tissue culture is that many of the normal restraints on metabolism are absent or at least may be over-ruled by hormonal or light treatment, so that theoretically it is possible to obtain clones which make massive amounts of a particular phenolic. Caffeic acid ester (e.g., chlorogenic acid) levels in plant leaves, for example, rarely exceed 1% dry wt. while caffeic acid derivatives may be formed in vast excess of this – up to 10% dry wt. – in cell cultures of *Machaeranthera* (HARBORNE, 1964b) and *Coleus* (RAZZAQUE and ELLIS, 1977). Such clones are ideal for studying the biosynthesis of these phenolic components and would also be valuable for commercial production, if these end products were of industrial value. Yields of phenolics normally vary in callus cultures and stable high-yielding strains may only be obtained after repeated selection and subculturing (MIZUKAMI et al., 1978).

The major success in producing large quantities of medicinally useful phenolics in tissue culture has, however, been with hydroxyquinones (TABATA, 1977). A number of anthraquinone derivatives are the active principles of plant drugs used pharmaceutically as laxatives and it is interesting that callus tissue of one such plant *Cassia tora* will produce ten times the normal amount of these quinones – up to 6% dry wt. Suspension cultures of *Morinda citrifolia* are even more rewarding and yield 10% dry wt. of total anthraquinone. Finally, callus of *Lithospermum erythrorhizon* produce the record amount of 12% dry wt. naphthoquinones, compounds used medicinally in the treatment of skin diseases and burns.

One other way of utilising tissue culture to produce medicinally useful phenolics is to feed in "unnatural" precursors and obtain modified metabolites as a result. The feasibility of this approach has been demonstrated by SCHRALL and BECKER (1977) with *Silybum marianum* (Compositae) cultures. When supplied with taxifolin (dihydroquercetin) and coniferyl alcohol, these cultures produced silybin, one of the main flavonolignans of the *Silybum* plant. The same cultures supplied with luteolin instead of dihydroquercetin were able to produce hydnocarpin, a flavonolignan previously only known as a plant product in a different family, in *Hydnocarpus wightiana* (Flacourtiaceae). Medicinally, flavonolignans are of value in treating liver ailments (WAGNER, 1978) and feeding yet other flavones to these cells might conceivably yield new structures with enhanced pharmacological activities.

The ability of tissue cultures generally to take up and metabolise exogenously supplied phenolics has been discussed by BARZ (1977) in some detail. As is the case with whole plants, one of the first reactions is conjugation with glucose and phenolic acids supplied to tissue cultures are converted to the corresponding glucose esters. During feeding experiments it seems likely that such conjugation may prevent the incorporation of fed molecules into "new" biosynthetic routes. Undoubtedly, also, substances fed into cell cultures will eventually undergo turnover and degradation and many reactions of this type have been observed in such experiments. The catabolism of exogenously supplied phenolics is similar to that of endogenous substances, although sometimes polymerisation reactions are more in evidence.

Much use has been made of tissue cultures for biosynthetic studies in the assumption that the pathways followed are the same as in the plants from which the cultures have been obtained. Whether this assumption always holds true is open to question, since it has not often been deliberately tested. It does seem to be true for flavone synthesis in parsley cells and leaves, and also for anthraquinone synthesis in suspension cultures and seedlings of *Morinda citrifolia* (LEISTNER, 1973) but further comparisons could usefully be carried out in other cases.

6 Functions of Phenolics in Plants

6.1 Physiological Interactions

There is still considerable uncertainty as to whether phenolic compounds have a physiological role in plant growth and metabolism. Many phenols are clearly able to exert significant effects on growth processes when applied to plant tissues at physiological concentrations, but this does not necessarily imply they have an endogenous role. With so much variation in structure, it is unlikely that phenolics as a group of substances have one particular universal role in regulating growth and development. Rather, it is possible that individual classes or individual substances may possess significant activities in certain physiological processes. Many of the observed physiological interactions may be incidental to the wider function of phenolics as protective agents in plants (see SWAIN, 1977). The putative physiological role of phenols in plants has been discussed by a number of authors, including GALSTON (1969), GROSS (1975), KEFELI and KUTACEK (1977), MCCLURE (1975) and STENLID (1970). Only a brief summary of these physiological interactions need be given here.

Do phenolics have a role as endogenous hormones in the plant kingdom? From most considerations, the answer must be no for the common phenolics, and particularly for the flavonoids, since they do not have the requisite properties expected of hormonal material. Nevertheless, it is possible that individual phenols actually function in this way. A case has been made that lunularic acid is a dormin-like hormone in liverworts, replacing the more widespread abscisic acid which is absent from these plants (PRYCE, 1972b). Growth experiments with *Lunularia cruciata* support this hormonal role. The structural resemblance between lunularic and abscisic acids has already been pointed out. Whether other related stilbene-like structures such as the batatasins in *Dioscorea* have a role as dormancy hormones in higher plants is less certain, but further experiments may establish these dibenzyl derivatives as regulators in their own right.

Even if most phenolics are not hormones themselves, they may affect plant growth by interaction with one or other of the major classes of plant hormone such as the auxins. Much attention has been paid to the effect of phenolic compounds on indoleacetic acid oxidase, a peroxidase-type enzyme which is capable of auxin destruction. It is well established that, in vitro, flavonoids with a catechol B-ring have a sparing effect on IAA by inhibiting oxidase activity and hence theoretically they have a stimulating effect on plant growth. By contrast, flavonoids with a monohydroxy phenol B-ring uniformly augment enzyme activity and thus have a potentially inhibiting effect on growth (STENLID, 1968, 1976a). Related hydroxycinnamic acids have similar effects. How important this is in vivo has never been fully examined and indeed the question of whether IAA oxidase has a regulating role in auxin activity has not been satisfactorily answered. An in vitro effect of phenolics on the pathway of auxin biosynthesis from anthranilic acid and tryptophan has also been demonstrated, but again the in vivo significance of such interactions has yet to be established (KEFELI and KUTACEK, 1977).

Another point of hormone control that phenolics might affect is the biosynthesis of ethylene. Thus, it is known that a *p*-coumaric acid ester is a necessary co-factor for ethylene biosynthesis from methionine in cauliflower florets (MAPSON, 1970). It is a cofactor to a peroxidase-like enzyme on the pathway and it is interesting that caffeic acid can inhibit the enzyme activity. Thus, the balance of *p*-coumaric to caffeic acid at the site of the synthesis could theoretically provide a regulation of ethylene synthesis.

Phenolics may react with other hormones by synergism or inhibition and both situations have been recorded in the case of plant growth stimulated by gibberellic acid. There is evidence that dihydroconiferyl alcohol in lettuce has a synergistic effect on the GA_3-stimulated elongation of hypocotyl (KAMISAKA and SHIBATA, 1977); by contrast substitution of dihydroconiferyl alcohol by any one of several common hydroxycinnamic acids reverses this effect. It may be noted also that tannins have generally been shown in other plant systems to have an antagonist effect on GA_3 activity (CORCORAN et al., 1972).

In the above examples, phenolics appear to interact specifically with plant hormones to produce effects on growth. It is also clear that phenolics may have indirect effects on physiological processes, through more non-specific effects on intermediary metabolism. For example, many phenolics are capable of inhibiting ATP synthesis in mitochondria, of uncoupling respiration and of inhibiting ion absorption in roots (STENLID, 1970). Flavonoids may also affect the polar transport of auxins (STENLID, 1976b) and protoplasmic streaming in root hairs (POPOVICI and REZNIK, 1976). There are also a variety of enzymic activities which may be inhibited in the presence of compounds such as quercetin (VAN SUMERE et al., 1975). How significant these effects are in the normal growth pattern of the plant has yet to be determined.

The relatively recent discovery that certain phenolics, particularly caffeic acid esters and flavonoids, occur in plant chloroplasts (see e.g., SAUNDERS and MCCLURE, 1976) in small amounts raises the question of a further possible function in relationship to photosynthesis or to the effect of light on plant processes. Certainly, it is conceivable that phenolic constituents, because of their intense UV absorption, can provide protection from damaging UV radiation in the atmosphere and may be valuable in absorbing this radiation which might otherwise interfere with the more vital processes in the chloroplast. Their occurrence in the chloroplast may equally well be accidental and further studies are needed to see whether they are in fact of importance in the overall metabolism of these organelles.

6.2 Ecological Significance

While it has been difficult to assign a physiological role for the majority of plant phenolics (Sect 6.1), there is increasing evidence that a considerable number of these substances play an ecological role in plants (Table 7). The importance of flavonoid pigments (together with carotenoids) in contributing to flower and fruit colour for purposes of attracting animals to plants for pollination and seed dispersal was recognised by CHARLES DARWIN and by many naturalists before and since his time. The relationship between flavonoid structure and plant colour has recently been reviewed (HARBORNE, 1976b). The contribution of the coloured flavonoids in the cell vacuoles to flower colour is self-evident, but it has taken longer to recognise that the "colourless" flavones and flavonols universally present in flower tissues are essential as co-pigments to the anthocyanins and are also occasionally concerned as "hidden" UV honey guides for attracting bees to flowers.

Other ecological roles for phenolics (Table 7) are based on more recent observations in nature. A series of experiments in both field and laboratory have indicated during the last decade a role for a number of phenolic derivatives as allelopathic agents. These are chemicals excreted by the plant, which may be autotoxic or affect the growth of other plants in the environment (RICE, 1974). It has also been found that flavonoids, and especially tannins, have a role as feeding deterrents, protecting plants from overgrazing by many animal species (SWAIN, 1977; HARBORNE, 1977a).

Table 7. Ecological importance of phenolics in plants

Role	Phenolic class	Examples and plant source[a]
Flower pigments	Anthocyanins	Cyanidin 3,5-diglucoside in *Rosa*
	Chalcones	Coreopsin in *Coreopsis tinctoria*
	Aurones	Aureusin in *Antirrhinum majus*
	Yellow Flavonols	Gossypetin 7-glucoside in *Gossypium*
	Flavones	Apigenin 7-glucoside in *Bellis perennis*
Fruit pigments	Anthocyanins	Petunidin glycoside in *Atropa belladonna*
	Isoflavones	Osajin in *Maclura pomifera*
	Chalcones	Okanin in *Kyllingia brevifolia*
Allelopathic Agents	Quinones	Juglone in *Juglans regia*
	Phenols	Hydroquinone in *Arctostaphylos*
	Phenolic acids	Salicylic acid in *Quercus falcata*
	Hydroxycinnamic acids	Ferulic acid in *Adenostoma*
Feeding deterrents	Quinones	Juglone in *Carya ovata*
	Tannins	Gallotannins in *Quercus robur*
	Flavonols	Quercetin glycosides in *Gossypium*
Anti-fungal agents	Isoflavones	Luteone in *Lupinus*
	Phenolic acids	Protocatechuic acid in *Allium*
	Dihydrochalcones	Phloridzin in *Malus pumila*
Phytoalexins	Stilbenes	Resveratrol in *Arachis hypogaea*
	Phenanthrenes	Orchinol in *Orchis militaris*
	Isoflavans	Vestitol in *Lotus corniculatus*
	Pterocarpans	Pisatin in *Pisum sativum*
	Phenylpropanoids	Coniferyl alcohol in *Linum usitatissimum*
	Furocoumarins	Psoralen in *Petroselinum crispum*

[a] In most cases, many other examples could be quoted. References are given in the text or may be found in SWAIN (1977) and HARBORNE (1976b, 1977a)

Finally, yet another important function for certain classes of phenol is as antimicrobial agents, in providing resistance to various fungal, bacterial, and viral infections. Phenolics are significant here not only as preformed antifungal compounds (SCHÖNBECK and SCHLÖSSER, 1976) but also as phytoalexins formed post-infectionally. Phenolic phytoalexins of one type or another (Table 7) have been recognised in being formed on fungal infection in at least six different plant families (HARBORNE and INGHAM, 1978). See also the chapter on phytoalexins in Vol. 4 (pp. 636–652) of this Encyclopedia (1976).

The ecological importance of phenolics, together with that of other secondary constituents, has been reviewed at length elsewhere (see e.g., SWAIN, 1977 and references therein). A few selected examples are illustrated in Table 7; the structures of many of the compounds listed here may be found in earlier sections of this chapter. In conclusion, it may be pointed out that the ecological effectiveness of many phenolics may well lie in their ability to modify growth processes in other organisms through hormonal interactions. Thus progress in understanding the ecological functions of phenolics will undoubtedly illuminate at the same time our comprehension of their physiological importance in the plant.

References

Abe, H., Uchiyama, M., Sato, R.: Isolation of phenylacetic acid and its *p*-hydroxyderivative as auxin-like substances from *Undaria pinnatifida*. Agric. Biol. Chem. *38*, 897–898 (1974)

Arcamone, F., Camerino, B., Franceschi, G., Penco, S.: Structure and synthesis of new phenolic carotenoids from *Streptomyces mediolani*. Gazz. Chim. Ital. *100*, 581–590 (1970)

Ardenne, R., von: Bestimmung der Natur der Anthocyane in Gewebekulturen von *Happlopappus gracilis*. Z. Naturforsch. *20*b, 186–187 (1965)

Arihari, S., Ruedi, P., Eugster, C.H.: Dopaldehyde – isolation from a plant in the form of its (Z) enol (E) caffeic ester. Helv. Chim. Acta *58*, 447–453 (1975)

Asakawi, Y., Takemoto, T., Wollenweber, E., Aratani, T.: Lasiocarpin A, B and C, three novel phenolic triglycerides from *Populus lasiocarpa*. Phytochemistry *16*, 1791–1796 (1977)

Bailey, J.A.: Pisatin production by tissue cultures of *Pisum sativum*. J. Gen. Microbiol. *61*, 409–415 (1970)

Ball, E.A., Harborne, J.B., Arditti, J.: Anthocyanins of *Dimorphotheca*. I. Identity of pigments in flowers, stems and callus cultures. Am. J. Bot. *59*, 924–930 (1972)

Barz, W.: Catabolism of endogenous and exogenous compounds by plant cell cultures. In: Plant tissue culture and its bio-technological applications. Barz, W., Reinhard, E., Zenk, M.H. (eds.), pp. 153–171. Berlin, Heidelberg, New York: Springer 1977

Barz, W., Hösel, W.: Über den Umsatz von Flavonolen und Isoflavonen in *Cicer arietinum*. Phytochemistry *10*, 335–341 (1971)

Barz, W., Hösel, W.: Metabolism of flavonoids. In: The flavonoids. Harborne, J.B., Mabry, T.J., Mabry, H. (eds.), pp. 916–969. London: Chapman & Hall 1975

Barz, W., Reinhard, E., Zenk, M.H.: Plant tissue culture and its biotechnological application. Berlin, Heidelberg, New York: Springer 1977

Baskin, J.M., Ludlow, C.J., Harris, T.M., Wolf, F.T.: Psoralen, an inhibitor in the seeds of *Psoralea subacaulis*. Phytochemistry *6*, 1209–1213 (1967)

Bate-Smith, E.C.: Chemistry and phylogeny of the angiosperms. Nature (London). *236*, 353–4 (1972)

Bentley, R.: Biosynthesis of quinones. In: Biosynthesis, Geissmann, T.A. (eds.), Vol. III, pp. 181–246. London: The Chemical Society 1975

Berlin, J., Barz, W.: Stoffwechsel von Isoflavonen und Cumostanen in Zell- und Callussuspensions–Kulturen von *Phaseolus aureus*. Planta *98*, 300–314 (1971)

Bhat, S.V., Kalyanarame, P.S., Kohl, H., De Sousa, N.J., Fehlhaber, H.W.: Inuroyleanol and 7-ketoroyleanine, diterpenoids of *Inula royleana* roots. Tetrahedron *31*, 1001 (1975)

Billek, G., Kindl, H.: Über die phenolischen Inhaltsstoffe der Familie Saxifragaceae. Monatschr. Chem. *93*, 85–98 (1962)

Birkofer, L., Kaiser, C., Thomas, V.: Acteoside and neoacteoside, sugar esters of *Syringa vulgaris* flowers. Z. Naturforsch. *23*b, 1051–1059 (1968)

Bjeldanes, L.F., Chang, G.W.: Mutagenic activity of quercetin and related compounds. Science *197*, 577 (1977)

Bleichert, E., Ibrahim, R.K.: Flavonoids of *Parthenocissus* tissue culture. Experientia *30*, 104–105 (1974)

Bohm, B.A., Collins, F.W.: Flavonoids of *Philydrum lanuginosum*. Phytochemistry *14*, 315–316 (1975)

Boudet, A.M., Ranjeva, R., Gadal, P.: Propriétés allostériques specifiques des deux isoenzymes de la phenylalanine–ammoniaque lyase chez *Quercus pedunculata*. Phytochemistry *10*, 997–1005 (1971)

Bouillant, M.L., Wollenweber, E., Chopin, J.: Mise en evidence des hydroxy-2′ et dihydroxy-2′, 5 flavones dans la primevere *Primula florindae*. C.R. Acad. Sci. Paris Ser. D *273*, 1629–1631 (1971)

Brunet, G., Ibrahim, R.K.: Tissue culture of *Citrus* peel and its potential for flavonoid synthesis. Z. Pflanzenphysiol. *69*, 152–162 (1973)

Butcher, D.N.: Secondary products in tissue cultures. In: Applied and fundamental aspects of plant cell, tissue and organ culture. Reinert, J., Bajaj, Y.P.S. (eds.), pp. 668–693. Berlin, Heidelberg, New York: Springer 1977

Camm, E.L., Towers, G.H.N.: Phenylalanine ammonia lyase. Prog. Phytochem. *4*, 169–188 (1977)

Canononica, L., Manitto, P., Monti, D., Sanchez, M.: Biosynthesis of allylphenols in *Ocimum basilicum*. Chem. Commun. 1108 (1971)
Carpenter, I., Locksley, H.D., Scheinmann, F.: Xanthones in higher plants. Phytochemistry *8*, 2013–2026 (1969)
Castledine, R.M., Harborne, J.B.: Identification of the glycosylflavones of *Ephedra* and *Briza* by mass spectrometry of their permethyl ethers. Phytochemistry *15*, 803–804 (1976)
Chatterjee, A., Dhara, K.P., Rej, R.N., Ghosh, P.C.: Hexacosyl ferulate, a phenolic constituent of *Pinus roxburghii*. Phytochemistry *16*, 397–398 (1977)
Chopin, J., Bouillant, M.L.: *C*-Glycosylflavonoids. In: The flavonoids. Harborne, J.B., Mabry, T.J., Mabry, H. (eds.), pp. 632–691. London: Chapman & Hall 1975
Cooper, R., Levy, E.C., Lavie, D.: Novel germination inhibitors from *Aegilops ovata*. Chem. Commun. 794–795 (1977)
Cooperdriver, G., Swain, T.: Sulphate esters of caffeyl and *p*-coumaryl glucose in ferns. Phytochemistry *14*, 2506–2507 (1975)
Corcoran, M.R., Geissman, T.A., Phinney, B.O.: Tannins as gibberellin antagonists. Plant Physiol, *49*, 323–330 (1972)
Creasy, L.L., Swain, T.: Flavan production in strawberry leaves. Phytochemistry *5*, 501–509 (1966)
Crombie, L., Dewick, P.M., Whiting, D.A.: Biosynthesis of the rotenoid amorphigenin in germinating *Amorpha fruticosa* seeds. Chem. Commun 1182–1185 (1971)
Culberson, C.F.: Chemical and botanical guide to lichen products. Chapel Hill: Univ. North Carolina Press 1969
Dahmen, J., Leander, K., Rosenblom, J.: A new glycoside 2-glucosyloxy-4,5-dimethoxycinnamic acid from *Dendrobium densiflorum*. Acta Chem. Scand. *29*, 627 (1975)
Dean, F.M.: Naturally occurring oxygen ring compounds. London: Butterworth 1963
Dean, F.M., Costa, A.M., Harborne, J.B., Smith, D.M.: Leptodactylone, a yellow coumarin from *Leptodactylon* and *Linanthus* species. Phytochemistry *17*, 505–510 (1978)
Deverall, B.J.: Current perspectives in research on phytoalexins. In: Biochemical aspects of plant-parasite relationships. Friend, J., Threlfall, D.R. (eds.), pp. 208–224. London: Academic Press 1976
Dhillon, D.S., Brown, S.A.: Dimethylallyl pyrophosphate: umbelliferone dimethylallyl transferase from *Ruta graveolens*. Arch. Biochem. Biophys. *177*, 74–83 (1976)
Donnelly, D.M.X.: Neoflavanoids. In: The flavonoids. Harborne, J.B., Mabry, T.J., Mabry, H. (eds.), pp. 801–865. London: Chapman & Hall 1975
Dubois, J.A., Harborne, J.B.: Anthocyanin inheritance in petals of flax, *Linum usitatissimum*. Phytochemistry *14*, 2491–2494 (1975)
Dukhovlinova, L.I., Sklyar, Y.E., Sdobina, L.I.: Borneol *trans-p*-hydroxycinnamate from *Seseli mucronatum* and *S. asperulum*. Khim. Prir. Soedin. *11*, 99–100 (1975)
Durand, R., Zenk, M.H.: Plumbagin biosynthesis in *Plumbago europeaus*. Tetrahedron Lett. 3009–3010 (1971)
Ebel, J., Hahlbrock, K., Grisebach, H.: *O*-Dihydric phenol meta-*O*-methyl transferase from cell suspensions of parsley and its relation to flavonoid synthesis. Biochim. Biophys. Acta *268*, 313–320 (1972)
Ehmann, A.: *N-p*-coumaryltryptamine and *N*-ferulyltryptamine in kernels of *Zea mays*. Phytochemistry *13*, 1979–1984 (1974)
Ellis, B.E.: Degradation of aromatic compounds in plants. Lloydia *37*, 168–184 (1974)
Feucht, W., Nachit, M.: Flavolans and growth-promoting catechins in young shoot tips of *Prunus* species and hybrids. Plant Physiol. *40*, 230–234 (1977)
Fisch, M.H., Flick, B.H., Arditti, J.: Structure and antifungal activity of hircinol, loroglossol and orchinol. Phytochemistry *12*, 437–441 (1973)
Forrest, G.I.: Polyphenol metabolism of tissue cultures derived from the tea plant. Biochem. J. *113*, 765–772 (1969)
French, C.J., Vance, C.P., Towers, G.H.N.: Conversion of *p*-coumaric to *p*-hydroxybenzoic acid by cell free extracts of potato tubers and *Polyporus hispidus*. Phytochemistry *15*, 564–566 (1976)
Fritig, B., Hirth, L., Ourisson, G.: Biosynthesis of the coumarins: scopoletin formation in tobacco tissue cultures. Phytochemistry *9*, 1963–1975 (1970)
Fritsch, H., Grisebach, H.: Biosynthesis of cyanidin in cell cultures of *Haplopappus gracilis*. Phytochemistry *14*, 2437–2442 (1975)

Fröst, S., Harborne, J.B., King, L.: Identification of the flavonoids in five chemical races of cultivated barley. Hereditas *85*, 163–168 (1977)
Fujita, M., Inoue, T.: Biosynthesis of mangiferin in *Anemarrhena asphodeloides*. Tetrahedron Lett. 4503–4506 (1977)
Fujiwara, H., Nonaka, G., Yagi, A.: Flavonoids of the leaves of *Coptis japonica*. Chem. Pharm. Bull. *24*, 407–413 (1976)
Furuya, T., Kojima, H., Katsuta, T.: 3-Methylpurpurin and other anthraquinones from callus tissue of *Digitalis lanata*. Phytochemistry *11*, 1073–1076 (1972)
Galston, A.W.: Flavonoids and photomorphogenesis in peas. In: Perspectives in phytochemistry. Harborne, J.B., Swain, T. (eds.), pp. 193–204. London: Academic Press 1969
Gardner, P.D., Park, G.J., Albers, C.C.: α- and β-verbesinol, sesquiterpene alcohols of the *cis*-decalin series. J. Am. Chem. Soc. *83*, 1511–1512 (1961)
Gautier, J., Cave, A., Kunesch, G., Polonsky, J.: On the biosynthesis of neoflavonoids. Experientia *28*, 759–761 (1972)
Geiger, H., Quinn, C.: Biflavonoids. In: The flavonoids. Harborne, J.B., Mabry, T.J., Mabry, H. (eds.), pp. 692–742. London: Chapman & Hall 1975
Gestetner, B., Conn, E.E.: The 2-hydroxylation of *trans*-cinnamate by chloroplasts from *Melilotus alba*. Arch. Biochem. Biophys. *163*, 617–619 (1974)
Giannasi, D.E.: Flavonoid chemistry and evolution in *Dahlia*. Bull. Torrey Bot. Club *102*, 404–412 (1975)
Gorham, J.: Lunularic acid and related compounds in liverworts, algae and *Hydrangea*. Phytochemistry *16*, 249–253 (1977)
Gottlieb, O.R.: Chemosystematics of the Lauraceae. Phytochemistry *11*, 1537–1570 (1972)
Gottlieb, O.R.: Flavonols. In: The flavonoids. Harborne, J.B., Mabry, T.J., Mabry, H. (eds.), pp. 296–375. London: Chapman & Hall 1975
Gray, A.I., Waterman, P.G.: Coumarins in the Rutaceae. Phytochemistry *17*, 845–864 (1978)
Gray, R., Bonner, J.: An inhibitor of plant growth from the leaves of *Encelia farinosa*. Am. J. Bot. *34*, 52–57 (1948)
Griffiths, L.A., Barrow, A.: Metabolism of flavonoid compounds in germ-free rats. Biochem. J. *130*, 1161–1162 (1972)
Gross, D.: Growth regulating substances of plant origin. Phytochemistry *14*, 2105–2112 (1975)
Gross, G.G.: Recent advances in the chemistry and biochemistry of lignin. Recent Adv. Phytochem. *12*, 177–220 (1979)
Gupta, P., Lewis, J.R.: Biogenesis of xanthones in *Gentiana lutea*. J. Chem. Soc. C, 629–631 (1971)
Hagen, C.W.: Changes in pigments during development of buds in *Impatiens balsamina*. Am. J. Bot. *53*, 54–60 (1966)
Hahlbrock, K., Grisebach, H.: Biosynthesis of flavonoids. In: The flavonoids. Harborne, J.B., Mabry, T.J., Mabry, H. (eds.), pp. 866–915. London: Chapman & Hall 1975
Harborne, J.B.: Phenolic glycosides and their natural distribution. In: Biochemistry of phenolic compounds. Harborne, J.B. (eds.), pp. 129–170. London: Academic Press 1964a
Harborne, J.B.: Production of phenols in plant tissue culture. John Innes Annu. Rep. 45–46 (1964b)
Harborne, J.B.: Flavonols as yellow flower pigments. Phytochemistry *4*, 647–657 (1965)
Harborne, J.B.: Caffeic acid ester distribution in higher plants. Z. Naturforsch. *21*b, 604–605 (1966)
Harborne, J.B.: Comparative biochemistry of the flavonoids. London: Academic Press 1967a
Harborne, J.B.: Flavonoid patterns in the Bignoniaceae and the Gesneriaceae. Phytochemistry *6*, 1643–1651 (1967b)
Harborne, J.B.: Flavonoid and phenylpropanoid patterns in the Umbelliferae. Bot. J. Linn. Soc. *64*, Suppl. 1, 293–314 (1971)
Harborne, J.B.: A unique pattern of anthocyanins in *Daucus carota* and other Umbelliferae. Biochem. Syst. Ecol. *4*, 31–35 (1976a)
Harborne, J.B.: Functions of flavonoids in plants. In: Chemistry and biochemistry of plant pigments, 2nd ed. Goodwin, T.W. (ed.), Vol. I, pp. 736–779. London: Academic Press 1976b
Harborne, J.B.: Introduction to ecological biochemistry. London: Academic Press 1977a

Harborne, J.B.: Flavonoid sulphates – a new class of natural product of ecological significance in plants. Prog. Phytochem. *4*, 189–208 (1977b)
Harborne, J.B.: Phenolic compounds derived from shikimate. In: Biosynthesis. Bu'Lock, J.D. (ed.), Vol. 5, pp. 34–55. London: The Chemical Society 1977c
Harborne, J.B.: Flavonoid profiles in the Compositae. In: The biology and chemistry of the Compositae. Heywood, V.H., Harborne, J.B., Turner, B.L., (eds.), pp. 341–366. London: Academic Press 1978a
Harborne, J.B.: The rare flavone isoetin as a yellow flower pigment in *Heywoodiella oligocephala* and in other Cichorieae. Phytochemistry *17*, 915–917 (1978b)
Harborne, J.B., Ingham, J.L.: Biochemical aspects of the co-evolution of higher plants with their fungal parasites. In: Biochemical aspects of plant and animal co-evolution. Harborne, J.B. (ed.), pp. 343–405. London: Academic Press 1978
Harborne, J.B., Mokhtari, N.: Two sulphated anthraquinone derivatives in *Rumex pulcher*. Phytochemistry *16*, 1314–1315 (1977)
Harborne, J.B., Simmonds, N.W.: Potato cell and tissue culture. John Innes Inst. 53rd Annu. Rep. 29–30 (1962)
Harborne, J.B., Williams, C.A.: The identification of orcinol in higher plants in the family Ericaceae. Phytochemistry *8*, 2223–2226 (1969)
Harborne, J.B., Williams, C.A.: A chemotaxonomic survey of flavonoids and simple phenols in leaves of the Ericaceae. Bot J. Linn. Soc. *66*, 37–54 (1973)
Harborne, J.B., Williams, C.A.: Flavone and flavonol glycosides. In: The flavonoids. Harborne, J.B., Mabry, T.J., Mabry, H. (eds.), pp. 376–441. London: Chapman & Hall 1975
Harborne, J.B., Williams, C.A.: Sulphated flavones and caffeic acid esters in members of the Fluviales. Biochem. Syst. Ecol. *4*, 37–41 (1976)
Harborne, J.B., Heywood, V.H., Williams, C.A.: Distribution of myristicin in seeds of the Umbelliferae. Phytochemistry *8*, 1729–1732 (1969)
Harborne, J.B., Williams, C.A., Smith, D.M.: Species-specific kaempferol derivatives in ferns of the Appalachian *Asplenium* complex. Biochem. Syst. Ecol. *1*, 51–54 (1973)
Harborne, J.B., Mabry, T.J., Mabry, H. (eds.).: The flavonoids. London: Chapman & Hall 1975
Harborne, J.B., Ingham, J.L., King, L., Payne, M.: The isopentenyl isoflavone luteone as preinfectional antifungal agent in the genus *Lupinus*. Phytochemistry *15*, 1485–1487 (1976)
Hartley, R.D., Jones, E.C.: Diferulic acid as a component of cell walls of *Lolium multiflorum*. Phytochemistry *16*, 1157–1160 (1976)
Hashimoto, T., Hasegawa, K., Yamaguchi, H., Saito, M., Ishimoto, S.: Structure and synthesis of batatasins, dormancy inducing substances of yam bulbils. Phytochemistry *13*, 2849–2852 (1974)
Haslam, E.: The shikimate pathway. New York: John Wiley & Sons 1974
Haslam, E.: Natural proanthocyanidins. In: The flavonoids. Harborne, J.B., Mabry, T.J., Mabry, H. (eds.), pp. 505–550. London: Chapman & Hall 1975
Hathway, D.E.: The stilbenes. In: Wood extractives. Hillis, W.E. (ed.), pp. 159–188. New York: Academic Press 1962
Heble, M.R., Narayanaswamy, S., Chadha, M.S.: Tissue differentiation and plumbagin synthesis in variant cell strains of *Plumbago zeylanica*. Plant Sci. Lett. *2*, 405–409 (1974)
Heywood, V.H. (ed.): The biology and chemistry of the Umbelliferae. London: Academic Press 1971
Hostettmann, K., Wagner, H.: Xanthone glycosides. Phytochemistry *16*, 821–830 (1977)
Hrazdina, G., Kreuzaler, F., Hahlbrock, K., Grisebach, H.: Substrate specificity of flavanone synthase from cell suspension cultures of parsley. Arch. Biochem. Biophys. *175*, 392–399 (1976)
Hrazdina, G., Wagner, G.J., Siegelman, H.W.: Subcellular localization of enzymes of anthocyanin biosynthesis in protoplasts. Phytochemistry *17*, 53–56 (1978)
Ibrahim, R.K., Towers, G.H.N., Gibbs, R.D.: Syringic and sinapic acids as indicators of differences between major groups of vascular plants. J. Linn. Soc. Bot. *58*, 223–230 (1962)
Ibrahim, R.K., Thakur, M.L., Permanand, B.: Formation of anthocyanins in callus tissue cultures. Lloydia *34*, 175–182 (1971)
Imperato, F.: 1-(3-glucosyloxy-4-hydroxycinnamyl)glucose from *Lantana hybrida*. Phytochemistry *15*, 1786 (1976)

Imperato, F., Dileo, C., Trovato, P.: 1-Caffeylrhamnose from *Lantana hybrida*. Phytochemistry *14*, 2725 (1975)
Ingram, D.S.: Applications in plant pathology. In: Plant tissue and cell culture, 2nd ed. Street, A.E. (ed.), pp. 463–500. Oxford: Blackwell Scientific Pub 1977
Jalal, M.A.F., Collin, H.A.: Polyphenols of mature plant, seedling and tissue cultures of *Theobroma cacao*. Phytochemistry *16*, 1377–1380 (1977)
Kamisaka, S., Shibata, K.: Anticotyledon factor: competitive inhibitor of dihydroconiferyl alcohol in stimulating GA_3 induced lettuce hypocotyl elongation. Plant Cell Physiol. *18*, 1057–1066 (1977)
Kaul, B., Staba, E.J.: *Ammi visnaga* tissue cultures: multilitre suspension growth and examination for furanochromones. Planta Med. *15*, 145–156 (1967)
Keen, N.T., Ingham, J.L.: New stilbene phytoalexins from American cultivars of *Arachis hypogaea*. Phytochemistry *15*, 1794–1795 (1976)
Kefeli, V.I., Kutacek, M.: Phenolic substances and their possible role in plant growth regulation. In: Plant growth regulation. Pilet, P.E. (ed.), pp. 181–188. Berlin, Heidelberg, New York: Springer 1977
Klischies, M., Stockigt, J., Zenk, M.H.: Biosynthesis of the allylphenols eugenol and methyleugenol in *Ocimum basilicum*. Chem. Commun. 879–880 (1975)
Koukol, J., Conn, E.E.: Purification and properties of the phenylalanine deaminase of *Hordeum vulgare*. J. Biol. Chem. *236*, 2692–2698 (1961)
Kreuzaler, F., Hahlbrock, K.: Flavonoid glycosides from illuminated cell suspension cultures of *Petroselinum crispum*. Phytochemistry *12*, 1149–1153 (1973)
Langcake, P., Pryce, R.J.: A new class of phytoalexins from grapevines. Experientia *33*, 151 (1977)
Leistner, E.: A second pathway leading to anthraquinones in higher plants. Phytochemistry *10*, 3015–3020 (1971)
Leistner, E.: Biosynthesis of morindone and alizarin in intact plants and cell suspension cultures of *Morinda citrifolia*. Phytochemistry *12*, 1669–1674 (1973)
Leistner, E., Zenk, M.H.: Biosynthesis of alizarin in *Rubia tinctorum*. Tetrahedron Lett. 1395–1398 (1968)
Letcher, R.M.: Structure and synthesis of the growth inhibitor batatasin I from *Dioscorea batatas*. Phytochemistry *12*, 2789–2790 (1973)
Letcher, R.M., Nhamo, L.R.M., Gumiro, I.T.: Substituted phenanthrenes and 9,10-dihydrophenanthrenes and a substituted bibenzyl from heartwood of *Combretum molle*. J. Chem. Soc. Perkin *I*, 206–210 (1972)
Mansell, R.L., Gross, G.G., Stöckigt, J., Franke, H., Zenk, M.H.: Purification and properties of cinnamyl alcohol dehydrogenase from higher plants involved in lignin biosynthesis. Phytochemistry, *13*, 2427–2436 (1974)
Mapson, L.W.: Biosynthesis of ethylene and the ripening of fruit. Endeavour *39*, 29–33 (1970)
Margna, V.: Control at the level of substrate supply – an alternative in the regulation of phenylpropanoid accumulation in plant cells. Phytochemistry *16*, 419–426 (1977)
McClure, J.W.: Photocontrol of *Spirodela intermedia* flavonoids. Plant Physiol. *43*, 193–200 (1968)
McClure, J.: Physiology and functions of flavonoids. In: The flavonoids. Harborne, J.B., Mabry, T.J., Mabry, H. (eds.), pp. 970–1055. London: Chapman & Hall 1975
Mechoulam, R.: Marihuana chemistry. Science *168*, 1159–1166 (1970)
Mizukami, H., Konoshima, M., Tabata, M.: Variation in pigment production in *Lithospermum erythrorhizon* callus cultures. Phytochemistry *17*, 95–97 (1978)
Mizusaki, S., Tanabe, T., Noguichi, M., Tamaki, E.: *p*-Coumarylputrescine, caffeylputrescine and ferulylputrescine from callus tissue culture of *Nicotiana tabacum*. Phytochemistry *10*, 1347–1350 (1971)
Momose, T., Abe, K., Yoshitama, K.: 5-*O*-Methylcyanidin 3-glucoside from leaves of *Egeria densa*. Phytochemistry *16*, 1321 (1977)
Morita, N., Arisawa, M., Yoshikawa, A.: Glycoflavones in leaves of *Polygonatum odoratum*. Yakugaku Zasshi *96*, 1180–1183 (1976)
Murakoshi, I., Kakegawa, F., Toriizuka, K., Haginiwa, J., Ohmiya, S., Otomasu, H.: 4′-Rhamnosyloxycinnamyl-lupinine, a new lupin alkaloid in *Lupinus luteus*. Phytochemistry *16*, 2046–2047 (1977)

Nair, P.M., Vining, L.C.: Cinnamic acid hydroxylase in spinach. Phytochemistry *4*, 161–168 (1965)

Nakamura, Y., Fushiki, H., Higuchi, T.: Metabolic differences between gymnosperms and angiosperms in the formation of syringyl lignin. Phytochemistry *13*, 1777–1784 (1974)

Nicolaus, R.A., Piatelli, M.: Progress in the chemistry of natural black pigments. Rend. Accad. Sci. Fis. Mat. Naples *32*, 1–17 (1965)

Nimz, H., Ebel, J., Grisebach, H.: The structure of lignin from soybean cell suspension cultures. Z. Naturforsch. *30*c, 442–444 (1975)

Österdahl, B.G., Lindberg, G.: Luteolin 7-neohesperidoside-4′-sophoroside, another new tetraglycoside from *Hedwigia ciliata*. Acta Chem. Scand. *31*B, 293–296 (1977)

Panizzi, L., Scarpati, O.: 1,4- and 1,5-Dicaffeylquinic acids. Gazz. Chim. Ital. *95*, 71–82 (1965)

Paul, S.M., Skolnick, P.: Catechol oestrogens inhibit oestrogen-elicited accumulations of hypothalamic cyclic AMP suggesting a role as endogenous anti-oestrogens. Nature (London) *266*, 559–561 (1977)

Pendse, R., Rao, A.V.R., Venkataraman, K.: 5,7-Dihydroxychromone from *Arachis hypogaea* shells. Phytochemistry *12*, 2033–2034 (1973)

Phillips, I.D.J.: Some interactions of GA_3 with naringenin in the control of dormancy and growth in plants. J. Exp. Bot. *13*, 213–226 (1962)

Popovici, G., Reznik, H.: The effect of some flavonoids on protoplasmic streaming in oat. Z. Pflanzenphysiol. *80*, 417–425 (1976)

Popovici, G., Weissenbock, G., Bouillant, M.L., Dellamonica, G., Chopin, J.: Flavonoids from *Avena sativa*. Z. Pflanzenphysiol *85*, 103–115 (1977)

Poulton, J., Grisebach, H., Ebel, J., Hekelek, B.A., Hahlbrock, K.: Two distinct 3-*O*-methyltransferases in soyabean cell suspension cultures. Arch Biochem. Biophys. *173*, 301–305 (1976)

Pryce, R.J.: Metabolism of lunularic acid to a new plant stilbene by *Lunularia cruciata*. Phytochemistry *11*, 1355–1364 (1972a)

Pryce, R.J.: The occurrence of lunularic and abscisic acids in plants. Phytochemistry *11*, 1759–1762 (1972b)

Pryce, R.J.: Gallic acid as a natural inhibitor of flowering in *Kalanchoe blossfeldiana*. Phytochemistry *11*, 1911–1918 (1972c)

Racker, E.: Reconstitution, mechanism of action and control of ion pumps. Biochem. Soc. Trans. *3*, 785–802 (1975)

Razzaque, A., Ellis, B.E.: Rosmarinic acid production in *Coleus* cell cultures. Planta *137*, 287–291 (1977)

Reisch, J., Bathory, M., Szendrei, K., Novak, I., Minker, E.: Weitere Phenanthrene aus dem Rhizom von *Tamus communis*. Phytochemistry *12*, 228–229 (1973)

Rice, E.L.: Allelopathy. New York: Academic Press 1974

Russell, D.W., Conn, E.E.: The cinnamic acid 4-hydroxylase of pea seedlings. Arch. Biochem. Biophys. *122*, 256–258, (1967)

Saleh, N.A.M., Fritsch, H., Kreuzaler, F., Grisebach, H.: Flavanone synthase from cell suspension cultures of *Happlopappus gracilis* and comparison with the synthase from parsley. Phytochemistry *17*, 183–186 (1978)

Saunders, J.A., McClure, J.W.: The localization of flavonoids and phenolic enzymes in barley plastids. Am. J. Bot. *59*, 673 (1972)

Saunders, J.A., McClure, J.W.: The distribution of flavonoids in chloroplasts of 25 species of vascular plants. Phytochemistry *15*, 809–810 (1976)

Saylor, M.H., Mansell, R.L.: Hydroxycinnamyl: CoA transferase in the biosynthesis of kaempferol 3-*p*-coumaryltriglucoside in *Pisum sativum*. Z. Naturforsch. *32*C, 765–768 (1977)

Schill, L., Grisebach, H.: Properties of a phenolase preparation from cell suspension cultures of parsley. Hoppe Seyler's Z. Physiol. Chem. *354*, 1555–1562 (1973)

Schmitz, M., Seitz, U.: Hemmung der Anthocyansynthase durch Gibberellinsäure A_3 bei Kalluskulturen von *Daucus carota*. Z. Pflanzenzuecht. *68*, 259–265 (1972)

Schönbeck, F., Schlösser, E.: Preformed substances as potential protectants. In: Physiological plant pathology. Heitefuss, R., Williams, P.H. (eds.), pp. 653–678. Berlin, Heidelberg, New York: Springer 1976

Schrall, R., Becker, H.: Tissue cultures of *Silybum marianum*. II flavonolignans formed from flavonoids and coniferyl alcohol. Planta Med. *32*, 27–32 (1977)
Shulgin, A.T.: Possible implications of myristicin as a psychotropic substance. Nature (London) *210*, 380–384 (1966)
Shutt, D.A.: The effects of plant oestrogens on animal reproduction. Endeavour *35*, 110–113 (1976)
Sikurai, N., Shibata, K., Kakisaka, S.: Stimulation of cucumber hypocotyl elongation by dihydroconiferyl alcohol. Plant Cell Physiol. *15*, 709–716 (1974)
Smith, H.: The photocontrol of flavonoid biosynthesis. In: Phytochrome. Mitrakos, K., Shropshire, W., (eds.), pp. 433–484. London: Academic Press 1972
Smith, H., Billett, F.E., Giles, A.B.: The photocontrol of gene expression in higher plants. In: Regulation of enzyme synthesis and activity in higher plants. Smith, H. (ed.), pp. 93–128. London: Academic Press 1977
Smith, T.A.: Recent advances in the biochemistry of plant amines. Prog. Phytochem. *4*, 27–82 (1977)
Statham, C.M., Crowden, R.K.: Anthocyanin biosynthesis in *Pisum* – sequence studies in pigment production. Phytochemistry *13*, 1835–1840 (1974)
Steck, W., Bailey, B.K., Shyluk, J.P., Gamborg, O.L.: Coumarins and alkaloids from cell cultures of *Ruta graveolens*. Phytochemistry *10*, 191–198 (1971)
Steiner, A.M.: Turnover of anthocyanin 3-monoglucosides in petals of *Petunia hybrida*. Z. Pflanzenphysiol. *65*, 210–222 (1971)
Stenlid, G.: The physiological effects of phloidzin, phloretin and related substances upon higher plants. Physiol. Plant *21*, 882–894 (1968)
Stenlid, G.: Flavonoids as inhibitors of the formation of adenosine triphosphate in plant mitochondria. Phytochemistry *9*, 2251–2256 (1970)
Stenlid, G.: Effects of substituents in the A-ring on the physiological activity of flavones. Phytochemistry *15*, 911–912 (1976a)
Stenlid, G.: Effects of flavonoids on the polar transport of auxins. Physiol. Plant. *38*, 262–266 (1976b)
Steward, F.C.: Interpretations of the growth from free cell to carrot plant. Am. J. Bot. *45*, 709–713 (1958)
Stickland, R.G., Sunderland, N.: Production of anthocyanins, flavonols and chlorogenic acids by cultured callus tissues of *Happlopappus gracilis*. Ann. Bot. *36*, 443–457 (1972)
Stocker, M., Pohl, R.: Postmortale Bildung von 5,7-Dihydroxychromon-7-rutinosid in *Mentha longifolia*. Phytochemistry *16*, 571–572 (1976)
Street, H.E. (ed.): Plant tissue and cell culture, 2nd ed. Oxford: Blackwell Scientific Pub. 1977
Sutter, A., Poulton, J., Grisebach, H.: Oxidation of flavanone to flavone with cell free extracts from young parsley leaves. Arch. Biochem. Biophys. *170*, 547–555 (1975)
Swain, T.: The tannins. In: Plant biochemistry, 2nd ed. Bonner, J., Varner, J.E. (eds.), pp. 552–580. New York: Academic Press 1965
Swain, T.: Secondary compounds as protective agents. Annu. Rev. Plant Physiol. *28*, 479–501 (1977)
Swain, T., Bate-Smith, E.C.: Flavonoid compounds. In: Comparative biochemistry. Florkin, M., Mason, H.S. (eds.), Vol. III, pp. 755–809. New York: Academic Press 1962
Swain, T., Cooper-Driver, G.: Biochemical systematics in the Filicopsida. Bot. J. Linn. Soc. Suppl. 1, *67*, 111–134 (1973)
Tabata, M.: Recent advances in the production of medicinal substances by plant cell cultures. In: Plant tissue culture and its bio-technological application. Barz, W., Reinhard, E., Zenk, M.H. (eds.), pp. 3–16. Berlin, Heidelberg, New York: Springer 1977
Tabata, M., Hiraokan, N., Ikenoue, M., Sano, Y., Konoshima, M.: The production of anthraquinones in callus cultures of *Cassia tora*. Lloydia *38*, 131–134 (1975)
Takata, A., Scheuer, P.J.: Glyceryl esters of caffeic and *p*-coumaric from pineapple stems. Lloydia *39*, 409–411 (1976)
Tanguy, J.M., Cabanne, F., Perdrizet, E., Martin, C.: The occurrence of aromatic amides in various flowering plants. Phytochemistry *17*, 1927–1928 (1978)
Terashima, N., Mori, I., Kanda, T.: Biosynthesis of *p*-hydroxybenzoic acid in poplar lignin. Phytochemistry *14*, 1991–1992 (1975)

Thakur, M.L., Ibrahim, R.K.: Biogenesis of flavonoids in flax seedlings. Z. Pflanzenphysiol. *71*, 391–397 (1974)

Thompson, W.R., Meinwald, J., Aneshansley, D., Eisner, T.: Flavonol pigments responsible for UV Absorption in nectar guide of flowers. Science *177*, 528 (1972)

Thomson, R.H.: Naturally occurring quinones. London: Academic Press 1971

Thomson, R.H.: Quinones, nature, distribution and biosynthesis. In: Chemistry and biochemistry of plant pigments, 2nd ed. Goodwin, T.W. (ed.), Vol. I, pp. 527–559. London: Academic Press 1976

Thomson, R.H. (ed.): Marine natural products. Oxford: Pergamon Press 1977

Thomson, R.H.: Recent advances in the chemistry and biochemistry of quinone pigments. Recent. Adv. Phytochem. *12*, 287–312 (1979)

Timberlake, C.F., Bridle, P.: Anthocyanins. In: The flavonoids. Harborne, J.B., Mabry, T.J., Mabry, H. (eds.), pp. 214–266. London: Chapman & Hall 1975

Towers, G.H.N., Tso, A., Maass, W.S.G.: Phenolic acids and phenolic glycosides of *Gaultheria* species. Phytochemistry *5*, 677–681 (1966)

Turner, W.B.: Fungal metabolites. London: Academic Press 1971

Van Sumere, C.F., Cottenie, J., De Greef, J., Kint, J.: Biochemical studies in relation to the possible germination regulatory role of naturally occurring coumarin and phenolics. Recent Adv. Phytochem. *4*, 165–221 (1972)

Van Sumere, C.F., De Pooter, H., Ali, H., Van Bussel, M.D.: *N*-Ferulylglycylphenylalanine – a sequence in barley proteins. Phytochemistry *12*, 407–411 (1973)

Van Sumere, C.F., Albrecht, J., Dedonder, A., De Pooter, H., Pé, I.: Plant proteins and phenolics. In: The Chemistry and Biochemistry of Plant Proteins (Harborne, J.B. ed.), pp. 211–264. London: Academic Press 1975

Vaughan, P.F.T., Butt, V.S., Grisebach, H., Schill, L.: Hydroxylation of flavonoids by a phenolase preparation from leaves of spinach beet. Phytochemistry *8*, 1373–1378 (1969)

Venkataraman, K.: Flavones. In: The flavonoids. Harborne, J.B., Mabry, T.J., Mabry, H. (eds.), pp. 267–295. London: Chapman & Hall 1975

Wagner, H.: Pharmacology of plant phenolics. Recent Adv. Phytochem. *12*, 589–616 (1979)

Wagner, H., Bladt, S.: Cumarine aus südafrikanischen *Pelargonium*arten. Phytochemistry *14*, 2061–2064 (1975)

Weinges, K., Kunstler, K., Schilling, G., Jaggy, H.: 10-Cinnamyl and 10-*p*-hydroxy-cinnamylcatalpol from *Scutellaria altissima*. Liebig's Ann. 2190–2194 (1975)

Williams, C.A.: The leaf flavonoids of the Orchidaceae. Phytochemistry *18*, 803–814 (1979)

Williams, C.A., Harborne, J.B.: Luteolin and daphnetin derivatives in the Juncaceae and their systematic significance. Biochem. Syst. Ecol. *3*, 181–190 (1975)

Williams, C.A., Harborne, J.B.: The leaf flavonoids of the Zingiberales. Biochem. Syst. Ecol. *5*, 221–229 (1977)

Williams, C.A., Harborne, J.B., Crosby, T.S.: Tricetin, diosmetin and luteolin sulphates in leaves of *Lachenalia unifolia*. Phytochemistry *15*, 349–350 (1976)

Williams, R.T.: Metabolism of phenolics in animals. In: The biochemistry of phenolic compounds. Harborne, J.B. (ed.), pp. 205–248. London: Academic Press 1964

Wollenweber, E.: Rare methoxyflavonoids from buds of *Betula nigra*. Phytochemistry *15*, 438–439 (1976)

Yoshihara, T., Yoshikawa, H., Kunimatsu, S.: New amino acid derivatives of caffeic acid and DOPA from red clover. Agric. Biol. Chem. *41*, 1679–1684 (1977)

Yoshitama, K.: An acylated delphinidin 3-rutinoside-5,3′,5′-triglucoside from *Lobelia erinus*. Phytochemistry *16*, 1857–1858 (1977)

Zenk, M.H.: Recent work on Cinnamyl CoA derivatives. Recent Adv. Phytochem. *12*, 139–176 (1979)

Zenk, M.H., El-Shage, H., Schulte, U.: Anthraquinone production by cell suspension cultures of *Morinda citrifolia*. Planta Med. Suppl. 79–101 (1975)

7 Non-Protein Amino Acids in Plants

E.A. BELL

1 Introduction

The term non-protein amino acid is used here to include those naturally occurring amino acids, amino acid amides and imino acids which are not usually found as protein constituents. A few non-protein amino acids such as homoserine, ornithine, and saccharopine may be involved in the primary metabolic pathways of plants, but the great majority of these compounds can be fairly regarded as secondary products.

Most of our knowledge of non-protein amino acids dates from 1948 when DENT first applied paper chromatography to the separation of amino acids. This technique and later developments such as thin-layer chromatography, ion-exchange chromatography and high-voltage paper electrophoresis have made it possible to identify "known" amino acids and detect "unknown" amino acids in plant extracts with relative ease. Almost all of the "non-protein" amino acids which have been isolated from plants (and there are over 220 of them) were first detected as ninhydrin-reacting spots occupying unusual positions on chromatograms and/or electrophoresis papers.

Most of the non-protein amino acids of higher plants occur in the free state or as condensation compounds of low molecular weight such as the γ-glutamyl, oxalyl, and acetyl derivatives which can be readily extracted with water or 80% ethanol. In the fungi, however, non-protein amino acids are known to occur as constituents of small polypeptides such as the cyclic phalloidins and amanitins which are responsible for the toxicity of certain species of the mushroom genus *Amanita* (WIELAND, 1968) and islanditoxin, the toxin produced by the mould *Penicillium islandicum* which infects rice (BAMBURG et al., 1969).

A number of reviews concerned with non-protein amino acids in plants have appeared in recent years (FOWDEN, 1970, 1974; BELL, 1976; BELL, 1977) and these have dealt with the biosynthesis, distribution and the possible roles of these compounds in the plants which synthesise them.

While it is clearly not possible to present detailed information on all the non-protein amino acids which have been isolated from plants in a single chapter, I have tried to draw the reader's attention to both their chemical variety and their biological interest.

2 Chemistry and Occurrence

The great majority of non-protein amino acids are simple molecules, most of which contain fewer than twelve carbon atoms. In spite of this limitation of molecular size the group includes compounds showing a wide range of structural types. It is probably easiest to remember the structural types if we can relate them wherever possible to the more familiar protein amino acids and I have sub-divided the group on this basis into neutral aliphatic acids (cf. alanine and serine), sulphur-containing amino acids (cf. cysteine), aromatic amino acids (cf. phenylalanine), heterocyclic amino acids (cf. histidine), acidic amino acids and amides (cf. glutamic acid and asparagine), basic amino acids (cf. lysine), and imino acids (cf. proline). Arbitrary decisions have been made when a non-protein amino acid can be logically placed in more than one category. I have chosen for example to place 3-carboxy-phenylalanine with the aromatic rather than the acidic amino acids.

2.1 Neutral Aliphatic Amino Acids

While 2-aminobutyric acid (I), which is found in *Salvia officinalis* (BRIESKORN and GLASZ, 1964), homoisoleucine (II), which is found in *Aesculus californica* (FOWDEN and SMITH, 1968) and 2-aminooctanoic acid (III), a constituent of a fungal peptide (STARON et al., 1965), are characterised by saturated hydrocarbon chains, most of the neutral aliphatic amino acids are hydroxylated, unsaturated, or both. The legumes have proved a rich source of non-protein amino acids and hydroxyamino acids isolated from species of this family include 4-hydroxyisoleucine (IV) from *Trigonella foenum-graecum* (FOWDEN et al., 1973), 5-hydroxynorleucine (V) from *Crotalaria juncea* (PANT and FALES, 1974), 4-hydroxynorvaline (VI) from *Lathyrus odoratus* (FOWDEN, 1966a) and 5-hydroxynorvaline (VII) from *Canavalia ensiformis* (THOMPSON et al., 1964). It may well be, however, that the apparent richness of the Leguminosae merely reflects a greater interest shown by phytochem-

CH_3–CH_2–CH(NH_2)–CO_2H
2–Aminobutyric acid
(I)

CH_3–CH_2–CH(CH_3)–CH_2–CH(NH_2)–CO_2H
Homoisoleucine
(II)

CH_3–CH_2–CH_2–CH_2–CH_2–CH_2–CH(NH_2)–CO_2H
2–Aminooctanoic acid
(III)

CH_3 – CH(OH) – CH(CH_3) – CH(NH_2) – CO_2H
4–Hydroxyisoleucine
(IV)

CH_3–CH(OH)–CH_2–CH_2–CH(NH_2)–CO_2H
5–Hydroxynorleucine
(V)

CH_3 – CH(OH) – CH_2CH(NH_2) – CO_2H
4–Hydroxynorvaline
(VI)

HOCH_2–CH_2–CH_2–CH(NH_2)–CO_2H
5–Hydroxynorvaline
(VII)

CH_3–CH=C(CH_3)–CH_2–CH(NH_2)–CO_2H
2–Amino–4–methylhex–4–enoic acid
(VIII)

$HOCH_2-CH=C(CH_3)-CH_2-CH(NH_2)-CO_2H$

2-Amino-6-hydroxy-4-methylhex-4-enoic acid

(IX)

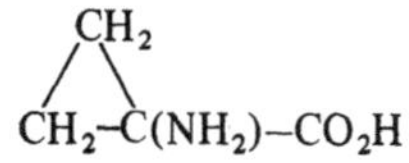

1-Aminocyclopropane-1-carboxylic acid

(X)

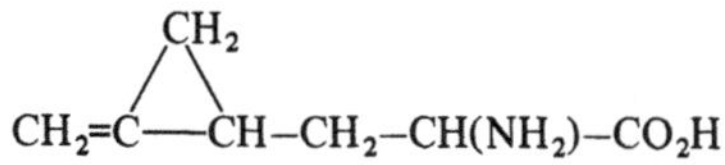

3-(Methylenecyclopropyl)alanine [Hypoglycine A]

(XI)

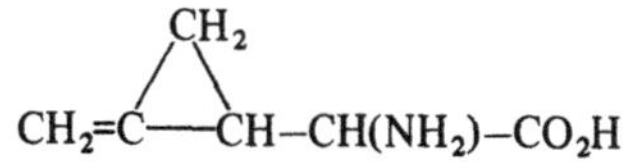

2-(Methylenecyclopropyl)glycine

(XII)

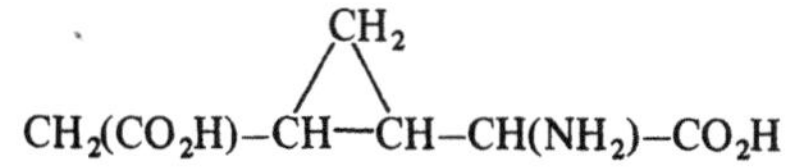

2-(2-Carboxymethylcyclopropyl)glycine

(XIII)

$CH_3-C(CH_2OH)=CH-CH_2-CH(NH_2)-CO_2H$

2-Amino-5-methyl-6-hydroxyhex-4-enoic acid

(XIV)

ists in this economically important family and it is worth noting that *Aesculus californica* (Sapindaceae), as well as being the source of homoisoleucine, has also been found to contain the unsaturated amino acid 2-amino-4-methylhex-4-enoic acid (VIII) and the corresponding 6-hydroxy compound (IX) (FOWDEN and SMITH, 1968). Neutral amino acids containing a cyclopropane ring include 1-aminocyclopropane-1-carboxylic acid (X) from pears and apples (BURROUGHS, 1957), 3-(methylenecyclopropyl)alanine (hypoglycine A) (XI) from *Blighia sapida* (HASSALL and REYLE, 1955), 2-(methylenecyclopropyl)glycine (XII) from *Litchi chinensis* (GRAY and FOWDEN, 1962a) and 2-(2-carboxymethylcyclopropyl)glycine (XIII) from *Blighia unijugata* (FOWDEN et al., 1972). This last species also contains L-2-amino-5-methyl-6-hydroxyhex-4-enoic acid (XIV, FOWDEN et al., 1972). Cyclopent-2-enyl-1-glycine (XV) is another naturally occurring alicyclic amino acid and this has been found in *Hydnocarpus anthelminthica* (CRAMER and SPENER, 1977), while 3-cyanoalanine (XVI), the only cyano amino acid yet found in plants, has been isolated from *Vicia sativa* (RESSLER, 1962). O-Acyl derivatives of neutral hydroxyamino acids are not infrequently met with in plants, the O-acetyl derivative of homoserine (XVII) occurs in *Pisum sativum* for example (GROBBELAAR and STEWARD, 1958) and the corresponding O-oxalyl derivative (XVIII) in *Lathyrus sativus* (PRZYBYLSKA and PAWELKIEWICZ, 1965). N-Methylated derivatives of both protein and non-protein amino acids also occur, N-methylalanine (XIX) and N-methylserine (XX) have been isolated from *Dichapetalum* species (ELOFF and GROBBELAAR, 1967, 1969) and N-(2-hydroxyethyl)glycine (XXI) has been identified in the alga *Petalonia fascia* (TAKAGI et al., 1970).

The fungi have yielded some 20 hydroxylated and unsaturated neutral amino acids, usually containing 5, 6, and 7 carbon atoms. From the peptides of *Amanita* species have been obtained 4-hydroxyleucine (XXII), 4,5-dihydroxyleucine (XXIII, WIELAND and SCHÖN, 1955), 3-methyl-4,5-dihydroxyisoleucine (XXIV, WIELAND, 1968), 3-methylene-L-norleucine (XXV, VERVIER and CASIMIR, 1970) and 2-amino-4,5-hexadienoic acid (XXVI, CHILTON et al., 1968). From *Tricholomopsis rutilans* has been obtained the acetylenic amino acid L-2-aminohex-4-ynoic acid (XXVII, HATANAKA et al., 1972), and both the threo and erythro forms of L-2-amino-3-hydroxyhex-4-ynoic acid (XXVIII, NIIMURA and HATANAKA, 1974). γ-Glutamyl

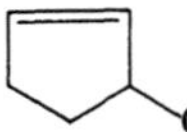$CH(NH_2)\text{–}CO_2H$

Cyclopent–2–enyl–1–glycine

(XV)

$HOOC\text{–}CO\text{–}OCH_2\text{–}CH_2\text{–}CH(NH_2)\text{–}CO_2H$

O–Oxalylhomoserine

(XVIII)

$NC\text{–}CH_2\text{–}CH(NH_2)\text{–}CO_2H$

3–Cyanoalanine

(XVI)

$CH_3\text{–}CH[NH\text{–}(CH_3)]\text{–}CO_2H$

N–Methylalanine

(XIX)

$CH_3CO\text{–}OCH_2\text{–}CH_2\text{–}CH(NH_2)\text{–}CO_2H$

O–Acetylhomoserine

(XVII)

$HOCH_2\text{–}CH[NH\text{–}(CH_3)]\text{–}CO_2H$

N–Methylserine

(XX)

$CH_2[NH\text{–}(C_2H_4OH)]\text{–}CO_2H$

N–(2–Hydroxyethyl)glycine

(XXI)

$CH_3\text{–}C(OH)\text{–}CH_2\text{–}CH(NH_2)\text{–}CO_2H$ (with CH_3 on C(OH))

4–Hydroxyleucine

(XXII)

$CH_3\text{–}CH_2\text{–}CH_2\text{–}C(=CH_2)\text{–}CH(NH_2)\text{–}CO_2H$

3–Methylenenorleucine

(XXV)

$HOCH_2\text{–}C(OH)\text{–}CH_2\text{–}CH(NH_2)\text{–}CO_2H$ (with CH_3 on C(OH))

4,5–Dihydroxyleucine

(XXIII)

$CH_2{=}C{=}CH\text{–}CH_2\text{–}CH(NH_2)\text{–}CO_2H$

2–Amino–4,5–hexadienoic acid

(XXVI)

$HOCH_2\text{–}CH(OH)\text{–}C(CH_3)_2\text{–}CH(NH_2)\text{–}CO_2H$

3–Methyl–4,5–dihydroxyisoleucine

(XXIV)

$CH_3\text{–}C{\equiv}C\text{–}CH_2\text{–}CH(NH_2)\text{–}CO_2H$

2–Aminohex–4–ynoic acid

(XXVII)

$CH_3\text{–}C{\equiv}C\text{–}CH(OH)\text{–}CH(NH_2)\text{–}CO_2H$

2–Amino–3–hydroxyhex–4–ynoic acid

(XXVIII)

derivatives of the first of these compounds and of the erythro-3-hydroxy acid have also been isolated from the same source recently (NIIMURA and HATANAKA, 1977).

2.2 Sulphur and Selenium-Containing Amino Acids

Cysteine and methionine are the two sulphur-containing amino acids normally found in proteins and many of the sulphur-containing non-protein amino acids

$CH_3{-}S{-}CH_2{-}CH(NH_2){-}CO_2H$

S–Methylcysteine

(XXIX)

$CH_3{-}CH{=}CH{-}S{-}CH_2{-}CH(NH_2){-}CO_2H$

S–(Prop–1–enyl)cysteine

(XXX)

$HO_2C{-}CH_2{-}S{-}CH_2{-}CH(NH_2){-}CO_2H$

S–(Carboxymethyl)cysteine

(XXXI)

$HO_2C{-}CH(NH_2){-}CH_2{-}S{-}CH_2{-}CH(CH_3){-}S{-}CH_2{-}CH(NH_2){-}CO_2H$

3,3′–(2Methylethylene–1,2–dithio)dialanine

(XXXII)

$CH_3{-}S{-}CH_2{-}CH_2{-}CH(NH_2){-}CO_2H$

Homomethionine

(XXXIII)

$CH_3{-}CH{=}CH{-}\underset{\downarrow \atop O}{S}{-}CH_2{-}CH(NH_2){-}CO_2H$

S–(Prop–1–enyl)cysteine sulphoxide

(XXXIV)

$CH_3{-}CH{=}CH{-}S{-}CH_2{-}CH(NH{-}OC{-}CH_2{-}CH_2{-}CH(NH_2){-}CO_2H){-}CO_2H$

γ–Glutamyl–S–(prop–1–enyl)cysteine

(XXXV)

$CH_3{-}\overset{+}{\underset{| \atop CH_3}{S}}{-}CH_2{-}CH(NH_2){-}CO_2H$

S–Methylmethionine

(XXXVI)

found in plants are structurally related to them. The Crucifereae and the Alliaceae have been rich sources of these compounds and mention can be made of S-methyl-L-cysteine (XXIX) in *Phaseolus vulgaris* (THOMPSON et al., 1959), S-(prop-1-enyl)-cysteine (XXX) from garlic (SUGII et al., 1963), S-(carboxymethyl)cysteine (XXXI) from radish (BUZIASSY and MAZELIS, 1974), 3,3′-(2-methylethylene-1,2-dithio)dialanine (XXXII) as its γ-glutamyl derivative from chives (MATIKKALA and VIRTANEN, 1964) and homomethionine (XXXIII) from cabbage (SUKETA et al., 1970). These and other sulphur containing amino acids are frequently accompanied in plants by the corresponding sulphoxides, such as S-(prop-1-enyl)-L-cysteine sulphoxide

$HO_2C{-}CH(OH){-}CH_2{-}SO_2{-}CH_2{-}S{-}CH_2{-}CH(NH_2){-}CO_2H$

Dichrostachinic acid

(XXXVII)

$HO_2C{-}CH(NH_2){-}CH_2{-}CH_2{-}S{-}CH_2{-}CH(NH_2){-}CO_2H$

Cystathionine

(XXXVIII)

$HO_2C{-}CH(NH_2){-}CH_2{-}S{-}CH_2{-}S{-}CH_2{-}CH(NH_2){-}CO_2H$

Djenkolic acid

(XXXIX)

$HO_2C{-}CH_2{-}CH(CH_3){-}S{-}CH_2{-}CH(NH_2){-}CO_2H$

S–(2–Carboxyisopropyl)cysteine

(XL)

$HO_2C{-}CH(OH){-}CH_2{-}S{-}CH_2{-}S{-}CH_2{-}CH(NH_2){-}CO_2H$

S–(2–Hydroxy–2–carboxyethanethiomethyl)cysteine

(XLI)

$CH_3{-}Se{-}CH_2{-}CH(NH_2){-}CO_2H$

Se–Methylselenocysteine

(XLII)

$HO_2Se{-}CH_2{-}CH(NH_2){-}CO_2H$

Selenocysteineselenic acid

(XLIII)

$$CH_3{-}CH{=}CH{-}\underset{\substack{\downarrow\\ O}}{Se}{-}CH_2{-}CH(NH_2){-}CO_2H$$

Se–Propen-1-ylselenocysteine selenoxide

(XLIV)

(XXXIV) which was first characterised from onion (SPARE and VIRTANEN, 1963) and as simple derivatives such as γ-glutamyl-S-(prop-1-enyl)cysteine (XXXV) which is found in chives (MATIKKALA and VIRTANEN, 1962).

S-Methylmethionine (XXXVI) was first isolated from the legume *Canavalia ensiformis*, a member of the Papilionoideae (GREEN and DAVIS, 1963); within this family, however, most sulphur amino acids have been found in species of the Mimosoideae. From *Dichrostachys* has been isolated dichrostachinic acid (XXXVII, GMELIN, 1962) while from species of *Neptunia* have been isolated L-cystathionine (XXXVIII, PETERSON and ROBINSON, 1972). *Acacia* species have received special attention however, and amino acids from these include djenkolic acid (XXXIX, GMELIN et al., 1957), N-acetyldjenkolic acid (GMELIN et al., 1962), S-(2-carboxyisopropyl)-cysteine (XL, GMELIN and HIETALA, 1960) and S-(2-hydroxy-2-carboxyethanethiomethyl)-cysteine (XLI, ITO and FOWDEN, 1972). No fewer than 8 sulphur-containing non-protein amino acids or their derivatives accumulate in major concentrations in the seeds of *Acacia* species and distribution patterns of these and other non-protein amino acids divide the genus into four biochemically defined sub-genera (EVANS et al., 1977).

When grown on soil containing selenium, many plants appear capable of absorbing this element, synthesising selenium analogues of methionine and cysteine and incorporating these into their proteins. The tolerance of most of these plants for selenium is not high however. In contrast a few such as *Stanleya pinnata, Lecythis ollaria, Neptunia amplexicaulis*, and a number of *Astragalus* species are tolerant of high selenium levels. These species accumulate the element primarily as amino acids other than selenomethionine and selenocystine and do not incorporate these into their proteins.

In a comprehensive review SHRIFT (1973) refers to the occurrence of Se-methylselenocysteine (XLII), Selenocysteineselenic acid (XLIII), Se-propen-1-ylselenocysteine selenoxide (XLIV), selenohomocystine, selenocystathione, selenomethionine selenoxide, Se-methylselenomethione and γ-L-glutamyl-Se-methylseleno-L-cysteine as occurring in higher plants and to this list have recently been added 2-γ-glutamylamino-4-(2-amino-2-carboxyethylselenyl)butyric acid and 2-amino-4-(2-γ-glutamylamino-2-carboxyethylselenyl)butyric acid, both of which have been isolated from *Astragalus pectinatus* (NIGAM and MCCONNELL, 1976).

2.3 Imino Acids

Proline is universally distributed as a protein constituent and is invariably considered together with the protein amino acids. It is appropriate therefore to consider other naturally occurring imino acids in this chapter.

The great majority of plant imino acids are structurally related to proline itself or to the higher homologue pipecolic acid (XLV). The lower homologue of proline, azetidine-2-carboxylic acid (XLVI) which is known to accumulate in some species of the Lilliaceae (FOWDEN, 1956) and the Leguminosae (WATSON and FOWDEN, 1973; EVANS and BELL, 1978) also occurs in very low concentrations in *Beta vulgaris* (FOWDEN, 1972). Other related compounds are nicotianamine (XLVII) first isolated from tobacco leaves (NOMA et al., 1971) and assigned a structure which was later revised when isolated again from *Fagus* together with 2(S),3′(S)-N-(3-amino-3-carboxypropyl) azetidine-2-carboxylic acid (XLVIII, KRISTENSEN et al., 1974). Naturally occurring polycyclic imino acids include 3-carboxy-6,7-dihydroxy-1,2,3,4-tetrahydroisoquinoline (XLIX, BELL, et al., 1971) and the corresponding 1-methyl compound (L) (DAXENBICHLER et al., 1972), both of which occur in seeds of *Mucuna* species and the tricyclic L-3-carboxy-1,2,3,4-tetrahydro-β-carboline (LI) from *Aleurites fordii* (OKUDA et al., 1975). Both trans-3-hydroxy-L-proline (LII, SUNG and FOWDEN, 1968), and trans-4-hydroxy-L-proline (LIII, WELTER et al., 1978) are known, but most plant prolines are substituted in the 4 position; examples are trans-4-methyl proline (LIV) which was first identified in young apple fruits (HULME and ARTHINGTON, 1954). HULME (1954) isolated 4-hydroxymethylproline (LV) from apple peel and this was later shown to be the cis

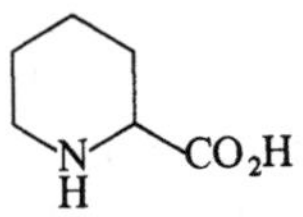

Pipecolic acid
(XLV)

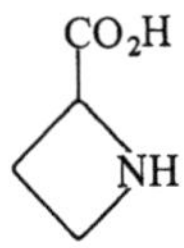

Azetidine–2–carboxylic acid
(XLVI)

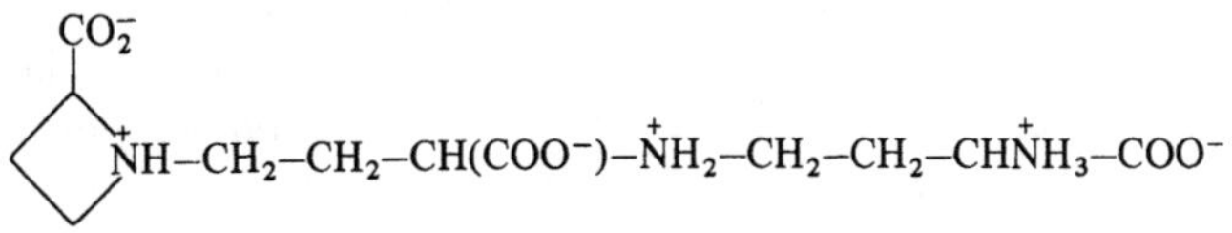

Nicotianamine
(XLVII)

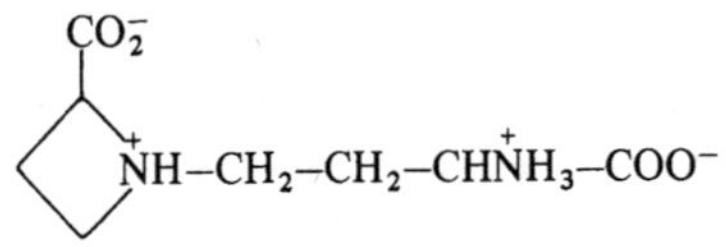

N–(3–Amino–3–carboxypropyl) azetidine–2–carboxylic acid
(XLVIII)

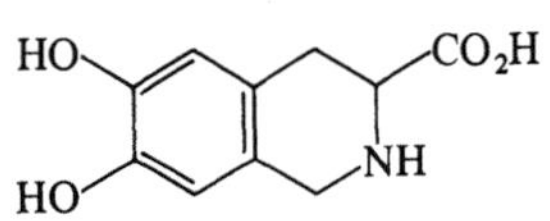

3–Carboxy–6,7–dihydroxy–1, 2, 3, 4–tetrahydroisoquinoline
(XLIX)

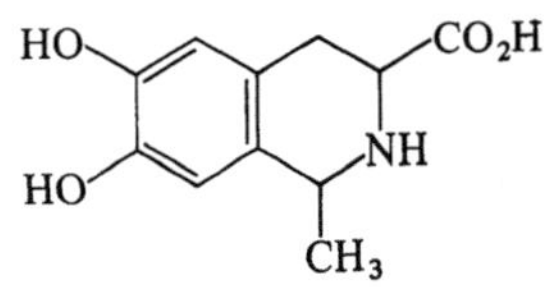

1–Methyl–3–carboxy–6,7–dihydroxy–1, 2, 3, 4–tetrahydroisoquinoline
(L)

L isomer (UNTCH and GIBBON, 1964). The same isomer is however accompanied by both the trans L and trans D forms in *Eriobotrya japonica* (GRAY, 1972). 4-Methylene-DL-proline (LVI) comes from the same source (GRAY and FOWDEN, 1962b) and 4-hydroxy-N-methyl-L-proline (LVII) from *Afrormosia elata* heartwood (MORGAN, 1964). Trans-4-Carboxy-L-proline (LVIII) has been reported in the legume *Afzelia bella* (WELTER et al., 1978) and also in an alga *Chondria coerulescens* (IMPELLIZZERI et al., 1977). Cis-3-Amino-L-proline (LIX) has been isolated from *Morchella esculenta* by HATANAKA (1969) while N^1-amino-D-proline (LX) as its γ-glutamyl derivative linatine has been found in *Linum* (flax) (KLOSTERMAN et al., 1967).

Pipecolic acid itself is found very commonly in plants while the γ-glutamyl derivative (LXI) has recently been isolated from a species of *Gleditsia* (Leguminosae) by DARDENNE et al. (1974a). Baikiain (LXII) (4,5-dehydro-L-pipecolic acid) originally isolated from *Baikiaea plurijuga* (KING et al., 1950) also occurs in the algae (MADGWICK et al., 1970). Both trans-4-hydroxypipecolic acid (LXIII) and trans-5-hydroxypipecolic acid (LXIV) were identified in *Armeria maritima* by VIRTANEN and KARI (1954) and the cis-5-hydroxypipecolic acid has now been found in *Gymnocladus dioicus* (DESPONTIN et al., 1977) *Morus alba* and *Lathyrus japonicus* (HATA-

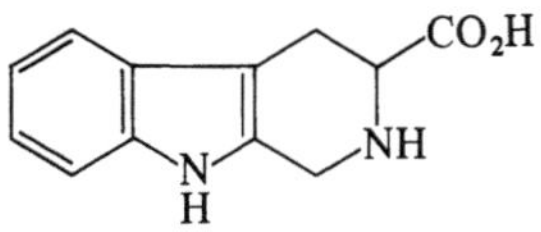

3–Carboxy–1, 2, 3, 4–tetrahydro–β–carbolıne
(LI)

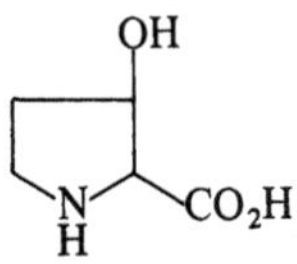

3–Hydroxyprolıne
(LII)

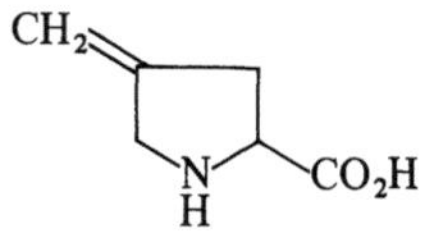

4–Methyleneprolıne
(LVI)

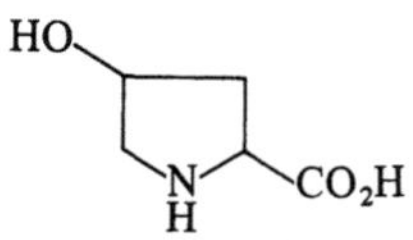

4–Hydroxyproline
(LIII)

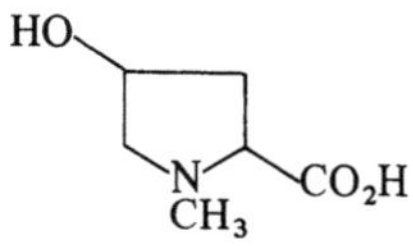

4–Hydroxy–N–methylprolıne
(LVII)

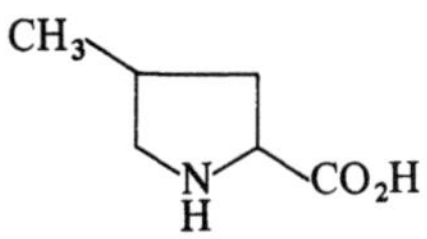

4–Methylproline
(LIV)

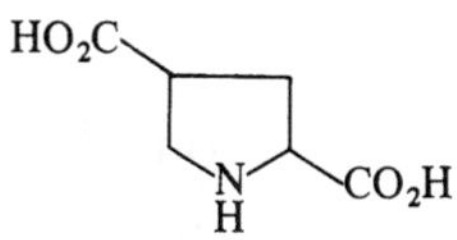

4–Carboxyprolıne
(LVIII)

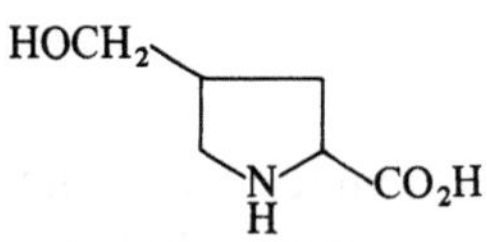

4–Hydroxymethylprolıne
(LV)

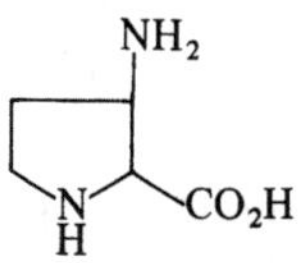

3–Amınoproline
(LIX)

NAKA and KANEKO, 1977). Two stereoisomers of 5-hydroxy-6-methylpipecolic acid (LXV) are found in *Fagus silvatica* (KRISTENSEN et al., 1974), while no fewer than three stereoisomers of 4,5-dihydroxy-pipecolic acid (LXVI) have been characterised from legumes (MARLIER et al., 1972; DARDENNE, 1975; MARLIER et al., 1976; SHEWRY and FOWDEN, 1976).

2.4 Acidic Amino Acids and Amides

Although aspartic acid and glutamic acid are both universally distributed as protein constituents, the majority of non-protein acidic amino acids and amides isolated

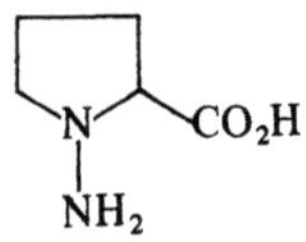

N^1–Aminoproline
(LX)

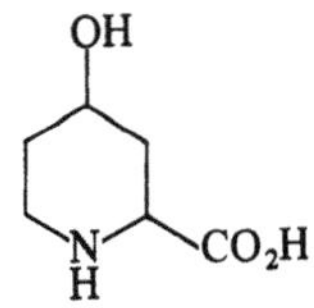

4–Hydroxypipecolic acid
(LXIII)

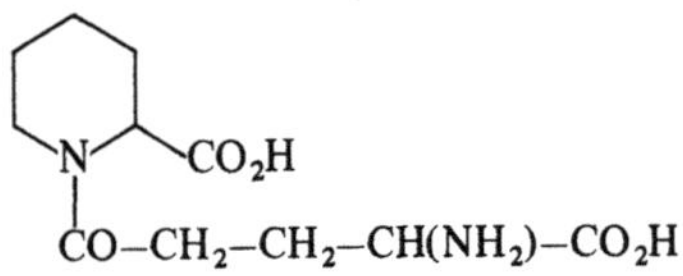

γ–Glutamylpipecolic acid
(LXI)

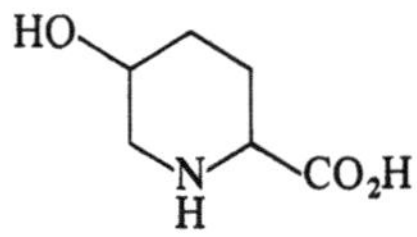

5–Hydroxypipecolic acid
(LXIV)

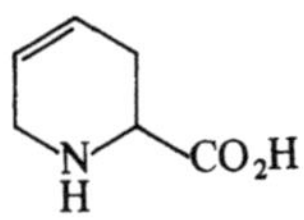

Baikiain
(LXII)

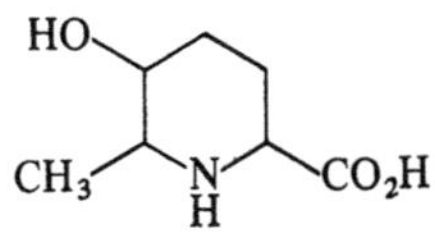

5–Hydroxy–6–methylpipecolic acid
(LXV)

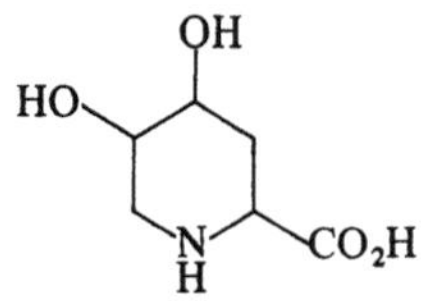

4,5–Dihydroxypipecolic acid
(LXVI)

from plants are structurally related to glutamic acid or glutamine. These include 4-methylglutamic acid (LXVII) isolated from *Lathyrus maritimus* (PRZYBYLSKA and STRONG, 1968) and the 4-methyleneglutamic acid (LXVIII) first isolated from *Arachis hypogaea* (DONE and FOWDEN, 1952). These two substituted glutamic acids are not uncommon in plants, the cis-4-ethylideneglutamic acid (LXIX), however, has only been isolated so far from three plant species, these being *Tulipa gesneriana* (FOWDEN, 1966b). *Tetrapleura tetraptera* (GMELIN and LARSEN, 1967) and *Guilandina crista* (NULU and BELL, 1972). The compound however has been detected chromatographically as a major constituent of the free amino acid pool of at least eleven other species of *Guilandina* (EVANS and BELL, 1978), a character which separates these species from other members of the Caesalpinoideae. 4-Propylideneglutamic acid (LXX) has been reported from one source, the fungus *Mycena pura* (HATANAKA and KATAYAMA, 1975). Among the hydroxylated glutamic acids which occur in plants are threo-4-hydroxy-glutamic acid (LXXI) first reported in *Phlox* (VIRTANEN and HIETALA, 1955) and two stereoisomeric forms of 4-hydroxy-4-methylglutamic acid (LXXII), the 2(S), 4(R) isomer being isolated from *Ledenber-*

$HO_2C{-}CH(CH_3){-}CH_2{-}CH(NH_2){-}CO_2H$

4–Methylglutamic acid

(LXVII)

$HO_2C{-}C(=CH_2){-}CH_2{-}CH(NH_2){-}CO_2H$

4–Methyleneglutamic acid

(LXVIII)

$HO_2C{-}C(=CH{-}CH_3){-}CH_2{-}CH(NH_2){-}CO_2H$

4–Ethylideneglutamic acid

(LXIX)

$HO_2C{-}C(=CH{-}CH_2{-}CH_3){-}CH_2{-}CH(NH_2){-}CO_2H$

4–Propylideneglutamic acid

(LXX)

$HO_2C{-}CH(OH){-}CH_2{-}CH(NH_2){-}CO_2H$

4–Hydroxyglutamic acid

(LXXI)

$HO_2C{-}C(OH)(CH_3){-}CH_2{-}CH(NH_2){-}CO_2H$

4–Hydroxy–4–methylglutamic acid

(LXXII)

$HO_2C{-}CH(CH_3){-}CH(OH){-}CH(NH_2){-}CO_2H$

3–Hydroxy–4–methylglutamic acid

(LXXIII)

$HO_2C{-}CH(=CH_2){-}CH(OH){-}CH(NH_2){-}CO_2H$

3–Hydroxy–4–methyleneglutamic acid

(LXXIV)

gia rosea-aenea (JADOT et al., 1967) and the 2(S), 4(S), form from *Phyllitis scolopendrium* (BLAKE and FOWDEN, 1964) and *Pandanus veitchii* (JADOT et al., 1967). 2-Stereoisomeric forms, 2(S), 3(S), 4(R) and 2(S), 3(R), 4(S) of 3-hydroxy-4-methylglutamic acid (LXXIII) have been isolated from seeds of *Gymnocladus dioicus* (DARDENNE et al., 1972a, b; EVRARD et al., 1974). The corresponding 2(S), 3(S)-3-hydroxy-4-methyleneglutamic acid (LXXIV) has also been isolated from a legume *Gleditsia caspica* (DARDENNE et al., 1974b). 3,4-Dihydroxyglutamic acid (LXXV) occurs in *Lepidium* and *Rheum* (VIRTANEN and ETTALA, 1957) and 2(S), 4(R)-4-(β-D-galactopyranosyloxy)-4-isobutylglutamic acid (LXXVI) in *Reseda odorata* (LARSEN et al., 1973). Hydroxyaspartic acid (LXXVII) has been identified in *Medicago* and *Trifolium* species by WILDING and STAHMANN (1962a). 2-Aminoadipic acid

$HO_2C{-}CH(OH){-}CH(OH){-}CH(NH_2){-}CO_2H$

3,4–Dihydroxyglutamic acid

(LXXV)

$HO_2C{-}C(CH_2{-}CH(CH_3)_2){-}CH_2{-}CH(NH_2){-}CO_2H$
O
Gal

4–(β–D–Galactopyranosyloxy)–4–isobutylglutamic acid

(LXXVI)

$HO_2C{-}CH(OH){-}CH(NH_2){-}CO_2H$

Hydroxyaspartic acid

(LXXVII)

$HO_2C{-}CH_2{-}CH_2{-}CH_2{-}CH(NH_2){-}COOH$

2–Aminoadipic acid

(LXXVIII)

$HO_2C{-}CH_2{-}CH_2{-}CH_2{-}CH_2{-}CH(NH_2){-}CO_2H$

2–Aminopimelic acid

(LXXIX)

$HO_2C{-}CH_2{-}CH_2{-}CH(CH_3){-}CH_2{-}CH(NH_2){-}CO_2H$

2–Amino–4–methylpimelic acid

(LXXX)

$H_2N{-}CH_2{-}CH_2{-}CH_2{-}CH{-}CO_2H$
NH
$HO_2C{-}CH_2{-}CH_2{-}CH{-}CO_2H$

N^2–(1,3–Dicarboxypropyl)ornithine

(LXXXI)

$HO_2C{-}CH(NH_2){-}CH{-}CH_2{-}CH{-}CO_2H$
O——NH

Penmaric acid

(LXXXII)

(LXXVIII) is widespread and possibly universal in higher plants (NAWAZ and SØRENSEN, 1977), while 2-aminopimelic acid (LXXIX) is known to occur in ferns (VIRTANEN and BERG, 1954). HATANAKA et al. (1975) have isolated L-2-amino-4-methylpimelic acid (LXXX) from a fungus, while N^2-(1,3-dicarboxypropyl)ornithine (LXXXI) is reported in crown gall tumours (FIRMIN and FENWICK, 1977), the corresponding arginine compound being nopaline (GOLDMANN et al., 1969) which is found in the same tumours. An unusual acidic amino acid with a pyrroli-

$(CH_3)HN{-}OC{-}CH_2{-}CH(NH_2){-}CO_2H$

N^4–Methylasparagine

(LXXXIII)

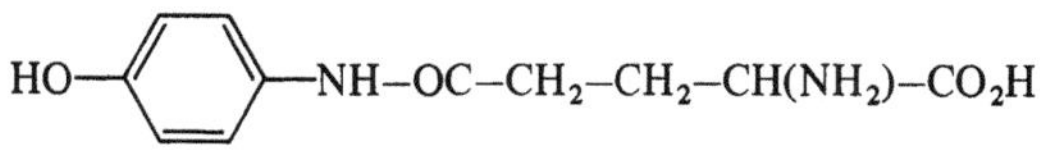

Agaratine

(LXXXIV)

done ring has been isolated independently from *Pentaclethra macrophylla* by Mbadiwe (1975) and Dardenne (1975) and named penmaric acid (LXXXII) by the former.

4-Methyleneglutamine from peanuts (Done and Fowden, 1952) and 4-hydroxyglutamine in *Phlox decussata* (Brandner and Virtanen, 1963) provide examples of amides corresponding to acidic non-protein amino acids. Not all the amides which have been isolated correspond to known free non-protein amino acids, however, and among these may be listed N^4-methylasparagine (LXXXIII), isolated from seeds of *Corallocarpus epigaeus* (Dunnill and Fowden, 1965), N^4-ethylasparagine isolated from *Ecballium elaterium* (Gray and Fowden, 1961), N^4-hydroxyethylglutamine isolated from *Bryonia dioica* (Fowden, 1961), N^5-hydroxy-ethylglutamine from *Lunaria annua* (Larsen, 1965), and agaratine (LXXXIV), an aromatic derivative of glutamine which was found in the fungus *Agaricus bisporus* (Levenberg, 1964).

The 4-ethyl amide of glutamic acid (L-ethanine) has also been detected in the leaves of the tea plant (Furuyama et al., 1964).

2.5 Basic Amino Acids and Related Compounds

The two basic amino acids normally found in proteins are lysine and arginine, while ornithine, although not a protein amino acid, occurs commonly as a metabolic intermediate. 4-Hydroxylysine (LXXXV) has been identified in *Salvia* (Brieskorn and Glasz, 1964b); N^6-acetyl-L-lysine and N^6-acetyl-allo-5-hydroxylysine have been both identified as minor components of *Beta vulgaris* (Fowden, 1972) while chromatographic evidence suggests that the free 5-hydroxylysine itself occurs in *Medicago sativa* (Wilding and Stahmann, 1962b). The quaternary ammonium derivative of lysine, laminine (LXXXVI), has been found both in *Laminaria* (Takemoto et al., 1964a) and in the higher plant *Reseda luteola* (Larsen, 1968). A number of naturally occurring derivatives of ornithine are also known. N^5-Acetylornithine has been obtained from *Corydalis ochotensis* (Manske, 1937) and subsequently from many other species, while other ornithine derivatives include 4-hydroxyornithine, identified in *Vicia* spp. (Bell and Tirimanna, 1964), and fusarinine (LXXXVII) and gigartinine (LXXXVIII) which have been isolated from algae by Emery (1965) and by Ito and Hashimoto (1966), respectively.

2,4-Diaminobutyric acid (LXXXIX) was first reported in *Polygonatum* (Fowden and Bryant, 1958); this amino acid is not uncommon in higher plants where

$H_2N-CH_2-CH_2-CH(OH)-CH_2-CH(NH_2)-CO_2H$
4-Hydroxylysine
(LXXXV)

$(CH_3)_3\overset{+}{N}-CH_2-CH_2-CH_2-CH_2-CH(NH_2)-CO_2H$
Laminine
(LXXXVI)

$HO-CH_2-CH_2-C(CH_3)=CH-CO-N(OH)-(CH_2)_3-CH(NH_2)CO_2H$
Fusarinine
(LXXXVII)

$H_2N-C(=NH)-NH-CO-NHCH_2-CH_2-CH_2-CH(NH_2)-CO_2H$
Gigartinine
(LXXXVIII)

$H_2N-CH_2-CH_2-CH(NH_2)-CO_2H$
2,4-Diaminobutyric acid
(LXXXIX)

$H_2N-CH_2-CH(NH_2)-CO_2H$
2,3-Diaminopropionic acid
(XC)

$CH_3NH-CH_2-CH(NH_2)-CO_2H$
2-Amino-3-methylaminopropionic acid
(XCI)

$H_2N-C(=NH)-NH-CH_2-CH_2-CH_2-CH[NH-CH(CO_2H)-CH_3]-CO_2H$
Octopine
(XCII)

it occurs both in the free state and as derivatives such as the N^4-acetyl compound which LISS (1962) found in *Euphorbia* and the N^4-lactyl compound which occurs in *Beta vulgaris* (FOWDEN, 1972). In addition to these, BELL and O'DONOVAN (1966) found both the N^2- and N^4-oxalyl derivatives of 2,4-diaminobutyric acid in *Lathyrus latifolius*. The lowest member of the homologous series containing lysine is 2,3-diaminopropionic acid (XC) which was first found as a naturally occurring amino acid by GMELIN et al. (1959) in the Mimosaceae. The N^3-acetyl derivative accumulates in the seeds of many *Acacia* species (SENEVIRATNE and FOWDEN, 1968), while the N^2-oxalyldiaminopropionic acid, N^3-oxalyldiamino-propionic acid have been reported in seeds of *Lathyrus* species (RAO et al., 1964; MURTI et al., 1964; BELL and O'DONOVAN, 1966). 2-Amino-3-methylaminopropionic (XCI) acid has

$H_2N{-}C({=}NH){-}NH{-}CH_2{-}CH_2{-}CH_2{-}CH_2{-}CH(NH_2){-}CO_2H$

Homoarginine

(XCIII)

$H_2N{-}C({=}NH){-}NH{-}CH_2{-}CH_2{-}CH(OH){-}CH_2{-}CH(NH_2){-}CO_2H$

4-Hydroxyhomoarginine

(XCIV)

$H_2N{-}C({=}NH){-}NH{-}CH_2{-}CH(OH){-}CH_2{-}CH(NH_2){-}CO_2H$

4-Hydroxyarginine

(XCV)

$H_2N{-}CO{-}NH{-}CH_2{-}CH_2{-}CH_2{-}CH_2{-}CH(NH_2){-}CO_2H$

Homocitrulline

(XCVI)

$H_2N{-}CO{-}NH{-}CH_2{-}CH(OH){-}CH_2{-}CH(NH_2){-}CO_2H$

4-Hydroxycitrulline

(XCVII)

$H_2N{-}CO{-}NH{-}CH_2{-}CH(NH_2){-}CO_2H$

Albizziine

(XCVIII)

$H_2N{-}CO{-}NH{-}CH_2{-}CH_2{-}CH_2{-}CH(NH_2){-}CO_2H$

Citrulline

(XCIX)

$H_2N{-}C({=}NH){-}NH{-}O{-}CH_2{-}CH_2{-}CH(NH_2){-}CO_2H$

Canavanine

(C)

been isolated from seeds of *Cycas circinalis* (VEGA and BELL, 1967). Non-protein amino acids related structurally to arginine include octopine (XCII) from crown gall (MENAGE and MOREL, 1964), homoarginine (XCIII) and 4-hydroxyhomoarginine (XCIV) originally isolated from *Lathyrus* species (BELL, 1962; BELL, 1963) and 4-hydroxyarginine (XCV) isolated from *Vicia sativa* (BELL and TIRIMANNA, 1963). Corresponding ureido compounds have also been identified, homocitrulline (XCVI) in *Vicia* (INATOMI et al., 1970) and 4-hydroxycitrulline (XCVII) in the same genus (BELL and TIRIMANNA, 1965; INATOMI et al., 1970). Albizziine (XCVIII), a lower homologue of citrulline (XCIX), was first found in *Albizzia* (GMELIN et al., 1959). Canavanine (C), originally isolated from *Canavalia ensiformis* (KITAGAWA and TOMIYAMA, 1929) and now known to occur in many species of the

$H_2N{-}C({=}NH){-}CH_2{-}CH_2{-}CH_2{-}CH_2{-}CH(NH_2){-}CO_2H$
Indospicine
(CI)

$H_2N{-}O{-}CH_2{-}CH_2{-}CH(NH_2){-}CO_2H$
Canalin
(CII)

leguminous sub-family Papilionoideae (BELL et al., 1978), is the only example of an amino acid containing a guanidoxy function, while indospicine (CI), which was isolated from *Indigofera spicata* (HEGARTY and POUND, 1968) is also unique in containing an amidino function. Canalin (CII), which bears the same relationship to canavanine as ornithine does to arginine has been found to occur in *Astragalus* species (INATOMI et al., 1968).

2.6 Heterocyclic Amino Acids

Both oxygen and nitrogen occur singly and together in the heterocyclic rings of various non-protein amino acids. 3-(2-Furoyl)-L-alanine (CIII) has been found in *Fagopyrum esculentum* by ICHIHARA et al. (1973) and 3-(3-carboxylfuran-4-yl)-L-alanine (CIV) was found in the fungus *Phyllotopsis nidulans* by DOYLE and LEVENBERG (1974), while stizolobic acid (CV) and stizolobinic acid (CVI) have been isolated from *Stizolobium hassjoo* (HATTORI and KOMAMINE, 1959; SENOH et al., 1964). Derivatives of pyrrolidine are represented by 3-amino-3-carboxypyrrolidine (CVII) from the Cucurbitaceae (FANG et al., 1961). 3-Pyrazol-1-ylalanine (CVIII) has also been isolated from the Cucurbitaceae (NOE and FOWDEN, 1960), dimethylhistidine (CIX) from a seaweed (MADGWICK et al., 1970) and enduracididine (CX) from *Lonchocarpus sericeus* (FELLOWS et al., 1977). Amino acids with six-membered nitrogen-containing rings include mimosine (CXI) from *Leucaena* and *Mimosa* species (RENZ, 1936), willardine (CXII) and isowillardine (CXIII) from *Acacia* and *Pisum* species (GMELIN, 1959; LAMBEIN and VAN PARIJS, 1968), nicotianine (CXIV) from tobacco leaves (NOGUCHI et al., 1968), lathyrine (CXV) from *Lathyrus* species (BELL and FOSTER, 1962) and 3-(2,6-dihydroxypyrimidin-5-yl)alanine (CXVI) from pea seedlings (BROWN and MANGAT, 1969). Indoles such as 5-hydroxy-L-tryptophan (CXVII) from *Griffonia* species (BELL and FELLOWS, 1966; BELL et al., 1976), N-methyltryptophan (CXVIII) and N,N,dimethyltryptophan (CXIX) both from *Abrus precatorius* are also known (MANDAVA et al., 1974).

Amino acids with both oxygen and nitrogen in their heterocyclic rings are also found. Tricholomic acid (CXX) is a constitutent of a small peptide from the fungus *Tricholoma muscarium* (TAKEMOTO and NAKAJIMA) while ibotenic acid (CXXI), amarin the corresponding amine (CXXII) and muscazone (CXXIII) have all been isolated from species of *Amanita* (TAKEMOTO et al., 1964b; BOWDEN et al., 1965; EUGSTER et al., 1965; FRITZ et al., 1965). The higher plants have also yielded a number of isoxazoline derivatives, these include 3-(isoxazolin-5-one-2-yl)-alanine (CXXIV) from *Pisum sativum*, 3-(2-β-D-glucopyranosyl-isoxazolin-5-one-4-yl)alanine (CXXV) from the same plant and 2-amino-4-(isoxalin-5-one-2-yl)butyric acid (CXXVI) from *Lathyrus odoratus*. The occurrence of these amino acids and other heterocyclic compounds in legume seedlings has been reviewed by LAMBEIN et al. (1976).

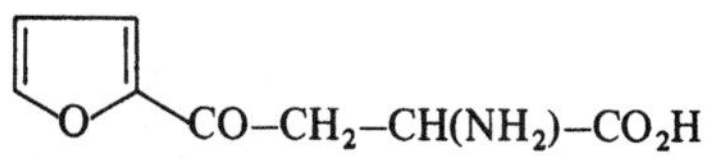

3–(2–Furoyl)alanine
(CIII)

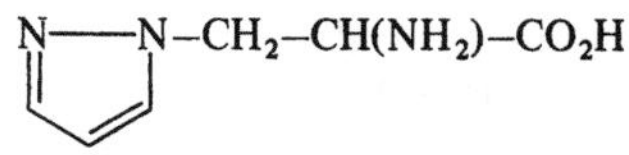

3–Pyrazol–1–ylalanine
(CVIII)

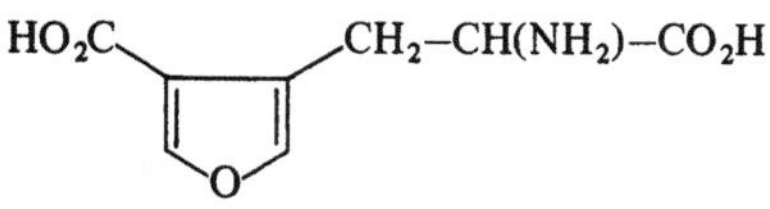

3–(3–Carboxyfuran–4–yl)alanine
(CIV)

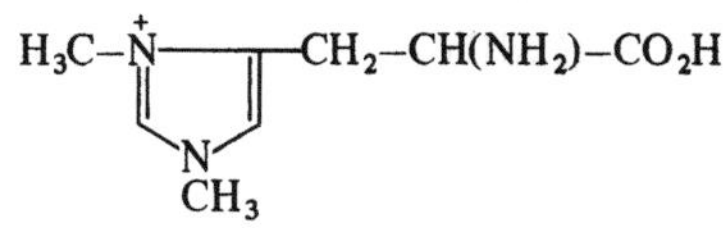

Dimethylhistidine
(CIX)

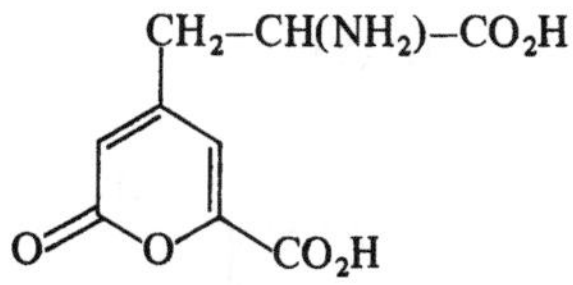

Stizolobic acid
(CV)

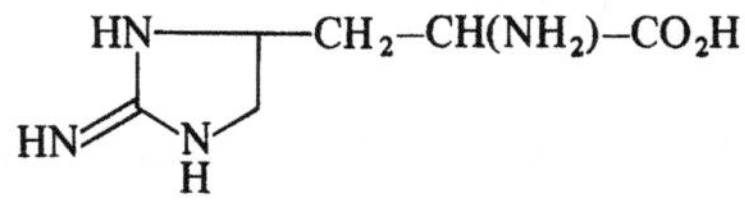

Enduracididine
(CX)

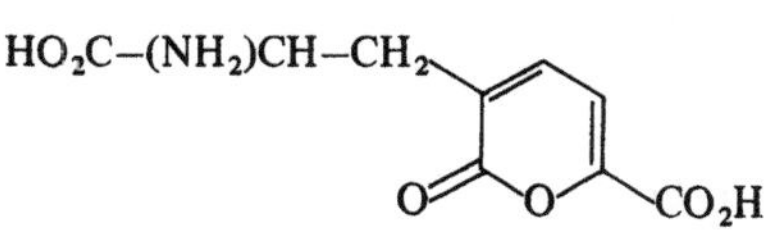

Stizolobinic acid
(CVI)

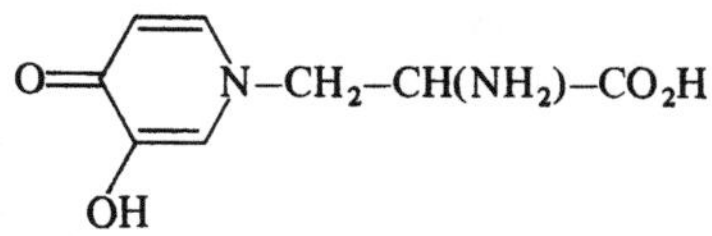

Mimosine
(CXI)

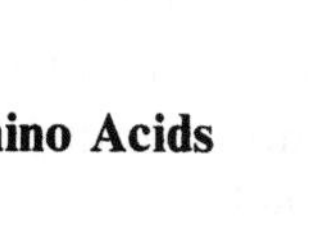

3–Amino–3–carboxypyrrolidine
(CVII)

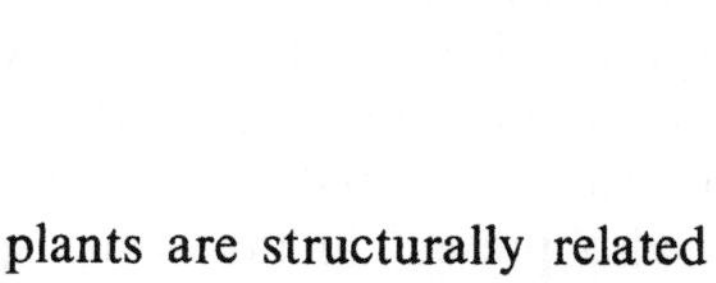

Willardine
(CXII)

2.7 Aromatic Amino Acids

Most of the aromatic non-protein amino acids in plants are structurally related to the protein amino acids, phenylalanine and tyrosine or to phenylglycine (CXXVII), which was first shown to occur naturally in the phloem sap of *Fagus* (DIETRICHS and FUNKE, 1967), but is not known as a protein constituent. Among the hydroxylated derivatives of phenylalanine are L-m-tyrosine (CXXVIII) from *Euphorbia myrsinites* (MOTHES et al., 1964), L-3,4-dihydroxyphenylalanine (CXXIX) from various plants (DAXENBICHLER et al., 1971; BELL and JANZEN, 1971), and 3-orcylalanine (CXXX) from *Agrostemma githago* (SCHNEIDER, 1958). 3-Hydroxymethyl-phenylalanine (CXXXI), L-4-hydroxy-3-hydroxymethylphenylalanine (CXXXII) and L-3-carboxyphenylalanine (CXXXIII) have all been found in *Caesal-*

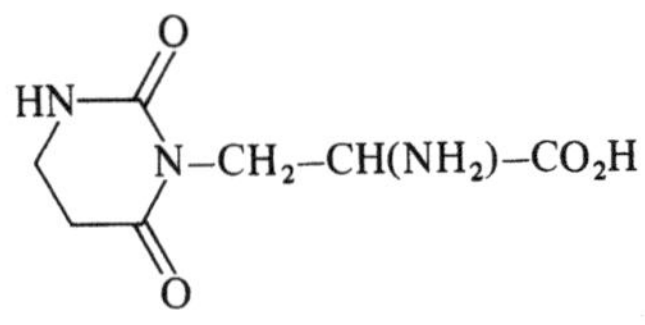

Isowillardine
(CXIII)

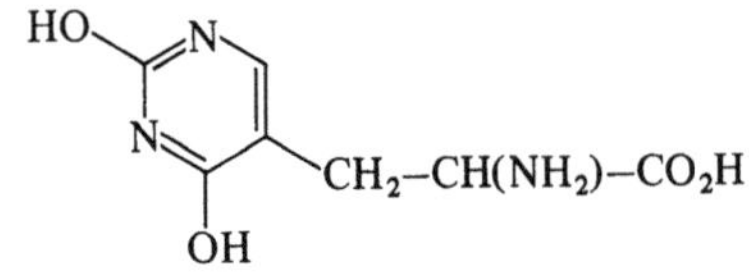

3-(2,6-Dihydroxypyrimidin-5-yl)alanine
(CXVI)

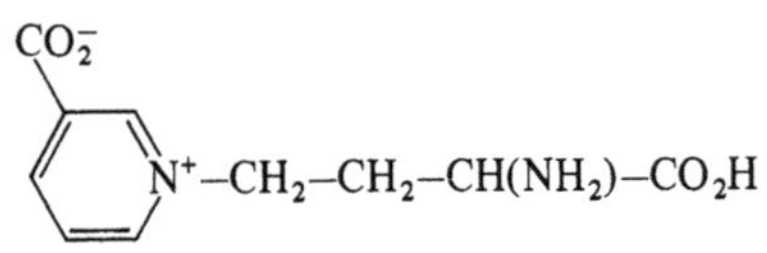

Nicotianine
(CXIV)

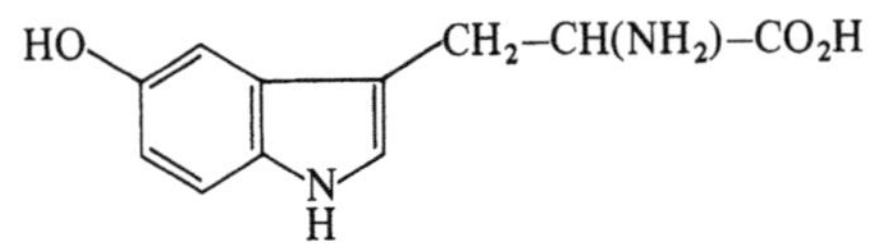

5-Hydroxytryptophan
(CXVII)

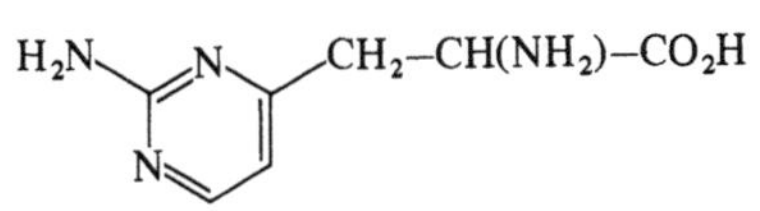

Lathyrine
(CXV)

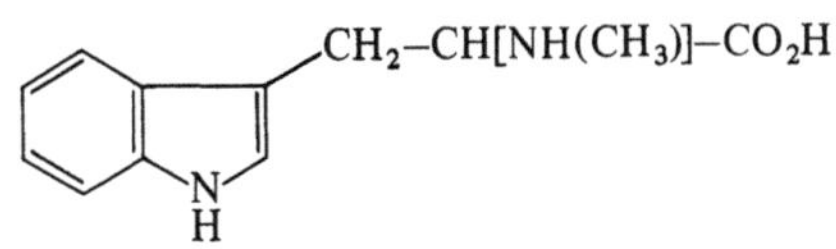

N-Methyltryptophan
(CXVIII)

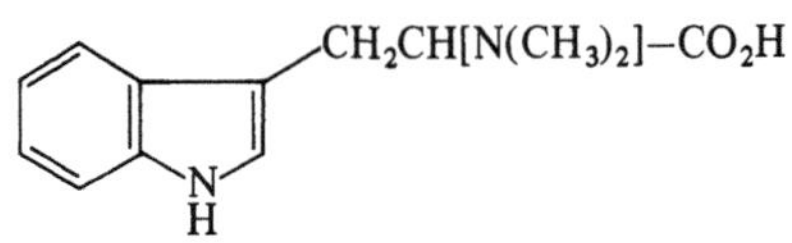

N,N-Dimethyltryptophan
(CXIX)

pinia tinctoria (WATSON and FOWDEN, 1973), the last of these three having previously been found in the Iridaceae (THOMPSON et al., 1961), the Cucurbitaceae (DUNNILL and FOWDEN, 1965) and the Resedaceae, where it was shown by KJAER and LARSEN (1963) to be accompanied by L-3-carboxytyrosine (CXXXIV). The finding of 3-aminomethylphenylalanine (CXXXV) by MWAULUKA et al. (1975b) in *Combretum zeyheri* where 3-carboxy-phenylalanine and 3-hydroxymethylphenylalanine have also been identified suggests that the 3-substituted aromatic amino acids are probably widely distributed in plants. It is of interest that N-methyl-L-tyrosine (CXXXVI) is also present in this same species of *Combretum* (MWAULUKA et al., 1975a). 4-Aminophenylalanine (CXXXVII) itself has been discovered recently in the seeds of *Vigna* species (DARDENNE et al., 1972c).

Phenylglycine has already been referred to; 3-hydroxyphenylglycine (CXXXVIII) and 3,5-dihydroxyphenylglycine (CXXXIX) both occur in *Euphorbia helioscopia* (MÜLLER and SCHÜTTE, 1967); the N-carbamoyl derivative of the racemic 4-hydroxyphenylglycine (CXL) has been reported in *Vicia faba* (EAGLES et al., 1971) while D-3-carboxyphenylglycine has been reported in Iridaceae (MORRIS et al., 1959) and in Resedaceae (KJAER and LARSEN, 1963). D-3-Carboxy-4-hydroxyphenylglycine was also found by these workers in the same species.

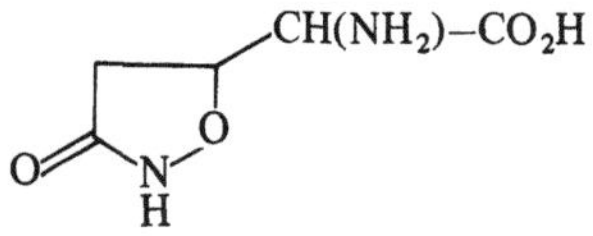

Tricholomic acid
(CXX)

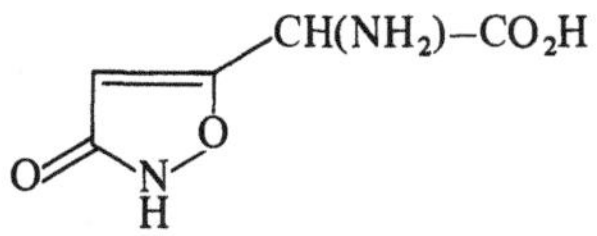

Ibotenic acid
(CXXI)

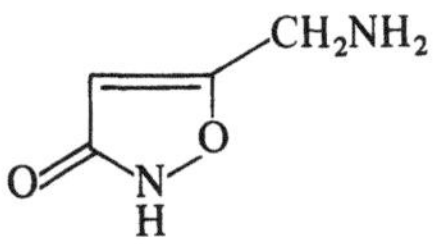

Amarin
(CXXII)

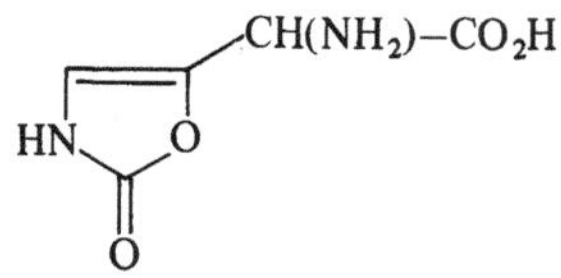

Muscazone
(CXXIII)

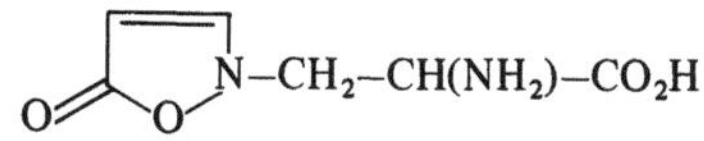

3–(Isoxazolin–5–one–2–yl)alanine
(CXXIV)

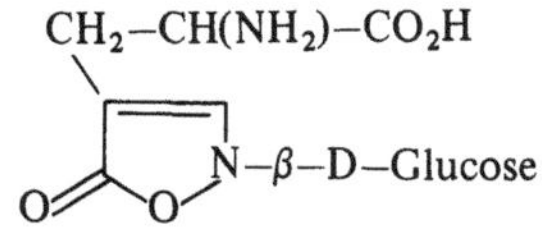

3–(2–β–D–Glucopyranosyl–
isoxazolin–5–one–4–yl)alanine
(CXXV)

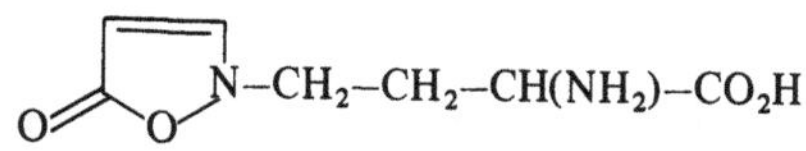

2–Amino–4–(isoxalin–5–one–2–yl)
butyric acid
(CXXVI)

2.8 Miscellaneous Amino Acids

Among the non-protein amino acids are some which are not close analogues of any of the protein amino acids. These include the commonly occurring 4-amino-butyric acid (CXLI) a decarboxylation product of glutamic acid. Other examples of non-α-amino acids are provided by 3-aminoglutaric acid (CXLII) which has been found in the seaweed *Chondria* (Takemoto et al., 1965) and β-phenyl-β-alanine (CXLIII) which occurs as the γ-glutamyl derivative in *Phaseolus* species (Koyama and Obata, 1966).

Saccharopine (CXLIV) originally isolated from yeast (Darling and Larsen, 1961; Kjaer and Larsen, 1961) and lysopine (CXLV) from crown gall tumours (Biemann et al., 1960) are uncyclised imino acids. The chlorinated amino acids trans-2-amino-5-chloro-4-hexenoic acid (CXLVI) and 2-amino-4-chloro-4-pente-noic acid (CXLVII) which have been isolated from an *Amanita* species by Chilton and Tsou (1972) and by Hatanaka et al. (1974) represent a third group, while pinnatanine (CXLVIII) from *Staphylea pinnata* (Grove et al., 1971) is an example of an unsaturated amino acid with an unusual structure.

These examples serve to illustrate the structural diversity found within this particular group of secondary compounds.

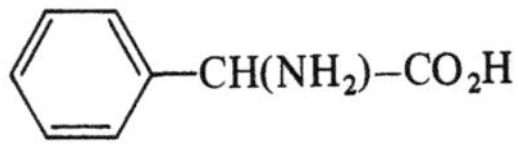

Phenylglycine
(CXXVII)

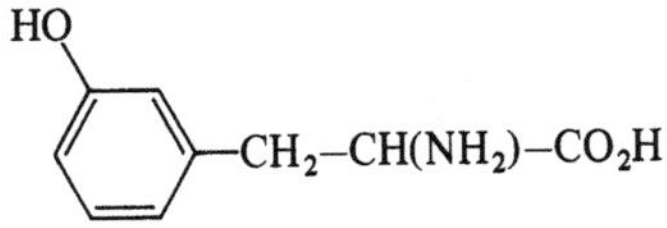

m–Tyrosine
(CXXVIII)

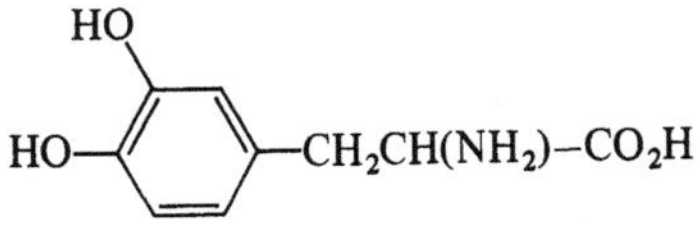

3,4–Dihydroxyphenylalanine
(CXXIX)

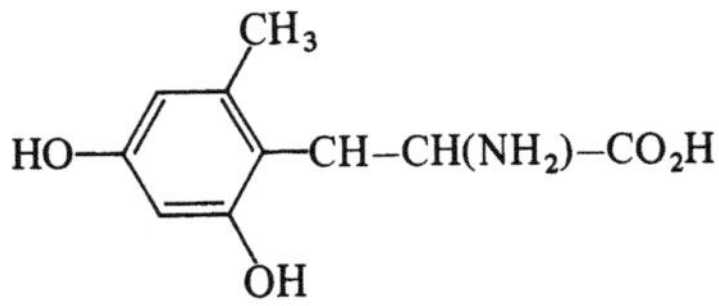

3–Orcylalanine
(CXXX)

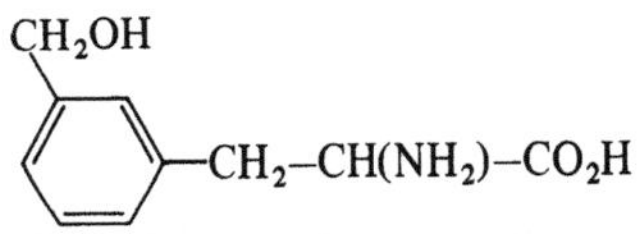

3–Hydroxymethylphenylalanine
(CXXXI)

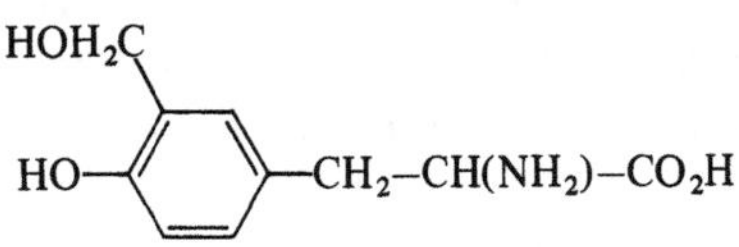

4–Hydroxy–3–hydroxymethylphenylalanine
(CXXXII)

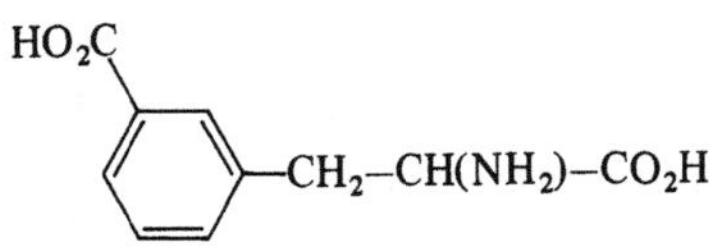

3–Carboxyphenylalanine
(CXXXIII)

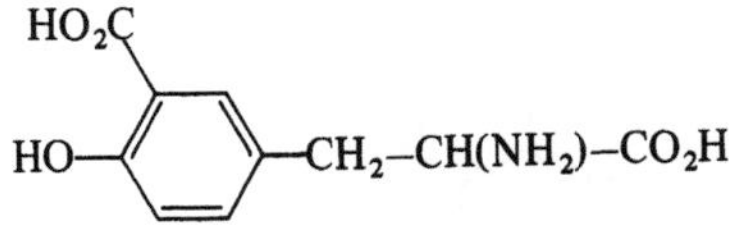

3–Carboxytyrosine
(CXXXIV)

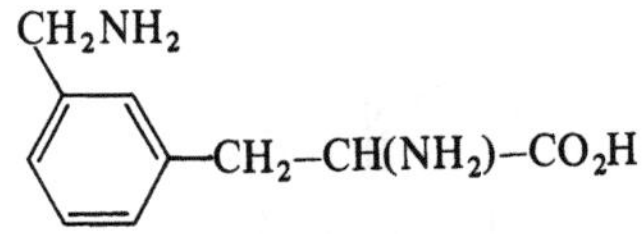

3–Aminomethylphenylalanine
(CXXXV)

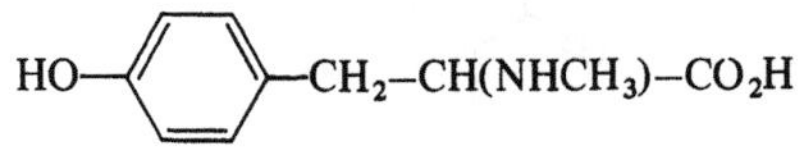

N–Methyltyrosine
(CXXXVI)

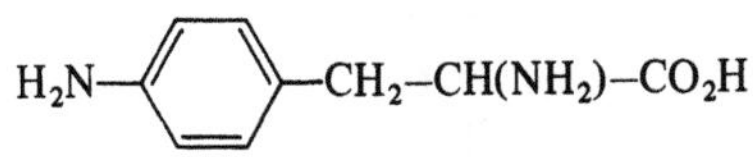

4–Aminophenylalanine
(CXXXVII)

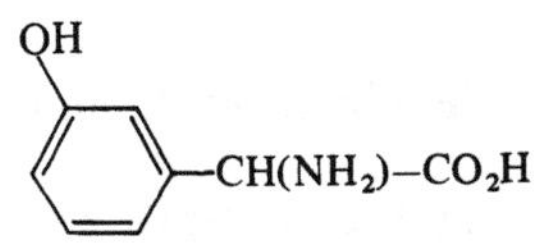

3–Hydroxyphenylglycine
(CXXXVIII)

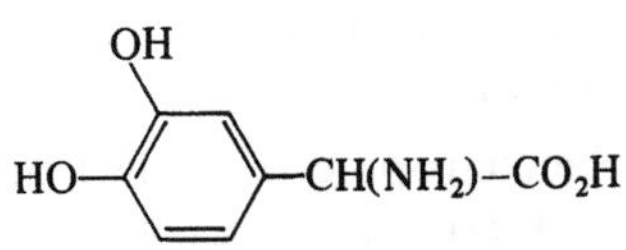

3,5–Dihydroxyphenylglycine
(CXXXIX)

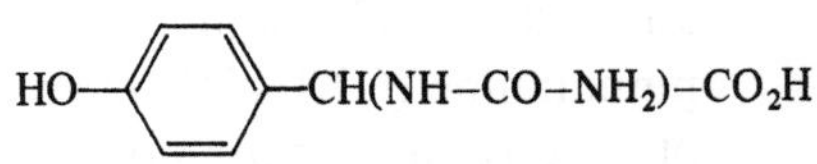

N–Carbamoyl–4–hydroxyphenylglycine
(CXL)

$H_2N-CH_2-CH_2-CH_2-CO_2H$
4–Aminobutyric acid
(CXLI)

$HO_2C-CH_2-CH(NH_2)-CH_2-CO_2H$
3–Aminoglutaric acid
(CXLII)

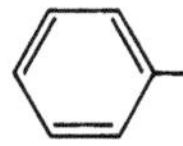$-CH(NH_2)-CH_2-CO_2H$

β–Phenyl–β–alanine
(CXLIII)

$HO_2C-CH(NH_2)-CH_2-CH_2-CH_2-CH_2-NH-CH(CO_2H)-CH_2-CH_2-CO_2H$
Saccharopine
(CXLIV)

$H_2N-CH_2-CH_2-CH_2-CH_2-CH(CO_2H)-NH-CH(CH_3)-CO_2H$
Lysopine
(CXLV)

$CH_3-C(Cl)=CH-CH_2-CH(NH_2)-CO_2H$
2–Amino–5–chloro–4–hexenoic acid
(CXLVI)

$CH_2=C(Cl)-CH_2-CH(NH_2)-CO_2H$
2–Amino–4–chloro–4–pentenoic acid
(CXLVII)

$CH_2=CH-C(CH_2OH)=CH-NH-CO-CHOH-CH_2-CH(NH_2)-CO_2H$
Pinnatanine
(CXLVIII)

3 Accumulation of Non-Protein Amino Acids

It is not possible to generalise about the accumulation of non-protein amino acids in plants. High concentrations of azetidine-2-carboxylic acid are stored in the seeds of the legume *Bussea massaiensis* (WATSON and FOWDEN, 1973) while the same compound is absent from the seeds of another legume *Delonix regia*, yet is present in the seedlings (SUNG and FOWDEN, 1968).

Canavanine, which is a major constituent of the free amino acid pool of the seeds of *Medicago sativa*, is rapidly metabolised during germination (BELL, 1960). In contrast albizziine, which accumulates in equally high concentrations in the seeds of *Albizzia julibrissin* is also a major component of the free amino acid pool of the developing seedlings.

Perhaps the most remarkable feature of non-protein amino acid accumulation is the way in which a single compound may represent a major part of the plant's soluble amino acid pool or stored nitrogen reserve. L-3,4-Dihydroxyphenylalanine accounts for 8% of the dry seed weight of *Mucuna mutisiana* (BELL and JANZEN, 1971), while 5-hydroxy-L-tryptophan may account for as much as 14% of the seed weight of *Griffonia simpliciolia* (BELL et al., 1976).

4 Biosynthesis of Non-Protein Amino Acids

Relatively little work has been reported on the biosynthesis of non-protein amino acids. Clearly, however, these compounds are formed in a variety of different ways. 5-Hydroxy-L-tryptophan (FELLOWS and BELL, 1970) and 5-acetylornithine (BROWN and FOWDEN, 1966) are formed by the hydroxylation and acetylation of tryptophan and ornithine respectively while 4-methylglutamic acid is formed by the oxidation of L-leucine (PETERSON and FOWDEN, 1972).

Another way in which non-protein amino acids may arise is by modification of biosynthetic pathways normally concerned in the synthesis of protein amino acids or amino acids found as primary metabolic intermediates. In a series of exemplary publications LARSEN et al. (1975) showed that the 3-substituted aromatic amino acids in *Reseda* and *Iris* species are derived from shikimic acid by a modification of the normal route to tyrosine and/or phenylalanine and that this modified route permits the formation of an aromatic ring without loss of the carboxyl carbon of shikimic acid.

A number of the heterocyclic amino acids found in plants are probably synthesised by a pathway analagous to the synthesis of tryptophan which involves the enzymic condensation of indole with serine. An example of such a synthesis is provided by 3-pyrazol-l-ylalanine, which is formed by the condensation of pyrazole and serine in the *Cucumis sativus* (DUNNHILL and FOWDEN, 1963). It is not clear that all heterocyclic amino acids arise this way, however, for BELL and PRZYBYLSKA (1965) have shown that radioactivity from [^{14}C]-homoarginine and [^{14}C]-4-hydroxyhomoargine is incorporated into lathyrine in *Lathyrus tingitanus*, suggesting that the pyrimidinyl nucleus is formed by the cyclisation of 4-hydroxyhomoarginine and subsequent loss of hydrogen atoms. BROWN and AL-BALDAWI (1977) have also shown, however, that radioactivity from [^{14}C]orotate and [^{14}C]uracil is incorporated into lathyrine in the same plant and these authors suggest that alternative pathways to lathyrine exist.

From the foregoing then it is clear that non-protein amino acids may be formed by the modification of pre-formed protein amino acids, by modification of pathways to protein amino acids or by biosynthetic routes which show little similarity to any concerned in the synthesis of protein amino acids.

5 The Distribution of Non-Protein Amino Acids

The distribution of non-protein amino acids and their possible taxonomic significance has been reviewed previously (BELL and FOWDEN, 1964; BELL, 1966, 1971; FOWDEN, 1970) and it is clear that valuable information concerning relationships within and between different genera, tribes and families of plants can be obtained from the study of these compounds.

If a non-protein amino acid is restricted to a limited number of plant species which are related in other respects, then it is probable that these species are all derived from a common ancestor in which the biosynthetic pathway to that non-protein amino acid first arose. If more than one non-protein amino acid is common to these same species and these amino acids are unrelated biosynthetically then the presumption of common origin is greatly strengthened.

Specific examples of the way in which distribution studies of non-protein amino acids have contributed to our understanding of plant phylogeny are given in Chapter 2 of this volume.

6 Ecological Significance of Non-Protein Amino Acids

The general question of whether secondary compounds in plants have an ecological role is discussed elsewhere in this volume (BELL, Chap. 2) and the reader is referred to that chapter for detailed information on the possible ecological significance of certain specific non-protein amino acids. Suffice to state here that many of the non-protein amino acids from plants are known to be toxic to vertebrate and invertebrate animals, to higher plants and to microorganisms. They can also act as feeding deterrents in insects (BELL, 1977).

It is possible that the presence of non-protein amino acids in a plant may protect that plant from predation, it is also possible that it may reduce competition from other plant species. Evidence for such roles is discussed in Chapter 2, this volume.

References

Bamburg, J.R., Strong, F.M., Smaley, E.B.: Toxins from moldy cereals. J. Agric. Food Chem. *17*, 443–450 (1969)

Bell, E.A.: Canavanine in the Leguminosae. Biochem. J. *75*, 618–620 (1960)

Bell, E.A.: The isolation of L-homoarginine from seeds of *Lathyrus cicera*. Biochem. J. *85*, 91–93 (1962)

Bell, E.A.: A new amino acid, γ-hydroxyhomoarginine in *Lathyrus*. Nature (London) *199*, 70–71 (1963)

Bell, E.A.: In: Comparative phytochemistry. Swain, T. (ed.), pp. 195–209. London: Academic Press 1966

Bell, E.A.: In: Chemotaxonomy of the Leguminosae. Harborne, J.B., Boulter, D., Turner, B.L. (eds.), pp. 179–206. London: Academic Press 1971

Bell, E.A.: Uncommon amino acids in plants. FEBS Lett. *64*, 29–25 (1976)

Bell, E.A.: The possible significance of uncommon amino acids in plant–vertebrate, plant–insect and plant–plant relationships. In: Natural products and the protection of plants. Marini-Bettolo, G.B. (ed.), Scripta Varia 41, pp. 571–595. Vatican City: Pontifical Academy of Science 1977

Bell, E.A., Fellows, L.E.: Occurrence of 5-hydroxy-L-tryptophan as a free plant amino acid. Nature (London) *210*, 529 (1966)

Bell, E.A., Foster, R.G.: Structure of lathyrine (a new amino acid). Nature (London) *194*, 91–92 (1962)

Bell, E.A., Fowden, L.: Studies on amino acid distribution and their possible value in plant classification. In: Taxonomic biochemistry and serology. Leone, C.A. (ed.), pp. 203–223. New York: Ronald Press 1964

Bell, E.A., Janzen, D.H.: Medical and ecological considerations of L-DOPA and 5-HTP in seeds. Nature (London) *229*, 136–137 (1971)

Bell, E.A., O'Donovan, J.P.: The isolation of α and γ-oxalyl derivatives of α,γ-diaminobutyric acid from seeds of *Lathyrus latifolius*, and the detection of the α-oxalyl isomer of the neurotoxin α-amino-β-oxalylaminopropionic acid which occurs together with the neurotoxin in this and other species. Phytochemistry *5*, 1211–1219 (1966)

Bell, E.A., Przybylska, J.: The origin and site of synthesis of the pyrimidine ring in the amino acid lathyrine. Biochem. J. *94*, 35p (1965)

Bell, E.A., Tirimanna, A.S.L.: Occurrence of γ-hydroxyarginine in plants. Nature (London) *197*, 901–902 (1963)

Bell, E.A., Tirimanna, A.S.L.: The isolation of γ-hydroxyarginine from seeds of *Vicia sativa* and the identification of γ-hydroxyornithine as a naturally occurring amino acid. Biochem. J. *91*, 356–358 (1964)

Bell, E.A., Nulu, J.R., Cone, C.: L-DOPA and L-3-carboxy-6,7-dihydroxy-1,2,3,4-tetrahydroisoquinoline, a new imino acid from seeds of *Mucuna mutisiana*. Phytochemistry *10*, 2191–2194 (1971)

Bell, E.A., Fellows, L.E., Qureshi, Y.: 5-Hydroxy-L-tryptophan: taxonomic character and chemical defense in *Griffonia*. Phytochemistry *15*, 823 (1976)

Bell, E.A., Lackey, J.A., Polhill, R.M.: The systematic significance of canavanine in the Papilionoideae (Faboideae). Biochem. Syst. Ecol. *6*, 201–212 (1978)

Biemann, K., Lioret, C., Asselineau, J., Lederer, E., Polonsky, J.: Structure of lysopine, a new amino acid isolated from crown gall tissue. Biochim. Biophys. Acta *40*, 369–370 (1960)

Blake, J., Fowden, L.: γ-Methyleneglutamic acid and related compounds in plants. Biochem. J. *92*, 136–142 (1964)

Bowden, K., Drysdale, A.C., Mogey, G.A.: Constituents of *Amanita muscaria*. Nature (London) *206*, 1359–1360 (1965)

Brandner, G., Virtanen, A.I.: Isolierung und Synthese von γ-Hydroxyl-L-glutamin. Acta Chem. Scand. *17*, 2563–2565 (1963)

Brieskorn, C.H., Glasz, J.: L(+)α-Amino-n-buttersäure als Baustein der Eiweißstoffe der Samen von *Salvia officinalis* L. Naturwissenschaften *51*, p. 216 (1964a)

Brieskorn, C.H., Glasz, J.: Zum Vorkommen der α-ε-Diamino-α-hydroxycapronsäure in den Samen von *Salvia officinalis* L. Naturwissenschaften *51*, p. 216 (1964b)

Brown, D.H., Fowden, L.: Metabolism of δ-acetylornithine in two leguminous species. Phytochemistry *5*, 887–893 (1966)

Brown, E.G., Al-Baldawi, F.: Biosynthesis of the pyrimidinyl amino acid lathyrine by *Lathyrus tingitanus* L. Biochem. J. *164*, 589–594 (1977)

Brown, E.G., Mangat, B.S.: Structure of a pyrimidine amino acid from pea seedlings. Biochem. Biophys. Acta *177*, 427–433 (1969)

Burroughs, L.F.: 1-Aminocyclopropane-1-carboxylic acid: a new amino-acid in perry pears and cider apples. Nature (London) *179*, 360–361 (1957)

Buziassy, C., Mazelis, M.: The formation of S-carboxymethyl-L-cysteine by radish (*Raphanus sativus*) seedling homogenates. Biochim. Biophys. Acta *86*, 185–186 (1974)

Chilton, W.S., Tsou, G.: A chloro amino acid from *Amanita solitaria*. Phytochemistry *11*, 2853–2857 (1972)

Chilton, W.S., Tsou, G., Kirk, L., Benedict, R.G.: A naturally occurring allenic amino acid. Tetrahedon Lett. 6283–6284 (1968)

Cramer, U., Spener, F.: Biosynthesis of cyclopentenyl fatty acids. Eur. J. Biochem. *74*, 495–500 (1977)

Dardenne, G.: Free amino acids from Leguminosae. Phytochemistry *14*, 860 (1975)

Dardenne, G., Bell, E.A., Nulu, J.R., Cone, C.: Absolute configuration of β-hydroxy-γ-methylglutamic acids from *Gymnocladus dioicus*. Phytochemistry *11*, 791–797 (1972a)

Dardenne, G., Casimir, J., Bell, E.A., Nulu, J.R.: Two stereoisomers of β-hydroxy-γ-methylglutamic acid from seeds of *Gymnocladus dioicus*. Phytochemistry *11*, 787–790 (1972b)

Dardenne, G., Marlier, M., Casimir, J.: Un nouvel acide amine dans les grains de *Vigna vexillata*. Phytochemistry *11*, 2567–2570 (1972c)

Dardenne, G., Casimir, J., Sørensen, H.: γ-L-Glutamyl-L-pipecolic acid in *Gleditsia capsica*. Phytochemistry *13*, 1515–1517 (1974a)
Dardenne, G., Casimir, J., Sørensen, H.: 2(S),3(S)-Hydroxy-4-methyleneglutamic acid from *Gleditsia capsica*. Phytochemistry *13*, 2195–2199 (1974b)
Darling, S., Larsen, P.O.: Saccharopine, a new amino acid in baker's and brewer's yeast. I. Isolation and properties. Acta Chem. Scand. *15*, 743–749 (1961)
Daxenbichler, M.E., Vanetten, C.H., Hallinan, E.A., Earle, F.R.: Seeds as sources of L-dopa. J. Med. Chem. *14*, 463–465 (1971)
Daxenbichler, M.E., Klieman, R., Weisleder, D., van Etten, C.H., Carlson, K.D.: A new amino acid, (−)-1-methyl-3-carboxy-6,7-dihydroxy-1,2,3,4-tetrahydroisoquionoline. Tetrahedron Lett. 1801–1802 (1972)
Dent, C.E.: Behaviour of sixty amino acids and other ninhydrin-reacting substances on phenol-collidine filter paper chromatograms, with notes as to the occurrence of some of them in biological fluids. Biochem. J. *43*, 169–180 (1948)
Despontin, J., Marlier, M., Dardenne, G.: L-*Cis*-5-Hydroxypipecolic acid from seeds of *Gymnocladus dioicus*. Phytochemistry *16*, 387–388 (1977)
Dietrichs, H.H., Funke, H.: Free amino acids in phloem and spring sap of red beeches (*Fagus sylvatica*). Holzforschung *21*, 102–107 (1967)
Done, J., Fowden, L.: A new amino acid amide in the groundnut plant (*Arachis hypogaea*): evidence of the occurrence of γ-methyleneglutamine and γ-methyleneglutamic acid. Biochem. J. *51*, 451–458 (1952)
Doyle, R.R., Levenberg, B.: L-3-(3-carboxyfuran-4-yl)alanine, a new amino acid from the mushroom *Phyllotopsis nidulans*. Phytochemistry *13*, 2813–2813 (1974)
Dunnill, P.M., Fowden, L.: The biosynthesis of β-pyrazol-1-ylalanine. J. Exp. Bot. *14*, 237–248 (1963)
Dunnill, P.M., Fowden, L.: The amino acids of seeds of the Cucurbitaceae. Phytochemistry *4*, 933–944 (1965)
Eagles, J., Laird, W.M., Matai, S., Self, R., Synge, R.L.M.: N-Carbamoyl-2-(p-hydroxyphenyl)-glycine from leaves of broad bean (*Vicia faba* L.). Biochem. J. *121*, 425–430 (1971)
Eloff, J.N., Grobbelaar, N.: The isolation and characterization of N-methyl-L-alanine, a novel amino acid from gifblaar (*Dichapetalum cymosum*). J. S. Afr. Chem. Inst. *20*, 190–193 (1967)
Eloff, J.N., Grobbelaar, N.: Isolation and characterization of N-methyl-L-serine from *Dichapetalum cymosum*. Phytochemistry *8*, 2201–2204 (1969)
Emery, T.: Isolation, characterization and properties of fusarainine, a δ-hydroxamic acid derivative of ornithine. Biochemistry *4*, 1410–1417 (1965)
Eugster, C.H., Mueller, G.F., Good, R.: Wirkstoffe aus *Amanita muscaria*: Ibotensäure und Muscazon. Tetrahedron Lett. 1813–1815 (1965)
Evans, C.S., Bell, E.A.: "Uncommon" amino acids in 64 species of Caesalpinieae. Phytochemistry *17*, 1127–1129 (1978)
Evans, C.S., Qureshı, Y., Bell, E.A.: Free amino acids in the seeds of *Acacia* species. Phytochemistry *16*, 565–670 (1977)
Evrard, G., Durant, F., Dardenne, G.A.: 2(S) 3(R) 4(S)-β-hydroxy-γ-methylglutamic acid. Cryst. Struct. Commun. *3*, 65–67 (1974)
Fang, S.T., Li, L.C., Ching, I.N., Tseng, K.-F.: Chemical studies on *Cucurbita moschata*. I. The isolation and structural studies of cucurbitine, a new amino acid. Sci. Sin. *10*, 845–851 (1961)
Fellows, L.E., Bell, E.A.: 5-Hydroxytryptophan, 5-hydroxytryptamine and L-tryptophan-5-hydroxylase in *Griffonia simplicifolia*. Phytochemistry *9*, 2389–2396 (1970)
Fellows, L.E., Hider, R., Bell, E.A.: 3[2-Amino-2-imidazolin-4(5)-yl]alanine (enduracididine) and 2-[2-amino-2-imidazolin-4(5)-yl]acetic acid in seeds of *Lonchocarpus sericeus*. Phytochemistry *16*, 1957–1959 (1977)
Firmin, J.L., Fenwick, R.G.: N^2-(1,3-Dicarboxypropyl)ornithine in crown gall tumours. Phytochemistry *16*, 761–762 (1977)
Fowden, L.: Azetidine-2-carboxylic acid: a new cyclic imino acid occurring in plants. Biochem. J. *64*, 323–332 (1956)
Fowden, L.: New amino acids from plants. Annu. Rep. Prog. Chem. *56*, 359–372 (1959)
Fowden, L.: A new asparagine derivative, N^4-(2-Hydroxyethyl)-L-asparagine from Bryony (*Bryonia dioica*). Biochem. J. *81*, 154–156 (1961)

Fowden, L.: Isolation of γ-hydroxynorvaline from *Lathyrus odoratus* seed. Nature (London) *209*, 807–808 (1966a)
Fowden, L.: The acidic amino acids of Tulip: Isolation of γ-ethylideneglutamic acid. Biochem. J. *98*, 57–61 (1966b)
Fowden, L.: The non-protein amino acids of plants. Prog. Phytochem. *2*, 203–266 (1970)
Fowden, L.: The amino acid complement of plants. Phytochemistry *11*, 2271–2276 (1972)
Fowden, L.: Non-protein amino acids from plants, distribution, biosynthesis and analog functions. Recent Adv. Phytochem. *8*, 95–122 (1974)
Fowden, L., Bryant, M.: Nitrogenous compounds and nitrogen metabolism in the Lilaceae. 4. Isolation of azeitidine-2-carboxylic acid and evidence for the occurrence of α,γ-diaminobutyric acid in *Polygonatum*. Biochem. J. *70*, 626–629 (1958)
Fowden, L., Smith, A.: Newly characterised amino acids from *Aesculus californica*. Phytochemistry *7*, 809–819 (1968)
Fowden, L., MacGibbon, C.M., Mellon, F.A., Sheppard, R.C.: Newly characterized amino acids from *Blighia unijugata*. Phytochemistry *11*, 1105–1110 (1972)
Fowden, L., Pratt, H.M., Smith, A.: 4-Hydroxyisoleucine fròm seed of *Trigonella foenum-graecum*. Phytochemistry *12*, 1707–1711 (1973)
Fritz, H., Gagneux, A.R., Zbinden, R., Eugster, C.H.: Structure of muscazone. Tetrahedron Lett. 2075–2076 (1965)
Furuyama, Lt., Yamashita, T., Senoh, S.: The synthesis of L-theanine. Bull. Chem. Soc. Jpn. *37*, p. 1078 (1964)
Ghosal, S., Dutta, S.K.: Alkaloids of *Abrus precatorius*. Phytochemistry *10*, 195–198 (1971)
Gmelin, R.: Die freien Aminosäuren der Samen von *Acacia willardiana* (Mimosaceae). Isolierung von Willardiin, einer neuen pflanzlichen Aminosäure vermutlich L-Uracil-β[-(a-Amino-Propionsäure)]-(3). Z. Physiol. Chem. *316*, 164–169 (1959)
Gmelin, R.: Dichrostachinsäure, eine neue schwefelhaltige Aminosäure aus den Samen von *Dichrostachys glomerata* (Forsk.) Hatch und Dalz. (Mimosaceae). Z. Physiol. Chem. *327*, 186–194 (1962)
Gmelin, R., Hietala, P.K.: S-[β-Carboxy-isopropyl]-L-Cysteine, eine neue Aminosäure aus den Samen von *Acacia millefolia* und *Acacia willardiana* (Mimosaceae). Z. Physiol. Chem. *322*, 278–282 (1960)
Gmelin, R., Larsen, P.O.: L-γ-Ethylideneglutamic acid and L-γ-methyleneglutamic acid in seeds of *Tetrapleura tetraptera* (Schum. et Thonn.) Taub. (Mimosaceae). Biochim. Biophys. Acta *136*, 572–573 (1967)
Gmelin, R., Hasenmaier, G., Strauss, G.: Über das Vorkommen von Djenkolsäure und einer c-s-Lyase in den Samen von *Albizzia lophantha* Benth. (Mimosaceae). Z. Naturforsch. *126*, 687–697 (1957)
Gmelin, R., Strauss, G., Hasenmaier, G.: Über neue Aminosäuren aus Mimosaceen. Z. Physiol. Chem. *341*, 28–32 (1959)
Gmelin, R., Kjaer, A., Larsen, P.O.: N-Acetyl-L-djenkolic acid a novel amino acid isolated from *Acacia farnesiana* Willd. Phytochemistry *1*, 233–236 (1962)
Goldmann, A., Thomas, D.W., Morel, G.: Sur la structure de la nopaline métabolite anormal de certaines tumeurs de crown-gall. C. R. Acad. Sci. *268*, 852–854 (1969)
Gray, D.O.: *Trans*-4-Hydroxymethyl-D-proline from *Eriobotrya japonica*. Phytochemistry *11*, 751–756 (1972)
Gray, D.O., Fowden, L.: N-Ethyl-L-asparagine: a new amino acid amide from *Ecballium*. Nature (London) *189*, 401–402 (1961)
Gray, D.O., Fowden, L.: α-(Methylenecylopropyl)glycine from *Litchi* seeds. Biochem. J. *82*, 385–389 (1962a)
Gray, D.O., Fowden, L.: 4-Methyleneprolıne: a new naturally occurrıng prolıne derivative. Nature (London) *193*, 1285–1286 (1962b)
Green, R.G., Davis, N.B.: Biosynthesis of S-methylmethionine in the jack bean. Biochim. Biophys. Acta *43*, 360–362 (1963)
Grobbelaar, N., Steward, F.C.: O-Acetylhomoserine in *Pisum*. Nature (London) *182*, 1358–1359 (1958)
Grove, M.D., Daxenbichler, M.E., Weisleder, D., van Etten, C.H.: The structure of pinnatinine, a new amino acid from *Staphylea pinnata* L. Tetrahedron Lett. *47*, 4477–4480 (1971)

Hassall, C.H., Reyle, K.: Hypoglycin A and B, two biologically active polypeptides from *Blighia sapida*. Biochem. J. *60*, 334–339 (1955)
Hatanaka, S.-I.: A new amino acid isolated from *Morchella esculenta* and related species. Phytochemistry *8*, 1305–1308 (1969)
Hatanaka, S.-I., Katayama, H.: L-γ-Propylideneglutamic acid and related compounds from *Mycena pura*. Phytochemistry *14*, 1434–1436 (1975)
Hatanaka, S.-I., Kaneko, S.: *Cis*-5-Hydroxy-L-pipecolic acid from *Morus alba* and *Lathyrus japonicus*. Phytochemistry *16*, 1041–1042 (1977)
Hatanaka, S.-I., Niimura, Y., Taniguchi, K.: L-2-Aminohex-4-ynoic acid: a new amino acid from *Tricholomopsis rutilans*. Phytochemistry *11*, 3327–3329 (1972)
Hatanaka, S.-I., Kaneko, S., Niimura, Y., Kinoshita, F., Soma, G.: L-2-Amino-4-chloro-4-pentenoic acid. A new natural amino acid from *Amanita pseudoporphyria* Hongo. Tetrahedron Lett. 3931–3932 (1974)
Hatanaka, S.-I., Iizumi, H., Atsuko, T.: L-2-Amino-4-methylpimelic acid: a new amino acid from *Lactarius* species. Phytochemistry *14*, 1559–1560 (1975)
Hattori, S., Komamine, A.: Stizolobic acid: a new amino acid in *Stizolobium hassjoo*. Nature (London) *183*, 1116–1117 (1959)
Hegarty, M.P., Pound, A.W.: Indospicine, a new hepatotoxic amino acid from *Indigofera spicata*. Nature (London) *217*, 354–355 (1968)
Honkanen, E., Moisio, T., Virtanen, A.I., Melera, A.: Structure of the new amino acid A, 2-methylene-cycloheptene-1,3-diglycine, isolated from the mushroom *Lactarius helvus*. Acta Chem. Scand. *18*, 1319–1320 (1964)
Hulme, A.C.: A new amino acid from the peel of apple fruits. Nature (London) *174*, 1055–1056 (1965)
Hulme, A.C., Arthington, W.: Methyl proline in young apple fruits. Nature (London) *173*, 588–589 (1954)
Ichihara, A., Hasegawa, H., Sato, H., Koyama, M., Sakamura, S.: The structure of a new amino acid from *Fagopyrum esculentum* Moench. Tetrahedron Lett. 37–38 (1973)
Impellizzeri, G., Piattelli, M., Sciuto, S., Fattorrusso, E.: Pyrolidine-2,4-dicarboxylic acid, a new naturally occurring imino acid. Phytochemistry *16*, 1601–1602 (1977)
Inatomi, H., Inugai, F., Murakami, T.: Isolierung und Identifizierung von Canalin in den unreifen Samen von *Astragalus sinicus* L. Chem. Pharm. Bull. *16*, p. 2521 (1968)
Inatomi, H., Morita, Y., Ozawa, M., Suyama, Y., Inugai, F.: Free amino acids in plants. VII. Isolation of γ-hydroxy-citrulline from unripe seeds of *Vicia faba*. Chem. Abstr. 18541y (1970)
Ito, K., Fowden, L.: New characterisations of amino acids and γ-glutamyl peptides from *Acacia georginae* seed. Phytochemistry *11*, 2541–2545 (1972)
Ito, K., Hashimoto, Y.: Gigartinine: a new amino acid in red algae. Nature (London) *211*, 417 (1966)
Jadot, J., Casimir, J., Loffet, A.: Séparation et charactérisation de l'acide 2(S), 4(R)-γ-hydroxy-γ-methylglutamique à partir de *Ledenbergia rosea-aenea* et de l'acid 2(S), 4(S)-γ-hydroxy-γ-methylglutamique à partir de *Pandanus veitchii*. Biochim. Biophys. Acta *136*, 79–88 (1967)
King, F.F., King, T.J., Warwick, A.J.: The chemistry of extractives from hardwoods. Pt. III. Baikiain an amino acid present in *Baikiaea plurijuga*. J. Chem. Soc. 3590–3597 (1950)
Kitagawa, M., Tomiyama, T.: A new amino compound in the Jack bean and a corresponding new enzyme. J. Biochem. (Tokyo) *11*, 265–271 (1929)
Kjaer, A., Larsen, P.O.: Saccharopine, a new amino acid in baker's and brewer's yeast. II. Structure and synthesis. Acta. Chem. Scand. *15*, 750–759 (1961)
Kjaer, A., Larsen, P.O.: D-(3-Carboxy-4-hydroxyphenyl)-glycine and related amino acids in higher plants. Acta Chem. Scand. *17*, 2397–2409 (1963)
Klosterman, H.J., Lameroux, G.L., Parsons, J.L.: Isolation, characterization and synthesis of linatine. A vitamin B_6 antagonist from flax seed (*Linum usitatissimum*). Biochemistry *6*, 170–177 (1967)
Koyama, M., Obata, Y.: Isolation and structure of γ-L-glutamyl-L-β-phenyl-β-alanine, a new γ-glutamyl peptide from *Phaseolus angularis*. Agric. Biol. Chem. Jpn. *30*, 472–477 (1966)
Kristensen, I., Larsen, P.O., Sørensen, H.: Free amino acids and γ-glutamyl peptides in seeds of *Fagus silvatica*. Phytochemistry *13*, 2803–2811 (1974)
Lambein, F., van Parijs, R.: Isolation and characterization of 1-alanyl-uracil (willardine) and

3-alanyl-uracil (isowillardine) from *Pisum sativum*. Biochem. Biophys. Res. Commun. *32*, 474–479 (1968)
Lambein, F., Kuo, Y.-H., van Parijs, R.: Isoxazolin-5-ones. Chemistry and biology of a new class of plant products. Heterocycles *4*, 567–593 (1976)
Larsen, P.O.: Amino acids and γ-glutamyl derivatives in seeds of *Lunaria annua* L. Part II. Acta Chem. Scand. *19*, 1071–1078 (1965)
Larsen, P.O.: N^6-Trimethyl-L-lysine betaine from seeds of *Reseda luteola* L. Acta Chem. Scand. *22*, p. 1369 (1968)
Larsen, P.O., Sørensen, H., Cochran, D.W., Hagaman, E.W., Wenkert, E.: 2(S), 4(R)-4-(β-D-Galactopyranosyloxy)-4-isobutylglutamic acid: a new amino acid in *Reseda odorata*. Phytochemistry *12*, 1713–1719 (1973)
Larsen, P.O., Onderka, D.K., Floss, H.G.: Biosynthesis of phenylalanine, tyrosine, 3-(3-carboxyphenyl)alanine and 3-(3-carboxy-4-hydroxyphenyl)alanine in higher plants. Examples of the transformation possibilities for chorismic acid. Biochim. Biophys. Acta *381*, 397–408 (1975)
Levenberg, B.: Isolation and structure of agaritine, a γ-glutamyl-substituted arylhydrazine derivative from Agaricaceae. J. Biol. Chem. *239*, 2267–2273 (1964)
Liss, I.: N-Acetyldiaminobuttersäure, eine neue Aminosäure aus dem Latex von *Euphorbia pulcherrima*. Willd. ex Klotzsch. Phytochemistry *1*, 87–88 (1962)
Madgwick, J.C., Ralph, B.J., Shannon, J.S., Simes, J.J.: Non-protein amino acids in Australian seaweeds. Arch. Biochem. Biophys. *141*, 766–767 (1970)
Mandava, N., Anderson, J.D., Dutky, S.R.: Indole plant-growth inhibitor from *Abrus precatorius* seeds. Phytochemistry *13*, 2853–2856 (1974)
Manske, R.H.F.: The natural occurrence of acetyl-ornithine. Can. J. Res. *15 B*, 84–87 (1937)
Marlier, M., Dardenne, G., Casimir, J.: 4,5-Dihydroxy-L-pipecolique à partir de *Calliandra haematocephala*. Phytochemistry *11*, 2597–2599 (1972)
Marlier, M., Dardenne, G., Casimir, J.: 2S-Carboxy-4R,5S-dihydroxy-piperidine et 2S-carboxy-4S,5S-dihydroxypiperidine à partir de *Derris elliptica*. Phytochemistry *15*, 183–185 (1976)
Matikkala, E.J., Virtanen, A.I.: A new γ-glutamyl peptide, γ-L-glutamyl-S-(prop-1-enyl)-L-cysteine in seeds of chives (*Allium schoenoprasum*). Acta Chem. Scand. *16*, 2461–2462 (1962)
Matikkala, E.J., Virtanen, A.I.: Synthesis of 3,3′-(2-methylethylene-1,2-dithio)-dialanine, an amino acid found as γ-glutamylpeptide in seeds of chive (*Allium schoenoprasum*). Acta Chem. Scand. *18*, 2009–2011 (1964)
Mbadiwe, E.I.: A new dicarboxylic acid from seeds of *Pentaclethra macrophylla*. Phytochemistry *14*, 1351–1354 (1975)
Menage, A., Morel, G.: Sur la présence d'octopine dans les tissus de crown-gall. C. R. *259*, 4795–4796 (1964)
Morgan, J.W.W.: 4-Hydroxy-N-methyl-L-proline from *Afrormosia elata* heatwood. Chem. Ind. 542–543 (1964)
Morris, C.J., Thompson, J.F., Asen, S., Irreverre, F.: Isolation of a new acidic aromatic amino acid (m-carboxy-α-phenylglycine) from Iris bulb. J. Am. Chem. Soc. *81*, 6069–6070 (1959)
Mothes, von K., Schütte, H.R., Müller, P., Ardenne, M.V., Tümmler, R.: m-Tyrosin, eine neue Aminosäure aus dem Milchsaft von *Euphorbia myrsinites* L. Z. Naturforsch. 19b, 1161–1162 (1964)
Müller, L.P., Schütte, H.R.: Zur Biosynthese des m-Tyrosine aus dem Latex von *Euphorbia myrsinites* (L.). Flora (Jena) *158*, 421–432 (1967)
Murti, V.V.S., Seshadri, T.R., Venkitasubramanian, T.A.: Neurotoxic compounds of the seeds of *Lathyrus sativus*. Phytochemistry *3*, 73–78 (1964)
Mwauluka, K., Bell, E.A., Charlwood, B.V., Briggs, J.M.: N-Methyl-L-tyrosine (L-surinamine) from seeds of *Combretum zeyheri*. Phytochemistry *14*, 1657–1658 (1975a)
Mwauluka, K., Charlwood, B.V., Briggs, J.M., Bell, E.A.: L-3-(3-Amino-methylphenyl)alanine, a new acid from seeds of *Combretum zeyheri*. Biochem. Physiol. Pflanz. *168*, 15–18 (1975b)
Navon, A., Bernays, E.A.: Inhibition of feeding in acridids by non-protein amino acids. Comp. Biochem. Physiol. 59A, 161–164 (1978)
Nawaz, R., Sørensen, H.: Saccharopine and 2-aminoadipic acid in higher plants. Phytochemistry *16*, 599–600 (1977)

Nigam, S.N., McConnell, W.B.: Isolation and identification of two isomeric glutamylselenocystathionines from the seeds of *Astragalus pectinatus*. Biochim. Biophys. Acta *437*, 116–121 (1976)
Niimura, Y., Hatanaka, S.-I.: L-*Threo* and L-*erythro* 2-amino-3-hydroxyhex-4-ynoic acids: new amino acids from *Tricholomopsis rutilans*. Phytochemistry *13*, 175–178 (1974)
Niimura, Y., Hatanaka, S.-I.: Two γ-glutamylpeptides of acetylenic amino acids in *Tricholomopsis rutilans*. Phytochemistry *16*, 1435–1436 (1977)
Noe, F.F., Fowden, L.: β-Pyrazol-1-ylalanine, an amino acid from water-melon seeds (*Citrullus vulgaris*). Biochem. J. *77*, 543–547 (1960)
Noguchi, M., Sakuma, H., Tamaki, E.: The isolation and identification of nicotianine: a new amino acid from tobacco leaves. Phytochemistry *7*, 1861–1866 (1968)
Noma, M., Noguchi, M., Tamaki, E.: A new amino acid nicotianamine, from tobacco leaves. Tetrahedron Lett. 2017–2020 (1971)
Nulu, J.R., Bell, E.A.: Configuration of L-γ-ethylideneglutamic acid from *Guilandina crista*. Phytochemistry *11*, 2573–2576 (1972)
Okuda, T., Yoshida, T., Shiota, N., Nobuhara, J.: L-3-Carboxy-1,2,3,4-tetrahydro-β-carboline, a new amino acid from seeds of *Aleurites fordii*. Phytochemistry *14*, 2304–2305 (1975)
Pant, R., Fales, H.M: Occurrence of a new amino acid in *Crotalaria* seeds. Phytochemistry *13*, 1626–1627 (1974)
Peterson, P.J., Fowden, L.: The biosynthesis of γ-substituted glutamic acids in *Gleditsia triacanthos*. Phytochemistry *11*, 663–667 (1972)
Peterson, P.J., Robinson, P.J.: L-Cystathionine and its selenium analogue in *Neptunia amplexicaulis*. Phytochemistry *11*, 1837–1839 (1972)
Przybylska, J., Pawelkiewicz, J.: O-Oxalylhomoserine, a new homoserine derivative in young pods of *Lathyrus sativus*. Bull. Acad. Pol. Sci. Ser. Biol. *13*, 327–329 (1965)
Przybylska, J., Strong, F.M: Identification of γ-methylglutamic acid in *Lathyrus maritimus*. Phytochemistry *7*, 471–475 (1968)
Rao, S.L.N., Adiga, P.R., Sarma, P.S.: The isolation and characterization of β-N-oxalyl-L-α,β-diaminopropionic acid: a neurotoxin from the seeds of *Lathyrus sativus*. Biochemistry *3*, 432–436 (1964)
Renz, J.: Über das Mimosin. Z. Physiol. Chem. *244*, 153–158 (1936)
Ressler, C.: Isolation and identification from common vetch of the neurotoxin β-cyano-L-alanine, a possible factor in neurolathyrism. J. Biol. Chem. *237*, 733–735 (1962)
Schneider, G.: 2,4-Dihydroxy-6-methylphenylalanin, eine neue Aminosäure aus den Samen von *Agrostemma githago*. Biochem. Z. *330*, 428–432 (1958)
Senoh, S., Imamoto, S., Maeno, Y.: Structure of stizolobic acid and stizolobinic acid; two novel amino acids in *Stizolobium hassjoo*. Tetrahedron Lett. 3431–3438 (1964)
Seneviratne, A.S., Fowden, L.: The amino acids of the genus *Acacia*. Phytochemistry *7*, 1039–1045 (1968)
Shewry, P.R., Fowden, L.: 4,5-Dihydroxypipecolic acids in the seed of *Julbernardia*, *Isoberlinia* and *Brachystegia*. Phytochemistry *15*, 1981–1983 (1976)
Shrift, A.: Metabolism of selenium by plants and microorganisms. In: Organic selenium compounds, their chemistry and biology. Klayman, L., Gunther, W.H.H. (eds.), pp. 763–814. Wiley-Interscience 1973
Spare, C.-G., Virtanen, A.I.: On the lachrymatory factor in onion (*Allium cepa*) vapours and its precursor. Acta Chem. Scand. *17*, 641–650 (1963)
Staron, T., Allard, C., Dat Xuong, N.: Isolement de l'acide α-aminooctanoique à partir d'un hétéropeptide d'origine fongique. C. R. *260*, 3502–3504 (1965)
Sugii, M., Suzuki, T., Nagasawa, S.: Isolation of (−)-(S)-propenyl-L-cysteine from garlic. Chem. Pharm. Bull. *11*, 548–549 (1963)
Suketa, Y., Sugii, M., Suzuki, T.: Isolation and biosynthesis of L-homomethionine (L-5-methylthionorvaline). Chem. Pharm. Bull. *18*, 249–254 (1970)
Sung, M., Fowden, L.: Trans-3-hydroxy-L-proline: a constituent of *Delonix regia*. Phytochemistry *7*, 2061–2063 (1968)
Sung, M.-L., Fowden, L.: Azetidine-2-carboxylic acid from the legume *Delonix regia*. Phytochemistry *8*, 2095–2096 (1969)
Takagi, N., Hsu, H.Y., Takemoto, T.: Hypotensive constituents of marine algae. 5. Amino acid constituents of *Petalonia fascia*. Yakugaku Zasshi *90*, 899–902 (1970)

Takemoto, T., Nakajima, T.: Structure of tricholomic acid, a novel amino acid isolated from *Tricholoma muscarium*. Yakugaku Zasshi *84*, 1230–1232 (1964)
Takemoto, T., Daigo, K., Takagi, N.: Hypotensive constituents of marine algae (i) basic amino acid laminine, and the basic constituents from *Laminaria augustata*, (ii) synthesis of laminine and related compounds. Yakugaku Zasshi *84*, 1176–1179 (1964a)
Takemoto, T., Nakajima, T., Yokobe, T.: Structure of ibotenic acid. Yakugaku Zasshi *84*, 1232–1233 (1964b)
Takemoto, T., Daigo, K., Sai, T.: Constituents of *Chondria armata*. VII. Isolation of 3-aminoglutaric acid. Yakugaku Zasshi *85*, 83–85 (1965)
Thompson, J.F., Morris, C.J., Zacharius, R.M.: Isolation of (−) S-methyl-L-cysteine from beans (*Phaseolus vulgaris*). Nature (London) *178*, 593 (1959)
Thompson, J.F., Morris, C.F., Asen, S., Irreverre, F.: m-Carboxy-phenyl-L-alanine: a dicarboxylic aromatic amino acid from Iris bulbs. J. Biol. Chem. *236*, 1183–1185 (1961)
Thompson, J.F., Morris, C.J., Hunt, G.E.: The identification of L-α-amino-δ-hydroxyvaleric acid and L-homoserine in Jack Bean seeds (*Canavalia ensiformis*). J. Biol. Chem. *239*, 1122–1125 (1964)
Untch, K.G., Gibbon, G.A.: The stereochemistry of 4-hydroxymethylproline. Tetrahedron Lett. 3259–3264 (1964)
Vega, A., Bell, E.A.: α-Amino-β-methylaminopropionic acid, a new amino acid from seeds of *Cycas circinalis*. Phytochemistry *6*, 759–762 (1967)
Vervier, R., Casimir, J.: Isolement et characterisation d'un nouvel acid aminé, la β-méthylene-1-(+)-norleucine, dans les carpophores d'*Amanita vaginata* var *Fulva*. Phytochemistry *9*, 2059–2060 (1970)
Virtanen, A.I., Berg, A.M.: A new α-aminodicarboxylic acid, α-aminopimelic acid in green plants. Acta Chem. Scand. *8*, 1085–1086 (1954)
Virtanen, A.I., Ettala, T.: Dihydroxyglutamic acid in plants. Acta Chem. Scand. *11*, 182–184 (1957)
Virtanen, A.I., Hietala, P.K.: γ-Hydroxyglutamic acid in green plants. Acta Chem. Scand. *9*, 175–176 (1955)
Virtanen, A.I., Kari, S.: 5-Hydroxy-piperidine-2-carboxylic acid in green plants. Acta Chem. Scand. *8*, 1290–1291 (1954)
Watson, R., Fowden, L.: Amino acids of *Caesalpinia tinctoria* and some allied species. Phytochemistry. *12*, 617–622 (1973)
Welter, A., Jadot, J., Dardenne, G., Marlier, M., Casimir, J.: Acide 3(R)[1′(S)-aminocarboxymethyl]2-pyrrolidone-5(S)-carboxylique dans les graines de *Pentaclethra macrophylla*. Phytochemistry *14*, 1347–1350 (1975)
Welter, A., Marlier, M., Dardenne, G.: Nouveau acids amines libres de *Afzelia bella*: *Trans*-hydroxy-4-L-proline et *trans*-carboxy-4-L-proline. Phytochemistry *17*, 131–134 (1978)
Wieland, T.: Poisonous principles of mushrooms of the genus *Amanita*. Science *159*, 946–952 (1968)
Wieland, T., Fahrmeir, A.: Oxydation und Reduction an der γ,δ-Dihydroxy-Isoleucine-Seitenkette des O-Methyl-α-Amatines, Methyl-aldoamanitin, ein ungiftiges Abbauprodukt. Ann. *736*, 95–99 (1970)
Wieland, T., Schön, W.: Über die Giftstoffe des grünen Knollenblätterpilzes. Die Konstitution des Phalloidins. Ann. *593*, 157–178 (1955)
Wilding, M.D., Stahmann, M.A.: Hydroxyaspartic acid from alfalfa and clover roots. Phytochemistry *1*, 241–244 (1962a)
Wilding, M.D., Stahmann, M.A.: Hydroxylysine from alfalfa roots. Phytochemistry *1*, 263–265 (1962b)

8 Plant Amines

T.A. SMITH

1 Introduction

Amines are related structurally to ammonia by substitution. The fundamental classes of the amines are: NH_2R primary amine, NHRR′ secondary amine, NRR′R″ tertiary amine, $N^+RR'R''R'''$ (OH^-) quaternary amine. Secondary amines may react with nitrous acid to give carcinogenic nitrosamines, e.g., dimethylnitrosamine $[(CH_3)_2N-N=O]$ which occurs naturally in *Solanum incanum* (DU PLESSIS et al., 1969). Tertiary amines may form *N*-oxides, e.g., $(CH_3)_3N\rightarrow O$, (trimethylamine *N*-oxide) found in marine algae (FUJIWARA-ARASAKI and MINO, 1972). Quaternary amines are unstable, in the case of fully methylated amines decomposing to give trimethylamine (Hofmann degradation).

A wide variety of amines occurs in higher plants. These are often derived from amino acids by decarboxylation (Table 1), though aldehyde transamination is an important source for some, especially the simple aliphatic amines (see Sect. 2). The distinction between alkaloids and amines is sometimes not well defined and some true amines are often referred to as alkaloids (e.g., mescaline). However, alkaloids almost invariably have the N atoms in heterocyclic groups, and they usually have a higher molecular weight than the amines. Amines are precursors for several alkaloids. The group termed the biogenic amines excludes compounds bearing carboxyl groups, substances which may even be quite basic, like arginine. Also excluded are the purines and pyrimidines, which are best considered with the nucleic acids. The plant amines have been reviewed earlier by WERLE (1955), GUGGENHEIM (1958), SCHÜTTE (1969) and SMITH (1971, 1977d).

Table 1. Amino acids as precursors of amines in plants. The evidence cited for decarboxylation is sometimes very indirect, and in other cases results from the extraction and characterisation of the enzyme responsible

Amino acid	Product	Source	Reference
Glycine	Methylamine	*Limonium vulgare*	LARHER and BERNARD (1971)
		Lichens	BERNARD and LARHER (1971)
Alanine	Ethylamine	*Cucumis sativus* *Ecballium elaterium*	CROCOMO and FOWDEN (1970)
		Camellia sinensis	TAKEO (1978)
		Zea mays	PREUSSER (1975)
2-Aminobutyric acid	*n*-Propylamine	Rhodophyceae	HARTMANN (1972a, b, c)

Table 1 (continued)

Amino acid	Product	Source	Reference
Norvaline	*n*-Butylamine	*Claviceps*	HARTMANN (1965)
		Rhodophyceae	HARTMANN (1972a, b, c)
Valine	iso-Butylamine	*Arum maculatum*	SIMON (1962)
		Crataegus monogyna *Sorbus aucuparia*	RICHARDSON (1966)
		Rhodophyceae	HARTMANN (1972a, b, c)
		Lychnis coronaria	HARTMANN et al. (1972b)
		Vicia faba	PREUSSER (1975)
Norleucine	*n*-Amylamine	*Claviceps*	HARTMANN (1965)
		Rhodophyceae	HARTMANN (1972a, b, c)
Leucine	iso-Amylamine	*Crataegus monogyna* *Sorbus aucuparia*	RICHARDSON (1966)
		Claviceps purpurea	HARTMANN (1965)
		Rhodophyceae	HARTMANN (1972a, b, c) HARTMANN and AUFERMANN (1973)
Isoleucine	2-Methyl-butylamine	Rhodophyceae	HARTMANN (1972a, b, c)
1-Aminoheptanoic acid	*n*-Hexylamine	*Claviceps purpurea*	HARTMANN (1965)
1-Aminocaprylic acid	*n*-Heptylamine	*Claviceps purpurea*	HARTMANN (1965)
Serine	Ethanolamine	*Cucumis sativus* *Ecballium elaterium*	CROCOMO and FOWDEN (1970)
Ornithine	Putrescine	*Aspergillus nidulans*	STEVENS and MCKINNON (1977)
		Blastocladiella	MENNUCCI et al. (1975)
		Euglena gracilis	LAFARGE-FRAYSSINET et al.(1978)
		Limonium vulgare	LARHER (1973)
		Atropa belladonna *Datura stramonium* *Hordeum vulgare* *Lycopersicon esculentum* *Nicotiana tabacum*	MIZUSAKI et al. (1973)
		Pisum sativum	HASSE et al. (1967)
Citrulline	*N*-Carbamyl-putrescine	*Saccharum officinarum*	MARETZKI et al. (1969)
		Sesamum indicum	CROCOMO and BASSO (1974)
Arginine	Agmatine	*Glycine max*	LE RUDULIER and GOAS (1975a)
		Hordeum vulgare	SMITH (1963)
		Lathyrus sativus	RAMAKRISHNA and ADIGA (1975)
		Limonium vulgare	LARHER (1973)
		Nicotiana tabacum	YOSHIDA (1969)
		Oryza sativa	LEE et al. (1972)
		Panus tigrinus	BOLDT et al. (1971)
		Scorzonera hispanica	SPERANZA and BAGNI (1977)
		Sesamum indicum	CROCOMO and BASSO (1974)
Homoarginine	Homoagmatine	*Lathyrus sativus*	RAMAKRISHNA and ADIGA (1976)

Table 1 (continued)

Amino acid	Product	Source	Reference
Lysine	Cadaverine	*Lathyrus sativus*	RAMAKRISHNA and ADIGA (1976)
		Lupinus luteus	SCHÜTTE and KNÖFEL (1968)
		Pisum sativum	HASSE et al. (1967)
		Sarothamnus scoparius	SCHÜTTE and KNÖFEL (1968)
Cysteine	2-Thioethylamine	Rhodophyceae	HARTMANN (1972a)
Homocysteine	3-Thiopropylamine	Rhodophyceae	HARTMANN (1972a)
Methionine	3-Methylthio-propylamine	Rhodophyceae	HARTMANN (1972a, c)
		Coprinus lagopus	WILSON (1970)
S-Adenosyl-methionine	*S*-Methyladenosyl-homocysteamine	*Saccharomyces cerevisiae*	COPPOC et al. (1971) PÖSÖ et al. (1975)
		Brassica chinensis *Daucus carota* *Phaseolus aureus*	COPPOC et al. (1971)
		Lathyrus sativus	SURESH and ADIGA (1977)
		Vinca rosea	BAXTER and COSCIA (1973)
Histidine	Histamine	*Dolichothele sphaerica*	ROSENBERG and STOHS (1976)
		Gossypium hirsutum	LLOYD and NICHOLLS (1965)
		Spinacia oleracea	HAARTMANN et al. (1966)
Tryptophan	Tryptamine	*Citrullus vulgaris*	DANNENBURG and LIVERMAN (1957)
		Cucumis sativus	SHERWIN (1970) ELKINAWY and RAA (1974)
		Juglans regia	BERGMANN et al. (1970)
		Lycopersicon esculentum *Hordeum vulgare*	GIBSON et al. (1972a) SCHNEIDER and WIGHTMAN (1974)
		Nicotiana tabacum	PHELPS and SEQUEIRA (1967)
		Piptadenia peregrina	FELLOWS and BELL (1971)
		Phalaris tuberosa	BAXTER and SLAYTOR (1972)
		Solanum tuberosum	GIEBEL and JACKOWIAK (1976)
		Zea mays	KEFELI et al. (1969)
5-Hydroxy-tryptophan	5-Hydroxytrypt-amine	*Lycopersicon esculentum*	GIBSON et al. (1972a)
Phenylalanine	Phenethylamine	Rhodophyceae	HARTMANN (1972a, b)
Tyrosine	Tyramine	*Hordeum vulgare*	GALLON and BUTT (1971) HOSOI (1974)
		Lolium multiflorum	O'DONOVAN and HORAN (1971)
		Lophophora williamsii	LUNDSTRÖM (1971) LUNDSTRÖM and AGURELL (1971)
		Trichocereus pachanoi	LUNDSTRÖM (1970)

Table 1 (continued)

Amino acid	Product	Source	Reference
Dihydroxyphenyl-alanine	Dopamine	*Cytisus scoparius*	TOCHER and TOCHER (1972)
		Lophophora williamsii	LUNDSTRÖM (1971)
		Monostroma fuscum	TOCHER and TOCHER (1969)
		Trichocereus pachanoi	LUNDSTRÖM (1970)

2 Aliphatic Monoamines

The lower aliphatic monoamines are widely distributed in the plant kingdom (Table 2). They are often produced in flowers at anthesis or by the fruiting bodies of fungi. Insects which carry pollen or spores are attracted by the smell of these amines, which may simulate rotting meat. Methylamine and ethanolamine are found ubiquitously in distillates of plant material, and in many cases they may be formed as artefacts by the degradation of larger molecules (e.g., phosphatides) in the course of isolation (HARTMANN et al., 1972b; LERCH and STEGEMANN, 1966). However, LERCH and STEGEMANN (1966), using mild extraction procedures, have found free ethanolamine in 41 species of higher plants. Trimethylamine occurs at high levels especially in plants from maritime environments, such as *Limonium* (LARHER and BERNARD, 1971), *Chenopodium* (CROMWELL and RICHARDSON, 1966) and marine algae (FUJIWARA-ARASAKI and MINO, 1972). The occurrence of trimethylamine in marine fish and in these plants may be related to the common saline habitat, possibly in osmotic regulation.

In the Rhodophyceae, a wide range of amines is formed by a non-specific particulate amino acid decarboxylase (HARTMANN, 1972a, b, c; HARTMANN and AUFERMANN, 1973). Leucine was the best substrate (Scheme 1), but the L-forms of norleucine, isoleucine, valine, norvaline, 2-aminobutyric acid, phenylalanine, methionine, cysteine, and homocysteine were also attacked. Pyridoxal phosphate could be displaced from the enzyme by incubation with substrate, and almost pure apodecarboxylase could be obtained in this way. The endogenous amino acids are not available to the enzyme, and it appears that free amino acids in the

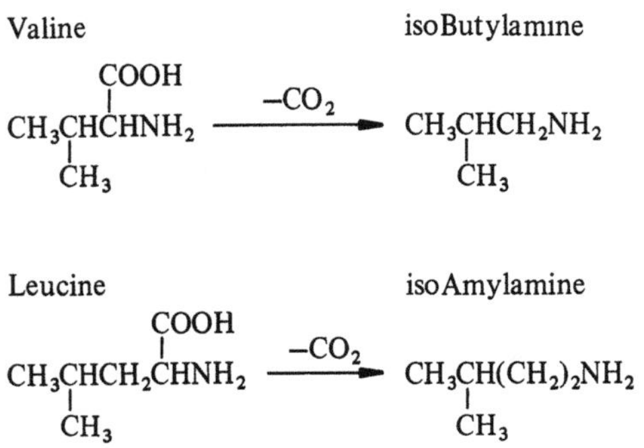

Scheme 1. Formation of isobutylamine and isoamylamine by decarboxylation of valine and leucine respectively

Scheme 2. Transamination of propanal with alanine as amino group donor

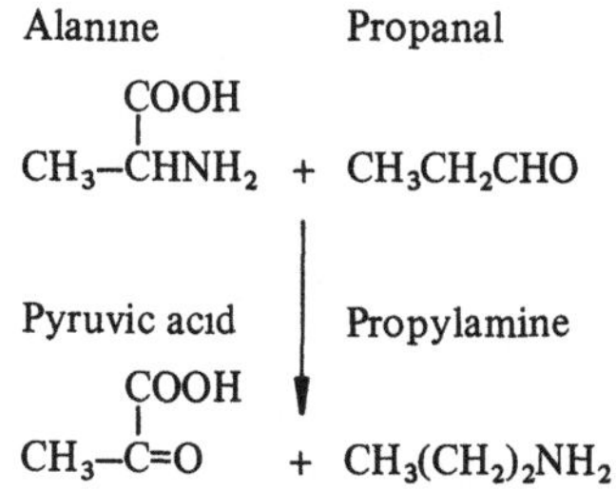

sea water are the natural substrates. The amines formed are adsorbed to the acidic polysaccharides of the cell walls, and HARTMANN and AUFERMANN (1973) suggest that they serve to protect the algae against micro-organisms.

In higher plants, although some simple amines are known to be produced by amino acid decarboxylation, aldehyde amination appears to be a more common biosynthetic pathway (HARTMANN et al., 1972a, b; PREUSSER, 1975). The amino acid aldehyde transaminase of *Mercurialis* has a pH optimum of 8.5. Out of 29 amino acids investigated only L-alanine, 5-aminocaproic acid, 4-aminobutyric acid, and L-glutamic acid acted as donors. Amines and D-amino acids were ineffective. However, a wide range of aldehydes was aminated, including a homologous series from ethanal to undecanal, with hexanal as the best substrate (Scheme 2). The enzyme was inhibited by several keto acids, notably pyruvate, oxaloacetate, 2-ketoglutarate, and glyoxylate. The equilibrium for amine:aldehyde was 10:1. At least four of the six amines found in *Mercurialis* could be formed in vitro by transamination. Aldehyde amination probably occurs through the activity of a normal amino acid-keto acid transaminase, which also reacts with aldehydes (UNGER and HARTMANN, 1976).

Several amines are formed by aldehyde transamination in apple fruits. Infiltration of a wide range of aldehydes gave rise to the corresponding amines, but activity could not be detected in apple homogenates (HARTMANN, 1967).

In *Zea mays*, ethylamine and isoamylamine may be formed by transamination, though alanine decarboxylation is an additional source of ethylamine. In *Vicia faba*, methylamine, propylamine and isoamylamine are formed by transamination, while isobutylamine is produced by decarboxylation (PREUSSER, 1975).

4-Glutamylethylamide (theanine) (I) and 4-glutamylmethylamide (II) are formed in the roots of the tea plant by condensation of ethylamine or methylamine respectively with glutamic acid (SUZUKI, 1973; TAKEO, 1974).

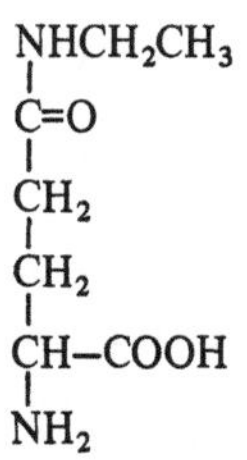

Theanine
(I)

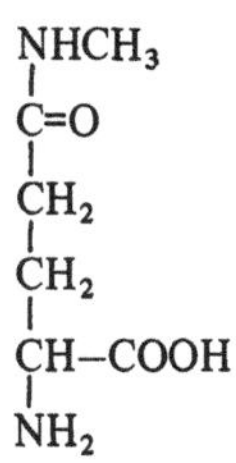

4–Glutamyl-
methylamide
(II)

In the Araceae, several aliphatic and aromatic amines are produced as part of a complex mechanism to attract flies which effect pollination. In *Sauromatum guttatum*, indole is evolved for 24 h at flowering. Amine release is stimulated by an inducer named calorigen, which may be separated into two components (Chen and Meeuse, 1975).

Isobutylamine is found as amides of various unsaturated aliphatic and aromatic organic acids in the Rutaceae, Compositae, and Piperaceae (Guggenheim, 1958; Boit, 1961; Tackie et al., 1975). Some of these derivatives are insecticidal. Sylvatine isolated from *Piper sylvaticum* is an amide formed by condensation of 10-methyl-trans-5-undecenylamine with piperic acid (Vig et al., 1975).

Table 2. Aliphatic monoamines found in plants. References to the occurrence of methylamine and ethanolamine are not included, since these amines are very widespread and may be artefacts of extraction (see Sect. 2). Additional references to the occurrence of amines in fungi may be found in Stein von Kamienski (1957)

Amine	Structure	Abbreviation
Dimethylamine	$(CH_3)_2NH$	DM
Trimethylamine	$(CH_3)_3N$	TM
Ethylamine	$CH_3CH_2NH_2$	E
Ethylmethylamine	$CH_3NHCH_2CH_3$	EM
Diethylamine	$CH_3CH_2NHCH_2CH_3$	DE
n-Propylamine	$CH_3(CH_2)_2NH_2$	P
n-Butylamine	$CH_3(CH_2)_3NH_2$	B
Dibutylamine	$CH_3(CH_2)_3NH(CH_2)_3CH_3$	DB
Amylamine	$CH_3(CH_2)_4NH_2$	A
n-Hexylamine	$CH_3(CH_2)_5NH_2$	HX
n-Heptylamine	$CH_3(CH_2)_6NH_2$	HP
n-Octylamine	$CH_3(CH_2)_7NH_2$	OC
isoPropylamine	$(CH_3)_2CHNH_2$	iP
isoButylamine	$(CH_3)_2CHCH_2NH_2$	iB
isoAmylamine	$(CH_3)_2CH(CH_2)_2NH_2$	iA
2-Methylbutylamine	$CH_3CH_2CH(CH_3)CH_2NH_2$	MB
3-Methylthiopropylamine	$CH_3S(CH_2)_3NH_2$	MTP
2-Propanolamine	$CH_3CH(OH)CH_2NH_2$	PR

Group or family	Species	Amines	Reference
Lower plants	Freshwater algae	DM, TM, E, P, iP	Rolle et al. (1977) Herrmann and Jüttner (1977)
	Marine algae	DM, TM, E, P, iB, iA, MTP	Steiner and Hartmann (1968) Fujiwara-Arasaki and Mino (1972)
Fungi	*Claviceps purpurea*	E, P, HX, iB, iA	Steiner (1966)
	Coprinus spp.	DM, P, B, iA	List and Hetzel (1960) List and Reith (1960)
	Inocybe patouillardii	DM, E, P, iA	List and Müller (1959)
	Phallus impudicus	DM, E, iA	List and Reinhard (1962)
	Polyporus sulphureus	DM, E, P, iA	List and Menssen (1959)

Table 2 (continued)

Group or family	Species	Amines	Reference
	Ustilago maydis	E, iA	LIST and WAGNER (1963)
Aceraceae	*Acer pseudoplatanus* (L)	iA	KLEIN and STEINER (1928)
Araceae	*Amorphophallus rivieri* (Sx)	TM, iB	KLEIN and STEINER (1928) STEINER and LÖFFLER (1929) STEIN VON KAMIENSKI (1957)
	Arum spp. (L, Sx, Se)	DM, E, DE, iB, iA	STAHL and KALTENBACH (1965)
	Dracunculus spp. (Sx)	DM, TM, E, iA	SMITH and MEEUSE (1966)
	Hydrosme spp. (Sx)	E, iB, iA	HARTMANN et al. (1972b)
	Sauromatum spp. (Sx)	DM, TM, E, DE, iB, iA	KOLEVA and AKHTARDZHIEV (1976)
Aristolochiaceae	*Aristolochia gigas* (Fl)	TM	KLEIN and STEINER (1928)
Asclepiadaceae	*Cynanchum* spp. (Fl)	iA	KLEIN and STEINER (1928) STEINER and LÖFFLER (1929)
	Gymnema sylvestre	TM	WILLAMAN and LI (1970)
	Vincetoxicum officinale (Fl)	iB, iA	STEIN VON KAMIENSKI (1957)
Berberidaceae	*Berberis vulgaris* (Fl)	iB	KLEIN and STEINER (1928)
	Mahonia aquifolium (Fl)	iB, iA	KLEIN and STEINER (1928)
Caprifoliaceae	*Viburnum* spp. (Fl)	TM, HX, OC, iB, iA	KLEIN and STEINER (1928)
	Sambucus nigra (Fl)	HX, HP, OC, iB, iA	STEINER and LÖFFLER (1929) STEIN VON KAMIENSKI (1957) HARTMANN et al. (1972b)
Caryophyllaceae	*Lychnis coronaria* (Fl)	E, HX, iA, MB	HARTMANN et al. (1972b)
Chenopodiaceae	*Chenopodium vulvaria* (L)	TM	CROMWELL and RICHARDSON (1966)
Compositae	*Hieracium pilosella* (Fl)	E	HARTMANN et al. (1972b)
Cornaceae	*Cornus sanguinea* (Fl)	TM, iA	KLEIN and STEINER (1928) STEINER and LÖFFLER (1929) STEIN VON KAMIENSKI (1957)
Cruciferae	*Iberis amara*	MTP	DALGAARD et al. (1977)
	Sinapis alba (Fl)	HX, iA, MB	HARTMANN et al. (1972b)
Cucurbitaceae	*Bryonia dioica* (Fl)	E	STEIN VON KAMIENSKI (1957)
Euphorbiaceae	*Mercurialis perennis* (L)	E, HX, iP, iB, iA	STEIN VON KAMIENSKI (1957) HARTMANN et al. (1972a)

Table 2 (continued)

Group or family	Species	Amines	Reference
Geraniaceae	*Erodium cicutarium* (L)	E	EIJK (1951)
Gramineae	*Hordeum vulgare* (L, Sd)	DM, E, i(?)B, i(?)A	DREWS et al. (1957) NEURATH and SCHREIBER (1973)
	Zea mays (L)	DM, E, P, HX, iB, iA	PREUSSER (1966) HARTMANN et al. (1972b)
Grossulariaceae	*Ribes grossularia* (Fl)	iA	STEINER and LÖFFLER (1929)
Iridaceae	*Iris germanica* (L)	iB, MB	HARTMANN et al. (1972b)
Leguminosae	*Vicia faba* (L)	DM, E, DE, P, HX, iB, iA	PREUSSER (1966)
Moraceae	*Humulus lupulus* (Fl)	DM, E	NEURATH and SCHREIBER (1973)
Musaceae	*Musa sapientum* (Fr)	DM, E, iB, iA, PR	ASKAR et al. (1972)
Nymphaeaceae	*Nuphar lutea* (Fl)	iB, iA	STEIN VON KAMIENSKI (1957)
Onagraceae	*Oenothera lamarckiana* (Fl)	iA	STEIN VON KAMIENSKI (1957)
Polygonaceae	*Polygonum cuspidatum* (Fl)	iA	STEIN VON KAMIENSKI (1957)
Plumbaginaceae	*Limonium vulgare* (L)	DM, TM	LARHER and BERNARD (1971)
Rosaceae	*Amelanchier rotundifolia* (Fl)	iA	NEU and FIEDLER (1954) KLEIN and STEINER (1928) STEINER and LÖFFLER (1929) STEIN VON KAMIENSKI (1957) HARTMANN (1967) HARTMANN et al. (1972b)
	Aruncus sylvester (Fl)	iA	
	Chaenomeles spp. (Fl)	HX, HP, OC, iB, iA	
	Cotoneaster spp. (Fl)	iA	
	Crataegomespilus grandiflorus (Fl)	iA	
	Crataegus spp. (Fl, Fr)	DM, E, TM, iB, iA	
	Filipendula spp. (Fl)	iB, iA	
	Malus spp. (Fl, Fr)	E, P, B, HX, iP, iB, iA	
	Prunus spp. (Fl)	TM, iA	
	Pyracantha coccinea (Fl)	iA	
	Pyrus spp. (Fl)	TM, iB, iA	
	Rosa canina (Fl)	iB	
	Sanguisorba officinalis (Fl)	iA	
	Sorbaria sorbifolia (Fl)	TM, iA	
	Sorbus spp. (Fl)	TM, iB, iA	
	Spiraea spp. (Fl)	iA	
Rubiaceae	*Coffea arabica* (Sd)	DM, EM, E, P, iB, A, iA	NEURATH and SCHREIBER (1973)
	Galium spp. (Fl)	iA	STEIN VON KAMIENSKI (1957)
Rutaceae	*Phellodendron amurense* (Fl)	iA	STEIN VON KAMIENSKI (1957)
Saxifragaceae	*Chrysosplenium alternifolium* (Fl)	TM, iA	STEINER and LÖFFLER (1929)
Scrophulariaceae	*Digitalis purpurea* (Fl)	E	HARTMANN et al. (1972b)

Table 2 (continued)

Group or family	Species	Amines	Reference
Solanaceae	*Atropa belladonna* (Fl)	iA	{ STEIN VON KAMIENSKI (1957)
	Datura stramonium (L)	iA, MB	HARTMANN et al. (1972b) }
	Nicotiana tabacum (L)	many	IRVINE and SAXBY (1969)
	Solanum tuberosum (Tb)	DE, P, iB	SCHORMÜLLER and WEDER (1966)
Staphyleaceae	*Staphylea colchica* (Fl)	iB, iA, MB	HARTMANN et al. (1972b)
Sterculiaceae	*Theobroma cacao* (Sd)	iB, iA	NEURATH and SCHREIBER (1973)
Theaceae	*Thea sinensis* (Sd)	TM, E, DE, P, DB	SERENKOV (1959)
Umbelliferae	*Anthriscus sylvestris* (Fl)	P, iP, iA	
	Chaerophyllum spp. (Fl)	TM, iA	{ STEINER and LÖFFLER (1929)
	Conium maculatum (Fl)	iP, iB	STEIN VON KAMIENSKI (1957)
	Heracleum sphondylium (Fl)	DM, TM, iA	HARTMANN et al. (1972b) }
	Pimpinella maior (FL)	iA	

Key: Fl = flowers, Fr = fruit, L = leaf, Tb = tuber, Sx = spadix, Se = spathe, Sd = seed

3 Di- and Polyamines

3.1 Growth Effects

The diamine putrescine and the polyamines spermidine and spermine (Scheme 3) have been demonstrated in many plants and probably occur universally in eukaryotes. These amines interact with nucleic acids in the double helical form, their structure allowing association of the amino groups with the phosphate groups on the nucleic acid. Polyamines may be implicated in protein synthesis through this association, and enzymes for their formation and breakdown may be concerned in mechanisms for the regulation of growth.

Putrescine, cadaverine, spermidine, and spermine at 10^{-4} to 10^{-6} M stimulated growth (fresh and dry wt.) in explants of *Helianthus tuberosus* (BAGNI, 1966). Protein synthesis by ribosomes from dormant tissue and tissue activated with 2×10^{-6} M IAA for 18 h was also studied. In the absence of polyamines, at 5×10^{-3} to 4×10^{-2} M Mg^{2+}, activity was low in the ribosomes from the dormant tissue. At 10^{-2} M Mg^{2+}, spermidine and putrescine applied together stimulated activity about 14-fold. Activity of the ribosomes from the IAA-activated tissue was much higher in the absence of polyamines, and only a 2–3-fold stimulation was found at 10^{-2} M Mg^{2+} on adding spermidine and putrescine. No polyamine stimulation was found at the optimum Mg^{2+} concentration (2×10^{-2} M). Total spermidine was increased 14-fold and spermine 21-fold in terms of fresh weight in the IAA-activated tissue. By comparison with ribosomes from dormant tissue, there was about twice as much spermidine and spermine in the ribosomal fraction of the IAA-activated tissues. As a consequence of their work, COCUCCI and BAGNI (1968)

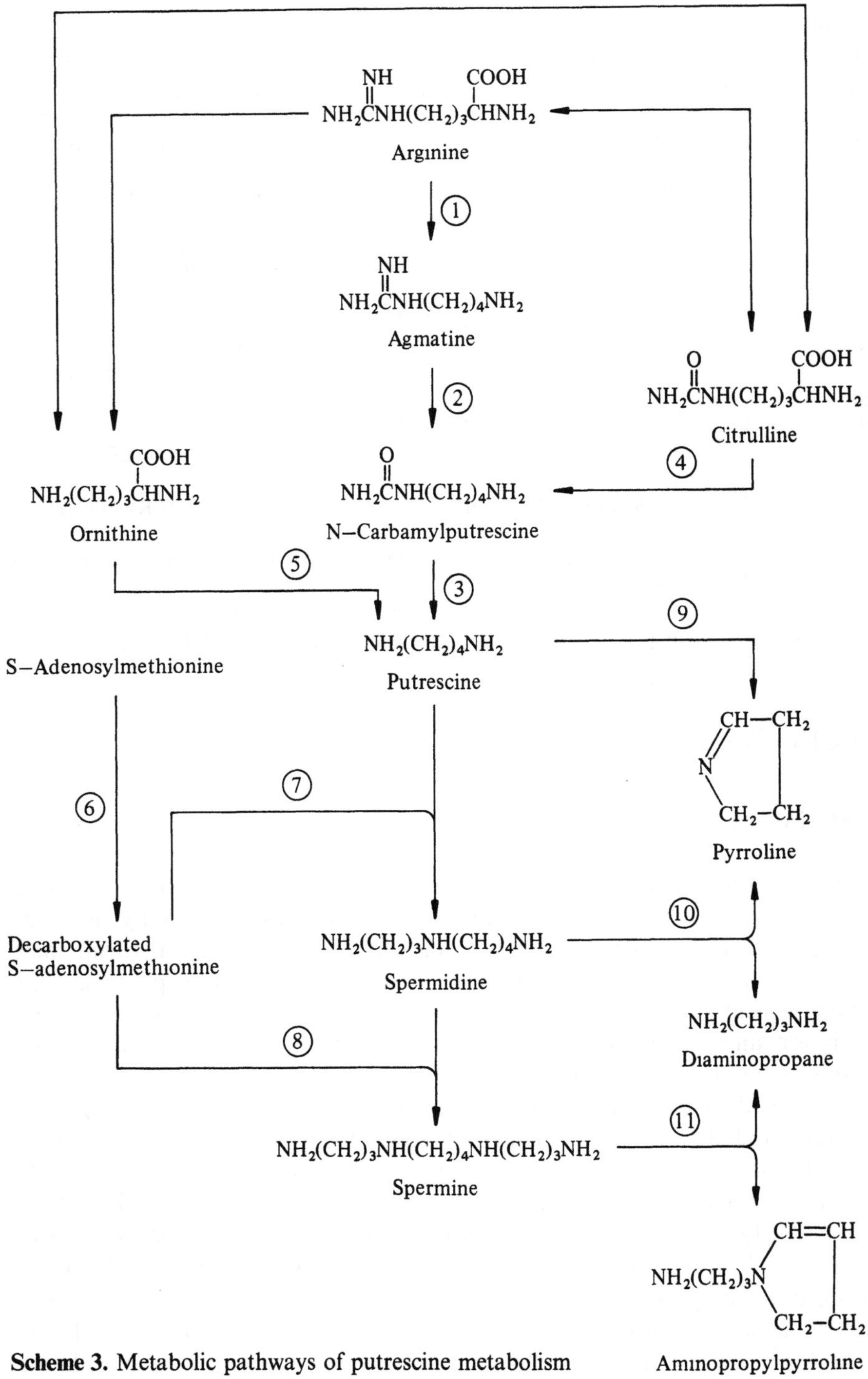

Scheme 3. Metabolic pathways of putrescine metabolism

suggest that at least part of the effect of IAA in stimulating growth in *Helianthus* depends on the auxin-induced activation of polyamine synthesis.

In both *Scorzonera hispanica* and *Nicotiana glauca* tumours, putrescine was considerably increased (up to 100 times) by comparison with the normal tissues (BAGNI et al., 1972; AUDISIO et al., 1976).

3.2 Effect of Mineral Nutrition on Amine Metabolism

Putrescine accumulation in conditions of K^+ deficiency is now established for many plant species and is probably universal amongst higher plants (MURTY et al., 1971). In barley, putrescine is formed from arginine with the intermediate production of agmatine and *N*-carbamylputrescine. In K^+-deficient barley increases in the activities of arginine carboxy-lyase (Scheme 3, step 1) and *N*-carbamylputrescine amidohydrolase (Scheme 3, step 3) may provide the mechanism for putrescine accumulation (SMITH, 1968).

Since acid feeding induces the formation of arginine carboxy-lyase and *N*-carbamylputrescine amidohydrolase in barley seedlings, it was suggested that putrescine, a strong organic base, is produced by the plant in response to a reduced cytoplasmic pH. Deficiency of K^+ may cause a reduction of pH in the tissue which induces putrescine formation, thereby correcting the acidity (SMITH and SINCLAIR, 1967). Putrescine is also accumulated in Mg^{2+} deficiency (BASSO and SMITH, 1974) indicating that this diamine may substitute for Mg^{2+}, perhaps in protein synthesis.

By comparison with plants grown in a nitrate medium, those grown with ammonium salts as nitrogen source accumulate large amounts of putrescine. The arginine carboxy-lyase of *Glycine* plants was 7–12 times greater with ammonium than with nitrate as N source and putrescine was 25–100 times greater on a dry weight basis (LE RUDULIER and GOAS, 1975a). Potassium levels were almost unchanged in plants grown in the two media. It was unlikely that reduction in pH of the medium caused the putrescine accumulation, but it was suggested that the ammonium ion may be competing with potassium at specific sites within the cell (LE RUDULIER and GOAS, 1975b).

3.3 Biosynthesis of the Di- and Polyamines

The pathway to putrescine from arginine by decarboxylation probably occurs widely in higher plants (Scheme 3), since agmatine is frequently found in plant extracts (SMITH, 1963). Moreover, agmatine iminohydrolase (Scheme 3, step 2) has been detected in extracts of *Zea* and *Helianthus*, and *N*-carbamylputrescine is formed readily on feeding agmatine to barley (SMITH, 1969). However in *Sesamum*, citrulline is decarboxylated to *N*-carbamylputrescine (Scheme 3, step 4) (CROCOMO and BASSO, 1974) and in *Nicotiana* putrescine is formed directly from ornithine (Scheme 3, step 5) (MIZUSAKI et al., 1973).

The polyamines spermidine and spermine are probably synthesised in plants by the same route which is found in bacteria and animals (Scheme 3, steps 6, 7 and 8) (COPPOC et al., 1971; BAXTER and COSCIA, 1973).

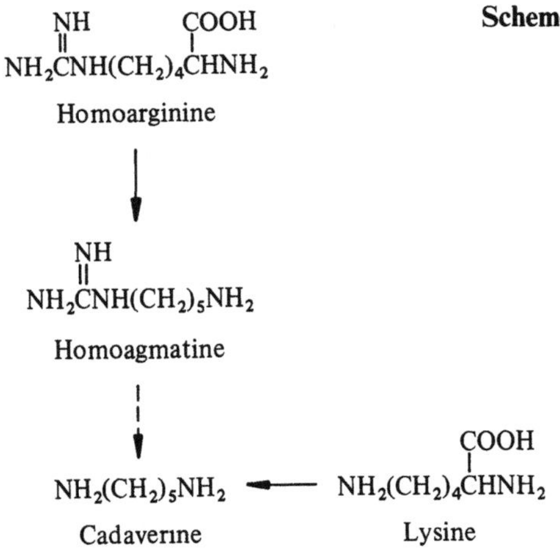

Scheme 4. Formation of cadaverine in *Lathyrus sativus*

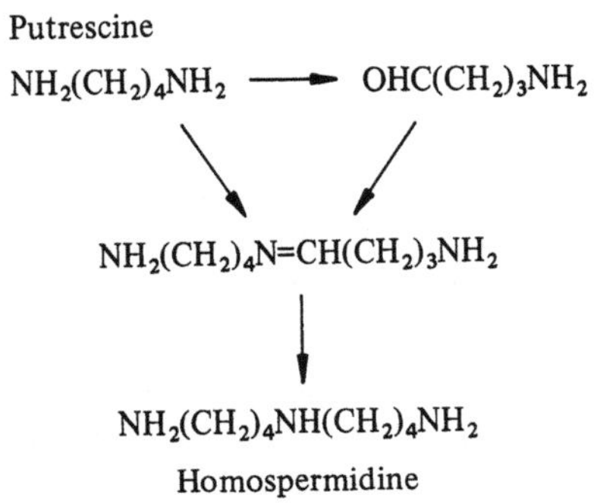

Scheme 5. Formation of homospermidine in leaves of *Santalum album*

Cadaverine has now been found in several plants, especially in the Leguminosae (SMITH and WILSHIRE, 1975). In *Lathyrus sativus* cadaverine may be formed from homoarginine with homoagmatine as an intermediate. However, lysine is a more efficient precursor of cadaverine than homoarginine (Scheme 4) (RAMAKRISHNA and ADIGA, 1974).

In the leaves of the sandalwood tree (*Santalum album*) a spermidine homologue, symhomospermidine, occurs at a concentration of up to 1.5% of the dry wt., and is probably formed from putrescine by the route shown in Scheme 5 (RADHAKRISHNAN, 1974). This amine is also found in *Rhizobium* spp., which occur in legume root nodules (SMITH, 1977c), and together with diaminodipropylamine, in the alga *Scenedesmus acutus* (ROLLE et al., 1977).

3.4 Polyamine Conjugates

Many conjugates of putrescine, spermidine, and spermine with cinnamic acid and its derivatives, or with fatty acids, are now known in plants. The occurrence of these has been reviewed by BADAWI et al. (1973) and SMITH (1977d). The hordatines (III) which are dimers of coumarylagmatine found in barley seedlings have antifungal properties (STOESSL and UNWIN, 1970) and 2-hydroxyputrescine amides of ferulic acid and *p*-coumaric acid are formed in wheat in response to rust infection

Hordatines { A : R = –H
B : R = –OCH_3

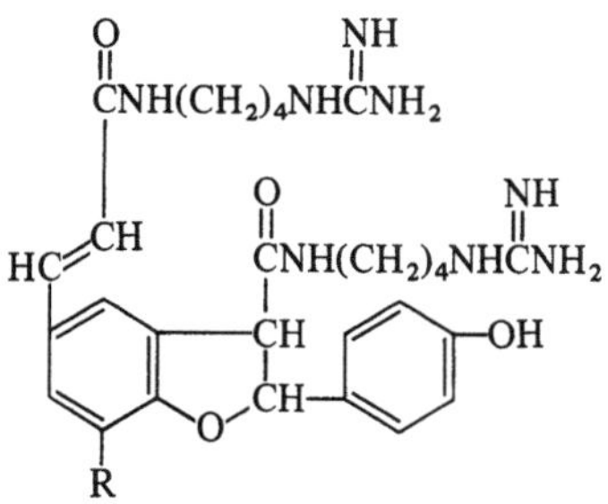

(III)

(Samborski and Rohringer, 1970). Moreover, aromatic amides of putrescine are formed in tobacco in response to virus infection and inhibit the multiplication of the virus. Caffeyl-putrescine may also be concerned in floral induction in this plant (Martin-Tanguy et al., 1976).

4 Amine Oxidases

The mono- and diamine oxidases may be separated on the basis of their substrate specificity. Both classes usually attack a wide range of substrates, and activity with a single substrate is insufficient for categorisation. For this reason the classification shown in Table 3 may not be correct for some of the enzymes. However, the polyamine oxidases of the Gramineae are quite distinct from the mono- and diamine oxidases in their absolute specificity for polyamines.

All amine oxidases appear to produce hydrogen peroxide, together with an aldehyde, which may cyclize with the elimination of water (Scheme 3). The function of the amine oxidases is at present obscure, though the di- and polyamine oxidases may control the level of the polyamines and thus influence the growth rate. Moreover, the growth substance IAA may be formed in at least some plants by enzymic oxidation of tryptamine (Gibson et al., 1972b).

Enzymes capable of oxidising amines occur in several families. Diamine oxidases occur widely in the Leguminosae and the pea seedling enzyme has been studied intensively, since it is very active. Putrescine is oxidised to pyrroline by this enzyme (Scheme 3, step 9). This enzyme contains copper which is easily removed by dialysis against chelating agents (Hill and Mann, 1968). One mol of reactive carbonyl is present per mol enzyme, though pyridoxal phosphate has not been found. The amino acid composition is characterised by a high content of ornithine (Nylén and Szybek, 1974). The enzyme is absent from ungerminated seed but activity per seedling increases to a peak about 14 days after germination, with most activity in the cotyledons (Ružička and Minář, 1974). The enzyme in the cotyledons, which differs from the embryo enzyme, is induced on feeding putrescine, spermidine, or ornithine (Srivastava et al., 1977; Srivastava and Prakash, 1977).

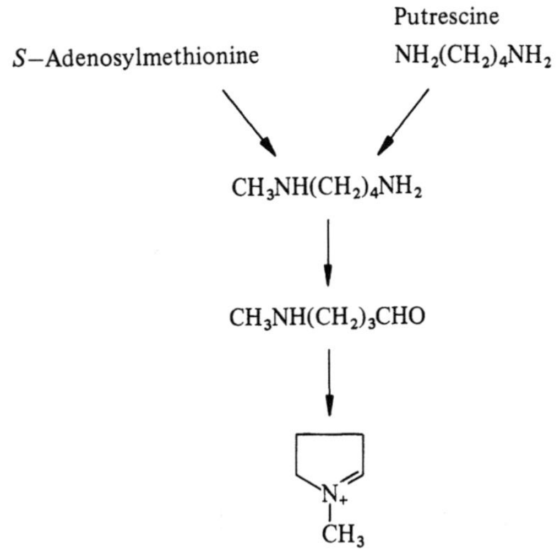

Scheme 6. Formation of *N*-methylpyrrolinium from putrescine in the biosynthesis of nicotine in tobacco roots

Both mono- and diamine oxidases occur in cucumber seedlings (PERCIVAL and PURVES, 1974). *N*-methylputrescine oxidase, found in the roots of tobacco plants and now purified 150-fold, is concerned in the biosynthesis of nicotine (Scheme 6) (MIZUSAKI et al., 1972, 1973).

The particulate polyamine oxidase from the leaves of oats, barley and maize oxidises spermidine and spermine at the secondary amino group (Scheme 3, steps 10 and 11). This enzyme is especially active in oat seedlings from which it has been purified over 1000-fold (SMITH, 1976).

Table 3. Amine oxidising systems found in plants

Family	Species	Best substrate known	Reference
Monoamine oxidases			
Ascomycetae	*Aspergillus niger*	Benzylamine	SUZUKI and OGURA (1972)
	Trichosporon	Ethylamine	YAMADA et al. (1966)
Bromeliaceae	*Ananas sativus*	*n*-Butylamine	WERLE and ROEWER (1952)
Chenopodiaceae	*Beta vulgaris*	Ethanolamine	RICHARDSON et al. (1966)
	Spinacia oleracea	iso-Propylamine	RICHARDSON et al. (1966)
Convolvulaceae	*Ipomoea batatas*	Ethanolamine	RICHARDSON et al. (1966)
Cruciferae	*Brassica rapa*	Ethanolamine	RICHARDSON et al. (1966)
	Raphanus sativus	Ethanolamine	RICHARDSON et al. (1966)
Cucurbitaceae	*Cucumis sativus*	Benzylamine	PERCIVAL and PURVES (1974)
	Momordica balsamina	*n*-Butylamine	WERLE and ROEWER (1952)
Gramineae	*Avena sativa*	Ethanolamine	RICHARDSON et al. (1966)
	Triticum vulgare	Ethanolamine	RICHARDSON et al. (1966)
Labiatae	*Salvia uliginosa*	*n*-Butylamine	WERLE and ROEWER (1950)
Liliaceae	*Allium cepa*	Ethanolamine	RICHARDSON et al. (1966)
	Colchicum bornmülleri	*n*-Butylamine	WERLE and ROEWER (1952)
Moraceae	*Cannabis indica*	*n*-Butylamine	WERLE and ROEWER (1952)

Table 3 (continued)

Family	Species	Best substrate known	Reference
Papaveraceae	*Chelidonium majus*	*n*-Butylamine	WERLE and ROEWER (1952)
	Papaver rhoeas	*n*-Butylamine	WERLE and ROEWER (1952)
Ranunculaceae	*Aconitum napellus*	*n*-Butylamine	WERLE and ROEWER (1952)
Solanaceae	*Solanum tuberosum*	Ethanolamine	RICHARDSON et al. (1966)
Theaceae	*Thea sinensis*	Methylamine	SUZUKI (1973)
Umbelliferae	*Daucus carota*	Ethanolamine	RICHARDSON et al. (1966)
	Pastinaca sativa	Ethanolamine	RICHARDSON et al. (1966)
Diamine oxidases			
Basidiomycetae	*Panus tigrinus*	Putrescine	BOLDT et al. (1971)
Cucurbitaceae	*Cucumis sativus*	Putrescine	PERCIVAL and PURVES (1974)
Compositae	*Artemisia*	Histamine	WERLE and ZABEL (1948)
	Taraxacum officinalis	Histamine	WERLE and ZABEL (1948)
Labiatae	*Lavandula* sp.	Putrescine	WERLE and PECHMANN (1949)
	Lavandula spica	Histamine	WERLE and ZABEL (1948)
	Lavandula vera	Histamine	WERLE and ZABEL (1948)
	Rosmarinus officinalis	Histamine	WERLE and ZABEL (1948)
	Salvia officinalis	Histamine	WERLE and ZABEL (1948)
	Thymus vulgaris	Histamine	WERLE and ZABEL (1948)
Leguminosae	*Glycine max*	Cadaverine	LE RUDULIER and GOAS (1977)
	Lathyrus sativus	Putrescine	SURESH et al. (1976)
	Lens culinaris	Dopamine	ROTHE (1974)
	Lupinus luteus	Putrescine	SUZUKI (1959a) SCHÜTTE et al. (1966)
	Medicago sativa	Cadaverine	WERLE and PECHMANN (1949)
	Phaseolus vulgaris	Cadaverine	WERLE and PECHMANN (1949)
	Pisum sativum	Putrescine	HILL and MANN (1968) NYLÉN and SZYBEK (1974)
	Trifolium incarnatum	Histamine	WERLE and ZABEL (1948)
	Trifolium pratense	Putrescine	HILL (1973)
	Trifolium repens	Cadaverine	WERLE and PECHMANN (1949)
	Vicia faba	Cadaverine	WERLE and PECHMANN (1949) MATSUDA and SUZUKI (1977)
Oleaceae	*Jasminum*	Histamine	WERLE and ZABEL (1948)
Rosaceae	*Fragaria*	Histamine	WERLE and ZABEL (1948)
	Rubus	Histamine	WERLE and ZABEL (1948)
Saxifragaceae	*Hortensia*	Histamine	WERLE and ZABEL (1948)
Solanaceae	*Atropa belladonna*	Putrescine	CROMWELL (1943)
	Nicotiana tabacum	*N*-Methyl putrescine	MIZUSAKI et al. (1972)
	Scopolia japonica	Putrescine	SUZUKI (1959b)
Umbelliferae	*Aegopodium podagraria*	Histamine	WERLE and ZABEL (1948)
Urticaceae	*Urtica dioica*	5-Hydroxy-tryptamine	COLLIER and CHESHER (1956)
Polyamine oxidases			
Gramineae	*Avena sativa*	Spermidine	SMITH (1976)
	Hordeum vulgare	Spermine	SMITH (1976)
	Triticum vulgare	Spermine	SMITH (1976)
	Secale cereale	Spermine	SMITH (1976)
	Zea mays	Spermidine	HIRASAWA and SUZUKI (1975)

5 Histamine

This amine (IV) occurs at 2–3 μmol/g fr. wt. in spinach leaves and up to 10 μ mol/g fr.wt. in the flowers (WERLE, 1955). *N*-methylated and acetylated derivatives are also found in spinach (APPEL and WERLE, 1959) and *N,N*-dimethylhistamine has been found in *Casimiroa edulis* (MAJOR and DÜRSCH, 1958). Histamine also occurs in association with serotonin in the hairs of the stinging nettle (VIALLI et al., 1973). Plants containing free histamine are listed in Table 4. Histamine also occurs as the amides of various organic acids (RAMAN et al., 1962; POUPAT and SÉVENET, 1975; REPKE, 1975; ROSENBERG and STOHS, 1976).

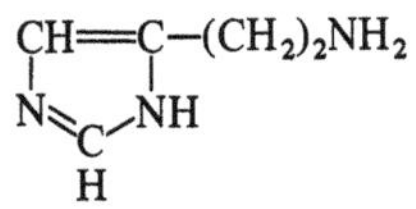

Histamine
(IV)

Table 4. Occurrence of histamine in plants

Family	Species	Reference
Basidiomycetae	*Phallus impudicus*	LIST and REINHARD (1962)
Araceae	*Hydrosme rivieri* (Sx)	SMITH and MEEUSE (1966)
Asclepiadaceae	*Calotropis gigantea* (Lx)	SAHA and KASINATHAN (1963)
Chenopodiaceae	*Beta vulgaris* var *rapa* (L)	WERLE (1955)
	Beta trigyna (L)	WERLE (1955)
	Chenopodium album (L, Fr)	WERLE (1955)
	Chenopodium bonus-Henricus (L, R)	WERLE (1955)
	Chenopodium quinoa (Fr)	WERLE (1955)
	Kochia childsii (L)	WERLE (1955)
	Salsola kali (L)	WERLE (1955)
	Spinacia oleracea (L, Sd, Fl)	APPEL and WERLE (1959) HAARTMANN et al. (1966)
Cruciferae	*Lepidium sativum* (Sl)	HAARTMANN et al. (1966)
	Sinapis alba (Sl)	HAARTMANN et al. (1966)
Droseraceae	*Drosera* (L)	WERLE (1955)
Euphorbiaceae	*Jatropha urens* (L)	VILLALOBOS et al. (1974)
Geraniaceae	*Erodium cicutarium*	EIJK (1957)
Gramineae	*Avena sativa* (Sl)	HAARTMANN et al. (1966)
	Dactylis glomerata (L)	FOWLER (1962)
	Lolium perenne (L)	FOWLER (1962)
Labiatae	*Lamium album*	KWASNIEWSKI (1959)
	Salvia (L)	WERLE (1955)
Leguminosae	*Mimosa* (L)	WERLE (1955)
	Trifolium pratense (L)	WERLE (1955)
	Trifolium repens (l, Sl)	FOWLER (1962) HAARTMANN et al. (1966)
Lentibulariaceae	*Pinguicula* (L)	WERLE (1955)
Loranthaceae	*Viscum album*	SAJNER and VĚŘÍŠ (1958a)
Malvaceae	*Gossypium hirsutum* (L)	LLOYD and NICHOLLS (1964)
Musaceae	*Musa sapientum* (Fr)	ASKAR et al. (1972)
Nepenthaceae	*Nepenthes* (L)	WERLE (1955)
Papaveraceae	*Chelidonium majus*	KWASNIEWSKI (1958)
Plantaginaceae	*Plantago lanceolata* (L)	WERLE (1955)

Table 4 (continued)

Family	Species	Reference
Primulaceae	*Cyclamen* (L, R, Fl)	WERLE (1955)
Ranunculaceae	*Delphinium* (Fl)	WERLE (1955)
Rosaceae	*Crataegus* (Fl)	SAJNER and VĚŘÍŠ (1958b)
Rutaceae	*Citrus vulgaris* (L)	WERLE (1955)
Sarraceniaceae	*Sarracenia* (L)	WERLE (1955)
Solanaceae	*Lycopersicon* (Fr, Sl)	HOLTZ and JANISCH (1937) HAARTMANN et al. (1966)
Urticaceae	*Girardinia heterophylla* (L)	SAXENA et al. (1966)
	Laportea (L, Fr)	WERLE (1955)
	Urera (L)	WERLE (1955)
	Urtica dioica (L)	EMMELIN and FELDBERG (1949)
	Urtica parviflora (L)	WERLE (1955)
	Urtica urens (L)	SAXENA et al. (1965)
		VIALLI et al. (1973)

Key: L=leaf, R=root, Fl=flowers, Sd=seed, Sl=seedling, Fr=fruit, Sx=spadix, Lx=latex

6 Tryptamines

In tomato and barley the hormone indol-3-yl-acetic acid is formed from tryptophan by the pathways shown in Scheme 7 (GIBSON et al., 1972b). In barley shoots gramine is also derived from tryptophan by the route shown in Scheme 8 (GROSS et al., 1974).

Bufotenine, *N,N*-dimethyltryptamine and psilocybin and some of their derivatives are hallucinogens. Some of the plants containing these amines are used as intoxicating snuffs and decoctions by various ethnic groups (SCHULTES, 1976). The hallucinogen bufotenine is formed in *Piptadenia peregrina* by the route shown in Scheme 9 (FELLOWS and BELL, 1971). The ability to form tryptamine derivatives which are physiologically active in animals may have evolved in plants as a defensive mechanism against predators. Several members of the Urticaceae (nettles) have stinging hairs containing serotonin.

The occurrence of the tryptamines has been reviewed by SMITH (1977b): see also RIVIER and LINDGREN (1972).

7 Phenethylamines

A wide variety of phenethylamines related metabolically to phenylalanine are found in plants (SMITH, 1977a). The hallucinogen mescaline occurs in several species of cacti, especially in *Lophophora williamsii*, the peyote cactus, in which it comprises 1% of the dry wt. This amine is formed from tyrosine, either via tyramine or via DOPA (Scheme 10) (PAUL, 1973). The occurrence of mescaline and of related phenolic amines in cacti has been reviewed by KAPADIA and FAYEZ (1970). *N*-Methylmescaline has been found in the legume *Alhagi pseudalhagi* (GHOSAL et al., 1974).

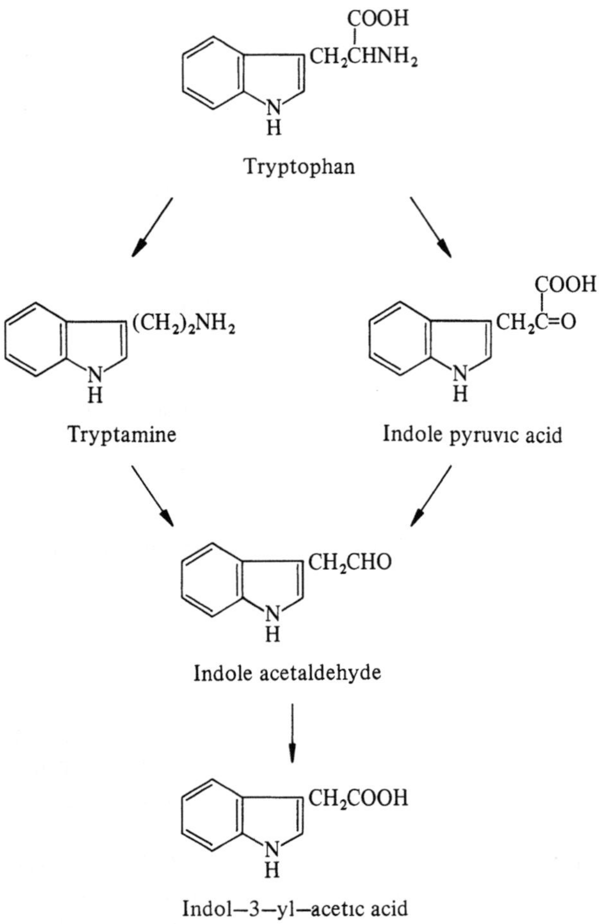

Scheme 7. Probable pathway of IAA formation in higher plants

The genus *Acacia* has high levels of phenethylamine (CAMP and NORVELL, 1966) in addition to a variety of tryptamines. Several phenethylamines are found in *Citrus* species. Tangerine juice contains about 0.8 μmol per ml of synephrine (STEWART and WHEATON, 1964). The biosynthetic route for synephrine (WHEATON and STEWART, 1969) is shown in Scheme 11.

Ephedra, which contains the vasoconstrictor and bronchodilator ephedrine, has been used in medicine for thousands of years. In *Ephedra distachya* ephedrine is biosynthesised from benzoic acid, benzaldehyde or cinnamic acid. The ephedrine appears to be formed by condensation of a C_6-C_1 portion derived from phenylalanine via cinnamic acid, with an unidentified C_2N fragment (YAMASAKI et al., 1973).

In the roots of barley seedlings tyrosine is decarboxylated to tyramine (HOSOI, 1974) which is successively methylated with *S*-adenosylmethionine as donor (MUDD, 1960) to give *N*-methyltyramine and hordenine. The quaternary base of hordenine, candicine, has also been isolated from barley roots. Demethylation of hordenine fed to barley seedlings suggested that hordenine is an active metabolite (RABITZSCH, 1959).

Phenethylamine, tyramine, dihydroxyphenethylamine (dopamine) and noradrenaline occur in the banana fruit, together with serotonin and histamine, and

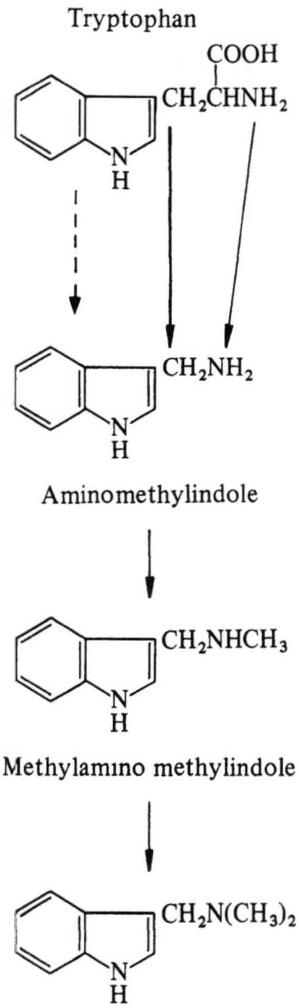

8 Gramine

Scheme 8. Biosynthesis of gramine from tryptophan in barley leaves

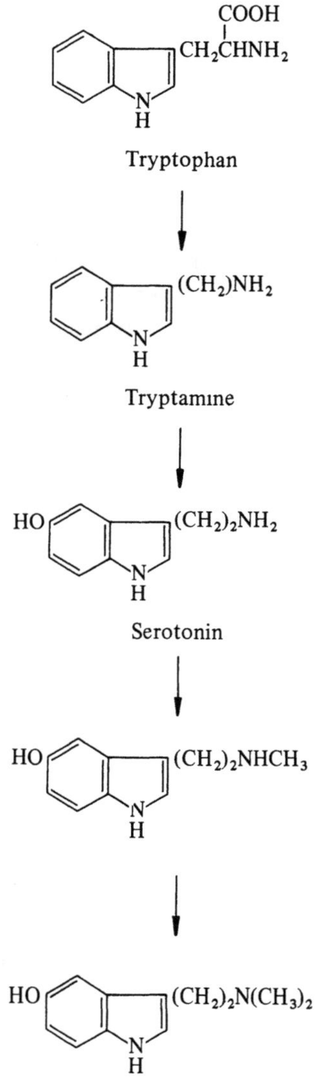

9 Bufotenine

Scheme 9. Formation of bufotenine in *Piptadenia peregrina*

other simple aliphatic amines (UDENFRIEND et al., 1959; ASKAR et al., 1972). Dopamine, formed by the hydroxylation of tyramine (DEACON and MARSH, 1971), is further hydroxylated to noradrenaline by an enzyme found in all parts of the banana fruit (SMITH and KIRSCHNER, 1960) (Scheme 12). By contrast, dopamine is formed by decarboxylation of dihydroxyphenylalanine in *Monostroma fuscum* (a marine alga) and in *Cytisus scoparius* (Scots broom) (TOCHER and TOCHER, 1969, 1972).

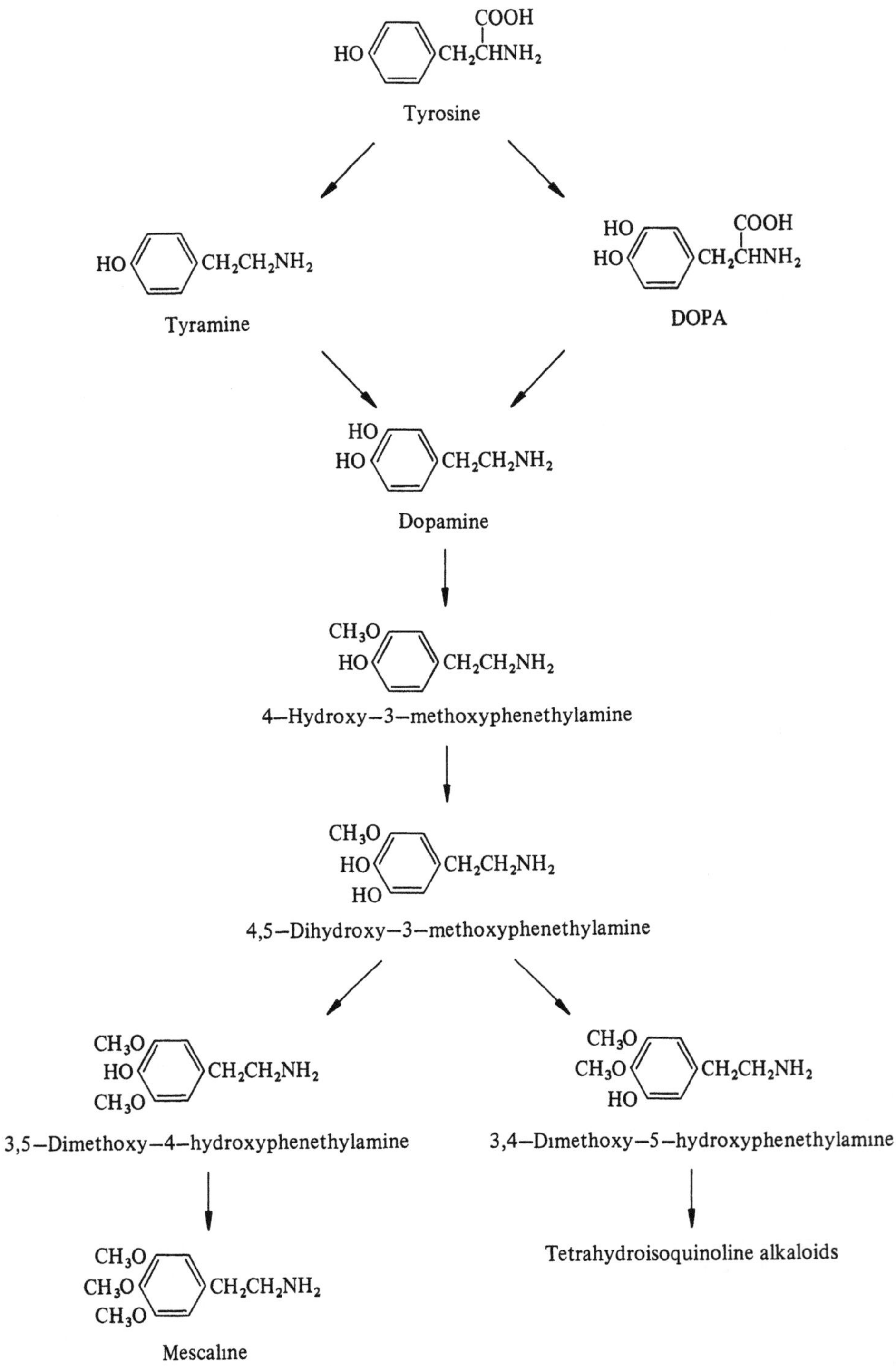

Scheme 10. Formation of mescaline in the cactus *Lophophora williamsii*

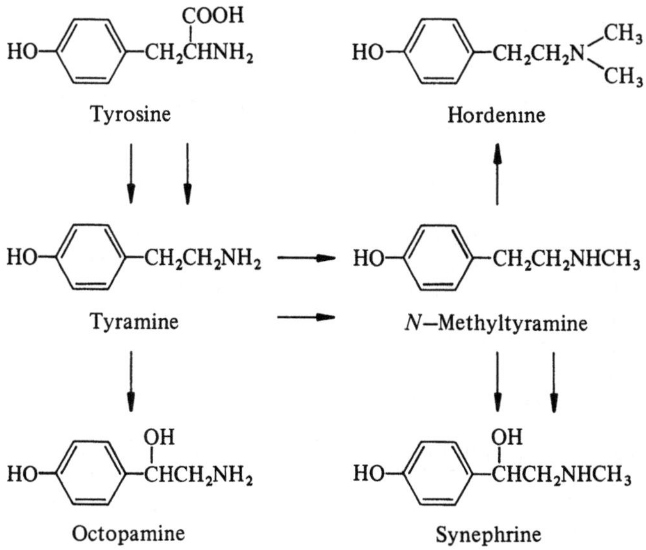

Scheme 11. Formation of synephrine and other phenolic amines in *Citrus* seedlings

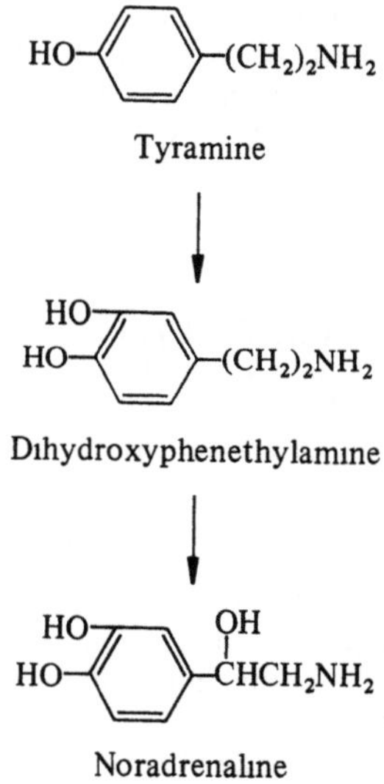

Scheme 12. Biosynthesis of noradrenaline in banana fruit

References

Appel, W., Werle, E.: Nachweis von Histamin, *N*-Dimethylhistamin, *N*-Acetylhistamin und Acetylcholin in *Spinacia oleracea*. Arzneimittelforschung *9*, 22–26 (1959)

Askar, A., Rubach, K., Schormüller, J.: Dünnschichtchromatographische Trennung der in Bananen vorkommenden Amin-Fraktion. Chem. Mikrobiol. Technol. Lebensm. *1*, 187–190 (1972)

Audisio, S., Bagni, N., Serafini Fracassini, D.: Polyamines during the growth in vitro of *Nicotiana glauca* R. Grah. habituated tissue. Z. Pflanzenphysiol. *77*, 146–151 (1976)

Badawi, M.M., Bernauer, K., van den Broek, P., Gröger, D., Guggisberg, A., Johne, S., Kompiš, I., Schneider, F., Veith, H.-J., Hesse, M., Schmid, H.: Macrocyclic spermidine and spermine alkaloids. Pure Appl. Chem. *33*, 81–108 (1973)

Bagni, N.: Aliphatic amines and a growth factor of coconut milk as stimulants of cellular proliferation of *Helianthus tuberosus* (Jerusalem artichoke) in vitro. Experientia *22*, 732 (1966)

Bagni, N., Serafini Fracassini, D., Corsini, E.: Tumours of *Scorzonera hispanica*; their polyamine content. Z. Pflanzenphysiol. *67*, 19–23 (1972)

Basso, L.C., Smith, T.A.: The effect of mineral deficiency on amine formation in higher plants. Phytochemistry *13*, 875–883 (1974)

Baxter, C., Coscia, C.J.: In vitro synthesis of spermidine in the higher plant, *Vinca rosea*. Biochem. Biophys. Res. Commun. *54*, 147–154 (1973)

Baxter, C., Slaytor, M.: Partial purification and some properties of tryptophan decarboxylase from *Phalaris tuberosa*. Phytochemistry *11*, 2763–2766 (1972)

Bergmann, L., Grosse, W., Ruppel, H.G.: Die Bildung von Serotonin in *Juglans regia* L. Planta *94*, 47–59 (1970)

Bernard, T., Larher, F.: Contribution à l'étude du métabolisme azoté des Lichens: rôle de la glycine ^{14}C-2 dans la formation des méthylamines chez *Lobaria laetevirens*. C.R. Acad. Sci. Paris Ser D *272*, 568–571 (1971)

Boit, H.G.: Ergebnisse der Alkaloid-Chemie bis 1960 pp. 1–2. Berlin: Akademie Verlag 1961

Boldt, A., Miersch, J., Reinbothe, H.: Metabolism of agmatine in fruit bodies of the fungus *Panus tigrinus* (Bull. ex Fr.) Sing. Phytochemistry *10*, 731–738 (1971)

Camp, B.J., Norvell, M.J.: The phenylethylamine alkaloids of native range plants. Econ. Bot. *20*, 274–278 (1966)

Chen, J., Meeuse, B.J.D.: Purification and partial characterization of two biologically active compounds from the inflorescence of *Sauromatum guttatum* Schott. (Araceae). Plant Cell Physiol. (Tokyo) *16*, 1–11 (1975)

Cocucci, S., Bagni, N.: Polyamine-induced activation of protein synthesis in ribosome preparations from *Helianthus tuberosus* tissue. Life Sci. *7*, 113–120 (1968)

Collier, H.O.J., Chesher, G.B.: Identification of 5-hydroxytryptamine in the sting of the nettle (*Urtica dioica*). Br. J. Pharmacol. Chemother. *11*, 186–189 (1956)

Coppoc, G.L., Kallio, P., Williams-Ashman, H.G.: Characteristics of *S*-adenosyl-L-methionine decarboxylase from various organisms. Int. J. Biochem. *2*, 673–681 (1971)

Crocomo, O.J., Basso, L.C.: Accumulation of putrescine and related amino acids in potassium deficient *Sesamum*. Phytochemistry *13*, 2659–2665 (1974)

Crocomo, O.J., Fowden, L.: Amino acid decarboxylases of higher plants. The formation of ethylamine. Phytochemistry *9*, 537–540 (1970)

Cromwell, B.T.: The role of putrescine in the synthesis of hyoscyamine. Biochem. J. *37*, 722–726 (1943)

Cromwell, B.T., Richardson, M.: Studies on the biogenesis of some simple amines and quaternary ammonium compounds in higher plants. Trimethylamine in *Chenopodium vulvaria* L. Phytochemistry *5*, 735–746 (1966)

Dalgaard, L., Nawaz, R., Sørensen, H.: 3-Methylthiopropylamine and (R)-3-methylsulphinylpropylamine in *Iberis amara*. Phytochemistry *16*, 931–932 (1977)

Dannenburg, W.N., Liverman, J.L.: Conversion of tryptophan-2-^{14}C to indoleacetic acid by watermelon tissue slices. Plant Physiol. *32*, 263–269 (1957)

Deacon, W., Marsh, H.V.: Properties of an enzyme from bananas (*Musa sapientum*) which hydroxylates tyramine to dopamine. Phytochemistry *10*, 2915–2924 (1971)

Drews, B., Just, F., Drews, H.: Vorkommen und Bildung von Aminen in Malz, Würze und Bier. Proc. 6th Eur. Brew. Conv. Copenhagen. pp. 167–172 (1957)
Du Plessis, L.S., Nunn, J.R., Roach, W.A.: Carcinogen in a Transkeian Bantu food additive. Nature (London) *222*, 1198–1199 (1969)
Eijk, J.L.: Phytochemical investigations of *Erodium cicutarium*. Thesis, Univ Utrecht (1951) cited in Chem. Abstr. *46*, 9262a
Eijk, J.L.: Fytochemisch onderzoek von *Erodium cicutarium* III. Pharm. Weekbl. Ned. *92*, 581–587 (1957)
Elkinawy, M., Raa, J.: Tryptophan decarboxylase activity in developing cucumber seedlings. Physiol. Plant. *30*, 172–175 (1974)
Emmelin, N., Feldberg, W.: Distribution of acetylcholine and histamine in nettle plants. New Phytol. *48*, 143–148 (1949)
Fellows, L.E., Bell, E.A.: Indole metabolism in *Piptadenia peregrina*. Phytochemistry *10*, 2083–2091 (1971)
Fowler, H.D.: Histamine in grasses and clovers. Nature (London) *193*, 582–583 (1962)
Fujiwara-Arasaki, T., Mino, N.: The distribution of trimethylamine and trimethylamine oxide in marine algae. Proc. 7th Int. Sea Weed Symp. pp. 506–510 (1972)
Gallon, J.R., Butt, V.S.: L-Tyrosine decarboxylase from barley roots. Biochem. J. *123*, 5P (1971)
Ghosal, S., Srivastava, R.S., Bhattacharya, S.K., Debnath, P.K.. The active principles of *Alhagi pseudalhagi*, β-phenethylamine and tetrahydroisoquinoline bases. Planta Med. *26*, 318–326 (1974)
Gibson, R.A., Barrett, G., Wightman, F.: Biosynthesis and metabolism of indol-3-yl-acetic acid. III. Partial purification and properties of a tryptamine-forming L-tryptophan decarboxylase from tomato shoots. J. Exp. Bot. *23*, 775–786 (1972a)
Gibson, R.A., Schneider, E.A., Wightman, F.: Biosynthesis and metabolism of indol-3yl-acetic acid. II. In vivo experiments with ^{14}C labelled precursors of IAA in tomato and barley shoots. J. Exp. Bot. *23*, 381–399 (1972b)
Giebel, J., Jackowiak, N.: Tryptophan decarboxylase in resistant and susceptible potato roots infected with *Heterodera rostochiensis*. Nematologica *22*, 462–466 (1976)
Gross, D., Lehmann, H., Schütte, H.R.: Zur Biosynthese des Gramins. Biochem. Physiol. Pflanz. *166*, 281–287 (1974)
Guggenheim, M.: Die biogenen Amine in der Pflanzenwelt. In: Handbuch der Pflanzenphysiologie. Ruhland, W. (ed.), Vol. 8, pp. 889–988 Berlin-Göttingen-Heidelberg: Springer 1958
Haartmann, U., Kahlson, G., Steinhardt, C.: Histamine formation in germinating seeds. Life Sci. *5*, 1–9 (1966)
Hartmann, T.: Zur Biogenese flüchtiger Amine beim Mutterkorn-Pilz *Claviceps purpurea*. Planta *66*, 191–206 (1965)
Hartmann, T : Biosynthese flüchtiger Amine in Äpfeln durch Aldehydaminierung. Z. Pflanzenphysiol. *57*, 368–375 (1967)
Hartmann, T.: Leucin-carboxy-lyase aus marinen Rhodophyceae. Vorkommen, Verbreitung und einige Eigenschaften. Phytochemistry *11*, 1327–1336 (1972a)
Hartmann, T.: Leucin-carboxy-lyase aus marinen Rhodophyceae. 2. Eigenschaften des Enzyms aus *Polysiphonia urceolata*. Biochem. Physiol. Pflanz. *163*, 1–13 (1972b)
Hartmann, T.: Leucin-carboxy-lyase aus marinen Rhodophyceae. 3. Coenzymspezifität und Aktivierung durch Carbonylverbindungen. Biochem. Physiol. Pflanz. *163*, 14–29 (1972c)
Hartmann, T., Aufermann, B.: Zur Physiologie der Aminbildung bei der marinen Rotalge *Polysiphonia urceolata*. Mar. Biol. *21*, 70–74 (1973)
Hartmann, T., Dönges, D., Steiner, M.: Biosynthese aliphatischer Monoamine in *Mercurialis perennis* durch Aminosäure-Aldehyd-Transaminierung. Z. Pflanzenphysiol. *67*, 404–417 (1972a)
Hartmann, T., Ilert, H.-I., Steiner, M.: Aldehydaminierung, der bevorzugte Biosyntheseweg für primäre aliphatische Monoamine in Blütenpflanzen. Z. Pflanzenphysiol. *68*, 11–18 (1972b)
Hasse, K., Ratych, O.T., Salnikow, J.: Transaminierung und Decarboxylierung von Ornithin und Lysin in höheren Pflanzen. Hoppe-Seylers Z. Physiol. Chem. *348*, 843–851 (1967)
Herrmann, V., Jüttner, F.: Excretion products of algae. Identification of biogenic amines by gas-liquid chromatography and mass spectrometry of their trifluoroacetamides. Anal. Biochem. *78*, 365–373 (1977)

Hill, J.M.: The changes with age in the distribution of copper and some copper-containing oxidases in red clover (*Trifolium pratense* L. cv. Dorset Marlgrass). J. Exp. Bot. *24*, 525–536 (1973)
Hill, J.M., Mann, P.J.G.: Some properties of plant diamine oxidase, a copper-containing enzyme. In: Recent aspects of nitrogen metabolism in plants. Hewitt, E.J., Cutting, C.V. (eds.), pp. 149–161. London: Academic Press 1968
Hirasawa, E., Suzuki, Y.: Purification and properties of a polyamine oxidase from *Zea mays*. Phytochemistry *14*, 99–101 (1975)
Holtz, P., Janisch, H.: Über das Vorkommen von Azetylcholin, Histamin und Adenosin in Pflanzen. Arch. Exp. Pathol. Pharmakol. *187*, 336–343 (1937)
Hosoi, K.: Purification and some properties of L-tyrosine carboxy-lyase from barley roots. Plant Cell Physiol. (Tokyo) *15*, 429–440 (1974)
Irvine, W.J., Saxby, M.J.: Steam volatile amines of Latakia tobacco leaf. Phytochemistry *8*, 473–476 (1969)
Kapadia, G.V., Fayez, M.B.E.: Peyote constituents. Chemistry, biogenesis and biological effects. J. Pharm. Sci. *59*, 1699–1727 (1970)
Kefeli, V.I., Turetskaya, R.K., Kutacek, M., Vackova, K.: Action of natural phenolic compounds and metabolic inhibitors on the biosynthesis of indoles in etiolated tissues of higher plants. Biokhim. Immun. Pokoya Rast. pp. 193–206 (1969) cited in Chem. Abstr. *74*, 20467
Klein, G., Steiner, M.: Stickstoffbasen im Eiweißabbau höherer Pflanzen. 1. Ammoniak und flüchtige Amine. Jahrb. Wiss. Bot. *68*, 602–710 (1928)
Koleva, M., Akhtardzhiev, Kh: Study of basic substances in the leaves of *Arum orientale*. Farmatsiya (Sofia) *26*, 25–27 (1976)
Kwasniewski, von V.: Untersuchungen über die nichtalkaloidischen Inhaltstoffe des Schöllkrauts (*Chelidonium majus* L.). Pharmazie *13*, 363–364 (1958)
Kwasniewski, von V.: Zur Kenntnis der Inhaltstoffe der Blüten der Taubnessel. Planta Med. *7*, 35–40 (1959)
Lafarge-Frayssinet, C., Bertaux, O., Valencia, R., Frayssinet, C.: Evolution of ornithine decarboxylase activity during the cell cycle of *Euglena gracilis* Z in synchronous culture. Influence of vitamin B-12. Biochim. Biophys. Acta *539*, 435–444 (1978)
Larher, F.: Métabolisme azoté des halophytes. Recherche des voies de formation de la putrescine dans les jeunes rameaux de *Limonium vulgare* Mill. C.R. Acad. Sci. Paris Ser D *277*, 1333–1336 (1973)
Larher, F., Bernard, T.: Contribution à l'étude du métabolisme azoté de *Limonium vulgare* Mill. Formation des méthylamines à partir de glycine ^{14}C-2. C.R. Acad. Sci. Paris Ser D *272*, 715–718 (1971)
Lee, K.K., Furuhashi, K., Yatazawa, M.: Utilization of arginine by rice callus tissue. Soil Sci. Plant Nutr. (Tokyo) *18*, 179–184 (1972)
Lerch, B., Stegemann, H.: Äthanolamine, ein verbreiteter Inhaltsstoff der Blätter höherer Pflanzen. Z. Naturforsch. *21B*, 216–218 (1966)
Le Rudulier, D., Goas, G.: Étude de l'activité de l'arginine décarboxylase dans les jeunes plantes de *Glycine max* privées de leurs cotylédons. Phytochemistry *14*, 1723–1725 (1975a)
Le Rudulier, D., Goas, G.: Influence des ions ammonium et potassium sur l'accumulation de la putrescine chez les jeunes plantes de *Soja hispida* Moench, privées de leurs cotylédons. Physiol. Vég. *13*, 125–136 (1975b)
Le Rudulier, D., Goas, G.: La diamine oxydase dans les jeunes plantes de *Glycine max*. Phytochemistry *16*, 509–511 (1977)
List, P.H., Hetzel, H.: Basische Pilzinhaltsstoffe 8. Planta Med. *8*, 105–106 (1960)
List, P.H., Menssen, H.G.: Basische Pilzinhaltsstoffe. 3. Arch. Pharm. *292*, 21–28 (1959)
List, P.H., Müller, H.: Basische Pilzinhaltsstoffe. 7. Arch. Pharm. *64*, 777–787 (1959)
List, P.H., Reinhard, C.: Pilzinhaltsstoffe. 11. Eine schwefelhaltige Diaminodicarbonsäure im Hexenei von *Phallus impudicus* L. Arch. Pharm. *295*, 564–571 (1962)
List, P.H., Reith, H.: Der Faltentintling, *Coprinus atramentarius* Bull., und seine dem Tetraäthylthiuramdisulfid ähnliche Wirkung. Arzneimittelforschung *10*, 34–40 (1960)
List, P.H., Wagner, K.: Ustilagin und andere Inhaltsstoffe des Maisbrandes *Ustilago maydis* Corda. Arzneimittelforschung *13*, 36–41 (1963)

Lloyd, G.R., Nicholls, P.J.: The presence of histamine in the cotton plant. J. Physiol. (London) *172*, 56P–57P (1964)

Lloyd, G.R., Nicholls, P.J.: Formation of histamine in the cotton plant. Nature (London) *206*, 298 (1965)

Lundström, J.: Biosynthesis of mescaline and 3,4-dimethoxy-phenethylamine in *Trichocereus pachanoi*. Acta. Pharm. Suec. *7*, 651–666 (1970)

Lundström, J.: Biosynthesis of mescaline and tetrahydroisoquinoline alkaloids in *Lophophora williamsii* (Lem). Coult. Acta. Chem. Scand. *25*, 3489–3499 (1971)

Lundström, J., Agurell, S.: Biosynthesis of mescaline and tetrahydroisoquinoline alkaloids in *Lophophora williamsii* (Lem). Coult. Acta. Pharm. Suec. *8*, 261–274 (1971)

Major, R.T., Dürsch, F.: N^{α},N^{α}-Dimethylhistamine, a hypotensive principle in *Casimiroa edulis* Llave et Lex. J. Org. Chem. *23*, 1564–1565 (1958)

Maretzki, A., Thom, M., Nickell, L.G.: Products of arginine catabolism in growing cells of sugar cane. Phytochemistry *8*, 811–818 (1969)

Martin-Tanguy, J., Martin, C., Gallet, M., Vernoy, R.: Sur de puissants inhibiteurs naturels de multiplication du virus de la mosaique du Tabac. C.R. Acad. Sci. Paris Ser D *282*, 2231–2234 (1976)

Matsuda, H., Suzuki, Y.: Some properties of the amine oxidase in *Vicia faba* seedlings. Plant Cell Physiol. (Tokyo) *18*, 1131–1137 (1977)

Mennucci, L., Rojas, S., Camargo, E.P.: Polyamines and ornithine decarboxylase activity during growth and differentiation in *Blastocladiella emersonii*. Biochim. Biophys. Acta *404*, 249–256 (1975)

Mizusaki, S., Tanabe, Y., Noguchi, M., Tamaki, E.: *N*-methylputrescine oxidase from tobacco roots. Phytochemistry *11*, 2757–2762 (1972)

Mizusaki, S., Tanabe, Y., Noguchi, M., Tamaki, E.: Changes in the activities of ornithine decarboxylase, putrescine *N*-methyltransferase and *N*-methylputrescine oxidase in tobacco root in relation to nicotine biosynthesis. Plant Cell Physiol (Tokyo) *14*, 103–110 (1973)

Mudd, S.H.: *S*-Adenosylmethionine requirement for plant transmethylations. Biochim. Biophys. Acta *37*, 164–165 (1960)

Murty, K.S., Smith, T.A., Bould, C.: The relation between the putrescine content and potassium status of black currant leaves. Ann. Bot. (London) *35*, 687–695 (1971)

Neu, R., Fiedler, U.: Über neue basische Inhaltsstoffe aus Weissdorn. Naturwissenschaften *41*, 259–260 (1954)

Neurath, G.B., Schreiber, O.: Investigations on amines in the human environment. In: *N*-Nitroso compounds in the environment. Bogovski, P., Walker, E.A. (eds.), pp. 211–214. International agency for research on cancer. Scientific Publication No. 9 1973

Nylén, U., Szybek, P.: Kinetics and other characteristics of diamine oxidase of pea seedlings. Acta Chem. Scand. *28 B*, 1153–1160 (1974)

O'Donovan, D.G., Horan, H.: The biosynthesis of annuloline, a unique oxazole alkaloid. J. Chem. Soc. C 331–334 (1971)

Paul, A.G.: Biosynthesis of the peyote alkaloids. Lloydia *36*, 36–45 (1973)

Percival, F.W., Purves, W.K.: Multiple amine oxidases in cucumber seedlings. Plant Physiol. *54*, 601–607 (1974)

Phelps, R.H., Sequeira, L.: Synthesis of IAA via tryptamine by a cell free system from tobacco terminal buds. Plant Physiol. *42*, 1161–1163 (1967)

Pösö, H., Sinervirta, R., Jänne, J.: *S*-Adenosylmethionine decarboxylase from baker's yeast. Biochem. J. *151*, 67–73 (1975)

Poupat, C., Sévenet, T.: Cinnamoylhistamine et hordenine, alcaloides d'*Acacia spirorbis*. Phytochemistry *14*, 1881–1882 (1975)

Preusser, von E.: Aliphatische Amine in Samen beziehungsweise Früchten und in Keimpflanzen von *Vicia faba* und *Zea mays*. Biol. Zentralbl. *85*, 19–27 (1966)

Preusser, von E.: Zur Biosynthese von flüchtigen Stickstoffverbindungen bei Pflanzen. Biol. Zentralbl. *94*, 75–86 (1975)

Rabitzsch, G.: Zur Analytik und Biochemie der *p*-Hydroxy-*β*-phenyl-alkylamine in *Hordeum vulgare* L. Planta Med. *7*, 268–297 (1959)

Radhakrishnan, A.N.: Some unusual amino compounds from sandal, *Santalum album* L. J. Sci. Ind. Res. *33*, 461–466 (1974)

Ramakrishna, S., Adiga, P.R.: Amine biosynthesis in *Lathyrus sativus* seedlings. Phytochemistry *13*, 2161–2166 (1974)
Ramakrishna, S., Adiga, P.R.: Arginine decarboxylase from *Lathyrus sativus* seedlings. Eur. J. Biochem. *59*, 377–386 (1975)
Ramakrishna, S., Adiga, P.R.: Decarboxylation of homoarginine and lysine by an enzyme from *Lathyrus sativus* seedlings. Phytochemistry *15*, 83–86 (1976)
Raman, S., Reddy, J., Lipscomb, W.N., Kapoor, A.L., Djerassi, C.: Structure of the alkaloid casimiroedine. Tetrahedron Lett. 357–360 (1962)
Repke, D.B.: The histamine amides of *Acacia longifolia*. Lloydia *38*, 101–105 (1975)
Richardson, K.E., Miedema, E., Vandor, S.L.: The oxidation of ethanolamine by plant tissue preparations. Can. J. Biochem. *44*, 1271–1275 (1966)
Richardson, M.: Studies on the biogenesis of some simple amines and quaternary ammonium compounds in higher plants. Phytochemistry *5*, 23–30 (1966)
Rivier, L., Lindgren, J-E.: "Ayahuasca", the South American hallucinogenic drink: an ethnobotanical and chemical investigation. Econ. Bot. *26*, 101–129 (1972)
Rolle, I., Hobucher, H-E., Kneifel, H., Paschold, B., Riepe, W., Soeder, C.J.: Amines in unicellular green algae – 2. Amines in *Scenedesmus acutus*. Anal. Biochem. *77*, 103–109 (1977)
Rosenberg, H., Stohs, S.J.: Effects of histidine decarboxylase inhibitors on the production of an aberrant alkaloid in *Dolichothele sphaerica*. Phytochemistry *15*, 501–503 (1976)
Rothe, G.M.: Das Vorkommen von Dopaminoxidase, Polyphenoloxidase, Peroxidase und Monoamineoxidase in *Lens culinaris*. Beitr. Biol. Pflanz. *50*, 113–120 (1974)
Ruzicka, V., Minář, J.: The significance of the cotyledons for the formation of diamine oxidase in pea plants. Biol. Plant. *16*, 215–223 (1974)
Saha, J.C., Kasinathan, S.: Histamine action of the latex of *Calotropis gigantea* Ait. Arch. Int. Pharmacodyn. Thér. *143*, 78–89 (1963)
Sajner, J., Věříš, O.: Über das Vorkommen von Histamin in *Viscum album*. Pharmazie *13*, 170 (1958a)
Sajner, J., Věříš, O.: Histaminwirkungen von Weißdorninfus. Pharmazie *13*, 52–54 (1958b)
Samborski, D.J., Rohringer, R.: Abnormal metabolites of wheat: occurrence, isolation and biogenesis of 2-hydroxyputrescine amides. Phytochemistry *9*, 1939–1945 (1970)
Saxena, P.R., Pant, M.C., Kishor, K., Bhargava, K.P.: Identification of pharmacologically active substances in the Indian stinging nettle *Urtica parviflora* Roxb. Can. J. Physiol. Pharmacol. *43*, 869–876 (1965)
Saxena, P.R., Tangri, K.K., Bhargava, K.P.: Identification of acetylcholine, histamine and 5-hydroxytryptamine in *Girardinia heterophylla* (Decne.). Can. J. Physiol. Pharmacol. *44*, 621–627 (1966)
Schneider, E.A., Wightman, F.: Amino acid metabolism in plants V. Can. J. Biochem. *52*, 698–705 (1974)
Schormüller, J., Weder, J.: Untersuchungen über Aromastoffe von Lebensmitteln. 5. Z. Lebensmittelforsch. *130*, 213–220 (1966)
Schultes, R.E.: Indole alkaloids in plant hallucinogens. Planta Med. *29*, 330–342 (1976)
Schütte, H.R.: Einfache Amine. In: Biosynthese der Alkaloide. Mothes, K., Schütte, H.R. (eds.) pp. 168–182. Berlin: VEB Deutscher Verlag der Wissenschaften 1969
Schütte, H.R., Knöfel, D.: Über Zwischenprodukte der Lupinin und Sparteinbildung. Z. Pflanzenphysiol. *59*, 80–86 (1968)
Schütte, H.R., Knöfel, D., Heyer, O.: Biosynthese von Lupinenalkaloiden. 19. Mitt. Anreicherung und Charakterisierung der Diaminoxydase aus Keimlingen von *Lupinus luteus*. Z. Pflanzenphysiol. *55*, 110–118 (1966)
Serenkov, G.P.: Ammonia and volatile amines in tea seeds. Nauch. Dokl. Vses. Skh. Biol. Nauk. *2*, 175–179 (1959)
Sherwin, J.E.: A tryptophan decarboxylase from cucumber seedlings. Plant Cell Physiol. (Tokyo) *11*, 865–872 (1970)
Simon, E.W.: Valine decarboxylation in *Arum* spadix. J. Exp. Bot. *13*, 1–4 (1962)
Smith, B.N., Meeuse, B.J.D.: Production of volatile amines and skatole at anthesis in some *Arum* lily species. Plant Physiol. *41*, 343–347 (1966)
Smith, T.A.: L-Arginine carboxy-lyase of higher plants and its relation to potassium nutrition. Phytochemistry *2*, 241–252 (1963)

Smith, T.A.: The biosynthesis of putrescine in higher plants and its relation to potassium nutrition In: Recent aspects of nitrogen metabolism in plants. Hewitt, E.J., Cutting, C.V. (eds.). pp. 139–146. London: Academic Press 1968
Smith, T.A.: Agmatine iminohydrolase in maize. Phytochemistry *8*, 2111–2117 (1969)
Smith, T.A.: The occurrence, metabolism and functions of amines in plants. Biol. Rev. *46*, 201–241 (1971)
Smith, T.A.: Polyamine oxidase from barley and oats. Phytochemistry *15*, 633–636 (1976)
Smith, T.A.: Phenethylamine and related compounds in plants. Phytochemistry *16*, 9–18 (1977a)
Smith, T.A.: Tryptamine and related compounds in plants. Phytochemistry *16*, 171–175 (1977b)
Smith, T.A.: Homospermidine in *Rhizobium* and legume root nodules. Phytochemistry *16*, 278–279 (1977c)
Smith, T.A.: Recent advances in the biochemistry of plant amines. In: Progress in phytochemistry. Reinhold, L., Harborne, J.B., Swain, T. (eds.). *4*, 27–82 (1977d)
Smith, T.A., Sinclair, C.: The effect of acid feeding on amine formation in barley. Ann. Bot. (London) *31*, 103–111 (1967)
Smith, T.A., Wilshire, G.: Distribution of cadaverine and other amines in higher plants. Phytochemistry *14*, 2341–2346 (1975)
Smith, W.J., Kirshner, N.: Enzymatic formation of noradrenaline by the banana plant. J. Biol. Chem. *235*, 3589–3591 (1960)
Speranza, A., Bagni, N.: Putrescine biosynthesis in *Agrobacterium tumefaciens* and in normal and crown gall tissues of *Scorzonera hispanica* L. Z. Pflanzenphysiol. *81*, 226–233 (1977)
Srivastava, S.K., Prakash, V.: Purification and properties of pea cotyledon and embryo diamine oxidase. Phytochemistry *16*, 189–190 (1977)
Srivastava, S.K., Prakash, V. Naik, B.I.: Regulation of diamine oxidase activity in germinating pea seeds. Phytochemistry *16*, 185–187 (1977)
Stahl, E., Kaltenbach, U.: Die basischen Inhaltsstoffe des Aronstabes (*Arum maculatum* L.). Arch. Pharm. *298*, 599–604 (1965)
Steiner, M.: Flüchtige Amine in Pflanzen. Arbeitsgem. Forsch. Landes Nordrhein Westfalen *161*, 7–46 (1966)
Steiner, M., Hartmann, T.: Über Vorkommen und Verbreitung flüchtiger Amine bei Meeresalgen. Planta *79*, 113–121 (1968)
Steiner, M., Löffler, H.: Stickstoffbasen im Eiweißabbau höherer Pflanzen. 2. Histochemische Studien über Verbreitung, Verteilung und Wandel des Ammoniaks und der flüchtigen Amine. Jahrb. Wiss. Bot. *71*, 463–532 (1929)
Stein von Kamienski, E.: Untersuchungen über die flüchtigen Amine der Pflanzen, Mitt. 2 und 3. Planta *50*, 315–352 (1957)
Stevens, L., McKinnon, I.M.: The effect of 1,4-diaminobutanone on the stability of ornithine decarboxylase from *Aspergillus nidulans*. Biochem. J. *166*, 635–637 (1977)
Stewart, I., Wheaton, T.A.: Phenolic amines in *Citrus* juice. Proc. Fl. State Hortic. Soc. *77*, 318–320 (1964)
Stoessl, A., Unwin, C.H.: The antifungal factors in barley. 5. Antifungal activity of the hordatines. Can. J. Bot. *48*, 465–470 (1970)
Suresh, M.R., Adiga, P.R.: Putrescine-sensitive (artifactual) and insensitive (biosynthetic) *S*-adenosyl-L-methionine decarboxylase activities of *Lathyrus sativus* seedlings. Eur. J. Biochem. *79*, 511–518 (1977)
Suresh, M.R., Ramakrishna, S., Adiga, P.R.: Diamine oxidase of *Lathyrus sativus* seedlings. Phytochemistry *15*, 483–485 (1976)
Suzuki, H., Ogura, Y.: Mode of inhibition of the *Aspergillus* amine oxidase by 8-hydroxyquinoline. J. Biochem. *72*, 833–839 (1972)
Suzuki, T.: Metabolism of methylamine in the tea plant (*Thea sinensis* L.). Biochem. J. *132*, 753–763 (1973)
Suzuki, Y.: Nitrogen metabolism in the seedling of *Lupinus luteus*. the absence of L-lysine decarboxylase and the presence of cadaverine oxidase. Sci. Rep. Tôkohu Univ. Ser. IV (Biol.) *25*, 139–144 (1959a)
Suzuki, Y.: Studies on a polyphenolase in *Scopolia japonica*. 6. Deamination of putrescine. Sci. Rep. Tôkohu Univ. Ser. 4 *25*, 125–129 (1959b)
Tackie, A.N., Dwuma-Badu, D., Ayim, J.S.K., El Sohly, H.N., Knapp, J.E., Slatkin, D.J.,

Schiff, P.L.: *N-iso*-Butyloctadeca-*trans*-2-*trans*-4-dienamide: a new constituent of *Piper guineense*. Phytochemistry *14*, 1888–1889 (1975)
Takeo, T.: L-Alanine decarboxylase in *Camellia sinensis*. Phytochemistry *17*, 313–314 (1978)
Tocher, R.D., Tocher, C.S.: Biosynthesis of 3-hydroxytyramine in plants. Abstr. 11th Int. Bot. Congr. 219 (1969)
Tocher, R.D., Tocher, C.S.: DOPA decarboxylase in *Cytisus scoparius*. Phytochemistry *11*, 1661–1667 (1972)
Udenfriend, S., Lovenberg, W., Sjoerdsma, A.: Physiologically active amines in common fruits and vegetables. Arch. Biochem. Biophys. *85*, 487–490 (1959)
Unger, W., Hartmann, T.: Biosynthese aliphatischer Monoamine in Blütenpflanzen durch Nebenaktivität einer Glutamate-Oxalacetat-Aminotransferase? Z. Pflanzenphysiol. *77*, 255–267 (1976)
Vialli, D.M., Barbetta, F., Zanotti, L., Mihályi, K.: Estendibilità del concetto di systema cellulare enterocromaffine ai vegetali. Arch. Histochem. *45*, 270–282 (1973)
Vig, O.P., Vig, A.K., Handa, V.K., Sharma, S.D.: Synthesis of sylvatine. Indian J. Chem. *13*, 225–227 (1975)
Villalobos, J., Ramirez, F., Moussatché, H.: Some observations on the presence of acetylcholine and histamine in plants. Ciênc. Cult. *26*, 690–693 (1974)
Werle, E.: Amine und Betaine. In: Modern methods of plant analysis. Paech, K., Tracey, M.V. (eds.), Vol. 4, pp. 517–623. Berlin: Springer 1955
Werle, E., Pechmann, E.: Über die Diaminoxydase der Pflanzen und ihre adaptive Bildung durch Bakterien. Ann. Chem. *562*, 44–60 (1949)
Werle, E., Roewer, F.: Monaminoxydase in Pflanzen. Biochem. Z. *320*, 298–301 (1950)
Werle, E., Roewer, F.: Über tierische und pflanzliche Monaminoxydasen. Biochem. Z. *322*, 320–326 (1952)
Werle, E., Zabel, A.: Über die Verbreitung der Histaminase im Pflanzenreich. Biochem. Z. *318*, 554–559 (1948)
Wheaton, T.A., Stewart, I.: Biosynthesis of synephrine in *Citrus*. Phytochemistry *8*, 85–92 (1969)
Willaman, J.J., Li, H.-L.: Alkaloid-bearing plants and their contained alkaloids (1957–1968). Lloydia (Suppl.) *33*, No. 3A (1970)
Wilson, R.H.: The regulation by methionine of three inducible enzymes in *Coprinus lagopus*. Biochem. J. *118*, 16P (1970)
Yamada, H., Kumagai, H., Uwajima, T., Ogata, K.: Trimethylamine metabolism. I. Monomethylamine oxidase. Mem. Res. Inst. Food Sci. Kyoto Univ. *27*, 1–14 (1966)
Yamasaki, K., Tamaki, T., Uzawa, S., Sankawa, U., Shibata, S.: Participation of C_6-C_1 unit in the biosynthesis of ephedrine in *Ephedra*. Phytochemistry *12*, 2877–2882 (1973)
Yoshida, D.: Effect of agmatine on putrescine formation in tobacco plants. Plant Cell Physiol. (Tokyo) *10*, 923–924 (1969)

9 Cyanogenic Glycosides [1]

E.E. CONN

1 Introduction

The ability of living organisms to produce hydrocyanic acid, HCN, is termed *cyanogenesis*. Organisms which exhibit this phenomenon usually contain one or more compounds as precursors which liberate the HCN on hydrolysis. The *cyanogenic substances* in plants are usually carbohydrate derivatives, specifically, β-glycosides of 2-hydroxynitriles (cyanohydrins). Cyanogenic glycosides will release HCN on treatment with dilute acid at elevated temperatures. However, the cyanogenesis observed when cyanophoric plant tissues are crushed or their cellular structure otherwise disrupted is the result of the action of enzymes in the plant acting on the cyanogenic glycosides which the plant contains.

An alternate source of HCN in plants has recently been described; these are the cyanolipids whose occurrence appears limited to the Sapindaceae (MIKOLAJCZAK, 1977). Certain bacteria and fungi also produce HCN by a mechanism that apparently does not involve cyanogenic glycosides (or lipids) as intermediates. The reader is referred to KNOWLES (1976) for a thorough discussion of cyanogenesis in bacteria and fungi. HCN is also produced by millipedes (EISNER, 1970) and by burnet moths (JONES et al., 1962); the mechanism of HCN production in the former is being examined (DUFFEY and TOWERS, 1978). The present review will be limited to cyanogenic glycosides which occur in higher plants.

2 Structure and General Chemical Properties

2.1 Structure

2.1.1 General Structural Features

A total of 23 cyanogenic glycosides has been reported in higher plants. They have the general formula shown in Figure 1 and as such are glycosidic derivatives of 2-hydroxynitriles. The structural variations are of three types involving (1) the nature of the sugar moiety, (2) the nature of the groups R_1 and R_2, and (3) the chirality of the carbinol carbon atom.

---

1 The present review is based on literature published through December 1977

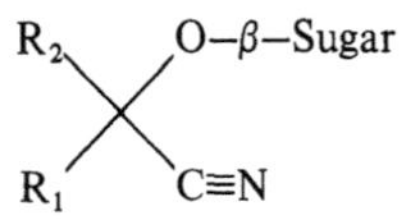

Fig. 1. General formula for cyanogenic glycosides

The sugar moiety in all but three of the known cyanogenic glycosides is the monosaccharide, D-glucose. This sugar is linked to the carbinol carbon by an O-β-glucosyl bond. The three exceptions are amygdalin, vicianin, and lucumin which have disaccharides as their carbohydrate moiety (see Sect. 2.1.4).

R_1 is either an aliphatic or aromatic group in a majority of the cyanogenic glycosides; the exceptions are triglochinin and its methyl ester and the glycosides having a cyclopentene ring structure (Sect. 2.1.6). In the latter, R_1 and R_2 are the elements of the cyclopentene ring. R_2 in a majority of the cyanogens is H; the exceptions, besides triglochinin, methyl triglochinin and the cyclopentene glycosides already mentioned, are linamarin (I) and lotaustralin (II). In these two compounds R_2 is $-CH_3$.

Since R_1 is usually not the same as R_2 in these compounds, the carbinol carbon atom is chiral. This introduces the possibility of epimeric pairs of glycosides which would differ only in configuration at that carbon. Several such epimeric pairs do occur in nature, usually not in the same plant or even in related families.

It is possible to classify the cyanogenic glycosides based on the chemical (i.e., aromatic, aliphatic, alicyclic) nature of the aglycone, i.e., the group to which the sugar is attached in glycosidic linkage (EYJOLFSSON, 1970). Since this method tends to deemphasize the biosynthetic origin of the aglycones, others (CONN, 1973a; SEIGLER, 1975) have classified the cyanogenic glycosides on the basis of their origin, either demonstrated or presumed. This will be the basis for the classification used in this review.

2.1.2 Glycosides Derived from Valine and Isoleucine

Linamarin and lotaustralin are two of the cyanogenic glycosides most commonly encountered in higher plants (see Sect. 4 on distribution). Their aglycones are formed (see Sect. 5 and CONN, 1973a) from L-valine and L-isoleucine by a pathway involving loss of the carboxyl carbon of the amino acids, oxidation of the amino nitrogen and 2-carbon atoms to a nitrile, oxygenation of the 3-carbon atom and glycosylation. These two glycosides, without apparent exception, always occur in the same plant although not necessarily in the same ratio. Thus, BUTLER (1965) found that linamarin predominated over lotaustralin in the ratio 19:1 in *Phaseolus lunatus* while the reversed ratio of 1:19 was observed in *Lotus tenuis*. The occurrence of both compounds within a single species is attributed to existence of a single

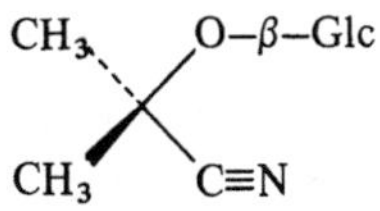

Linamarin
(I)

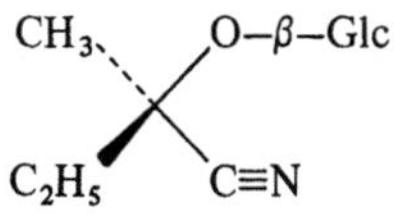

(*R*)-Lotaustralin
(II)

Glc = D-glucose

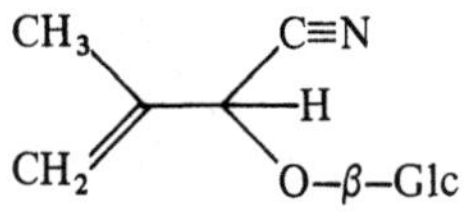

Proacacipetalin
(III)

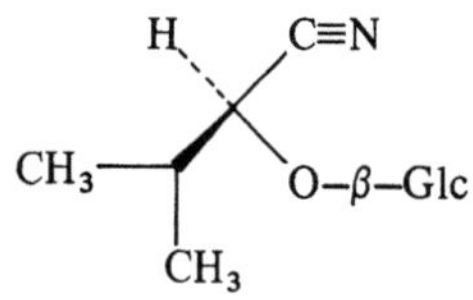

(*S*)-Heterodendrin
(Dihydroacacipetalin)
(IV)

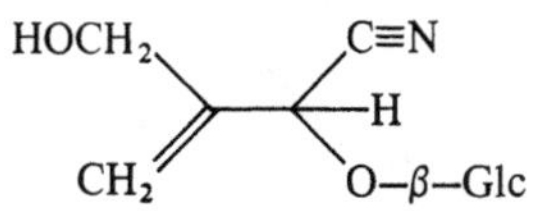

Cardiospermin
(V)

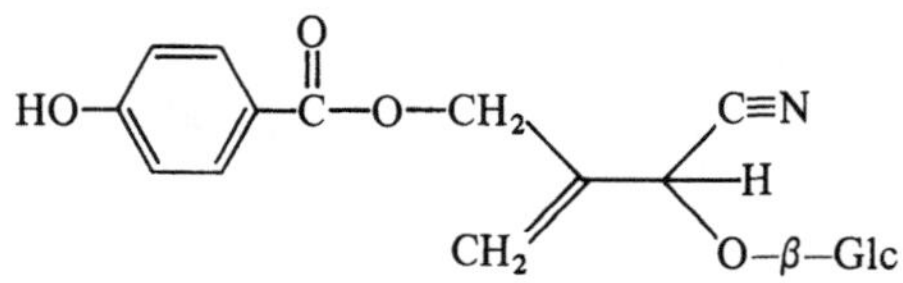

Cardiospermin-*p*-hydroxybenzoate
(VI)

set of biosynthetic enzymes that can act on either valine or isoleucine and the related intermediates (CONN, 1973a). In support of this claim, only a single glucosyl transferase, which catalyzes the last step of the biosynthetic pathway, is required for synthesizing either linamarin or lotaustralin (HAHLBROCK and CONN, 1971).

2.1.3 Glycosides Derived from Leucine

Although four such glycosides are known, data indicating the biosynthetic origin of the aglycone from L-leucine are available only in the case of proacacipetalin (III) (BUTTERFIELD et al., 1975) in *Acacia sieberiana* var. *woodii*. For a discussion of the relationship between acacipetalin and proacacipetalin see ETTLINGER et al. (1977). Definitive studies on the configuration of heterodendrin (IV), presumed to be identical with dihydroacacipetalin, have recently established that it has the (*S*)-configuration at the carbinol carbon (NAHRSTEDT and HUBEL, 1978). Cardiospermin (V) was isolated from *Cardiospermum hirsutum*, a Sapindaceous species known to contain one or more cyanolipids of closely related structures (SEIGLER et al., 1974). A novel ester of cardiospermin, in which the hydroxymethyl group is esterified with p-hydroxybenzoic acid (VI), has also been isolated from *Sorbaria arborea* (NAHRSTEDT, 1976).

2.1.4 Glycosides Derived from Phenylalanine

L-Phenylalanine would appear to be the precursor of the aglycone of seven cyanogenic glycosides, and among these are the three cyanogens amygdalin (VII), vicianin (VIII), and lucumin (IX) which have disaccharides as the sugar moiety. This group contains two epimeric pairs, (*R*)-prunasin (X)/(*S*)-sambunigrin (XI) and (*R*)-holocalin (XII)/(*S*)-zierin (XIII). Thus (*R*)-prunasin differs from (*S*)-sambunigrin in its configuration at only a single chiral center, namely the carbinol carbon atom. The (*S*)-configuration was originally assigned to holocalin by GMELIN et al. (1973) on the basis of specific rotation. However, a comparative study of the NMR

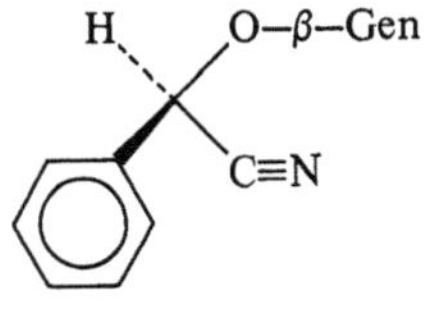

(*R*)-Amygdalin*
(VII)

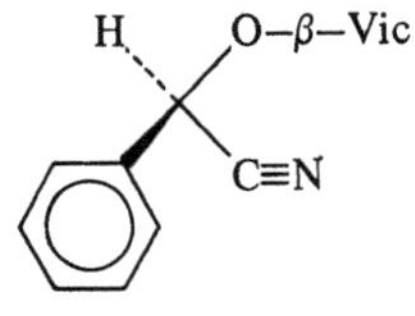

(*R*)-Vicianin*
(VIII)

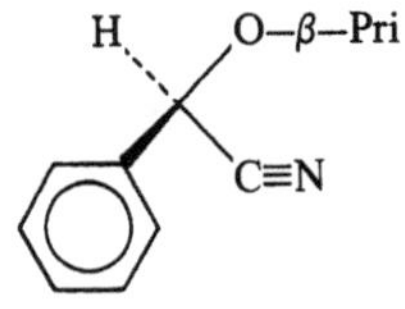

(*R*)-Lucumin*
(IX)

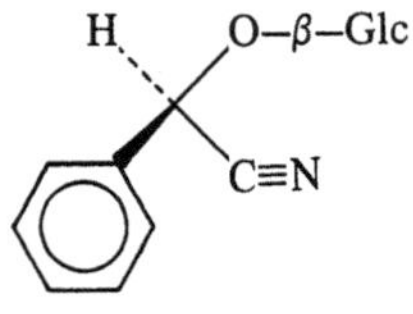

(*R*)-Prunasin
(X)

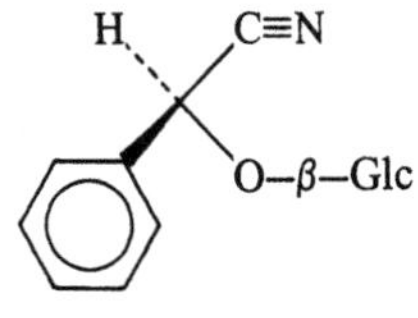

(*S*)-Sambunigrin
(XI)

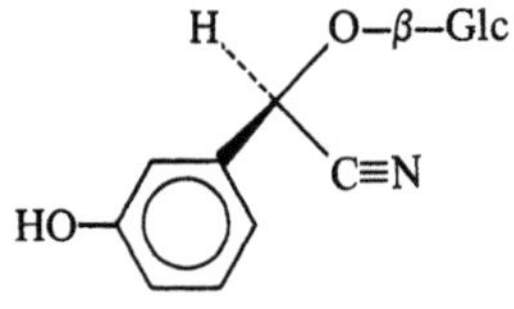

(*R*)-Holocalin
(XII)

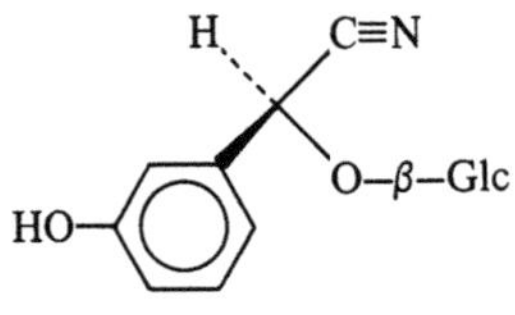

(*S*)-Zierin
(XIII)

* The abbreviations Gen, Vic, and Pri stand for the disaccharides gentiobiose, vicianose, and primaverose respectively

spectra of these compounds and the closely related glycosides derived from tyrosine [(*R*)-taxiphyllin and (*S*)-dhurrin] has led SEIGLER (1975) to argue for the (*R*)-configuration for holocalin. His conclusion is supported by independent work by NAHRSTEDT (1973b) on the separation of epimeric mixtures of these glycosides by GLC. Therefore the long-known zierin (FINNEMORE and COOPER, 1936) is assigned the (*S*)-configuration.

The older literature has references to the isolation by HERISSEY (1906) of an epimeric mixture of (*R*)-prunasin and (*S*)-sambunigrin from leaves of the cherry laurel, *Prunus lauracerasus*. This mixture, given the name "prulaurasin" was indeed obtained by HERISSEY, but some years later (PLOUVIER, 1935) was shown to be an artifact of isolation resulting from the addition of $CaCO_3$ to the crude plant extracts. Glycosides of the amygdalin type having a chiral carbon at the point of attachment of the sugar epimerize readily at alkaline pH (see Sect. 2.2.1.1) and the $CaCO_3$ provided the alkaline environment necessary for epimerization.

2.1.5 Glycosides Derived from Tyrosine

Six cyanogens are known which derive their aglycone from L-tyrosine. These include another epimeric pair, (*R*)-taxiphyllin (XIV) and (*S*)-dhurrin (XV) which, like such pairs previously discussed, do not necessarily occur in closely related species.

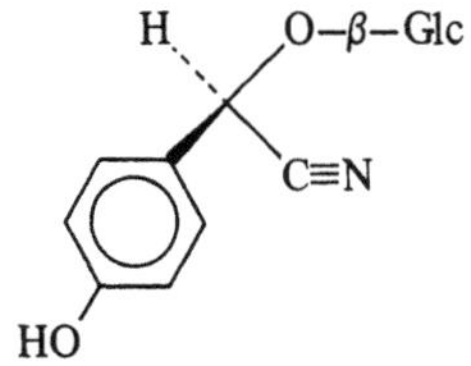

(*R*)-Taxiphyllin
(XIV)

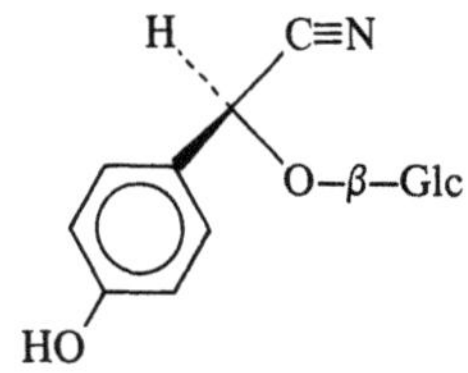

(*S*)-Dhurrin
(XV)

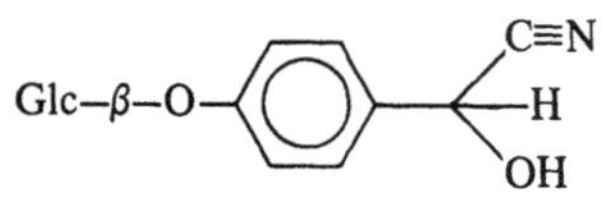

p–Glucosyloxymandelonitrile
(XVI)

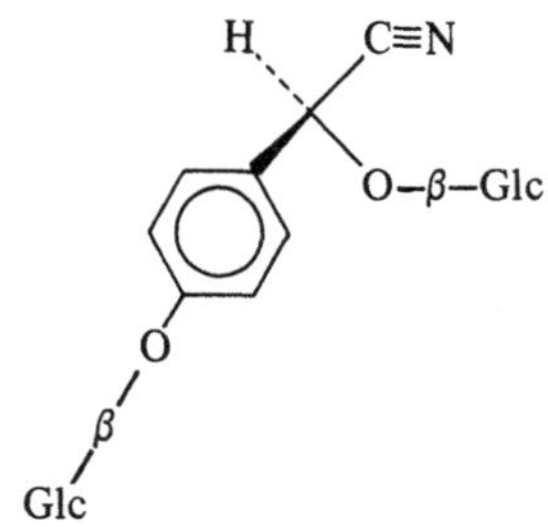

(*S*)-Proteacin
(XVII)

This group also includes the unstable *p*-glucosyloxymandelonitrile (XVI) which occurs in *Nandina domestica* (ABROL et al., 1966), *Thalictrum aquilegifolium* (SHARPLES and STOKER, 1969) and *Goodia lotifolia* (FINNEMORE and LARGE, 1936). In this glucoside the labile cyanohydrin is not stabilized in glycosidic linkage and consequently dissociates during isolation, making it difficult to study. The *p*-glucosyloxybenzaldehyde formed when HCN is released is well characterized. The diglucosyl proteacin (XVII) is believed, from NMR studies, to have the (*S*)-configuration (YOUNG and HAMILTON, 1966).

Triglochinin (XVIII) and its methylester (XIX) complete the glycosides presumably derived from tyrosine. Direct biosynthetic evidence on this point for the methylester has been provided by SHARPLES et al. (1972). The relationship of these two compounds, which lack the aromatic ring found in the other glycosides, to tyrosine itself may not be immediately obvious. However, hydroxylation of dhurrin (or taxiphyllin) on the aromatic ring at the carbon adjacent to the existing hydroxyl followed by oxidative ring cleavage of the 3,4-dihydroxymandelonitrile glucoside so formed would yield triglochinin (ETTLINGER and EYJOLFSSON, 1972).

2.1.6 Glycosides with Cyclopentene Rings

A total of four cyanogenic glycosides with cyclopentenyl aglycones is known. Gynocardin (XX) is the only one for which the structural assignment is complete. The problem of assignment of two different names to the same compound also exists in this group with tetraphyllin A (XXI) presumably being identical with deidaclin, and tetraphyllin B (XXII) being the same as barterin (see SEIGLER, 1975 for discussion). An East African species has also yielded evidence for the existence of *epi*-tetraphyllin B (XXIII) and its occurrence together with tetraphyllin B in the same species (GONDWE et al., 1978).

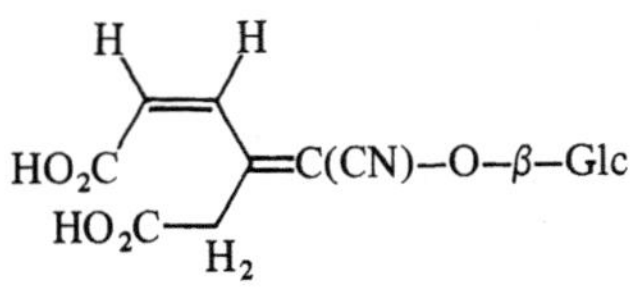

Triglochinin
(XVIII)

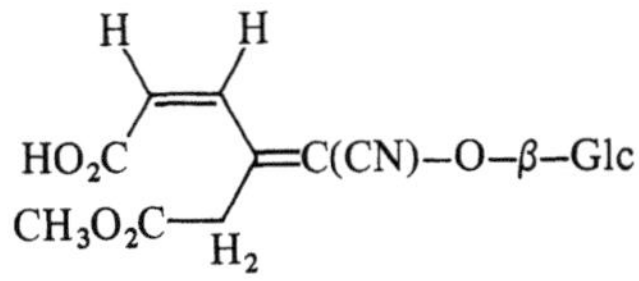

Triglochinin, methylester
(XIX)

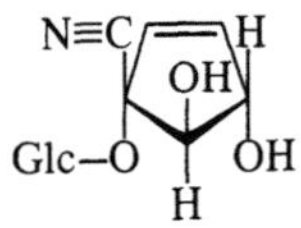

Gynocardin
(XX)

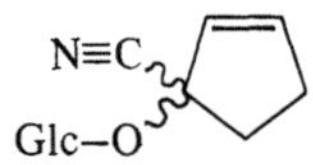

Tetraphyllin A
(Deidaclin)
(XXI)

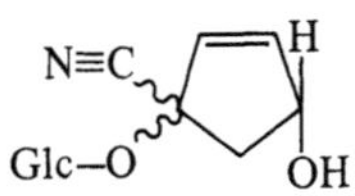

Tetraphyllin B
(Barterin)
(XXII)

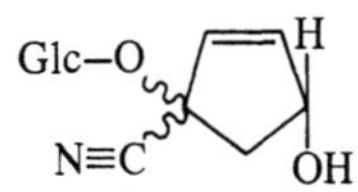

epi–Tetraphyllin B
(XXIII)

2.2 Properties of Cyanogenic Glycosides

A detailed description of the chemical properties of the known cyanogenic glycosides is beyond the purpose of this review. For the interested reader, the reviews of EYJOLFSSON (1970) and SEIGLER (1975) contain much information. Alternatively, the original literature, which usually contains physical constants, spectral data, and other properties can be consulted. When this is done it will be observed that the cyanogenic glycoside most thoroughly studied is amygdalin, discovered in 1830 by ROBIQUET and BOUTRON-CHARLARD. This compound, readily isolated from bitter almonds and apricot kernels, has long been available commercially. Within the past decade linamarin has also become available commercially and we can expect that further studies on this compound will be forthcoming. Since none of the other glycosides are available except by their isolation from original or new sources, additional information over that reported in the original paper(s) describing their isolation and characterization is meager.

The properties of cyanogenic glycosides of interest in plant physiology will be summarized briefly below.

2.2.1 Hydrolysis

2.2.1.1 Chemical Hydrolysis

The cyanogenic glycosides react in a predictable manner to acid and alkaline hydrolysis. Being glycosides they are cleaved in dilute acid at elevated temperatures

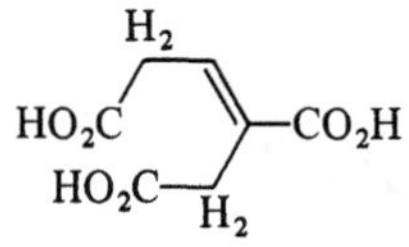

Triglochinic acid
(XXIV)

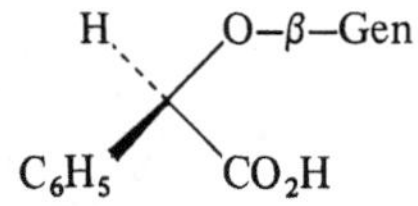

Amygdalinic acid*
(XXV)

* Gen stands for gentiobiose

at the β-glycosidic bond to yield their sugar component and a cyanohydrin. The latter is unstable dissociating to HCN and an aldehyde or ketone. Thus, acetone and 2-butanone are formed from linamarin and lotaustralin, respectively, while the glycosides derived from phenylalanine yield benzaldehyde or, in the case of holocalin and zierin, *m*-hydroxybenzaldehyde. The cyclopentene glycosides have yielded cyclopentenones, and those tyrosine-derived glycosides which retain the aromatic ring in their structure yield *p*-hydroxybenzaldehyde. On the other hand, triglochinin, also tyrosine-derived, on dissociation of its cyanohydrin produces a ketene which reacts with water to produce triglochinic acid (XXIV).

Hydrolysis by concentrated acids can yield the corresponding 2-hydroxy acid and NH_3. These products would be expected if the rate constant for hydrolysis of the nitrile group is greater than that for hydrolysis of the β-glycosidic bond.

Hydrolysis under mild alkaline conditions [saturated $Ba(OH)_2$] typically hydrolyzes the nitrile group in several cyanogenic glycosides to a carboxyl function without cleavage of any other bonds. The resultant product is the corresponding glycosidic acid, i.e., amygdalinic acid (XXV) is formed from amygdalin. On the other hand, dhurrin is labile in dilute alkaline solution and decomposes to D-glucose, HCN, and *p*-hydroxybenzaldehyde (MAO and ANDERSON, 1965).

Alkaline conditions facilitate epimerization at the carbinol carbon of cyanogenic glycosides having electron withdrawing groups (e.g., an aromatic ring) adjacent to that carbon (NAHRSTEDT, 1975). As reported in Section 2.1.4, this erroneously led to the report of a naturally occurring epimeric mixture of (*R*,*S*)-mandelonitrile glucoside in cherry laurel (HERRISEY, 1906). The epimerization of amygdalin will even occur at significant rates at neutral pH (pH 7.0) and elevated temperatures (NAHRSTEDT, 1975). This finding requires that care be taken with isolation procedures involving boiling aqueous solvents in order to avoid epimerization. In those few cases (JENSEN and NIELSEN, 1973; GONDWE et al., 1978) where epimeric mixtures have been isolated, the authors, aware that epimerization could occur, have concluded that the conditions employed would keep the process at a minimum.

2.2.1.2 Enzymatic Hydrolysis

Enzymatic hydrolysis of cyanogenic glycosides was discovered at the same time as the compounds themselves because plant tissues usually contain both the glycosides and the enzymes. This process is represented for prunasin in Figure 2. The hydrolytic reaction is catalyzed by a β-glucosidase yielding D-glucose and (*R*)-mandelonitrile, a cyanohydrin. The cyanohydrin in turn will dissociate nonenzymically to form benzaldehyde and HCN. However, plant tissues usually contain a second enzyme, a hydroxynitrile lyase, which catalyzes the cyanohydrin dissociation.

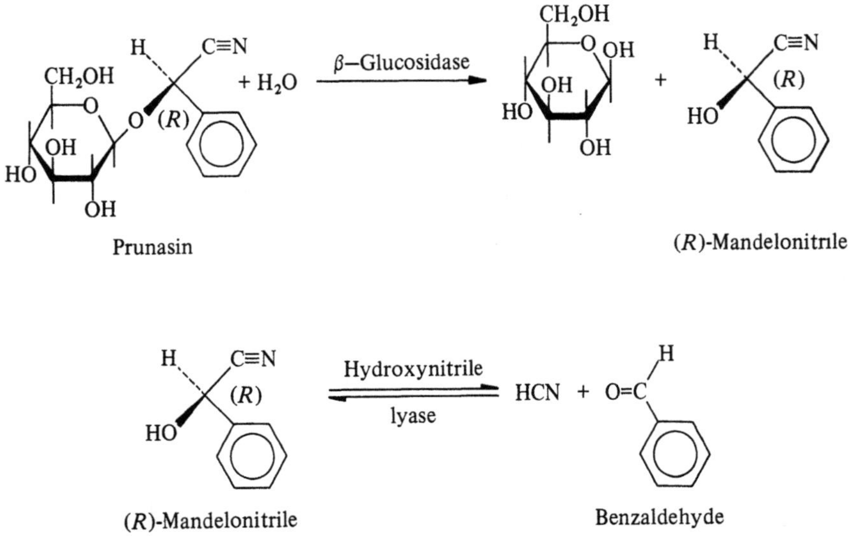

Fig. 2. Enzymatic hydrolysis of (*R*)-prunasin [reproduced from Effects of Poisonous Plants on Livestock, p. 302. Keeler, R.F., von Kampen, K.R., James, L.F. (eds.). New York: Academic Press 1978]

The two reactions indicated in Figure 2 are responsible for the release of HCN from a cyanogenic plant when that plant tissue is crushed or otherwise disrupted and the substrate (cyanogenic glycoside) is brought in contact with the enzyme. It is generally believed that substrate and enzyme are spatially separated in the intact cell, thereby preventing their interaction until the cell structure is destroyed. Actual evidence as to the cellular localization of these two components is only now forthcoming. Using autoradiographic techniques SAUNDERS et al. (1977) have shown that the cyanogenic glucoside in dark-grown sorghum seedlings is localized in the vacuoles of that tissue. Similarly, the glucoside in green sorghum leaves is also confined to the vacuoles (SAUNDERS and CONN, 1978). Supporting the concept of separation of substrate and enzyme is the recent study by BUTCHER et al. (1977) showing that β-glucosidases were not present in vacuoles isolated from *Hippeastrum* petals but were present in the soluble cytoplasmic fraction.

The activity of β-glycosidases on cyanogenic glycosides can be measured directly by following the formation of the sugar as it is released. However, if the hydroxynitrile lyase is present in excess, or if the experimental conditions are such that nonenzymic dissociation of the cyanohydrin is not rate-limiting, then the production of HCN or the aldehyde (or ketone) may also be followed. In the specific case of prunasin (Fig. 2) the production of benzaldehyde may be followed spectrophotometrically due to its significant absorption at 250 nm ($a_M = 1.4 \times 10^4$).

There is a substantial amount of literature on the occurrence of β-glycosidases in plants (NISIZAWA and HASHIMOTO, 1970). However, the literature is regrettably inadequate as regards the action of such enzymes on the cyanogenic glycosides, mustard oil glycosides, and other glycosides that occur naturally in plants. For example, HELFERICH and KLEINSCHMIDT (1968) studied the activity of two β-glucopyranosidases obtained in a pure state from sweet almonds towards a number of phenolic glycosides but apparently did not study their action towards amygdalin and prunasin, the two cyanogenic glycosides found in abundance in *Prunus amygdalus*. Even more numerous are the studies of the action of β-glycosidases on entire families of chromogenic substrates such as *p*-nitrophenyl-β-D-glucopyranoside or 2,4-dinitrophenyl-β-D-glucopyranoside. The ease and convenience of measuring enzymatic activities with such substrates has to some degree distracted workers in this field.

Studies (HAISMAN and KNIGHT, 1967; BUTLER et al., 1965; MAO and ANDERSON, 1967) on the hydrolysis of cyanogenic glycosides by specific enzymes isolated from the cyanogenic plants that contain these compounds have been summarized by EYJOLFSSON (1970). This work indicates that the β-glycosidase(s) of almonds hydrolyze the glycosides derived from phenylalanine and tyrosine, as well as proacacipetalin, but have little activity against linamarin and lotaustralin. Similarly, a β-glucosidase in *Sorghum vulgare*, which contains dhurrin, is active towards that cyanogen and its epimer taxiphyllin, but will also hydrolyze amygdalin and prunasin. Linseed, in turn, contains a β-glucosidase active towards linamarin which will also hydrolyze prunasin and taxiphyllin but not dhurrin or amygdalin.

HUGHES (1968) has observed two β-glucosidases in *Trifolium repens*, a species which is polymorphic for cyanogenesis (Sect. 5.1.4). However, only the enzyme found in leaf tissue appears to be regulated by the *Li* gene and therefore involved in cyanogenesis in this species. These results, interesting as they are, must be qualified by the fact that the substrate employed was *p*-nitrophenyl-β-D-glucopyranoside instead of the cyanogenic glycoside linamarin.

HÖSEL and NAHRSTEDT (1975) have purified a β-glucosidase(s) from *Alocasia macrorrhiza* which exhibits maximum catalytic activity towards triglochinin, the cyanogen of this species. The enzyme will also hydrolyze several chromogenic substrates (at much lower rates) and dhurrin, at about 1% that observed for triglochinin. It shows little or no detectable activity towards seven other cyanogenic glycosides and therefore is remarkably specific for triglochinin. The enzyme was electrophoretically homogeneous after a ninefold purification and thus must represent a significant part of the soluble protein in the extract.

2.2.2 Spectral Properties

Data on the ultraviolet (UV), infrared (IR), and nuclear magnetic resonance (NMR) spectra of the cyanogenic glycosides have been gathered and evaluated in an especially useful review by SEIGLER (1975). On study it is apparent that NMR spectra are of greatest value and general application in studies on structure elucidation and identification. However, UV spectra have seen use in the case of the cyanogenic compounds derived from phenylalanine and tyrosine.

In the case of NMR spectra, the compounds may be examined as the free glycosides in D_2O since they are all readily soluble in water. However, the occurrence of an HDO peak near 4.6 δ often obscures peaks of interest in the cyanogenic glycosides. For this and other reasons, the TMS-ether derivatives and nonpolar solvents such as CCl_4 or $CDCl_3$ are used in structural elucidation and the identification of unknown glycosides. The TMS-derivatives, which are easily prepared, have found additional use in the separation of cyanogenic glycosides by GLC (BISSETT et al., 1969; NAHRSTEDT, 1973a; SEIGLER, 1975) and in mass spectrometry to determine molecular weights and to identify aglycones by interpretation of the fragmentation patterns (BISSETT et al., 1969; ZILG et al., 1972).

Fully acetylated derivatives of the cyanogenic glycosides have also found use in this field (SEIGLER, 1975). The acetates are more difficult to prepare than the TMS-ethers, but they are crystalline, give sharp and reproducible melting points, and can be used in NMR and MS studies. They are also stable to aqueous environments and therefore are easier to preserve.

2.2.3 Other Properties

The cyanogenic glycosides show the expected physical properties of this class of carbohydrate derivatives. That is, they are difficult to crystallize, and if a solid is obtained, it may exhibit different melting points depending on the degree of hydration. Thus a range of melting points and specific rotations will usually be encountered if several papers are consulted on a specific compound. The acetylated derivatives have the advantage of solubility in less polar solvents and show sharper melting points. A table of such data, gleaned from the literature, is available in SEIGLER'S review (1975).

3 Detection, Isolation, and Characterization

3.1 Detection of Cyanogenic Glycosides

3.1.1 Detection in Plant Material

The presence of a cyanogenic glycoside in fresh plant material will be indicated by the release of HCN when the plant material is crushed or otherwise disrupted. This action presumably brings the glycoside into contact with the β-glycosidase and hydroxynitrile lyase referred to in Section 2.2.1.2. The HCN can then be detected qualitatively or measured quantitatively by several different techniques.

Three qualitative tests commonly employed are the picrate paper test, the nitrobenzaldehyde-dinitrobenzene test, and the benzidine-cupric acetate test. All three have been described by EYJOLFSSON (1970) and the reader is referred to that author for details. The principle is the same in all three tests in that a small sample of plant tissue is placed in a microtest-tube or vial, the tissue is then crushed with a stirring rod (a small stick will work in the field), and autolysis is allowed to proceed while the tube (vial) is tightly stoppered. In the picrate test, a piece of filter paper impregnated with picric acid-sodium carbonate is suspended at the top of the tube or vial and held in place as the tube is tightly stoppered. (In the other tests, the paper is either buffered at pH 8.5 and suspended to trap any HCN released, or paper impregnated with benzidine acetate and cupric acetate is added after autolysis is completed. In the former, the buffered pH paper is then removed and wetted with a solution of p-nitrobenzaldehyde and o-dinitrobenzene.) The presence of HCN is indicated if the picrate paper turns from its initial bright-yellow color to a reddish-brown. The more rapidly the paper changes, the higher the concentration of HCN in the tissue.

The picrate test will detect 1 µg or less of HCN and therefore is relatively sensitive. However, the following precautions should be noted. Firstly, a positive picrate test, even if rapidly obtained, does not necessarily prove that a cyanogenic glycoside is present. One family of plants, the Sapindaceae, contains cyanogenic lipids in which the β-D-glucosyl moiety is replaced with a long chain fatty acid (MIKOLAJCZAK, 1977). Such lipids are presumably hydrolyzed by lipases in the plant when the plant tissue is disrupted and the HCN released. Secondly, a weakly positive test may develop over a period of 16–24 h. In this case other evidence for the presence of HCN should be obtained as other volatiles (e.g., H_2S, volatile aldehydes) may give a reaction over an extended time period. The ability of certain organic thiocyanates and nitriles to react in tests for HCN may be responsible for reports (GIBBS, 1974; HONEYMAN, 1966) that certain Crucifereae are cyanogenic. STONE (1977) has recently shown that benzylthiocyanate will react in the modified Lambert method for HCN and is sufficiently volatile to distill into the center well traps used for HCN analysis.

Thirdly, whenever the picrate test is applied as described in the preceding paragraph it should be compared with an identical sample to which has been added a few drops of a dilute (1 mg/ml) almond emulsin solution or preferably a mixture of emulsin and flax linamarase. These two enzymes will hydrolyze glycosides having aromatic and aliphatic aglycones, respectively, and therefore insure the rapid release of HCN from any one of a number of cyanogens that might be present. While enzymes capable of degrading the cyanogenic glycosides are usually present in cyanophoric plants, there also are species which contain a cyanogen but apparently lack the β-glucosidase (FINNEMORE and GLEDHILL, 1928; SECOR et al., 1976). In some instances, e.g., *Trifolium repens* and *Lotus corniculatus*, this is due to the fact that these species are polymorphic with respect to cyanogenesis and that the plant may have the cyanogen but lack the degradative enzymes. The addition of enzymes to the second vial will ensure a positive test from such plants.

3.1.2 Detection on Chromatograms

A number of methods are available for detecting cyanogenic glycosides on paper or TL chromatograms (see SEIGLER, 1975 for review). These methods depend on the enzymatic release of HCN from the chromatogram by spraying with the appropriate enzyme and subsequent detection with picrate paper or other HCN-specific reagents. Since, however, the cyanogens can vary in their stability in different chromatographic solvents, several different conditions may have to be employed to identify the cyanogen.

More recently, procedures have been developed which are relatively specific for the mandelonitrile glycosides (SCHWARZMAIER, 1975a) and for linamarin (ZITNAK et al., 1977). In the latter case, the method can be adapted for quantitative analysis. SCHWARZMAIER (1975b) has also described a color reagent (hydroxylamine/ferric chloride) which distinguishes between the (*R*) and (*S*) forms of the mandelonitrile glycosides.

3.1.3 Quantitative Analysis

Research on cyanogenic plants has required methods for measuring the amount of HCN released by known amounts of plant tissues. Older methods involved steam distillation of the HCN released on autolysis, but these procedures required comparatively large amounts of material. Micromethods involving diffusion of the HCN into 1*N* NaOH during a period of autolysis and shaking in a water bath have been described (HAHLBROCK and CONN, 1970; REAY and CONN, 1970). The HCN can then be determined quantitatively by the micromethod of ALDRIDGE (1944), EPSTEIN (1947) or LAMBERT et al. (1975). The latter two are preferred in our laboratory as the Aldridge procedure requires the use of bromine water, sodium arsenite, and benzidine, all of which are hazardous reagents.

WISSING (1974) has described an apparatus for removing micromole quantities from reaction mixtures by rapid stirring and aeration. The HCN is then trapped in alkali for analysis. This apparatus can be used not only for removing preformed HCN but also for monitoring reaction mixtures in which HCN is being continuously produced. Rather surprisingly, enzymes do not appear to be inactivated by the combined stirring and aeration which the reaction mixture undergoes.

For quantitative determination of the intact cyanogenic glycosides gas liquid chromatography can be utilized. This technique has been successfully applied to the TMS-ethers of a number of the cyanogenic glycosides (BISSETT et al., 1969; NAHRSTEDT, 1970; see SEIGLER, 1975 for tabulation) including the separation of epimeric pairs such as prunasin/sambunigrin (NAHRSTEDT, 1973a).

Application of these quantitative techniques has shown that the cyanogenic glycosides can occur in any of several organs and parts of a cyanogenic plant, i.e., flowers, stems, petioles, leaves, root, bark. The cyanogen may also be present at one point in the life cycle of the plant and not at others. Thus, sorghum seed contains no dhurrin, but the dark-grown seedling may contain 5% dhurrin (dry wt.) with lesser quantities in green leaves. On the other hand, the seed of the wild lima bean contains 3–4% (fr. wt.) linamarin which apparently is not metabolized during germination and growth of the seedling (CLEGG, 1977). In general, leaves contain the highest concentration, but flowers and seeds may also be rich sources. Thus bitter almonds and apricot pits are sources of the amygdalin of commerce.

The amount of cyanogen within a cyanogenic species can also be greatly influenced by seasonal, environmental, nutritional, and genetic factors. Thus, the levels of prunasin vary with the season in the California toyon, *Heteromeles arbutifolia*, being higher in the spring and in the late fall (DEMENT AND MOONEY, 1974). The amount of dhurrin in *Sorghum* species is stated to increase when plants undergo moisture stress (NELSON, 1953). There are numerous studies on the effect of fertilizers on the cyanide content of sorghum (see, for example, NELSON, 1953; GILLINGHAM et al., 1969; HARMS and TUCKER, 1973). The influence of genetic factors on cyanogenesis is discussed in Section 5.1.4.

3.2 Isolation

In order to isolate a cyanogenic glycoside from its natural plant source, care must be taken to inactivate the degradative enzymes that accomplish the breakdown of the cyanogen. This is especially true if fresh plant material is being extracted. However, inactivation is also important in the case of material collected and dried at a distance from where the extraction is performed, since β-glucosidase may be active in dried specimens. Indeed, significant β-glucosidase activity has been detected in herbarium species collected in the previous century.

The extraction process usually involves grinding of the plant tissue followed by extraction in solvents such as ethanol, methanol, H_2O, or mixtures thereof. Since the hydrolysis of the glycoside is set in action with grinding of the tissue, there is a need to inactivate the β-glucosidase as soon as possible. One convenient technique for fresh plant material (HAHLBROCK and CONN, 1971) is to freeze the tissue in liquid nitrogen and grind it as soon as the liquid N_2 has evaporated. The brittle, frozen tissue is extremely easy to grind into a fine powder which then can be transferred, while still frozen, into boiling solvent (e.g., 80% ethanol) for 3–5 min. The temperature and alcohol will inactivate the degradative enzymes; at the same time, efficient extraction of the finely ground tissue is accomplished, and the solid residue can be filtered, reextracted, and discarded. In our experience, these conditions are sufficiently mild so that loss of cyanogenic material will be minimized except for the thermolabile taxiphyllin (SCHWARZMAIER, 1976). Moreover, these extraction conditions do not epimerize glycosides of the benzaldehyde-cyanohydrin type (Sect. 2.2.1.1), since epimeric mixtures were not encountered when this technique was employed in the extraction of prunasin and sambunigrin from *Acacias* (SECOR et al., 1976).

Following extraction, the volume is reduced by removal of solvent at low temperatures; if required, undesirable pigments (e.g., chlorophyll) and lipids can be removed by extraction with a non-polar solvent such as petroleum ether or chloroform. At this point a number of techniques can be employed. Polyvinylpyrrolidone has been employed to remove phenolic impurities (REAY, 1969; RUSSELL and REAY, 1971). If acidic impurities are present, they can be removed with lead acetate or ion-exchange treatment. The latter should be cautiously employed since some glycosides (triglochinin and its methylester) possess carboxyl groups. Alkaline conditions should be avoided if dhurrin or taxiphyllin are being isolated because of their alkaline lability. Eventually, column chromatography is required with selective elution of the desired products (or impurities) with appropriate solvents. SEIGLER (1975) has tabulated the procedures which have been used in recent years. To this list may be added the isolation procedures employed by NAHRSTEDT (1976) with a new glycoside (cardiospermin-*p*-hydroxybenzoate) from *Sorbaria arborea* by SCHWARZMAIER (1976) who isolated taxiphyllin from bamboo shoots, and by SECOR et al. (1976) and SEIGLER et al. (1976) who have extracted proacacipetalin from several *Acacia* species.

In addition to column chromatography, paper chromatography has been used extensively in the isolation of cyanogenic glycosides (SEIGLER, 1975). In view of the success which high pressure liquid chromatography (HPLC) has enjoyed in the preparation of certain natural products, it will be only a matter of time before this technique is applied to the cyanogenic glycosides.

3.3 Characterization

The cyanogenic glycosides can be characterized by the usual physical constants (e.g., by melting points if sufficiently purified to yield crystals, and by optical rotation). Being carbohydrates, they can be easily acetylated and the melting points of these derivatives noted. Spectral (UV, IR, and NMR) properties can also be compared; of the three, the NMR spectra are most informative. SEIGLER (1975) is the authority on this type of data and his laboratory has made extensive use of this technique in elucidating the structure of proacacipetalin (BUTTERFIELD et al., 1975), cardiospermin (SEIGLER et al., 1974) and *epi*-tetraphyllin B (GONDWE et al., 1978).

The aldehyde or ketone formed on dissociation of the cyanohydrin aglycone also serves to identify the general class of cyanogen. Thus the aromatic glycosides derived from tyrosine yield *p*-hydroxybenzaldehyde, while those from phenylalanine yield benzaldehyde. Although acetone and 2-butanone are derived from linamarin and lotaustralin, respectively, the volatility of these compounds has prevented much use centering on their detection and identification. Basic hydrolysis will frequently yield the glycosidic acid, from which the sugar can be removed to yield an α-hydroxy acid. For comparison the same hydroxy acid can be obtained by treatment with concentrated acids.

The sugar moieties derived on hydrolysis of cyanogenic glycosides can be identified by PC, TLC and GLC, by preparation of phenylosazones and methylated derivatives, and by specific rotation. Since the sugar produced on hydrolysis is frequently D-glucose, the highly substrate-specific oxidation of glucose by glucose oxidase has been useful both in identifying the sugar (RUSSELL and REAY, 1971; GMELIN et al., 1973) and in quantitating that product (SEIGLER et al., 1976).

4 Distribution

4.1 Occurrence of Cyanogenesis

Cyanogenesis is a property exhibited widely within the plant kingdom. A recent tabulation by GIBBS (1974) lists approximately 1,000 cyanogenic species in 500 genera comprising 100 families. In a vast majority of these species, however, only the cyanogenic potential has been observed with little or no information as to the nature of the cyanogenic precursor. This is, of course, due to the ease with which the picrate or other qualitative test can be performed on live specimens, often in the field, as compared with the effort to isolate and characterize any cyanogenic substance in the plant.

Although the number of cyanogenic plant species is large, the number of known cyanogenic compounds is not, numbering only 32 if one includes the cyanolipids and pseudocyanogenic glycosides (SEIGLER, 1975). The list of cyanogenic glycosides (Sect. 2.1) includes 23 compounds, twice the number known when the author reviewed the subject a decade earlier (CONN, 1969). This is the result of the increased research on these compounds in the past decade, and should encourage continued work in this area. However, we cannot expect to see the list of cyanogenic glycosides increase tenfold as (ideally) the 800–900 species whose cyanogens remain unidentified are studied. Because the cyanogen in a genus, or even a family, is usually the same compound(s), workers hoping to discover new cyanogenic substances would be advised to examine unstudied families and genera, or alternatively to examine thoroughly those few genera where more than one kind of cyanogen has been identified.

4.2 Natural Distribution

A list of the families and genera in which the 21 known cyanogenic glycosides have been shown to occur (Table 1) gives some indication of how widespread cyanogenesis is in plants. Families noted for their production of cyanogenic glycosides are the Rosaceae (150 species), Leguminosae (125), Gramineae (100), Araceae (50), Compositae (50), Euphorbiaceae (50) and Passifloraceae (30) (GIBBS, 1974).

Examination of Table 1 discloses certain relationships of possible interest in chemotaxonomy. Thus, five glycosides occur in only one family each, i.e., acacipetalin in Leguminosae (in the genus *Acacia* only), amygdalin in Rosaceae, cardiospermin in Sapindaceae, gynocardin in Flacourticeae, and the closely related tetraphyllin A, tetraphyllin B, and deidaclin in Passifloraceae. The grasses (Gramineae) with one exception (bamboo) contain only dhurrin [see however, TJON SIE FAT (1977)]; the Rosaceae contain the closely related amygdalin and prunasin, but not sambunigrin, the epimer of prunasin. Tyrosine-derived compounds (proteacin, *p*-hydroxymandelonitrile derivatives and triglochinin) appear to be associated with monocotyledonous plants and members of the subclass Magnoliidae.

There has been a continuing interest in cyanogenesis as a taxonomic tool (ALSTON and TURNER, 1963; GIBBS, 1974). However, the subject is somewhat at a disadvantage because of the large number of cyanophoric species in which the source of the HCN remains unknown. Still, certain general relationships seem clear, and problem areas have been described (HEGNAUER, 1971). HEGNAUER and his school have been major contributors to our knowledge of this subject and his treatise (HEGNAUER, 1963–73), reviews (e.g., HEGNAUER, 1971, 1973, 1977) and continuing series of research papers (e.g., FIKENSCHER and HEGNAUER, 1977), must be consulted by those who wish to be well informed on this aspect of cyanogenesis.

Table 1 contains some findings that deserve comment and further study. The epimeric pairs (*S*)-sambunigrin/(*R*)-prunasin and (*S*)-zierin/(*R*)-holocalin are listed as occurring in the genus *Sambucus*, i.e., *Sambucus nigra* (JENSEN and NIELSEN, 1973). The occurrence of epimeric pairs in this single species is reminiscent of the claim for "prulaurasin" in cherry laurel (HERISSEY, 1906). However, JENSEN and NIELSEN, who were aware of the ease of epimerization, cite, as evidence it did not occur, the isolation of sambunigrin only using the same techniques with plants collected in different localities in Denmark.

Table 1 also shows that three different cyanogens have been found in the genus *Acacia*, i.e., sambunigrin in the phyllodinous *A. cunninghamii*, prunasin in the Australian bipinnate *A. paucijuga* and proacacipetalin in African and American species (SECOR et al., 1976). This occurrence of several cyanogens of different biosynthetic origin in the same genus is unique and deserves the additional study it is receiving in our laboratory. The study not only may have chemotaxonomic value but offers the hope that new cyanogens may occur in unstudied members of the genus.

The occurrence of the *p*-hydroxybenzoyl ester of cardiospermin in the Rosaceae genus *Sorbaria* is intriguing. This is the first instance of a glycoside, other than those derived from phenylalanine, occurring in this highly cyanogenic family.

Table 1. Distribution of cyanogenic glycosides (adapted from Seigler, 1977)

Glycoside	Family	Genera
Acacipetalin (= proacacipetalin)	Leguminosae	*Acacia* (African and American)
Amygdalin	Rosaceae	*Cotoneaster* spp., *Cydonia*, *Eriobotrya*, *Malus* spp., *Photinia*, *Prunus* spp., *Sorbus*
Cardiospermin	Sapindaceae	*Cardiospermum*, *Heterodendron*
Cardiospermin, p-hydroxybenzoylester	Rosaceae	*Sorbaria*
Deidaclin	Passifloraceae	*Deidamia*
Dhurrin	Gramineae	*Sorghum*
	Proteaceae	*Macadamia*, *Stenocarpus*
Dihydroacacipetalin	Leguminoseae	*Acacia* (African)
(= heterodendrin)	Sapindaceae	*Heterodendron*
Gynocardin	Flacourtiaceae	*Gynocardia*, *Pangium*
Holocalin	Caprifoliaceae	*Sambucus*
	Leguminoseae	*Holocalyx*
p-Glucosyloxymandelonitrile	Berberidaceae	*Nandina*
	Leguminosae	*Goodia*
	Ranunculaceae	*Thalictrum* spp.
Linamarin	Compositae	*Dimorphotheca* spp., *Osteospermum*
	Euphorbiaceae	*Cnidoscolus*, *Hevea*, *Manihot* spp.
	Leguminosae	*Lotus* spp., *Phaseolus*, *Trifolium*
	Linaceae	*Linum* spp.
Lotaustralin	Compositae	*Dimorphotheca* spp.
	Euphorbiaceae	*Manihot*
	Leguminosae	*Lotus* spp., *Phaseolus*, *Trifolium*
	Linaceae	*Linum* spp.
Lucumin	Sapotaceae	*Lucuma*
Proteacin	Proteaceae	*Macadamia*
	Ranunculaceae	*Thalictrum*
Prunasin	Caprifoliaceae	*Sambucus*
	Leguminosae	*Acacia* sp. (Australian)
	Myoporaceae	*Eremophila*
	Myrtaceae	*Eucalyptus*
	Polypodiaceae	*Cystopteris*, *Pteridium*
	Rosaceae	*Cotoneaster* spp., *Cydonia* *Prunus* spp.
	Saxifragaceae	*Jamesia*
	Scrophulariaceae	*Linaria* spp.
Sambunigrin	Caprifoliaceae	*Sambucus*
	Leguminosae	*Acacia* (Australian)
	Olacaceae	*Ximenia*
Taxiphyllin	Cupressaceae	*Juniperus*
	Euphorbiaceae	*Phyllanthus*
	Taxaceae	*Taxus* spp.
	Taxodiaceae	*Metasequoia*
Tetraphyllin A	Passifloraceae	*Tetrapathaea*
Tetraphyllin B	Passifloraceae	*Barteria*, *Tetrapathaea*
Triglochinin	Araceae	*Alocasia*
	Juncaginaceae	*Scheuzeria*, *Triglochin*
	Liliaceae	*Lilaea*
Triglochinin methylester	Ranunculaceae	*Thalictrum*
Vicianin	Leguminosae	*Vicia* spp.
	Polypodiaceae	*Davallia* spp.
Zierin	Caprifoliaceae	*Sambucus*
	Rutaceae	*Zieria*

4.3 Literature on Cyanogens in Plants

Two useful lists of cyanogenic plants found in the northeastern United States (SEIGLER, 1976a) and in Oklahoma and Texas (SEIGLER, 1976b) have been published. Such lists are valuable not only to those concerned about the potential toxicity of these plants to animals including man, but also to the scientist desiring a list of all cyanogenic species in a geographical region. Eventually, similar lists should be prepared for the rest of the United States, and for other countries and regions of the world for they would complement the information already available on poisonous plants. Three excellent sources of the latter type are KINGSBURY (1964), WATT and BREYER-BRANDWIJK (1962), and EVERIST (1974). These authors are of course concerned with all poisonous plants, not just those which produce toxic levels of HCN. On the other hand, if one wants to know if a given species has ever been described as being cyanogenic, one should consult HEGNAUER'S treatise (HEGNAUER, 1963–1973) or that of GIBBS (1974) on chemotaxonomy.

5 Metabolism of Cyanogenic Glycosides

5.1 Biosynthesis

5.1.1 Early Studies

The initial studies on the biosynthesis of cyanogenic glycosides demonstrated that the aglycone of dhurrin, the cyanogen of sorghum, was derived from L-tyrosine (GANDER, 1960; AKAZAWA et al., 1960). Subsequent experiments in several laboratories in the next few years then established the additional precursor-product relationships indicated in Fig. 3 which shows that four protein amino acids give rise to the aglycones of dhurrin, prunasin, linamarin, and lotaustralin. These early experiments, which involved feeding amino acids labeled with ^{14}C in one or more carbon atoms to young seedlings of flax, sorghum, and peach and measuring the incorporation of radioactivity into the aglycones of the glycosides, have been reviewed (CONN and BUTLER, 1969; CONN, 1973a).

In order to learn more about the nature of the biosynthetic conversion, amino acid precursors doubly labeled with ^{14}C and ^{15}N were then employed in several laboratories (BUTLER and CONN, 1964; BLEICHERT et al., 1966; URIBE and CONN, 1966). This work showed that there was little change in the isotopic ratios of $^{14}C:^{15}N$ as the amino acid was converted. Taking the conversion of tyrosine to dhurrin as an example (Fig. 4) this meant that the bond between the 2-carbon of the amino acid and the amino nitrogen was not broken as the conversion occurred. This important observation in turn indicated that all intermediates in the pathway were nitrogenous. Further, in the case of dhurrin, tyrosine double-labeled in the 2 and 3 position was employed and the ratio of specific activities for the corresponding atoms in dhurrin examined (KOUKOL et al., 1962). As the ratio remained unchanged, it was concluded that the carbon–carbon bond linking carbon-2 and carbon-3 was not severed during biosynthesis.

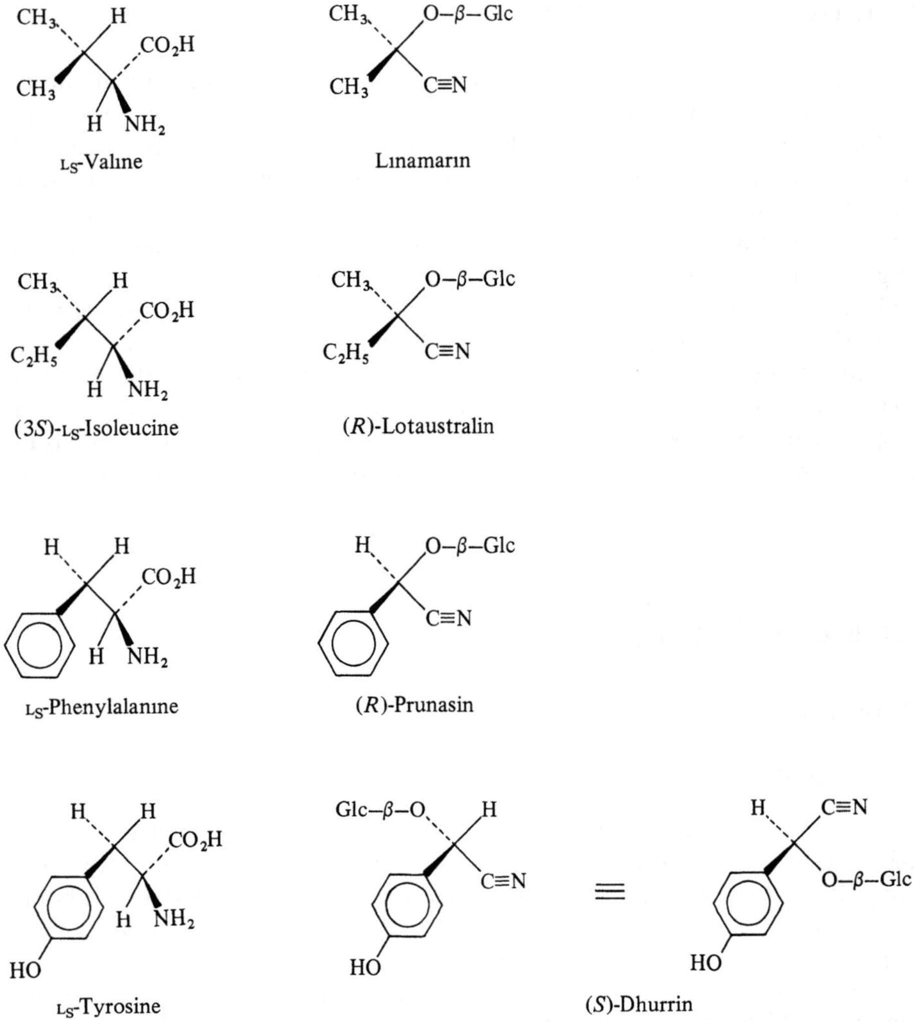

Fig. 3. Precursor-product relationship between four amino acids and related cyanogenic glycosides

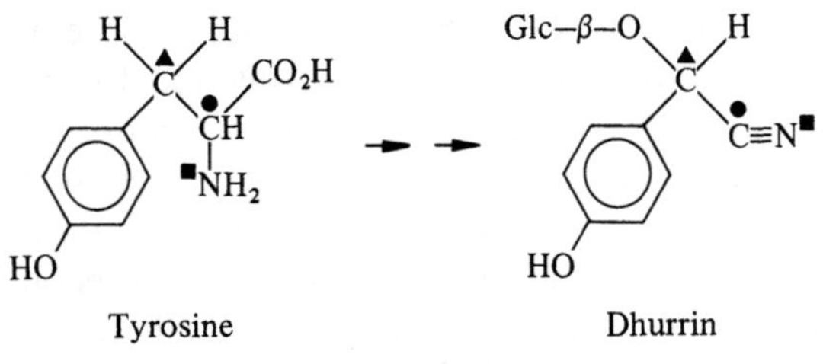

Fig. 4. Precursor-product relationship between certain atoms in L-tyrosine and the aglycone of dhurrin

These results were interpreted in terms of a biosynthetic process in which the carboxyl carbon of tyrosine is lost while the 2-carbon is oxidized to the level of a nitrile (a 4-electron oxidation) with retention of nitrogen atom. In addition carbon-3 also undergoes a 2-electron oxidation to form the hydroxyl group to which glucose is attached. Subsequent work (ZILG et al., 1972) eventually demonstrated that the oxygen atom to which the sugar is attached is derived in biosynthesis from oxygen (O_2) of the atmosphere. Consequently, the oxidation process involves an oxygenation (hydroxylation) reaction.

The indirect evidence that intermediates in the biosynthetic pathway are nitrogenous, referred to above, coincided with the early testing of tyramine, the one obvious nitrogenous metabolite of tyrosine (KOUKOL et al., 1962). When it did not serve as a precursor, there was an indication that the metabolic pathway might well involve unusual intermediates not normally encountered in amino acid metabolism in animals and bacteria. Attempts were also made to identify labeled intermediates in flax and sorghum seedlings which had been administered ^{14}C-labeled valine and tyrosine, respectively. Although extensive conversions (20%–50%) of the labeled amino acids into the glycosides were observed, it was impossible to detect any ^{14}C-labeled intermediates in alcoholic extracts of the seedlings. This finding was surprising in view of the extensive conversions observed, and only recently has our knowledge of the process provided a possible explanation (see Sect. 5.1.3).

5.1.2 Experiments with Labeled Intermediates

A major breakthrough was provided by TAPPER and BUTLER (1972) who observed that the conversion of ^{14}C-labeled valine to linamarin in flax seedlings was inhibited by DL-O-methylthreonine, DL-*allo*-O-methylthreonine and DL-2-methoxypropionaldoxime. Moreover, isobutyraldoxime and its O-β-glucoside accumulated in flax seedling shoots treated with these inhibitors. Their observation, made several years earlier and generously communicated to other workers in the field before it was published in detail, provided the impetus for examining oximes and their dehydration products, nitriles, as intermediates in linamarin, prunasin, and dhurrin biosynthesis. Table 2 contains early data showing the efficient incorporation of isotope from ^{14}C-labeled isobutyraldoxime, isobutyronitrile, and 2-hydroxyisobutyronitrile (acetone cyanohydrin) into the cyanogenic glycoside linamarin in excised shoots

Table 2. Comparison of precursors of linamarin

	Compound administered	% Precursor converted into linamarin
Expt. 1	L-[U-^{14}C]-Valine[a]	25
	[U-^{14}C]-Isobutyraldoxime	21
	[U-^{14}C]-Isobutyraldehyde	0.7
Expt. 2	L-[U-^{14}C]-Valine[a]	23
	[1-^{14}C]-Isobutyronitrile	11
	α-Hydroxy[1-^{14}C]-Isobutyronitrile	28

[a] Corrected for presumed loss of the carboxyl carbon atom. Data are those of Tapper et al. (1967) and Hahlbrock et al. (1968).

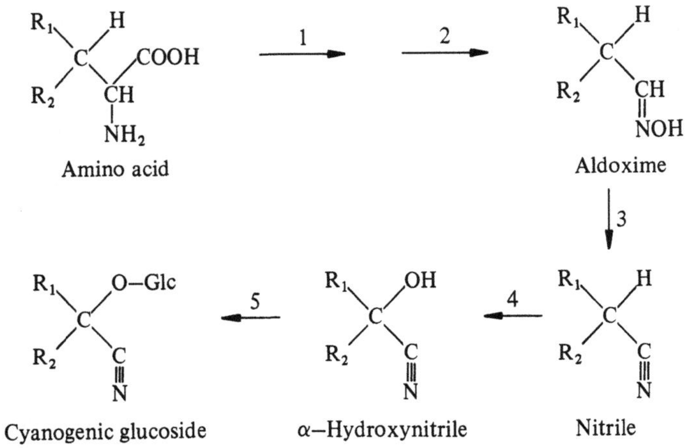

Fig. 5. A biosynthetic pathway for the formation of cyanogenic glucosides

of flax seedlings. To account for these observations a biosynthetic pathway (Fig. 5) was proposed which involved these compounds as intermediates (HAHLBROCK et al., 1968).

The sequence indicated in Fig. 5 for the aldoxime, nitrile, and cyanohydrin was based primarily on the fact that oximes are known to dehydrate readily forming nitriles, and that a cyanohydrin, possibly formed by hydroxylation, might be glucosylated enzymically. ^{14}C-Labeled oximes, nitriles, and cyanohydrins (2-hydroxynitriles) that could be obtained by the metabolism of valine and phenylalanine were administered to flax seedlings and cherry laurel leaves, respectively (TAPPER et al., 1967; HAHLBROCK et al., 1968; TAPPER and BUTLER, 1971). The corresponding derivatives for L-tyrosine were also labeled and administered to sorghum (FARNDEN et al., 1973). These collective studies clearly demonstrated that oximes, nitriles, and 2-hydroxynitriles were efficiently converted to cyanogenic glycosides.

Such experiments were of course subject to the criticism that compounds administered in feeding experiments may be incorporated into the product being studied because they closely resemble but are not actually the true intermediates. Besides this criticism, it was also disturbing that investigators had never observed these postulated intermediates, or any other possible intermediates, in extracts of untreated plants. The failure to observe such intermediates could possibly be due to the analytical methods employed not being sufficiently sensitive to detect very small amounts. An alternative possibility was that the intermediates remained bound on the surface of the enzyme system(s) that carried out the biosynthesis. Still another possibility was that oximes, nitriles, and cyanohydrins might in fact not be the correct intermediates.

Additional experiments therefore were performed to test the validity of the proposed pathway. These involved "trapping experiments" in which the proposed intermediates (aldoximes and nitriles but not the labile cyanohydrins) were administered together with the ^{14}C-labeled precursor amino acid to cyanogenic plants (shoots, leaves, seedlings). After a period of metabolism, the plant tissues were extracted and the intermediates isolated from the plant, purified and examined for radioactivity. When this was done in flax seedlings for example, TAPPER and BUTLER (1972) showed that isobutyraldoxime and isobutyronitrile acquired radioac-

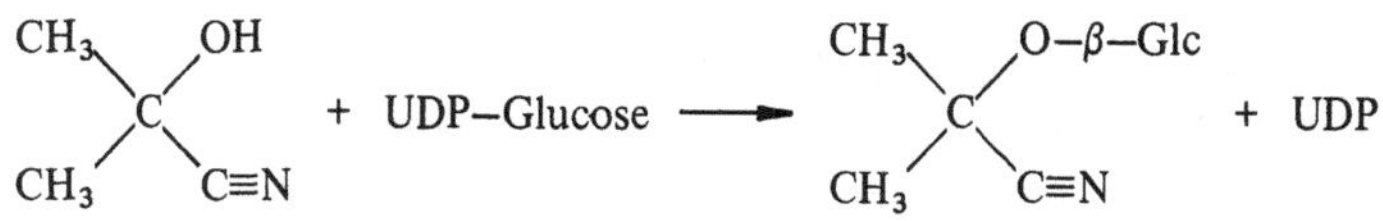

Fig. 6. Enzymatic glucosylation of α-hydroxyisobutyronitrile

tivity from the [^{14}C]-valine fed. Similar data supporting the pathway were obtained by FARNDEN et al. (1973), who examined dhurrin biosynthesis in sorghum seedlings.

During this work, some attention was paid to the possibility that the aldoxime might first be hydroxylated to yield an 2-hydroxyaldoxime which then would be dehydrated to form the cyanohydrin. (This sequence of hydroxylation followed by dehydration would be reversed over that shown in Fig. 5). This possibility was difficult to test because of the compounds involved, but recent work with a cell-free enzyme system (SHIMADA and CONN, 1977) from sorghum seedlings has supported the earlier conclusion that 2-hydroxyaldoximes are not involved in the pathway.

To the biochemist, the most satisfactory evidence for the pathway proposed in Fig. 5 would be the demonstration, isolation, and characterization of individual enzymes catalyzing the pathway. Studies of this sort are described in the next section.

5.1.3 Enzymatic Studies

The first enzymatic studies on cyanogenic glucoside biosynthesis involved the last step in the pathway in which the 2-hydroxynitrile is converted to the cyanogen. HAHLBROCK and CONN (1970) partly purified and studied an enzyme from flax seedlings that catalyzed the glucosylation of a series of aliphatic 2-hydroxynitriles by UDPG (Fig. 6). The enzyme exhibited maximum rates with the two 2-hydroxynitriles that yield linamarin and lotaustralin on glucosylation, but was inactive on aromatic 2-hydroxynitriles such as (*R*,*S*)-mandelonitrile.

Some interesting stereochemical relationships exist here. Since 2-hydroxy-2-methylbutyronitrile contains a chiral carbon atom, this molecule exists in two enantiomeric forms, both of which would be present in an aqueous solution of the cyanohydrin. Enzymatic glucosylation of the enantiomeric mixture could yield either (*R*)-lotaustralin or (*S*)-*epi*-lotaustralin, depending on the stereospecificity of the flax UDPG-glucosyl transferase. However, when this was tested experimentally, ZILG and CONN (1974) found that both enantiomers were glucosylated in the enzyme mixture. Since only (*R*)-lotaustralin is found in the intact flax plant (ZILG and CONN, 1974), and also in *Manihot esculenta* (BISSETT et al., 1969), this means that in vivo the glucosyltransferase is presented with only one of the two enantiomeric cyanohydrins. Presumably this is determined by the configuration of the 3-carbon of the (3*S*)-L_S-isoleucine which serves as precursor. These and related stereochemical studies have been reviewed in detail by CONN (1973a).

REAY and CONN (1974) have partly purified and studied the glucosyl transferase which catalyzes the synthesis of dhurrin in sorghum. The stereospecificity of the sorghum enzyme contrasts sharply with that of the flax enzyme because, when presented with the enantiomeric mixture of its substrate, i.e., *p*-hydroxy-(*R*,*S*)-mandelonitrile, only dhurrin was produced. This means that only one of the enantiomeric

2-hydroxynitriles was glucosylated. As expected, the sorghum enzyme also utilizes aromatic instead of aliphatic cyanohydrins. It will also glucosylate the aromatic hydroxyl groups of *p*-hydroxybenzaldehyde and *p*-hydroxybenzoic acid but at much lower rates. In other respects, the flax and sorghum enzymes are similar. Both form β-glucosides, and both use only UDP-glucose as the glucosyl donor; they have similar pH optima and have K_M values for UDP-glucose of less than 1 mM.

The enzymatic studies on biosynthesis were facilitated by the demonstration (McFARLANE et al., 1975) that a high-speed particulate fraction (microsomal fraction), isolated from dark-grown sorghum seedlings, can catalyze the conversion of L-tyrosine to *p*-hydroxymandelonitrile, i.e., steps 1 through 4 of Fig. 5. The particles require only NADPH (or an NADPH-generating system) in addition to the substrate and the reaction can be followed by measuring one or both of the compounds formed (HCN or *p*-hydroxybenzaldehyde) on dissociation of the cyanohydrin. Evidence that the conversion of L-tyrosine to the 2-hydroxynitrile proceeds by the pathway shown in Fig. 5 was presented by McFARLANE et al. (1975) who showed the stepwise conversion of L-tyrosine to *p*-hydroxyphenylacetaldoxime (steps 1 and 2), the conversion of the aldoxime to *p*-hydroxyphenylacetonitrile (step 3), and the conversion of the nitrile to the 2-hydroxynitrile, *p*-hydroxymandelonitrile (step 4). Moreover, when a soluble protein fraction from sorghum seedlings containing the UDP-glucosyl transferase was added to the microsomal system, the cell-free synthesis of dhurrin occurred (McFARLANE et al., 1975).

The sorghum microsomal system has been utilized to clarify the role of 2-hydroxyaldoximes in the biosynthetic pathway. As pointed out in the preceding section the conversion of the aldoxime to the 2-hydroxynitrile could occur by two different routes. It could occur by dehydration and hydroxylation or it could occur by hydroxylation followed by dehydration (Fig. 7). Earlier studies with intact plant seedlings or shoots had not clarified this matter. Using sorghum microsomes, SHIMADA and CONN (1977) performed "trapping" experiments using L-[U-^{14}C]-tyrosine and either *p*-hydroxyphenylacetonitrile or (2-hydroxy)*p*-hydroxyphenylacetaldoxime as traps. The data showed that the nitrile served effectively as a trap while the 2-hydroxyaldoxime did not. These results then permit the conclusion that 2-hydroxyaldoximes are not intermediates in the pathway.

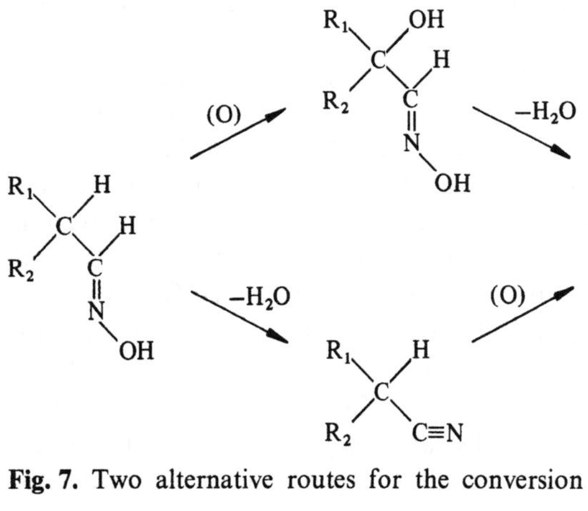

Fig. 7. Two alternative routes for the conversion of an aldoxime into an α-hydroxynitrile

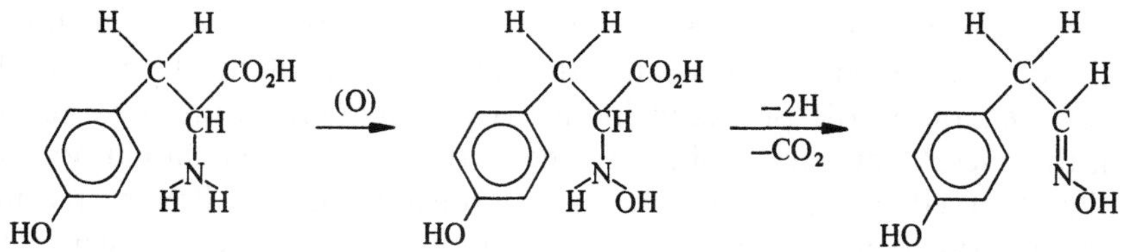

Fig. 8. The conversion of L-tyrosine to p-hydroxyphenylacetaldoxime

The microsomal preparation has also been used to clarify the early steps in the biosynthetic pathway. The conversion of L-tyrosine to *p*-hydroxyphenylacetaldoxime represents a 4-electron oxidative decarboxylation. This is a process unlikely to occur without at least one intermediate, although the oxidative decarboxylation of amino acids to the homologous isoatomic amides does occur (OHNISHI et al., 1976). KINDL and UNDERHILL (1968), in related studies on the biosynthesis of mustard oil glycosides (glucosinolates), had suggested that N-hydroxyphenylalanine might be an intermediate in the conversion of L-phenylalanine to phenylacetaldoxime. Supporting this proposal, these authors obtained an enzyme which catalyzed the conversion of N-hydroxyphenylalanine to the homologous aldoxime. No attempt to demonstrate the N-hydroxyamino acid as occurring naturally in their plants or in in vitro systems was reported.

In view of this work, it seemed necessary to determine whether N-hydroxytyrosine was an intermediate in the early steps of dhurrin biosynthesis. This required developing a method for synthesis of N-hydroxytyrosine, both labeled and unlabeled (MØLLER et al., 1977), as well as a suitable method for detecting N-hydroxytyrosine and separating it from other intermediates in the pathway. The method of choice was separation by GLC of the trimethylsilyl derivatives of the individual compounds (MØLLER, 1977).

Using this method, MØLLER and CONN (1977) have demonstrated [^{14}C]-N-hydroxytyrosine formation from L-[U-^{14}C]-tyrosine by sorghum microsomes provided a pool of unlabeled N-hydroxytyrosine is present in the reaction mixture as a trap. Moreover, the particles will convert N-hydroxytyrosine to the homologous aldoxime, nitrile and cyanohydrin. Thus N-hydroxytyrosine appears to be an intermediate in the formation of *p*-hydroxyphenylacetaldoxime and we may write steps 1 and 2 of Figure 5 as shown in Figure 8.

In the course of these studies, it was observed that the sorghum particles utilized the proposed intermediates at different and inconsistent rates. Thus, MCFARLANE et al. (1975) observed high rates of oxidation with L-tyrosine and *p*-hydroxyphenylacetaldoxime in comparison with *p*-hydroxyphenylacetonitrile. Later SHIMADA and CONN (1977) determined the optimum concentrations for oxidation of these substrates by the particles and confirmed that *p*-hydroxyphenylacetaldoxime was oxidized about three times as rapidly as *p*-hydroxyphenylacetonitrile. This difference in rate appears inconsistent with the order in which these compounds occur in the biosynthetic pathway (Fig. 5). If the oxidation of the nitrile is the last step in the sequence, then the oxidation of *p*-hydroxyphenylacetaldoxime, or L-tyrosine for that matter, should not proceed more rapidly than that of nitrile.

Such reasoning should apply if the rate of oxidation of exogenous nitrile added to the particulate enzyme system is the same as that of nitrile produced in situ. If however, the nitrile produced in situ is preferentially utilized by the particles, then the overall conversion of L-tyrosine to the end product should not be limited by the observed, slower oxidation rate of exogenous nitrile. Preferential utilization of the intermediate produced in situ in a two-step reaction sequence has been observed by Czichi and Kindl (1975) in the conversion of L-phenylalanine to *p*-coumaric acid by thylakoids of *Dunaliella marina*. These authors have proposed a mechanism involving compartmentation of the reactants.

The results observed with the sorghum particles resembles those obtained by Gaertner et al. (1970) with two soluble multienzyme complexes in *Neurospora crassa*. In describing their research, these authors termed the phenomenon "catalytic facilitation" and pointed out that it is an expected property of multienzyme complexes. Such complexes should exhibit increased catalytic efficiency as they channel the flow of carbon atoms through a series of intermediates tightly bound to the complex. In turn, these complexes can account for metabolic pool separation or channeling that is observed in vivo. In the specific case of sorghum seedlings, dhurrin biosynthesis occurs at a remarkable rate because the 4-day-old dark-grown seedling contains about 15% dhurrin (dry wt.). Since the seedling requires tyrosine for protein synthesis and the synthesis of phenolic compounds as well (Stafford, 1974), there must be some mechanism for channeling a major portion of the available tyrosine into dhurrin formation. The membrane-bound enzyme system which catalyzes steps 1 through 4 of the biosynthetic pathway for dhurrin may well be an example of a multi-enzyme complex which produces and uses the intermediates indicated without release from the surface of the enzyme.

5.1.4 Genetic Control of Cyanogenesis

The genetic polymorphism for cyanogenesis has been extensively studied in *Trifolium repens* and *Lotus corniculatus* (see Nass, 1972 for review). *Sorghum bicolor*, *Prunus amygdalus*, *Manihot esculenta*, and *Macadamia ternifolia* undoubtedly are also polymorphic since there are individual plants within these species which are either strongly cyanogenic or only weakly so (Jones, 1972; Seigler, 1976a).

Study of the cyanogenic potential of *T. repens* by Corkill (1942) and by Atwood and Sullivan (1943) has established that two independent alleles are involved. As shown in Figure 9, the *Ac* gene controls the biosynthesis of the cyanogenic glycoside from its amino acid precursor. The second allele, *Li*, controls the amount of *β*-glucosidase in the plant. Thus, for a plant to release HCN, it must contain at least one dominant allele of both genes. The four phenotypes *AcAcLiLi*, *AcAclili*, *acacLiLi*, and *acaclili* can readily be obtained by selective breeding and studied.

Hughes and Conn (1976) have compared the biosynthesis of linamarin in specimens of *T. repens* which produce cyanogenic glucosides with other plants

Amino acid ⟶⟶⟶⟶ Cyanogenic glycoside —Glycosidase⟶ HCN + Sugar and aldehyde or ketone

Ac gene   *Li* gene

Fig. 9. Genetics of cyanogenesis in *Trifolium repens*

which lack the cyanogen. For practical purposes, the plants were selected from polymorphic commercial varieties and in both cases lacked linamarase activity. Thus, although the genetic crosses were not performed to prove the genotype, the plants used corresponded to the genotypes *AcAclili* and *acaclili*. The results clearly showed that plants lacking the cyanogenic glucoside were unable to convert L-valine to isobutyraldoxime, and therefore presumably lack the enzyme(s) for catalyzing the synthesis of N-hydroxyvaline and/or the aldoxime. Indirect evidence was also obtained suggesting that the plants lacking the glucoside also lacked the UDP-glucosyl transferase which catalyzes the last step in the pathway. Unfortunately, direct comparison of the levels of this enzyme in the two types of plant cannot be made since the latter enzyme has not yet been demonstrated in this species. However, these results indicate that plants lacking the cyanogenic glucoside may lack the enzymes required for *two* reactions in the biosynthetic pathway. This must be either the result of mutation in two genes, each of which caused the loss of one function, or the result of a mutation in a single-gene resulting in the loss of both activities. Although the cyanoglucoside polymorphism in *T. repens* has commonly been used as an example of a single-gene controlled biochemical difference, the loss of two activities in the plants lacking the cyanoglucoside means that the single gene nature of the polymorphism must be questioned.

5.2 Catabolism in Plants

The major catabolic process for cyanogenic glycosides in plants is initiated when, after crushing or maceration, the glycosides are brought in contact with the β-glycosidases and hydroxynitrile lyase. The action of these enzymes, described in Section 2.2.1.2, results in the formation of one or two mol of sugar, HCN and, usually, an aldehyde or ketone. As pointed out elsewhere, the enzymes and substrates are apparently physically separated in the intact cell, the vacuole being the site of glucoside accumulation in at least one plant (SAUNDERS et al., 1977; SAUNDERS and CONN, 1978).

There is evidence indicating that the cyanogenic glycosides undergo metabolic turnover in the intact plant. This was realized when ABROL and CONN (1966) and ABROL et al. (1966) fed ^{14}C-labeled amino acids to cyanogenic plants (*Lotus* sp. and *Nandina domestica*) and, after a period of metabolism, observed that radioactivity had been incorporated into the amide carbon atom of asparagine. Apparently a β-glucosidase, not necessarily the one involved in cyanogenesis, has limited access to the cyanogens in these plants and thus released HCN. The HCN was converted to asparagine by way of β-cyanoalanine, as discussed below. Similar conclusions regarding the active turnover of cyanogenic glucosides may be drawn from the work of BOUGH and GANDER (1970) who carried out pulse-labeling experiments of dhurrin in etiolated sorghum seedlings. These authors estimated a turnover rate of about 0.05 μmol/h in seedlings containing 1.0 μmol dhurrin/shoot. This corresponds to a metabolic half-life of approximately 10 h.

BLUMENTHAL et al. (1963) were the first to observe that [^{14}C]-HCN, administered as a gas to young sorghum seedlings, is incorporated in remarkable yield (i.e., between 20%–50%) into the amide carbon atom of asparagine (Fig. 10).

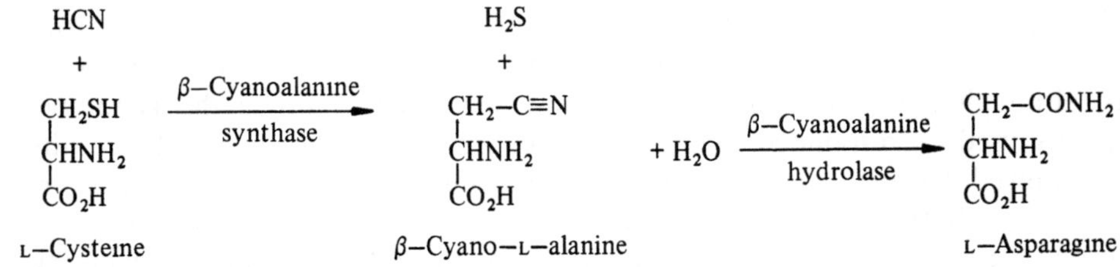

Fig. 10. Metabolism of HCN in plants

This process not only occurred in cyanogenic flax and white clover seedlings but also in the noncyanogenic plants barley, pea and red clover. In only one species, *Vicia sativa*, did asparagine fail to become labeled (TSCHIERSCH, 1964), another compound being labeled instead. This compound was subsequently identified as γ-glutamyl-β-cyanoalanine (RESSLER et al., 1963); therefore β-cyanoalanine was postulated to be the initial product formed by the reaction of HCN. FOWDEN and BELL (1965) then went on to demonstrate that extracts of *Vicia sativa* contained an active 4-glutamyl transferase which catalyzed the formation of the dipeptide.

Initial studies on the enzymatic assimilation of HCN indicated that serine reacted with HCN to form β-cyanoalanine in extracts of *Lotus tenuis* (FLOSS et al., 1965) *E. coli* (DUNNILL and FOWDEN, 1965) and *Vicia sativa* (TSCHIERSCH, 1965). Another legume, *Lupinus angustifolius*, was used as an enzyme source for purification and characterization because of the activity in crude extracts (BLUMENTHAL et al., 1968). HENDRICKSON and CONN (1969), who achieved a 250-fold purification of this enzyme, then showed that the enzyme utilizes cysteine and not serine as its 3-carbon substrate. Since the enzyme will also utilize O-acetylserine as an alternate substrate, this latter reaction resembles cysteine synthase (i.e., O-acetylserine sulfhydrase). However, the two enzymes clearly are not identical since β-cyanoalanine synthase in blue lupine is located in the mitochondria while cysteine synthase is a cytoplasmic enzyme. The β-cyanoalanine synthase of blue lupine has recently been purified to homogeneity, and its mechanism studied (AKOPYAN et al., 1975). It utilizes pyridoxal phosphate as a coenzyme.

To complete our understanding of the metabolism of HCN in plants, CASTRIC et al. (1972) detected and partially purified an enzyme (β-cyanoalanine hydrolase) from *Lupinus angustifolius* that catalyzes the hydration of β-cyanoalanine forming asparagine. Other workers have reported that asparaginase II of *E. coli*, which catalyzes the hydrolysis of asparagine forming aspartic acid and ammonia, will also catalyze the hydration of β-cyanoalanine (LAUINGER and RESSLER, 1970; JACKSON and HANDSCHUMACHER, 1970). Although it could be inferred that plant asparaginase might be similarly involved in the utilization of β-cyanoalanine in plants, this is not the case in *Lupinus angustifolius*.

The reactions constituting HCN metabolism are summarized in Figure 10. The pertinent question remains unanswered, however. Why should higher plants have this organized series of reactions for the metabolism of HCN? In the cyanogenic plant, the reactions could constitute a mechanism for the detoxification of HCN if it should be released within the intact cell. However, most plants, and certainly *Lupinus angustifolius*, do not have a source of HCN that requires detoxification.

6 Physiological Role of Cyanogenic Glycosides

The function of cyanogenic glycosides is frequently said to be that assigned to secondary plant compounds as a group generally. It is the role of repelling or inhibiting herbivores, pathogens, or competitors, with subsequent selection favoring the better-protected individuals. A representative list of papers which discuss this problem are those by FRAENKEL (1959), EHRLICH and RAVEN (1965), and WHITTAKER and FEENEY (1971). During the past decade numerous symposia have treated this subject; see, for example the volumes edited by HARBORNE (1972); HEYWOOD (1973), and LUCKNER et al. (1976).

The cyanogenic glycosides have received special attention when this subject is discussed, however. This is because of their widespread distribution in plants and some evidence indicating that these compounds may play a role in the primary metabolism of the plants. JONES has thoroughly presented arguments suggesting that the cyanogens may be primary or essential metabolic compounds which have been selected in some plants (i.e., "cyanogenic species") because of their defensive role. He has been joined in this argument by SEIGLER and PRICE (1976) who also argue for a possible primary metabolic role of these substances.

The cyanogenic glycosides have attracted attention, not so much because of their physiological role in the plant, but rather because of their physiological effects on man and his animals. Little has been said in this review of the toxicity of cyanogenic glycosides; that subject has been reviewed elsewhere (CONN, 1973b). In addition, one cyanogenic glycoside, amygdalin, has attracted attention because of claims made for its effectiveness in the treatment of cancer. The controversy surrounding this matter has also been reviewed (LEWIS, 1977).

Acknowledgments. The author's research in cyanogenic glycosides described in this review has been supported in part by a continuing grant GM-05301 from the National Institute of General Medical Sciences, U.S. Public Health Service and by National Science Foundation grants GB-30319 and BMS 74-11997.

References

Abrol, Y.P., Conn, E.E.: Studies on cyanide metabolism in *Lotus arabicus* L. Phytochemistry *5*, 237–242 (1966)

Abrol, Y.P., Conn, E.E., Stoker, J.R.: Studies on the identification, biosynthesis and metabolism of a cyanogenic glucoside in *Nandina domestica* Thunb. Phytochemistry *5*, 1021–1027 (1966)

Akazawa, T., Miljanich, P., Conn, E.E.: Studies on cyanogenic glycoside of *Sorghum vulgare*. Plant Physiol. *35*, 535–538 (1960)

Akopyan, T.N., Braunstein, A.E., Goryachenkova, E.V.: Beta-cyanoalanine synthase: purification and characterization. Proc. Natl. Acad. Sci. USA *72*, 1617–1621 (1975)

Aldridge, W.N.: New methods for estimation for microquantities of cyanide and thiocyanate. Analyst *69*, 262–265 (1944)

Alston, R.E., Turner, B.L.: Biochemical systematics. p. 404. New Jersey: Prentice Hall 1963

Atwood, S.S., Sullivan, J.T.: Inheritance of a cyanogenic glucoside and its hydrolyzing enzyme in *Trifolium repens*. J. Hered. *34*, 311–320 (1943)

Bissett, F.H., Clapp, R.C., Coburn, R.A., Ettlinger, M.G., Long, L.: Cyanogenesis in manioc: concerning lotaustralin. Phytochemistry *8*, 2235–2247 (1969)
Bleichert, E.F., Neish, A.C., Towers, G.H.N.: Biosynthesis of taxiphyllin in *Taxus*. In: Biosynthesis of aromatic compounds. Billek, G. (ed.), Vol. 3. Proc. 2nd Meet. Fed. Eur. Biochem. Soc. Oxford: Pergamon Press 1966
Blumenthal, S.G., Butler, G W., Conn, E.E.: Incorporation of hydrocyanic acid labelled with carbon-14 in asparagine in seedlings. Nature (London) *197*, 718–719 (1963)
Blumenthal, S.G., Hendrickson, H.R., Abrol, Y.P., Conn, E.E.: The biosynthesis of β-cyanoalanine. J. Biol. Chem. *243*, 5302–5307 (1968)
Bough, W.A., Gander, J.E.: Exogenous L-tyrosine metabolism and dhurrin turnover in sorghum seedlings. Phytochemistry *10*, 67–77 (1971)
Butcher, H.C., Wagner, G.J., Siegelman, H.W.: Localization of acid hydrolases in protoplasts. Plant Physiol. *59*, 1098–1103 (1977)
Butler, G.W.: The distribution of the cyanoglucosides linamarin and lotaustralin in higher plants. Phytochemistry *4*, 127–131 (1965)
Butler, G.W., Conn, E.E.: Biosynthesis of the cyanogenic glucosides linamarin and lotaustralin. J. Biol. Chem. *239*, 1674–1679 (1964)
Butler, G.W., Bailey, R.W., Kennedy, L.D.: Studies on the glucosidase "linamarase". Phytochemistry *4*, 369–381 (1965)
Butterfield, C.S., Conn, E.E., Seigler, D.S.: Elucidation of structure and biosynthesis of acacipetalin. Phytochemistry *14*, 993–997 (1975)
Castric, P.A., Farnden, K.J.F., Conn, E.E.: The formation of asparagine from β-cyanoalanine. Arch. Biochem. Biophys. *152*, 62–69 (1972)
Clegg, D.O.: Unpublished observations (1977)
Conn, E.E.: Cyanogenic glycosides. Agric. Food Chem. *17*, 519–526 (1969)
Conn, E.E.: Biosynthesis of cyanogenic glycosides. Biochem. Soc. Symp. *38*, 277–302 (1973a)
Conn, E.E.: Cyanogenic glycosides. In: Toxicants occurring naturally in foods. pp. 299–307. Washington: National Acad. Sci. 1973b
Conn, E.E., Butler, G.W.: The biosynthesis of cyanogenic glycosides and other simple nitrogen compounds. In: Perspectives in phytochemistry. Harbone, J.B., Swain, T. (eds.), pp. 47–74. London: Academic Press 1969
Corkill, L.: The inheritance of cyanogenesis. N.Z. J. Sci. Technol. Sect. B *23*, 178–193 (1942)
Czichi, U., Kindl, H.: Formation of hydroxycinnamic acids from L-phenylalanine on thylakoids of *Dunaliella marina*. Hoppe-Seylers Z. Physiol. Chem. *356*, 475–485 (1975)
Dement, W.A., Mooney, H.A.: Seasonal variation in the production of tannins and cyanogenic glucosides in the chaparral shrub *Heteromeles arbutifolia*. Oecologia *15*, 65–76 (1974)
Duffey, S.S., Towers, G.H.N.: On the biochemical basis of HCN production in the millipede *Harpaphe haydeniane*. Can. J. Zool. *56*, 7–16 (1978)
Dunnill, P.M., Fowden, L.: Enzymic formation of β-cyano-alanine from cyanide by *Escherichia coli* extracts. Nature (London) *208*, 1206–1207 (1965)
Ehrlich, P.R., Raven, P.H.: Butterflies and plants: a study in coevolution. Evolution *18*, 586–608 (1965)
Eisner, T.: Arthropod defences. In: Chemical ecology. Sondheimer, E., Simeone, J.B. (eds.), pp. 157–217. New York: Academic Press 1970
Epstein, J.: Estimation of microquantities of cyanide. Anal. Chem. *19*, 272–274 (1947)
Ettlinger, M., Eyjolfsson, R.: Revision of the structure of the cyanogenic glucoside triglochinin. J. Chem. Soc. Chem. Commun. 572–573 (1972)
Ettlinger, M.G., Jaroszewski, J.W., Jensen, S.R., Nielsen, B.J., Nartey, F.: Proacacipetalin and acacipetalin. J. Chem. Soc. Chem. Commun. 952–953 (1977)
Everist, S.L.: Poisonous plants of Australia. Sydney: Angus and Robertson 1974
Eyjolfsson, R.: Recent advances in the chemistry of cyanogenic glycosides. Fortschr. Chem. Org. Naturst. *27*, 74–108 (1970)
Farnden, K.J.F., Rosen, M.A., Liljegren, D.R.: Aldoximes and nitriles as intermediates in the biosynthesis of cyanogenic glucosides. Phytochemistry *12*, 2673–2677 (1973)
Fikenscher, L.H., Hegnauer, R.: Über die cyanogenen Verbindungen bei einigen *Compositae*, bei den *Olinaceae* und in der Rutaceen-Gattung, *Zieria*. Pharm. Weekblad. *112*, 11–20 (1977)

Finnemore, H., Cooper, J.M.: Cyanogenetic glucosides in Australian plants, pt. 4. *Zieria laevigata*. J. Proc. R. Soc. N. S. W. *70*, 175–182 (1936)

Finnemore, H., Gledhill, W.C.: The presence of cyanogenetic glucosides in certain species of *Acacia*. Australas. J. Pharm. N.S. *9*, 174–178 (1928)

Finnemore, H., Large, D.K.: An unstable cyanogenetic constituent in *Goodia lotifolia*. J. Proc. R. Soc. N. S. W. *70*, 440 (1936)

Floss, H.G., Hadwiger, L., Conn, E.E.: Enzymatic formation of β-cyanoalanine from cyanide. Nature (London) *208*, 1207–1208 (1965)

Fowden, L., Bell, E.A.: Cyanide metabolism by seedlings. Nature (London) *206*, 110–112 (1965)

Fraenkel, G.S.: The raison d'etre of secondary plant substances. Science *129*, 1466–1470 (1959)

Gaertner, F.H., Ericson, M.C., DeMoss, J.A.: Catalytic facilitation in vitro by two multienzyme complexes from *Neurospora crassa*. J. Biol. Chem. *245*, 595–600 (1970)

Gander, J.E.: Factors influencing *in vivo* formation of p-hydroxymandelonitrile-β-D-glucoside. Plant Physiol. *35*, 767–771 (1960)

Gibbs, R.D.: Chemotaxonomy of flowering plants. Vol. I-IV. Montreal: McGill-Queens Univ. Press 1974

Gillingham, J.T., Shirer, M.M., Starnes, J.J., Page, N.R., McClain, E.F.: Relative occurrence of toxic concentrations of cyanide and nitrate in varieties of sudan grass and sorghum sudan grass hybrids. Agron. J. *61*, 727–730 (1969)

Gmelin, R., Schüler, Bordan, E.: Holocalin: ein neues cyanogenes Glykosid aus *Holocalyx balansae*. Phytochemistry *12*, 457–461 (1973)

Gondwe, A., Seigler, D.S., Dunn, J.E.: Tetraphyllin B and *epi*-tetraphyllin B from *Adenia volkensii*. Phytochemistry *17*, 271–274 (1978)

Hahlbrock, K., Conn, E.E.: Purification and properties of a UDP-glucose:ketone cyanohydrin β-glucosyltransferase from *Linum usitatissimum* L. J. Biol. Chem. *245*, 917–922 (1970)

Hahlbrock, K., Conn, E.E.: Evidence for the formation of linamarin and lotaustralin in flax seedlings by the same glucosyl transferase. Phytochemistry *10*, 1019–1023 (1971)

Hahlbrock, K., Tapper, B.A., Butler, G.W., Conn, E.E.: Conversion of nitriles and α-hydroxynitriles to cyanogenic glucosides in flax seedlings and cherry laurel leaves. Phytochemistry *10*, 1019–1023 (1968)

Haisman, D.R., Knight, D.J.: The enzymic hydrolysis of amygdalin. Biochem. J. *103*, 528–534 (1967)

Harborne, J.B. (ed.): Phytochemical ecology. p. 272. London, New York: Academic Press 1972

Harms, C.L., Tucker, B.B.: Influence of nitrogen fertilization and other factors on yield, prussic acid, nitrate and total nitrogen concentration of sudan grass cultivars. Agron. J. *65*, 21–26 (1973)

Hegnauer, R.: Chemotaxonomie der Pflanzen. Vol. I-VI. Basel: Birkhäuser-Verlag 1963–1973

Hegnauer, R.: Probleme der Chemotaxonomie, erläutert am Beispiel der cyanogenen Pflanzenstoffe. Pharm. Acta Helv. *46*, 585–601 (1971)

Hegnauer, R.: Die cyanogenen Verbindungen der *Liliatae* und *Magnoliatae-Magnoliidae*: Zur systematischen Bedeutung der Merkmale der Cyanogenese. Biochem. Syst. *1*, 191–197 (1973)

Hegnauer, R.: Cyanogenic compounds as systematic markers in *Tracheophyta*, Plant Syst. Evol. Suppl. 1, 191–209 (1977)

Helferich, B., Kleinschmidt, T.: Zur Kenntnis des Süssmandelemulsins. Hoppe-Seylers Z. Physiol. Chem. *349*, 25–28 (1968)

Hendrickson, H.R., Conn, E.E.: Purification and properties of the β-cyanoalanine synthase of blue lupine. J. Biol. Chem. *244*, 2632–2640 (1969)

Herissey, H.: Sur la "prulaurasine", glucoside cyanhydrique cristallisé retiré des feuilles de laurier-cerise. J. Pharm. Ch. *23*, 5–14 (1906)

Heywood, V.H. (ed.): Taxonomy and ecology. New York: Academic Press 1973

Honeyman, J.M.: On the occurrence of cyanogenetic glycosides in the order Rhoeadales. Taxon *5*, 33–34 (1956)

Hösel, W., Nahrstedt, A.: Specifische Glucosidasen für das Cyanglucosid Triglochinin: Reinigung und Charakterisierung von β-Glucosidasen aus *Alocasia macrorrhiza* Schott. Hoppe-Seylers Z. Physiol. Chem. *356*, 1265–1275 (1975)

Hughes, M.A.: Studies on the β-glucosidase system of *Trifolium repens*. J. Exp. Bot. *19*, 427–434 (1968)
Hughes, M.A., Conn, E.E.: Cyanoglucoside biosynthesis in white clover (*Trifolium repens*). Phytochemistry *15*, 697–701 (1976)
Jackson, R.C., Handschumacher, R.E.: *Escherichia coli* L-asparaginase. Catalytic activity and subunit nature. Biochemistry *9*, 3585–3590 (1970)
Jensen, S.R., Nielsen, B.J.: Cyanogenic glucosides in *Sambucus nigra* L. Acta Chem. Scand. *27*, 2661–2662 (1973)
Jones, D.A.: Cyanogenic glycosides and their function. In: Phytochemical ecology. Harborne, J. (ed.), pp. 103–124. New York: Academic Press 1972
Jones, D.A.: Co-evolution and cyanogenesis. In: Taxonomy and ecology. Heywood, V.H. (ed.), pp. 213–242. New York: Academic Press 1973
Jones, D.A., Parsons, J., Rothschild, M.: Release of hydrocyanic acid from crushed tissues of all stages in the life-cycle of species of the *Zygaeninae* (Lepidoptera). Nature (London) *193*, 52–53 (1962)
Kindl, H., Underhill, E.W.: Biosynthesis of mustard oil glycosides: N-hydroxy-phenylalanine, a precursor of glucotropaeolin and a substrate for the enzymatic and nonenzymatic formation of phenylacetaldoxime. Phytochemistry *7*, 745–756 (1968)
Kingsbury, J.M.: Poisonous plants of the United States and Canada: p. 626. Englewood Cliffs: Prentice-Hall 1964
Koukol, J., Miljanich, P., Conn, E.E.: Studies on the biosynthesis of dhurrin, the cyanogenic glucoside of *Sorghum vulgare*. J. Biol. Chem. *237*, 3223–3228 (1962)
Knowles, C.J.: Microorganisms and cyanide. Bacteriol. Rev. *40*, 652–680 (1976)
Lambert, J.L., Ramasamy, J., Paukstelis, J.V.: Stable reagents for the colorimetric determination of cyanide by modified Koenig reactions. Anal. Chem. *47*, 916–918 (1975)
Lauinger, C., Ressler, C.: β-Cyanoalanine as a substrate for asparaginase. Stoichiometry, kinetics, and inhibition. Biochim. Biophys. Acta *198*, 316–323 (1970)
Lewis, J.P.: Laetrile. West. J. Med. *127*, 55–62 (1977)
Luckner, M., Mothes, K., Nover, L. (eds.): Secondary metabolism and coevolution. p. 614. Halle (Saale): Deutsche Akademie der Naturforscher Leopoldina 1976
Mao, C.-H., Anderson, L.: Mechanism of the alkaline hydrolysis of dhurrin. J. Org. Chem. *30*, 603–607 (1965)
Mao, C.-H., Anderson, L.: Partial purification and characterization of two β-glucosidases from *Sorghum* tissues. Phytochemistry *6*, 473–483 (1967)
McFarlane, I.J., Lees, E.M., Conn, E.E.: The in vitro biosynthesis of dhurrin, the cyanogenic glycoside of *Sorghum bicolor*. J. Biol. Chem. *250*, 4708–4713 (1975)
Mikolajczak, K.L.: Cyanolipids. Prog. Chem. Lipids Rel. Compds. *15*, 97–130 (1977)
Møller, B.L.: Intermediates in the biosynthesis of cyanogenic glucosides determined by use of a gas chromatograph coupled with a gas proportional counter. Anal. Biochem. *81*, 292–304 (1977)
Møller, B.L., Conn, E.E.: N-Hydroxytyrosine, an intermediate in the biosynthesis of the cyanogenic glucoside dhurrin in *Sorghum vulgare*. Plant Physiol. Suppl. *59*, 82 (1977)
Møller, B.L., McFarlane, I.J., Conn, E.E.: Chemical synthesis and disproportionation of N-hydroxytyrosine. Acta Chem. Scand. B *31*, 343–344 (1977)
Nahrstedt, A.: Zur Cyanogenese in *Prunus avium*. Phytochemistry *9*, 2085–2089 (1970)
Nahrstedt, A.: Gaschromatographie diastereomerer Benzaldehydcyanhydrin-Glycoside. Planta Med. *24*, 83–89 (1973a)
Nahrstedt, A.: Zur Configuration des Cyanglycosides Holocalin. Phytochemistry *12*, 2799–2800 (1973b)
Nahrstedt, A.: Die Isomerisierung von Amygdalin und Homologen. Arch. Pharm. *308*, 903–910 (1975)
Nahrstedt, A.: Ein neues cyanogenes Glykosid aus der Rosacee *Sorbaria arborea*. Z. Naturforsch. *31 C*, 397–400 (1976)
Nahrstedt, A., Hubel, W.: Die cyanogenen Verbindungen von *Heterodendron oleaefolium* Deaf. Phytochemistry *17*, 314–315 (1978)
Nass, H.G.: Cyanogenesis: Its inheritance in *Sorghum bicolor, Sorghum sudanense, Lotus,* and *Trifolium repens*, a review. Crop Sci. *12*, 503–506 (1972)
Nelson, C.E.: HCN content of certain sorghums under irrigation as affected by nitrogen fertilizer and soil moisture stress. Agron. J. *45*, 615–617 (1953)

Nisizawa, K., Hashimoto, Y.: Glycoside hydrolases and glycosyl transferases. In: The carbohydrates. Ch. 33, 2nd ed. Pigman, W., Horton, D. (eds.), Vol. IIA, pp. 241–300. New York: Academic Press 1970

Ohnishi, T., Yamamoto, S., Hayaishi, O., Izumi, T., Shiba, T.: Studies on the reaction specificity of the flavoprotein lysine monooxygenase with modified substrates. Arch. Biochem. Biophys. *176*, 358–365 (1976)

Plouvier, V.: Recherches sur l'isomérisation d'hétérosides cyanogénétiques. C.R. Acad. Sci. *200*, 1985 (1935)

Reay, P.F.: An improved procedure for the isolation of dhurrin. Phytochemistry *8*, 2259–2260 (1969)

Reay, P.F., Conn, E.E.: Dhurrin synthesis in excised shoots and roots of sorghum seedlings. Phytochemistry *9*, 1825–1827 (1970)

Reay, P.F., Conn, E.E.: Purification and properties of a uridine diphosphate glucose: aldehyde cyanohydrin β-glucosyltransferase from sorghum seedlings. J. Biol. Chem. *249*, 5826–5830 (1974)

Ressler, C., Giza, Y.H., Nigam, S.N.: Biosynthesis and metabolism in species of vetch and lathyrus of γ-glutamyl-β-cyanoalanine: relation to the biosynthesis of asparagine. J. Am. Chem. Soc. *85*, 2874–2875 (1963)

Robiquet, P.J., Boutron-Charlard, A.F.: Nouvelles experiences sur les amandes amères et sur l'huile volatile qu'elles fournissent. Ann. Chim. Phys. Ser. 2 *44*, 352–382 (1830)

Russell, G.B., Reay, P.R.: The structures of tetraphyllin A and B, two new cyanoglucosides from *Tetrapathaea tetrandra*. Phytochemistry *10*, 1373–1377 (1971)

Saunders, J.A., Conn, E.E.: Presence of the cyanogenic glucoside dhurrin in isolated vacuoles of *Sorghum*. Plant Physiol. *61*, 154–157 (1978)

Saunders, J.A., Conn, E.E., Lin, C.H., Stocking, C.R.: Subcellular localization of the cyanogenic glucoside of *Sorghum* by autoradiography. Plant Physiol. *59*, 647–652 (1977)

Schwarzmaier, U.: 2,3,5-Triphenyltetrazoliumhydroxid – ein Farbreagenz für Mandelsäurenitrilglycoside. J. Chromatogr. *114*, 235–236 (1975a)

Schwarzmaier, U.: Hydroxylamin/eisen (III) Chlorid, ein stereospezifisches Farbreagenz für Mandelsäurenitrilglycoside. J. Chromatogr. *111*, 430–433 (1975b)

Schwarzmaier, U.: Über die Cyanogenese von *Bambusa vulgaris* and *B. guadua*. Chem. Ber. *109*, 3379–3389 (1976)

Secor, J.B., Conn, E.E., Dunn, J.E., Seigler, D.S.: Detection and identification of cyanogenic glucosides in six species of Acacia. Phytochemistry *15*, 1703–1706 (1976)

Seigler, D.S.: Isolation and characterization of naturally occurring cyanogenic compounds. Phytochemistry *14*, 9–29 (1975)

Seigler, D.S.: Plants of the northeastern United States that produce cyanogenic compounds. Econ. Bot. *30*, 395–407 (1976a)

Seigler, D.S.: Plants of Oklahoma and Texas capable of producing cyanogenic compounds. Proc. Okla. Acad. Sci. *56*, 95–110 (1976b)

Seigler, D.S.: Cyanolipids in *Cordia verbenacea* – a correction. Biochem. Syst. Ecol. *4*, 235–236 (1976c)

Seigler, D.S.: The naturally occurring cyanogenic glycosides. In: Progress in phytochemistry. Reinhold, L., Harborne, J.B., Swain, T. (eds.), Vol. IV, pp. 83–120. 1977

Seigler, D.S., Price, P.W.: Secondary compounds in plants: primary functions. Am. Nat. *110*, 101–104 (1976)

Seigler, D.S., Eggerding, C., Butterfield, C.: A new cyanogenic glycoside from *Cardiospermum hirsutum*. Phytochemistry *13*, 2330–2332 (1974)

Seigler, D.S., Dunn, J.E., Conn, E.E.: Acacipetalin in *Acacia constricta* from North America. Phytochemistry *15*, 219–220 (1976)

Sharples, D., Stoker, J.R.: The identification and biosynthesis of two cyanogenic glucosides in *Thalictrum aquilegifolium*. Phytochemistry *8*, 597–601 (1969)

Sharples, D., Spring, M.S., Stoker, J.R.: Biosynthesis of the major cyanogenic glycoside of *Thalictrum aquilegifolium*. Phytochemistry *11*, 2999–3002 (1972)

Shimada, M., Conn, E.E.: The enzymatic conversion of *p*-hydroxyphenylacetaldoxime to *p*-hydroxyphenylacetonitrile. Arch. Biochem. Biophys. *180*, 199–207 (1977)

Stafford, H.A.: Possible multi-enzyme complexes regulating the formation of C_6-C_3 phenolic compounds and lignins in higher plants. Recent Adv. Phytochem. *8*, 53–79 (1974)

Stone, M.P.: Unpublished observations (1977)

Tapper, B.A., Butler, G.W.: Oximes, nitriles and 2-hydroxynitriles as precursors in the biosynthesis of cyanogenic glucosides. Biochem. J. *124*, 935–941 (1971)
Tapper, B.A., Butler, G.W.: Intermediates in the biosynthesis of linamarin. Phytochemistry *11*, 1041–1046 (1972)
Tapper, B.A., Conn, E.E., Butler, G.W.: Conversion of α-keto-isovaleric acid oxime and isobutyraldoxime to linamarin in flax seedlings. Arch. Biochem. Biophys. *119*, 593–595 (1967)
Tjon Sie Fat, L.: Cyanogenesis in some grasses. Proc. Koninkl. Nederl. Akad. Wetensch. *80*, 227–237 (1977)
Tschiersch, B.: Assimilation of $H^{14}CN$ by *Lathyrus odoratus* L., *Vicia sativa* L. and *Ricinus communis* L. Phytochemistry *3*, 365–367 (1964)
Tschiersch, B.: β-Cyanoalanine-Bildung durch zellfreie Extrakte höherer Pflanzen. Flora (Jena) Abt. A *156*, 363–366 (1965)
Uribe, E., Conn, E.E.: The origin of the nitrile nitrogen atom of dhurrin (β-D-glucopyranosyloxy-L-*p*-hydroxymandelonitrile). J. Biol. Chem. *241*, 92–94 (1966)
Watt, J.M., Breyer-Brandwijk, M.G.: The medicinal and poisonous plants of Southern and East Africa, 2nd ed. Edinburgh-London: Livingstone 1962
Whittaker, R.H., Feeny, P.P.: Allelochemics: Chemical interactions between species. Science *171*, 757–770 (1971)
Wissing, F.: Cyanide formation from oxidation of glycine by a *Pseudomonas* species. J. Bacteriol. *117*, 1289–1294 (1974)
Young, R.L., Hamilton, R.A.: A bitter principle in macadamia nuts. Proc. VI. Annu. Meet. Hawaii Macadamia Prod. Assoc. p. 27 (1966)
Zilg, H., Conn, E.E.: Stereochemical aspects of lotaustralin biosynthesis. J. Biol. Chem. *249*, 3112–3115 (1974)
Zilg, H., Tapper, B.A., Conn, E.E.: The origin of the glucosidic linkage oxygen of the cyanogenic glucosides linamarin and lotaustralin. J. Biol. Chem. *247*, 2384–2386 (1972)
Zitnak, A., Hill, D.C., Alexander, J.C.: Determination of linamarin in biological tissues. Anal. Biochem. *77*, 310–314 (1977)

10 Glucosinolates

E.W. UNDERHILL

1 Introduction

The glucosinolates are a uniform class of naturally occurring anions which have, so far, been found exclusively in the plant kingdom, and only in a limited number of dicotyledon families. Although all are derivatives of a common general structure, there is a great variation in the chemistry of their aglycones.

Glucosinolates are characterized by their ability to be hydrolysed by the enzyme myrosinase to yield glucose and a labile aglucone which spontaneously rearranges with the loss of sulfate to give an isothiocyanate (mustard oil) as the major product. Thus the glucosinolates have been termed thioglucosides, mustard oil glucosides and isothiocyanate producing glucosides. Many of the isothiocyanate fission products, which may be present in the intact plant in minute amounts, are released when the plant tissues are crushed. These volatile, fission products (mustard oils) are responsible for the distinctive, pungent flavor and odor with which we associate some of the common mustards (Cruciferae species) used as vegetables and condiments (VAUGHAN and HEMINGWAY, 1959).

2 General Structure and Nomenclature

The first glucosinolates were isolated more than a century ago at a time when the enzymatic formation of the mustard oils was being studied (BOUTRON and FREMY, 1840). A mustard oil-producing compound, sinalbin, was isolated from white mustard seeds (*Sinapis alba* L.) (ROBIQUET and BOUTRON, 1831) and BUSSY (1840) recovered the potassium salt of the "acid myronique", later to be known as sinigrin, from black mustard seeds (*Brassica nigra* Koch.). HOFMANN (1868) showed the mustard oils to be isothiocyanates. The first structural formula for these two classical glucosides was proposed by GADAMER (1897). Although his formula accounted for the principal enzymic hydrolytic products of sinigrin, viz allylisothiocyanate, glucose and bisulfate, a number of the other known reactions of sinigrin were not reconcilable with his formula. In spite of this, the GADAMER structure remained essentially unchallenged until ETTLINGER and LUNDEEN (1956) presented the correct formula for sinigrin and sinalbin. A detailed discussion of the early work on the properties and structure of the mustard oil glucosides and isothiocyanates is given in a monograph by CHALLENGER (1959).

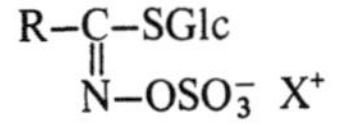

(I)

(Glc–OH = β–D–glucose)

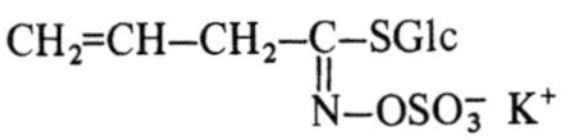

(II)

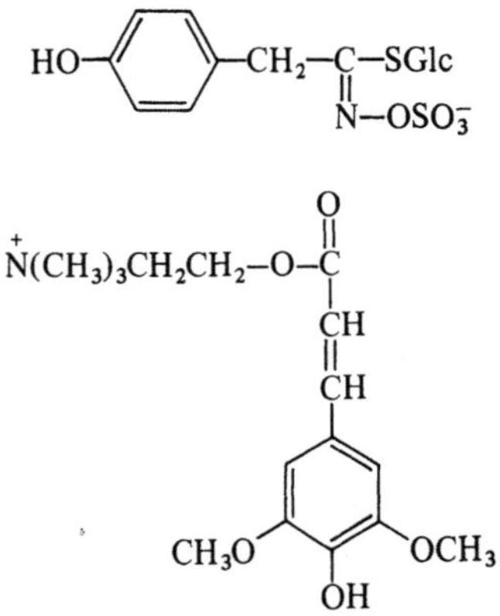

(III)

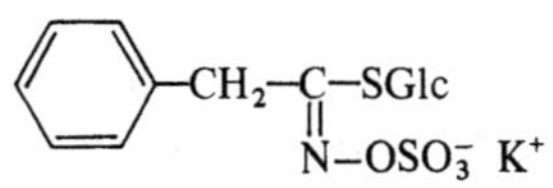

(IV)

The revised structure, verified by synthesis (ETTLINGER and LUNDEEN, 1957) and now extended by analogy to the general formula (I), has been accepted as the correct formula for all glucosinolates known to date. Glucose has invariably been found as the sugar moiety attached through a β-thioglucosidic linkage. X-Ray crystallographic studies (WASER and WATSON, 1963) confirmed the structure and established that the configuration around the C=N is with the side-chain "R" *anti* to the sulfate. The glucosinolates have been isolated most frequently as their potassium salts, although sinapine (the choline ester of sinapic acid) is the cation of the mustard oil glucoside sinalbin.

Prior to 1948 only nine glucosinolates were known; by 1960 the number had increased to thirty (KJAER, 1961), and now exceeds seventy. Apart from the classical names for sinigrin and sinalbin, the nomenclature employed to designate individual members of this class was based on a system of trivial names. These names were derived from the Latin name of the plant from which the glucoside was first isolated, prefixed with "gluco" and suffixed with "in", e.g., glucotropaeolin from *Tropaeolum* species. Such a system of nomenclature, besides failing to designate the nature of the cation, and having little functional value, became most unwieldy as the number of new glucosinolates isolated increased. On the other hand the structures are too complicated to permit a fully systematic scheme to be workable. To overcome these difficulties ETTLINGER and DATEO (1961) introduced a semi-systematic method of nomenclature in which the anion of the salt (I; R=H) is designated by the term "glucosinolate" and the chemical name of the radical "R" (the glucosinolate side chain) is used as a prefix; thus, the mustard oil glucosides sinigrin (II), sinalbin (III) and glucotropaeolin (IV) are now designated as potassium allylglucosinolate, sinapine 4-hydroxybenzylglucosinolate and potassium benzylglucosinolate, respectively.

3 Glucosinolate Side Chains

Glucosinolates are difficult to crystallize and less than half of those presently recognized have been characterized as crystalline glucosides. The structures of those which have not been recovered as crystalline entities have been deduced on the basis of the products of enzymic hydrolysis and chemical degradation.

A great variety of complex and interesting structures comprise the side chains of the known glucosinolates. The structures include those with simple and branched alkyl and alkenyl side chains, as well as ω-methylthioalkyl, monoketoalkyl, aralkyl, and heterocyclic side chains. There are thirteen aralkyl glucosinolates and of these nine have benzyl side chains with the ring variously substituted with hydroxyl, methoxyl (including some *m*-substituted derivatives) and *o*-(α-L-rhamnopyranosyl) groups while the remaining four have phenylethyl side chains. Two unusual features not generally found in secondary plant products but which occur amongst the glucosinolates are the large number of homologues and the high frequency of β-hydroxylated analogues. Most unusual is the number of homologous series of glucosinolates, for example the extensive series of ω-methylthioalkyl side chains ranging from $MeS(CH_2)_3$ to $MeS(CH_2)_8$ and (R)-$MeS(O)(CH_2)_3$ to (R)-$MeS(O)(CH_2)_{11}$. For a detailed discussion of the individual glucosinolates with references to their discoveries, the following should be consulted (KJAER, 1960; ETTLINGER and KJAER, 1968; KJAER and OLESEN LARSEN, 1973, 1976).

4 Occurrence

The glucosinolates are now known to occur in eleven families of dicotyledons. They have been detected in practically every species studied of the Cruciferae, Capparaceae, Tovariaceae, Resedaceae, and Moringaceae (these five families constitute the well-defined order Capparales sensu CRONQUIST, 1968). Other families in which they occur in one or more taxa include Limnanthaceae, Caricaceae, Tropaeolaceae, Gryostemonaceae, Salvadoraceae, and Euphorbiaceae (KJAER, 1974).

There is a wide variation in the distribution of glucosinolates within these eleven plant families, as indicated by the occurrence of methylglucosinolate and benzylglucosinolate. Methylglucosinolate rarely occurs outside the Capparaceae and is the closest to a family-specific glucosinolate, whereas benzylglucosinolate has a broad family distribution. Three keto-alkylglucosinolates found in South American species of the genus *Capparis* have not been detected in "Old-World" species of the same genus (KJAER and THOMSEN, 1963). When considered on the basis of their biosynthetic origin, (see below) glucosinolates derived from chain-extended amino acids are found primarily in the Cruciferae and Resedaceae. Studies on the occurrence of the indole glucosinolates indicate that they are characteristic of the first four listed families of the order Capparales (SCHRAUDOLF, 1967). Investigation at the genus and species level has shown certain distribution patterns and, although of a preliminary nature, has been of significant chemosystematic value (ETTLINGER and THOMPSON, 1962; AL-SHEHBAZ, 1973).

Generally a plant species contains more than one glucosinolate and often there are both qualitative and quantitative differences between the roots, leaves, and seeds of a plant (ELLIOTT and STOWE, 1971). Variations in the total glucosinolate content between different varieties and lines as well as between individual plants of a line have also been noted and are the basis of breeding programs to eliminate glucosinolates from oil seed crops (JOSEFSSON and JÖNSSON, 1969).

5 Myrosinase and Glucosinolate Hydrolytic Products

Glucosinolates, although considered non-toxic themselves, are important because of the multiplicity of physiologically active products derived from them upon enzymic cleavage (see TAPPER and REAY, 1973; BENN, 1977 for review). Almost without exception they are accompanied within the same plant by the hydrolytic enzyme myrosinase, a thioglucosidase (thioglucoside glucohydrolase, EC 3.2.3.1) which is released when the plant tissues are crushed or when autolysis within the plant occurs. It has also been reported that hydrolysis may occur at a slow rate in normal plant tissues (TANG, 1971). Enzymes from other sources with hydrolytic activity resembling plant myrosinase have been found in fungi (REESE et al., 1958; OHTSURU et al., 1969), bacteria of the lower mammalian intestine (OGINSKY et al., 1965), and the cabbage aphid (MCGIBBON and ALLISON, 1968).

SCHWEIDLER (1905) studied the location of myrosinase in plants using light microscopic techniques and reported it to be present in cells called idioblasts, specialized cells present in parenchymatous tissue of the green parts of crucifer plants. Recent detailed studies on the localization of myrosinase in *S. alba* roots indicate that the enzyme is present in the majority of the root-tip cells and is not confined to specific cells (IVERSEN, 1970). Electron microscopical cytochemistry and differential centrifugation of subcellular organelles have demonstrated that it is mainly present in the mitochondria, dictyosome, and smooth endoplasmic reticulum fractions (IVERSEN, 1970; PIHAKASKI and IVERSEN, 1976). It was concluded that the activity of the enzyme was membrane-associated and that the dictyosomes might also be loci for glucosinolate hydrolysis.

For several years there has been a controversy as to whether myrosinase was one or two enzymes (a thioglucosidase and a sulfatase) but it is now considered to be one enzyme often existing in isoenzymic forms. The action of the plant enzyme on glucosinolates is that of a thioglucoside hydrolase (BJÖRKMAN, 1976). NAGASHIMA and UCHIYAMA (1959) found that myrosinase from *S. alba* was strongly activated by ascorbic acid, while ETTLINGER et al. (1961) separated two myrosinase isoenzymes from *S. alba* and found that one was strongly activated by ascorbic acid and the other not. Similar results were obtained by VOSE (1972). The existence of several myrosinase isoenzymes in plants has been demonstrated by gel electrophoresis (VAUGHAN et al., 1968), isoelectric focusing (VOSE, 1972) and immunoelectrophoretic experiments (VAUGHAN and GORDON, 1969).

Purified myrosinase isoenzymes have been obtained preparatively, three from both *S. alba* (BJÖRKMAN and JANSON, 1972) and *Brassica napus* (LÖNNERDAL and JANSON, 1973), and four from *B. juncea* (OHTSURU and HATA, 1972). The isoenzymes from *S. alba* and *B. napus* had similar properties with respect to temperature maximum, stability, and hydrolytic pattern with different glucosinolates; the main differences were in respect to pH optima, degree of activation with ascorbic acid and some physicochemical properties such as molecular weight and isoelectric point (BJÖRKMAN and LÖNNERDAL, 1973). All are glycoproteins of two to four subunits with a molecular weight between 125,000 and 153,000. Screening experiments on the myrosinase activity in various cultivars of rapeseed (*B. napus* and *B. campestris*) and white mustard showed large differences both between and within species. The activity of myrosinase derived from white mustard was five to ten times greater than that derived from rapeseed. However, there were no differences between the myrosinases obtained from these three species on their rates of hydrolysis of six different glucosinolates, signifying that myrosinases from different species are not specifically adapted to a particular set of glucosinolates (BJÖRKMAN, 1976).

Myrosinase when added to a glucosinolate solution yields D-glucose and an aglucone (Fig. 1). The aglucone (V), under neutral conditions, gives rise to sulfate and by a Lossen-type rearrangement, an isothiocyanate (VI); under weakly acidic conditions or in the presence of ferrous ion, a nitrile (VII) and elemental sulfur are formed. Isothiocyanates possessing a *β*-hydroxyl (VIII) spontaneously cyclize to form substituted oxazolidine-2-thiones (IX). The isothiocyanates formed from the indole glucosinolates and *p*-hydroxybenzylglucosinolate are unstable in neutral or alkaline conditions and give rise to thiocyanate ion (see SCHLÜTER and GMELIN, 1972). At lower pH, enzymic fission of 3-indolylmethylglucosinolate leads to the formation of the growth-promoting compound, 3-indolylacetonitrile.

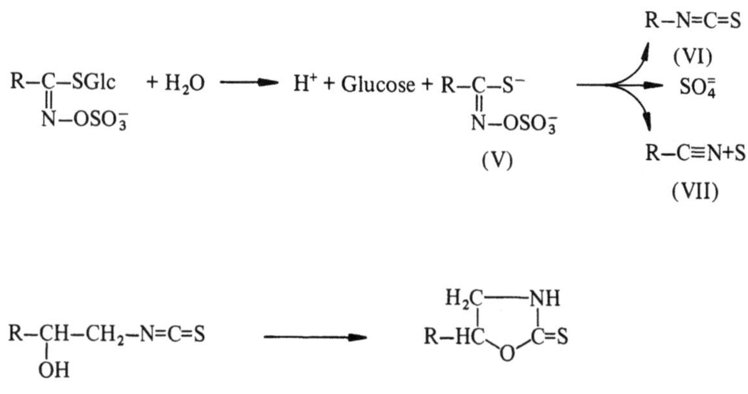

Fig. 1. Some enzymic hydrolysis products of glucosinolates

The foregoing products are routinely observed when exogenous myrosinase is supplied to aqueous extracts of glucosinolates or to moistened seed meals or other plant parts which have been previously heat-treated to destroy the endogenous enzyme. Still other products are formed from the hydrolysis of glucosinolates by endogenous enzymes (autolysis) such as often occurs when seeds are crushed and moistened. GMELIN and VIRTANEN (1959) found organic thiocyanates were primarily formed rather than isothiocyanates, when fresh plant material was crushed with water; from *Thlaspe arvense* L. (containing allylglucosinolate) and *Lepidium ruderale* L. (benzylglucosinolate) they isolated allylthiocyanate and benzylthiocyanate, respectively. Although their data supported the enzymic formation of the products, they and others have been unable to isolate the enzyme(s) and it has not been ascertained whether the products were formed enzymatically or whether they were formed secondarily after the action of myrosinase (SAARIVIRTA, 1973).

It has been observed frequently that formation of isothiocyanates and 5-vinyloxazolidine-2-thione is favored when rapeseed or *Crambe abyssinica* meal is heat-treated prior to hydrolysis using exogenous enzyme, whereas nitriles are the major products formed under autolytic conditions. That the exogenous source of myrosinase is not responsible for this difference in products formed has been clarified by SRIVASTAVA and HILL (1974). The nitriles formed from ω-alkenyl glucosinolates under conditions of autolysis include 1-cyano-2-hydroxy-3-butene and 1-cyano-2-hydroxy-3,4-epithiobutane both from 2-hydroxy-3-butenylglucosinolate (VAN ETTEN et al., 1969), 1-cyano-3,4-epithiobutane from 3-butenylglucosinolate (KIRK and MACDONALD, 1974), 1-cyano-2,3-epithiopropane from allylglucosinolate and 1-cyano-4,5-epithiopentane from 4-pentenylglucosinolate (COLE, 1975).

A protein factor which greatly alters the nature of the products formed during hydrolysis has been isolated from *C. abyssinica* (TOOKEY, 1973). The factor, termed epithio specifier protein, is by itself devoid of thioglucosidase activity. When the thioglucosidase fraction obtained from *C. abyssinica* was combined with the specifier protein, the product formed from the parent glucosinolate was 1-cyano-2-hydroxy-3,4-epithiobutane, whereas the enzyme alone yielded only 1-cyano-2-hydroxy-3-butene from the same substrate. The specifier is unique in that it specifies the product rather than the substrate. Notwithstanding the foregoing, the factors which determine product formation on autolysis of glucosinolates are not clearly understood, and a great need exists for further studies in this area.

Many plants of the Cruciferae are used as foods and condiments, examples of which include cabbage and related plants, rutabaga, turnip, salad greens from *Lepidium* species, radish, horseradish, and white and yellow mustard seed. Often there is a considerable variation in the flavor volatiles of these plants due to the different kinds and amounts of hydrolytic products formed. These products depend on the plant variety and conditions of harvest, storage, and processing (VAN ETTEN et al., 1969; DAXENBICHLER et al., 1977).

Generally, isothiocyanates have, at low concentration, a pleasant appetite-stimulating flavor whereas nitriles have a garlic-like flavor. The less pleasing flavor of cooked frozen Brussels sprouts compared with the flavor of the cooked fresh vegetable is related to the low concentration of isothiocyanate in the frozen material as a result of myrosinase inactivation during blanching, and to the high concentration of allylcyanide in the frozen vegetable due to autolysis or non-enzymic degradation of allylglucosinolate (MACLEOD, 1976).

An important by-product after the extraction of the edible and industrial oils from rapeseed and crambe is a protein-rich meal which may be used as a supplement in livestock and poultry feeds (APPELQVIST, 1972). However, the oilseed meal has been known for many years to have detrimental effects on animal growth and reproduction and this has limited the use of rapeseed meal as a feed supplement. These deleterious effects are the result of the hydrolytic products of glucosinolates, most of which are goitrogenic, namely isothiocyanates, organic thiocyanates, thiocyanate ion, and oxazolidine-2-thiones. Although these compounds can produce goiter in animals when large amounts of Cruciferae plants are fed, it has not been firmly established that endemic benign goiter in humans has resulted, even in times of food shortages when large amounts of cruciferous plants have been consumed. A detailed review on goitrogens in cruciferous plants used for human and animal consumption has been prepared by VAN ETTEN (1969) and on the goitrogenic properties of processed rapeseed meal by BELL and BELZILE (1965).

6 Glucosinolates and Myrosinase in Plant Cell Cultures

The presence of glucosinolates and myrosinase has been studied in tissue cultures derived from plants known to accumulate these compounds. KRIKORIAN and STEWARD (1969) and AFZALPURKAR (1974) were unable to detect glucosinolates in cultures of *Iberis sempervirens* and *Brassica napus* whereas KIRKLAND et al. (1971) reported their presence in cell suspension cultures of two of seven species examined. The glucosinolates found in the cells, but not in the culture media, included benzylglucosinolate and 2-phenylethylglucosinolate in cultures of *Tropaeolum majus* and *Nasturtium officinale*. Although no glucosinolates could be detected in cells of *S. alba*, one of the enzymes involved in glucosinolate biosynthesis, UDP-glucose: thiohydroximate glucosyltransferase, was isolated. Thus the capability for glucosinolate synthesis was demonstrable in some but not all cultures.

Recent studies (UNDERHILL, unpublished) suggest this apparent lack of glucosinolate synthetic capability may have been a result of a nutritional deficiency in the cultures, as has been observed in whole plants cultured under conditions of low sulfur (FREEMAN and MOSSADEGHI, 1972). Further, a functional relationship between glucosinolates, myrosinase, and plant nutrition is suggested from the following results. Although the growth of the culture increased almost linearly over fourteen days, the amount of 2-phenylethylglucosinolate was found to increase rapidly, reaching a maximum at day 5 or 6, and then to decrease precipitously almost to zero over the next two days. During this period the amount of inorganic sulfate in the media decreased rapidly from the first day and was exhausted by day 5; myrosinase activity, barely detectable during the first three days, increased dramatically during day 4, reached a maximum at day 5, and decreased thereafter. It was evident that as the concentration of sulfate in the media decreased the myrosinase became activated with a concomitant loss of glucosinolate within the cells. When the inorganic sulfate in the media was maintained at its original level the amount of glucosinolates reached a maximum at eight days and remained almost constant during the next week while the myrosinase activity remained at a low level. It appears likely that glucosinolates were not detected previously because of a deficiency of inorganic sulfate in the media with the resulting hydrolysis of the glucosinolate by myrosinase. It is also evident that destruction of plant tissues is not the only requirement for glucosinolate hydrolysis by myrosinase.

7 Quantitative Analysis

Numerous methods have been reported for the quantitative analysis of glucosinolates (LEIN, 1970) and while a detailed discussion of these methods is outside the scope of this presentation, brief mention will be made of some of the more important methods. Most of the quantitative methods are based upon the analysis of one or more of the enzymic hydrolytic products and therefore assays for both total and individual glucosinolates are available. Assays for total glucosinolates in seeds or plant extracts are based on the measurement of either the D-glucose or sulfate liberated on enzymic hydrolysis. A simple, rapid method, which has been used to screen for low glucosinolate concentrations in individual seeds, is based on the analysis of the total liberated D-glucose using a commercially available test paper containing glucose oxidase, peroxidase, and a chromogen. Here, it has been found necessary to pass the crude aqueous plant extract through charcoal prior to assay in order to remove colored and other interfering compounds (VAN ETTEN et al., 1974). Glucose formed by myrosinase has also been measured by GLC (OLSSON et al., 1976). Total glucosinolates can also be determined by measurement of the sulfate released either gravimetrically (MCGHEE et al., 1965) or titrimetrically. A number of indirect methods have been developed to quantitate particular glucosinolates or groups of glucosinolates. Colorimetric (JOHNSTON and JONES, 1966) and titrimetric (RAGHAVAN et al., 1971) assays of thiocyanate ion have been adapted for the estimation of the indole glucosinolates and *p*-hydroxybenzylglucosinolate; here, too, colored materials found in plant extracts often interfere with the assay. A method of analysis in rapeseed meal for the total substituted oxazolidine-2-thiones and alkali stable isothiocyanates, following conversion of the latter to thioureas, is based upon their specific UV absorbance (WETTER and YOUNGS, 1976). Gas chromatography is the only available method which can reliably be used for both quantitative and qualitative analysis of complex mixtures containing isothiocyanates, nitriles, organic thiocyanates, and oxazolidine-2-thiones formed during autolysis of glucosinolates in foods (VAN ETTEN et al., 1976). Most glucosinolates can be gas chromatographed without prior hydrolysis as their trimethylsilyl derivatives with good resolution and quantitation (UNDERHILL and KIRKLAND, 1971; THIES, 1976).

8 Biosynthesis

Biosynthetic studies of glucosinolates have revealed that all are derived from amino acids and that most glucosinolates are formed by a common biosynthetic pathway. These studies have involved administration of labeled compounds to plants, isolation of intermediates and, in a few cases, the detection of some of the enzymes involved in the pathway.

8.1 Glucosinolates from Amino Acids

Assuming the loss of the carboxyl carbon, a remarkable similarity exists between the carbon-nitrogen structures of a number of common amino acids and some glucosinolates. This relationship between amino acids and glucosinolates was clearly demonstrated by the facile conversion of labeled tryptophan into 3-indolylmethylglucosinolate in *Brassica oleracea* L. (KUTÁČEK et al., 1962) and of phenylalanine into benzylglucosinolate in *Tropaeolum majus* L. (BENN, 1962; UNDERHILL et al., 1962). It was established that the carbon skeleton, except for the carboxyl, and the amino nitrogen of L-phenylalanine were incorporated intact into benzylglucosinolate (X, Fig. 2) (UNDERHILL and CHISHOLM, 1964). Similarly it has been demonstrated that other glucosinolates, including *p*-hydroxybenzyl-, *p*-methoxybenzyl- and isopropyl-glucosinolates are formed from their corresponding amino acids with loss of the carboxyl carbon (see UNDERHILL and WETTER, 1973).

The majority of glucosinolates comprising the extensive series of homologues possess aglycone structures which are not related to protein amino acids. It is generally accepted, however, that these glucosinolates are also derived from amino acids formed in turn from protein amino acids via condensation of their 2-oxo-acids with acetate involving a chain elongation pathway as outlined below (Fig. 3). Studies on the biosynthesis of 2-phenylethylglucosinolate (I; $R=C_6H_5-CH_2-CH_2$) in *N. officinale* L. have shown it is formed from phenylalanine (XI; $R=C_6H_5-CH_2$) and acetate by chain extension (UNDERHILL, 1965). The natural occurrence of the intermediates 2-benzylmalic acid (XII; $R=C_6H_5-CH_2$) and 3-benzylmalic acid (XIII; $R=C_6H_5-CH_2$) and of the chain extended 2-amino-4-phenylbutyric acid (XIV; $R=C_6H_5-CH_2$), the amino acid precursor of the glucosinolate, was also demonstrated (UNDERHILL, 1968; DÖRNEMANN et al., 1974).

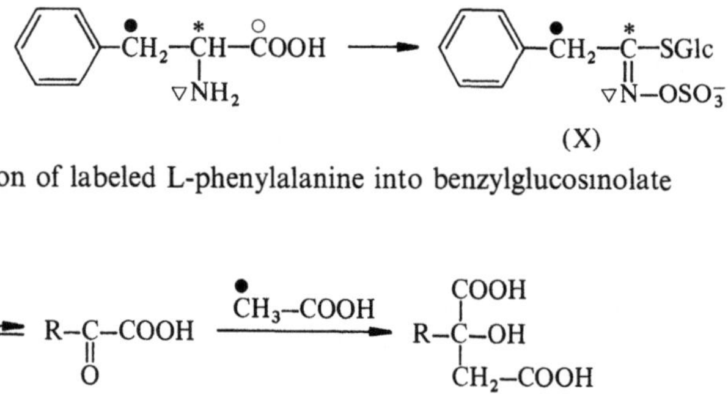

(X)

Fig. 2. Incorporation of labeled L-phenylalanine into benzylglucosinolate

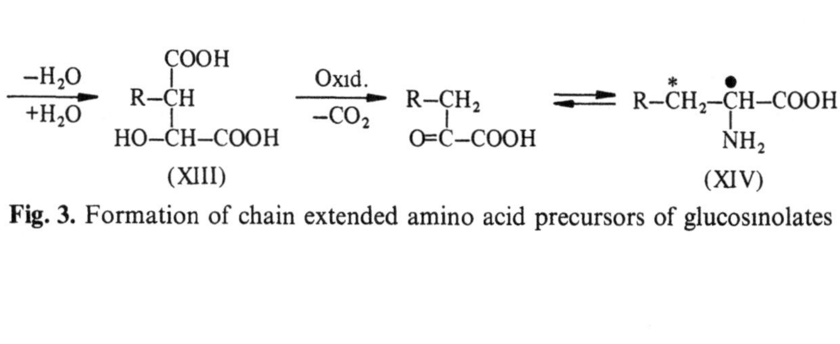

Fig. 3. Formation of chain extended amino acid precursors of glucosinolates

Although intermediates involved in the chain elongation pathway have been isolated only in the formation of 2-amino-4-phenylbutyric acid, the labeling patterns observed in the formation of other glucosinolates are entirely consistent with such a pathway. The chain extension of amino acids receives further support from the work on the biosynthesis of allylglucosinolate (I; $R=CH_2=CH\text{–}CH_2$) (CHISHOLM and WETTER, 1964, 1966; MATSUO and YAMAZAKI, 1968), 3-butenylglucosinolate [I; $R=CH_2=CH\text{–}(CH_2)_2$] and 4-pentenylglucosinolate [I; $R=CH_2=CH\text{–}(CH_2)_3$] and their 2-hydroxylated analogues (CHISHOLM and WETTER, 1967; SERIF and SCHMOTZER, 1968; LEE and SERIF, 1970). All are derived from C_2, C_3, and C_4 of methionine and the methyl carbon of acetate, and the labeling pattern in the glucosinolate aglycone was in each case consistent with the operation of the route involving repeated chain extension of methionine. The natural occurrence of 2-amino-5-methylthiovaleric acid [XIV; $R=CH_3S(CH_2)_2$], homomethionine, in horseradish and its formation from methionine and acetate have been demonstrated (CHISHOLM and WETTER, 1966). The entire homologous series of glucosinolates as in [I; $R=CH_3S(CH_2)_n$, n=3 to 8] is considered to be derived by repeated chain extensions starting with methionine; each time the sequence is traversed a new higher amino acid homologue is formed. Studies on the biosynthesis of 4-methylsulfinyl-3-butenylglucosinolate [I; $R=CH_3S(O)\text{–}CH=CH\text{–}(CH_2)_2$] have shown it is formed from 2-amino-6-methylthiocaproic acid [XIV; $R=CH_3S(CH_2)_3$] a naturally occurring compound in *Matthiola* species, which is derived from methionine via homomethionine by two successive chain elongation steps (LÖFFELHARDT and KINDL, 1975). Nothing is known concerning the substrate specificity of the enzymes involved in chain extension nor about the number of times the sequence can be repeated and many of the amino acid equivalents of glucosinolates have yet to be isolated.

8.2 Intermediates Between Amino Acids and Glucosinolates

The conversion of amino acids to glucosinolates without the severance of the C=N bond has been confirmed by several workers, a result that implies that the intermediates between amino acids and glucosinolates are nitrogenous. A general biosynthetic pathway starting from α-amino acids (Fig. 4) may now be deduced mainly on the basis of in vivo studies on the formation of benzylglucosinolate in *T. majus* and supported by work on other glucosinolates.

The first nitrogenous intermediates found were aldoximes, independently reported by TAPPER and BUTLER (1967) and UNDERHILL (1967). It was demonstrated that isobutyraldoxime [XVII; $R=(CH_3)_2\text{–}CH$], phenylacetaldoxime (XVII; $R=C_6H_5\text{–}CH_2$) and phenylpropionaldoxime (XVII; $R=C_6H_5\text{–}CH_2\text{–}CH_2$) were each incorporated into isopropyl-, benzyl-, and phenylethyl-glucosinolates with an efficiency equal to or greater than that from their natural amino acid precursors. Further, phenylacetaldoxime was shown to be a naturally occurring aldoxime derived from phenylalanine in *T. majus*. The number of aldoximes known to be intermediates in the biosynthesis of glucosinolates has been appreciably extended through subsequent studies and several have been shown as natural plant components (see KJAER and OLESEN LARSEN, 1973; UNDERHILL and WETTER, 1973). Aldoximes are not only involved in the formation of glucosinolates, but are important interme-

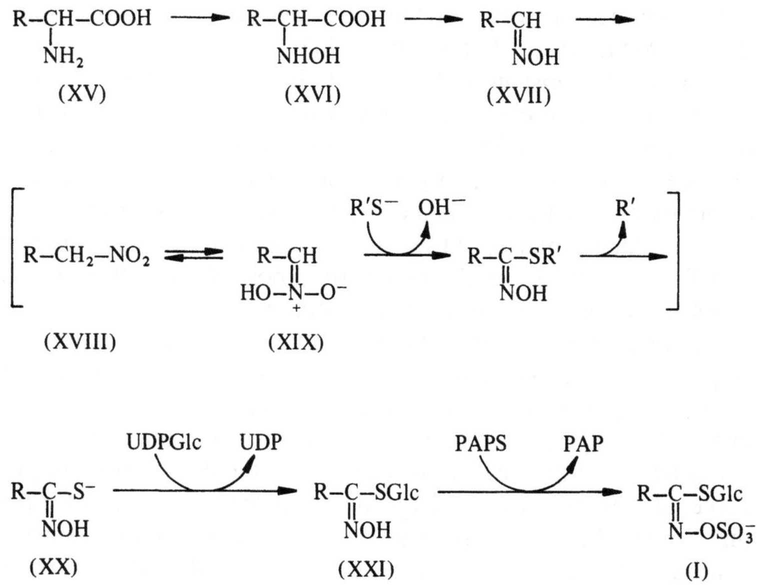

Fig. 4. Pathway for the biosynthesis of glucosinolates

diates in the biosynthesis of cyanogenic glucosides (cf. CONN, Chap. 9, this vol.) nitriles, amides, aldehydes, and alcohols and have been the subject of another review (MAHADEVAN, 1973).

In a search for intermediates between amino acids and aldoximes en route to glucosinolates a number of 2-oximino acids, which in theory could be progenitors of aldoximes by decarboxylation, have been fed to plants (UNDERHILL and CHISHOLM, 1964; TAPPER and BUTLER, 1967; KINDL and UNDERHILL, 1968) but all were found to be incorporated into glucosinolates to a substantially lower extent than were the amino acid precursors. Another possible intermediate of phenylacetaldoxime, N-hydroxyphenylalanine (XVI; $R=C_6H_5-CH_2$) has been fed to *T. majus* and found to be incorporated into benzylglucosinolate to a somewhat greater extent than phenylalanine (KINDL and UNDERHILL, 1968). Enzymic studies, conducted at the same time, demonstrated the conversion of N-hydroxyphenylalanine into phenylacetaldoxime, but attempts to demonstrate in vivo formation of the former from labeled phenylalanine in *T. majus* were unsuccessful. That N-hydroxyamino acids may be derived from amino acids has subsequently been demonstrated by MOELLER and CONN (1977), who recovered radioactive N-hydroxytyrosine from sorghum microsomal preparations incubated with tyrosine-^{14}C, carrier N-hydroxytyrosine and a NADPH regenerating system. The reaction was specific for L-tyrosine and required oxygen and NADPH.

An integral part of the biosynthetic sequence leading from aldoximes to glucosinolates is the incorporation of sulfur and the first well-defined sulfur intermediate to be isolated was phenylacetothiohydroximic acid (XX; $R=C_6H_5-CH_2$). Experiments with [^{14}C]- and [^{35}S]- labeled phenylacetothiohydroximate established that both carbon and sulfur were incorporated into the glucosinolate intact (UNDERHILL and WETTER, 1969). Evidence supporting the true intermediacy of this thiohydroximate came from the demonstration of its formation from phenylacetaldoxime in vivo (MATSUO and UNDERHILL, 1969).

One of the most important points in the biosynthesis of glucosinolates which still remains to be clarified concerns the nature of the reduced sulfur donor and the oxime-derived sulfur acceptor ultimately leading to the formation of the thiohydroximic acid. Several inorganic and organic sulfur compounds have been fed to plants and of these cysteine is the most efficient source of the reduced sulfur atom in glucosinolates, but no further details have been reported (see UNDERHILL and WETTER, 1973). With respect to a possible acceptor, ETTLINGER and KJAER (1968) proposed a mechanism of sulfur incorporation in which the *aci* tautomer (XIX) of a primary nitro compound (XVIII) functioned as the thiol acceptor. A plant feeding experiment in which [^{14}C] 1-nitro-2-phenylethane (XVIII; R= C_6H_5–CH_2) was fed to *T. majus* provided data consistent with this hypothesis in that the nitro precursor was specifically incorporated into benzylglucosinolate aglycone, albeit at a level only slightly greater than that from phenylalanine (MATSUO et al., 1972). Further, it was demonstrated by means of a trapping experiment that 1-nitro-2-phenylethane was formed in vivo from labeled phenylacetaldoxime. Although its formation from the aldoxime could be demonstrated it is not known whether, in fact, it functions in the plant as a sulfur acceptor. In this respect, investigations involving cell-free systems will be required before either cysteine or the *aci*-tautomer of primary nitro compounds can be considered intermediates in the general biosynthetic pathway.

The next, and last, sulfur intermediate in the sequence was recognized as a result of plant feeding experiments in which [^{14}C]-labeled desulfobenzylglucosinolate [S-(β-D-glucopyranosyl) phenylacetothiohydroximic acid] (XXI; R=C_6H_5–CH_2) was converted into benzylglucosinolate. Following this, an enzyme was isolated from four plant species which catalysed the formation of desulfobenzylglucosinolate by glucosyl transfer from UDP-glucose to phenylacetothiohydroximate (MATSUO and UNDERHILL, 1971). It was found that out of seven thiohydroximic acids studied as glucosyl acceptors only one failed to function as a substrate, namely acetothiohydroximic acid, the intermediate of methylglucosinolate whose presence is confined to the family Capparidaceae. The final step in the sequence, sulfation of the desulfoglucosinolate, is mediated by a sulfotransferase which has been detected in cell-free systems obtained from *T. majus* (UNDERHILL, unpublished). The enzyme catalysed the transfer of sulfate from 3′-phosphoadenosine 5′-phosphosulfate (PAPS) to a number of desulfoglucosinolates, showing little substrate specificity with respect to variations in the aglycone moiety.

Although a number of important details remain to be clarified, a single anabolic route appears to be operative leading from α-amino acid→N-hydroxyamino acid→aldoxime→[*aci* tautomer of primary nitro compound+cysteine?]→thiohydroximic acid→desulfoglucosinolate→glucosinolate. Thus, the distribution of glucosinolates in nature is often considered in terms of their amino acid precursors, whether they are those derived by chain extension or otherwise.

8.3 Glucosinolate Interconversions

It is known that a number of glucosinolates are derived from a single amino acid precursor by side chain modification of previously formed glucosinolates including oxidation, cleavage of CH_3S (or oxidized equivalent), acylation and N-sulfonation. Glucosinolates whose aglycones contain an aliphatic hydroxyl are derived from their desoxy analogues (UNDERHILL and KIRKLAND, 1972a, b; JOSEFSSON, 1973) and in this respect differ from those possessing phenolic hydroxylation which are formed from the corresponding phenolic amino acids (KINDL and SCHIEFER, 1971). Feeding experiments have demonstrated that 3-indolylmethylglucosinolate is the progenitor of both the 1-sulfo- and 1-methyl-derivatives in *Isatis tinctoria* (ELLIOTT and STOWE, 1971; MAHADEVAN and STOWE, 1972). The coexistence in plants of glucosinolates containing ω-methylthioalkyl side chains with sulfur at three levels of oxidation suggests their interconversion through oxidation and reduction; since the stereochemistry of the sulfoxide sulfur of all known ω-methylsulfinylalkyl-glucosinolates is (**R**), a common enzyme associated with their derivation seems apparent. Biosynthetic studies conducted on wallflower leaves (CHISHOLM, 1972) have demonstrated the reversible formation of 3-methylthiopropylglucosinolate and 3-methylsulfinylpropylglucosinolate. The formation of ω-alkenylglucosinolates from ω-methylthioalkylglucosinolates has been experimentally verified in horseradish (CHISHOLM and MATSUO, 1972). However, both 3-methylthiopropyl- and 3-methylsulfinylpropylglucosinolate when fed to horseradish leaves were converted with high efficiency into allylglucosinolate and it was not established with certainty which was the immediate precursor.

Of the more than seventy glucosinolates known, the formation of most can be envisaged according to the general pathway described above. However, nothing is known concerning the biosynthesis of the three oxo glucosinolates [I; R= $EtCO(CH_2)_4$-, $PrCO(CH_2)_3$-, $PrCO(CH_2)_4$-] and of 4,5,6,7-tetrahydroxydecylisothiocyanate, reported to be of glucosinolate origin (GAIND et al., 1975), all of which occur in the genus *Capparis* (Capparaceae); indeed their formation may be notable exceptions to the established biosynthetic pathway.

9 Elimination of Glucosinolates Through Plant Breeding

The discovery in 1967 that the *B. napus* cultivar Bronowski contained very low levels of glucosinolates, and hence possessed low goitrogenic properties, was a major step in improving the feeding value of rapeseed meal for livestock and poultry (see FINLAYSON et al., 1973). The high yielding, low erucic acid cultivars Tower (STEFANSSON and KONDRA, 1975) and Erglu (RÖBBELEN, 1976) both of which contain less than one-tenth the glucosinolate level of normal rapeseed cultivars have been developed from crosses with Bronowski.

Extensive feeding trials using commercially processed Tower and Bronowski meals indicate that the amount of these low glucosinolate meals which can be safely included in non-ruminant rations may be increased between 33% and 100% over that recommended previously for rapeseed meals (BELL, 1977). Glucosinolate levels lower than those presently attained may be required to eliminate completely the reduced feed intake and growth depression associated with high glucosinolate levels (JOSEFSSON and MUNCK, 1972). Recently, the first low glucosinolate, low erucic acid cultivar of *B. campestris*, Candle, has been released (R.K. DOWNEY, personal communication) and commercially processed meal from this cultivar is undergoing nutritional evaluation.

The genetic control of glucosinolate formation in rapeseed is still only partially understood even though most rapeseed breeders are engaged in the development of low glucosinolate cultivars. Not only is the inheritance pattern complex but the methodology employed by many plant breeders for glucosinolate determination has frequently lacked precision and has often failed to assay all of the glucosinolates present in the breeding material (KONDRA and STEFANSSON, 1970). For example, in selecting for low glucosinolate levels in both *B. napus* and *B. campestris*, the presence of the indole glucosinolates which occur in minor but significant amounts was not detected by the assay system used. As a result, the selection had little or no effect on the amount of these glucosinolates present, whereas the alkenyl glucosinolates were substantially reduced. Although it is known that the amount of glucosinolate present in the seed depends primarily on the genotype of the maternal parent, researchers examining data from segregating populations of Bronowski crosses have not been able to distinguish discrete genetic classes, other than the low glucosinolate parental class. KONDRA and STEFANSSON (1970) have proposed the involvement of at least 11 recessive alleles conditioning the low values of 3-butenyl-, 4-pentenyl-, and 2-hydroxy-3-butenyl glucosinolates. While the understanding of gene action in controlling glucosinolate levels in Bronowski crosses is sufficient to develop effective breeding strategies, it is, as yet, not complete enough to determine at what points in the biosynthetic pathway these genes are acting. JOSEFSSON (1973) has attempted to determine the location of the biochemical block(s) in the biosynthesis of glucosinolates in Bronowski by precursor feeding experiments. His data indicates the existence of a block prior to the formation of 5-methylthiopentanaldoxime and another at the stage of glucosinolate hydroxylation. Inheritance studies using the newly developed, low glucosinolate strains of diploid *B. campestris* may provide a clearer picture of gene action than is now possible in the amphidiploid *B. napus* species.

There are a number of biological effects associated with glucosinolates and their isothiocyanate hydrolytic products. Many of the isothiocyanates are highly toxic to several species of insects and are powerful antibacterial and antifungal agents. The glucosides themselves serve as gustatory stimulants for insect pests of cruciferous plants while their isothiocyanate fission products serve as attractants for many of these pests (see BECK and REESE, 1976). Little is known about the functions of glucosinolates in the plants which produce them and of the ultimate effects their elimination through plant breeding will have.

References

Afzalpurkar, A.B.: Tissue cultures of rapeseed. Glucosinolates and fatty acid composition. Fette, Seifen, Anstrichm. *76*, 550–552 (1974)

Al-Shehbaz, I.A.: The biosystematics of the genus *Thelypodium* (Cruciferae). Contribu. Gray Herb. Harv. Univ. *204*, 3–148 (1973)

Appelqvist, L.-A.: Glucosinolates. In: Rapeseed. Cultivation, composition, processing and utilization. Appelqvist, L.-A., Ohlson, R. (eds.), pp. 149–154. Amsterdam-New York. Elsevier Publ. Co. 1972

Beck, S.D., Reese, J.C.: Insect-plant interactions: nutrition and metabolism. Recent Adv. Phytochem. *10*, 41–92 (1976)

Bell, J.M.: Rapeseed oil, meal and by-product utilization. Publication No. 45, Vancouver, Rapeseed Association of Canada 1977

Bell, J.M., Belzile, R.J.: Goitrogenic properties. In: Rapeseed meal for livestock and poultry – a review. Bowland, J.P., Clandinin, D.R., Wetter, L.R. (eds.), pp. 45–60. Ottawa: Can. Dep. Agric. 1965

Benn, M.H.: Biosynthesis of mustard oils. Chem. Ind. (London) 1907 (1962)

Benn, M.: Glucosinolates. Pure and Appl. Chem. *49*, 197–210 (1977)

Björkman, R.: Plant myrosinases. In: The biology and chemistry of the cruciferae. Vaughan, J.G., MacLeod, A.J., Jones, B.M.G. (eds.), pp. 191–205. London: Academic Press 1976

Björkman, R., Janson, J-C.: Studies on myrosinases. 1. Purification and characterization of a myrosinase from white mustard seed (*Sinapis alba* L.). Biochim. Biophys. Acta *276*, 508–518 (1972)

Björkman, R., Lönnerdal, B.: Studies on myrosinases. III. Enzymatic properties of myrosinases from *Sinapis alba* and *Brassica napus* seeds. Biochim. Biophys. Acta *327*, 121–131 (1973)

Boutron, F., Fremy, E.: Untersuchung des schwarzen und weißen Senfsamens. Annalen *34*, 230–232 (1840)

Bussy, A.: Untersuchungen über die Bildung des ätherischen Senföls. Annalen *34*, 223–230 (1840)

Challenger, F.: Aspects of the organic chemistry of sulfur. London: Butterworths 1959

Chisholm, M.D.: Biosynthesis of 3-methylthiopropylglucosinolate and 3-methylsulfinylpropylglucosinolate in wallflower *Cheiranthus kewensis*. Phytochemistry *11*, 197–202 (1972)

Chisholm, M.D., Matsuo, M.: Biosynthesis of allylglucosinolate and 3-methylthiopropylglucosinolate in horseradish, *Armoracia lapathifolia*. Phytochemistry *11*, 203–207 (1972)

Chisholm, M.D., Wetter, L.R.: Biosynthesis of mustard oil glucosides. IV. The administration of methionine-C^{14} and related compounds to horseradish. Can. J. Biochem. *42*, 1033–1040 (1964)

Chisholm, M.D., Wetter, L.R.: Biosynthesis of mustard oil glucosides. VII. Formation of sinigrin in horseradish from homomethionine-2-^{14}C and homoserine-2-^{14}C. Can. J. Biochem. *44*, 1625–1632 (1966)

Chisholm, M.D., Wetter, L.R.: The biosynthesis of some isothiocyanates and oxazolidinethiones in rape (*Brassica campestris* L.). Plant Physiol. *42*, 1726–1730 (1967)

Cole, R.: 1-Cyanoepithioalkanes: major products of alkenylglucosinolate hydrolysis in certain cruciferae. Phytochemistry *14*, 2293–2294 (1975)

Cronquist, A.: The evolution and classification of flowering plants. London: Nelson 1968

Daxenbichler, M.E., Van Etten, C.H., Spencer, G.F.: Glucosinolates and derived products in cruciferous vegetables. Identification of organic nitriles from cabbage. J. Agric. Food Chem. *25*, 121–124 (1977)

Dörnemann, D., Löffelhardt, W., Kindl, H.: Chain elongation of aromatic amino acids: the role of 2-benzylmalic acid in the biosynthesis of a C_6C_4 amino acid and a C_6C_3 mustard oil glucoside. Can. J. Biochem. *52*, 916–921 (1974)

Elliott, M.C., Stowe, B.B.: Distribution and variation of indole glucosinolates in woad (*Isatis tinctoria* L.). Plant Physiol. *48*, 498–503 (1971)

Etten, C.H. van: Goitrogens. In: Toxic constituents of plant foodstuffs. Leiner, I.E. (ed.), pp. 103–142. London: Academic Press 1969

Etten, C.H. van, Daxenbichler, M.E., Wolff, I.A.: Natural glucosinolates (thioglucosides) in foods and feeds. J. Agric. Food Chem. *17*, 483–491 (1969)

Etten, C.H. van, McGrew, C.E., Daxenbichler, M.E.: Glucosinolate determination in cruciferous seeds and meals by means of enzymatically released glucose. J. Agric. Food Chem. *22*, 483–487 (1974)

Etten, C.H. van, Daxenbichler, M.E., Williams, P.H., Kwolek, W.F.: Glucosinolates and derived products in cruciferous vegetables. Analysis of the edible part from twenty-two varieties of cabbage. J. Agric. Food Chem. *24*, 452–455 (1976)

Ettlinger, M.G., Dateo, G.P., Jr.: Studies of mustard oil glucosides, 12, Final report. Contract DA19-129-QM-1059 Natick, Mass.: U.S. Army Natick Laboratories 1961

Ettlinger, M.G., Kjaer, A.: Sulfur compounds in plants. In: Recent advances in phytochemistry. Mabry, T.J., Alston, R.E., Runeckles, V.C. (eds.), Vol. I, pp. 60–144. New York: Appleton-Century-Crofts 1968

Ettlinger, M.G., Lundeen, A.J.: The structure of sinigrin and sinalbin; an enzymatic rearrangement. J. Am. Chem. Soc. *78*, 4172–4173 (1956)

Ettlinger, M.G., Lundeen, A.J.: First synthesis of a mustard oil glucoside: the enzymatic Lossen rearrangement. J. Am. Chem. Soc. *79*, 1764–1765 (1957)

Ettlinger, M.G., Thompson, C.P.: Studies of mustard oil glucosides (II), AD No. 290747, Office of Technical Services. Washington D.C.: U.S. Dep. of Commerce 1962

Ettlinger, M.G., Dateo, G.P., Harrison, B.W., Mabry, T.J., Thompson, C.P.: Vitamin C as a coenzyme: the hydrolysis of mustard oil glucosides. Proc. Natl. Acad. Sci. U.S.A. *47*, 1875–1880 (1961)

Finlayson, A.J., Krzymanski, J., Downey, R.K.: Comparison of chemical and agronomic characteristics of two *Brassica napus* L. cultivars, Bronowski and Target. J. Am. Oil Chem. Soc. *50*, 407–410 (1973)

Freeman, G.G., Mossadeghi, N.: Studies on sulphur nutrition, flavour and allyl isothiocyanate formation in *Brassica juncea* (L.) Coss and Czern. (brown mustard). J. Sci. Food Agric. *23*, 1335–1345 (1972)

Gadamer, J.: Über das Sinigrin. Ber. Dtsch. Chem. Ges. *30*, 2322–2326 (1897)

Gaind, K.N., Gandhi, K.S., Juneja, T.R., Kjaer, A., Nielsen, B.J.: 4,5,6,7-Tetrahydroxydecyl isothiocyanate derived from a glucosinolate in *Capparis grandis*. Phytochemistry *14*, 1415–1418 (1975)

Gmelin, R., Virtanen, A.I.: A new type of enzymatic cleavage of mustard oil glucosides. Formation of allylthiocyanate in *Thlaspe arvense* L. and benzylthiocyanate in *Lepidium ruderale* L. and *Lepidium sativum* L. Acta Chem. Scand. *13*, 1474–1475 (1959)

Hofmann, A.W.: Über die dem Senföl entsprechenden Isomeren der Schwefelcyanwasserstoffäther. Ber. Dtsch. Chem. Ges. *1*, 169–182 (1868)

Iversen, T.-H : Cytochemical localization of myrosinase (β-thioglucosidase) in root tips of *Sinapis alba*. Protoplasma *71*, 451–466 (1970)

Johnston, T.D., Jones, D.I.H.: Variations in the thiocyanate content of kale varieties. J. Sci. Food Agric. *17*, 70–71 (1966)

Josefsson, E.: Studies of the biochemical background to differences in glucosinolate content in *Brassica napus* L. III. Further studies to localize metabolic blocks. Physiol. Plant. *29*, 28–32 (1973)

Josefsson, E., Jönsson, R.: Studies of variation in glucosinolate content of seed of *Cruciferae* plants, especially in material with high erucic acid content. Z. Pflanzenzücht. *62*, 272–283 (1969)

Josefsson, E., Munck, L.: Influence of genetically determined differences in glucosinolate content in marrow stem kale. Z. Pflanzenzücht. *67*, 353–359 (1972)

Kindl, H., Schiefer, S.: Aldoximes as intermediates in the biosynthesis of tyrosol and tyrosol derivatives. Phytochemistry *10*, 1795–1802 (1971)

Kindl, H., Underhill, E.W.: Biosynthesis of mustard oil glucosides: N-hydroxyphenylalanine, a precursor of glucotropaeolin and a substrate for the enzymatic and nonenzymatic formation of phenylacetaldehyde oxime. Phytochemistry *7*, 745–756 (1968)

Kirk, J.T.O., MacDonald, C.G.: 1-Cyano-3,4-epithiobutane: a major product of glucosinolate hydrolysis in seeds of certain varieties of *Brassica campestris*. Phytochemistry *13*, 2611–2615 (1974)

Kirkland, D.F., Matsuo, M., Underhill, E.W.: Detection of glucosinolates and myrosinase in plant tissue cultures. Lloydia *34*, 195–198 (1971)

Kjaer, A.: Naturally derived *iso* thiocyanates (mustard oils) and their parent glucosides. Fortschr. Chem. Org. Naturst. *18*, 122–176 (1960)

Kjaer, A.: Naturally occurring isothiocyanates and their parent glycosides. In: Organic sulphur compounds. Kharasch, N. (ed.), pp. 409–420. Oxford: Pergamon Press 1961

Kjaer, A.: The natural distribution of glucosinolates: a uniform group of sulfur-containing glucosides. In: Chemistry in botanical classification. Bendz, G., Santesson, J. (eds.), pp. 229–234. London: Academic Press 1974

Kjaer, A., Olesen Larsen, P.: Nonprotein amino acids. cyanogenic glycosides and glucosinolates. In: Biosynthesis. Geissman, T.A. (ed.), Vol. 2, pp. 71–105. Specialist periodical reports. London: The Chemical Society 1973

Kjaer, A., Olesen Larsen, P.: *Ibid.* Vol. 5, pp. 179–203, 1976

Kjaer, A., Thomsen, H.: Isothiocyanate-producing glucosides in species of Capparidaceae. Phytochemistry *2*, 29–32 (1963)

Kondra, Z.P., Stefansson, B.R.: Inheritance of the major glucosinolates of rapeseed (*Brassica napus*) meal. Can. J. Plant Sci. *50*, 643–647 (1970)

Krikorian, A.D., Stewart, F.C.: Biochemical differentiation: The biosynthetic potentialities of growing and quiescent tissue. In: Plant physiology. Steward, F.C. (ed.), Vol. VB, p. 278. London, New York: Academic Press 1969

Kutáček, M., Procházka, Ž., Vereš, K.: Biogenesis of glucobrassicin, the *in vitro* precursor of ascorbigen. Nature (London) *194*, 393–394 (1962)

Lee, C.J., Serif, G.S.: Precursor role of [^{14}C, ^{15}N]-2-amino-6-(methylthio)-caproic acid in progoitrin biosynthesis. Biochemistry *9*, 2068–2071 (1970)

Lein, K-A.: Quantitative Bestimmungsmethoden für Samenglucosinolate in *Brassica*-Arten und ihre Anwendung in der Züchtung von glucosinolatarmem Raps. Z. Pflanzenzücht. *63*, 137–154 (1970)

Löffelhardt, W., Kindl, H.: Biosynthesis of 4-methylsulfinyl-3-butenylglucosinolate (glucoraphenin) in *Matthiola* species. Evidence for homomethionine and 2-amino-6-methylthiocaproic acid as intermediates. Z. Naturforsch. C, Biosci. *30C*, 233–239 (1975)

Lönnerdal, B., Janson, J-C.: Myrosinases II. Purification and characterization of a myrosinase from rapeseed (*Brassica napus* L.). Biochim. Biophys. Acta *315*, 421–429 (1973)

MacLeod, A.J.: Volatile flavour compounds of the Cruciferae. In: The biology and chemistry of the cruciferae. Vaughan, J.G., MacLeod, A.J., Jones, B.M.G. (eds.), pp. 307–330. London: Academic Press 1976

Mahadevan, S.: Role of oximes in nitrogen metabolism in plants. Annu. Rev. Plant Physiol. *24*, 69–88 (1973)

Mahadevan, S., Stowe, B.B.: An intermediate in the synthesis of glucobrassicins from 3-indoleacetaldoxime by woad leaves. Plant Physiol. *50*, 43–50 (1972)

Matsuo, M., Underhill, E.W.: A UDP glucose: thiohydroximate glucosyltransferase from *Tropaeolum majus* L. Biochem. Biophys. Res. Commun. *36*, 18–23 (1969)

Matsuo, M., Underhill, E.W.: Purification and properties of a UDP glucose: thiohydroximate glucosyltransferase from higher plants. Phytochemistry *10*, 2279–2286 (1971)

Matsuo, M., Yamazaki, M.: Biosynthesis of sinigrin VI. Incorporation from homomethionine (2-^{14}C, ^{15}N) and some labelled compounds into sinigrin. Chem. Pharm. Bull. (Tokyo) *16*, 1034–1039 (1968)

Matsuo, M., Kirkland, D.F., Underhill, E.W.: 1-Nitro-2-phenylethane, a possible intermediate in the biosynthesis of benzylglucosinolate. Phytochemistry *11*, 697–701 (1972)

McGhee, J.E., Kirk, L.D., Mustakas, G.C.: Methods for determining thioglucosides in *Crambe absyssinica*. J. Am. Oil Chem. Soc. *42*, 889–891 (1965)

McGibbon, D.B., Allison, R.M.: Glucosinolase system in the aphid *Brevicoryne brassicae*. N.Z. J. Sci. *11*, 440–446 (1968)

Moeller, B.L., Conn, E.E.: Biological formation of N-hydroxyamino acids. Proc. West. Pharmacol. Soc. *20*, 103–107 (1977)

Nagashima, Z., Uchiyama, M.: Possibility that myrosinase is a single enzyme and mechanism of the decomposition of mustard oil glucoside by myrosinase. Bull. Agric. Chem. Soc. Jpn. *23*, 555–556 (1959)

Oginsky, E.L., Stein, A.E., Greer, M.A.: Myrosinase activity in bacteria as demonstrated by the conversion of progoitrin to goitrin. Proc. Soc. Exp. Biol. Med. *119*, 360–364 (1965)

Ohtsuru, M., Hata, T.: Molecular properties of multiple forms of plant myrosinase. Agric. Biol. Chem. *36*, 2495–2503 (1972)

Ohtsuru, M., Tsuruo, I., Hata, T.: Studies on fungus myrosinase. Part I. Production, purification and some characteristics. Agric. Biol. Chem. *33*, 1309–1314 (1969)

Olsson, K., Theander, O., Aman, P.: Determination of total glucosinolate content in rapeseed and turnip rapeseed meals by gas liquid chromatography. Swed. J. Agric. Res. *6*, 225–229 (1976)

Pihakaski, K., Iversen, T-H.: Myrosinase in Brassicaceae I. Localization of myrosinase in cell fractions of roots of *Sinapis alba* L. J. Exp. Bot. *27*, 242–258 (1976)

Raghavan, B., Shankaranarayana, M.L., Nagalakshmi, S., Natarajan, C.P.: Volumetric determination of *p*-hydroxybenzyl isothiocyanate in sinalbin (*p*-hydroxybenzylglucosinolate) and in white mustard seed (*Sinapis alba* L.). J. Sci. Food Agric. *22*, 523–525 (1971)

Reese, E.T., Clapp, R.C., Mandels, M.: A thioglucosidase in fungi. Arch. Biochem. Biophys. *75*, 228–242 (1958)

Röbbelen, G.: Züchtung und Erzeugung von Qualitätsraps in Europa. Fette, Seifen, Anstrichm. *78*, 10–17 (1976)
Robiquet, P.J., Boutron, F.: Sur la semence de moutarde. J. Pharm. Chim. *17*, 279–282 (1831)
Saarivirta, M.: The formation of benzylcyanide, benzylthiocyanate, benzylisothiocyanate and benzylamine from benzylglucosinolate in *Lepidium*. Planta Med. *24*, 112–119 (1973)
Schlüter, M., Gmelin, R.: Abnormal enzymic cleavage of 4-methylthiobutylglucosinolate in fresh plants, *Eruca sativa*. Phytochemistry *11*, 3427–3431 (1972)
Schraudolf, H.: Untersuchungen zur Verbreitung von Indolglucosinolaten in höheren Pflanzen: *Moringa oleifera* Lam. Experientia *23*, 103 (1967)
Schweidler, J.H.: Die systematische Bedeutung der Eiweiß- oder Myrosinzellen der Cruciferen nebst Beiträgen zu ihrer anatomisch-physiologischen Kenntnis. Ber. Dtsch. Bot. Ges. *23*, 274–280 (1905)
Serif, G.S., Schmotzer, L.A.: Biosynthesis of the aglycones of plant thioglucosides I. Precursor studies of the aglycone of progoitrin. Phytochemistry *7*, 1151–1157 (1968)
Srivastava, V.K., Hill, D.C.: Glucosinolate hydrolytic products given by *Sinapis alba* and *Brassica napus* thioglucosidases. Phytochemistry *13*, 1043–1046 (1974)
Stefansson, B.R., Kondra, Z.P.: Tower summer rape. Can. J. Plant Sci. *55*, 343–344 (1975)
Tang, C-S.: Benzyl iothiocyanate of papaya fruit. Phytochemistry *10*, 117–121 (1971)
Tapper, B.A., Butler, G.W.: Conversion of oximes to mustard oil glucosides (Glucosinolates). Arch. Biochem. Biophys. *120*, 719–721 (1967)
Tapper, B.A., Reay, P.F.: Cyanogenic glycosides and glucosinolates. In: Chemistry and Biochemistry of Herbage Vol. 1. Butler, G.W., Bailey, R.W. (eds.), pp. 447–478. London-New York: Academic Press, 1973
Thies, W.: Quantitative gas-liquid chromatography of glucosinolates on a microliter scale. Fette, Seifen, Anstrichm. *78*, 231–234 (1976)
Tookey, H.L.: Crambe thioglucoside glucohydrolase (EC 3.2.3.1): Separation of a protein required for epithiobutane formation. Can. J. Biochem. *51*, 1654–1660 (1973)
Underhill, E.W.: Biosynthesis of mustard oil glucosides V. Formation of gluconasturtiin from L-γ-phenylbutyrine-C^{14}-N^{15} in watercress. Can. J. Biochem. *43*, 179–187 (1965)
Underhill, E.W.: Biosynthesis of mustard oil glucosides: Conversion of phyenylacetaldehyde oxime and 3-phenylpropionaldehyde oxime to glucotropaeolin and gluconasturtiin. Eur. J. Biochem. *2*, 61–63 (1967)
Underhill, E.W.: Biosynthesis of mustard oil glucosides: 3-benzylmalic acid, a precursor of 2-amino-4-phenylbutyric acid and gluconasturtiin. Can. J. Biochem. *46*, 401–405 (1968)
Underhill, E.W., Chisholm, M.D.: Biosynthesis of mustard oil glucosides III. Formation of glucotropaeolin from L-phenylalanine-C^{14}-N^{15}. Biochem. Biophys. Res. Commun. *14*, 425–430 (1964)
Underhill, E.W., Kirkland, D.F.: Gas chromatography of trimethylsilyl derivatives of glucosinolates. J. Chromatogr. *57*, 47–54 (1971)
Underhill, E.W., Kirkland, D.F.: A new thioglucoside, 2-methylpropylglucosinolate. Phytochemistry *11*, 2085–2088 (1972a)
Underhill, E.W., Kirkland, D.F.: L-2-Amino-4-phenylbutyric acid and 2-phenylethylglucosinolate, precursors of 2-hydroxy-2-phenylethylglucosinolate. Phytochemistry *11*, 1973–1979 (1972b)
Underhill, E.W., Wetter, L.R.: Biosynthesis of mustard oil glucosides: sodium phenylacetothiohydroximate and desulfobenzylglucosinolate, precursors of benzylglucosinolate in *Tropaeolum majus* L. Plant Physiol. *44*, 584–590 (1969)
Underhill, E.W., Wetter, L.R.: Biosynthesis of glucosinolates. Biochem. Soc. Symp. *38*, 303–326 (1973)
Underhill, E.W., Chisholm, M.D., Wetter, L.R.: Biosynthesis of mustard oil glucosides. I. Administration of C^{14}-labelled compounds to horseradish, nasturtium and watercress. Can. J. Biochem. Physiol. *40*, 1505–1514 (1962)
Vaughan, J.G., Gordon, E.I.: Comparative serological studies of myrosinase from *Sinapis alba* and *Brassica juncea* seeds. Phytochemistry *8*, 883–887 (1969)
Vaughan, J.G., Hemingway, J.S.: The utilization of mustards. Econ. Bot. *13*, 196–204 (1959)

Vaughan, J.G., Gordon, E., Robinson, D.: The identification of myrosinase after the electrophoresis of *Brassica* and *Sinapis* seed proteins. Phytochemistry *7*, 1345–1348 (1968)
Vose, J.R.: The fractionation of two glucosinolases from *Sinapis alba* seed by isoelectric focussing. Phytochemistry *11*, 1649–1653 (1972)
Waser, J., Watson, W.H.: Crystal structure of sinigrin. Nature (London) *198*, 1297–1298 (1963)
Wetter, L.R., Youngs, C.G.: A thiourea UV assay for total glycosinolate content in rapeseed meals. J. Am. Oil Chem. Soc. *53*, 162–164 (1976)

11 Betalains

T.J. Mabry

1 Introduction

In 1961 it was found that the red-violet betacyanins, characteristic constituents of most members of the order Centrospermae, represented a new class of pigments structurally unrelated to the widespread anthocyanins (Mabry et al., 1962, 1967; Wyler et al., 1963). Later in 1966 when it became evident that the betaxanthins, the yellow vacuolar pigments also found in members of the Centrospermae, were structurally and biogenetically closely related to the betacyanins (Piattelli et al., 1964c; Wyler et al., 1965), the term "betalain" was introduced to describe this class of pigments to which both the betacyanins and betaxanthins belong (Mabry and Dreiding, 1968).

The present review deals with the structures of betacyanins and betaxanthins and betalain biogenesis, synthesis, function, and phylogenetic significance, all with respect to the Centrospermae. We also comment on the occurrence of betalains in mushrooms, notably in the bright red cap of the fly agaric, *Amanita muscaria*, (Döpp and Musso, 1973, 1974; von Ardenne et al., 1974; Musso, 1978). Betalains and related topics have been previously reviewed by Dreiding, 1961; Mabry, 1964, 1966, 1970, 1973, 1976, 1977; Mabry and Dreiding, 1968; Mabry et al., 1972; Mabry and Behnke, 1976a; Reznik, 1975; Piattelli, 1976.

2 Naturally Occurring Betalains in the Centrospermae

2.1 Betacyanins

Except for the betacyanins in a few cacti whose flowers exhibit bluish colors (see Barthlott and Rauh, 1976), the betacyanins, of which over 50 are known, tend to impart a red-violet color to the plants which contain them. Until the groups of Dreiding (Wyler and Dreiding, 1957) and Schmidt (Schmidt and Schönleben, 1957) prepared crystalline betanin (the betacyanin from *Beta vulgaris*), little evidence had been accumulated regarding the structural nature of the betacyanins. Indeed, these pigments were still widely referred to as "nitrogenous anthocyanins", a name which acknowledged their nitrogen content but incorrectly presumed a direct structural relationship to the anthocyanins.

In the years immediately following the isolation of betanin, extensive degradative investigations were carried out with the aglycone betanidin which could be crystallized as a salt from the hydrolysate of betanin.

2.1.1 Aglycones

Today we know that, with the exception of the single report of 2-descarboxybetanidin (from *Carpobrotus acinaciformus*; PIATTELLI and IMPELLIZZERI, 1970), all betacyanins are based on just two aglycones, betanidin and isobetanidin, which differ only in stereochemistry at C_{15}. Strong acid hydrolysis of betanin gives, along with glucose, a mixture of both aglycones (WYLER and DREIDING, 1959), while only betanidin is obtained under milder conditions (SCHMIDT et al., 1960) or from β-glucosidase hydrolysis (PIATTELLI et al., 1964a). Furthermore, treatment of either betanidin or isobetanidin under acid hydrolysis conditions yields an equilibrium mixture of 30% betanidin and 70% isobetanidin (WYLER and DREIDING, 1959; DUNKELBLUM et al., 1972).

Alkaline hydrolysis of betanidin yields three fragments which together account for all the carbon atoms in betanidin: 4-methylpyridine-2,6-dicarboxylic acid (I), 5,6-dihydroxy-2,3-dihydroindole-2-carboxylic acid (II) and formic acid (Fig. 1) (WYLER and DREIDING, 1959; 1962).

MABRY et al. (1962, 1967) deduced the nature of the betacyanins after establishing the structures of a series of derivatives of neobetanidin and neobetanin; these 14,15-dehydrobetanidin and 14,15-dehydrobetanin derivatives were all readily crystallizable, in part because they had lost the zwitterion character common to naturally occurring betalains. A typical neobetanidin can be formed by first esterifying betanidin with methanolic hydrochloric acid and then acetylating the trimethyl ester hydrochloride thus obtained; under the basic conditions of the latter reaction a 14,15-dehydrogenation also occurs to give 5,6-di-O-acetylneobetanidin trimethyl ester (III). When dissolved in an acidic medium, this yellow-colored neobetanidin (III) undergoes a bathochromic shift of 100 nm giving a bright pink solution (MABRY et al., 1962, 1967). The unique neobetanidin structures demonstrated for the first time that the betacyanins were unrelated to anthocyanins and indeed were members of a new class of plant pigments. The misleading expression "nitrogenous anthocyanins" was finally discarded. With the structures of several neobetanidin derivatives in hand, formulation of betanidin and isobetanidin as (IV) and (V), respectively, quickly followed.

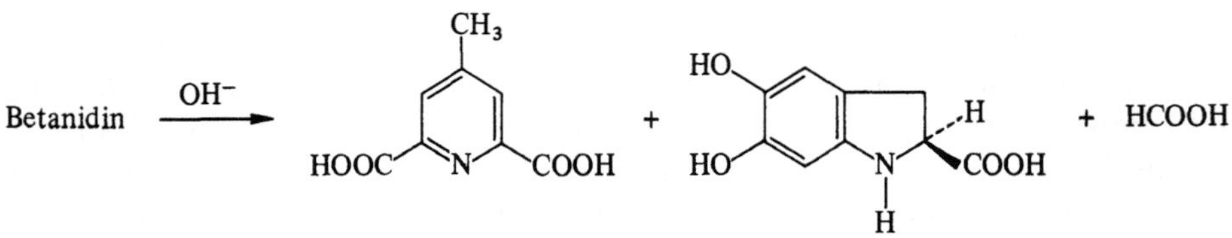

(I) (II)

Fig. 1. Alkaline degradation of betanidin yields 4-methylpyridine-2,6-dicarboxylic acid (I), 5,6-dihydroxy-2,3-dihydroindole-2-carboxylic acid (II) and formic acid (WYLER and DREIDING, 1962)

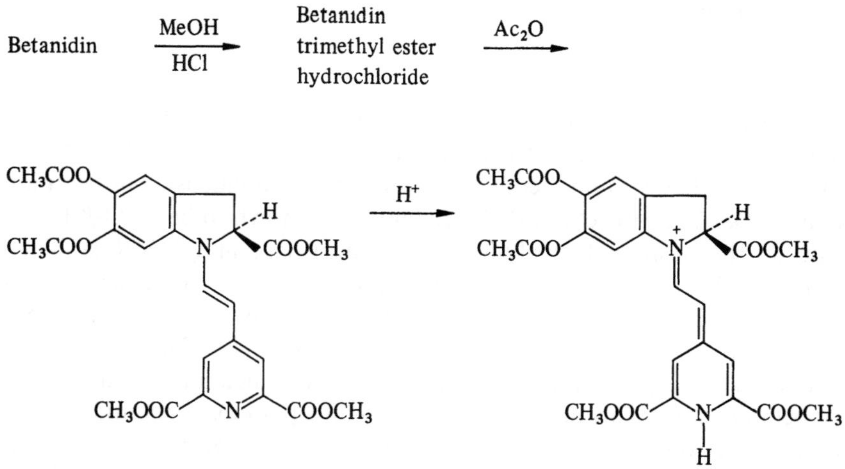

(III; λ_{max} 383 nm) (λ_{max} 483 nm)

Fig. 2. Betanidin can be converted into the neobetanidin derivative III, which upon protonation undergoes a bathochromic shift typical of cyanine dyes (MABRY et al., 1962, 1967)

The final structure for betanidin (IV) emerged from the analysis of 5,6-di-O-acetylbetanidin (WYLER et al., 1963). The absolute configuration at C-2 was shown to be [S] by the isolation of 5,6-dihydroxy-2,3-dihydro-2S-indole-2-carboxylic acid (II) from the alkaline degradation of betanidin (WYLER and DREIDING, 1962; WYLER et al., 1963). An [S] absolute configuration for the only other chiral center in betanidin, C-15, was established by a correlation with indicaxanthin which was known at the time to possess the [S] configuration at the corresponding carbon atom (C-11) (WILCOX et al., 1965b). That betanidin and isobetanidin differ only in stereochemistry at C-15 was confirmed by converting both compounds to the same product, namely, 5,6-di-O-methyl-2S-neobetanidin trimethyl ester (WYLER et al., 1963). Thus, the zwitterion structures for both betanidin (IV) and isobetanidin (V) contain dihydroindole and dihydropyridine rings and a polymethylene cyanine chromophore system.

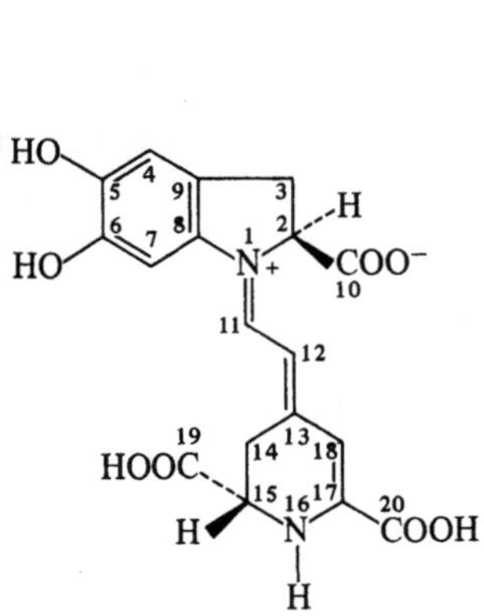

Betanidin
(IV)

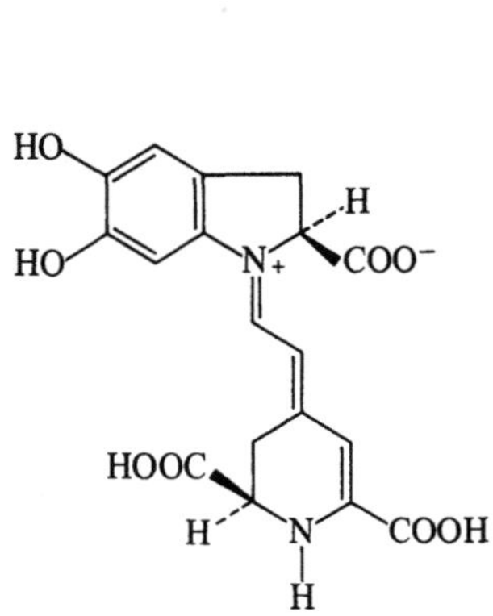

Isobetanidin
(V)

2.1.2 Glycosides

2.1.2.1 Nonacylated Glycosides

The structure of betanin, the red beet pigment, was readily elucidated once the aglycone betanidin had been determined, especially since it had been known for many years that betanin merely represents the monoglucoside of betanidin (PUCHER et al., 1938; SCHMIDT et al., 1960). Betanin was shown to be an O-β-D-glucopyranoside on the basis of hydrolysis with β-glucosidase and NMR studies (PIATTELLI et al., 1964a; WILCOX et al., 1965b). Using different procedures PIATTELLI et al. (1964a) and WILCOX et al. (1965b) obtained 5-hydroxy-6-methoxyindole-2-carboxylic acid or its methyl ester (VIII) from betanin (Fig. 3), thus establishing that betanin is a 5-O-glucoside of betanidin. Therefore, betanin (IX) and its C-15 epimer isobetanin can be formulated as 5-O-β-D-glucopyranosyl-betanidin and -isobetanidin, respectively. The corresponding 6-O-glucosyl derivatives, gomphrenin-I (X) and -II, were obtained from *Gomphrena globosa* (MINALE et al., 1967).

Three bioside pairs are known; they are amaranthin, bougainvillein-r-I, and betanidin 6-sophoroside and their iso forms. Amaranthin (XI) and isoamaranthin were obtained from the leaves of *Amaranthus tricolor* (PIATTELLI et al., 1964b; PIATTELLI and MINALE, 1966). Hydrolysis of amaranthin and isoamaranthin with β-glucoronidase followed by treatment with β-glucosidase transformed the glyco-

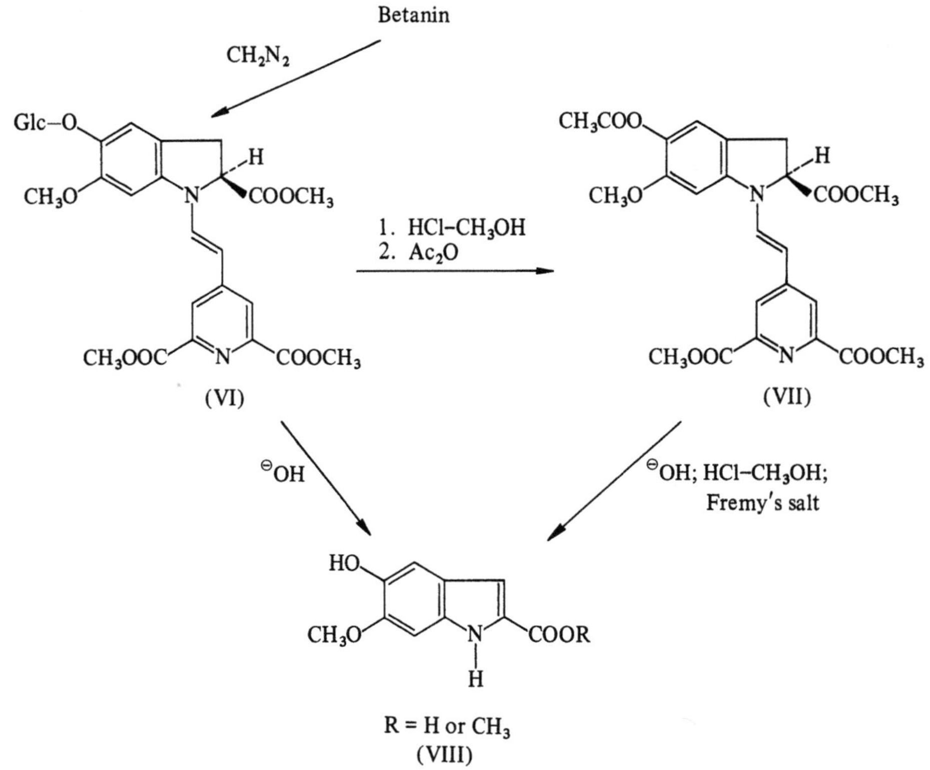

(VI) (VII)

R = H or CH_3
(VIII)

Fig. 3. The location of the glucosyl moiety in betanin was established by conversion of betanin to (VIII) from (VI) directly (PIATTELLI et al., 1964a) and via (VII) (WILCOX et al., 1965b)

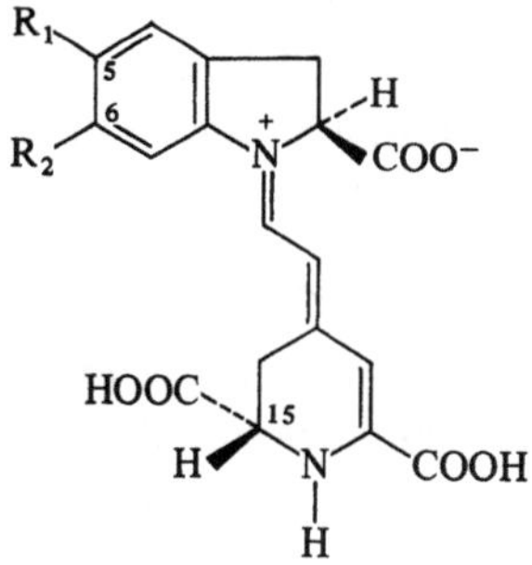

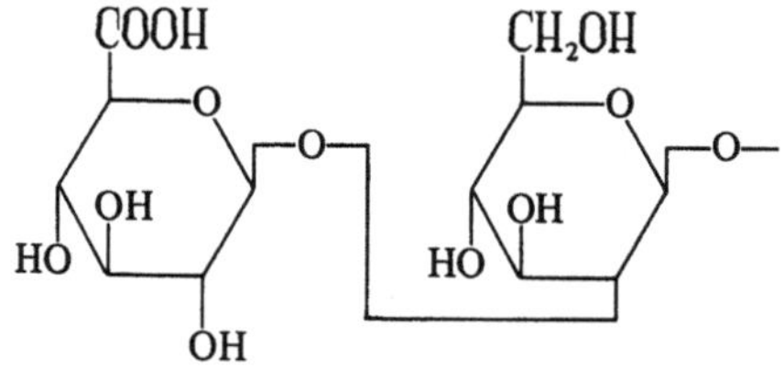

Amaranthin (XI),
$R_2 = OH$,
R_1 = glucuronic acid–glucosyl

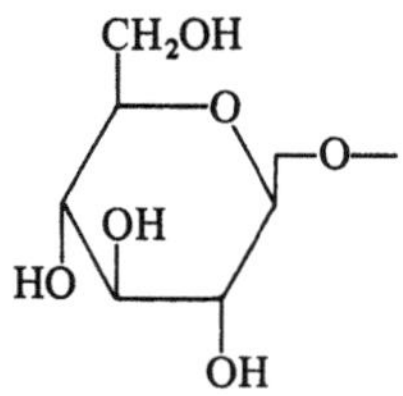

Betanin (IX),
$R_2 = OH$,
R_1 = glucosyl

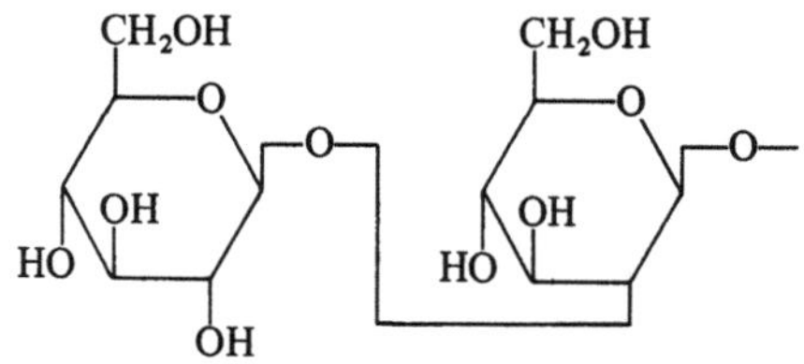

Bougainvillein–r–I (XII),
$R_2 = OH$,
R_1 = sophorosyl

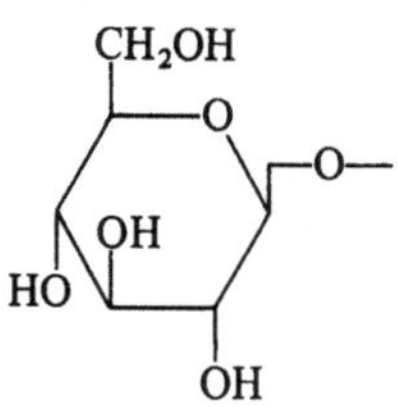

Gomphrenin I (X),
$R_1 = OH$,
R_2 = glucosyl

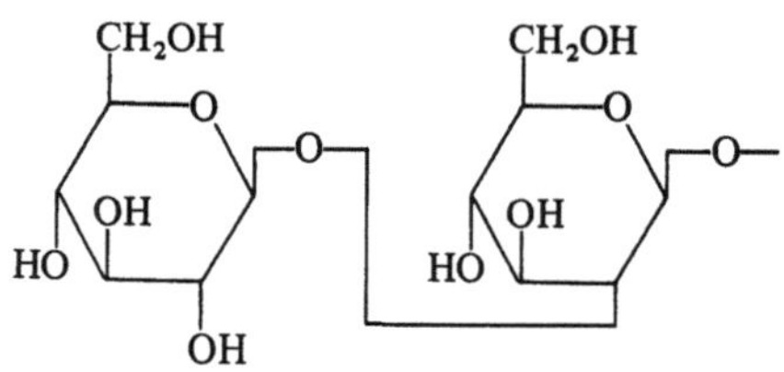

(XIII),
$R_1 = OH$,
R_2 = sophorosyl

sides into betanidin and isobetanidin, respectively. These two betacyanins could be formulated as O-(β-D-glucopyranosyluronic acid)-5-O-β-D-glucopyranosides of betanidin [see(XI)] and isobetanidin, respectively.

A horticultural variety of *Bougainvillea* yielded bougainvillein-r-I (XII) and its C-15 epimer, isobougainvillein-r-I; that is, the 5-O-β-sophorosides of betanidin and isobetanidin (PIATTELLI and IMPERATO, 1970a); *Bougainvillea glabra* var. *sanderiana* afforded the corresponding 6-sophorosides [see (XIII)] but in an acylated form (PIATTELLI and IMPERATO, 1970b). The latter species also contains incompletely characterized 6-O-rhamnosyl-sophorosides of betanidin and isobetanidin (PIATTELLI and IMPERATO, 1970b).

2.1.2.2 Acylated Glycosides

In all of the forty known acylated derivatives of betacyanin glycosides, the acyl group is always linked to the sugar moiety and only acyl derivatives of the glycosides discussed above have so far been found. The acylated monosides are prebetanin, a minor pigment from *Beta vulgaris*, the sulphuric acid half-ester of betanin (WYLER et al., 1967); phyllocactin from *Phyllocactus hybridus*, the malonic acid half-ester of betanin (MINALE et al., 1966); iresinin, the major pigment from leaves of *Iresine herbstii*, the 3-hydroxy-3-methylglutaric acid half-ester of amaranthin (MINALE et al., 1966); and gomphrenin-III and -V from *Gomphrena globosa*, the *p*-coumaroyl and feruloyl esters of gomphrenin-I (MINALE et al., 1967). The C-15 epimers of each of these pigments are also known. Other acylated betacyanin glycosides, many of which are not fully characterized, contain caffeic, citric, *p*-coumaric and sinapic acid moieties (PIATTELLI, 1976).

2.1.3 Summary

In summary there are three betacyanin agylcones (betanidin, isobetanidin, and the once reported 2-descarboxybetanidin), as well as 9 nonacylated and 40 acylated glycosides of betanidin and isobetanidin; thus there are over fifty betacyanins known at the present time.

2.2 Betaxanthins

Although it was long suspected that a direct structural relationship existed between the two classes of Centrospermae pigments, the red-violet betacyanins ($\lambda max \simeq 540$ nm) and the yellow betaxanthins ($\lambda max \simeq 480$ nm), confirmation came only with the analysis of indicaxanthin (XIV), the betaxanthin from *Opuntia ficus-indica* (PIATTELLI et al., 1964c). Acid and alkaline degradations of indicaxanthin afforded L-proline and 4-methylpyridine-2,6-dicarboxylic acid, the latter being a product typical of the dihydropyridine ring in betacyanins (Fig. 4). Moreover,

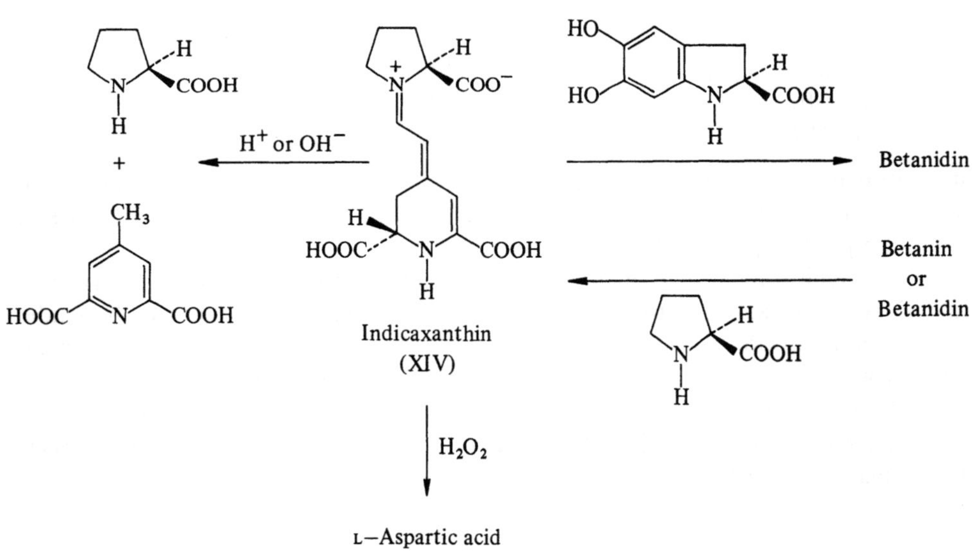

Fig. 4. Degradative reactions of indicaxanthin and the interconversion of betacyanins and indicaxanthin via the imino acid exchange reaction

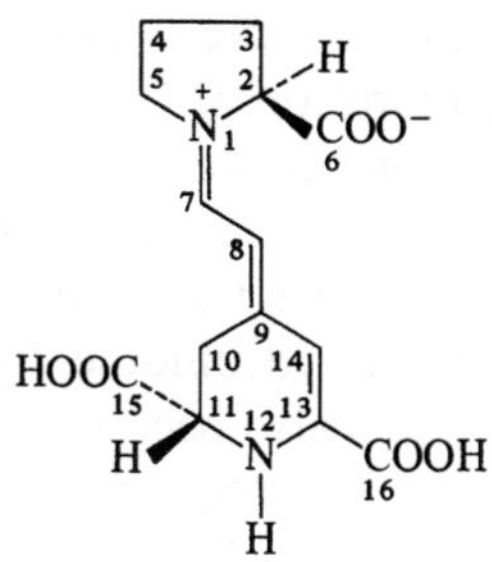

Indicaxanthin
(XIV)

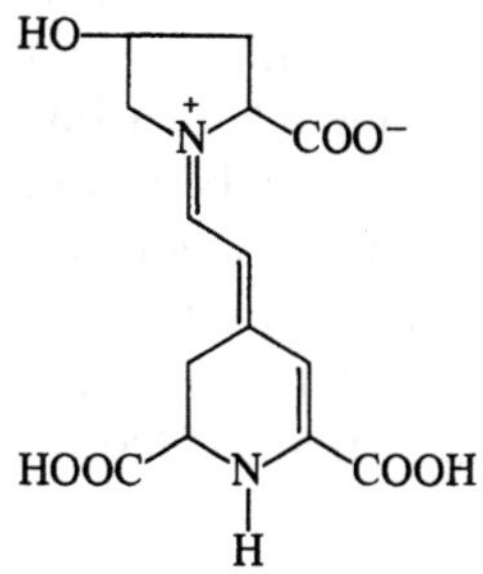

Portulaxanthin
(XV)

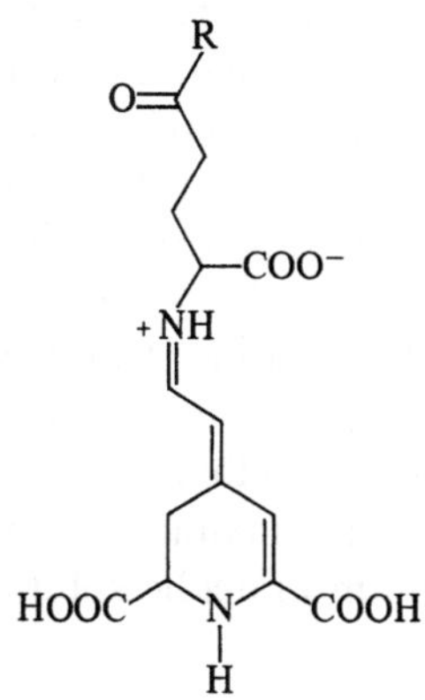

Vulgaxanthin–I
(XVI, R = NH_2);
Vulgaxanthin–II
(XVII, R = OH)

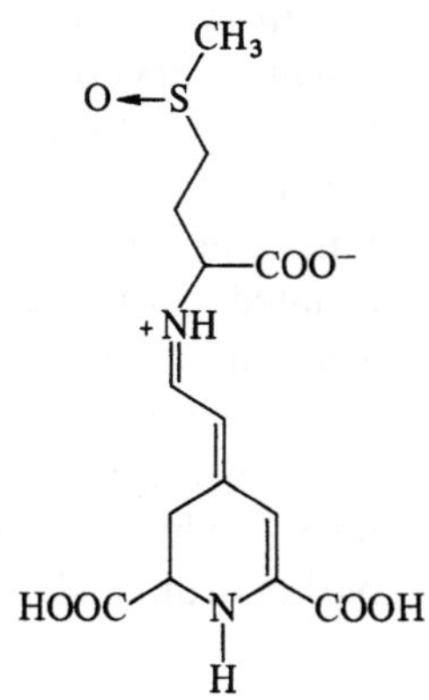

Miraxanthin–I
(XVIII)

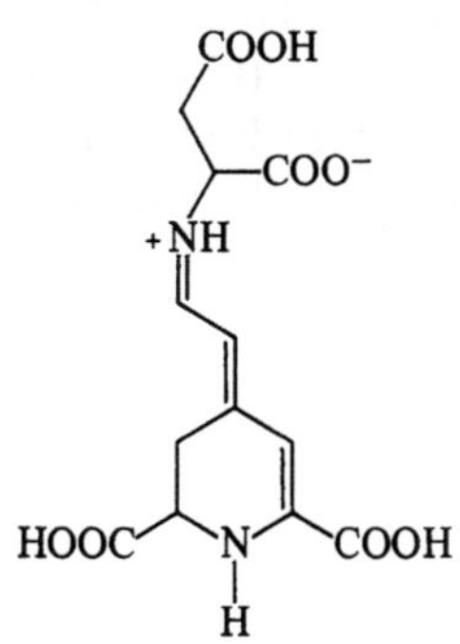

Miraxanthin–II
(XIX)

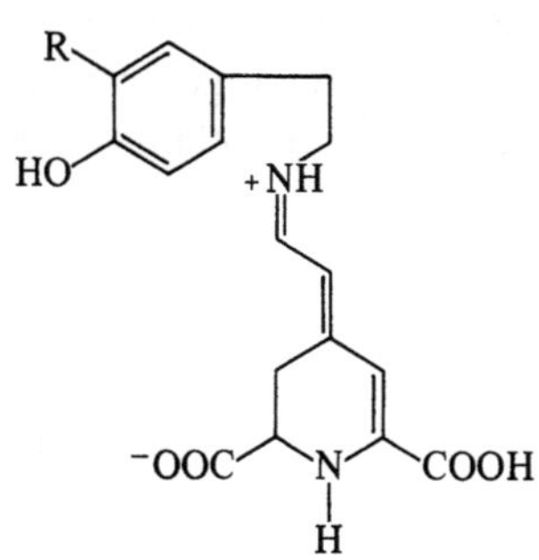

Miraxanthin–III
(XX, R = H);
Miraxanthin–IV
(XXI, R = OH)

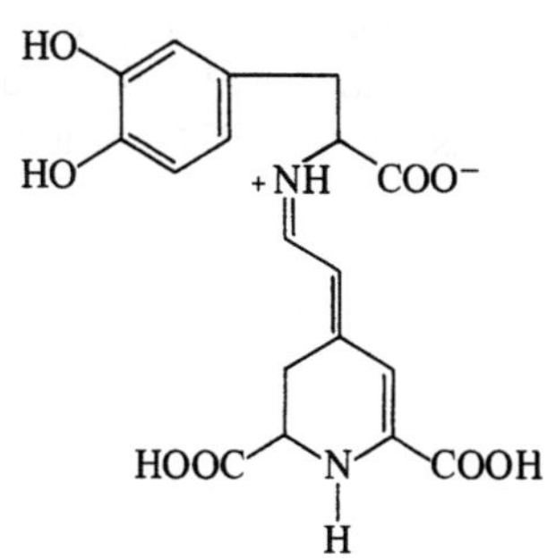

Dopaxanthin
(XXII)

the detection of L-aspartic acid after the oxidation of indicaxanthin by hydrogen peroxide indicated an [S] configuration at C-11.

A partial synthesis of indicaxanthin was effected by a novel imino acid exchange reaction involving treatment of betanin with L-proline in the presence of base. Furthermore, indicaxanthin could be transformed into betanidin by allowing it to react with 5,6-dihydroxy-2,3-dihydroindole-2-carboxylic acid (Fig. 4) (WYLER et al., 1965). This same method has also been used for the preparation of dopaxanthin (IMPELLIZZERI et al., 1973; PIATTELLI et al., 1965a) obtained indicaxanthin and other betaxanthins (vulgaxanthin-I and several miraxanthins) by dissolving betanin in water and then saturating the solution with sulfur dioxide and concentrating the solution in vacuo; addition of the appropriate amino acid or amine yielded the expected betaxanthin.

Several betaxanthins are now known: portulaxanthin (XV) from *Portulaca grandiflora* (PIATTELLI et al., 1965b); vulgaxanthin-I (XVI) and -II (XVII) from *Beta vulgaris* (PIATTELLI et al., 1965b); four miraxanthins ((XVIII)–(XXI)) from *Mirabilis jalapa* (PIATTELLI et al., 1965a), and dopaxanthin (XXII) from *Glottiphyllum longum* (IMPELLIZZERI et al., 1973). From the acid hydrolysate of all these betaxanthins the amine or amino acid attached to the dihydropyridine moiety can be isolated; moreover, alkaline fusion always yields 4-methylpyridine-2,6-dicarboxylic acid, indicating that a dihydropyridine group is indeed present in betaxanthins and moreover, that it corresponds to the same group found in betacyanins. Just as for betacyanins, diazomethane methylation converts betaxanthins into neo-pigments, which exhibit absorption maxima between 340 and 360 nm and show a bathochromic shift of 80–100 nm in acidic solutions (PIATTELLI et al., 1965a).

2.3 Betalamic Acid

Betalamic acid (XXIV), which in many respects may be the most important betalain, does not formally belong to either the betacyanin or betaxanthin type (XXIII), but represents an intermediate between L-DOPA and all other betalains. Betalamic acid occurs as a yellow pigment in many centrospermous plants (CHANG et al., 1974) and is also formed in the base-exchange procedures shown in Figure 4 (KIMLER et al., 1971; SCIUTO et al., 1972). Betalamic acid reacts readily with amino

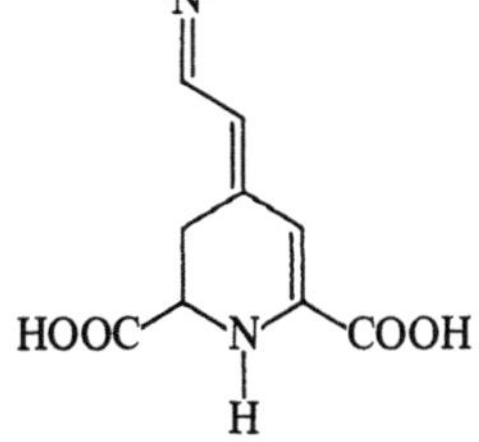

General formula for betacyanins and betaxanthins (XXIII)

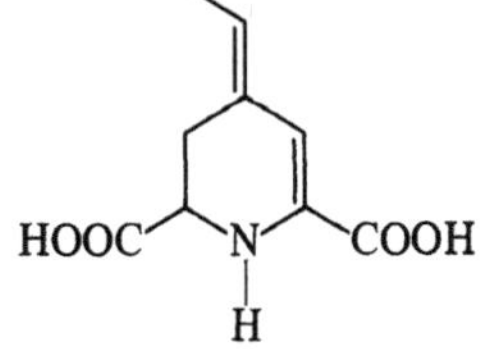

Betalamic acid (XXIV)

acids or amines to give not only naturally occurring betalains but also new pigments not known from nature. Both the dimethyl ester semicarbarbazone (HERMANN and DREIDING, 1975) and dimethyl ester (BUCHI et al., 1977) of betalamic acid have been synthesized.

3 Detection and Isolation of Betalains

Our current procedure (MABRY et al., 1975) for distinguishing betalains from anthocyanins (especially for small quantities) is as follows: The red pigmented part of a plant is ground with acidic MeOH (MeOH: 2 N HCl:100:1.5). The mixture is centrifuged and the supernatant evaporated to dryness in vacuo at 40°. About 1–2 ml of 0.01 N HCl are added to dissolve the pigment. The solution is absorbed onto the top of a small cation exchange column (0.5 × 3 cm Dowex 50 W-XI, 50–100 mesh, packed in H_2O). The column is first washed with 10 ml MeOH followed by 0.01 N HCl. The betalain pigments are then eluted with H_2O; anthocyanins are only eluted with acidic methanol (MeOH: conc HCl:100:4). The pigmented solution from the column, after concentration, is spotted on Whatman No. 1 paper with standard samples of betanin, malvidin 3,5-diglucoside, and rutin. Electrophoresis is conducted at pH 4.5 using a pyridine-formate buffer (0.5 M) and 300 V for 40 min. Under these conditions betanin migrates 3 cm towards the anode, and most anthocyanins migrate about 0.2 cm toward the cathode, while the nonionic rutin remains at the origin.

For both small and large-scale isolations of betalains PIATTELLI (1976) employs a preliminary separation of the crude aqueous plant extract by nonionic absorption on strongly acid resins and subsequent chromatography on a polyamide column, using increasing concentrations of methanol in aqueous citric acid as the developing solvent. The resolved betalains are separated from the citric acid by resin treatment.

MUSSO and co-workers (see MUSSO, 1978) have found that Sephadex chromatography gives good separation of betalains from *Amanita muscaria*. For example, 20 kg of pigmented material from 30 kg of caps were extracted with cold methanol and, after evaporation of the extract, 600 g of a dark brown powder were obtained. The powder was first chromatographed on CM-Sephadex C-25 (with $CO_2^- Na^+$ groups) to remove impurities. Next, chromatography over Sephadex G-10 in water yielded about 250 g of pigment material. The final separation was achieved on a DEAE-Sephadex A-25 column (which contains diethylamino groups in the chloride form). By gradient chromatography with increasing sodium chloride solutions (0.3 to 1.0 M), nine distinct pigments were eluted.

ADAMS and VON ELBE (1977) reported that chromatography on gel filtration supports (Sephadex or Bio-Gel polyacrylamide) is a rapid and efficient method for separating betalains from beet juice. When raw and fermented beet juice (*Beta vulgaris*) was directly applied to a column of polyacrylamide gel (Bio-Gel P-6), numerous pigment bands were observed. They suggest that the major mechanism for retention of the pigments on gel supports is adsorption rather than gel-filtration. Resolution of betanin and betanidin was greater on columns packed with Sephadex

G-25 compared to Bio-Gel P-6; however, loading capacity was greater on the latter. These workers also report excellent separations with Bio-Gel P-6 using a phosphate buffer at pH 3.0.

Detailed electrophoretic procedures for betalain separations are reported by WYLER (1969a, b) and KREMER (1975); HPLC procedures are given by VINCENT and SCHOLZ (1978).

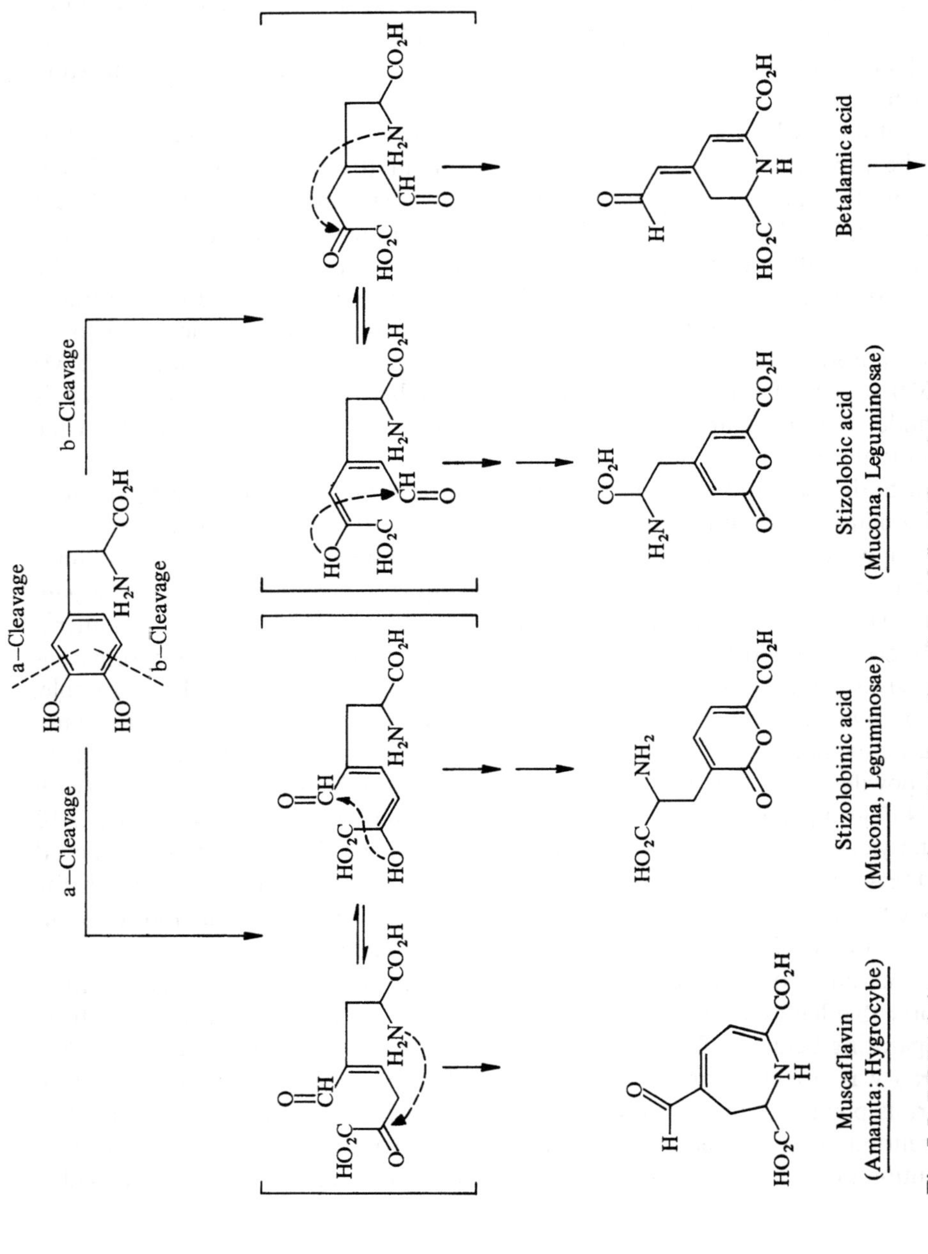

Fig. 5. L-DOPA is known to undergo 2,3-(a-pathway) or 4,5-(b-pathway) extra diol cleavages in different organisms. Only in the Centrospermae and in mushrooms is betalamic acid formed

4 Biogenesis of Betalains

4.1 Extra-Diol Cleavage of L-DOPA to Betalamic Acid

Before any experimental evidence was available, it was recognized that betanidin could theoretically be formed in vivo from two molecules of L-DOPA, one of which had to undergo ring cleavage and recyclization to form the dihydropyridine ring present in all betalains (WYLER et al., 1965; MABRY, 1964). This cleavage is now known to occur between the 4 and 5 positions in L-DOPA (FISCHER and DREIDING, 1972; IMPELLIZZERI and PIATTELLI, 1972; CHANG et al., 1974). Moreover it is likely that this same 4,5-extra-diol cleavage of L-DOPA leads to stizolobic acid (Fig. 5) both in mushrooms and members of the Leguminosae (ELLIS, 1976; SAITO et al., 1975, 1976; MUSSO, private communication).

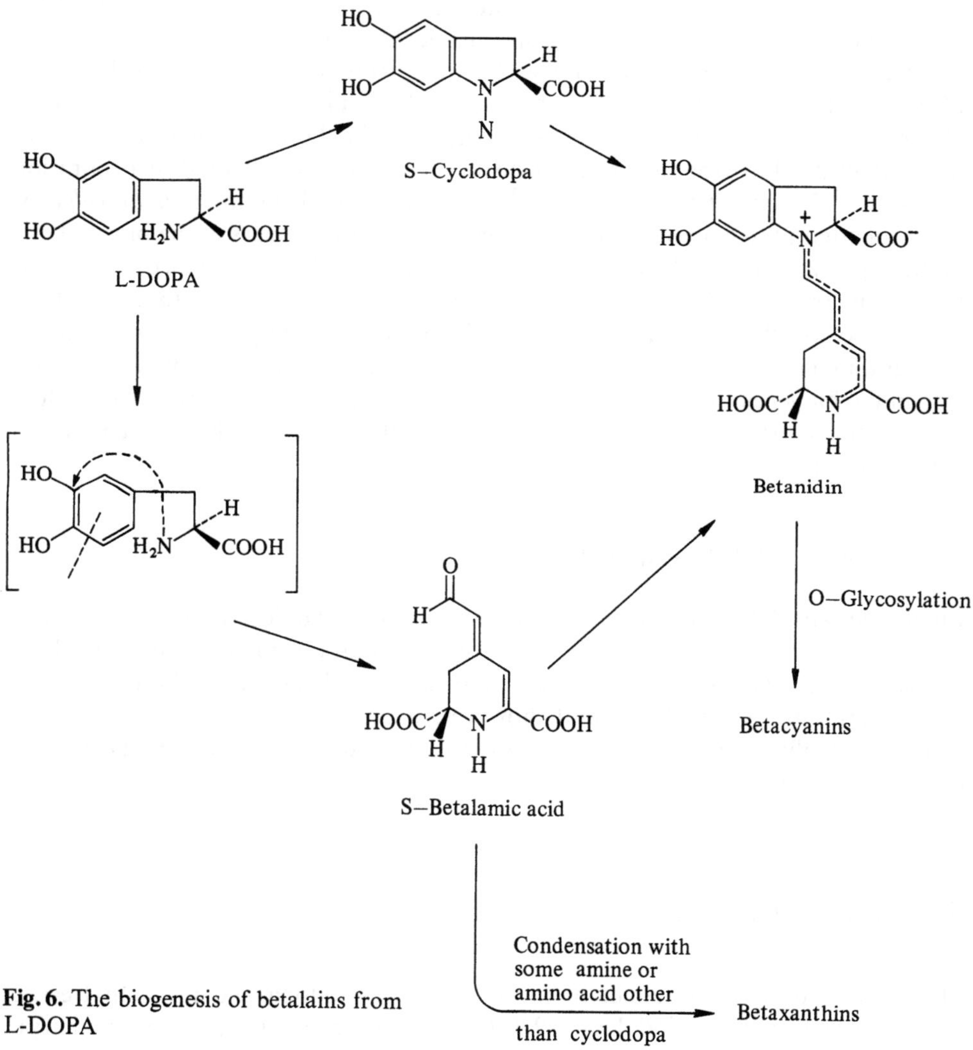

Fig. 6. The biogenesis of betalains from L-DOPA

In addition to betalamic acid, other compounds known or suspected to be derived either by 2,3- or 4,5-extra-diol cleavage of DOPA include the mushroom pigment muscaflavin (ARDENNE, VON et al., 1974) and stizolobinic acid (SAITO et al., 1975, 1976; ELLIS, 1976); these compounds are known from *Amanita* (CHILTON et al., 1974) and *Mucuna* (ELLIS, 1976 and references therein) (Fig. 5).

The conversion of L-DOPA to betalamic acid and its subsequent condensation with amines and amino acids to yield betalains is still known only for the Centrospermae and mushrooms (Fig. 6); it would be of interest to know if these two evolutionarily remote groups of organisms utilize the same enzymes for these transformations.

4.2 Incorporation of Labeled Precursors

In addition to the investigations which established the cleavage process whereby L-DOPA is converted to betalamic acid, feeding experiments had earlier shown that labeled L-DOPA is incorporated into betanin in *Beta vulgaris* (HÖRHAMMER et al., 1964), into amaranthin in *Amaranthus* seedlings (GARAY and TOWERS, 1966), and into indicaxanthin and betanin in *Opuntia ficus-indica* (MINALE et al., 1965). Moreover, MILLER et al. (1968) found that in the fruits of *Opuntia* species most of the labeled L-DOPA that appears in betanin is in the dihydropyridine moiety rather than the dihydroindole ring. LIEBISCH et al. (1969) found good incorporation of [1-^{14}C, ^{15}N]-tyrosine and [2-^{14}C, ^{15}N]-tyrosine into betanin in *Beta vulgaris*.

In an attempt to determine at what stage glycosylation occurs in the biosynthesis of betacyanins, labeled betanidin was found to be efficiently incorporated into betanin in fruits of *Opuntia dillenii* (SCIUTO et al., 1972). Later (SCIUTO et al., 1974) reported that cyclodopa 5-O-β-D-glucoside is an important precursor to amaranthin in *Celosia plumosa*, suggesting that glycosylation may be an early step in the biogenesis of betalain glycosides; however, this result may reflect factors such as transport of the precursor rather than the stage at which glycosylation actually occurs in the plant.

4.3 Control of Betalain Biogenesis

Light is an absolute requirement in some species for betalain synthesis and always mediates the amount of pigment synthesized; in addition, phytochrome and hormones such as kinetin have also been implicated in betalain synthesis in many centrospermous species. In general, however, details about the enzymes and cofactors controlling betalain biogenesis are not yet understood (see PIATTELLI, 1976 for a recent review).

5 Synthesis of Betalains

The total synthesis of a neobetanidin, as well as cyclodopa and its methyl ester, were achieved in the period 1964–1970 (see MABRY et al., 1972 for a summary).

However, it was not until 1975 when the syntheses of betanidin and indicaxanthin dimethyl ether were reported (HERMANN and DREIDING, 1975) that a naturally occurring betalain was prepared in the laboratory. These workers converted chelidamic acid into the dimethyl ester semicarbazone of betalamic acid, and this key intermediate was then subjected to the imino acid exchange reaction (Fig. 7).

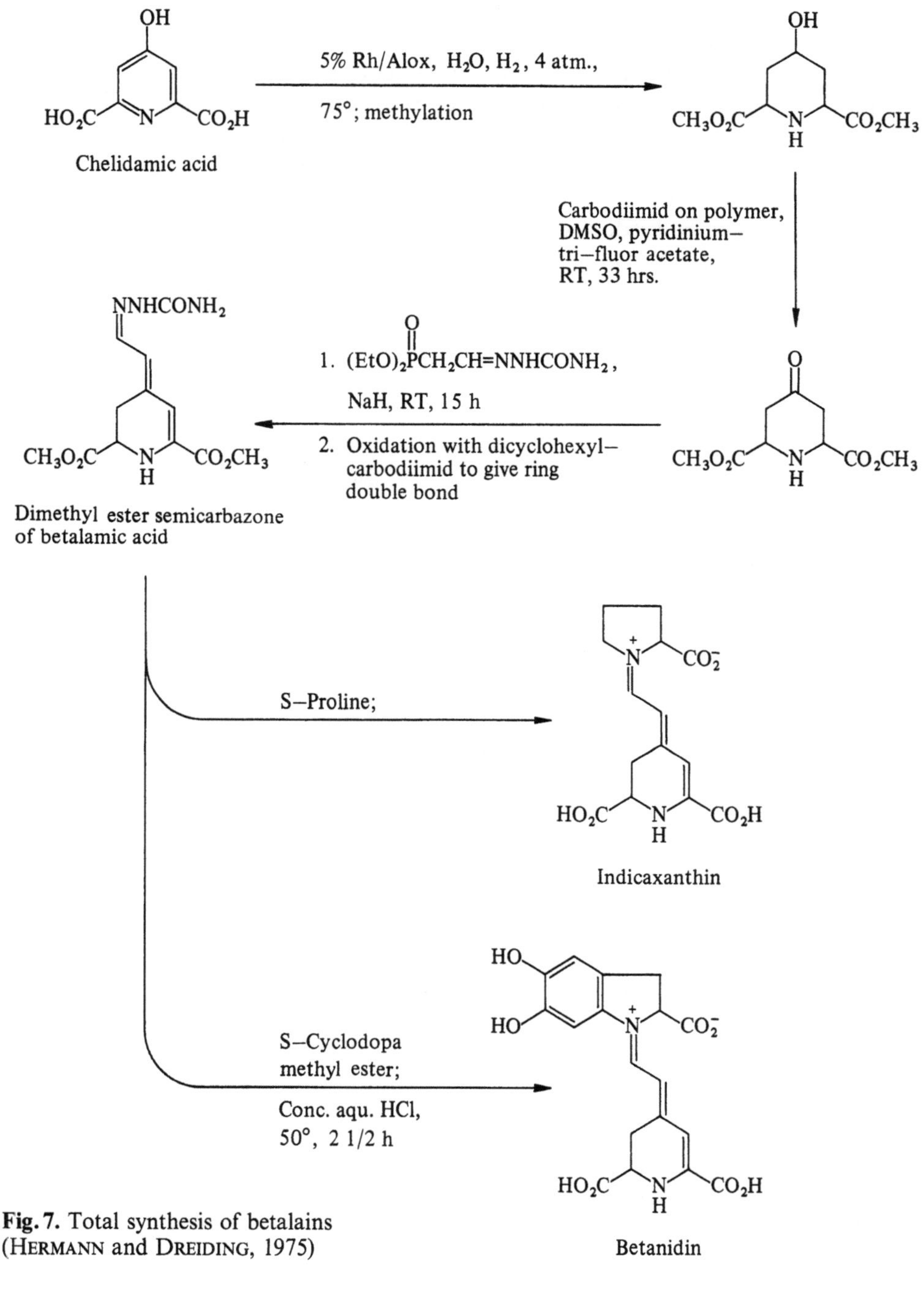

Fig. 7. Total synthesis of betalains (HERMANN and DREIDING, 1975)

6 Function of Betalains

Like the more widespread anthocyanins, betalains occur in cell vacuoles of flowers, fruits, and leaves and often accumulate in the stalk. Of course, in the beet they are found in high concentration in underground parts, but this occurrence represents selection by man in the middle ages. Comparison of wavelengths of the absorption maxima of betalains and anthocyanins indicates the importance of the chromophoric group, presumably as a pollinator attractant (Fig. 8); nevertheless, the high concentration of betalains in the stems and leaves of the plants which produce them suggests that these substances are polyfunctional.

Injury of centrospermous plants can cause accumulation of betalains in tissues which are not normally pigmented, indicating that these compounds may serve as a defense against viral infections (SOSNOVA, 1970). Betanin-containing beet seedlings have a higher resistance to the plant pathogenic fungus *Phythium debaryum* than do nonpigmented seedlings; moreover when this fungus, which is responsible for the plant disease known as "damping off", was grown in liquid cultures containing 50 ppm of betanin, a 50% reduction in fungal growth was observed (KIMLER, 1975).

The function of the betalains in mushrooms is not known.

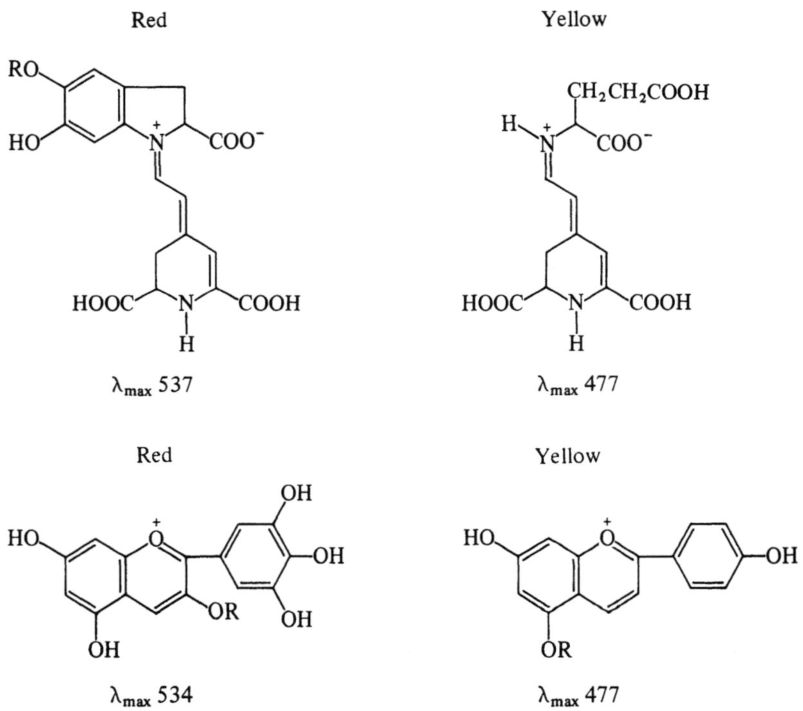

Fig. 8. The visible absorption maxima of typical red and yellow betalains (*top row*) which are found in phyletically related centrospermous families are similar to those for some red and yellow anthocyanins (*bottom row*), pigments which occur in most flowering families other than those which produce betalains

7 Phylogenetic Significance of Betalains with Respect to the Centrospermae

With the exception of a few mushrooms, betalains have been detected to date only in the red-violet, orange, and yellow-pigmented species which belong to a few closely related centrospermous families; moreover, on the basis of several lines of evidence, these betalain families are also phyletically close to two anthocyanin families, the Caryophyllaceae and Molluginaceae. Remarkably, the two classes of biogenetically and structurally distinct pigments, betalains and anthocyanins, are mutually exclusive, never having been detected together in the same species or even in different species of the same family (see KIMLER et al., 1970).

The long-held view that the anthocyanin families Caryophyllaceae and Molluginaceae are related to those which produce betalains is based upon morphological and anatomical characteristics including distinctive embryological features. Indeed, the classical data are sufficiently unique to suggest that the betalain families and the Caryophyllaceae and Molluginaceae have a common ancestor. Nevertheless, it is the molecular data which have contributed most to the circumscription of the Centrospermae as shown in Table 1.

Table 1. Order Centrospermae[a] or Caryophyllales (modified here from MABRY, 1977)

Suborder Chenopodiineae[a] (betalain families with P-III subtype sieve-element plastids)		
Aizoaceae	Kranz[b]	CAM
including Tetragoniaceae		
and possibly Gisekiaceae	Kranz	
Amaranthaceae	Kranz	
Basellaceae		
Cactaceae		CAM
Chenopodiaceae	Kranz	
including Dysphaniaceae		
Didiereaceae		
Halophytaceae		
Nyctaginaceae	Kranz	
Phytolaccaceae		
including Achatocarpaceae,		
Agdestidaceae,		
Petiveriaceae,		
Stegnospermaceae		
Portulacaceae	Kranz	
including Hectorellaceae		
Suborder Caryophyllineae (anthocyanin families with P-III subtype sieve-element plastids)		
Caryophyllaceae	Kranz	
Molluginaceae	Kranz	

[a] Whether or not such betalain taxa as *Petivera* and *Agdestis* (BEHNKE et al., 1974), *Halophytum* (HUNZIKER et al., 1974), *Gisekia* (MABRY et al., 1976), *Dysphania* (MABRY and BEHNKE, 1976b) and *Hectorella* (MABRY et al., 1978) should each be treated as core families is not resolved; however, they clearly belong to the suborder Chenopodiineae either as families or as genera in one of nine core families.

[b] Kranz = C_4 photosynthesis; CAM = crassulacean acid metabolism pathway. All families designated Kranz or CAM have many C_3 photosynthetic taxa; in contrast all that have no designation have only the C_3 pathway. Except for *Hectorella* (presumably C_3), all families have been examined.

7.1 Sieve-Element Plastids

The most striking piece of evidence supporting a close relationship between the two anthocyanin families Caryophyllaceae and Molluginaceae and the betalain families is the ultrastructural data for sieve-element plastids. The discovery that the sieve-element plastids in all the betalain families, as well as those in the Caryophyllaceae and Molluginaceae, contain ring-like inclusions composed of proteinaceous filaments of a type (see Fig. 9) not found elsewhere in the angiosperms, established that these eleven families are clearly related and represent the core centrospermous families (for recent review see BEHNKE, 1976). Thus today a number of families sometimes treated as being centrospermous but which contain neither the unique proteins in their sieve-element plastids (called P-III subtype) nor betalains are now excluded from the Centrospermae proper (see MABRY and BEHNKE,

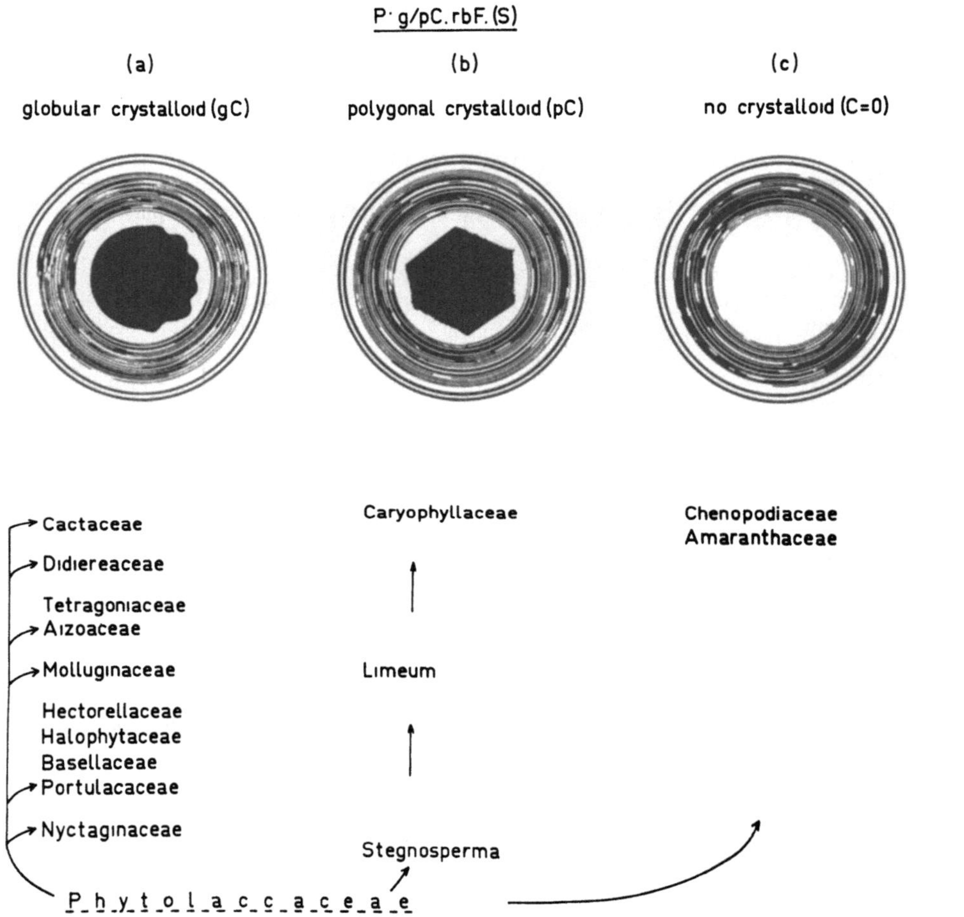

Fig. 9. The P-III subtype sieve-element plastids which are characteristic for the Centrospermae always contain a ring-shaped bundle of protein filaments with either globular (a) or polygonal (b) protein crystalloid or no crystalloid at all (*c*). We thank Prof. H.-D. BEHNKE for this figure (see BEHNKE, 1976), which illustrates the distribution of these three modifications of P-III subtype in the Centrospermae

1976a; BEHNKE and MABRY, 1976) these include Polygonaceae, Plumbaginaceae, Fouquieraceae, Frankeniaceae, Rhabdodendraceae, Vivianiaceae, Theligonaceae, Bataceae, and Gyrostemonaceae. So far neither anthocyanins nor betalains have been detected in the latter two families, both of which contain glucosinolates (GOLDBLATT et al., 1976).

7.2 C_4 Photosynthesis in the Centrospermae

It is remarkable that among the dicotyledons which have evolved the C_4 photosynthetic pathway (the Kranz syndrome), seven of the eleven families (see Table 1) and about 85% of the genera are from the Centrospermae (MABRY, 1977). This preadaptation for the Kranz syndrome has permitted repeated and independent evolution of the syndrome in numerous genera in seven core families of the order: Nyctaginaceae, Chenopodiaceae, Amaranthaceae, Portulacaceae, Aizoaceae, Molluginaceae, and Caryophyllaceae. The only other dicotyledonous families where a few species exhibit the C_4 pathway are the Boraginaceae, Compositae, Euphorbiaceae, and Zygophyllaceae. It should also be noted that the CAM (crassulacean acid metabolism) photosynthetic pathway occurs in members of the Cactaceae and Aizoaceae; this pathway, although anatomically, functionally, and phylogenetically distinct, does utilize enzymes of the Kranz syndrome.

The widespread occurrence of the Kranz syndrome in both betalain and anthocyanin centrospermous families and its sporadic and limited occurrence in other dicotyledons lends support to the view that the Centrospermae is an old evolutionary line containing two groups of families distinguished by their pigments.

7.3 DNA-RNA Hybridization Data

The DNA-RNA hybridization studies of CHANG and MABRY (1974) support a close relationship for the several betalain families tested. Significantly, the Caryophyllaceae was the family closest to the betalain group. These data also indicate that the Bataceae, often treated as a member of the Centrospermae, is misplaced there. Indeed, the Bataceae as well as the Gyrostemonaceae are probably best allied with the glucosinolate families.

7.4 Summary

It is our view that the nine centrospermous betalain families and the two centrospermous anthocyanin families are derived from a common ancestor which had evolved the unique protein depositions characteristic of the sieve-element plastids found in these families. In addition, the ancestor was probably preadapted for C_4 photosynthesis. The centrospermous line is now represented by a core of eleven families which can be treated in one order named either the Centrospermae or Caryophyllales; in any case, the order is probably best treated as consisting of a betalain-suborder, the Chenopodiineae, and an anthocyanin-suborder, the Caryophyllineae (Table 1).

Mushroom pigments
(Order Agaricinales)

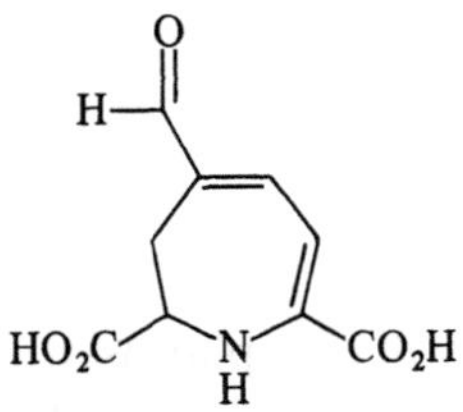

Muscaflavin
(Amanita, Hygrocybe)
(XXV)

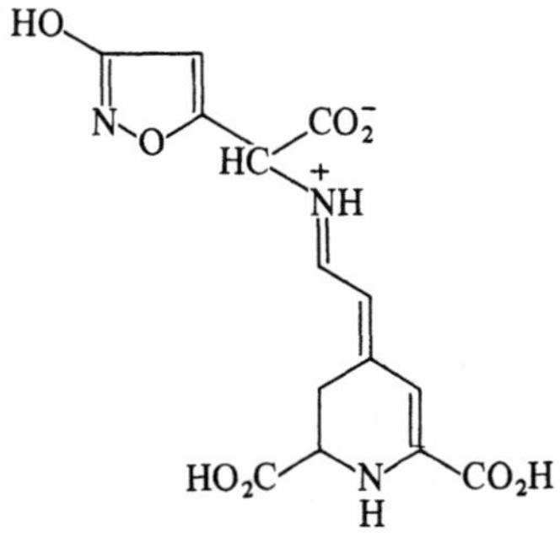

Musca–aurin–I
(Amanita)
(XXVI)

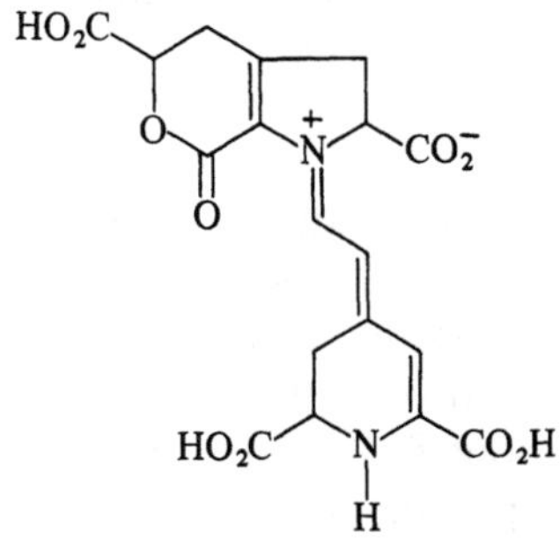

Musca–purpurin
(Amanita)
(XXVII)

Fig. 10. Three of the pigments found in the mushroom *Amanita muscaria* including two betalains, musca-aurin-I and musca-purpurin

8 Betalains in Mushrooms

One of the more intriguing discoveries relating to betalains has been the detection of this class of pigments in mushrooms; indeed several of the same pigments found in members of the Centrospermae, e.g., vulgaxanthin-I, II, miraxanthin-II and indicaxanthin, occur in *Amanita muscaria* (Order Agaricinales) (for reviews, see MUSSO, in press; STEGLICH, 1975).

Other betalains found in mushrooms include pigments which can be derived by condensation of betalamic acid with adipic acid and ibotenic acid. One of these, musca-aurin-I (XXVI), was readily prepared by condensation of ibotenic acid with betalamic acid (MUSSO, 1978). All of the seven known musca-aurins are yellow pigments, i.e., betaxanthins, with absorption maxima around 475 nm. Muscaflavin (XXV, Fig. 10), another yellow pigment found in both *Amanita* and *Hygrocybe* (λ max 425 nm), is not a betalain but is related to the betalains on the basis of its probable origin from L-DOPA (see Fig. 5). Structure (XXVII) (Fig. 10) has been proposed for the betacyanin musca-purpurin (λ max 540 nm) from *Amanita muscaria*.

Acknowledgments. Some of the research reviewed here was supported by the National Science Foundation (DEB 79-02703), the National Institutes of Health (HD-04488) and the Robert A. Welch Foundation (F-130).

References

Adams, J.P., Elbe, von, J.H.: Betanine separation and quantification by chromatography on gels. J. Food Sci. *42*, 410–414 (1977)

Ardenne, von, R., Döpp, H., Musso, H., Steglich, W.: Isolation of muscaflavin from *Hygrocybe* species (Agaricales) and its dihydroazepine structure. Z. Naturforsch. *29c*, 637–639 (1974)

Barthlott, W., Rauh, W.: *Wittia amazonica* K. Schumanh (syn. *Wittia panamensis* Britton und Rose) ein bemerkenswerter epiphytischer Kaktus. Kakteen Sukkulenten *27*, 145–150 (1976)
Behnke, H.-D.: Ultrastructure of sieve-element plastids in Caryophyllales (Centrospermae), evidence for the delimitation and classification of the Order. Plant Syst. Evol. *126*, 31–54 (1976)
Behnke, H.-D., Chang, C., Eifert, I.J., Mabry, T.J.: Betalains and P-type sieve-tube plastids in *Petiveria* and *Agdestis* (Phytolaccaceae). Taxon *23*, 541–542 (1974)
Behnke, H.-D., Mabry, T.J.: S-Type sieve-element plastids and anthocyanins in Vivianiaceae: evidence against its inclusion into Centrospermae. Plant Syst. Evol. *126*, 371–375 (1976)
Buchi, G., Fliri, H., Shapiro, R.: A synthesis of betalamic acid. J. Org. Chem. *42*, 2192–2194 (1977)
Chang, C., Mabry, T.J.: The constitution of the order Centrospermae: RNA-DNA hybridization studies among betalain- and anthocyanin-producing families. Biochem. Syst. *1*, 185–190 (1974)
Chang, C., Kimler, L., Mabry, T.J.: Biogenesis of betalamic acid. Phytochemistry *13*, 2771–2775 (1974)
Chilton, W.S., Hsu, C.P., Zdybak, W.T.: Stizolobic and stizolobinic acids in *Amanita pantheria*. Phytochemistry *13*, 1179–1181 (1974)
Dreiding, A.S.: The betacyanins, a class of red pigments in the Centrospermae. In: Recent developments in the chemistry of natural phenolic compounds. Ollis, W.D. (ed.), pp. 194–211. Oxford: Pergamon Press 1961
Döpp, H., Musso, H.: Fliegenpilz-Farbstoffe II. Isolierung und Chromophore der Farbstoffe aus *Amanita muscaria*. Chem. Ber. *106*, 3473–3482 (1973)
Döpp, H., Musso, H.: Chromatographic analysis of betalain pigments in toadstools and higher plants. Z. Naturforch. *29c*, 640–642 (1974)
Dunkelblum, E., Miller, H.E., Dreiding, A.S.: On the mechanism of decarboxylation of betanidin, a contribution to the interpretation of the biosynthesis of betalains. Helv. Chim. Acta *55*, 642–648 (1972)
Ellis, B.E.: Dopa ring-cleavage in the biogenesis of stizolobic acid in *Mucuna deeringiana*. Phytochemistry *15*, 489–491 (1976)
Fischer, N., Dreiding, A.S.: Biosynthesis of betalains. On the cleavage of the aromatic ring during the enzymatic transformation of dopa into betalamic acid. Helv. Chim. Acta *55*, 649–659 (1972)
Garay, A.S., Towers, G.H.N.: Studies on the biosynthesis of amaranthin. Can. J. Bot. *44*, 231–236 (1966)
Goldblatt, P., Nowicke, J.W., Mabry, T.J., Behnke, H.-D.: Gyrostemonaceae: status and affinity. Bot. Not. *129*, 201–206 (1976)
Hermann, K., Dreiding, A.S.: Totalsynthese von Betalainen. Helv. Chim. Acta *58*, 1805–1808 (1975)
Hörhammer, L., Wagner, H., Fritzsche, W.: Zur Biosynthese der Betacyane I. Biochem. Z. *339*, 398–400 (1964)
Hunziker, H., Behnke, H.-D., Eifert, I.J., Mabry, T.J.: *Halophytum ameghinoi*: a betalain-containing and P-type sieve-tube plastid species. Taxon *23*, 537–539 (1974)
Impellizzeri, G., Piattelli, M.: Biosynthesis of betalains: formation of indicaxanthin in *Opuntia ficus-indica* fruits. Phytochemistry *11*, 2499–2502 (1972)
Impellizzeri, G., Piattelli, M., Sciuto, S.: A new betaxanthin from *Glottiphyllum longum*. Phytochemistry *12*, 2293–2294 (1973)
Kimler, L.M.: Betanin, the red beet pigment, as an antifungal agent. Abstr. Bot. Soc. Am. 70th Annu. Meet. 36 (1975)
Kimler, L., Mears, J., Mabry, T.J., Rösler, H.: On the question of the mutual exclusiveness of betalains and anthocyanins. Taxon *19*, 875–878 (1970)
Kimler, L.M., Larson, R.A., Messenger, L., Moore, J.B., Mabry, T.J.: Betalamic acid, a new naturally occurring pigment. J. Chem. Soc. Chem. Commun. 1329–1330 (1971)
Kremer, B.P.: Das Experiment: Trennung von Anthocyanin und Betalainen. Biol. unserer Zeit *5*, 155–157 (1975)
Liebisch, H.W., Matschiner, B., Schuette, H.R.: Beiträge zur Physiologie und Biosynthese des Betanins. Z. Pflanzenphysiol. *61*, 269–278 (1969)

Mabry, T.J.: The betacyanins, a new class of red violet pigments, and their phylogenetic significance. In: Taxonomic biochemistry and serology. Leone, C.A. (ed.), pp. 239–253. New York: The Ronald Press 1964

Mabry, T.J.: The betacyanins and betaxanthins. In: Comparative phytochemistry. Swain, T. (ed.), pp. 231–244. London: Academic Press 1966

Mabry, T.J.: The betalains. In: The chemistry of alkaloids. Pelletier, S.W. (ed.), pp. 719–746. New York: Reinhold 1970

Mabry, T.J.: Is the order Centrospermae monophyletic? In: Chemistry in botanical classification. Bendz, G., Santesson, J. (eds.), pp 275–285. London: Academic Press 1973

Mabry, T.J.: Pigment dichotomy and DNA-RNA hybridization data for centrospermous families. Plant Syst. Evol. *126*, 79–94 (1976)

Mabry, T.J.: The order Centrospermae. Ann. M. Bot. Gard. *64*, 210–220 (1977)

Mabry, T.J., Behnke, H.-D. (eds.): Evolution of centrospermous families. Plant Syst. Evol. *126*, 1–105 (1976a)

Mabry, T.J., Behnke, H.-D.: Betalains and P-type sieve-element plastids: the systematic position of *Dysphania* R. Br. (Centrospermae). Taxon *25*, 109–111 (1976b)

Mabry, T.J., Dreiding, A.S.: The betalains. In: Recent advances in phytochemistry. Mabry, T.J., Alston, R.E., Runeckles, V.C. (eds.), Vol. 1, pp. 145–160. New York: Appleton-Century-Crofts 1968

Mabry, T.J., Wyler, H., Sassu, G., Mercier, M., Parikh, I., Dreiding, A.S.: Die Struktur des Neobetanidens: Über die Konstitution des Randenfarbstoffes Betanin. Helv. Chim. Acta *45*, 640–647 (1962)

Mabry, T.J., Wyler, H., Parikh, I., Dreiding, A.S.: The conversion of betanidin and betanin to neobetanidin derivatives. Tetrahedron *23*, 3111–3127 (1967)

Mabry, T.J., Kimler, L., Chang, C.: The betalains: structure, function and biogenesis and the plant order Centrospermae. In: Recent advances in phytochemistry. Runeckles, V.C., Tso, T.C. (eds.), Vol. V, pp. 105–134. New York: Appleton-Century-Crofts 1972

Mabry, T.J., Eifert, I.J., Chang, C., Mabry, H., Kidd, C., Behnke, H.-D.: Theligonaceae: pigment and ultrastructural evidence which exclude it from the order Centrospermae. Biochem. Syst. Evol. *3*, 53–55 (1975)

Mabry, T.J., Behnke, H.-D., Eifert, I.J.: Betalains and P-type sieve-element plastids in *Gisekia* L. (Centrospermae). Taxon *25*, 112–114 (1976)

Mabry, T.J., Neuman, P., Philipson, W.R.: *Hectorella:* a member of the betalain-suborder Chenopodiineae of the order Centrospermae. Plant Syst. Evol. *130*, 163–165 (1978)

Miller, H.E., Rösler, H., Wohlpart, A., Wyler, H., Wilcox, M.E., Frohofer, H., Mabry, T.J., Dreiding, A.S.: Biogenese der Betalaine: Biotransformation von Dopa und Tyrosin in den Betalaminsäureteil des Betanins. Helv. Chim. Acta *51*, 1470–1474 (1968)

Minale, L., Piattelli, M., Nicolaus, R.A.: Pigments of Centrospermae IV: On the biogenesis of indicaxanthin and betanin in *Opuntia ficus-indica* Mill. Phytochemistry *4*, 593–597 (1965)

Minale, L., Piattelli, M., De Stefano, S., Nicolaus, R.A.: Pigments of Centrospermae VI: acylated betacyanins. Phytochemistry *5*, 1037–1052 (1966)

Minale, L., Piattelli, M., De Stefano, S.: Pigments of Centrospermae VII: betacyanins from *Gomphrena globosa* L. Phytochemistry *6*, 703–709 (1967)

Musso, H.: Colouring matters of fly agaric (*Amanita muscaria*). Tetrahedron Lett. (in press)

Piattelli, M.: Betalains. In: Chemistry and biochemistry of plant pigments. Goodwin, T.W. (ed.), pp. 560–596. New York: Academic Press 1976

Piattelli, M., Impellizzeri, G.: 2-Descarboxybetanidin, a minor betacyanin from *Carpobrotus acinaciformis*. Phytochemistry *9*, 2553–2556 (1970)

Piattelli, M., Imperato, F.: Pigments of Centrospermae XI: betacyanins from *Bougainvillea*. Phytochemistry *9*, 455–458 (1970a)

Piattelli, M., Imperato, F.: Pigments of *Bougainvillea glabra*. Phytochemistry *9*, 2557–2560 (1970b)

Piattelli, M., Minale, L.: Struttura dell'amarantina e dell'isoamaranthina. II. Ann. Chim. *56*, 1060–1064 (1966)

Piattelli, M., Minale, L., Prota, G.: Struttura della betanina. Ann. Chim. *54*, 955–962 (1964a)

Piattelli, M., Minale, L., Prota, G.: Isolamento e struttura dell amarantina e dell isoamarantina. Ann. Chim. *54*, 963–968 (1964b)

Piattelli, M., Minale, L., Prota, G.: Isolation, structure and absolute configuration of indicaxanthin. Tetrahedron *22*, 2325–2329 (1964c)

Piattelli, M., Minale, L., Nicolaus, R.A.: Pigments of Centrospermae V: betaxanthins from *Mirabilis jalapa* L. Phytochemistry *4*, 817–823 (1965a)
Piattelli, M., Minale, L., Nicholaus, R.A.: Ulteriori ricerche sulle betaxanthine. Rend. Accad. Sci. Fis. Mat. Naples *32*, 55–56 (1965b)
Piattelli, M., Minale, L., Prota, G.: Pigments of Centrospermae III: Betaxanthins from *Beta vulgaris* L. Phytochemistry *4*, 121–125 (1965c)
Pucher, G.W., Curtis, L.C., Vickery, H.B.: The red pigment of the root of the beet (*Beta vulgaris*). J. Biol. Chem. *123*, 61–75 (1938)
Reznik, H.: Betalaine (Betalains). In: Pigments in plants. Czygan, F.-C. (ed.), pp. 179–190. Stuttgart: Gustav Fischer 1975
Saito, K., Komamine, A., Senoh, S.: Biosynthesis of stizolobinic acid and stizolobic acid in the etiolated seedlings of *Stizolobium hassjoo*. Z. Naturforsch. *30c*, 659–662 (1975)
Saito, K., Komamine, A., Senoh, S.: Further studies of the biosynthesis of stizolobinic acid and stizolobic acid in etiolated seedlings of *Stizolobium hassjoo*. Z. Naturforsch. *31c*, 15–17 (1976)
Schmidt, O. Th., Schönleben, W.: Zur Kenntnis der Farbstoffe der Roten Rübe, II. Z. Naturforsch. *12b*, 262–263 (1957)
Schmidt, O. Th., Becher, P., Hübner, M.: Zur Kenntnis der Farbstoffe der Roten Rübe, III. Chem. Ber. *93*, 1296–1304 (1960)
Sciuto, S., Oriente, G., Piattelli, M.: Betanidin glucosylation in *Opuntia dillenii*. Phytochemistry *11*, 2259–2262 (1972)
Sciuto, S., Oriente, G., Piattelli, M., Impellizzeri, G., Amico, V.: Biosynthesis of amaranthin in *Celosia plumosa*. Phytochemistry *13*, 947–951 (1974)
Sosnova, V.: Reproduction of sugar beet mosaic and tobacco mosaic viruses in anthocyanized beet plants. Biol. Plant. *12*, 424–427 (1970)
Steglich, W.: Pilzfarbstoffe. Chem. unserer Zeit *9*, 117–123 (1975)
Vincent, K.R., Scholz, R.G.: Separation and quantification of red beet betacyanins and betaxanthins by high-performance liquid chromatography. J. Agric. Food Chem. *26*, 812–816 (1978)
Wilcox, M.E., Wyler, H., Dreiding, A.S.: Stereochemie von Betanidin und Isobetanidin: Zur Konstitution des Randenfarbstoffes Betanin. Helv. Chim. Acta *48*, 1134–1147 (1965a)
Wilcox, M.E., Wyler, H., Mabry, T.J., Dreiding, A.S.: Über die Konstitution des Randenfarbstoffes Betanin: Die Struktur des Betanins. Helv. Chim. Acta 48, 252–258 (1965b)
Wohlpart, A., Mabry, T.J.: The distribution and phylogenetic significance of the betalains with respect to the Centrospermae. Taxon *17*, 148–152 (1968a)
Wyler, H.: Das Experiment: Papierelektrophorese. Chem. unserer Zeit *3*, 111–115 (1969a)
Wyler, H.: Die Betalaine. Chem. unserer Zeit *3*, 146–151 (1969b)
Wyler, H., Dreiding, A.S.: Kristallisiertes Betanin. Helv. Chim. Acta. *40*, 191–192 (1957)
Wyler, H., Dreiding, A.S.: Über die Konstitution des Randenfarbstoffes Betanin: Darstellung und Abbauprodukte des Betanidins. Helv. Chim. Acta *42*, 1699–1702 (1959)
Wyler, H., Dreiding, A.S.: Über die Konstitution des Randenfarbstoffes Betanin: Abbauprodukte des Betanidins. Helv. Chim. Acta *45*, 638–640 (1962)
Wyler, H., Mabry, T.J., Dreiding, A.S.: Über die Konstitution des Randenfarbstoffes Betanin: Zur Struktur des Betanidins. Helv. Chim. Acta *46*, 1745–1748 (1963)
Wyler, H., Wilcox, M.E., Dreiding, A.S.: Umwandlung eines Betacyanes in ein Betaxanthin. Synthese von Indicaxanthin aus Betanin. Helv. Chim. Acta *48*, 361–366 (1965)
Wyler, H., Rösler, H., Mercier, M., Dreiding, A.S.: Prabetanin, ein Schwefelsäure-Halbester des Betanins. Ein Beitrag zur Kenntnis der Betacyane. Helv. Chim. Acta *50*, 545–561 (1967)

12 Plant Lipids of Taxonomic Significance

G.A. Thompson, Jr.

1 Introduction

The designation of certain lipids as "secondary products" is necessarily rather arbitrary. Following the guideline of Luckner (1972), this review considers those lipids which apparently "are not necessarily vital, often differ from species to species, and may be considered as an expression of the chemical individuality of the organism." Following this orientation, the essential structural and metabolic functions of lipids have been omitted in order to emphasize their chemotaxonomic potential.

The literature on plant lipids is voluminous. A great deal of compositional data has been drawn together by Hilditch and Williams (1964). Hitchcock and Nichols (1971) have published a comprehensive account of the more biochemical and physiological aspects of plant lipids. An introductory general treatment of plant chemotaxonomy, including a brief discussion of lipids as taxonomic markers, has been written by Smith (1976).

With the advent of newer methods of chromatography and physical chemical analysis, the isolation and characterization of plant lipids has been transformed from an art to a science. The techniques needed for routine lipid analysis have been carefully outlined by Kates (1972) and Harborne (1973) and are, in most cases, well within the capabilities of the typically equipped laboratory.

2 Fatty Acids

2.1 Commonly Occurring Fatty Acids

The major lipids of most plant tissues derive their hydrophobic properties from their constituent fatty acids. Fatty acid-containing lipids of vegetative plant parts consist mainly of glycolipids and phospholipids (Hitchcock and Nichols, 1971). Both of these lipid classes function as structural components of membranes. On the other hand, reproductive tissues, such as seeds and, in some cases, fruits, accumulate large stores of triglycerides as a reservoir of metabolic energy. With some important exceptions, to be discussed later, the triglycerides and membrane lipids contain the same types of fatty acids. Table 1 lists the fatty acids more commonly found in plant (and animal) tissues.

Table 1. The major fatty acids of photosynthetic organisms

No. of carbons	Structural formula	Systematic name	Common name
16	$CH_3(CH_2)_{14}COOH$	hexadecanoic	palmitic
16	$CH_3(CH_2)_5CH{=}CH(CH_2)_7COOH$	cis-9-hexadecenoic	palmitoleic
18	$CH_3(CH_2)_7CH{=}CH(CH_2)_7COOH$	cis-9-octadecenoic	oleic
18	$CH_3(CH_2)_4[CH{=}CHCH_2]_2(CH_2)_6COOH$	cis, cis-9,12-octadeca-dienoic	linoleic
18	$CH_3CH_2[CH{=}CHCH_2]_3(CH_2)_6COOH$	all cis-9,12,15-octadeca-trienoic	α-linolenic
18	$CH_3(CH_2)_4[CH{=}CHCH_2]_3(CH_2)_3COOH$	all cis-6,9,12-octadeca-trienoic	γ-linolenic
20	$CH_3(CH_2)_4[CH{=}CHCH_2]_4(CH_2)_2COOH$	all cis-5,8,11,14-eicosa-tetraenoic	arachidonic

The physical properties of lipids are markedly dependent upon their fatty acid composition, with viscosity decreasing as the number of unsaturated bonds of the component fatty acids increases. Each organism maintains the viscosity of its membranes at a fairly constant level, despite the tendency of environmental changes to alter viscosity, by adjusting the relative proportions of these major fatty acids.

However, despite this regulatory pressure against variation of fatty acid composition, there does exist a degree of biochemically and taxonomically interesting diversity. Perhaps the best known example of this is found in the algae. There are significant differences among the algal classes with respect to the pathways of biosynthesis of polyunsaturated fatty acids. Two distinct metabolic schemes for these fatty acids are known, one leading mainly to 9,12,15-octadecatrienoic acid (α-linolenate) and the other to 6,9,12-octadecatrienoic acid (γ-linolenate) (Fig. 1). The formation of α-linolenate involves the placement of the second and third double bond on the methyl side of the initial double bond at carbon 9. This mechanism is characteristic of plants. The "animal system", on the other hand, is capable of introducing double bonds on both the methyl and the carboxyl side of the initial double bond, and this gives rise to products such as γ-linolenate.

The algae and protozoa have been found to contain one or the other or, uncommonly, both of the systems for desaturation. Accordingly, there has been considerable interest in comparing the polyunsaturated fatty acid distribution in different taxonomic groups of algae. Figure 2 summarizes a scheme in which ERWIN (1973) superimposed the major polyunsaturated fatty acids upon the phylogenetic relationships proposed by KLEIN and CRONQUIST (1967). The pattern conforms to the expected pattern of gradually increasing complication in biosynthetic patterns. The blue-green algae (Cyanophyceae) frequently contain no polyenoic fatty acids (example: *Anacystis nidulans*). When trienoic acids are found in the blue-greens, they may be either of the γ-linolenate pathway (*Spirulina platensis*) or the α-linolenate pathway (*Anabaena cylindrica*). Marine algae, including the green Chlorophyceae, the red Rhodophyceae, and the brown Phaeophyceae contain, in addition to the C_{18} polyenes, C_{20} and C_{22} polyenoic fatty acids containing 4–6 double bonds.

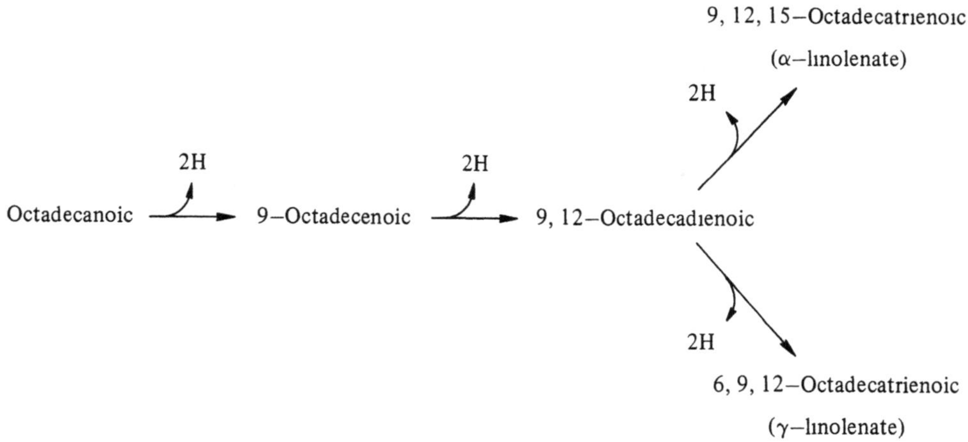

Fig. 1. Basic outline of C_{18} polyunsaturated fatty acid synthesis. For details of the biosynthetic pathway, which produces fatty acid species not usually considered to be "natural products", see GURR (1974)

Whereas the observed fatty acid patterns can be logically explained by the KLEIN and CRONQUIST scheme, they are not well correlated with the phylogenetic arrangement of MARGULIS (formerly SAGAN, see SAGAN, 1967) which is based on the endosymbiont theory. Ordering the algae according to this latter scheme yields no obvious sequence of progressive metabolic complexity from lower to higher levels of evolution.

However, interpretation of the available data is confounded by the probability that many of the variations in algal fatty acid patterns are caused, not by a gradual evolution of the basic desaturation pathway, but rather by the coexistence in certain species of two different pathways. Thus *Euglena gracilis* grown photo-autotrophically produced fatty acids mainly of the plant type, while cells grown on an organic medium in the absence of light contained animal-type products as their major polyunsaturated fatty acids.

As in the algae, the membrane structural lipids of higher plants contain primarily those fatty acids illustrated in Table 1. Variations from family to family are small, and although efforts have been made to employ compositional differences as a chemotaxonomic tool, e.g., in citrus species and cultivars (NORDBY et al., 1976), no general taxonomic use has been made of the structural lipid distributions.

On the other hand the storage lipids, primarily the seed triglycerides, exhibit considerably more variation. The major triglyceride fatty acids of most seed fats are also included among those listed in Table 1. Taxonomic groupings have been constructed based on the relative abundance of these major fatty acids (HILDITCH and WILLIAMS, 1964, p. 204). The value of such schemes is severely compromised by the fact that environmental factors, such as growth temperature, can have a striking effect on the degree of unsaturation in these fats. For example, seeds from a strain of sunflower grown in Africa contained 43% oleic acid and 44% linoleic acid (BARKER and HILDITCH, 1950a) whereas the same strain grown in the cooler climate of England produced 21% oleic acid and 70% linoleic acid (BARKER and HILDITCH, 1950b).

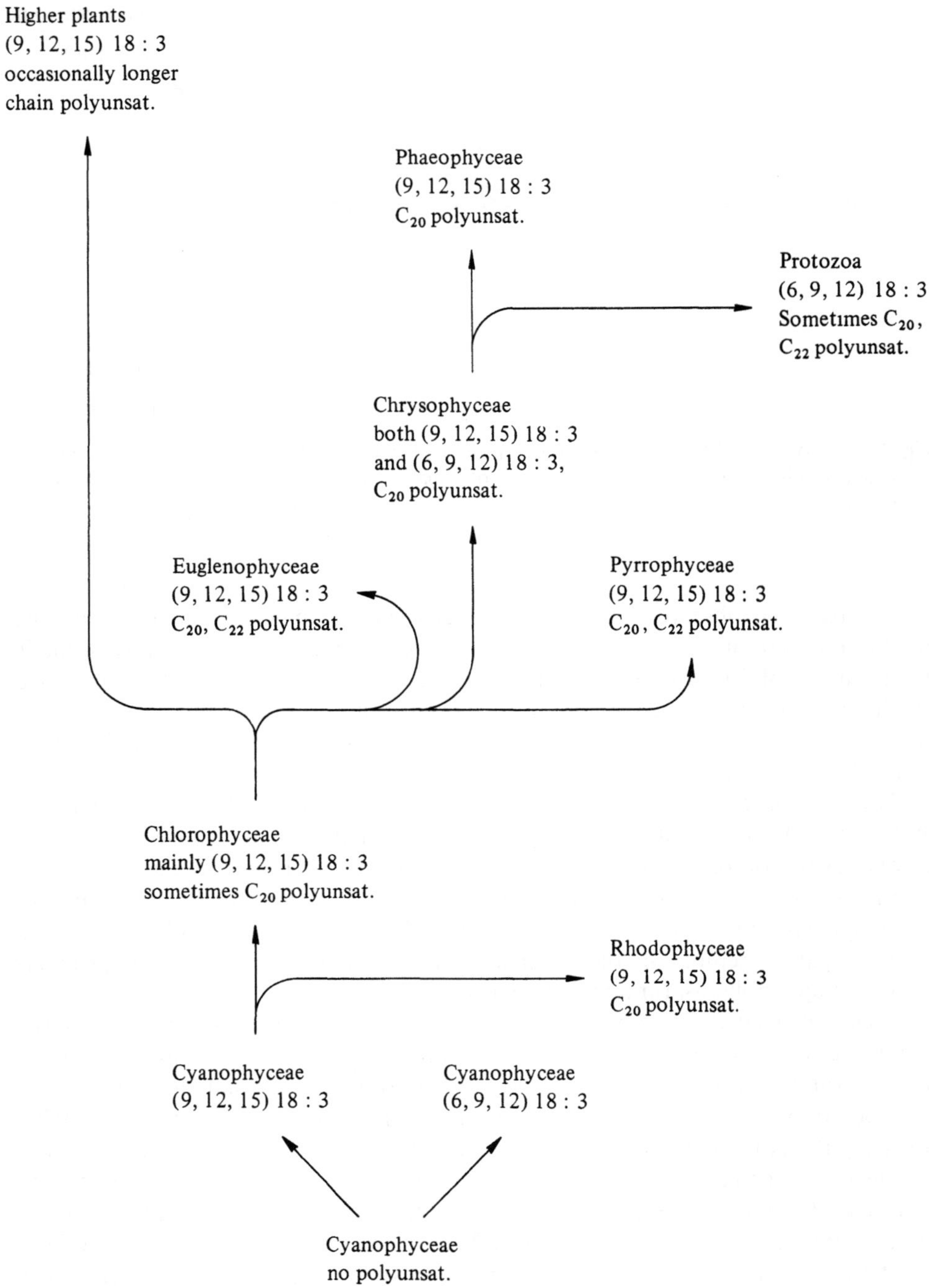

Fig. 2. Phylogenetic distribution of C_{18} fatty acids in algae. When C_{20} and longer chain polyunsaturates occur, they may have additional double bonds added on either side of the original Δ 9 unsaturation

2.2 Fatty Acids of Unusual Structure

Of perhaps more utility from a chemotaxonomic point of view are the many unusual fatty acids found in plants. Table 2 lists representative fatty acids having a more narrow distribution. The notable properties of these acids are that they have either longer or shorter chains than the more common length of 12–18 carbon atoms (Table 2A), they have unusual placement or numbers of unsaturated bonds (Table 2B, C, D), or they have characteristic substituent groups (Table 2E, F), such as ring structures. In most cases, no special function directly useful to the plant has yet been associated with these atypical fatty acids. Of course the toxic effects on animal predators exhibited by certain of the compounds, e.g., the inhibi-

Table 2. Some characteristic fatty acids prevalent only in certain plant species

No. of carbons	Structural formula	Systematic name	Common name	Present mainly in seed fats of
A. Saturated				
10	$CH_3(CH_2)_8COOH$	n-decanoic	capric	Ulmaceae, Lauraceae, Lythraceae
24	$CH_3(CH_2)_{22}COOH$	n-tetracosanoic	lignoceric	Leguminosae, Sapindaceae
B. Unsaturated				
18	$CH_3(CH_2)_{10}CH{=}CH(CH_2)_4COOH$	*cis*-6-octadecenoic	petroselenic	Umbelliferae, Araliaceae
22	$CH_3(CH_2)_7CH{=}CH(CH_2)_{11}COOH$	*cis*-13-docosenoic	erucic	Cruciferae, Tropaeolaceae
C. Unsaturated (conjugated)				
18	$CH_3CH_2[CH{=}CH]_4(CH_2)_7COOH$	*cis, trans, trans, cis*-9,11,13,15-octadecatetraenoic	parinaric	Rosaceae Balsaminaceae
D. Acetylenic				
18	$CH_3(CH_2)_{10}C{\equiv}C(CH_2)_4COOH$	octadec-6-ynoic	tariric	Simarubaceae
18	$CH_2{=}CH(CH_2)_4C{\equiv}C{-}C{\equiv}C(CH_2)_7COOH$	octadec-17-en-9,11-dienoic	isanic	Olacaceae
E. Cyclic				
18	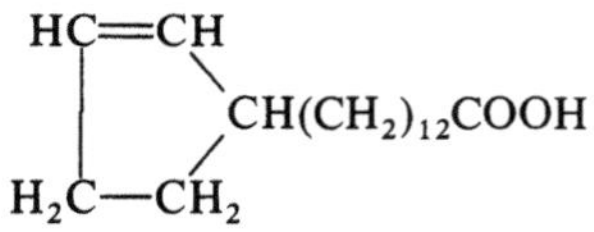	13-*cyclo* pent-2-enyl-n-tridecanoic	chalmoogric	Flacourtiaceae
19	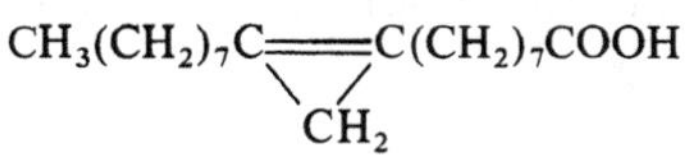	8-(2-n-octyl*cyclo*prop-1-enyl)octanoic	sterculic	Sterculiaceae
F. Hydroxy				
18	$CH_3(CH_2)_5CH(OH)(CH_2)_{10}COOH$	12-hydroxyoctadecanoic	ricinoleic	Euphorbiaceae

tion of fatty acid desaturation by sterculic acid (RAJU and REISER, 1967) are of great indirect value to the plants containing them.

A more thorough compilation of the types and natural distribution of unusual fatty acids may be found in HILDITCH and WILLIAMS (1964) and in various more recent references, e.g., SMITH (1970) and GIBBS (1974a). Not surprisingly, little information is available on the metabolism of these fatty acids. The mono-unsaturated acids which differ from their more common counterparts by having an atypical placement of double bonds are thought to arise through the action of a fatty acid desaturase having a positional specificity which is different from that of the ubiquitous Δ^9-desaturase. Alternatively, desaturation in the usual position followed by modification in chain length could place the double bond in an atypical position (HITCHCOCK and NICHOLS, 1971).

Naturally occurring conjugated fatty acids normally have three or more conjugated double bonds with one or more of the bonds having a *trans* configuration. Although very few firm data are available concerning their formation, it seems likely that a free radical-induced isomerization of a diene is followed by the addition of oxygen to yield a hydroxy-diene, perhaps via a hydroperoxide intermediate (HITCHCOCK and NICHOLS, 1971). Dehydration could then give rise to a conjugated triene.

The formation of the cyclopropene ring of sterculic acid (Table 2) involves the addition of a methyl group, probably via S-adenosyl methionine, to oleic acid to produce dihydrosterculic acid (9,10-methylene octadecanoic acid) (JOHNSON et al., 1967). This intermediate is then desaturated to sterculate.

In higher plants, seed oils may contain hydroxy acids which account for as much as 90% of the total fatty acids. Such is the case in the castor bean plant (*Ricinus communis*), which forms large amounts of ricinoleic acid during seed development. MORRIS (1967) demonstrated that ricinoleic acid arises in *R. communis* by hydroxylation of oleic acid (Fig. 3). Curiously enough, ricinoleic acid of ergot oil from the fungus *Claviceps purpurea* has been shown by MORRIS et al. (1966) to be derived, not as in higher plants, but through the hydration of linoleic acid (Fig. 3).

Further metabolic details of the examples of unusual fatty acids given above have been compiled by HITCHCOCK and NICHOLS (1971). As additional information concerning the biosynthetic mechanisms responsible for their formation becomes available, evolutionary relationships will surely emerge.

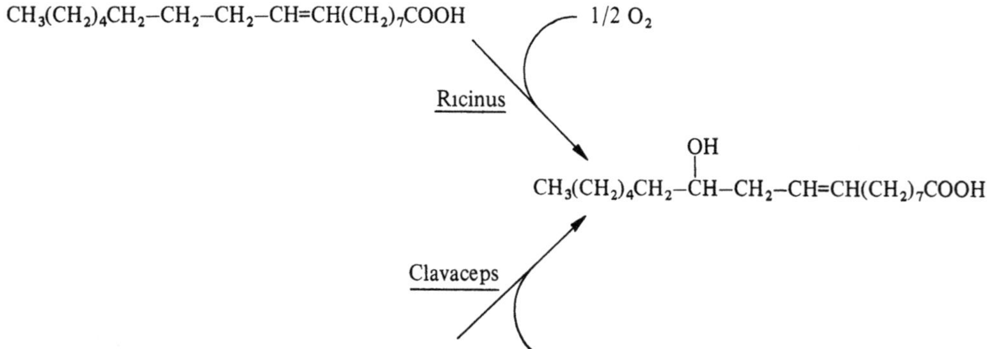

Fig. 3. The biosynthesis of ricinoleic acid in *Ricinus communis* and *Claviceps purpurea*

2.3 The Use of Fatty Acids for Taxonomic Purposes

Unfortunately, the number of plant species analyzed for seed lipid composition is still too limited to permit a thorough assessment of fatty acids as useful chemotaxonomic markers. Only rarely has the study of fatty acid composition been of real value in formulating phylogenetic relationships. In those cases where fatty acids have provided strong support for taxonomic assignments, the variation of significance has most often involved one or more of the unusual fatty acids (Table 2). There are noteworthy exceptions to this generalization: for example, the assignment of Proteaceae species into sub families and tribes was strengthened by analysis of their seed oil fatty acids, which contain unusually high concentrations of the 16 and 18 carbon monoenes (VICKERY, 1971). However, where possible, it is more comforting to rely on groupings based on the presence of absence of some unusual fatty acid, thus lessening the risk of environmentally produced artifacts.

The value of these characteristic fatty acids in systematics is well illustrated by studies of the genus *Lesquerella* (Cruciferae) (ROLLINS and SHAW, 1973, APPLEQVIST, 1976). The auriculate-leaved species of the genus contain the distinctive C_{18} monohydroxy acid, densipolic acid (12-hydroxy-octadeca-*cis*-9-*cis*-15-dienoic acid). In place of this component, the nonauriculate species contain a C_{20} monohydroxy compound, lesquerolic acid (14-hydroxy-*cis*-eicos-11-enoic acid) as a major component of their seed oils. Relationships based on these chemical data correlate well with independent measurements of factors such as interfertility and chromosome numbers. For example, two exceptional auriculate-leaved species, *L. grandiflora* and *L. lasiocarpa*, which, contrary to the above rule, contained lesquerolic acid were also anomalous in having haploid chromosome numbers of 9 and 7, respectively, instead of the usual n=8. Interestingly, *L. auriculata*, the species thought by some to be an evolutionary link between the two main groups, contains both densipolic acid and its C_{20} homolog, auricolic acid. The presence of the latter compound adds credence to the proposed relationship of *L. auriculata* to the lesquerolic acid-rich groups of species.

Because the unusually long chain compounds eicosenoic acid ($C_{20\ 1}$) and erucic acid ($C_{22\ 1}$) are common in cruciferous seed oils, taxonomic studies of their occurrence have been widespread (APPLEQVIST, 1976). Yet even now only a few hundred of the approximately 3000 species of Cruciferae have been analyzed, and a clear-cut pattern of fatty acid distribution cannot yet be established. In the species examined so far, erucic acid and/or eicosenoic acid almost always occur in the seed oil, varying in amount from 1% to 60% of the total fatty acids. As usual, environmental as well as genetic influences determine the exact amounts of $C_{20\ 1}$ and $C_{22\ 1}$ present, contributing to the confusion of chemotaxonomists. By using single plant selection techniques, it has been possible to develop erucic acid-free cultivars of some agriculturally important plants such as *Brassica napus* and *B. campestris*. Thus the undesirable physiological effects of high levels of erucic acid in commercial rapeseed oil can be avoided.

With the advent of recent advances in gas chromatography and other analytical techniques, novel, unusual plant fatty acids have continued to be discovered during the past few years (SMITH, 1970). Charting the distribution of these compounds among plant species offers an exciting challenge for the systematist.

Although many years of labor will be required before it is possible to compile a key to the higher plants featuring all their lipids (GIBBS, 1974b, has made a beginning), a surprisingly good sampling can be gleaned from the reviews of SHORLAND (1963), HILDITCH and WILLIAMS (1964), SMITH (1970), etc. It is immediately apparent from examining these sources that some families (including many of those listed in Table 2) contain a rich and potentially useful variety of fatty acids, while others, such as Palmae, Graminae, and Leguminosae, exhibit less diversity. Thus the chemotaxonomist unfamiliar with the lipid field can often assess rather quickly the likelihood that fatty acid analysis will be of interest in comparing members of a particular group of plants.

3 Waxes

Coating the surfaces of most higher plant leaves, fruits, and other parts is a mixture of very hydrophobic lipids known as cuticular wax. The most obvious benefits of this impervious layer are to reduce water loss and protect against invasion by fungi and other harmful organisms. The wax can be shown by scanning electron microscopy to exist as plates, rods, tubes or other forms projecting from the leaf surfaces. Although the microscopic appearance of the wax is often characteristic of the plant species, the morphology of the wax projections is greatly influenced by environmental factors, such as humidity and temperature (BAKER, 1974) and therefore can seldom be used as a reliable taxonomic marker.

Cuticular wax is generally a mixture of relatively simple hydrocarbons, wax esters, and fatty alcohols, ketones, and acids. There are some exceptions; for example, 70% of the wax of sultana grape skin is the triterpenoid oleanolic acid. A large proportion of the wax is present as a thin coat on the plant surface where it can be readily removed free of contaminating intracellular lipids merely by dipping the plant part in an organic solvent. Although convenient, this treatment is not guaranteed to effect a quantitative extraction of the waxes, particularly in those plant species where some wax components are tightly associated with the cutin layer.

KOLATTUKUDY and WALTON (1973) have reviewed the known components of cuticular waxes. Some of the more widespread components are listed in Table 3.

3.1 Hydrocarbons

Hydrocarbons are prevalent in most higher plant waxes. The most commonly found hydrocarbons are the C_{29} and C_{31} straight chain alkanes. Iso- and, less often, anteiso-alkanes are also found, with the former having mainly odd chain lengths and the latter even chain lengths. Alkenes of similar chain lengths are sometimes present.

Hydrocarbons, as well as the other wax components, are formed enzymatically in the epidermal layer of cells and extruded by an unknown mechanism to the

Table 3. Some common components of plant cuticular waxes

Compound type	Structural formula	Usual range of chain lengths
n-Alkanes	$CH_3(CH_2)_nCH_3$	C_{25}-C_{35}
Iso-alkanes	$CH_3\overset{CH_3}{\overset{\vert}{C}}H(CH_2)_nCH_3$	C_{25}-C_{35}
Alkenes	$CH_3(CH_2)_nCH{=}CH(CH_2)_mCH_3$	C_{17}-C_{33}
Monoketones	$CH_3(CH_2)_n\overset{O}{\overset{\Vert}{C}}(CH_2)_mCH_3$	C_{24}-C_{33}
β-Diketones	$CH_3(CH_2)_n\overset{O}{\overset{\Vert}{C}}CH_2\overset{O}{\overset{\Vert}{C}}(CH_2)_mCH_3$	C_{31}-C_{33}
Secondary alcohols	$CH_3(CH_2)_n\overset{OH}{\overset{\vert}{C}}H(CH_2)_mCH_3$	C_{20}-C_{33}
Wax esters	$CH_3(CH_2)_n\overset{O}{\overset{\Vert}{C}}{-}O(CH_2)_mCH_3$	C_{30}-C_{60}
Primary alcohols	$CH_3(CH_2)_nCH_2OH$	C_{12}-C_{36}
Normal fatty acids	$CH_3(CH_2)_n\overset{O}{\overset{\Vert}{C}}OH$	C_{12}-C_{36}
ω-hydroxy acids	$\overset{OH}{\overset{\vert}{C}}H_2(CH_2)_n\overset{O}{\overset{\Vert}{C}}OH$	C_{10}-C_{34}

tissue surface. At one time it was thought that the hydrocarbons were synthesized by the "head-to-head" condensation of two long-chain fatty acids. However, direct evidence obtained using higher plants is more in keeping with the continued elongation of fatty acids past the usual chain length of 16 or 18 carbon atoms (KOLATTUKUDY and WALTON, 1973). Small amounts of these very long-chain fatty acids are indeed found in the same tissues that contain the hydrocarbons. It appears that palmitic acid synthesized by the well-known fatty acid synthetase complex is elongated further by a different enzyme system located in the endoplasmic reticulum and using malonyl CoA as substrate for chain lengthening (KOLATTUKUDY and BUCKNER, 1972). It appears that a C_{29} alkane is formed by the decarboxylation of a C_{30} fatty acid.

3.2 Secondary Alcohols and Ketones

Most waxes contain secondary alcohols and ketones having chain lengths corresponding to those of the alkanes present. The hydroxy or carbonyl groups are located near the middle of the molecules. Isotope data (KOLATTUKUDY and LIU, 1970) indicate that in higher plants the secondary alcohols and ketones arise from hydrocarbons as shown in the following scheme (Fig. 4).

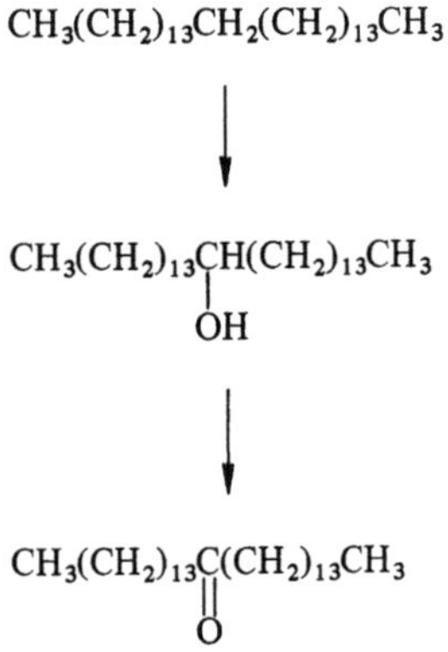

Fig. 4. The biosynthetic origin of secondary alcohols and ketones

3.3 Wax Esters

Wax esters are generally rather simple structures consisting of saturated primary alcohols, frequently C_{24}, C_{26}, and C_{28}, esterified to saturated fatty acids of the same size range. Many of the older compositional data (HILDITCH and WILLIAMS, 1964) were obtained by analyzing the free alcohols and acids produced by hydrolysis of the esters. The application of gas chromatography-mass spectrometry has enabled the molecular species of intact wax esters to be quantified (BRUUN, 1976), and newer developments in the field (BURLINGAME et al., 1976) promise to make molecular species analysis routine in the near future.

The mechanism of wax ester biosynthesis is not well understood. In broccoli leaf preparations KOLATTUKUDY (1967) found evidence for a direct transesterification of palmitic acid from phospholipid to a long-chain alcohol. Yet an acyl-CoA fatty alcohol transacylase was also demonstrated, leaving some uncertainty as to the preferred biosynthetic route.

3.4 Fatty Acids and Fatty Alcohols

The very long-chain fatty acids found in waxes have been shown experimentally to arise either de novo or from exogenous C_{16} and C_{18} acids (KOLATTUKUDY, 1968). This presumably indicates that the elongation step involves a fatty acid substrate bound to coenzyme A rather than acyl carrier protein.

Fatty alcohols are known to be derived from fatty acids in plant tissues. Enzymes from *Brassica oleracea* catalyzed the NADH-dependent reduction of [1-^{14}C]-palmitoyl-CoA to palmitaldehyde (KOLATTUKUDY, 1971). Another protein from the same preparation carried out further reduction of palmitaldehyde to hexadecanol with NADPH as cofactor.

3.5 The Use of Waxes as a Taxonomic Tool

The waxes, like the fatty acids of seed oils, are subject to considerable variation due to the influence of environmental factors. Furthermore, variation in wax com-

Table 4. ω-Hydroxy acids of leaf wax esters from Cupressaceae species

		C_{18}	C_{16}	C_{14}	C_{12}
Sub-family:	Cupressoideae				
Genus:	Cupressus				
Species:	*C. arizonica* Greene		19	13	68
	C. torulosa Don.		65	17	18
	C. funebris Endl.		60	30	10
	C. macrocarpa Hartw.		34	11	55
	C. lusitanica Mill.		19	8	73
	C. benthamii Endl.		22	7	71
	C. lindleyi Klotzsch		22	11	67
	C. nevadensis Abrams		25	14	61
	C. cashmeriana Royle		63	20	17
Sub-family:	Callitroideae				
Genus:	Callitris				
Species:	*C. columellaris* F. Muell.	15	34	42	9
	C. hugelii (Carr.) Franco	9	49	31	15
	C. preissii Miq.	11	59	10	20
	C. endlicheri Comb.	11	43	31	16
	C. rhomboidea R. Br.	26	45	18	11
Genus:	Widdringtonia				
Species:	*W. juniperoides* Endl.		25	3	72
	W. schwarzii Manl.		32	9	59
	W. cupressoides Endl.		40	8	52
	W. whytei Rendle		48	4	48

position can occur from one organ of a plant to another, and in *Juglans* quantitative differences were even observed between the lower and the upper surfaces of the same leaf (STRANSKY et al., 1967). For these reasons waxes are of dubious value as taxonomic markers. Qualitative variations in wax components among species would seem more useful for comparative studies than quantitative changes. A number of taxonomic analyses of waxes have been summarized by MARTIN and JUNIPER (1970). Because of their ease of extraction, the alkanes have been most widely compared. However, in some cases other wax components have been shown to be superior guides in aligning genera and even species groups within genera; for example, Table 4 indicates the close correlation between leaf wax ω-hydroxy acids and species arrangements in the Cupressaceae (HERBIN and ROBINS, 1968). Alkanes did not show nearly so striking a specificity in this study.

4 Cutin

4.1 The Nature and Origin of Cutin Monomers

Although the solubility properties of cutins in their natural form would not suggest their classification as lipids, they are in fact lipid polymers in which hydroxy

Table 5. Characteristic cutin monomers

Structural Formula	Name
$\underset{OH}{CH_2}-(CH_2)_4CH{=}CH-CH_2-CH{=}CH-(CH_2)_7COOH$	18-hydroxyoctadeca-9,12-dienoic
$\underset{OH}{CH_2}-(CH_2)_5\underset{OH}{CH}-(CH_2)_8COOH$	10,16-dihydroxyhexadecanoic
$\underset{OH}{CH_2}-(CH_2)_7\underset{OH}{CH}-\underset{OH}{CH}-(CH_2)_7COOH$	*threo*-9,10,18-trihydroxy-octadecanoic

fatty acids exist as an interesterified, rigid meshwork. The plant cuticle consists of this material embedded in a matrix of wax (KOLATTUKUDY and WALTON, 1973). Depolymerization of cutin by saponification or hydrogenolysis (WALTON and KOLATTUKUDY, 1972) followed by analysis using gas chromatography-mass spectrometry, has revealed dramatic differences in cutin monomers among various species. The cutin acids, primarily mono-, di-, and tri-hydroxy fatty acids (Table 5) fall into two principal families, depending upon whether they contain 16 or 18 carbon atoms. Cutin from papaya fruit is composed exclusively of C_{16} acids, while peach, grape, apple, and pear fruit cutins each contains a characteristic mixture of C_{16} and C_{18} monomers (WALTON and KOLATTUKUDY, 1972).

KOLATTUKUDY and WALTON (1973) proposed the following pathway for biosynthesis of the major cutin acids (Fig. 5), based on a variety of studies with isotopically labeled intermediates. Polymerization of the cutin acids to form cutin itself may be at least partly a nonenzymatic process. However, the rapidity of cutin polymerization at physiological temperatures suggests that enzymes are involved.

4.2 The Use of Cutin Monomers for Taxonomic Purposes

Little use has yet been made of cutin acids as a taxonomic tool. A survey of gymnosperm cutins by HUNNEMAN and EGLINTON (1972) revealed a large variety of components (some 155 cutin acids seem possible) which may be of taxonomic value. The authors felt that they could detect a pattern whereby plants higher on the evolutionary scale contained cutin acids having a higher degree of oxidation. This is obviously an area deserving extensive investigation.

A recent examination of cutins from *Sphagnum* mosses showed them to be rather similar to those found in higher plants, whereas two species of liverworts contained large amounts of bound 15-hydroxyhexadecanoic acid, a component hitherto unreported in higher plants (CALDICOTT and EGLINTON, 1976). Thus *Sphagnum* mosses may be more closely related to the higher plants than are other bryophytes, such as the liverworts.

As the study of cutin acids increases, the same precautions necessary when using seed fatty acids for chemotaxonomic purposes will need to be exercised. Significantly different cutin compositions were found in leaves and in embryonic

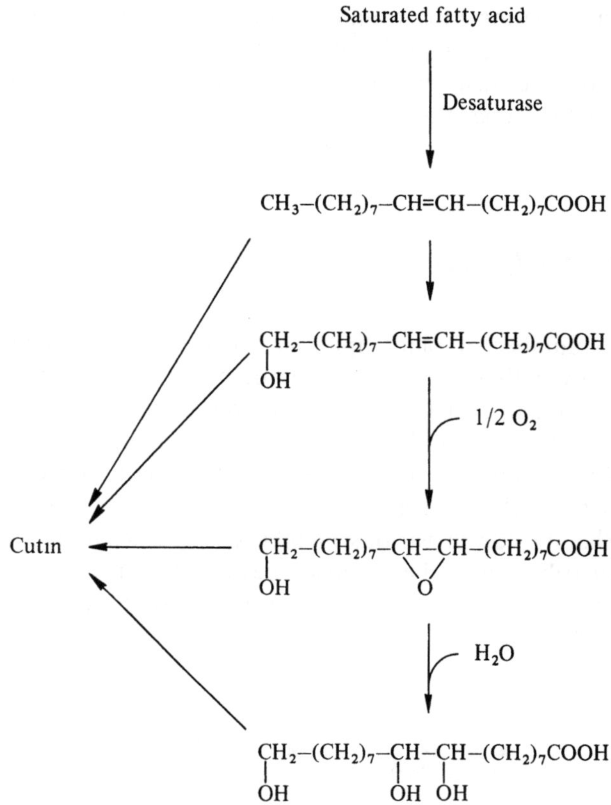

Fig. 5. The biosynthesis of various types of cutin monomer

shoots of *Vicia faba*, suggesting that the physiological state of the tissue may sometimes be of importance in determining cutin make-up (KOLATTUKUDY and WALTON, 1973).

5 Suberin

Whereas cutin is mainly associated with the aerial parts of plants, the corresponding lipid-derived polymer shielding underground parts (and periderms) is known as suberin (KOLATTUKUDY, 1975). Suberin is qualitatively rather similar in composition to cutin, but the proportions of the various constituents are quite different. Long-chain (C_{16} to C_{22}) dicarboxylic acids (predominantly octadec-9-ene-1,18-dioic acid) and very long-chain (C_{20} to C_{26}) acids and alcohols are much more in evidence in suberin. These aliphatic monomers account for only about one-third of the weight of suberin, with the remaining portion being composed of phenolic compounds.

Only a handful of plant species has been analyzed for suberin monomer composition. A consideration of suberin as a useful predictor of taxonomic or evolutionary interrelationships would certainly be premature at this time. However, since of the species examined to date each appears to have its own characteristic quantitative monomer distribution (KOLATTUKUDY et al., 1975), suberin cannot be ignored for phylogenetic studies.

6 Polyketides

A number of plant lipids are formed by the sequential condensation of acetyl CoA with several molecules of malonyl CoA in a mechanism which does not involve loss of the resulting carbonyl groups. The polyketo acids are themselves very unstable and probably exist only complexed with their biosynthetic enzymes. Cyclization can give rise to a variety of natural products, including anthraquinones such as the hydrophobic compound emodin (Fig. 6), which occurs mainly as the rhamnoside in various members of the Rhamnaceae and the Polygonaceae. The distribution of these polyketides has been reviewed by THOMSON (1976) and, since they are usually accumulated as water-soluble glycosides, they will not be considered further here.

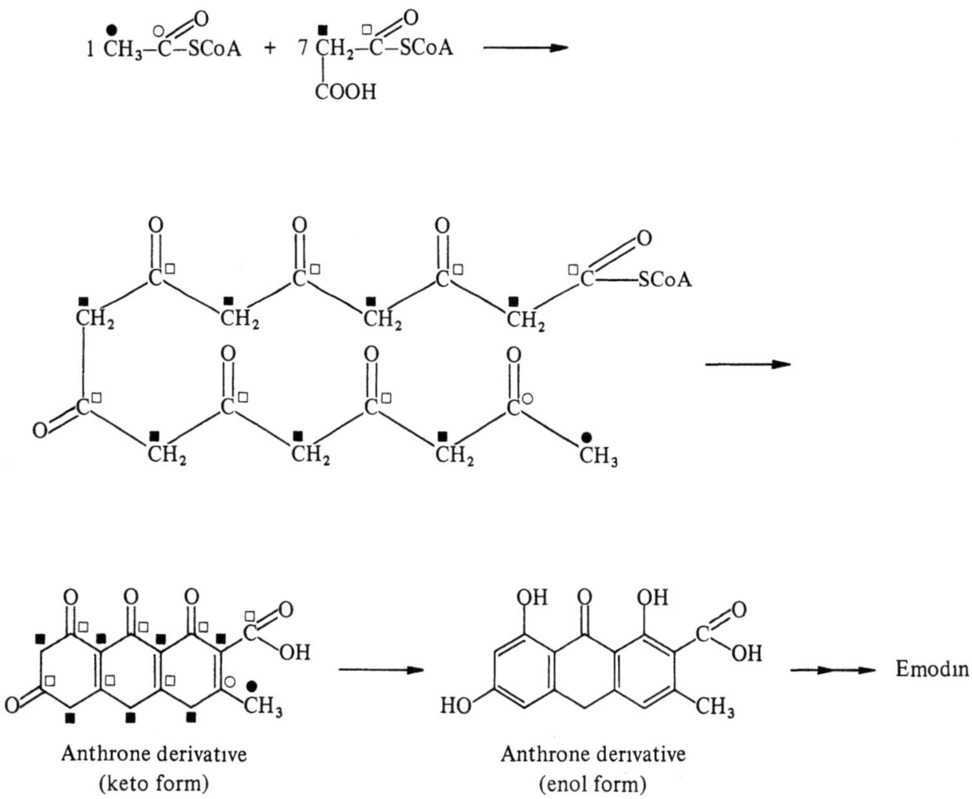

Fig. 6. Polyketide participation in the formation of emodin

7 Acetylenic Compounds

7.1 The Nature and Metabolism of Acetylenes

Analyses of many higher plant lipids have detected several hundred different compounds containing one or more acetylenic linkages. Still other acetylenes have been found in microorganisms. The structures of the more common acetylenes have been reviewed by BOHLMANN (1973), and typical examples of acetylenic compounds are given in Tables 2 and 6.

Many of the higher plant acetylenes contain 17 carbon atoms. These are known to be derived from linoleic acid by the pathway shown in Figure 7. A C_{17}-hydrocarbon arises through decarboxylation of one of the structures as depicted in Figure 8. Many variations on the above theme can yield a diversity of acetylenic compounds, including shorter chain length or hydroxy derivatives, amides, and a number of cyclic products.

Quite unlike the unusual fatty acids, which are confined mainly to seed fats and the waxes of the plant cuticle, the acetylenic compounds seem to occur throughout the plant. The dehydrogenation steps outlined in Figure 7 were shown to be catalyzed by enzymes of the large particulate fraction (the 1000 g pellet) of immature safflower seeds (ICHIHARA and NODA, 1977).

7.2 The Use of Acetylenes for Taxonomic Purposes

Most of the known acetylenes occur in the Compositae and Umbelliferae. The wide variety of compounds coupled with the fact that many of them are restricted to only a very few genera make for a taxonomically valuable asset. For example,

Table 6. Typical acetylenic compounds of higher plants (SORENSEN, 1963)

Compound	Genus
$CH_3[C{\equiv}C]_3CH{=}CH{-}CH{=}CH(CH_2)_3CH{=}CH_2$	*Chrysanthemum* (Compositae)
$CH_3[C{\equiv}C]_3CH{=}CH{-}CH{=}CH{-}CH_2{-}CH_2{-}O{-}C({=}O){-}CH_3$	*Chrysanthemum* (Compositae)
$CH_3(CH_2)_6CH{=}CH{-}CH(OH)[C{\equiv}C]_2C({=}O){-}CH{=}CH_2$	*Carum, Oenanthe, Sium* (Umbelliferae)
$CH_3[C{\equiv}C]_4CH{=}CH{-}CH(Cl){-}CH_2OH(OAC)$	*Centaurea* (Compositae)
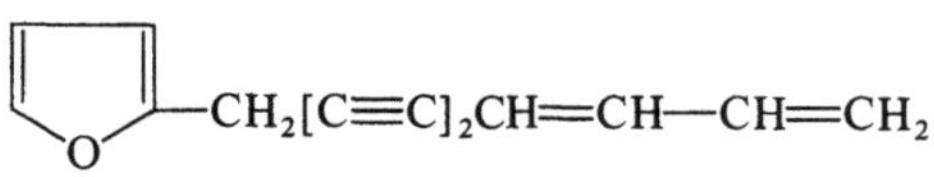	*Carlina* (Compositae)
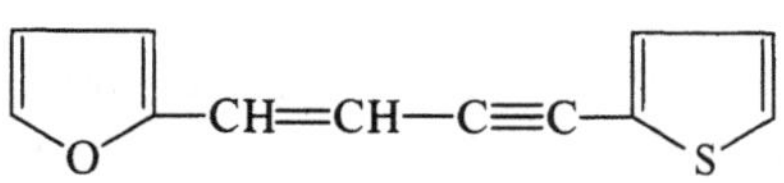	*Santolina* (Compositae)

$CH_3(CH_2)_4CH{=}CHCH_2CH{=}CH(CH_2)_7COOH$

↓

$CH_3(CH_2)_4C{\equiv}CCH_2CH{=}CH(CH_2)_7COOH$

↓

$CH_3CH_2CH_2CH{=}CHC{\equiv}CCH_2{-}CH{=}CH(CH_2)_7COOH$

↓

$CH_3[CH{=}CH]_2C{\equiv}CCH_2CH{=}CH(CH_2)_7COOH$

↓

$CH_3CH{=}CH[C{\equiv}C]_2CH_2CH{=}CH(CH_2)_7COOH$

↓

$CH_3[C{\equiv}C]_3CH_2CH{=}CH(CH_2)_7COOH$

Fig. 7. Formation of acetylenic compounds from linoleic acid

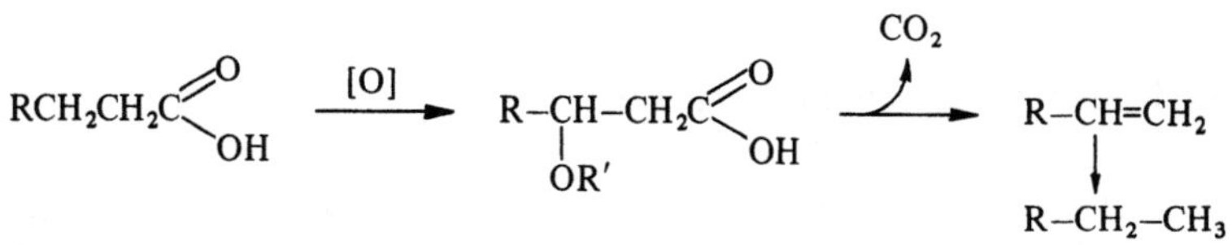

Fig. 8. The formation of acetylenic hydrocarbons by decarboxylation of acetylenic fatty acids

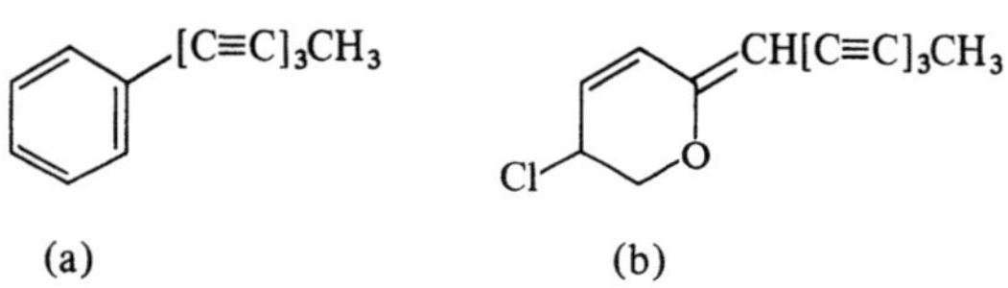

Fig. 9. Characteristic acetylenic compounds of **a** *Coreopsis* and **b** *Anaphalis*

compound (a) (Fig. 9), occurs only in *Coreopsis* species while (b) is found only in *Anaphalis* (BOHLMAN, 1973). Many other species can be grouped together on the basis of their acetylene distribution, and interrelationships among genera have also been based upon the occurrence of these compounds.

8 Cyanolipids

Since 1970 an entirely new class of seed oil components has been characterized. The compounds, termed cyanolipids, are of great practical interest because, like the cyanogenic glycosides, most of them give rise to hydrogen cyanide under

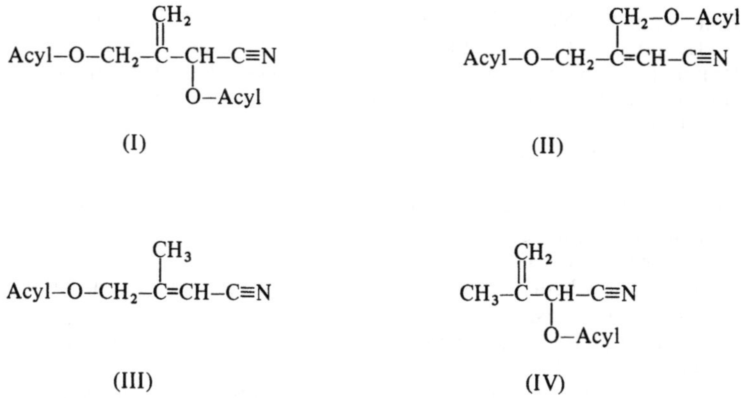

Fig. 10. Naturally occurring cyanolipids

mild hydrolysis, either enzymatically of nonenzymatically. Their distribution and properties have recently been reviewed (MIKOLAJCZAK, 1977).

Four cyanolipids have been identified (Fig. 10). In many plant species one compound occurs to the virtual exclusion of the others. Thus compound I is by far the predominant cyanolipid in *Allophyllus edulis*, *Cardiospermum halicacabum*, and *Paullina meliaefolia*, while compound II is the principal cyanolipid in *Nephelium lappaceum*, and *Sapindus mukorossi* (MIKOLAJCZAK et al., 1970). Compounds III and IV are found in *Stocksia brahuica* and *Ungnadia speciosa*, respectively. The acyl moieties consist mainly of oleic acid and both saturated and monounsaturated C_{20} fatty acids (MIKOLAJCZAK, 1977).

All authenticated sources of cyanolipids are in the family Sapindaceae, an early report of compound I in *Cordia verbenaceae* (Boraginaceae) probably having been erroneous (SEIGLER, 1976). SEIGLER and KAWAHARA (1976) have extended the search for cyanolipids to a wide variety of species in the Sapindaceae and in Hippocastanaceae, Meliomaceae, and Staphyleceae, all families in the Sapindales Order. Cyanolipids were found only in the Sapindaceae and not in all members of this family. Evidence sufficient for a meaningful chemotaxonomic evaluation should soon be available.

References

Appleqvist, L.-Å.: Lipids in the cruciferae. In: The biology and chemistry of the cruciferae. Vaughan, J.G., MacLeod, A.J., Jones, B.M.G. (eds.), pp. 221–277, New York: Academic Press 1976

Baker, E.A.: The influence of environment on leaf wax development in *Brassica oleracea* var. *gemmifera*. New Phytol. *73*, 955–966 (1974)

Barker, C., Hilditch, T.P.: The influence of environment upon the composition of sunflower seed oils. I. Individual varieties of sunflowers grown in different parts of Africa. J. Sci. Food Agric. *1*, 118–121 (1950a)

Barker, C., Hilditch, T.P.: The influence of environment upon the composition of sunflower seed oils. II. Composition of the seed oils of sunflowers grown in English gardens from five specimens of different African sunflower seed. J. Sci. Food Agric. *1*, 140–144 (1950b)

Bohlmann, F.: Naturally occurring acetylenes. In: Phytochemistry. Miller, L.P. (ed.), Vol. III, pp. 112–131, New York: Van Nostrand Reinhold 1973

Bruun, T.: Aliphatic compounds in some lichens. Phytochemistry *15*, 1261–1263 (1976)

Burlingame, A.L., Kimble, B.J., Derrick, P.J.: Mass spectrometry. Anal. Chem. *48*, 368R–403R (1976)

Caldicott, A.B., Eglinton, G.: Cutin acids from Bryophytes: an ω-1 hydroxyalkanoic acid in two liverwort species. Phytochemistry *15*, 1139–43 (1976)

Erwin, J.: Comparative biochemistry of fatty acids in eukaryotic microorganisms. In: Lipids and biomembranes of eukaryotic microorganisms. Erwin, J. (ed.), pp. 41–143, New York: Academic Press 1973

Gibbs, R.D.: Chemotaxonomy of flowering plants. Vol. 1, pp. 493–523. Montreal: McGill-Queen's University Press 1974a

Gibbs, R.D.: Chemotaxonomy of flowering plants, 4 Vols. Montreal: McGill-Queen's University Press 1974b

Gurr, M.I.: The biosynthesis of unsaturated fatty acids. In: MTP Int. Rev. Sci. Biochem. Ser. 1. Goodwin, T.W. (ed.), Vol. 4, pp. 181–235. London: Butterworths 1974

Harborne, J.B.: Phytochemical methods. pp. 132–165, London: Chapman and Hall 1973

Herbin, G.A., Robins, P.A.: Studies on plant cuticular waxes. III. The leaf wax alkanes and ω-hydroxy acids of some members of the Cupressaceae and Pinaceae. Phytochemistry *7*, 1325–1337 (1968)

Hilditch, T.P., Williams, P.N.: The chemical constitution of natural fats, 4th ed. London: Chapman and Hall 1964

Hitchcock, C., Nichols, B.W.: Plant lipid biochemistry. pp. 157–160. London: Academic Press 1971

Hunneman, D.H., Eglinton, G.: The constituent acids of Gymnosperm cutins. Phytochemistry *11*, 1989–2001 (1972)

Ichihara, K.-I., Noda, M.: Distribution and metabolism of polyacetylenes in safflower. Biochim. Biophys. Acta *487*, 249–260 (1977)

Johnson, A.R., Pearson, J.A., Shenstone, F.S., Fogerty, A.C., Giovanelli, J.: The biosynthesis of cyclopropane and cyclopropene fatty acids in plant tissues. Lipids *2*, 308–315 (1967)

Kates, M.: Techniques in lipidology. In: Laboratory techniques in biochemistry and molecular biology. Work, S.T., Work, E. (eds.), Vol. 3, pp. 269–610. Amsterdam: North-Holland 1972

Klein, R.M., Cronquist, A.: Consideration of the evolutionary and taxonomic significance of some biochemical, micromorphological, and physiological characters in the Thallophytes. Q. Rev. Biol. *42*, 105–296 (1967)

Kolattukudy, P.E.: Mechanisms of synthesis of waxy esters in broccoli (*Brassica oleracea*). Biochemistry *6*, 2705–2717 (1967)

Kolattukudy, P.E.: Tests whether a head to head condensation mechanism occurs in the biosynthesis of n-hentriacontane, the paraffin of spinach and pea leaves. Plant Physiol. *43*, 1466–1470 (1968)

Kolattukudy, P.E.: Enzymatic synthesis of fatty alcohols in *Brassica oleracea*. Arch. Biochem. Biophys. *142*, 701–709 (1971)

Kolattukudy, P.E.: Biochemistry of cutin, suberin, and waxes, the lipid barriers on plants. In: Recent advances in the chemistry and biochemistry of plant lipids. Galliard, T., Mercer, E.I. (eds.). pp. 203–246. London: Academic Press 1975

Kolattukudy, P.E., Buckner, J.S.: Chain elongation of fatty acids by cell-free extracts of epidermis from pea leaves (*Pisum sativum*). Biochem. Biophys. Res. Commun. *46*, 801–807 (1972)

Kolattukudy, P.E., Liu, T.-Y.J.: Direct evidence for biosynthetic relationships among hydrocarbons, secondary alcohols and ketones in *Brassica oleracea*. Biochem. Biophys. Res. Commun. *41*, 1369–1374 (1970)

Kolattukudy, P.E., Walton, T.J.: The biochemistry of plant cuticular lipids. Prog. Chem. Fats Other Lipids Part 3, *13*, 121–175 (1973)

Kolattukudy, P.E., Kronman, K., Poulose, A.J.: Determination of structure and composition of suberin from the roots of carrot, parsnip, rutabaga, turnip, red beet, and sweet potato by combined gas-liquid chromatography and mass spectrometry. Plant Physiol. *55*, 567–573 (1975)

Luckner, M.: Secondary metabolism in plants and animals (English translation). p. 3. London: Chapman and Hall 1972
Martin, J.T., Juniper, B.E.: The cuticles of plants. pp. 140–144. New York: St. Martin's Press 1970
Mikolajczak, K.L.: Cyanolipids. Prog. Chem. Fats Other Lipids *15*, 97–130 (1977)
Mikolajczak, K.L., Smith, C.R., Jr., Tjarks, L.W.: Cyanolipids of *Cardiospermum halicacabum L.* and other sapindaceous seed oils. Lipids *5*, 812–817 (1970)
Morris, L.J.: The mechanism of ricinoleic acid biosynthesis in *Ricinus communis* seeds. Biochem. Biophys. Res. Commun. *29*, 311–315 (1967)
Morris, L.J., Hall, S.W., James, A.T.: The biosynthesis of ricinoleic acid by *Claviceps purpurea.* Biochem. J. *100*, 29c-30c (1966)
Nordby, H.D., Nagy, S., Hall, N.: Fatty acids of monogalactosyl diglycerides from *Citrus.* Phytochemistry *15*, 957–960 (1976)
Raju, P.K., Reiser, R.: Inhibition of fatty acyl desaturase by cyclopropene fatty acids. J. Biol. Chem. *242*, 379–384 (1967)
Rollins, R.C., Shaw, E.A.: The genus *Lesquerella* (Cruciferae) in North America. pp. 8–9. Cambridge: Harvard Univ. Press 1973
Sagan, L.: On the origin of mitosing cells. J. Theor. Biol. *14*, 225–274 (1967)
Seigler, D.S.: Cyanolipids in *Cordia verbenacea*- a correction. Biochem. Syst. Ecol. *4*, 235–236 (1976)
Seigler, D.S., Kawahara, W.: New reports of cyanolipids from sapindaceous plants. Biochem. Syst. Ecol. *4*, 263–265 (1976)
Shorland, F.B.: The distribution of fatty acids in plant lipids. In: Chemical plant taxonomy. Swain, T. (ed.), pp. 253–311. London: Academic Press 1963
Smith, C R., Jr.: Occurrence of unusual fatty acids in plants. Prog. Chem. Fats Other Lipids Part 1, *11*, 139–177 (1970)
Smith, P M.: The chemotaxonomy of plants. pp. 70–76. Amsterdam: Elsevier 1976
Sorensen, N.A.: Chemical taxonomy of acetylenic compounds. In: Chemical plant taxonomy. Swain, T. (ed.), pp. 219–252. London: Academic Press 1963
Stransky, K.M., Streibl, M., Herout, V.: On natural waxes. VI. Distribution of wax hydrocarbons in plants at different evolutionary levels. Collect. Czech. Chem. Commun. Engl. ed. *32*, 3213–3220 (1967)
Thomson, R.H.: Quinones: nature, distribution and biosynthesis. In: Chemistry and biochemistry of plant pigments, 2nd ed. Goodwin, T.W. (ed.), Vol. 1, pp. 527–559. London: Academic Press 1976
Vickery, J.R.: The fatty acid composition of the seed oils of Proteaceae: a chemotaxonomic study. Phytochemistry *10*, 123–130 (1971)
Walton, T.J., Kolattukudy, P.E.: Determination of the structures of cutin monomers by a novel depolymerization procedure and combined gas chromatography and mass spectrometry. Biochemistry *11*, 1885–1896 (1972)

13 Plant Carbohydrates [1]

A.M. STEPHEN

1 Introduction

In reviewing plant carbohydrates of potential taxonomic interest it has been decided to consider only Gymnosperms and Angiosperms amongst land plants (Sects. 2–6), and certain algae (Sect. 7). The carbohydrate components of fungi, of which the branched hetero-mannans, -galactans and -glucans are particularly significant, have been the subject of recent comprehensive reviews (GORIN and SPENCER, 1968; LINDBERG and SVENSSON, 1974; STURGEON, 1974, 1975, 1976), and bacterial polysaccharides (LINDBERG and SVENSSON, 1974) constitute so large a field as to make it pointless to attempt to generalise adequately in the space available. The common monosaccharides that participate in normal metabolic processes, and their simple, water-soluble derivatives (e.g., sucrose, raffinose, stachyose, maltose, lactose), are so widely distributed (BAILEY, 1965; DARNLEY GIBBS, 1974) as to be without taxonomic value; however, even the fructose/glucose/sucrose ratios in the nectars of over thirty species of *Fritillaria* (Liliaceae) have been used to support the division, made on general taxonomic grounds, of these species into four groups (RIX and RAST, 1975). Volume 1 of this Encyclopedia includes a survey (ZIMMERMANN and ZIEGLER, 1975) of certain alditols and oligosaccharides in the exudates from cut bark taken from over 500 species in 100 dicotyledonous families. Distinct taxonomic relationships can be seen between some of the main components selected for the study and the families from which they were obtained. The distribution in plant taxa of alditols containing from four to eight carbon atoms (STACEY, 1974) and of certain heptuloses (McCOMB and RENDIG, 1977) has recently been reviewed. A detailed account of the alditols, cyclitols, and simple glycosides that occur in algae is available (CRAIGIE, 1974), and in this regard there has been an important investigation as to whether or not D-mannitol is a constituent of Rhodophyceae, where glycerol galactosides are characteristic (KREMER, 1976). A rather more general examination of algal constituents has also been started (IMPELLIZZERI et al., 1975). Less common derivatives of monosaccharides, and other oligosaccharides, are found widely in plants glycosidically bonded to flavonoids and isoprenoids, and are discussed appropriately in other chapters of this volume.

Other low molecular weight carbohydrates might be considered, among them the series of galactosyl derivatives of sucrose of higher molecular weight than stachyose, certain glucosyl- and fructosyl-sucroses (DARNLEY GIBBS, 1974) and the novel pinitol galactoside of legume seeds (BEVERIDGE et al., 1977), but it was concluded that this article should concentrate on the polysaccharide group. It has been recognized (PERCIVAL, 1968a; ANDERSON and DEA, 1969; BAILEY, 1973)

1 In the general physiological context plant carbohydrates will be treated in two forthcoming volumes of this Encyclopedia (TANNER, W., LOEWUS, F., eds.).

that heteroglycans, in particular, have taxonomic potential because of the numbers of constituent monosaccharide units that may be present (six or even more) and the numerous possibilities for inter-sugar linkage and the grouping of sugars in different sequences (TOWLE and WHISTLER, 1973). Furthermore, there is a logical means of subdividing heteroglycans on the basis of the main features of the interior or "core" chains of sugars upon which the polysaccharide structure is built (ASPINALL, 1967, 1969). The enzyme systems responsible for the many structural features present in polysaccharide molecules are of course genetically controlled, and these will, in general, be complementary to those that govern the morphological characteristics used in plant classification. It is also recognised that the physiological functions required of different polysaccharides are determined by conformations and inter-chain associations which in turn depend upon their primary structures (REES, 1972, 1974); thus molecular structure as expressed in terms of inter-sugar linkages and sequences is tied to function, and function to classification.

Several hundred different polysaccharides have been isolated from plants (TOWLE and WHISTLER, 1973), and although from the chemotaxonomic point of view the ubiquity of many glucans (cellulose, callose, and starches), fructosans of the inulin type, and many of the heteroglycans that comprise the pectic group of substances (ALBERSHEIM, 1974a) disqualifies these groups from serious consideration, certain correlations of the structures of plant exudates (gums), endosperm polysaccharides, and marine algal compounds with taxa of origin appear to be possible. Whereas there is good reason for describing algal polysaccharide constituents under the headings of the taxonomic groups with which they are associated, it is more practical to work from the accepted general classification of plant polysaccharides based upon interior molecular structure (ASPINALL, 1967, 1969, 1973; OVODOV, 1975; PERCIVAL, 1966; PERCIVAL and MCDOWELL, 1967) and to note the distribution of particular examples among the taxa of land plants.

A number of circumstances militate against the successful use of polysaccharide structures by the taxonomist, most of them relating to the difficulty of determining any such structure that can be said, uniquely, to represent a component that is characteristic of the living plant or part thereof. For instance, the quantity of a polysaccharide generated is often subject to seasonal and environmental variations, and to stages of growth (BAILEY, 1973; CRAIGIE, 1974; HAUG et al., 1974); polysaccharide components of different tissues may vary (HOGSETT and QUATRANO, 1975); the polysaccharide itself may undergo continuous, sometimes rapid, modifications in molecular structure (BUCHALA and WILKIE, 1973; IZUMI, 1971, 1972; JOSELEAU and BARNOUD, 1975; MORRISON, 1974; NORTHCOTE, 1974; PERCIVAL and YOUNG, 1974; REES, 1961); and the problems associated with isolating the polysaccharide for study may differ from one extreme (solvent extraction of the free substance) to another [where polysaccharides may be physically and covalently joined to each other and to protein (ANDERSON and HENDRIE, 1971; HENDERSON and HAY, 1972) or lignin (BAILEY, 1973)]. Sub-fractionation of isolated polysaccharide material is normally carried out to the point where all available physical and chemical criteria fail to give evidence of heterogeneity (ASPINALL, 1967), but the technical difficulties are such that small variations in structure may, on the one hand, remain undetected, or, if observed, be of questionable significance.

Nevertheless, the very complexity of many of the polysaccharides to be discussed is a point in favour of their being recognised as useful taxonomic markers; the more complicated the molecular structure, the more demanding are the genetic requirements for its synthesis. Taxonomic categories are, of course, based largely on resemblances in habit and floral morphology, and the gene complexes that govern morphological characteristics are not necessarily linked with those which lead to polysaccharide synthesis; nevertheless the degree of success attained for other classes of natural product in correlating chemical structure with sources of origin suggests that, given sufficient examples to work upon, valid conclusions may be reached for polysaccharides as well. At present the situation is that far fewer than one species of plant in a thousand has received anything like adequate study with respect to its carbohydrate composition, and any generalisations which follow must necessarily be regarded in this light. More often than not the reason for studying the carbohydrate constituents of plants has been a commercial one; only in exceptional cases have systematic chemotaxonomic programmes been initiated (ANDERSON and DEA, 1968, 1969; ANDERSON and HENDRIE, 1970; ANDERSON and MUNRO, 1969). For the purposes of this present review, a dendrogram showing probable relationships of the orders of flowering plants has been included (see diagram on p. 560, 561).

2 Arabinans, Galactans and Arabinogalactans

Few arabinans, which are highly branched, acid-sensitive polysaccharides, have been isolated under sufficiently mild conditions to keep the molecular structure intact. Present in the seeds of two species of *Brassica* (Cruciferae; Capparales), white mustard (*B. alba*) (REES and RICHARDSON, 1966), and rapeseed (*B. campestris*) (SIDDIQUI and WOOD, 1974), and of soybean (*Glycine max*; Phaseoleae; Papilionaceae; Fabales) (ASPINALL and COTTRELL, 1971) are polymers made up of 1→5-linked α-L-arabinofuranosyl residues, extensively substituted at C-3 and also at C-2. Arabinans are also generally associated with the polysaccharides of plant cell walls (NORTHCOTE, 1972; KEEGSTRA et al., 1973) and constitute part of what has been described as a matrix of covalently cross-linked polymers whose composition changes continuously with growth. Such arabinans necessarily pose additional problems in their isolation. Products similar in molecular constitution have been extracted from *Pinus pinaster* (Pinaceae; Coniferae; Gymnospermae) wood (ROUDIER and EBERHARD, 1965), from the bark of young twigs of *Salix alba* (KARACSONYI et al., 1975) and bark of another tree of the same family aspen (*Populus tremuloides*) (JIANG and TIMELL, 1972), and from inner bark of *Rosa glauca* (Rosaceae; Rosales) (JOSELEAU et al., 1977).

The types of linkage between L-arabinofuranosyl residues that are found in the arabinans are found also between the arabinose units covalently linked to D-galactan frameworks in arabinogalactans. Here again the extent to which these heteroglycans are bound to woody tissues (ASPINALL, 1973; KEEGSTRA et al., 1973) is reflected in their ease of extraction or otherwise. For example, hot water dissolves from the wood of several *Larix* spp. (Pinaceae; Coniferae) (ASPINALL, 1967; TOWLE and WHISTLER, 1973) a considerable proportion of the total carbohydrate present, which consists of a chain of 1→3-linked β-D-galactopyranosyl residues, substituted at C-6 by short galactose chains, to which L-arabinofuranosyl and 3-O-β-L-arabinopyranosyl-L-arabinofuranosyl groups are attached. The extract from *L. occidentalis* has been separated into two fractions, one comprising about 100 sugar units and the other six times this number (BOUVENG, 1959). The constitution of these arabinogalactans is of considerable importance in relation to the molecular structures of many plant exudates as described in detail below (Sect. 3), the more so as some larch arabinogalactans do include bound D-glucuronic acid residues in their structure (ASPINALL, 1967; PERCIVAL, 1966; SIMSON et al., 1968). Similarly constituted, neutral arabinogalactans may be extracted from the sap of the sugar maple (*Acer saccharum*; Aceraceae; Sapindales, from suspension-cultured sycamore (*Acer pseudoplatanus*) cells (ASPINALL et al., 1969), from young bamboo shoots (*Phyllostachys reticulata* and *edulis*; Gramineae; Poales) (MAEKAWA and KITAO, 1974; MAEKAWA, 1976) and, under alkaline, degradative conditions, from sugarbeet (*Beta vulgaris*; Chenopodiaceae; Caryophyllales) (ASPINALL, 1970; cf. ASPINALL, 1973). Arabinogalactans of this type are in fact widely distributed and have also been extracted from cereal sources, such as wheat flour (NEUKOM and MARKWALDER, 1975) and (in association with peptides) wheat endosperm (FINCHER et al., 1974), and from a moss *Fontinalis antipyretica* (Bryophyta) (GEDDES and WILKIE, 1971).

D-Galactans have been isolated from citrus pectin (LABAVITCH et al., 1976), from seeds (of *Lupinus albus*; Papilionaceae, and *Strychnos nux-vomica*; Loganiaceae; Gentianales), from tubers (of *Solanum tuberosum*; Solanaceae; Scrophulariales) and bulbs (of *Allium sativum*; Liliaceae; Liliales) (DAS et al., 1977), and, after appropriate delignification, from the compression wood of Norway spruce (*Picea excelsa*) (ASPINALL, 1970) and from the bark of *Salix alba* (TOMAN et al., 1972), but here the linkage between the β-D-galactopyranosyl residues is predominantly $1 \rightarrow 4$. Where, on the other hand, uronic acid residues as well as arabinose have been found joined to a galactan framework, as in the normal wood (and in compression wood) of *Larix laricina* (Pinaceae; Coniferae) (FU and TIMELL, 1972) and other conifers, the inter-sugar linkages are the same as in the neutral arabinogalactan extracts from *Larix* spp. A water-soluble polysaccharide present in cashew nut shells (from *Anacardium occidentale*; Anacardiaceae; Rutales) also contains a $1 \rightarrow 3$- and $1 \rightarrow 6$-linked β-D-galactan structure (BOSE and SONI, 1972), to which short arabinose chains linked to galacturonic acid are attached (BOSE and SONI, 1974). An acidic arabinogalactan from leaves of *Phaseolus atropurpureus* (Papilionaceae) differs from the neutral arabinogalactans of other legumes (*Glycine max, Medicago sativa* and *Centrosema plumieri*) and *Opuntia dillenii* (Cactaceae; Caryophyllales) (SRIVASTAVA and PANDE, 1974) which are based upon a $1 \rightarrow 4$-linked D-galactan, in having a $1 \rightarrow 3$- and $1 \rightarrow 6$-linked D-galactan framework and $1 \rightarrow 3$-linked L-arabinofuranosyl units in side chains (FORD, 1972).

ASPINALL (1973) has categorised the arabinogalactans on the basis of their galactan core structures as Type I (linear, $1 \rightarrow 4$) and Type II (branched).

3 Plant Gum Exudates Based on Arabinogalactan Structures

The main taxonomic interest in arabinogalactan-based polysaccharides lies in their occurrence as exudates from land plants of various species. On appropriate stimulation gum-producing plants exude, from stem, root, leaf or fruit, polysaccharides that are made up of a β-D-galactan core (linked 1 → 3 and 1 → 6), heavily substituted by outer chains containing L-arabinofuranosyl, D-glucopyranuronosyl, and L-rhamnopyranosyl groups. This common pattern is exhibited, for example, by gums of the genus *Acacia* (Mimosaceae; Fabales), the most widely studied group (SMITH and MONTGOMERY, 1959; ASPINALL, 1969, 1970; BAILEY, 1971; OVODOV, 1975). Other sugars, such as D-mannose, D-xylose, L-fucose, D-galacturonic acid and 3-O-methyl-L-rhamnose, are found linked as substituents in gums from other species of land plants (cf. CAFFINI and PRIOLO DE LUFRANO, 1974, 1975). The relative ease of purification of the gums for chemical study is a great advantage, and their use in the mining, foodstuffs, pharmaceutical and many other industries has provided a constant incentive for examination of their physical and chemical properties (WHISTLER, 1973).

3.1 *Acacia* Gums in Relation to Species of Origin

Gum exudates have been examined in detail for upwards of 40 genera representing some 30 families and 20 different orders of Angiosperms, and there are isolated examples from among Gymnosperms (STEPHEN and SCHELPE, 1964; SHAW and STEPHEN, 1965). Gums from the genus *Acacia* were among the first to be studied, possibly because of the extensive traditional use of "gum arabic" as an article of commerce, and by 1940 many of the features of the complex structure of gum samples, probably from *A. senegal*, had been elucidated using the painstaking methodology of the day. Newly developed chromatographic and spectroscopic methods opened the way to rapid examination of authenticated material from *Acacia* trees of different species, and by 1955 it was clear that while the same four constituent sugars (D-galactopyranose, L-arabinopyranose and -furanose, L-rhamnopyranose and D-glucopyranuronic acid) were present in *A. mearnsii* (then *A. mollissima*), *A. pycnantha* and *A. saligna* (then *A. cyanophylla*) gums, the proportions of the sugars and the total effects of the variously linked sugar residues on optical rotation varied considerably (SMITH and MONTGOMERY, 1959). There followed a series of investigations into the structure–species relationships of *Acacia* gums that has continued to the present time. The modes of linkage of the sugars in many examples from different species were found to be limited in number (KAPLAN and STEPHEN, 1967), and the basal polysaccharide chains were demonstrated in all cases to include 1→3-linked β-D-galactosyl and L-arabinosyl residues (as in the arabinogalactans of woody tissue, Sect. 2); in addition the outer chains include D-glucopyranuronic acid units often substituted at C-4 by methyl or L-rhamnopyranosyl groups. In all *Acacia* gums a β1 → 6 linkage joins the acid to D-galactose, and in many, an α1 → 4 linkage is present as well (ANDERSON and CREE, 1968). Details of variations of fine structure with species have been reviewed (ANDERSON and DEA, 1969; ASPINALL, 1969), and reference should be made to representative formulations for the gums of *A. pycnantha* (ASPINALL et al., 1959) and *A. senegal* (ANDERSON et al., 1966b).

It is clear, however, that although in these examples the inter-sugar linkages are accurately known, and they are present in the proportions implied by the formulae, there are in fact virtually limitless possibilities for the gum structures,

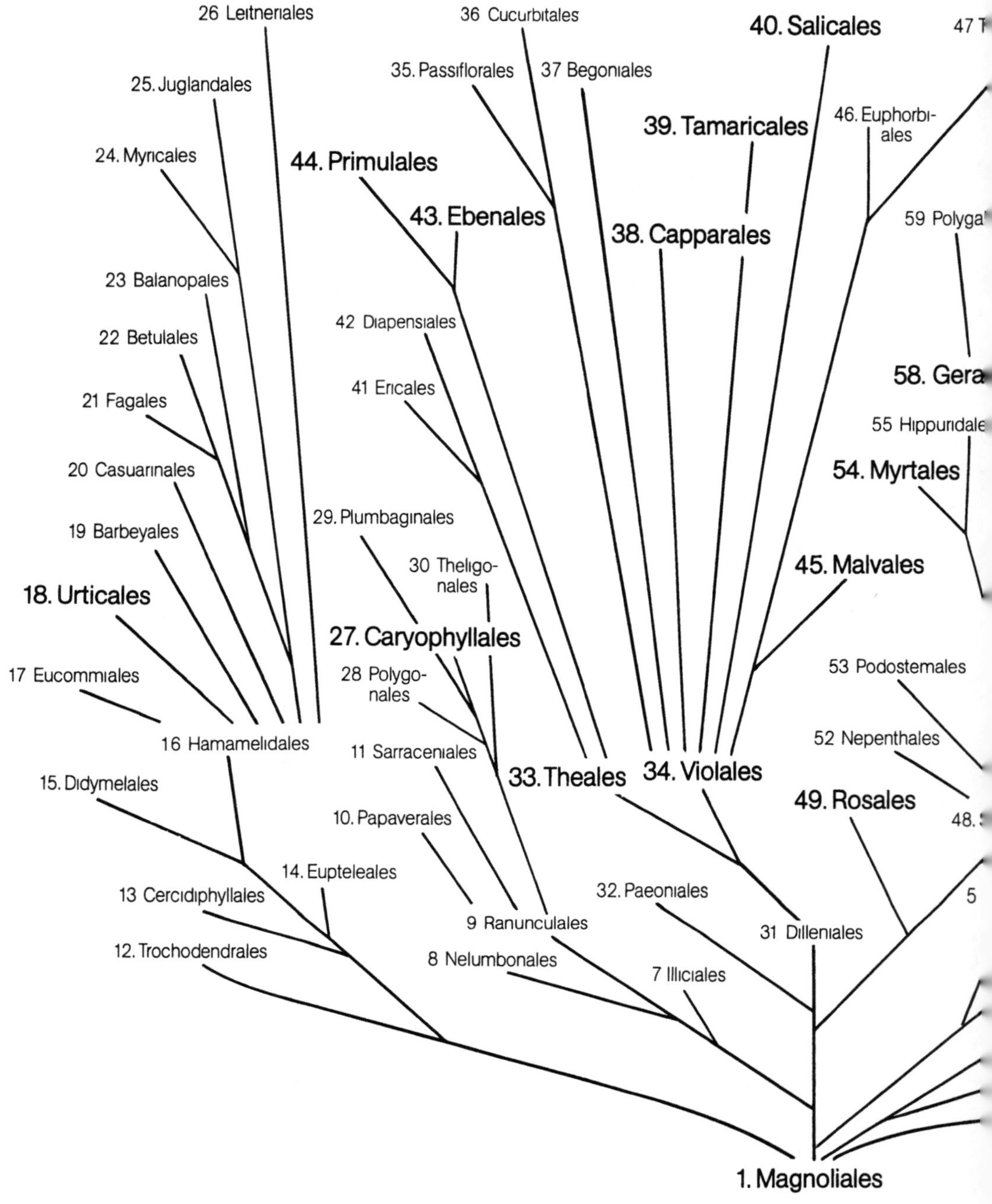
26 Leitneriales
36 Cucurbitales
40. Salicales
47 T
25. Juglandales
35. Passiflorales
37 Begoniales
24. Myricales
44. Primulales
39. Tamaricales
46. Euphorbi-
ales
43. Ebenales
38. Capparales
59 Polyga
23 Balanopales
22 Betulales
42 Diapensiales
58. Gera
21 Fagales
41 Ericales
55 Hippuridale
20 Casuarinales
54. Myrtales
19 Barbeyales
29. Plumbaginales
30 Theligo-
nales
45. Malvales
18. Urticales
27. Caryophyllales
17 Eucommiales
28 Polygo-
nales
53 Podostemales
16 Hamamelidales
11 Sarraceniales
52 Nepenthales
15. Didymelales
33. Theales
34. Violales
49. Rosales
10. Papaverales
48.
14. Eupteleales
13 Cercidiphyllales
32. Paeoniales
5
9 Ranunculales
31 Dilleniales
12. Trochodendrales
8 Nelumbonales
7 Illiciales
1. Magnoliales

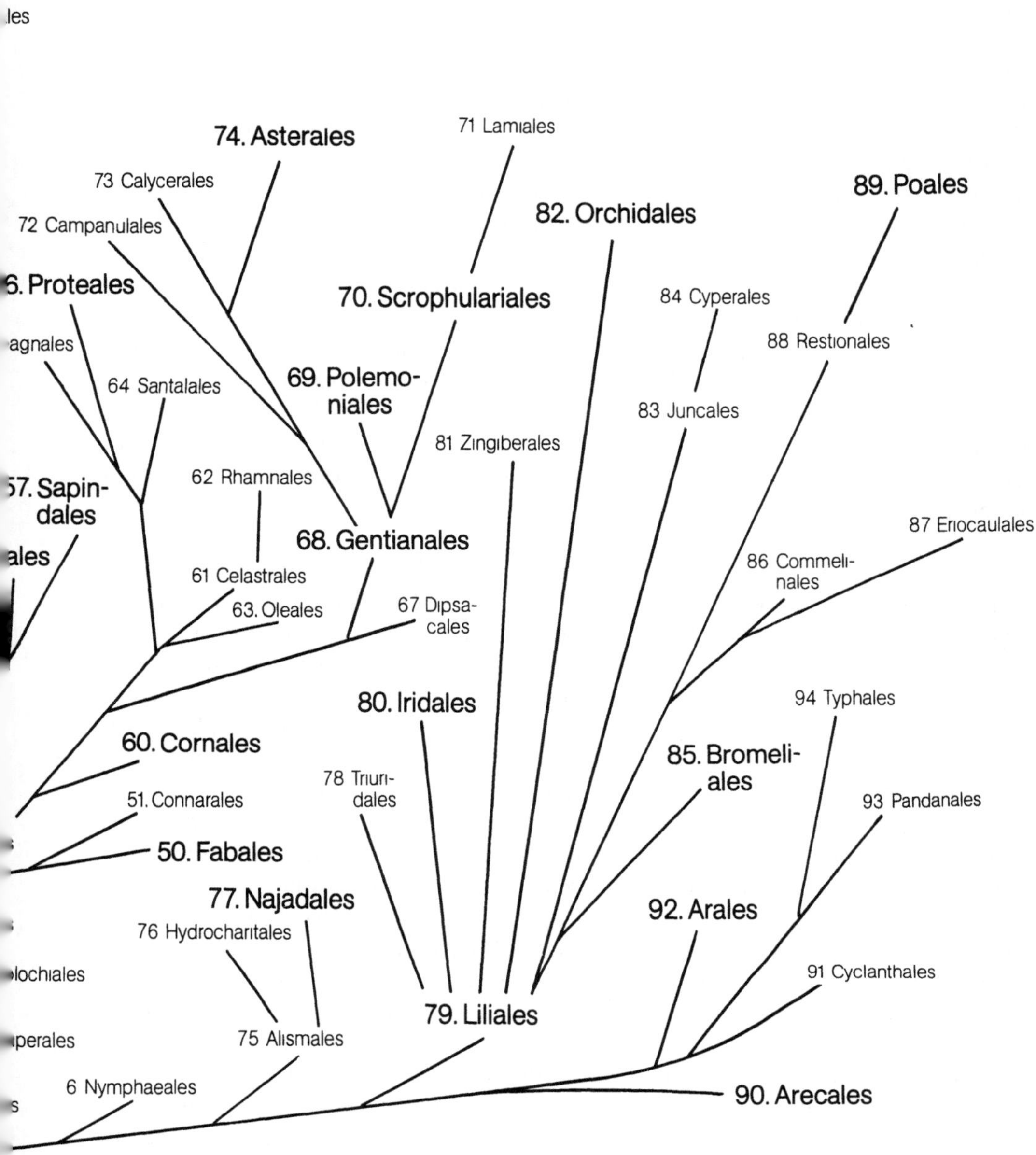

Dendrogram showing the probable relationships of the orders of flowering plants. (Modified after TAKHTAJAN, 1966). The orders referred to in this Chapter are in **bold face** so as to relate taxa of origin to the carbohydrate constituents discussed.

regular or otherwise, that fit the experimental facts obtained. The modes of biosynthesis of the gums are not yet known (cf. KAUSS, 1974), nor is the particular significance of the arrangements of peripheral sugars that are found, among which acidic groups, which possess the obvious property of binding cations, are prominent. However, it is certain that there are differences in the proportions of constituent sugars in different nodules of gums from the one species of tree and in gum samples taken from different varieties of the same species, and there are possible variations dependent upon circumstances of growth (ANDERSON and DEA, 1971). Gum specimens from a particular species exhibit different molecular weights, even when taken from different branches of a single tree, although in at least one instance this has been demonstrated to follow from the union of varying numbers of sub-units (see below). It may well be that each isolated gum sample represents stages in the transformation, by addition or removal of peripheral sugars, of the basal core structure, or even the joining together of quite large and complex sub-units, and so includes a "family" of related polysaccharide molecules differing from one to another in points of detail. This concept will be referred to again in connection with other polysaccharide groups (see Sects. 3.2.2, 6.3, 7.2, 7.3).

No group of workers has pursued more actively the chemical examination of gum exudates from within a genus, with a view to determining their taxonomic significance, than that of D.M.W. ANDERSON. A number of useful correlations between species and molecular composition of *Acacia* gums were recognised, and have been reviewed in detail with chemotaxonomy the main objective (ANDERSON and DEA, 1969; ANDERSON, 1978). Exudates from *Acacia* species classified by BENTHAM as series 4 (Gummiferae) and 5 (Vulgares) showed only minor differences in structure, even when examined in great detail; perhaps the most significant generalisation related to their β-D-galactan cores, where the series 4 gums were shown to contain blocks of 1→6-linked units in an otherwise 1→3-linked, branched framework (but see also ANDERSON and CREE, 1968). A more extensive examination of gums from series 1 (Phyllodineae) (ANDERSON et al., 1972; ANDERSON and GILL, 1975) and 2 (Botryocephalae) (ANDERSON et al., 1971) followed, and those shown to have below-average molecular weights were deliberately selected for further study on the grounds that where there was a reasonable limit to the number of sugar residues in the molecule (of the order of 100), the analytical and structural examination ought to be easier (CHURMS et al., 1977). It turned out that for at least six species the molecular weight distributions conformed to a general pattern of multiples of that of the lowest molecular-weight component, which was usually between 4,000 and 6,000. A single Smith-degradation (periodate oxidation; borohydride reduction; strictly controlled acid hydrolysis) invariably led to a monodisperse fragment consisting essentially of the 1→3-linked β-D-galactan core, the molecular weight of which corresponded to that of the smallest component of the gum from which the bulk of the peripheral sugars had been detached. This implies cleavages by periodate, at regular intervals, within the larger components, and the sub-unit hypothesis received further corroboration when the chemistry of the lowest molecular weight component was shown (ADAM et al., 1977) to be virtually identical with that of the rest of the gum. It is possible, therefore, that degrees of regularity will in due course be detected within the structures of gums in general, and that taxonomic correlations may thereby be assisted.

The production of gum from certain *Acacia* trees may be extremely erratic, and not all *Acacia* species yield gum exudates, but with one exception (*A. melanoxylon*) (cf. ASPINALL and STEPHEN, 1974) all that do give gums which have molecular structures that are remarkably alike, though it seems that interspecies variations outweigh differences arising from habitat, maturity, and so on. So far, in only two cases (*A. cyanophylla* and *A. pycnantha* – series 1, sub-series 6) does the classification of *Acacia* species based on gum chemistry conflict with the traditional classification (ANDERSON et al., 1972). Recently there has been an attempt to separate sub-species of *A. tortilis* on the basis of their gum structures and arising from this work there was evidence that in one of these (ssp. *heteracantha*), two types of molecule were present, which is not usual for *Acacia* gums (ANDERSON and BRENAN, 1975).

3.2 Gums, Mainly of Arabinogalactan Type, from Other Genera, Families, and Orders

In view of the relatively large numbers of types of leguminous trees that are gum-producing, this section is subdivided according to source – Fabales and others.

3.2.1 Order Fabales

Gum exudates from over sixty species of *Acacia* (ANDERSON et al., 1973) have been shown to conform to a general structural pattern, that of a branched heavily substituted β-D-galactan in which 1 → 3-linkages predominate over 1 → 6, to which single units or short chains of sugar residues comprising L-arabinofuranose (less usually pyranose), L-rhamnopyranose and D-glucuronic acid are attached. The gums from other genera within the order Fabales have not been as extensively studied, but they demonstrate the kinds of structural variation of the *Acacia* gum pattern that are shown by gums throughout all the other orders so far investigated, and exemplify in addition the types of exudate based upon D-galacturono-L-rhamnan and L-arabino-D-xylan frameworks that are described in Sections 4 and 5. An early attempt (STEPHEN and SCHELPE, 1964; SHAW and STEPHEN, 1965) at classifying gums on the basis of molecular structure used the designations A, B, and C to relate to the three distinct types of core structure, and the same will be done here. Types B and C are structurally related to the pectins and xylans found universally in plant tissues.

Reference to Table 1 indicates that among Mimosaceae there are general structural resemblances between the gums from *Prosopsis juliflora* (which is highly complex), and from several *Albizia* species, to the type A *Acacia* gums, though a significant amount of D-mannose is introduced in the case of *Albizia* (ANDERSON et al., 1966a). β-D-Glucuronic acid is linked 1 → 2 to this hexose, a structural feature that is found also in the exudates of *Virgilia oroboides* and *V. divaricata* of the Papilionaceae, and in gums from the orders Rosales, Proteales, Myrtales (SMITH and MONTGOMERY, 1959; ASPINALL, 1969), and (for this one example is available) Scrophulariales (SHAW and STEPHEN, 1965). There is a further significance in the occurrence of D-glucuronic acid and D-mannose linked in this way, as explained in Section 3.2.2. *Virgilia* gum contains β-D-xylopyranosyl end-groups

Table 1. As far as can be judged from their compositions and from structural studies the gum exudates conform to the generalised structures given in the key

Order	Family	Genus and species	Gum type
Angiosperms (after TAKHTAJAN, 1966)			
27 Caryophyllales	Cactaceae	*Opuntia fulgida*	A
		O. megacantha (= *O. ficus-indica*)	A
33 Theales	Guttiferae	*Garcinia* spp.	A
34 Violales	Cochlospermaceae	*Cochlospermum gossypium*	B
38 Capparales	Moringaceae	*Moringa oleifera*	A
39 Tamaricales	Tamaricaceae	*Tamarix manifera*	*
43 Ebenales	Sapotaceae	*Sapota achras* (= *Achras zapota*)	C
	Ebenaceae	*Euclea natalensis*	A
45 Malvales	Sterculiaceae	*Sterculia urens*	B
		S. setigera	B
		S. acerifolia	B
		Brachychiton diversifolium	B
		Firmiana simplex	A
	Malvaceae	*Salmalia malabarica*	A
49 Rosales	Rosaceae	*Prunus* (> 10 spp.)	A and A^+ ($A^{\ddagger}$)
50 Fabales	Mimosaceae	*Acacia* (> 60 spp.)	A
		Albizia (5 spp.)	A^+
		Prosopis juliflora	A
	Caesalpiniaceae	*Caesalpinia* spp.	
		Cercidium australe	C
	Papilionaceae	*Virgilia oroboides*	A^+
		V. divaricata	A^+
		Lasiobema retusum	B
		Astragalus gummifer (+ others)	* and B
54 Myrtales	Combretaceae	*Combretum leonense*	A and B
		C. verticillatum	A
		C. hartmannianum	A
		Anogeissus latifolia	$A^{\ddagger}$
		A. leiocarpus	$A^{\ddagger}$
		Terminalia tomentosa	
		T. sericea	
		T. superba	
	Rhizophoraceae	*Rhizophora mangle*	B
56 Rutales	Anacardiaceae	*Anacardium occidentale* (+ others)	A
		Lannea spp.	A
		Spondias cytheria	A
		Mangifera indica	A
		Rhus viminalis	A
		Loxopterigium huasango	A
		Odina wodier (= Lannea?)	A
	Burseraceae	*Boswellia carteri*	A^{Φ}
		B. serrata	A^{Φ}
		Commiphora myrrha	A^{Φ}
	Rutaceae	*Aegle marmelos*	A
		Citrus limonia	A
		C. maxima	A
		Feronia elephantum	A
		Fagara xanthoxyloides	A
		Chloroxylon swietenia	A

Table 1. (continued)

Order	Family	Genus and species	Gum type
	Meliaceae	*Azadirachta indica*	A
		Khaya senegalensis	A and B
		K. grandifoliola	B
		K. ivorensis	B
		Carapa procera	A
		Cedrela odorata	A
60 Cornales	Araliaceae	*Cussonia spicata*	A
	Umbelliferae	*Dorema ammoniacum*	A
		Ferula asafoetida	A^{Φ}
66 Proteales	Proteaceae	*Grevillea robusta*	A^{+}
		Hakea acicularis	A^{+}
		Brabejum stellatifolium	A^{+}
70 Scrophulariales	Pedaliaceae	*Sigmatosiphon guerichii* (=*Sesamothamnus guerichii*)	A^{+} and A^{Φ}
79 Liliales	Agavaceae	*Agave americana*	A
	Liliaceae	*Phormium tenax*	C
80 Iridales	Iridaceae	*Watsonia pyramidata*	C
		W. versveldii	C
85 Bromeliales	Bromeliaceae	*Puya* spp.	C
		Bromelia spp.	C
89 Poales	Gramineae		C and *
90 Arecales	Arecaceae	*Phoenix reclinata*	A
Gymnosperms (after LAWRENCE, 1951)			
10 Cycadales	Zamiaceae	*Encephalartos longifolius* (+ others)	$A^{\ddagger}$
12 Coniferae	Araucariaceae	*Araucaria bidwillii* (+ others)	A^{Φ}
13 Gnetales	Welwitschiaceae	*Welwitschia mirabilis*	A^{+}

Key.
Gum type

A = Arabinogalactan. Sugar residues from: rhamnose, xylose, fucose, glucuronic and galacturonic acids; B = Galacturonorhamnan; arabinose, galactose, fucose and glucuronic acid residues; C = Xylan; arabinose, galactose and glucuronic acid residues; * = Neutral sugar residues only; A^{+} = Includes $Glu_pA\beta1\rightarrow2M$ as a constituent; $A^{\ddagger}$ = Includes glucuronomannan chains; A^{Φ} = Occurs in conjunction with water-insoluble resin.

The sources upon which botanical classifications have been based are:
TAKHTAJAN, A.: Flowering Plants. Edinburgh: Oliver & Boyd 1969. (Translation from Russian Edition of 1966)
LAWRENCE, G.H.: Taxonomy of vascular plants. London: Macmillan 1951

to a limited extent, linked 1 → 5 to arabinose, which contribute to the complex pattern of peripheral sugar residues attached to a predominantly 1 → 6-linked β-D-galactan core. Within the exudates from *Astragalus* spp. (Papilionaceae) more than one structural type is represented, among which there is evidence for the occurrence as a minor component of a *neutral* arabinogalactan. The bulk of the polysaccharide, however, is of quite different structure, based upon a chain of 1 → 4-linked D-galactopyranuronic acid units, most of which are substituted at

C-3 by D-xylopyranosyl groups. Two out of every three of the xylose residues carry a further substituent at C-2, either a β-D-galactopyranosyl or an α-L-fucopyranosyl end-group; and there are further, minor structural components. This interesting molecular structure is directly related to pectic acid in respect of the interior chain, and represents an extreme form of a type B gum polysaccharide with minimal rhamnose content. To complete a picture of great diversity of structural types amongst gums from the order Fabales, *Cercidium australe* (Caesalpiniaceae) yields a polysaccharide gum based on a chain of $1 \rightarrow 4$-linked β-D-xylopyranose units with D-glucopyranuronosyl or D-xylopyranosyl groups or both joined at C-2 of most of them (CEREZO et al., 1969). It must be emphasised that wherever D-glucuronic acid occurs as a component of polysaccharide gums, there is a likelihood of its being methylated at O-4 (ANDERSON and CREE, 1968; KAUSS, 1974); this applies to gums of all three structural types A, B, and C.

3.2.2 Other Orders Derived from Magnoliales, and from Gymnosperms

Representative species of several plant orders have been found to give exudates whose composition and structures correspond in general to the gums of type A described in the previous Section (Table 1). Reference to the comprehensive reviews available should suffice to illustrate the diversity of arrangements of sugars in the outer chains for each polysaccharide gum. It appears that the general structural pattern is followed for exudates from plants in widely different orders, for example from *Agave americana* (Agavaceae; Liliales, a monocotyledonous order), from *Opuntia* spp. (Cactaceae; Caryophyllales), from seven genera of Anacardiaceae, two of Burseraceae, five of Rutaceae and four of Meliaceae (all order Rutales), from two families (Araliaceae and Umbelliferae) of order Cornales, and from three orders of Gymnosperms (*Encephalartos* spp., Zamiaceae; Cycadales; *Araucaria* spp., Araucariaceae; Coniferae; and *Welwitschia mirabilis*, Welwitschiaceae; Gnetales) (LANGHAMMER et al., 1977). *Khaya senegalensis* exudates contain two quite different types of polysaccharide (A and B) as *Astragalus* gum has also been shown to do, and certain of the gums of type A are accompanied by substantial proportions of water-insoluble resin [from *Boswellia carteri* and *B. serrata* and *Commiphora myrrha*, both Burseraceae, from *Ferula asafoetida* (Umbelliferae; Cornales) and from *Araucaria* spp., for example], or by protein (particularly in the case of *Azadirachta indica* gum). This last observation provides a link with the role of arabinogalactans in the hemicellulosic complex of wood tissue (ALBERSHEIM, 1974a); there may well be covalent linkage to peptide which in turn constitutes part of the layered protein in the cell wall.

Gums from the families Combretaceae and Rhizophoraceae (Myrtales) show a diversity of molecular structures, including a variant which has not been commonly encountered elsewhere. *Anogeissus leiocarpus* (Combretaceae) gum, for example, has been fractionated to yield a major polysaccharide component with a backbone of alternating 4-linked β-D-glucopyranuronic acid and 2-linked α-D-mannopyranose residues to which are attached side chains including $1 \rightarrow 3$-linked L-arabinofuranosyl residues, $1 \rightarrow 6$-linked β-D-galactopyranose sequences terminating in a $1 \rightarrow 3$ link to L-arabinopyranose, and D-xylopyranosyl groups (ASPINALL and CARLYLE, 1969). On structural grounds, therefore, this polysaccharide merits classification as a D-glucurono-D-mannan. *Anogeissus latifolia* gum is of the same general type, with greater variety in the linkages between the L-arabinofuranose residues, and longer side-chains. Related material has also been identified as an extracellular constituent of suspension-cultured tobacco (*Nicotiana tabacum*; Solanaceae; Scrophulariales) cells (KATO et al., 1976, 1977). Other plants, from many different orders (Table 1), contain the 1→2 linkage between β-D-glucuronic acid and D-mannose, and there is some indication that *Prunus americana* (Rosaceae; Rosales) and *Encephalartos* (Zamiaceae; Cycadales) gums include repetition of this biouronic acid grouping within their molecular structures. *Combretum leonense* gum (ASPINALL and BHAVANANDAN, 1965) contains a number of sugar sequences in common with that of *Anogeissus latifolia* (and of *A. schimperi*), but the D-glucopyranuronosyl-D-mannose chains are absent and there is instead evidence for repeated combinations of galactose with α-D-galacturonic acid linked 1→2 to L-rhamnopyranose (cf. Type B gums). Furthermore, this gum may be fractionated into components that vary in composition while retaining similar structural units. In the apparently homogeneous *Combretum hartmannianum* gum (ANDERSON and BELL, 1976) mannose and both glucuronic and galacturonic acids are present, though the modes of linkage of the latter are not yet clear; there are appreciable differences in the manner whereby sugars common to the two *Combretum* gums are located within their respective structures. Analytical comparisons have been made of a number of other *Combretum* gums (ANDERSON and BELL, 1977). *Terminalia* spp. (Combretaceae) gums contain such a complexity of different sugar units that as yet they cannot be assigned to any particular type based on arabinogalactan, galacturonorhamnan or even glucuronomannan core structures (ANDERSON and BELL, 1974). By contrast, *Rhizophora mangle* (Rhizophoraceae) gum (SARKAR and RAO, 1975), also order Myrtales, is clearly of the B type discussed below.

Gums from four representative families of order Rutales have been examined, some of them in depth, and in all of them type A molecular structures are found. More than one species of *Lannea* (ANDERSON and HENDRIE, 1973), *Anacardium* (Anacardiaceae), *Boswellia* (Burseraceae), *Citrus* (Rutaceae) and *Khaya* (Meliaceae) have been studied, and no very unusual modifications are apparent. However, all three *Khaya* gums (*K. senegalensis*, *K. grandifoliola* and *K. ivorensis*) are predominantly of the B type, the first of these containing also a type A component; and glucose has been detected for the first time as a gum constituent in *Anacardium occidentale* (ANDERSON and BELL, 1975). Regarding gums from other plant orders, reference to Table 1 will show that at least one genus in each of Caryophyllales, Ebenales, Rosales, Cornales, Proteales, Scrophulariales and the monocotyledonous Liliales and Arecales has been shown to yield a type A product, and the wide distribution of arabinogalactan-based gums is thus evident. Rather fewer examples have been studied of the similarly structured, though usually simpler, constituent polysaccharides of wood (cf. Sect. 2). Although gums from only one genus of Rosales have been investigated, a great deal of research has been carried out on the exudates of many *Prunus* spp. from the 1930's to the present time (ASPINALL, 1969; SMITH and MONTGOMERY, 1959; TOWLE and WHISTLER, 1973), and important, detailed structural work is continuing; even within a species quite wide compositional variations have been found (RÓSIK and WILKIE, 1975).

Earlier in this Section reference was made to gums from *Encephalartos* spp., *Welwitschia* and *Araucaria* spp., representing three different Gymnosperm orders, as being of type A. They contain the unusual sugar component 3-O-methyl-L-rhamnose, and taxonomic implications arising from their structures have recently been discussed (ANDERSON and MUNRO, 1969; LANGHAMMER et al., 1977). Furthermore, a closely related polysaccharide has been isolated from the stigmatic exudate of *Lilium longiflorum*; here the acidic arabinogalactan plays a nutritional role in pollen tube wall formation (ASPINALL and ROSELL, 1978).

4 Plant Polysaccharide Gums of the D-Galacturono-L-Rhamnan Type

Gums of type B consist of highly branched molecules based upon D-galacturono-L-rhamnan main chains in which the 1→4 -linked acid is interrupted at intervals by 1→2-linked L-rhamnopyranose units (ASPINALL, 1969). They are clearly related to the pectins of hemicellulosic character (REES, 1969; ASPINALL, 1973) which are found widespread throughout the plant kingdom in young tissue, bark and fruit, but are of taxonomic interest only in respect of the peripheral sugars. Various assemblages of sugar units are connected to the main chains of the gums, which themselves show variation, particularly in the distribution of rhamnose units and in the incorporation of, for example, D-galactopyranose. It was seen in Section 3.2.1 that a gum component ("tragacanthic acid") from *Astragalus* spp. is of this type, though in point of fact it does represent an extreme in that rhamnose is virtually absent from the galacturonan chains. In the major component of *Khaya senegalensis* gum exudate rhamnose residues occur more frequently, and to them are attached at C-4 units of D-galactopyranose; 4-O-methyl-α-D-glucuronic acid is linked 1→4 to one-half of these. There are nevertheless sequences of α-D-galacturonic acid residues mutually joined by 1→4-linkages in this gum polysaccharide and in exudates of *K. grandifoliola* and *K. ivorensis*, and there is therefore an obvious structural similarity between gums from the three species of this genus. In *Sterculia* (Sterculiaceae; Malvales) gums, that of *S. urens* for example, D-galactose is included in the main chains, adjacent D-galacturonic acid residues occur only infrequently, and β-D-glucuronic acid is found linked 1→3 to the D-galacturonic acid of the main chain. Partial acetylation of the *Khaya* and *Sterculia* polysaccharides is a characteristic.

Two type B gums have been found in other orders, and there are many points of similarity between the structures of the examples just given and those of the gums of *Cochlospermum gossypium* (Cochlospermaceae; Violales) and *Rhizophora mangle* (Rhizophoraceae; Myrtales) (SARKAR and RAO, 1975). These pectin-like exudates are therefore to be found in widely separated orders in the same manner as type A gums, though the number of available examples is small by comparison.

4.1 Apiogalacturonans

The branched-chain sugar component apiose is found in numerous linear 1→4-linked galacturonans which contain several other combined sugars (OVODOV, 1975; WATSON and ORENSTEIN, 1975; WOOLARD and JONES, 1978). From two species of

Zosteraceae (order Najadales), *Zostera marina* and *Z. pacifica*, and *Phyllospadix* (Potamogetonaceae), a polysaccharide (zosterine) of molecular weight 40,000–45,000 was extracted with hot aqueous ammonium oxalate using suitably prepared plant tissue. Partial acid hydrolysis liberated apiose, arabinose, galactose, and an O-methylxylose, yielding the galacturonan, while pectinase action on the polysaccharide left the apiose, attached (at C-2, C-3, or both) to the main chain. It has been suggested that the presence of apiose in this pectic substance affords protection towards microbial attack, particularly by the pectinases of pathogenic fungi. Polysaccharide isolated from the cell walls of *Lemna minor* (Lemnaceae; Arales) has been shown to include apiose units joined to one another, the disaccharide O-β-D-apiofuranosyl-(1→3)-D-apiose being liberated from apiogalacturonan by careful acid hydrolysis.

Apiose-containing polysaccharides, in which the proportions of apiose vary with species of origin and even within a species, are known to occur in *Lemna gibba* as well as *L. minor*, in the seagrasses of the Zosteraceae and *Posidonia australis* (Posidoniaceae; Najadales), and also in the unrelated *Tilia* spp. (Tiliaceae; Malvales) and *Taraxacum kok-saghyz* (Compositae; Asterales). The sugar, glycosidically linked, is ubiquitous in nature (WATSON and ORENSTEIN, 1975), as is free hamamelose (SELLMAIR et al., 1977); hamamelitol, on the other hand, is confined to *Primula* (Primulaceae; Primulales).

5 Gum Exudates of D-Xylan Type

Polysaccharide gums based upon chains of 1→4-linked β-D-xylopyranosyl units are found distributed widely amongst Angiosperm orders. One example (see Sect. 3.2.1) is unusual among many genera of Fabales that have been investigated, viz. *Cercidium australe* gum, in which the inner xylan chain is sheathed (through O-2) by 4-O-methyl-α-D-glucopyranuronosyl or occasional xylosyl groups, to which in turn are joined D-xylopyranosyl end-groups. L-Arabinopyranose partially replaces the exterior xylose residues. The gum of *Sapota achras* (=*Achras zapota*) (Sapotaceae; Ebenales) is similar (DUTTON and KABIR, 1973). Three monocotyledonous orders, Iridales, Liliales and Bromeliales, give xylan-based gums (Table 1). The gums from both *Puya* and *Bromelia* (Bromeliaceae) species have been studied and compared (BONGIORNO DE PFIRTER et al., 1973), and material from different *Watsonia* (Iridaceae) species has been investigated not only by chemical means, but also by X-ray analysis, as the polysaccharide exuded into channels within *Watsonia* corms shows sufficient regularity of orientation to permit excellent diffraction patterns to be obtained and the molecular conformation to be deduced (ASPINALL and STEPHEN, 1974; LELLIOTT et al., 1978). The gum from *Watsonia pyramidata* carries substituent L-arabinofuranosyl groups, either as single units or in the form of short chains, attached to *both* O-2 and O-3 of the xylan chain, and therefore shows an unusually high degree of substitution for this type of polysaccharide (SHAW and STEPHEN, 1966). α-D-Galactopyranosyl end-groups occur frequently, bound 1→3 to arabinose. Experiments have indicated that this polysaccharide acts as a food reserve (cf. p.568). The seed-boxes of *Watsonia versveldii* yield a gum similar to the corm-sac product just described, but with D-glucuronic acid residues present in addition.

The structural relationship of the type C gums to hemicellulosic L-arabino-D-xylans, D-glucurono-D-xylans and 4-O-methyl-D-glucurono-D-xylans is apparent, and as with the other major types of exudate gum structure, the differences between exudate and woody tissue component are most clearly seen in the elaboration of relatively complex sugar patterns in the periphery of the gum molecules. Hemicellulosic xylan components are of course widespread and of limited taxonomic significance though various generalisations have been made (TOWLE and WHISTLER, 1974; OVODOV, 1975). However, the fact that particular examples taken from among Gramineae (order Poales) show mutual similarities of structure is of some interest (BAILEY, 1973; BUCHALA, 1973). For many of these there are variations in the sugar composition at different stages of growth (BUCHALA and WILKIE, 1973; MORRISON, 1974; JOSELEAU and BARNOUD, 1975; cf. WILKIE and WOO, 1977), and material that is isolated may well represent a polydisperse range of molecular species as found in the cell wall. In addition to D-galactose (cf. WOOLARD et al., 1977), a constituent that has occasionally been shown present is L-galactose, found, for instance, as a terminal group in corn-cob (*Zea mays*; Gramineae; Poales) xylan (WHISTLER and CORBETT, 1955; ROBERTS and HARRER, 1973). The occurrence of this sugar as a polysaccharide component within sea plants is referred to in Section 7.2.

6 Polysaccharide Components of Bark and Seeds

A number of different types of polysaccharide, some acidic and some neutral, have been isolated from bark, leaves and seeds of a variety of plants (SMITH and MONTGOMERY, 1959; ASPINALL, 1969; TOWLE and WHISTLER, 1973). Some, the xyloglucans (or "amyloids"), merge into the common glucans by virtue of their having an inner core of glucose units to which other sugars are attached as short chains; mucilaginous substances, the galactomannans, have been extracted from numerous plant sources; and acidic polysaccharides of complex structure, related in many respects to the gum exudates, have long been known and exploited (WHISTLER, 1973).

6.1 Xyloglucans from Cotyledon Cell Walls and from Endosperm

Substances of this polysaccharide group have been known for well over a century, being easily recognised by the blue coloration imparted to them by application of iodine-potassium iodide solution. Extensive investigations of their occurrence have been made with a view to making taxonomic correlations (KOOIMAN, 1960), and a number of useful generalisations have emerged, e.g., that amyloids, which are a major constituent of seeds from the Caesalpiniaceae, are obtained from genera of three tribes only, and not at all from six others. Amyloid-containing species were found in sixteen out of 208 dicotyledonous families, but none amongst 25 families of the monocotyledones.

Detailed structural investigations of these gel-forming polysaccharides are limited in number (ASPINALL, 1969; COURTOIS and LE DIZET, 1974); that from *Tamarindus indica* (Caesalpiniaceae) has a cellulose-like D-glucose main-chain, to most units of which are linked α-D-xylopyranosyl groups at O-6. Occasional β-D-galactopyranosyl units extend the length of the side-chains to two. From seeds of other taxa, those from *Annona muricata* (Annonaceae, of the most primitive order, Magnoliales) yield a xyloglucan which is intermediate in structure between that of the tamarind seed and that of cellulose, while the product from *Tropaeolum majus* (Tropaeolaceae; Geraniales) seed resembles that of the tamarind. The seeds of *Brassica alba* (GOULD et al., 1971) and *B. campestris* (SIDDIQUI and WOOD, 1971, 1977) contain L-fucose as an additional, terminal side-chain constituent; compare the galactose- and fucose-containing xyloglucans of suspension-cultured sycamore cells, and, possibly, dicotyledonous cell walls generally (ALBERSHEIM, 1974b, 1975; ASPINALL et al., 1969, 1977). Examples of xyloglucan from barks are also known, e.g., from *Picea engelmannii*, as well as from red spruce (*Picea rubens*) compression wood (ASPINALL, 1969).

6.2 Galactomannans

Galactomannans are known to occur widely as mucilages, many of which are of considerable industrial importance (WHISTLER, 1973), in the endosperm of seeds of the Leguminosae (Fabales) [70 species were listed by DEA and MORRISON (1975), following BAILEY, (1971)] and of the Annonaceae, Convolvulaceae (order Polemoniales), Rubiaceae (Gentianales) and Palmae (Arecales) families. The usual molecular structure consists of a 1→4-linked chain of β-D-mannopyranosyl units to O-6 of which varying proportions of α-D-galactopyranosyl groups are joined (MCCLEARY et al., 1976) and variously distributed (BAKER and WHISTLER, 1975), though many deviations involving other linkages and the relative positions of the sugar constituents have been reported (ASPINALL and STEPHEN, 1974; COURTOIS, 1974). Sometimes the ratio of galactose to mannose residues is greater than unity, 2:1 being reported for a preparation from alfalfa (*Medicago sativa*) and about 12:1 for *Centrosema plumieri*, where the structure is obviously quite different. It has been noted (Sect. 2) that the seeds of the leguminous plant *Lupinus albus* and of *Strychos nux-vomica* (Loganaceae; Gentianales) contain a 1→4 galactan.

Wide differences in specific rotation are listed for different preparations of galactomannans (DEA and MORRISON, 1975), which may reflect corresponding variations in molecular structure and conformation (REES, 1972). Structural changes have been shown to be only slight, however, during germination, in several typical seed galactomannans, suggesting that once the process does begin it proceeds very rapidly to completion (MCCLEARY and MATHESON, 1974). This supports the contention (DEA and MORRISON, 1975) that mannose: galactose ratios may, sometimes, be useful in taxonomy. As an example, the unusually high ratio of 1.6 for the product from *Trigonella cretica*, which is out of keeping with values of 1.1 to 1.2 for eight other species of *Trigonella* (of the tribe Trifolieae), suggests the possible re-classification of this particular species. Seeds of only certain species of *Sophora* (tribe Sophoreae) appear to contain galactomannan, and taxonomic significance has been attributed to this (BAILEY, 1974).

The galactose content may be very low in galactomannans from non-leguminous plants, this being especially noticeable within divisions (Phytelephantoideae, Phoenicoideae) of the Palmae. The product from endosperm of *Phytelephas macrocarpa*, for example, is known as ivory nut mannan.

6.3 Glucomannans

Linear polymers resembling cellulose in composition but with mannose residues taking the place of glucose to differing extents are associated with hemicellulosic material in Gymnosperms, many examples from *Pinus, Larix,* and *Picea* species (all Pinaceae) being available. Branching of the chains (HOFFMANN and TIMELL, 1970), inclusion of 1→3 linkages (KATO et al., 1973), partial acetylation (KENNE et al., 1975) and the attachment of pendant D-galactopyranosyl groups are common structural variations (ASPINALL and STEPHEN, 1974). Soluble polysaccharides of similar type occur particularly in seeds, tubers or leaves of many species of Amaryllidaceae, Liliaceae and Asparagaceae (order Liliales) (JAKIMOW-BARRAS, 1973; PAULSEN et al., 1978), Iridaceae (Iridales), Araceae (Arales) and Orchidaceae (Orchidales) though it is clear that other, widely differing plant orders may also serve as sources of glucomannan and galactoglucomannan; *Populus* and *Salix* species (Salicaceae; Salicales) and *Stylosanthes* (Fabales) (ALAM and RICHARDS, 1971), for example. The galactoglucomannan of *Cercis siliquastrum* has been singled out for comment as being apparently unique amongst legumes in that glucose residues are present (cf. Sect. 6.2) (MCCLEARY et al., 1976), whereas D-xylopyranosyl as well as D-galactopyranosyl end-groups have been found in a branched glucomannan from *Cassia tora* seeds which has been shown (VARSHNEY et al., 1976) to include xyloglucan- and galactomannan-type moieties within its structure (see also KOOIMAN, 1960; DEA and MORRISON, 1975). In this connection the unusual (1→2) mode of linkage of D-galactopyranose units to the main, mannose chain in a polysaccharide from *Cassia absus* seeds (KAPOOR and MUKHERJEE, 1969, 1971), together with a xylose component, should be noted. A modified glucomannan (including ~1% galactose) occurs in the bracken fern *Pteridium aquilinum* (Dennstaedtiaceae; Filicales; Pteridophyta) (BREMNER and WILKIE, 1971), and galactoglucomannan in the aquatic moss *Fontinalis antipyretica* (GEDDES and WILKIE, 1971) and the fern *Osmunda cinnamonea* (Osmundaceae; Filicales).

6.4 Acidic Polysaccharides

In addition to the remarkable galacturonorhamnan-based mucilage from inner bark of *Ulmus fulva* (Ulmaceae; Urticales), which includes the unusual feature of chains of 3-O-methyl-D-galactose linked to rhamnose, related acidic products (SMITH and MONTGOMERY, 1959; ASPINALL, 1970; TOWLE and WHISTLER, 1973; CAFFINI and PRIOLO DE LUFRANO, 1974) have been extracted from seeds of *Lepidium sativum* and *Brassica campestris* (Cruciferae; Capparales), from species of *Althaea* and *Abelmoschus (Hibiscus)* (Malvaceae; Malvales) and *Plantago* (Plantaginaceae; Scrophulariales), and from *Linum usitatissimum* (Linaceae; Geraniales); L-galactose is a constituent end-group of both acidic polysaccharides isolated from the last-named and L-fucose of one of them. A further characteristic of many of these mucilages is that they are accompanied by complex, highly branched neutral polymers involving combinations of L-arabinose and D-xylose (ASPINALL, 1969).

7 Carbohydrates of Marine Algae

The three major divisions of marine algae, Phaeophyta, Rhodophyta, and Chlorophyta, have their respective pigmentations as their distinguishing characteristic, that is to say, the basis of classification depends on the proportions and molecular constitutions of coloured substances present (PERCIVAL and MCDOWELL, 1967). Their differentiation is primarily, therefore, on chemical grounds. It is also true that the shape, form, degree of rigidity, physiological requirements (water retention and cation exchange) and other characteristics of these plants that are adapted, more or less completely, to a marine environment are in no small measure dependent on the polysaccharides which make up the bulk of the dry weight and whose molecules are so structured as to fulfil special functions, many of which are not required of terrestrial plants (MACKIE and PRESTON, 1974). Sugar units which have not been found in components of land plant orders (D-mannuronic acid and its C-5 epimer L-guluronic acid) are incorporated in polysaccharides of marine algae, in this case the browns, and this same group is characterized by the occurrence of free D-mannitol and of its β-D-glucopyranosyl derivatives at O-1 and both O-1 and O-6 (CRAIGIE, 1974). There is no parallel elsewhere for the existence of long chains of mutually joined L-fucopyranose units as is found in this group of seaweeds. All three divisions utilize sulphate half-ester groups, in association with polysaccharide, and there are some instances, among the reds, of pyruvic acid joined as hemi-acetal (YOUNG et al., 1971). In this way additional acidic groupings of different strengths are added to the molecular structures, conveying properties, for example of ion-exchange, that are necessary for correct function. A consequence of sulphate esterifying primary hydroxyl groups of D- or L-galactose residues in the polysaccharides that are distributed widely amongst the Rhodophyceae is that these may be eliminated, at polysaccharide level (HAUG and LARSEN, 1974), by interaction of C-6 with the 3-OH as the sugar conformation allows. This means that the polysaccharides from many different sources within this division include residues of the 3,6-anhydrosugar which in turn have an influence on conformation and physical characteristics.

These and many other clear examples of the correlation that exists between carbohydrate structure and botanical source have attracted much research interest, and many excellent reviews have been written which describe the unique characteristics of algal polysaccharides in relation to their taxa of origin (SMITH and MONTGOMERY, 1959; PERCIVAL and MCDOWELL, 1967; PERCIVAL, 1966, 1968a, 1978a; TOWLE and WHISTLER, 1973). The following sections, which because of the occurrence in the algal divisions of chemical components that so clearly typify the divisions are grouped on a taxonomic basis, will accordingly be brief, recent work being given preference. Such structural information as is presently available on polysaccharides from algal sources other than marine has been reviewed elsewhere (MACKIE and PRESTON, 1974).

7.1 Polysaccharides of Phaeophyta

The polysaccharides found in members of this division, as in the others, belong to several structural types, which may be extracted simultaneously; in any event, there are often problems of fractionation and purification which generally offset the relative ease with which algal polysaccharides can be separated from the plant tissue.

Polymers (fucans) in which the principal units consist of 1→2 linked α-L-fucopyranosyl groups sulphated at C-4 occur widely in species of *Laminaria* (ELYAKOVA and ZYVAGINTSEVA, 1974) and *Macrocystis* (Laminariales), *Fucus* (HOGSETT and QUATRANO, 1975), *Ascophyllum* (PERCIVAL, 1968b; LARSEN, 1967), *Pelvetia* (EVANS et al., 1973), *Himanthalia* and *Bifurcaria* (Fucales) (MIAN and PERCIVAL, 1973a), *Padina pavonia* (Dictyotales) (MIAN and PERCIVAL, 1973b; ABDEL-FATTAH and EDRESS, 1977), *Colpomenia sinuosa* (Scytosiphonales) (HUSSEIN, 1975) and *Desmarestia* (Desmarestiales) (ELYAKOVA and ZYVAGINTSEVA, 1974; PERCIVAL and YOUNG, 1974). Small amounts of galactose, xylose, and glucuronic acid accompany the material first extracted, however, and extensive purification may be needed to remove most of these (PERCIVAL, 1968a). Studies involving the sequential extraction of polysaccharides from *Fucus vesiculosus*, into which $^{14}CO_2$ had been incorporated, suggest that "fucoidan" results from a series of transformations at polysaccharide level during which the other constituent sugar and uronic acid units are detached (BIDWELL et al., 1972); for examples of the complex heteroglycans that have been isolated reference should be made to the fractionation of *Ascophyllum nodosum* (LARSEN et al., 1966, 1970; MEDCALF and LARSEN, 1977a) and *Sargassum linifolium* extracts (ABDEL-FATTAH et al., 1973, 1974) and to the review by OVODOV (1975) of detailed structural investigations of the sulphated glucuronoglycans from *Sargassum pallidum* and *Pelvetia wrightii*. The main chains of these two polysaccharides consist of glucuronic acid and mannose residues variously linked and distributed, the side chains of xylose, fucose, and galactose (cf. the glucuronomannan gums of land plants, Sect. 3.2.2). Fucose residues linked in ways other than as 1→2 chain units are present in the "fucoidan" (TOWLE and WHISTLER, 1973); the matter is complicated further by the covalent attachment of fucose to protein (MEDCALF and LARSEN, 1977b). The sites and mechanism of sulphate incorporation have been studied for several species (EVANS et al., 1973). It follows from the complexity of the transformation that the polysaccharide constituents undergo, and the further likelihood of chemical decomposition during isolation (LARSEN et al., 1970), that it is not possible to define the structural features that differentiate species within the Phaeophyta, though it does seem that sulphated fucose polymers in general are characteristic of this division.

Alginic acid, the term used to designate linear polymers containing 1→4-linked β-D-mannuronic and α-L-guluronic acid residues (PERCIVAL and MCDOWELL, 1967; PERCIVAL, 1968a; MACKIE and PRESTON, 1974), is a typical component of members of the Phaeophyta, often being present (appropriately neutralised, though not esterified) in large amounts, and having numerous commercial outlets. Young tissue has been found to be rich in blocks of mannuronic acid units (HAUG et al., 1974), enzymatic conversion (HAUG and LARSEN, 1974) to the epimeric guluronic acid taking place with increasing age. The older, stronger portions of plants of different species show appreciable variations in the ratio of the two uronic acid components.

Both physical properties and chemical reactivity are greatly affected by the different conformations adopted by the polymannuronic acid, polyguluronic acid and mixed-unit segments of the chains, and the distribution of the acid units present (MACKIE and PRESTON, 1974; REES, 1972). Thus the real situation is far more complicated than would appear from consideration of the primary structure alone, and the investigation of physical, chemical, and biochemical relationships among alginates is retarded by the extreme difficulty of separating from one another, even grossly, components that differ in mannuronic and guluronic acid content and distribution (FUJIBAYASHI et al., 1970). The problems associated with chemical investigations of polysaccharides that are rich in uronic acid residues are thus compounded.

The reserve polysaccharide of Phaeophyta is a 1→3-linked β-D-glucan, laminaran, samples of which from different orders have been submitted to detailed chemical investigation with the objective of establishing minor structural variations (CRAIGIE, 1974; PERCIVAL, 1968a; STARK, 1977). A unique feature is that the relatively short chains of between ten and thirty units (all pyranose) are in part terminated by β-D-glucosidic bonding to C-1 of D-mannitol, so there is a distinction between those chains that have reducing terminals and those that do not (ELYAKOVA and ZYVAGINTSEVA, 1974), and separations can be effected on this basis (STARK, 1976). Certain β1→6 inter-glucose linkages occur, and these may be contiguous, as in *Eisenia bicyclis*. There may also be branching at C-6 of some units in the 1→3-linked chains. No meaningful correlations of structure with species of origin may, however, be made.

7.2 Polysaccharides of Rhodophyta

The distinctive groups of polysaccharides found widespread among the Rhodophyta consist of modified, sulphated D-galactans with α1→3 and β1→4 linkages, or similarly constituted molecules in which the 4-linked units have the L-configuration (MACKIE and PRESTON, 1974; PERCIVAL, 1966). More often than not representative species within a family are found to contain one *or* other of these two groups (fundamentally, D- and DL-galactans) which have been termed carrageenans and agars ("agaroids") respectively, and are readily extractable; many are known for their exceptional gelling characteristics. Arising from extensive researches into the molecular structures of these compounds (DUCKWORTH and YAPHE, 1971; REES, 1969; TURVEY and WILLIAMS, 1976), it is clear that not only are both types of sugar residue (3- and 4-linked) liable to modification in different species by conversion to sulphate half-esters and methyl ethers and the 4-linked residues of both D- and L-galactose to their 3,6-anhydro derivatives, but that variations in polysaccharide structure are to be expected, for a given species, at different stages of growth, seasons or localities (MACKIE and PRESTON, 1974). There may also be branching: for example, 4-O-methyl-α-L-galactosyl groups are joined to O-6 of 4-linked D-galactose chain units in the polysaccharide of *Aeodes ulvoidea* (ALLSOBROOK et al., 1974).

As to the alternation (contrast Sect. 2) of the galactose units linked 1→3 or 1→4, this has been demonstrated in some cases; but the D- to L-galactose ratio is not always unity. D-Xylose and D-glucuronic acid residues are present in quantity in sulphated polysaccharides of even greater complexity elaborated by the unicellular red algae of order Porphyridiales, *Porphyridium cruentum* (HEANEY-KIERAS and CHAPMAN, 1976; MEDCALF et al., 1975b) and *Rhodella maculata* (EVANS et al., 1974; FAREED and PERCIVAL, 1977); the latter polysaccharide includes rhamnose and the unusual component 3-O-methylxylose. There is evidence for the linkage of D-xylose to D-galactose within the galactan from *Laingia pacifica* (KOCHETKOV et al., 1973), and possibly for 3-O-substituted D-galacturonic acid in that from *Laurencia pinnatifida* (BOWKER and TURVEY, 1968).

The agars, for example those from numerous *Gelidium* and *Gracilaria* spp., are clearly mixed polysaccharides (IZUMI, 1971, 1972) in which the extent of sulphate ester formation varies from virtually nothing to above 30%; and residues of pyruvic acid are found linked as ketal to the 4,6-positions of certain galactose units (NUNN et al., 1973; YOUNG et al., 1971). The highly sulphated carrageenans are subject to similar extremes of structural variation (PERCIVAL, 1978b), and can be fractionated; their classification on a molecular basis is correspondingly difficult. Agaroid and carrageenan-like molecules have not yet been positively shown to occur in different species within any one genus (PERCIVAL, 1978a), though there are genera which metabolise polysaccharides that show certain characteristics of both (PERCIVAL, 1978b).

Some members of Rhodophyta contain neutral, 1→4 linked β-D-mannans and β-D-xylans which are skeletal in function, and in addition water-soluble xylans with mixed linkages, 1→3 as well as 1→4, so differing constitutionally from those typical of land plants; furthermore, branching may occur. There is also present in *Porphyra umbilicalis* an insoluble, and presumably structural, xylan containing 1→3 linkages only (ASPINALL and STEPHEN, 1974), and in *Nemalion vermiculare* a sulphated, 1→3-linked α-D-mannan (USOV et al., 1975). As little has been done on the structural examination of polysaccharides from fresh water members of the Rhodophyta, work reported (TURVEY and GRIFFITHS, 1973) on the non-sulphated polysaccharide from *Batrachospermum* spp. is of special interest; the 3-O-methyl derivatives of both D-galactose and L-rhamnose are present as constituent sugars.

7.3 Polysaccharides of Chlorophyta

A number of different polysaccharide types are found in extracts from members of the Chlorophyta (MACKIE and PRESTON, 1974; PERCIVAL, 1966, 1968a), some of them exceedingly complex in structure and containing groups of sugar residues unlike those found in any other taxonomic group. Thus water-soluble mucilaginous extracts from *Ulva lactuca* (MEDCALF et al., 1972; MEDCALF et al., 1975a) and *Enteromorpha compressa* (both Ulvaceae, order Ulvales) (MEDCALF et al., 1972) and from *Acrosiphonia centralis* (Cladophoraceae; Cladophorales) are all sulphated highly branched heteroglycans, probably hetero-polydisperse, with chains of D-xylose, L-rhamnose and D-glucuronic acid residues as well as sequences of D-xylose and of D-glucose units joined by mixed 1→3 and 1→4 linkages (cf. CARLBERG and PERCIVAL, 1977; PERCIVAL, 1978c). Mixtures of related polymers based upon a sulphated L-arabino-D-galactan framework (PERCIVAL, 1966), comprising certain linkages similar to those found in the arabinogalactans from various land plants (Sects. 2 and 3), and including xylose residues, are found in *Cladophora rupestris* and *Chaetomorpha capillaris* (both Cladophoraceae); *Codium fragile* (Codiaceae, Siphonales) and *Caulerpa filiformis* (Caulerpaceae, another family of Siphonales) contain polysaccharides of similar general type, though they are distinguished by differences in certain of their inter-sugar linkages and positions of sulphation. Extracts from the latter two species contain bound L-rhamnose, and that from *Caulerpa* mannose and traces of D-glucose and fucose as well (PERCIVAL, 1978c). Part of the structural material of many green seaweeds consists of 1→4-linked β-D-mannan (as found in higher plants) or 1→3-linked β-D-xylan (ASPINALL, 1973; ATKINS et al., 1969; MACKIE and PRESTON, 1974; REES, 1974) (cf. Sect. 7.2).

The fructosan (2→1-linked) found as a food reserve in Dasycladales and Cladophorales specimens is rather unusual among algae (CRAIGIE, 1974), though such polysaccharide material is widely spread among orders of land plants (DARNLEY GIBBS, 1974).

8 Conclusion

In view of the variability in production by different plants of the complex heteroglycans considered here, dependent very often upon the state and conditions of growth, it is the *capability* of a species to yield particular products that is relevant to chemotaxonomy. Complete descriptions of molecular structure are not available, and in view of the molecular weight variations which are common, the structural criteria outlined in this chapter have been the inter-sugar linkages occurring in blocks or groupings of sugars present in partial structures.

Correlations have been observed at species level, e.g., a species yields an exudate whose composition typifies the genus, but possesses its own distinguishing characteristics, which fluctuate relatively little from sample to sample. Sub-species differentiation is rare. Land plants generally may exude gums, the diverse chemical types of which occur widely distributed through the orders; all three types may be found in one order, and two have been found in more than one example of a genus. Among marine algae the grouping of divisions based on pigmentation is supported by the occurrence of distinctive polysaccharides (alginates, laminarans, and complex sulphated fucans in one division, modified sulphated galactans in another), and among land plants there are seed and tuber components (galactomannans, glucomannans and xyloglucans) which are concentrated within certain orders and families rather than others.

The incorporation of modified, derived, or unusual sugars (methyl ethers, acetates, deoxysugars, branched-chain sugars, sugars of uncommon configuration or ring-size) is also widespread in the plant kingdom, and, although interesting from many other points of view, cannot yet be seen to follow any particular taxonomic pattern.

Acknowledgements. It is a pleasure to record my appreciation of the advice and comment given by numerous colleagues, Professor E.A.C.L.E. Schelpe, Dr. H. Glen and Dr. D.T. Mitchell (Cape Town), Professors J.R. Nunn and H. Parolis (Grahamstown), Dr. E. Percival and Professor J.B. Pridham (London), Professor G.O. Aspinall (York, Downsville), Professor G.N. Richards (Queensland) and Professor R.L. Whistler (Purdue, W. Lafayette) among them.

References

Reference is made to reviews wherever possible as primary sources can unfortunately not be quoted in full in the space available.

Abdel-Fattah, A.F., Edrees, M.: Carbohydrates of the brown seaweed *Padina pavonia*. Phytochemistry *16*, 939–941 (1977)

Abdel-Fattah, A.F., Hussein, M.M., Salem, H.M.: Sargassan: a sulphated heteropolysaccharide from *Sargassum linifolium*. Phytochemistry *12*, 1995–1998 (1973)

Abdel-Fattah, A.F., Hussein, M.M., Salem, H.M.: Constitution of sargassan, a sulphated heteropolysaccharide from *Sargassum linifolium*. Carbohydr. Res. *33*, 209–215 (1974)

Adam, J.W.H., Churms, S.C., Stephen, A.M., Streefkerk, D.G., Williams, E.H.: Structural relationships of whole *Acacia pycnantha* gum and a component of low molecular-weight. Carbohydr. Res. *54*, 304–307 (1977)

Alam, M., Richards, G.N.: Polysaccharides of tropical pasture herbage. V. A galactoglucomannan from Townsville lucerne (*Stylosanthes humilis*). Aust. J. Chem. *24*, 2411–2416 (1971)

Albersheim, P.: The primary cell wall and control of elongation growth. In: Plant carbohydrate biochemistry. Pridham, J.B. (ed.), pp. 145–164. London: Academic 1974a

Albersheim, P.: Structure and growth of walls of cells in culture. In: Tissue culture and plant science. Street, H.E. (ed.), pp. 379–404. London: Academic 1974b

Albersheim, P.: The walls of growing plant cells. Sci. Am. *232*, 81–94 (1975)

Allsobrook, A.J.R., Nunn, J.R., Parolis, H.: The linkage of 4-*O*-methyl-L-galactose in the sulphated polysaccharide of *Aeodes ulvoidea*. Carbohydr. Res. *36*, 139–145 (1974)

Anderson, D.M.W.: Chemotaxonomic aspects of the chemistry of *Acacia* gum exudates. Kew Bulletin *32*, 529–536 (1978)

Anderson, D.M.W., Bell, P.C.: The composition and properties of the gum exudates from *Terminalia sericea* and *T. superba*. Phytochemistry *13*, 1871–1874 (1974)

Anderson, D.M.W., Bell, P.C.: Structural analysis of the gum polysaccharide from *Anacardium occidentale*. Anal. Chim. Acta *79*, 185–197 (1975)

Anderson, D.M.W., Bell, P.C.: Analytical and structural features of the gum exudate from *Combretum hartmannianum* Schweinf. Carbohydr. Res. *49*, 341–349 (1976)

Anderson, D.M.W., Bell, P.C.: The composition of the gum exudates from some *Combretum* species; the botanical nomenclature and systematics of the Combretaceae. Carbohydr. Res. *57*, 215–221 (1977)

Anderson, D.M.W., Brenan, J.P.M.: Chemotaxonomic aspects of the gum exudates from some sub-species of *Acacia tortilis*. Boissiera *24*, 307–309 (1975)

Anderson, D.M.W., Cree, G.M.: The aldobiouronic acids in gums from *Acacia* species. Carbohydr. Res. *6*, 214–219 (1968)

Anderson, D.M.W., Dea, I.C.M.: Chemotaxonomic aspects of the chemistry of *Albizia* gum exudates. Adv. Front. Plant Sci. *23*, 169–175 (1968)

Anderson, D.M.W., Dea, I.C.M.: Chemotaxonomic aspects of the chemistry of *Acacia* gum exudates. Review. Phytochemistry *8*, 167–176 (1969)

Anderson, D.M.W., Dea, I.C.M.: Recent advances in the chemistry of *Acacia* gums. J. Soc. Cosmet. Chem. *22*, 61–76 (1971)

Anderson, D.M.W., Gill, M.C.L.: The composition of *Acacia* gum exudates from species of the subseries *Juliflorae*. Phytochemistry *14*, 739–741 (1975)

Anderson, D.M.W., Hendrie, A.: An analytical study of gum exudates from some species of the genus *Lannea* A. Rich. Phytochemistry *9*, 1585–1588 (1970)

Anderson, D.M.W., Hendrie, A.: The proteinaceous, gum polysaccharide from *Azadirachta indica*. Carbohydr. Res. *20*, 259–268 (1971)

Anderson, D.M.W., Hendrie, A.: The structure of *Lannea coromandelica* gum. Carbohydr. Res. *26*, 105–115 (1973)

Anderson, D.M.W., Munro, A.C.: An analytical study of gum exudates from the genus *Araucaria* Jussieu (Gymnospermae). Carbohydr. Res. *11*, 43–51 (1969)

Anderson, D.M.W., Cree, G.M., Marshall, J.J., Rahman, S.: Studies on uronic acid materials. Part XIII. The composition of gum exudates from *Albizia sericocephala* and *A. glaberrima* with an appendix on botanical nomenclature in the genus *Albizia*. Carbohydr. Res. *2*, 63–69 (1966a)

Anderson, D.M.W., Hirst, Sir Edmund, Stoddart, J.F.: Studies on uronic acid materials. Part XVII. Some structural features of *Acacia senegal* gum (Gum arabic). J. Chem. Soc. (C), 1959–1966 (1966b)
Anderson, D.M.W., Bell, P.C., McNab, C.G.A.: An analytical study of some *Acacia* gum exudates of the Series Botryocephalae. Carbohydr. Res. *20*, 269–274 (1971)
Anderson, D.M.W., Bell, P.C., McNab, C.G.A.: Analysis of six *Acacia* gum exudates of the Series Phyllodineae. Phytochemistry *11*, 1751–1754 (1972)
Anderson, D.M.W., Bell, P.C., Conant, G.H., McNab, C.G.A.: The gum exudates from *Acacia dealbata* and *Acacia sieberana*, corrections to previous analytical data. Carbohydr. Res. *26*, 99–104 (1973)
Aspinall, G.O.: The exudate gums and their structural relationship to other groups of plant polysaccharides. Pure Appl. Chem. *14*, 43–55 (1967)
Aspinall, G.O.: Gums and mucilages. Adv. Carbohydr. Chem. Biochem. *24*, 333–379 (1969)
Aspinall, G.O.: Pectins, plant gums, and other plant polysaccharides. In: The carbohydrates. Pigman, W., Horton, D. (eds.), Vol. IIB, pp. 515–536. New York: Academic 1970
Aspinall, G.O.: Carbohydrate polymers of plant cell walls. In: Biogenesis of plant cell wall polysaccharides. Loewus, F. (ed.), pp. 95–115. New York: Academic 1973
Aspinall, G.O., Bhavanandan, V.P.: *Combretum leonense* gum, part I. Partial hydrolysis of the gum. J. Chem. Soc., 2685–2692 (1965)
Aspinall, G.O., Carlyle, J.J.: *Anogeissus leiocarpus* gum, part IV. Exterior chains of leiocarpan A. J. Chem. Soc. (C), 851–856 (1969)
Aspinall, G.O., Cottrell, I.W.: Polysaccharides of soybeans. VI. Neutral polysaccharides from cotyledon meal. Can. J. Chem. *49*, 1019–1022 (1971)
Aspinall, G.O., Rosell, K-G.: Polysaccharide component in the stigmatic exudate from *Lilium longiflorum*. Phytochemistry *17*, 919–922 (1978)
Aspinall, G.O., Stephen, A.M.: Polysaccharide methodology and plant polysaccharides. In: MTP Int. Rev. Sci. Org. Chem. Ser. 1 Aspinall, G.O. (ed.), *7*, 285–317 (1974)
Aspinall, G.O., Hirst, E.L., Nicolson, A.: The structure of *Acacia pycnantha* gum. J. Chem. Soc., 1697–1706 (1959)
Aspinall, G.O., Molloy, J.A., Craig, J.W.T.: Extracellular polysaccharides from suspension-cultured sycamore cells. Can. J. Biochem. *47*, 1063–1070 (1969)
Aspinall, G.O., Krishnamurthy, T.N., Rosell, K-G.: A fucogalactoxyloglucan from rapeseed hulls. Carbohydr. Res. *55*, 11–19 (1977)
Atkins, E.D.T., Parker, K.D., Preston, R.D.: The helical structure of the β-1→3-linked xylan in some siphoneous green algae. Proc. R. Soc. London Ser. B. *173*, 209–221 (1969)
Bailey, R.W.: Oligosaccharides. Oxford: Pergamon 1965
Bailey, R.W.: Polysaccharides in the Leguminosae. In: Chemotaxonomy of the Leguminosae. Harborne, J.B., Boulter, D. and Turner, B.L. (eds.), pp. 503–541. London: Academic 1971
Bailey, R.W.: Structural carbohydrates. In: Chemistry and biochemistry of herbage. Butler, G.W., Bailey, R.W. (eds.), pp. 157–211. London: Academic 1973
Bailey, R.W.: Galactomannans and other soluble polysaccharides in *Sophora* seed. N. Z. J. Bot. *12*, 131–136 (1974)
Baker, C.W., Whistler, R.L.: Distribution of D-galactosyl groups in guaran and locust-bean gum. Carbohydr. Res. *45*, 237–243 (1975)
Beveridge, R.J., Ford, C.W., Richards, G.N.: Polysaccharides of tropical pasture herbage. VII. Identification of a new pinitol galactoside from seeds of *Trifolium subterraneum* (Subterranean clover) and an analysis of several pasture legume seeds for cyclohexitols and their galactosides. Aust. J. Chem. *30*, 1583–1590 (1977)
Bidwell, R.G.S., Percival, E., Smestad, B.: Photosynthesis and metabolism of marine algae. VIII. Incorporation of $^{14}CO_2$ into the polysaccharides metabolized by *Fucus vesiculosus* during pulse labelling experiments. Can. J. Bot. *50*, 191–197 (1972)
Bongiorno de Pfirter, G.M., Caffini, N.O., Buttazzoni de Cozzarin, M.S.: Gum exudates in an Argentinian species of the genus Bromelia. II. Gum from *Bromelia hieronymi* (Bromeliaceae). Rev. Farm. (Buenos Aires) *115*, 48–50 (1973)
Bose, S., Soni, P.L.: Graded hydrolysis of cashew nut shell polysaccharide: II. Structures of trisaccharide and tetrasaccharides. J. Indian Chem. Soc. *49*, 593–597 (1972)
Bose, S., Soni, P.L.: Structure of the shell polysaccharides from cashew nuts (*Anacardium occidentale* nuts). Indian J. Chem. *12*, 680–683 (1974)

Bouveng, H.O.: Studies on arabogalactans, parts III and IV. Acta Chem. Scand. *13*, 1869–1883 (1959)
Bowker, D.M., Turvey, J.R.: Water-soluble polysaccharides of the red alga *Laurencia pinnatifida*. Part II. Methylation analysis of the galactan sulphate. J. Chem. Soc. (C), 989–992 (1968)
Bremner, I., Wilkie, K.C.B.: The hemicelluloses of bracken. Part II. A galactoglucomannan. Carbohydr. Res. *20*, 193–203 (1971)
Buchala, A.J.: An arabinogalacto(4-*O*-methylglucurono)xylan from the leaves of *Hordeum vulgare*. Phytochemistry *12*, 1373–1376 (1973)
Buchala, A.J., Wilkie, K.C.B.: Uronic acid residues in the total hemicelluloses of oats. Phytochemistry *12*, 655–659 (1973)
Caffini, N.O., Priolo de Lufrano, N.S.: Composition of gums and mucilages of higher plants. Rev. Farm. (Buenos Aires) 95–105 (1974); 38–45 (1975)
Carlberg, G.E., Percival, E.: The carbohydrates of the green seaweeds *Urospora wormskioldii* and *Codiolum pusillum*. Carbohydr. Res. *57*, 223–234 (1977)
Cerezo, A.S., Stacey, M., Webber, J.M.: Some structural studies of brea gum (an exudate from *Cercidium australe* Johnst.). Carbohydr. Res. *9*, 505–517 (1969)
Churms, S.C., Merrifield, E.H., Stephen, A.M.: Structural features of the gum exudates from some *Acacia* species of the Series Phyllodineae and Botryocephalae. Carbohydr. Res. *55*, 3–10 (1977)
Courtois, J-E.: Introductory address. In: Plant carbohydrate biochemistry. Pridham, J.B. (ed.), pp. 1–6. London: Academic 1974
Courtois, J-E., Le Dizet, P.: Comparative study of the structure of three galactoxyloglucans (amyloids) of seeds. An. Quim. *70*, 1067–1072 (1974)
Craigie, J.S.: Storage products. In: Algal physiology and biochemistry. Stewart, W.D.P. (ed.), pp. 206–235. Oxford: Blackwell 1974
Darnley Gibbs, R.: Chemotaxonomy of flowering plants. Vol. 1, pp. 394–420. Montreal: McGill-Queen's University 1974
Das, N.N., Das, A., Mukherjee, A.K.: Structure of the D-galactan isolated from garlic (*Allium sativum*) bulbs. Carbohydr. Res. *56*, 337–349 (1977)
Dea, I.C.M., Morrison, A.: Chemistry and interactions of seed galactomannans. Adv. Carbohydr. Chem. Biochem. *31*, 241–312 (1975)
Duckworth, M., Yaphe, W.: The structure of agar, part II. The use of a bacterial agarase to elucidate structural features of the charged polysaccharides in agar. Carbohydr. Res. *16*, 435–445 (1971)
Dutton, G.G.S., Kabir, S.: Structural studies of sapote (*Sapota achras*) gum. Carbohydr. Res. *28*, 187–200 (1973)
Elyakova, L.A., Zyvagintseva, T.N.: A study of the laminarans of some Far-Eastern, brown seaweeds. Carbohydr Res. *34*, 241–248 (1974)
Evans, L.V., Simpson, M., Callow, M.E.: Sulphated polysaccharide synthesis in brown algae. Planta *110*, 237–252 (1973)
Evans, L.V., Callow, M.E., Percival, E., Fareed, V.: Synthesis and composition of extracellular mucilage in the unicellular red alga *Rhodella*. J. Cell Sci. *16*, 1–21 (1974)
Fareed, V.S., Percival, E.: The presence of rhamnose and 3-*O*-methylxylose in the extracellular mucilage from the red alga *Rhodella maculata*. Carbohydr. Res. *53*, 276–277 (1977)
Fincher, G.B., Sawyer, W.H., Stone, B.A.: Chemical and physical properties of an arabinogalactan-peptide from wheat endosperm. Biochem. J. *139*, 535–545 (1974)
Ford, C.W.: Arabinogalactan from *Phaseolus atropurpureus* leaves. Phytochemistry *11*, 2559–2562 (1972)
Fu, Y., Timell, T.E.: Polysaccharides in compression wood of tamarack (*Larix laricina*). 5. Constitution of an acidic arabinogalactan. Svensk. Papperstidn. *75*, 680–682 (1972)
Fujibayashi, S., Habe, H., Nisizawa, K.: Heterogeneity of alginate in special reference to the enzymatic degradation. J. Biochem. (Tokyo) *67*, 37–45 (1970)
Geddes, D.S., Wilkie, K.C.B.: Hemicelluloses from the stem tissues of the aquatic moss *Fontinalis antipyretica*. Carbohydr. Res. *18*, 333–335 (1971)
Gorin, P.A.J., Spencer, J.F.T.: Structural chemistry of fungal polysaccharides. Adv. Carbohydr. Chem. *23*, 367–417 (1968)
Gould, S.E.B., Rees, D.A., Wight, N.J.: Polysaccharides in germination. Xyloglucans ('amyloids') from the cotyledons of white mustard. Biochem. J. *124*, 47–53 (1971)

Haug, A., Larsen, B.: Biosynthesis of algal polysaccharides. In: Plant carbohydrate biochemistry. Pridham, J.B. (ed.), pp. 209–218. London: Academic 1974
Haug, A., Larsen, B., Smidsrød, O.: Uronic acid sequence in alginate from different sources. Carbohydr. Res. *32*, 217–225 (1974)
Heaney-Kieras, J., Chapman, D.J.: Structural studies on the extracellular polysaccharide of the red alga *Porphyridium cruentum*. Carbohydr. Res. *52*, 169–177 (1976)
Henderson, G.A., Hay, G.W.: The carbohydrates of the leaves of common barberry (*Berberis vulgaris*). Carbohydr. Res. *23*, 379–398 (1972)
Hoffmann, G.C., Timell, T.E.: Isolation and characterisation of a galactoglucomannan from red pine (*Pinus resinosa*) wood. Tappi *53*, 1896–1899 (1970)
Hogsett, W.E., Quatrano, R.S.: Isolation of polysaccharides sulphated during early embryogenesis in *Fucus*. Plant Physiol. *55*, 25–29 (1975)
Hussein, M. M-D.: Two new polysaccharides from *Colpomenia sinuosa*. Phytochemistry *14*, 1866–1868 (1975)
Impellizzeri, G., Mangiafico, S., Oriente, G., Piartelli, M., Scinto, S., Fattorusso, E., Magno, S., Santacroce, C., Sica, D.: Amino-acids and low molecular-weight carbohydrates of some marine red algae. Phytochemistry *14*, 1549–1557 (1975)
Izumi, K.: Chemical heterogeneity of the agar from *Gelidium amansii*. Carbohydr. Res. *17*, 227–230 (1971)
Izumi, K.: Chemical heterogeneity of the agar from *Gracilaria verrucosa*. J. Biochem. (Tokyo) *72*, 135–140 (1972)
Jakimow-Barras, N.: Les polysaccharides des graines de quelques Liliacées et Iridacées. Phytochemistry *12*, 1331–1339 (1973)
Jiang, K.S., Timell, T.E.: Polysaccharides in the bark of aspen (*Populus tremuloides*) II. Isolation and structure of an arabinan. Cell. Chem. Technol. *6*, 499–502 (1972)
Joseleau, J-P., Barnoud, F.: Hemicelluloses of *Arundo donax* at different stages of maturity. Phytochemistry *14*, 71–75 (1975)
Joseleau, J-P., Chambat, G., Vignon, M., Barnoud, F.: Chemical and ^{13}C n.m.r. studies on two arabinans from the inner bark of *Rosa glauca*. Carbohydr. Res. *58*, 165–175 (1977)
Kaplan, M., Stephen, A.M.: Application of gas-liquid chromatography to the structural investigation of polysaccharides. II. The study of *Acacia* gums. Tetrahedron *23*, 193–198 (1967)
Kapoor, V.P., Mukherjee, S.: Galactomannan from *Cassia absus* seed. I. Nature of sugars present, methylation, and periodate oxidation studies. Can. J. Chem. *47*, 2883–2887 (1969)
Kapoor, V.P., Mukherjee, S.: Galactomannan of *Cassia absus* seed. II. Isolation of five oligosaccharides from the galactomannan of *Cassia absus* seed. Phytochemistry *10*, 655–659 (1971)
Karácsonyi, Š., Toman, R., Janaček, F., Kubačková, M.: Polysaccharides from the bark of the white willow (*Salix alba* L.): Structure of an arabinan. Carbohydr. Res. *44*, 285–290 (1975)
Kato, K., Kawaguchi, Y., Mizuno, T.: Structural analysis of Suisen glucomannan. Carbohydr. Res. *29*, 469–476 (1973)
Kato, K., Eda, S., Watanabe, F.: Glucuronomannan from extracellular polysaccharides of suspension-cultured tobacco cells. VIII. Int. Symp. Carbohydr. Chem. Kyoto, Aug. 16–20 1976, Abstr. p. 48 (1976)
Kato, K., Watanabe, F., Edy, S.: Interior chains of glucuronomannan from the extra-cellular polysaccharides of suspension-cultured tobacco cells. Agric. Biol. Chem. (Tokyo) *41*, 539–542 (1977)
Kauss, H.: Biosynthesis of pectin and hemicelluloses. In: Plant carbohydrate biochemistry. Pridham, J.B. (ed.), pp. 191–205. London: Academic 1974
Keegstra, K., Talmadge, K.W., Bauer, W.D., Albersheim, P.: The structure of plant cell walls. III. A model of the walls of suspension-cultured sycamore cells based on the interconnections of the macromolecular components. Plant Physiol. *51*, 188–196 (1973)
Kenne, L., Rosell, K-G., Svensson, S.: Studies on the distribution of the *O*-acetyl groups in pine glucomannan. Carbohydr. Res. *44*, 69–76 (1975)
Kochetkov, N.K., Usov, A.I., Miroshnikova, L.I., Chizhov, O.S.: The polysaccharides of algae. XII. Partial hydrolysis of a polysaccharide from *Laingia pacifica*. Zh. Obshch. Khim. *43*, 1832–1839 (1973)
Kooiman, P.: The occurrence of amyloids in plant seeds. Acta Bot. Neerl. *9*, 208–219 (1960)

Kremer, B.P.: Mannitol in the Rhodophyceae – a reappraisal. Phytochemistry *15*, 1135–1138 (1976)

Labavitch, J.M., Freeman, L.E., Albersheim, P.: Purification and characterisation of a β-1,4-galactanase which degrades a structural component of the primary cell walls of dicots. J. Biol. Chem. *251*, 5904–5910 (1976)

Langhammer, L., Horisberger, M., Horman, I., Strahm, A.: Partial structure of a polysaccharide from the leaves of *Welwitschia mirabilis* Hook. F. Phytochemistry *16*, 1575–1577 (1977)

Larsen, B.: Sulphated polysaccharides in brown algae. II. Isolation of 3-*O*-β-D-xylopyranosyl-L-fucose from ascophyllan. Acta Chem. Scand. *21*, 1395–1396 (1967)

Larsen, B., Haug, A., Painter, T.J.: Sulphated polysaccharides in brown algae. I. Isolation and preliminary characterisation of three sulphated polysaccharides from *Ascophyllum nodosum* (L.) le Jol. Acta Chem. Scand. *20*, 219–230 (1966)

Larsen, B., Haug, A., Painter, T.: Sulphated polysaccharides in brown algae. III. The native state of fucoidan in *Ascophyllum nodosum* and *Fucus vesiculosus*. Acta Chem. Scand. *24*, 3339–3352 (1970)

Lelliott, C., Atkins, E.D.T., Juritz, J.W.F., Stephen, A.M.: Conformation of the gummy polysaccharide from *Watsonia pyramidata* corms. Polymer *19*, 363–367 (1978)

Lindberg, B., Svensson, S.: Microbial polysaccharides. MTP Int. Rev. Sci. Org. Chem. Ser. 1. Aspinall, G.O. (ed.), *7*, 319–344 (1974)

Mackie, W., Preston, R.D.: Cell wall and intercellular region polysaccharides. In: Algal physiology and biochemistry. Stewart, W.D.P. (ed.), pp. 40–85. Oxford: Blackwell 1974

Maekawa, E.: An arabinogalactan from water-soluble polysaccharides of bamboo shoot. VIII. Int. Symp. Carbohydr. Chem. Kyoto, Aug. 16–20, 1976, Abstr. p. 46 (1976)

Maekawa, E., Kitao, K.: The structure of an arabinogalactan isolated from bamboo shoot. Agric. Biol. Chem. (Tokyo) *38*, 227–229 (1974)

McCleary, B.V., Matheson, N.K.: Galactomannan structure and β-mannanase and β-mannosidase activity in germinating legume seeds. Phytochemistry *14*, 1187–1194 (1974)

McCleary, B.V., Matheson, N.K., Small, D.M.: Galactomannans and β-mannanase. Phytochemistry *15*, 1111–1117 (1976)

McComb, E.A., Rendig, V.V.: Occurrence, biosynthesis and function of some heptuloses. In: Developments in food carbohydrate. Birch, G.G., Shallenberger, R.S. (eds.), Vol. 1, pp. 61–76. London: Applied Science 1977

Medcalf, D.G., Larsen, B.: Fucose-containing polysaccharides in the brown algae *Ascophyllum nodosum* and *Fucus vesiculosus*. Carbohydr. Res. *59*, 531–537 (1977a)

Medcalf, D.G., Larsen, B.: Structural studies on ascophyllan and the fucose-containing complexes from the brown alga *Ascophyllum nodosum*. Carbohydr. Res. *59*, 539–546 (1977b)

Medcalf, D.G., Root, C.F., Craney, C.L., Mukhopadhyay, D., Miller, C.K., Hopewell, W.D.: Chemical characterisation of mucilaginous polysaccharides from Ulvaceae species native to the Puget Sound. Proc. Int. Seaweed Symp. 7th, 1971, pp. 541–547. New York: Wiley 1972

Medcalf, D.G., Lionel, T., Brannon, J.H., Scott, J.R.: Seasonal variation in the mucilaginous polysaccharides from *Ulva lactuca*. Bot. mar. *18*, 67–70 (1975a)

Medcalf, D.G., Scott, J.R., Brannon, J.H., Hemerick, G.A., Cunningham, R.L., Chessen, J.H., Shah, J.: Some structural features and viscometric properties of the extracellular polysaccharide from *Porphyridium cruentum*. Carbohydr. Res. *44*, 87–96 (1975b)

Mian, A.J., Percival, E.: Carbohydrates of the brown seaweeds *Himanthalia lorea* and *Bifurcaria bifurcata*. Carbohydr. Res. *26*, 147–161 (1973a)

Mian, A.J., Percival, E.: Carbohydrates of the brown seaweeds *Himanthalia lorea*, *Bifurcaria bifurcata* and *Padina pavonia*. Carbohydr. Res. *26*, 133–146 (1973b)

Morrison, I.M.: Changes in the hemicellulosic polysaccharides of rye-grass with increasing maturity. Carbohydr Res. *36*, 45–51 (1974)

Neukom, H., Markwalder, H.: Isolation and characterisation of an arabinogalactan from wheat flour. Carbohydr. Res. *39*, 387–389 (1975)

Northcote, D.H.: Chemistry of the plant cell wall. Annu. Rev. Plant Physiol. *23*, 113–132 (1972)

Northcote, D.H.: Site of synthesis of the polysaccharides of the cell wall. In: Plant carbohydrate biochemistry. Pridham, J.B. (ed.), pp. 165–181. London: Academic 1974

Nunn, J.R., Parolis, H., Russell, I.: Sulphated polysaccharides of the Solieriaceae family, part II. The acidic components of the polysaccharide from the red alga *Anatheca dentata*. Carbohydr. Res. *29*, 281–289 (1973)

Ovodov, Y.S.: Structural chemistry of plant glycuronoglycans. Pure Appl. Chem. *42*, 351–369 (1975)

Paulsen, B.S., Fagerheim, E., Øverbye, E.: Structural studies of the polysaccharide from *Aloe plicatilis* Miller. Carbohydr. Res. *60*, 345–351 (1978)

Percival, E.: The natural distribution of plant polysaccharides. In: Comparative phytochemistry. Swain, T. (ed.), pp. 139–158. London: Academic 1966

Percival, E.: Marine algal carbohydrates. Oceanogr. Mar. Biol. Annu. Rev. *6*, 137–161 (1968a)

Percival, E.: Glucuronoxylofucan, a cell-wall component of *Ascophyllum nodosum*, part I. Carbohydr. Res. *7*, 272–283 (1968b)

Percival, E.: Do the polysaccharides of brown and red seaweeds ignore taxonomy? In: Systematics Association Special Vol. No. 10, Modern approaches to the taxonomy of red and brown algae. Irvine, D.E.G., Price, J.H. (eds.), pp. 47–62. London: Academic 1978a

Percival, E.: Sulphated polysaccharides of the Rhodophyceae. ACS Symp. Ser. Schweiger, R.G. (ed.) (1978b)

Percival, E.: Sulphated polysaccharides metabolized by the marine Chlorophyceae – a review. ACS Symp. Ser. Schweiger, R.G. (ed.), *77*, 203–212 (1978c)

Percival, E., McDowell, R.H.: Chemistry and enzymology of marine algal polysaccharides. London: Academic 1967

Percival, E., Young, M.: Carbohydrates of the brown sea weeds. Carbohydr. Res. *32*, 195–201 (1974)

Rees, D.A.: Enzymic synthesis of 3,6-anhydro-L-galactose within porphyran from L-galactose-6-sulphate units. Biochem. J. *81*, 347–352 (1961)

Rees, D.A.: Polysaccharide gels and networks. Adv. Carbohydr. Chem. Biochem. *24*, 267–332 (1969)

Rees, D.A.: Polysaccharide gels. A molecular review. Chem. Ind. (London), 630–636 (1972)

Rees, D.A.: Polysaccharide conformation. In: MTP Int. Rev. Sci. Org. Chem. Ser. 1. Aspinall, G.O. (ed.), *7*, 251–283 (1974)

Rees, D.A., Richardson, N.G.: Polysaccharides in germination. Occurrence, fine structure, and possible biological role of the pectic araban in white mustard cotyledons. Biochemistry *5*, 3099–3107 (1966)

Rix, E.M., Rast, D.: Nectar sugars and subgeneric classification in *Fritillaria*. Biochem. Syst. Ecol. *2*, 207–209 (1975)

Roberts, R.M., Harrer, E.: Determination of L-galactose in polysaccharide material. Phytochemistry *12*, 2679–2682 (1973)

Rosík, J., Wilkie, K C.B.: Quantitative estimation of sugar residues in acidic gums. Phytochemistry *14*, 1019–1021 (1975)

Roudier, A.J., Eberhard, L.. Recherches sur les hémicelluloses du bois de Pin maritime des Landes. Bull. Soc. Chim. Fr., 460–464 (1965)

Sarkar, M., Rao, C.V.N.: The structure of degraded mangle gum. Carbohydr. Res. *41*, 163–174 (1975)

Sellmair, J., Beck, E., Kandler, O., Kress, A.: Hamamelose and its derivatives as a chemotaxonomic marker in the genus *Primula*. Phytochemistry *16*, 1201–1204 (1977)

Shaw, D.H., Stephen, A.M.: Further observations on the structures of plant polysaccharide gums. S. Afr. Ind. Chem., 146–147 (1965)

Shaw, D.H., Stephen, A.M.: Structure of a polysaccharide found in the corm-sacs of *Watsonia pyramidata* (Andr.) Stapf. Carbohydr. Res. *1*, 400–413 (1966)

Siddiqui, I.R., Wood, P.J.: Structural investigation of water-soluble, rape-seed amyloid. Carbohydr. Res. *17*, 97–108 (1971)

Siddiqui, I.R., Wood, P.J.: Structural variation of oxalate-soluble rapeseed (*Brassica campestris*) polysaccharides, part III. An arabinan. Carbohydr. Res. *36*, 35–44 (1974)

Siddiqui, I.R., Wood, P. J.: Structural investigation of sodium hydroxide-soluble rapeseed (*Brassica campestris*) polysaccharides, part V. Fucoamyloid. Carbohydr. Res. *53*, 85–94 (1977)

Simson, B.W., Cote, W.A. Jr., Timell, T.E.: Larch arabinogalactans. IV. Molecular properties. Svensk Papperstidn. *71*, 699–710 (1968)

Smith, F., Montgomery, R.: The chemistry of plant gums and mucilages. New York: Reinhold 1959
Srivastava, B.K., Pande, C.S.: Arabinogalactan from the pods of *Opuntia dillenii*. Planta Med. *25*, 92–97 (1974)
Stacey, B.E.: Plant polyols. In: Plant carbohydrate biochemistry. Pridham, J.B. (ed.), pp. 47–59. London: Academic 1974
Stark, J.R.: A new method for the analysis of laminarans and for preparative-scale fractionation of their components. Carbohydr. Res. *47*, 176–178 (1976)
Stark, J.R.: Isolation of three water-soluble glucan components from *Pelvetia canaliculata*. Carbohydr. Res. *57*, C11–C12 (1977)
Stephen, A.M., Schelpe, E.A.C.L.E.: Plant polysaccharide gum structures in relation to the taxonomic affinities of their species of origin. S. Afr. Ind. Chem., 12–15 (1964)
Sturgeon, R.J.: Chemical and biochemical aspects of fungal cell walls. In: Plant carbohydrate biochemistry. Pridham, J.B. (ed.), pp. 219–233. London: Academic 1974
Sturgeon, R.J.: Microbial polysaccharides. In: Specialist periodical reports, carbohydrate chemistry. Vol. 7, pp. 282–293. London: The Chemical Society 1975
Sturgeon, R.J.: Microbial polysaccharides. In: Specialist periodical reports, carbohydrate chemistry. Vol. 8, pp. 225–261. London: The Chemical Society 1976
Toman, R., Karácsonyi, Š., Kováčik, V.: Polysaccharides from the bark of the white willow (*Salix alba* L.): structure of a galactan. Carbohydr. Res. *25*, 371–378 (1972)
Towle, G.A., Whistler, R.L.: Hemicelluloses and gums. Review. In: Phytochemistry. Miller, L.P. (ed.), Vol. I, pp. 198–248. New York: van Nostrand Reinhold 1973
Turvey, J.R., Griffiths, L.M.: Mucilage from a fresh-water red alga of the genus *Batrachospermum*. Phytochemistry *12*, 2901–2907 (1973)
Turvey, J.R., Williams, E.L.: The agar-type polysaccharide from the red alga *Ceramium rubrum*. Carbohydr. Res. *49*, 419–425 (1976)
Usov, A.I., Adamyants, K.S., Yarotskii, S.V., Anoshina, A.A.: Polysaccharides of algae. XVI. Study of the structure of the sulphated mannan from the red alga *Nemalion vermiculare* Sur. Zh. Obshch. Khim. *45*, 916–921 (1975)
Varshney, S.C., Rizvi, S.A.I., Gupta, P.C.: Structure of a polysaccharide from the seeds of *Cassia tora*, part II. Partial acidic hydrolysis. J. Chem. Soc. Perkin I, 1621–1628 (1976)
Watson, R.R., Orenstein, N.S.: Chemistry and biochemistry of apiose. Adv. Carbohydr. Chem. Biochem. *31*, 135–184 (1975)
Whistler, R.L.: Industrial gums, 2nd ed. New York: Academic 1973
Whistler, R.L., Corbett, W.M.: Oligosaccharides from partial acid hydrolysis of corn fiber hemicellulose. J. Am. Chem. Soc. *77*, 6328–6330 (1955)
Wilkie, K.C.B., Woo, S-L.: A heteroxylan and hemicellulosic materials from bamboo leaves, and a reconsideration of the general nature of commonly occuring xylans and other hemicelluloses. Carbohydr. Res. *57*, 145–162 (1977)
Woolard, G.R., Jones, J.K.N.: Polysaccharides of the sea-grass *Phyllospadix torreyi*. Carbohydr. Res. *63*, 327–332 (1978)
Woolard, G.R., Rathbone, E.B., Novellie, L.: A heteroxylan from the husks of *Sorghum* grain. Phytochemistry *16*, 957–959 (1977)
Young, K., Duckworth, M., Yaphe, W.: The structure of agar. Part III. Pyruvic acid, a common feature of agars from different agarophytes. Carbohydr. Res. *16*, 446–448 (1971)
Zimmermann, M.H., Ziegler, H.: List of sugars and sugar alcohols in sieve-tube exudates. In: Encyclopedia of plant physiology. New series. Zimmermann, M.H., Milburn, J.A. (eds.), Vol. 1, pp. 480–503. Berlin, Heidelberg, New York: Springer 1975

Author Index

Page numbers in *italics* refer to the references

Species Index

Subject Index

(Page numbers in *italics* refer to structures)

MIX
Papier aus verantwortungsvollen Quellen
Paper from responsible sources
FSC® C105338

If you have any concerns about our products,
you can contact us on
ProductSafety@springernature.com

In case Publisher is established outside the EU,
the EU authorized representative is:
Springer Nature Customer Service Center GmbH
Europaplatz 3, 69115 Heidelberg, Germany

Printed by Libri Plureos GmbH
in Hamburg, Germany